ADVANCES IN AGILE MANUFACTURING

Advances in Design and Manufacturing

Volume 4

ISSN: 0926-9622

Advances in Agile Manufacturing

Integrating Technology, Organization and People

Edited by

Paul T. Kidd

Manufacturing Knowledge™ Inc.

and

Waldemar Karwowski

University of Louisville

IOS Press

1994

Amsterdam • Oxford • Washington DC • Tokyo

ISBN 90 5199 176 2
Library of Congress Catalog Card Number: 94-077316

Publisher:
IOS Press
Van Diemenstraat 94
1013 CN Amsterdam
Netherlands

Sole distributor in the UK and Ireland:
IOS Press/Lavis Marketing
73 Lime Walk
Headington
Oxford OX3 7AD
England

Distributor in the USA and Canada:
IOS Press, Inc.
P.O. Box 10558
Burke, VA 2209-0558
U.S.A.

Distributor in Japan:
Kaigai Publications, Ltd.
21 Kanda Tsukasa-Cho 2-Chome
Chiyoda-Ku
Tokyo 101
Japan

Fourth International Conference on
Human Aspects of Advanced Manufacturing and Hybrid Automation
July 6-8, 1994, UMIST Conference Centre, Manchester, England

Jointly Organised by

Cheshire Henbury Research & Consultancy
Macclesfield, England

and

Centre for Industrial Ergonomics
University of Louisville, Kentucky, USA

Conference Theme - *In Search of Agility*

Conference Co-sponsored by:
The Ergonomics Society
The Human Factors and Ergonomics Society
The Institution of Electrical Engineers
The International Ergonomics Association
Manufacturing Knowledge[TM] Inc.
MONITOR/FAST Programme of the
Commission of the European Communities

International Scientific Advisory Board

Conference Chairman: Paul T. Kidd PhD, CEng

Honorary Chairman: Waldemar Karwowski PhD, PE

Foreword

There can now be little doubt that contemporary manufacturing industry is undergoing a major paradigm shift away from the mass production (Tayloristic) paradigm that has long dominated industrial corporations worldwide. The emerging paradigm, which is known as Agile Manufacturing [1], is forcing us to completely reassess our values, our culture and our approaches. A key element in this reassessment is the shift of focus towards a systematic examination of the organization and people issues in advanced manufacturing. This concern with organization and people issues manifests itself in at least three forms.

First, there is the commercial concern to build agile infrastructures for manufacturing enterprises by integrating technology, organization and people (TOP). This concern is reflected in asking fundamental questions about business precesses and the strategies required for TOP integration. Secondly, there is the need to achieve competitive advantage from organizational structures and practices and from the skills and knowledge of our people. This concern is seen through the widespread interest in concurrent engineering, team working, continuous improvement, empowerment, etc. Thirdly, there is an interest, often born from implementation failures and disappointments, to master the difficult process of designing and implementing advanced manufacturing systems, giving proper consideration to the importance of organization and people issues [2-4].

These three areas represent the dominant theme of this book. Of the 166 papers contained in this volume, about 70% are in someway connected with the issue of developing methods and tools to support TOP integration or to support the analysis, design and implementation of advanced manufacturing systems and concepts.

This book contains papers presented at the *Fourth International Conference on Human Aspects of Advanced Manufacturing and Hybrid Automation* held in Manchester, England on July 6-8 1994. The objective of this biennial conference is to provide an international forum to promote, exchange and disseminate information and research results on the organization and people aspects of advanced manufacturing systems, with particular emphasis on the enhancement of human skills and the integration of these skills with machine performance to achieve productivity, quality, flexibility and safety. Reflecting the nature of this field, the forum of the meeting was multidisciplinary with participation from all segments of the international scientific and engineering communities. The conference also provided industrial managers with an opportunity to familiarise themselves with the latest developments in this important field.

The book is organized into 10 parts. The first part consists of the four keynote papers given at the conference. The second part (9 papers) deals with the issue of *Manufacturing Paradigms*. Part III, consisting of 15 papers, addresses the issues surrounding *Concurrent Engineering and New Product Development*. Part IV groups 37 papers which are devoted to the *Design and Implementation of Advanced Manufacturing Systems* - a topic of increasing importance to industry. Part V (12 papers) is devoted to *Human Computer Interaction*. In Part VI, 12 papers address issues of *Reliability, Safety and Health Issues*. The emerging area of *Skill and Knowledge Enhancing Technologies* is considered in Part

VII (23 papers). In Part VIII (16 papers), traditional *Human Performance and Ergonomic Design Issues* are discussed. *Organizational and Cultural Change and Human Roles* (32 papers) are addressed in Part IX. The last 6 papers (Part X) discuss *Quality and Maintenance Strategies*.

We would like to acknowledge the advice of members of the International Scientific Advisory Board. We would also like to acknowledge the support and sponsorship of this conference by the Ergonomics Society in England, the Human Factors and Ergonomics Society in the United States, the Institution of Electrical Engineers in England, the International Ergonomics Association, Manufacturing Knowledge™ Inc. and the MONITOR-FAST Programme of the Commission of the European Communities.

References

[1]   Paul T. Kidd, Agile Manufacturing: Forging New Frontiers. Addison-Wesley, Wokingham, England, 1994.

[2]   A. Majchrzak, M. Fleischer, D. Roithman and J. Mokray, Reference Manual for Performing the HITOP Analysis. Industrial Technology Institute, Ann Arbor, MI, 1991.

[3]   Waldemar Karwowski and Gavriel Salvendy (Eds.), Organization and Management of Advanced Manufacturing. John Wiley & Sons, New York, 1994.

[4]   Gavriel Salvendy and Waldemar Karwowski (Eds.), Design of Work and Development of Personnel in Advanced Manufacturing. John Wiley & Sons, New York, 1994.

Paul T. Kidd, Manufacturing Knowledge™ Inc.
Louisville, Kentucky, USA.

Waldemar Karwowski, University of Louisville,
Louisville, Kentucky, USA.

July 1994

Contents

Part I
Keynote Papers

Advances in Agile Manufacturing
P.T. Kidd and W. Karwowski (Eds.)
IOS Press, 1994

Human Factors Collaborative Research Within Manufacturing

John D. GILLIS
Department of Trade & Industry
Technology Programmes & Services Division
London

Abstract. The paper provides background information on Human Factor collaborative applied research within manufacturing. It describes the elements of the UK - Manufacturing Organisation, People & Systems Programme "MOPS" its associated research projects, deliverables and awareness activities, together with that of the parallel European EUREKA project called Integration of Technology & Organisation for Quality Production "INTO".

1. Introduction

1.1 *Competitive Challenge*

UK manufacturing industry is facing major challenges to improve quality and response time, be flexible in its production and keep down costs. It needs to respond quickly to these challenges, so that it can meet the changing demands and competition from abroad.

During the 1980's, investment in Advanced Manufacturing Technology (AMT) was considered vital to industrial growth and competitiveness. However, most investments in such technology failed to reach full potential because industry did not perceive that *Organisational* structures, *Peoples* own skills and computer based information *Systems*, were interdependent, and as such, needed to be optimised together if the whole activity of a manufacturing unit was to achieve maximum performance and attended benefits.

1.2 *Past Experience*

Evaluation Reports on the past Flexible Manufacturing (FMS) and Advanced Manufacturing Systems (AMT) DTI Schemes identified "Human Factor" issues as one of the principal problems in the introduction of technology and achieving the expected business benefit. Some relevant paragraphs are given below:

"Skill problems were encountered by approximately one-third of firms in the sample. These ranged from project management skills in the actual implementation process to operational and management skills in running systems".

"Of particular note was the general underestimation of maintenance problems and the need for extra training to support the operation and maintenance of integrated systems".

"Problems of organisational fit included the lack of compatibility of the FMS with the physical layout of the rest of the plant, with the product range/families being made and with the functional inter-relationship such as between design and manufacturing or quality control".

UK industry has tended to think of manufacturing technology and its management as distinct entities. This was plausible when companies were simply replacing one machine tool with another. However, with new AMT systems which integrate with other parts of the business[#], this precept cannot be adopted and therefore, for their successful application, a change in attitude by people at all levels within an organisation is necessary.

2. Background to Human Factors Research

2.1 *DTI Study Reports*

In early 1990 DTI commissioned two reports which detailed the needs for research into the way that manufacturing should be organised, so as to optimise the full potential of people and modern technology systems. The reports were:

> a. A short collaborative project carried out with Rossmore Warwick and 40 industrial partners, to identify the "needs" of manufacturing industry in managing change.

and   b. A desk top study carried out by AMTRI, on "existing **human factors research**" in UK manufacturing and the current situation in academic institutes. One of the aims being to identify and promote a UK strategy on the way forward for the subject.

2.2 *Report Findings*

The reports indicated that existing research had concentrated on psychological and ergonomic aspects rather than practical implementation in the factory and indicated the "**fear factor**" as a major hurdle which management had to overcome for the implementation of change.

The reports reinforced the view that there is a need in all areas of an organisation for a "**change in attitude**" to change itself. The successful organisation of the future needed to see change as a continuing, welcome, and productive state, putting in place structures, attitudes and a culture which supports it, together with the people who are an intrinsic part of the success of such an attitude. Synergy between the three elements of *organisation, people* and *systems* was seen as an integral part of a company's manufacturing strategy.

2.3 *Study Conclusions*

The reports suggested that some academic research had deliverables but were sometimes hidden or not in a useable form, industry needed bridging tools to help them manage

---

[#] in this context, covering automated assembly and flexible manufacturing sytems (FAS, FMS); computer aided design, manufacturing and production management (CAD, CAM & CAPM)

change. The **"Manufacturing, Organisation People & Systems (MOPS) Programme"** wastherefore established to provide industry with some of the necessary background methodologies and tools to help facilitate this change, and test them in a working environment.

3. MOPS Programme

3.1 *Programme Objectives*

The MOPS programme through collaborative industrial led research will provide some of the necessary tools and methodologies to help all companies implement manufacturing strategies which achieve better management of operations through improved interaction of information systems and people. The research will also lead to demonstrators and case studies that will show industry the benefits of considering "MOPS" issues when implementing AMT.

The 3 year programme started in May 1991. The total cost of the programme being £5.1m with a DTI contribution of £1.515m. Programme elements included:

    a. identifying issues that managers should address when deploying a manufacturing strategy in the "MOPS" area.
    b. initiating UK industrial workshops on "MOPS" issues.
    c. establishment of collaborative projects to provide methodologies and tools for managers.
    d. production of suitable material (A4 leaflet) to support the "MOPS" programme.
    e. suitable coverage of the programme through national and technical media.
    f. generation of case studies and demonstration sites in the UK where industry can view "best practice".
    j. to maintain a watching brief on international developments in the "MOPS" area.

3.2 *Research Projects*

Research projects developed in the programme cover:

- *CONTINUOUS IMPROVEMENT FOR COMPETITIVE ADVANTAGE ("CIRCA")*
  Lucas Industries/Brighton Business School
  Continuous improvement ("CI") methodologies, which can improve manufacturing performance through constant systematic incremental change. Research will cover:-
      team working   organisation cell development
      multi-skilling communication   decentralised control

- *CONTINUOUS IMPROVEMENT - NORTH WEST (*"CIGNET"*)*
  NIMTECH - North West Technology Network (850 within network)
  Continuous improvement ("CI") methodologies, and supports/supplements the above "CIRCA" project, and the resulting experiences/best practice will be shared with a wider network of companies.
  Action research will cover:-
      problem finding/evaluation   problem solving

- *BEST PRACTICE CELLULAR MANUFACTURING ("BESTMAN")*
  Human Centred Systems Ltd
  Organisational and people issues connected with managing change associated with the implementation of cellular manufacturing. It builds on previous ESPRIT research (Project No 1199)

- *OBJECTIVE SYSTEMS TO IMPLEMENT SUCCESSFUL TEAMWORKING METHODS IN MANUFACTURING INDUSTRY ("OSISTEM")*
  Selection Research Ltd ("SRI")
  Team selection and review methodologies to assess the implementation of team-based organisations, and to review the performance of the whole team, as well as its individuals, on a continuous basis.

- *INTELLIGENT MANAGEMENT OF CHANGE IN MANUFACTURING ("IMOCIM")*
  Euristics - SME provider
  Management of Change methodologies and tools, to assist companies in the planning and implementation of advanced manufacturing technology systems

- *DEVELOPMENT OF A SOFTWARE TOOL FOR THE ANALYSIS AND IMPROVEMENT OF ORGANISATIONAL STRUCTURES IN MANUFACTURING INDUSTRY ("DESTINY")*
  Poly Enterprises Plymouth Ltd ("PEP")
  An integrated diagnostic PC software tool to graphically model organisational structures

- *METHODS AND TOOLS FOR REINFORCING & DEVELOPING TEAMWORK IN MANUFACTURING (TIM)*
  The Tavistock Institute of Human Relations
  Reinforcing & developing teamwork in manufacturing - inter-related technical, human and organisational issues connected to group-based work on the shop floor.

- *SUPPLY CHAIN MANAGEMENT TOOL*
  Glasgow Business School - Supply Chain Management Group
  Supply chain positioning tool techniques to improve relationships between manufacturing based customers and their suppliers

- *HUMAN FACTORS - THE MANAGEMENT OF CHANGE IN SME'S*
  The University of Nottingham - Institute for Occupational Ergonomics
  Investigate how "people considerations" will improve the management of change in SME's

4. Overall Benefits

Individual MOPS projects are researching the organisation and people issues associated with the integration of computer based manufacturing systems in various areas of the business   (eg. between sales, design, production, and supply) as well as the related management structures.

The results of the research will be of direct benefit to the participating companies, but will also made available to a wider audience through "awareness seminars" and conferences. In addition case studies will in 1994 be presented through "M90's".

5. European Collaboration

5.1 "INTO" Initiative

"Integration of Technology and Organisation for Quality Production - INTO (EU860) was originally initiated by the UK in March 1992, following a meeting with over 40 UK industrialists who unanimously agreed a need for a European programme to follow up the successful UK MOPS (Manufacturing Organisation, People and Systems) programme.

INTO (EU860) is the establishment of a EUREKA organisational and human-centred **network** to assist European industry to initiate innovative projects that help to increase the competitiveness, quality and reliability of manufacturing companies through improved organisation, people and technology systems, and their strategic interaction.

The UK industrial endorsement provided the mandate to circulate "a discussion document on the case and scope for collaboration under EUREKA" throughout Europe and to organise an international meeting in London in June 1992. Agreement was reached at the meeting to raise a EUREKA organisational and human-centred network umbrella; following which, a "Letter of Common Understanding" and "Eureka 18 Point Plan" was produced and circulated to the attending nations. As a result INTO was endorsed by the EUREKA HLG in October 1992.

5.2 Aim and Purpose of INTO

Within EUREKA, INTO aims to provide the coordination network and project management structure, enabling information on individual research and development projects to be communicated effectively throughout Europe. This network will bring together potential partners (eg. users, suppliers and academics) and help them to initiate innovative projects that increase the competitiveness of manufacturing companies through improved organisation, people and technology structures.

5.3 Participating members

The following EUREKA members are taking part in INTO: **Finland, France, Germany, Norway, Sweden, Switzerland and The United Kingdom.** In addition, collaborators in the projects endorsed to date also come from Hungary, Ireland, Italy, Portugal and Spain. It is anticipated that this involvement will lead to these nations formally joining INTO.

5.4 Work Programme

Overall progress has seen the registration of over 12 proposals within the INTO portfolio, 4 of which have attained EUREKA status.

TIME (EU824)   MAIRIE (EU1024)
RITMICS (EU1031)   CLEANTECH (EU1104)

Proposals are primarily aimed at products, processes and services based on advanced technologies/methodologies and directed to both private and public markets. Individual projects are initiated by the participants, who are free to choose their own research topics, partners and operation patterns ("bottom-up" approach).

5.4 *Awareness Activities*

During the initial stages of INTO, the national coordinators have concentrated their efforts on **creating awareness and assisting and encouraging the development of collaborative projects**. These activities have included:

* creation of INTO brochure/dissemination of information
* monitoring of cooperation and bilateral relationships
* coordination of the relationship between INTO to other European programmes
* dissemination of information and promotion of technology transfer.
* establishing cross-border training (eg. as a result of the Helsinki CI workshop, students from Sweden have attended a course in a UK university on the subject of Continuous Improvement)

6.   Conclusion/Justification For MOPS/INTO Research

6.1 *Industrial Awareness*

Changes which are introduced into the workplace without adequate consideration of the "MOPS/INTO" issues, are unlikely to fully benefit from the opportunities presented to them. The results being frequent union disputes, losses in productivity, quality, and competitiveness, or worst, total internal chaos.

The MOPS/INTO programmes aim to compensate for the market failure in the UK and Europe caused by the lack of appreciation and awareness of the methodology necessary to make the best optimised use of advanced manufacturing systems. In the past companies have not systematically recorded their experience of introducing change, therefore DTI co-ordination and support was required to bring this experience to the public domain, and disseminate the information more widely to UK industry.

6.2 *Benefits to All*

The major thrust within the programmes is aimed at companies in the 50-500 employee bracket, where it was perceived there was a real danger of departmental boundaries being established as companies grow.   The involvement of over 50 companies (25 SME's) in the direct research and over 1000 companies within the networks established around the individual projects demonstrate the industrial *interest* and *the need*.

If interested in collaborative research within INTO or require further information on MOPS then contact Gil Virgo on 071-215-1537 at the DTI.

Advances in Agile Manufacturing
P.T. Kidd and W. Karwowski (Eds.)
IOS Press, 1994

Human Oriented Manufacturing System

Hiromu Nakazawa

Waseda University, 3-4-1 Shinohkubo

Shinjuku-ku, Tokyo, Japan

Abstract. Fully automatic unmanned manuffacturing system have many problems. This paper present "Human–Oriented Manufacturing System" (HOMS), which aims to acheive high productivity and quality of production. Seven satisfaction factors are also shown from the survay of the foundry workers. These factors should be considered for the design of HOMS.

1. Problems of Unmanned Fullautomation

Manufacturing industries have employed fully automated and unmanned systems and excluded workers from the workshops. This way may have the following problems in the future:

(1) Since fully automatic unmanned manufacturing system is highly synchronised and adjusted, when a problem occurs in the system, the system falls down into a logical conflict and the system suffers serious damage.

(2) To cope with this conflict, the concept of autonomous distributed system has been proposed. But even with this idea, the functional level of the system will be very low and less flexible, because it is impossible to replace all human abilities with the machine systems.

(3) Techniques and skills are created and advanced by humans, and they are transferred from human to human. If fully automatic unmanned systems are realised, further progress of technique and manufacturing technology stops.

(4) Flexibility and productivity of the fully automatic unmanned manufacturing systems is still lower than that of the system with human interaction.

(5) The fully automatic unmanned manufacturing systems can be obtained at the sacrifice of the increasing depreciation.

(6) Fully automatic unmanned manufacturing systems may conflict with the creation of the employment chances.

2. New Manufacturing Systems

It is important to understand that the human role in manufacturing systems is neither to be an incorporated component of the system nor an excluded one. The alternative human role is to interact with the automated manufacturing systems to obtain high flexibility and productivity and to improve product quality.

Now we understand that the direction of fully automatic unmanned manufacturing system is wrong. Considering above discussion we proposed Human-OrientedManufacturing System (HOMS)[houmz] 1990 as an alternative manufacturing system.

Functional requirements of HOMS consist of the follwing three characteristics:

(1) Manufacturing systems in which humans have pride and pleasure in work;

(2) Manufacturing systems which are tender to humans;

(3) Manufacturing systems which make a profit.

Some ideas for realization of these functional requirements and introduction of some items of researchs are mentioned in other paper[1].

3. Satisfaction of Workers

3.1 Survey of Foundry

Next problem is how to design HOMS in view of the workers satisfaction. We can not design the human-oriented systems properly till we know the important factors in satisfaction of work. So first we have to know what the satisfaction of work is.

We, therefore, have made a survay of satisfaction of workers in a foundry which is considered as relatively bad working environment among manufacturing industries. The reason we chose a foundry is that the number of quitting workers is comparatively small in spite of bad working environment. Products of the surveyed foundry are large size ones such as beds of machine tools and dies for automobile bodies. The model workshop is located apart from a main foundry and 30 workers are there. On the other hand the main foundry has 50 workers. The average age of all the workers is 50, while ages range from 19 to 59. The manufacturing flow is devided into seven processes, which are as follows.

(1) Model: Styrofoam models are made closely to the drawings in an office-like bright and clean workshop.

(2) Model Coating: In this process the styrofoam models are coated with carbon-paint by a brush or spraygun and the coated models are dried. This workshop is very dirty and disgusting.

(3) Setup: Weak parts of the carbon-coated models are strengthened or holes are filled up or blocks are added, and carbon-coated. This workshop is also dirty.

(4) Model making: Finished models are buried in the sand inside the frame. The sand is fed by a machine into the frame and stamped. This work is physically hard.

(5) Melting: After having made sand models, melted metal is poured into it.

(6) Frame removing: After the melting process the frames are removed by a forklift, though the products are still hot. Environment of this workshop is not good.

(7) Finishing: Finishing the products is the last process. Products are put on the floor and worker must take difficult positions. They finish the products by a sander, hammer and so on.

3.2 First Questionnaire

In order to make clear the fundamental concept of pleasure and pride in manufacturing, first we made preliminary questionaire to the fifty workers except model workshop.

The answers to the questions are analysed, classified and summarized into some concepts by KJ method[2]. The following three motivations for work are made clear.

(1) Pleasure of manufacturing

(2) Good human relations

(3) Positive attitude toward work

It is also found that pleasure of manufacturing has seven factors. They are as follows.

(1) Creativity: The work is asking for the creativity.

(2) Feedback effect: The worker can see the results of what he has done by his effort and

ideas, and recongnize his abilities.

(3) Subjectivity: The work is done by his own ideas and subjective plan.

(4) Diversity: Products and contents of work change often.

(5) Technical skill: The worker can recongnize the progress of his skill under the accumulation of experiences and endevors.

(6) Interest: The worker is intersted in the work or the product personally.

(7) Social contribution: The worker can recongnize that the product he made contributes to the society.

Concequently we know that manufacturing system should be designed and constructed to fullfil these seven factors of satisfaction.

3.3 Second Questionnaire

Next pleasure of manufacturing was surveyed of more deeply by focusing on the above seven factors of satisfaction. This time all the eighty workers are requested to answer. Several questions are made for each factor changing the expression each other in order to check the reliability of the answers. As a result of the statistical analysis sufficient reliability was confirmed. Each question is answered by seven point method. For example, to the question "Is your work diversified?", the worker answers it giving a point (mark) from seven (the most diverse) to one (repetition of a same work).

3.4 Evaluation by Information Integration Method

All the data of the questionnaire are evaluated by information integration method[3]. Reason that this method is adopted is that the mean value of the marks is insufficient to evaluate the data because scatter of the data can not be considered. This method can evaluate the mean value as well as scatter range of the data.

Caluculation is done as follows. Obtain first the mean value of the all the data M. Design range is defined as a desired range, and in this case it is the range whose value is equal or larger than M as shown in Fig.1. Since the relative evaluation of satisfaction in the foundry is the main object, the design range defined above is reasonable.

System range L1 (shown in Fig.1) where scattering points exist is obtained as follows.

Standard deviation σ and mean value m are calulated from the points for a satisfaction factor

of a process, and width of σ of each side of m is defined as a system range.

Common range L2 is defined as a overlapping range of the system range and the design

range. Using above values we can get information value I by the following equation.

$$I = \ln \frac{\text{system range } (L_1)}{\text{common range } (L_2)}$$

Total satisfaction of a process Ip is obtained by summing up the seven information values of

the factors for the process based on the axion of this method.

The smaller the information is, the more the workers are satisfied with their work for the

item in this case. Therefore, zero is the best and the infinte value which is the case when the

system range comes out of the design range implies that the process or factor concerned is

dissatisfied.

3.5 Results of Second Questionnaire

One of the results is shown in Fig.2. Here pleasure of manufacturing is shown by the total

information value for the seven satisfaction factors of each process. This figure shows very

important facts. First, informations values are very large for the model, frame removing and

finishing processes. That is, the workers of those processes feel less pleasure. It is very

interesting to know that environment of the model and finishing processes are very good.

But they don't understand how the model is used or what results happen with it. This work

may not ask for any idea or matured skill or subjectivity. It is wrong to understand that

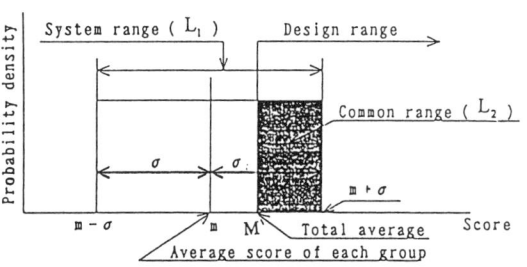

Fig.1. Design range and system range

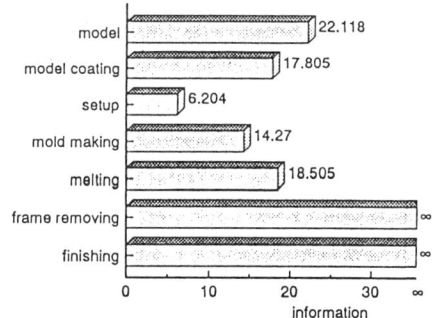

Fig.2. Total information for each process

when environment is comfortable every thing goes well.

Finishing process is provided also with good evironment, but workers don't satisfy with the work either. In this process technical skill factor is not satisfied enough. This process also forces the worker physically the difficult and uncomfortable position. In the frame removing process workers dissatisfy in view of feedback effect and social contribution factors.

On the other hand the setup, mold making annd model coating processes show high degree of satisfaction. Quality of the work of setup process effects quality of products.   The workers of these processes are asked ideas and subjectivity, and can see results of his work (feed back effect). Therefore almost all seven factors are satisfied in these process.

Another important conclusion of this survey is that regardless of dirty and disgusting environment if the seven or some factors are satisfied well, the workers feel pleasure of manufacturing.

4. Conclusion

Manufacturing system of next generation should aim for the fusion of automation technology and human abilities. Processes in which human can not have pleasure and pride should be fully automated and unmanned. On the other hand processes in which human have pride and pleasure should be designed for human to exhibit his abilities to the maximum.

Also the manufacturing system should be design so that the system can be evolved by human. Then with this design concept the system will be able to get high performances under the low investment cost. With this HOMS human can expect happy living while getting much profits. We must recognize again that the manufacturing systems exist for human and human doesn't exist for the manufacturing system.

Refarences

[1] Hiromu Nakazawa: Alternative Human Role in Manufacturing AI&Society (1993) 7,151-156

[2] Jiro Kawakita: Hasso-ho (Idea Creating Method in Japanese) Chuohkohron.1988.

[3] Hiromu Nakazawa: Information Integration Mathod (In Japanese) Corona 1987, or

   Hiromu Nakazawa: Principles of Precision Engineering (to be published) Oxford University Press.

Advances in Agile Manufacturing
P.T. Kidd and W. Karwowski (Eds.)
IOS Press, 1994

Automation in Manufacturing, Control versus Chaos?

Dietrich Brandt
HDZ/KDI, University of Technology, D-52068 Aachen, Germany

Abstract: In automation the ruling concept of the human role can be seen as follows: The control system -including its assisting systems - acts, the human observes and supervises, he/she only acts in case of emergency. But can this concept be sustained in view of increasing chaos around the world? This question is discussed in this presentation based on the work of different researchers in Europe and around the world. Alternative approaches to automation and control are described.

1. Introduction

There is increasing concern and criticism about technology, but there is also a strong belief in the potential of technology to solve our problems. The belief seems still be prevailing that automation should be developed further and further. Let us consider the design of systems which can be controlled by one human sitting in front of a control panel - many research groups worldwide are following this concept developing components for such systems, e.g.:

a) Data processing units which process all data measured, in real time; they display only those processed data which are at any moment really needed by the human to supervise the control process.

b) Control units which 'run' the manufacturing system at optimum performance, at any time.

c) Decision support systems allowing the human to call up information needed in case of trouble. They also suggest - or perform - emergency actions.

Technological concepts supporting these developments are, e.g.: expert systems and artificial intelligence, neural networks and fuzzy logic, automated speech recognition, virtual reality and visualization of processes, computer-supported cooperative work and multi-media, parallel processing of data, micro-systems and new sensor control devices etc.

Application areas of these developments are, e.g.: CAD/CAM/CNC, power plants and chemical engineering, textile industry, food and consumer goods production, aircraft and air traffic control, train and car automation.

In these approaches to automation, the ruling concept of the human role can be seen as follows:

The control system - including its assisting systems - acts, the human observes and supervises, he/she acts only in case of emergency.

However we may look at recent developments in European economy. Humans are largely the first 'sacrifices' in favour of further automation. It helps to cut down on expenses in times of economic difficulties and recessions. Despite rising unemployment the concept of increasing automation is seen by many as the way out of recession.

These developments towards increasing automation have to be considered in view of:

international and tribal warfare, social egotism, unemployment and poverty, environmental catastrophes and new kinds of crime and sabotage - just to mention a few of the symptons of increasing chaos.

Recent attempts seem to have failed to control this chaos according to rational reasoning or through imposing power structures - whether democratic or military power. The question may be asked: Can the present concept of automation be sustained in the long term and worldwide?

2.   Some Aspects of Chaos

The concept of Chaos came into public discussion through the work of E. Lorenz (USA). He calculated long-term weather changes. He discovered that small changes of start data were leading to completely different weather patterns. The behaviour of weather is, therefore, 'chaotic' because it is impossible to predict weather patterns beyond a certain time span. Sufficient data would never be available to calculate long-term weather patterns. However, complex systems - e.g. the weather - are likely to pass through a series of certain well-defined states ('attractors') while avoiding other states. Hence certain system states show higher probability than others.

Under certain conditions, more than one 'stable' state can exist in parallel. The system can be in any one of these different states at any time. It can change from one such state to any other without obvious 'reason'. It can be triggered by a minute disturbance, e.g. the wing movements of a butterfly.

It seems to be possible to cautiously transfer these conceps - among others - into considering a social system, e.g. a factory or a nation [1]. Under conditions of outside influence and inner uncertainty, the system may become destabilized. It may switch to a completely unexptected state, e.g. Eastern Europe, moving into tribal warfare; a factory bursting into fierce strikes or sabotage etc.

On the other hand, such 'switching' may contribute to increasing freedom of development. Let us consider systems which appear deterministic, or fixed in 'eternal ' patterns - suddenly they may become open for change. It may be seen as the task of society today:
a) to educate people to cope with such sudden changes,
b) to establish societal patterns which allow or encourage changes and flexibility
c) to develop technologies which do not become obsolete through changes but can easily be adapted to changes.

In the following two paragraphes, these concepts are applied to manufacturing and air traffic control.

3. An Example from Manufacturing

Today automation appears particularly strong in manufacturing. The un-manned factory is supposed to be unfeasible, but all control concepts referred to in the introduction, aim at increasing automation in production. However there are certain areas where people are still needed: feeding CAD, drafting CAM, maintaining CNC etc. But less and less people are needed for more and more productivity.

The social and cultural aspects of this development have been extensivly discussed e.g. by the IFAC Committees corresponding. This kind of automation only works in the long run if sufficient people remain in the factory who do exactly the jobs needed to keep the automated systems working. However the working population around this kind of production are being deskilled, and they unlearn their jobs if they are only challenged in cases of emergency or maintenance.

They loose - or never develop - "the willingness and ability to act in responsible ways." [2] Thus in a chaotic development, such factories may be the first to come to a standstill.

A different kind of control seems to be needed to counteract this danger. The workers need to be given control of the production process while making use of advantages of CNC technology. This concept has been followed by several European projects within ESPRIT, FAST, COMETT etc.

As an example, the proposal will be desribed here to use analogue input devices for CNC, e.g. conventional control systems such as manually controlled handwheels. Their movements more closely correspond to the production process, than digital/graphics input devices. The analogue input devices read data electronically and transfer them to the graphical representation of the workpiece and to the simulation of the production process [3]. The handwheel drives the simulation of the production process on the screen. Thus the worker can 'produce' the workpiece in all details as he is used to do on a conventional system. The computer stores and processes all data to subsequently control the 'real ' production process. But the control system can also be used through keyboard input like any other CNC system. Presently the system is being developed by a German-French consortium (Keller-NUM). Shopfloor users are fully involved into the devolopment process. For these systems, there is no automated CAD/CAM/CNC link, and no automated control of production flow. Once again, humans are responsible for production. They are not merely serving the machines.

The ESPRIT-Working Group HERMES (Helping Europe Revitalize Manufacturing: an Educational Strategy) has integrated this concept - among others - into a Europe-wide action plan. The core of this project is symbolized by the table following here [4].

4. Control of Large-Scale Systems - Networks of People and Technology

Today, it is necessary to optimize the human-machine system "air-traffic control" in such a way that the airspace capacities can be increased without decreasing safety and reliability. However should the human remain within the air traffic control system? As an example, automated direct data transfer ground-aircraft would be faster than personal communication between the (human) controller and the (human) pilot. Thus the humans would be outside the control process except to calculate the route, in advance. There would be nobody left to take charge in case of emergency. Those humans remaining in the system may un-learn how to respond to emergencies.

Therefore it is suggested to follow the Dual-Design Approach in order to design a human-centred ground-air communications system [5].

The Dual Design Approach is a set of principles to ensure appropriate development of both technical and human aspects of human-machine systems. The fully automated concept is represented by the left-hand triangle of the figure. Here, the major part of design efforts, creativity and research is used to obtain a fully automated system. However certain elements of the system cannot be fully automated. Humans have to become part of the concept. The-

Table: CIM Trends today

Old concepts	Currently dominant concepts at the end of their useful life	Emerging concepts likely to dominate the future
		clean manufacturing; green products
small operations	economy of scale, rigid production lines	flexible systems
	deep management hierarchies	team work; lean management
management by exception	management by objectives	management by learning, continuous improvement
	low cost products	innovative products; high value added products
	sequential product introduction process	concurrent product introduction process
	machine efficiency	lean manufacturing
	quality control	continuous quality improvement; TQM
technology-driven IT solutions	business-driven IT solutions	generic packages and systems
independent programs	centralised information systems	networks of PCs; distributed systems
islands of automation	hierarchical "CIM"	humans controlling "CIM"

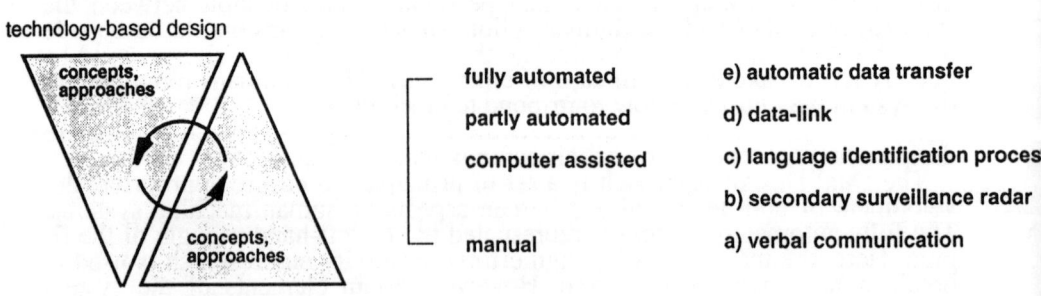

Figure: The Dual Design Approach applied to ground - air communication

refore, it is necessary to introduce a second approach, the working process based design in order to consider the human work as well. The working process based design raises the issue of how to solve the problem without or with a lower level of automation. This will result in a concept where tasks are performed by people. It means that the main part of design efforts, creativity, and ideas will be put into this approach (see figure, right-hand triangle).

Both the technology based design and the working process based design should be used in parallel to obtain an optimum. It is complementary in the technology based design and in the working process design. The weakness and the advantage of both concepts have to be considered.

This approach can be applied to ground-air communication. The level of "manual" corresponds to verbal communication. Verbal communication via radio-telephone is frequently the only available device. It is also the fallback option if other technical systems fail. Moreover, such verbal communication contains additional information that could not be substituted by automatic data link. It is carried by articulation and ways of speaking. The analysis of this Dual Design Approach has been performed in cooperation with Eurocontrol (NL). It leads to the following suggestion: Data-link with the possibility of verbal communication in parallel serves best in order to increase air traffic capacity without loss of safety. The data which are transferred via data-link, however, should not be transmitted directly to the auto-pilot of the aircraft. First, an examination and confirmation of data and of control strategy are to be performed by the pilot. This issue is the theme of several Europe-wide projects and networks [6].

Another problem becomes apparent when automated aircraft systems cause separation from the real world. This may happen, e.g., when a flight crew "flies" the small symbol of the aircraft on the navigation display along the coloured line representing the course, rather than flying the aircraft through the real airspace. The crew may not belief in the real mountain coming up in front of the aircraft window.

This experience has become a very recent issue: the accident of an Airbus A 320 (LH) at Warszawa (14.09.1993). It was admittedly caused by the computer system which did not allow the pilots to correct wrong data processing of the automated landing control devices. The computer had been fed with weather data which did not correspond to reality outside the landing aircraft. There was no way for the pilots to overrule these 'wrong' decisions of the control system. This problem of too much automation has been discussed among Airbus designers and experienced pilots, for several years.

On the basis of these and further experiences, the insight is slowly gaining relevance that the fully automated aircraft is unlikely to be feasible. Hence the question in designing control technologies is how to automate air transport. Such systems are to be designed to combine both people (e.g., the pilots) and control technology. It is shown by the Dual Design Approach that the maximum use of technology may mean full automation - but nobody would like to fly across Europe in an aircraft without a pilot controlling it. In the future, aircraft are likely to depend rather more than less on the flight crew because of their ability to deal with uncertainty - or what we may call 'chaos'[7].

5. Conclusions

These two examples from manufacturing and air traffic control show that
technology is very sensitive to chaos: in manufacturing, the danger of unlearning und deskilling is apparent in many societies; in air traffic control, the
weather is only one outside influence (among many others) which cannot be
incorporated into the systems' predictions. In view of these experiences, the
concept of automation may need to be reconsidered.

References

[1] I. Isenhardt, Change and Innovation in Complex Organisations. HDZ/KDI, RWTH
     Aachen 1994 (in German).
[2] T, Martin, J. Kivinen, J.E. Rijnsdorp, M.G. Rodd and W.B. Rouse, Appropriate Au-
     tomation - Integrating Technical, Human, Organizational, Economic and Cultural
     Factors. *Automatica* **27**, 6 (1991), 901-917.
[3] P. Fuchs-Frohnhofen, Worker-oriented Design of CNC Control. HDZ/KDI, RWTH
     Aachen 1994 (in German).
[4] G. Rzevski et al, CIM Trends. ESPRIT Project 9158: HERMES, 1993.
[5] K. Henning and B. Ochterbeck, Dualer Entwurf von Mensch-Maschine-Systemen.
     In: P. Meyer-Dohm et al (Eds), Der Mensch im Unternehmen. Bern, Stuttgart 1988,
     pp. 225-245.
[6] B. Harendt, Dual Design of Computer-based Air Traffic. 5th Int. Congress HCI, Or-
     lando (USA), 08.-13.08.1993, **19 A**, 1993, 398-403.
[7] T. Hancke and R.J. Braune, Human-centred Design of Human-Machine Systems and
     Examples from Air Transport. IFAC Congress, Sydney (Australia), 18.-23.07.1993,
     7, 343-346.

Advances in Agile Manufacturing
P.T. Kidd and W. Karwowski (Eds.)
IOS Press, 1994

Models of Design for Concurrent Engineering

Martin Helander
State University of New York at Buffalo
Buffalo, New York, USA

Abstract. The purpose of this paper is to present research relevant to the development of a model of human design behavior. Such a model can be useful in conceptualizing how individuals or groups can work productively with concurrent engineering. It is also useful for the construction of design aids. Below we will first discuss some of the assumptions made in engineering models of design. We then present descriptions of human design behavior. Finally, we characterize user requirements in concurrent engineering and propose some design aids.

1. Introduction

Engineering models view design work as a well ordered transformation from the formulation of a design problem to the selection of a solution. Typically, there are several distinct stages leading from problem formulation to the solution stage, for example, (1) problem structuring, (2) preliminary design, (3) design refinement, and (4) detailed design (e.g. [1] and [2]). Meister [3] acknowledged that many designers deviate from this process and that such deviations reduce the effectiveness of design. He also suggested that engineers' biases, such as the tendency to extrapolate from similar or parallel situations, intuitive thinking, and non-Bayesian decision making are some of the factors which may produce less effective design solutions.

Engineering models are not adequate. They are normative and suggest what should happen during design, but provide no insight to how designers actually design. Many studies have noted the great variability in design behavior between designers and the vast differences in design outcome. But this seemingly planless variability is typical of creative design, and to restrict it could mean to restrain creativity.

The common emphasis in present engineering research is to analyze deterministic design scenarios ([4] provides many examples). Many studies propose artificial intelligence and other computer routines which can generate design solutions. A common problem with these studies is that they address overly specialized or overly structured scenarios. But in real design work many goals remain unspecified and constraints are negotiated. Indeed Simon [5] characterized design problems as ill-structured and contrasted them with well-structured problems.

This paper summarizes some behavioral studies with the purpose of explaining the design process.

2. Behavioral Studies of Design

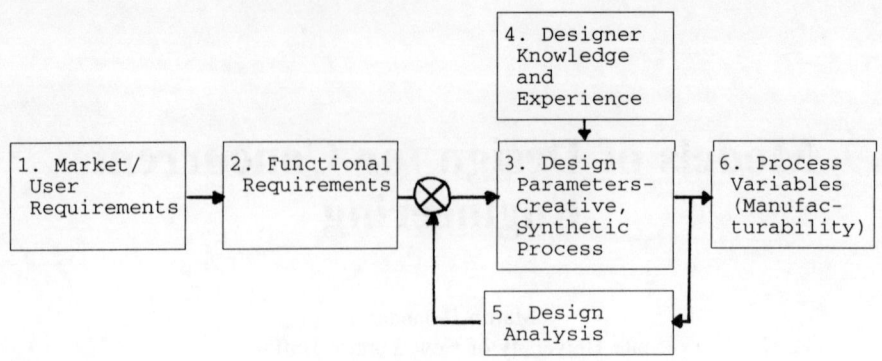

Figure 1. The gradual development of design.

The design process is a gradual development, see Figure 1. In the best case market and user requirements (1) are understood at the outset. This may be the case in the design of a new model of an existing product. Design goals are formulated, which are expressed as functional requirements (2). Design parameters (3) or design solutions are then found which satisfy the functional requirements. This is the actual design stage which results in a physical artifact. The design work is a synthetic, creative process, and the success thereof depends on the designers knowledge and experience (4).

The design is analyzed to understand to what extent the functional requirements are satisfied (5). The analysis can rely on design rules and constraints and can be automated [6]. The feedback from the analysis is used to modify the design.

The design process can be continued. In stage 6 process variables (for manufacturing) are sought that can satisfy the design parameters. Thereby concurrent engineering or life-cycle variables are introduced.

Much of the past behavioral research is based on the study of cryptarithmetic problems many of which were formulated by Newell and Simon [7] (e.g., Tower of Hanoi; DONALD + GERALD = ROBERT. Solve for D = 5). These are well-structured problems. There is a complete specification of the goal and one correct design solution, and hence they do not qualify as design problems.

Table 1 summarizes some characteristics of design activity. In real design the structure of the task and the goals are not well specified in advance. Many of the constraints are non-logical and the designer has the option to negotiate these constraints and formulate design goals that fit his/her knowledge and experience.

Because of lack of information on market requirements there are no right or wrong answers. Direct feedback during design is also lacking, and the rules for evaluation that the designer uses become personalized.

Goel [8] analyzed design behavior in architectural and mechanical design. He used a division of four major design phases: problem structuring, preliminary design, refinement, and detail design, and the design work was divided into six different activities: people, purpose, resources, behavior, function, and structure. During the first two design phases, designers have a strong interest in the people aspect, the purpose and the available resources. This is due to the loose definition of the design task and the need to establish goals and constraints. Therefore, the designer will initially seek information from the outside to clarify or negotiate design goals. At the last stage, "detailed design," the goals have been formulated (although they may still be the wrong goals). Behavior is then focused on the structure of design [9], [10].

It has been noted by a number of people that designers decompose design solutions into "leaky" or sparsely connected "modules." Goel [8] pointed out that designers had two main strategies for dealing with these interconnections: (a) they either made functional assumptions about the interconnected modules to disconnect the modules, or (b) put the current module on hold and attended to an interconnecting module.

Table 1. Design Problems are Characterized by the Following Factors.

A.	Design problems are generally large, complex and ill-structured. There is incompleteness of information including the goal state.
B.	There are two types of constraints: (1) Logical, such as natural laws. These are not negotiable. (2) Socio-economic rules and conventions. These are negotiable. Most task constraints can be negotiated.
C.	Due to the lack of feedback from the world during design, there are no right or wrong answers. Feedback occurs after the design is complete.
D.	The components of design problems are not logically connected. Due to the lack of structure the decomposition of the design is dictated by the experience of the designer.

Goel [8] also observed that only a very small percentage of design decisions (1.3%) were generated by deductive inference. Most decisions seemed to be the result of memory retrieval and modification of previous solutions (case-based reasoning). The designers response repertoire, knowledge and experience, are therefore instrumental.

Guindon's studies [11] can be interpreted with the same frame of reference. She investigated the design behavior of programmers writing a complex piece of software to control a set of elevators. The programmers defined several subtasks which they solved one after the other (depth first). Occasionally, they encountered problems that were not directly related to the subtask. They would then temporarily abandon the subtask at hand to solve the new problem. Guindon referred to this behavior as "opportunistic." With Goel's framework only a very small percentage of design decisions are based on deductive inference. Being side-tracked would, therefore, not disturb the designer's line of thought. In fact, Guindon's designers fit rather nicely into the case-based reasoning scenario, where problems are dealt with one at a time as they "present themselves."

The behavior is then a matter of random search of memory structures with rapid associations and evaluations. There is nothing opportune in this behavior, rather it is dictated by the ill-structured nature of the design task.

3. The Importance of Visualization

During the initial problem structuring designers solve most problems by reasoning and natural language. However, in the preliminary design phase sketches of design solutions become increasingly important. These sketches are refined during the following two stages of design refinement and detail design [12]. Goel [8] compared freehand sketches to sketches produced with Mac Draw. Most of his test subjects preferred freehand sketching over Mac Draw, since the computerized system imposed procedures that made the drawing task more difficult. In contrast, freehand drawings were found to be semantically and syntactically more dense and they were therefore more informative than sketches made by Mac Draw.

The level of abstraction in design is used to describe the gradual refinement in designing an artifact [13]. Typically there are four to five levels: functional purpose, abstract function, generalized function, physical function, and physical form. A design artifact can also be partitioned into up into its component parts ranging from the "whole system" to the "component level." To study designers' behavior, design actions can be obtained through verbal protocol data and a design trajectory is plotted in a two-dimensional diagram, see Figure 2. Typically, the point of origination is the upper left, where the goal or the purpose is related to the whole system. In most cases, the design trajectory jumps back and forth along the diagonal connecting the upper left and lower right corners. This has the implication that designs progress in an ordered fashion [14].

Much of the designer's knowledge is tacit knowledge which can be amplified by visual images rather than words. Miller [15] argued that, "thinking in images is an essential ingredient of scientific research of the highest creativity." Albert Einstein claimed that he rarely thought in words at all. His visual and "muscular" images had to be translated "laboriously" into conventional verbal and mathematical terms [16]. Visual aids such as those exemplified in Figure 3 are helpful, particularly if they are relevant to the abstraction level and the problem at hand. Guindon [17] suggested how software designers could profit from the use of similar aids that are made easily made available on the screen. Such design aids must be flexible to follow the designers path and allow a smooth transition between the different levels of abstraction.

It seems likely that the difficulty of a design could be classified using the abstraction matrix. For example, VLSI design has been greatly simplified since the elements of design are modular chips that perform generalized functions. Such design rarely involves components (transistors and resistors) at the lower level. Thereby the number of abstraction levels have been reduced.

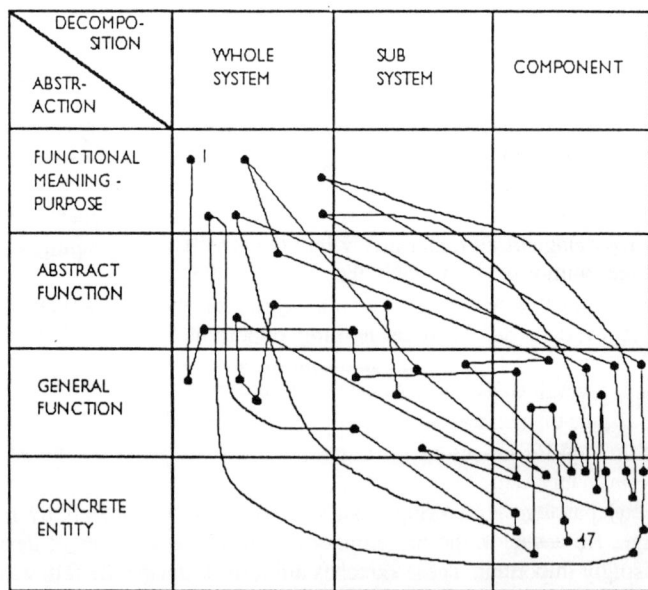

Figure 2. Design trajectory from designing a vehicle for the handicapped. The trajectory illustrate the predominance of diagonal elements [18].

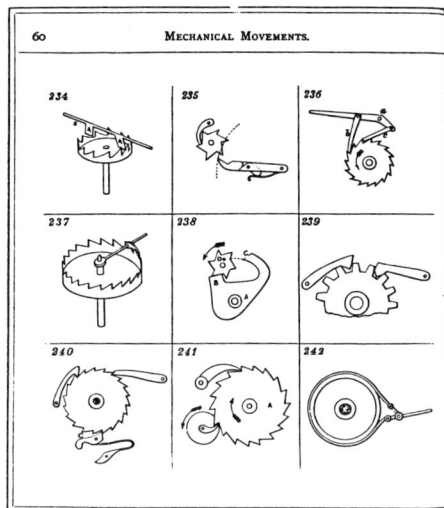

Figure 3. Visual aid for design at the component level. From the
book <u>Five Hundred and Seven Mechanical Movements</u>
by H.T. Brown [19].

Deviations from the diagonal can also be effortful. For example, a design that is
considered for its "functional purpose" at one or several a subsystem levels is likely to be
much more complex than a design where only the "whole system" has to be considered.

4. Implications for Concurrent Engineering

Concurrent engineering incorporates product life cycle requirements. For example:
manufacturability, safety, maintainability, and reuse. Figure 1 illustrates the design process
as composed of four steps: market requirements, functional requirements, design
parameters, and process variables. This figure incorporates concurrent requirements in that
the functional requirements are evaluated with respect to the physical design parameters as
well as the impact on the manufacturing processes. According to Suh [6], there should be
a one-to-one correspondence between functional requirements and design parameters and
between design parameters and process variables. In addition, the "information" in the
design should be minimized. For example physical parts with a large tolerances are
preferred over small tolerances which are more information-rich and difficult to
manufacture. These rules, which are referred to as design axioms, are helpful for the
analytical evaluation of a proposed design [20].
   There may be several ways to incorporate concurrency in the design process: 1. as
constraints, 2. as part of market/user requirements, 3. as functional requirements, or 4. by
adding a separate step following the design parameters. There is no formula for what is
appropriate. It probably depends on the type of life cycle parameter. For example to
"design for safety," many safety rules can be expressed as constraints (e.g., don't create
pinch points). But in a nuclear power plant one important functional requirement is: "no
unsafe design." This has ripple effects for the following design activity. In planning for
concurrent design, it is appropriate to bring up the design model (e.g., Figure 1) and
discuss how the constraints and concurrent requirements, can be dealt with. The
appropriateness of the design model depends on the social design scenario, such as if there
are several experts collaborating in a group, or if there is a single designer.

Design is a process of negotiation, where constraints and functional requirements can be redefined. Using the definition of design in Table 1 and assuming a negotiative process of design, we concur that design automation and artificial intelligence are of limited utility [21]. However, design aids based on the functionalities expressed by the abstraction hierarchy would be appropriate. These would serve as "look-up figures and tables," and could inspire the designer to seek better solutions.

References

[1]   J.M. Balley, An Experimental View of the Design Proces. In W.B. Rouse and K.R. Boff (Eds.), System Design: Behavioral Perspectives on Designers, Tools and Organizations. North-Holland, New York, 1987.
[2]   V. Hubka and W.E. Eder, Theory of Technical Systems. Springer-Verlag, New York, 1988.
[3]   D. Meister, Human Factors: Theory and Practice. Wiley, New York, 1971.
[4]   A. Kusiak (Ed), Intelligent Design and Manufacturing. Wiley, New York, 1992.
[5]   H.A. Simon, The Structure of Ill-Structured Problems, Artificial Intelligence 4 (1973) 181-201.
[6]   N. Suh, The Principles of Design. Oxford University Press, New York, 1990.
[7]   A. Newell and H.A. Simon, Human Problem Solving. Prentice-Hall, Englewood Cliffs, NJ, 1972.
[8]   V. Goel, Sketches of thought: A study of the role of sketching in design problem-solving and its implications for the computational theory of the mind. Dissertation. University of California, Berkeley, CA, 1991.
[9]   C. Alexander, Notes on the Synthesis of Form. Harvard University Press, 1974.
[10]   H.A. Simon, How Complex are Complex Systems? Proceedings of the 1976 Biennial Meeting of the Philosophy of Science Association, 2 (1977) 507-522.
[11]   R. Guindon, Designing the Design Process: Exploiting Opportunistic Thoughts. Human-Computer Interaction 5 (1990) 305-321.
[12]   D.G. Ullman, S. Wood and D. Craig, The Importance of Drawing in the Mechanical Design Process, Computer and Graphics 14 2 (1990) 263-274.
[13]   P. Vora and M. Helander, A Review of Design Methods and a Proposal for Taxonomy of Design. In: M. Helander (Ed.), Design for Manufacturability. Taylor and Francis, London, 1992.
[14]   J. Rasmussen, Personal Communication, University of Toronto, December 1993.
[15]   A.I. Miller, Imaging in Scientific Thought: Creating 20th Century Physics. MIT Press, Cambridge, MA, 1986.
[16]   M. Pines, Are We Left-Brained or Right-Brained? New York Times Magazine (September 9, 1973), 32.
[17]   R. Guindon, Requirement and Design of Design Vision, an Object-Oriented Graphical Interface to an Intelligent Software Design Assistant. Proceedings of CHI '92, 499-506. Association for Computing Machinery, New York.
[18]   C.A. Bussi, Designer's Behavior in Procedural and Creative Design. M.S. Thesis. Department of Industrial Engineering, State University of New York, Buffalo, New York, 1992.
[19]   E.S. Ferguson, Engineering and the Mind's Eye. MIT Press, Cambridge, MA, 1992.
[20]   M. Helander, Axiomatic Design with Human Factors Design Variables. Proceedings of International Ergonomics Association 1994 Congress. Human Factors Association of Canada, Toronto, 1994, in press.
[21]   T. Winograd and F. Flores, Understanding Computers and Cognition. Ablex, Norwood, NJ, 1986.

Part II
Manufacturing Paradigms

Part II
Manufacturing Paradigms

Agile Manufacturing: Key Issues

Paul T. KIDD
Manufacturing Knowledge Inc.
North American Office: 14407 Maple Ridge Place
Louisville, KY 40245, USA.
European Office: Tamworth House, PO Box 103,
Macclesfield, SK11 8UW, UK

Abstract. This paper outlines the concept of Agile Manufacturing. A definition is provided along with a description of basic concepts. A number of key issues in this new area are also explored.

1. Introduction

Manufacturing industry may well be on the verge of a major paradigm shift. This shift is likely to take us away from mass production, way beyond lean manufacturing, into a world of Agile Manufacturing. Agile Manufacturing, however, is a relatively new term, one which was first introduced with the publication of the Iacocca Institute report *21st Century Manufacturing Enterprise Strategy* [1]. Furthermore, at this point in time, Agile Manufacturing is not well understood and the conceptual aspects are still being defined. However, there is a tendency to view Agile Manufacturing as another programme of the month, and to use the term Agile Manufacturing as just another way of describing lean production, flexible manufacturing or CIM.

Many of our corporations today are under going massive transformations - reengineering business processes, flattening hierarchies, empowering people, implementing lean production concepts, etc. The list is almost endless. But none of these massive transformations, on their own or taken collectively, constitutes the implementation of Agile Manufacturing. What Agile Manufacturing really represents is the potential for a quantum leap forward in manufacturing. Instead of just chasing after the Japanese by copying their techniques in a prescriptive fashion, or implementing our own prescriptions such as CIM, in Agile Manufacturing we should be trying to achieve a competitive lead by doing something that our competitors are not doing.

Agile Manufacturing is something that many of our corporations have yet to fully comprehend, never mind implement. Agile Manufacturing is likely to be the way business will be conducted in the next century. It is not yet a reality. Our challenge is to make it a reality, first by more fully defining the conceptual aspects, and secondly by venturing into the frontier of implementation.

In this paper we will examine some of the key issues relevant to the development of Agile Manufacturing. Owing to space limitations we will only provide a very brief overview of Agile Manufacturing. The reader is referred to *21st Century Manufacturing*

Enterprise Strategy [1] and *Agile Manufacturing: Forging New Frontiers* [2] for more detailed information.

2. Definition and Concepts

The problem with a new idea such as Agile Manufacturing is the lack of a good sound definition and a set of concepts that most people would agree upon. References [1] and [2] have a reasonably common understanding of what constitutes Agile Manufacturing.

Agile Manufacturing should primarily be seen as a business concept. Its aim is quite simple - to put our enterprises way out in front of our primary competitors. In Agile Manufacturing our aim is to develop agile properties. We will then use this agility for competitive advantage, by being able to rapidly respond to changes occurring in the market environment and through our ability to use and exploit a fundamental resource - knowledge.

One fundamental idea in the exploitation of this resource is the idea of using technologies to lever the skills and knowledge of our people. We need to bring our people together, in dynamic teams formed around clearly identified market opportunities, so that it becomes possible to lever one another's knowledge. Through these processes we should seek to achieve the transformation of knowledge and ideas into new products and services, as well as improvements to our existing products and services.

The concept of Agile Manufacturing is also built around the synthesis of a number of enterprises that each have some core skills or competencies which they bring to a joint venturing operation, which is based on using each partners facilities and resources. For this reason, these joint venture enterprises are called virtual corporations, because they do not own significant capital resources of their own. This, it is believed, will help them to be agile, as they can be formed and changed very rapidly.

Central to the ability to form these joint ventures is the deployment of advanced information technologies and the development of highly nimble organisational structures to support highly skilled, knowledgeable and empowered people.

Agile Manufacturing enterprises are expected to be capable of rapidly responding to changes in customer demand. They should be able to take advantage of the windows of opportunities that, from time to time, appear in the market place. With Agile Manufacturing we should also develop new ways of interacting with our customers and suppliers. Our customers will not only be able to gain access to our products and services, but will also be able to easily assess and exploit our competencies, so enabling them to use these competencies to achieve the things that they are seeking.

3. Some Key Issues in Agile Manufacturing

3.1 The "I am a Horse" Syndrome

There is an old saying that hanging a sign on a cow that says "I am a horse" does not make it a horse. There is a real danger that Agile Manufacturing will fall prey to the unfortunate tendency in manufacturing circles to follow fashion and to relabel everything with a new fashionable label. The dangers in this are two fold. First, it will give Agile Manufacturing a bad reputation. Second, instead of getting to grips with the profound implications and issues raised by Agile Manufacturing, management will only acquire a superficial understanding, which leaves them vulnerable to those competitors that take Agile Manufacturing seriously. Of course this is good news for the competitors!

One sure way to fail with Agile Manufacturing is to hang a new sign up. Get smart, resist the temptation, and put the paint brush away.

3.2 The Existing Culture of Manufacturing

One of the important things that is likely to hold us back from making a quantum leap forward and exploring this new frontier of Agile Manufacturing, is the baggage of our traditions, conventions and our accepted values and beliefs. A key success factor is, without any doubt, the ability to master both the soft and hard issues in change management. However, if we are to achieve agility in our manufacturing enterprises, we should first try to fully understand the nature of our existing cultures, values, and traditions. We need to achieve this understanding, because we need to begin to recognise and come to terms with the fact that much of what we have taken for granted, probably no longer applies in the world of Agile Manufacturing. Achieving this understanding is the first step in facing up to the pain of consigning our existing culture to the garbage can of historically redundant ideas.

3.3 Understanding Agility

Agility is defined in dictionaries as quick moving, nimble and active. This is clearly not the same as flexibility which implies adaptability and versatility. Agility and flexibility are therefore different things.

Leanness (as in lean manufacturing [3]) is also a different concept to agility. Sometimes the terms lean and agile are used interchangeably, but this is not appropriate. The term lean is used because lean manufacturing is concerned with doing everything with less [4]. In other words, the excess of wasteful activities, unnecessary inventory, long lead times, etc are cut away through the application of just-in-time manufacturing, concurrent engineering, overhead cost reduction, improved supplier and customer relationships, total quality management, etc.

We can also consider CIM in the same light. When we link computers across applications, across functions and across enterprises we do not achieve agility. We might achieve a necessary condition for agility, that is, rapid communications and the exchange and reuse use of data, but we do not achieve agility.

Thus agility is not the same as flexibility, leanness or CIM. Understanding this point is very important. But if agility is none of these things, then what is it? This is a good question, and not one easily answered. Yet most of us would recognise agility if we saw it.

For example, we would not say the a Sumo wrestler was agile. Nor would we think that 50 Sumo wrestlers, tied together by a complex web of chains and ropes, all pulling in different directions, as agile. Quite the contrary. We would see them as lumbering, slow and unresponsive. However, we would all recognise a ballet dancer as agile. We would also think of a stage full of ballet dancers as agile, because what binds them together is something quite different.

This analogy between Sumo wrestlers and ballet dancers is very relevant to understanding the property of agility. Many of our corporations, to varying degrees, resemble Sumo wrestlers, tied together, but all pulling in different directions. If we want to develop agile properties, we need to understand what causes agility and what hinders agility. Only when we have developed this understanding can we begin to think about designing an agile enterprise. For, when we have such an understanding of the causes of agility, we can start to audit out current situation, and identify what needs to be changed.

4. Concluding Remarks

We have spent much time copying the Japanese. Now we may be about to teach the Japanese something. For a change, US manufacturing industry is realising that it has very little to gain, in the long term, by copying what other people are doing. There is now a growing realisation that global preeminence in manufacturing can only be achieved through innovation. We can learn from others, but in a highly competitive world we can only become world leaders if we develop new ideas that take us beyond the state-of-the-art. Basically, the issue is, should we adopt lean manufacturing in our own enterprises, i.e. should we mimic the Japanese, or should we do something different and better?

Without doubt there are a significant number of people who believe that we have to adopt lean manufacturing. But in adopting this approach we run the risk of forever chasing after a moving target. The Japanese will keep innovating. Thus, adopting lean manufacturing can only be a short term measure aimed doing something to close the competitive gap. In the longer term, if we want to catch up with and overtake the Japanese, lean manufacturing is not the answer. What we need to do, is something which the Japanese cannot do. That something may well be Agile Manufacturing.

References

[1]   Iacocca Institute, 21st Century Manufacturing Enterprise Strategy. An Industry-Led View. Volumes 1 & 2. Iacocca Institute, Bethlehem, PA, 1991.

[2]   Paul T. Kidd, Agile Manufacturing: Forging New Frontiers. Addison-Wesley, 1994.

[3]   J.P. Womack, D.T. Jones and D. Roos, The Machine that Changed the World. Rawson Associates, New York, 1990.

[4]   D.T. Jones, Beyond the Toyota Production System: The Era of Lean Production. In: C.A. Voss (Ed.), Manufacturing Strategy: Process and Content. Chapman & Hall, London, 1992, pp 189-210.

Advances in Agile Manufacturing
P.T. Kidd and W. Karwowski (Eds.)
IOS Press, 1994

Changing Manufacturing Paradigm: A Thematic Approach[1]

Stuart Smith, David Tranfield, Susan Whittle, Valerie Martin,
Sheffield Hallam University, Totley Campus, Totley Hall Lane, Sheffield S17 4AB, UK
Roger Maull, Stephen Childe, Plymouth University

Abstract. The paper reports the initial findings and the theoretical framework underpinning a current ACME funded research project into the approaches companies have adopted in attempting to regenerate their manufacturing competitiveness. It explores the extent to which the changes observed in manufacturing companies is attributable to shift in manufacturing paradigm and the usefulness of dissipative structures as an explanatory framework.

1. Introduction

This paper reports some of the initial findings and ideas emerging from a current research project[2] investigating companies' efforts to improve their manufacturing competitiveness. So far, the study has examined the change programmes of over 12 successfully "regenerated" companies, interviewing senior managers in both manufacturing and other functions. In all cases in depth interviews with managing director/CEO were conducted in order to begin to understand both what companies had been aiming to do in undertaking 'regenerative' activities i.e. what 'regeneration' meant for them, as well as how they had gone about it.

2. What had they done?

One area we have been exploring concerns the content of the changes companies have been seeking to implement and the labels they have been using to denote these activities. Much of the language used in describing changes is common, despite the fact that companies faced very different manufacturing situations (eg. engineer to order compared with assemble to stock). For example, of the 12 in our sample, all had used generic labels for their improvement programmes, either TQM, or a variation on World Class Manufacturing, usually with a bespoke label developed by the management team. These broad themes encompassed a range of specific improvement projects typically each one having its own three letter acronym label.

These similarities were even more remarkable in so far as the companies visited operated in differing industrial sectors, served different markets and deployed differing manufacturing processes. What are the factors underlying these commonalities? We

---

[1] This paper is informed by the Final Report of a Joint ESRC/SERC Research Project: Manufacturing Organisation for Computer Integrated Technologies by Stuart Smith, David Tranfield & John Bessant.

[2] The research reported in this paper was supported by the ACME Directorate of SERC.

considered a number of alternative propositions.

The most obvious explanation is that this is a reflection of the *fads and fashions* of manufacturing. The community of people working in manufacturing are subject to peer group pressures just like any other section of society and they are exposed to a plethora of persuasive material from consultants, conferences, journals, books etc.. From this perspective there are no fundamental forces underpinning these changes; the lurch from one *three letter acronym* to the next are merely manifestations of modish behaviour. Undoubtedly this partly explains some of what is going on; the manufacturing community are not immune to these normal social pressures. At its extreme this probably goes a long way to explain why there are so many widely reported disappointments and failures to deliver competitive edge from these change programmes.

A second possible explanation for these commonalities is that they reflect a change of *techno-economic paradigm* (Freeman and Perez [1], Dosi, [2] & Perez [3]) which itself involves a change in *manufacturing paradigm*. The paradigm defines the *common sense* rules which govern the workings, both structural and technical, of industrial society and set the pattern of best practice.

It is possible to see the currently dominant paradigm as the one which grew out of the early 20th century and took shape in the mass production factories typified by the nascent car industry. This blueprint emerged in a particular paradigm characterised by particular environmental and technological conditions and has served us well for over sixty years. But in the light of what many see as radical shifts in both environment (globalisation) and technology (integrated information and communications), it is beginning to look outdated and outmoded. The ground rules for 'best practice' in designing for a manufacturing organisation may need to be rethought.

For convenience the label 'Fordist' has been used to describe the old and increasingly inappropriate model based on mass production and rigid, bureaucratic organisational forms. The contrasting 'post-Fordist' paradigm is by no means clearly defined, but includes a number of features such as:
- greater emphasis on non price factors
- greater emphasis on flexibility in technology
- greater emphasis on flexibility in organisational structure
- changing relationships within and between organisations

Perhaps the key question for manufacturing enterprises as they move into the 1990s is how to effect the transition . Two elements would appear to be crucial for success, change based on the new technologies of information and communication, and organisational change towards new supporting structures.

The problem facing firms wishing to change is that although there is *'fuzzy' vision* of where to go to, there is no clear blueprint available defining what new organisational forms or technological configurations are needed - any more than earlier paradigms were clearly articulated at their start. Instead there is a need for experimentation, for innovation and learning. As new options are tried and evaluated both as models for 'best practice' and as alternatives to the increasingly inappropriate 'Fordist' model. One of the phrases beginning to enter the manufacturing language of the 1990s is the concept of *the learning organisation*. Given the challenge of trying to find and develop a new paradigm this ability to learn may become the critical skill determining competitiveness in the future.

Significantly the most successful post-war economy - that of Japan - is characterised by a fundamentally different model of factory organisation and operation. Arguably this arose out of the problems posed by immediate post-war conditions and represents a rethinking of the basic design rules, the creation, at least in embryo, of a new paradigm. The roots of this are undoubtedly in the old one, not least, the main sources of inspiration were writers like

Ford and Taylor. But the application and development of these took a different path and has produced, arguably, a first draft of a new blueprint better suited to the current and emerging technological and competitive environment.

Recent research on the motor vehicle industry has suggested that there is still considerable room for elaboration of the new paradigm, and that, although it was first developed in Japan, the elaboration and refinement of the model is now taking place in many locations around the world (see Womack, Jones and Roos, [4]). This mirrors the pattern of diffusion of Ford and Sloan's model for mass production which replaced craft production at the turn of the century.

In our opinion the data and findings from the companies we have visited support the argument about shifts in manufacturing paradigms. If the changes in organisation had been purely the result of changing technology or adapting to specific markets then it might be expected that the effects could be less diffuse and more specific to individual companies. This was not the case. The changes were similar across different companies, were organisation-wide, and were coherent as opposed to piecemeal. This was apparent in each picture obtained from talking to managers about their thinking and rationale for changing.

In our view what we are witnessing in many manufacturing firms' efforts to regenerate their competitiveness is an attempt to find their way out of the Fordist/mass-production paradigm. Although it is not fully articulated, the main values are defined and exemplars are available. However the 'recipe' still involves a fair degree of trial and error and it is inevitable that some 'cooks' will produce successes whilst others fail either by trying the wrong experiments or being unable to give up the ways embedded in their old paradigm.

3. How had they done it?

Although many of these companies have developed plans to enhance manufacturing competitiveness through building operational capabilities, not all have started from business strategy and cascaded this down through manufacturing and other functional strategies. Some have started with capability building and strategic consequences have emerged. Others have simply sought to address specific operational problems and have found capability emerging which in turn has had strategic implications. These initial findings illuminate some of the shortcomings of existing work on manufacturing strategy where generally it is portrayed as a process of rational planning rather than a more messy process of organisational learning where change and development can emerge from many positions from within the organisation.

Interesting theoretical perspectives to explain the regeneration phenomenon are offered in the study of transformational change in dissipative structures (Prigogine and Stengers [5]), self-organising systems and autopoeisis (Jantsch [6]). Essentially, they offer a systems perspective for understanding the transformation (emergence and self creation) of new structures by viewing human organisations as *dissipative structures* which maintain form or structure by a continuous dissipation or consumption of energy imported from the environment. By these means dissipative structures resist entropic processes. The organisation is best conceived as a 'flowing wholeness" which is the coherence provided by the dominant paradigm. This is similar to Miller & Friesen's [7] view of *quantum* changes in organisational configuration where change is not a continuous adaptive process to environmental contingencies, rather a step function shift when malalignment exceeds threshold levels.

To function as ongoing entities, organisations contain adaptive mechanisms which negate natural entropic forces. In *far from equilibrium conditions* such as those caused by environmental hyperturbulence, these are overwhelmed by internally and/or externally induced contingencies. The impact of these contingencies on the organisation is to induce chaos leading to *bifurcation point* or *point of singularity*. This is experienced by those inside the organisation as a period of potential crisis where the existing paradigm no longer meets the demands of external contingencies. This is similar to Morgan's [8] concept of *fracture lines* in explaining major change. Reaching a *bifurcation point* is a vital precursor either to transformation, or to greater perturbation and death.

Given the changes in the internal and external contexts of manufacturing companies, it is our view that the above theoretical concepts adequately describe the main processes which are underpinning the empirical observations described in our research. Contextual hyperturbulence over a protracted period on a wide variety of fronts is straining the adaptive capacity of the organisation of most manufacturing companies. Many are reaching the elastic limits of their current structural arrangements, and often experience the 'bifurcation point' as being attributable to one specific event which is experienced by management as the latest in a long line of changes which finally triggers crisis. John Parnaby of Lucas Industries has referred to this as *a significant emotional event.*

As this point is reached, the organisation finds itself in a particularly chaotic state. Davis [9] has argued that change comprises three states:-

i)   You don't know that you don't know   -   **IGNORANCE**
ii) You do know that you didn't know   -   **INSIGHT**
iii)   You know   -   **UNDERSTANDING**

The movement from i) to ii) can be described as the process of transformational change or breakthrough. It involves confusion and chaos, search, and intellectual exploration. It is a creative process, and, from the participants viewpoint, is a change of *mindset* or *weltanschauung*. The movement from ii) to iii) can be described as transitional change. It involves education, alternative generation, choice and implementation. It is a logical process, and often a process of putting into action that which is rendered obvious by the transformed mindset. The manufacturing companies which we observed could be interpreted as undergoing or having undergone transformational and transitional change.

4. References

[1] Freeman C & Perez C, Structural crises of adjustment, business cycles and investment behaviour', in Dosi G, (ed), Technical change and economic theory', Frances Pinter, London, 1989

[2] Dosi G, Technological paradigms and technological trajectories', Research Policy, 11, 3, 1982

[3] Perez C, Structural change and the assimilation of new technologies in the economic and social system', Futures, 15, 4, 1983

[4] Womack J, Jones D, & Roos D The machine that changed the world, Rawson Associates, New York, 1990

[5] Prigogine I & Stengers I, Order out of chaos, Bantam Books, Toronto, 1982

[6] Jantsch E, The self organising universe, George Braziller, New York

[7] Miller D & Friesen P, Organisations: a quantum view, Prentice Hall 1984

[8] Morgan G, Creative organisation theory: a resource book, Sage 1989

[9] Davis S, Transforming organisations: the key to strategy is context, Organisational Dynamics, 1982

Advances in Agile Manufacturing
P.T. Kidd and W. Karwowski (Eds.)
IOS Press, 1994

TAYLORISM AND LEAN PRODUCTION

D.G. Elton, Pall Europe Ltd., Walton Road, Portsmouth;
S.V. Madgwick and Dr. V. Newman, The CIM Institute, Cranfield University, Bedford

Abstract. If manufacturing films followed the conventions of cowboy films, Frederick Taylor would wear a black hat. To a Japanese audience, on the topic of manufacturing and lean production, Frederick Taylor's hat is clearly white. This paper presents a classic view of Taylor's work then examines why production guru's like Shigeo Shingo should regard Scientific Management with such esteem. A scenario of failure in the West looms as attempted lean implementations omit fundamental issues behind lean philosophy. The reasons for this failure are built into Western Management Culture, giving rise to myths which must be confronted.

1. What is Wrong With Taylorism?

Waterman (among others) views Taylor's work as "the starting point for the limited way we think about management", suggesting that "Taylor saw workers as robots" [1]. Until recently we have been unable to discuss Taylor's contribution to the way we think about work without ritually condemning him. In the light of lean production, our perception of work, and how it can be done, has changed. Only 6 years ago in 1988, Michael Rose could speak dismissively of what he termed "Productivism ... the conviction that the overriding business of life is to produce more goods more efficiently" [2]. Now, with the competitive evidence of Lean Production Systems, our discussion has changed. We no longer need to ask why should we go lean? but instead, how do we become lean? Even so, few observers can see, or would accept, the strong links between lean philosophy and Scientific Management.

2. Scientific Management

2.1. Mutual Benefit

According to Taylor "the principal object of management is to secure maximum prosperity for the employer, coupled with maximum prosperity for the employee" [3]; He insisted on the alignment of pay rises with productivity gained through Scientific Management. His philosophy was: the right man doing the right work in the best way, for a fair reward.

Taylor observed 'the workman trying to make his manager think that he is doing a large amount of work whilst actually doing the least' [4]. It was as common then as it is now. He termed this 'soldiering' and saw three main causes:

1. Redundancy - avoiding becoming too efficient and potentially making others redundant.
2. Reward - self preservation by not working harder without financial gain.
3. Inefficiency - through sticking to misguided rule of thumb methods.

2.2. Overcoming Soldiering

Today we know all to well that remaining inefficient, rather than becoming too efficient, is more likely to result in redundancy. Unfortunately reason alone will not remove the other aspects of soldiering; hence Taylor pioneered 'time-study' and 'method-study' to combat rule-of-thumb methods. Many believed that work-study alone could speed up production, but without financial gain the 'self preservation' aspect of soldiering proved difficult to overcome.

Taylor intended to bring the workers closer to management, to have a common goal, and developed the Four Principles of Scientific Management [5]:
1. Scientific development of work practices - to demolish old rules of thumb.
2. Scientific selection - to engage people in work that they are capable of and find fulfilling.
3. Individual training - by specialist foremen to provide workers with the best method.
4. Co-operation - between management and the workers.

Taylor included 'co-operation' because he knew how dangerous his revolutionary techniques could be if used with the wrong attitude. "The knowledge obtained from time study is a powerful implement, and can be used in one case to promote harmony between the workmen and the management, or, in the other case, as a club to drive the workmen into doing a larger day's work for approximately the same pay" [6].

Taylor's 'time and motion' ideas (similar, he notes, to Gilbreth's) were adopted throughout the Western world, without considering the true objectives of Scientific Management. Most managers saw their goal as achieving maximum prosperity for the employer, but gave little thought to the workforce. Implementations failed on the misconception that it was only necessary to measure and increase performance and, consequently, 'Taylorism' was rejected.

2.3. Taylorism and Lean Principles

Of those who have commented on the errors of Taylorism, few can have read his works. Shigeo Shingo, a key figure in the rise of Japanese industry writes:

> 'In 1931, I ran across a translation of Taylor's book (The Principles of Scientific Management). Thumbing through it, I found a most unusual statement. "Inexpensive goods", it said, "can be produced even when workers are paid high wages." The apparent impossibility of such a proposition aroused my suspicions, and as I continued to leaf through the book, I saw that Taylor claimed the feat was possible if efficiency was raised to a high level. For me, this argument was utterly novel, so I bought the book and did not sleep until I had read it from cover to cover. At that point I resolved to devote my life to scientific management" [7].

Shingo, among others, was influential in establishing The Toyota Production System, and hence the origins of lean production. It is of considerable significance that he should attribute such praise to a philosophy denounced in the West. How can the same concepts be viewed so differently and implemented with such contrasting results?

Even Taylor may have had difficulty in explaining the remarkable achievements of people such as Shingo. If 'Taylor saw workers as robots' can we put Japanese success down to a nation of Robots? In answer to this, two reasons why Japanese industry was forced to look for 'leaner', more flexible alternatives to Mass production were: Japanese domestic markets refused to accept 'mass produced' limited variety; and, their work force was not prepared to become an interchangeable resource, as at classic mass producing sites [8]. Some robots. So why did Taylorism succeed in Japan and fail in the West? The answer to this paradox lies not only in what the West omitted from Scientific Management, but equally in what Shingo and others added, perhaps almost unwittingly.

3. The Route to Lean Production

3.1. Adding Value to the Workforce

The Japanese have redefined the concept of waste and created various techniques to remove it. The question here, however, is how did they succeed in implementing Scientific Management and use it as a spring board for their 'Revolution in Manufacturing'.

Taylor attempted to increase skill, motivation and responsibility with regard to raising each individual to the highest state of efficiency. The Japanese saw further benefits through developing the workforce to monitor, analyse and maintain both the equipment they operate and the products make. They have added value not only to their products but also to their employees. The concept of 'ownership' becomes a key issue in motivating the workers to 'own' problems, and, transferring responsibility so they can make a difference.

3.2. Is History About to be Repeated?

If we fail to understand these fundamentals of Lean Production we will repeat the mistake made with Scientific Management. Konosuke Matsushita, founder of one of Japan's largest companies, is adamant that Western management style constitutes failure:

> "We are going to win and the industrial West is going to lose out; there's not much you can do about it because the reasons for failure are within yourselves. ... With your bosses doing the thinking while the workers wield the screwdrivers, you're convinced deep down that this is the right way to run a business. For you the essence of management is getting the ideas out of the heads of the bosses and into the hands of the labour. We are beyond your mindset. Business, we know, is now so complex and difficult ... that continued existence depends on the day-to-day mobilisation of every ounce of intelligence" [9].

Matsushita uses the abilities of every employee for continuous improvement. Although Taylor did not think that operators would be able to scientifically analyse the operations they performed, he was not opposed to them making suggestions:

> "It is true that with scientific management the workman is not allowed to use whatever implements and methods he sees fit in the daily practice of his work. Every encouragement, however, should be given him to suggest improvements, both in methods and in implements" [10].

3.3. Transferring 'Ownership' of Problems

The reserved nature of Western people, particularly in Britain, makes it difficult to enthuse employees so that they adopt 'ownership' and responsibility for their work. Japanese culture, being less reserved, has perhaps contributed toward their success, verses our failure with the same basic philosophy. However, enthusiasm or not, if you assume that management owns the work and the problems it generates, and that the workforce is just another component within the manufacturing process, then 'soldiering' will result and it's consequence is a management culture that demonstrates its power and ownership through what seem to be arbitrary decisions.

The route to 'ownership', through Employee Involvement, comes in many forms: Total Quality Management, Kaizen, Total Productive Maintenance, and other continuous improvement programmes. We can use these techniques for removing waste, much in the same way as Taylor's techniques were used, but the same problems will hinder success. Unless we provide for mutual benefit and focus on the individual, there will be much wasted effort and expense. Financial gains are a prime motivator, but Taylor's suggests other incentives including promotion, shorter hours, and better working conditions [11].

At some lean manufacturing plants the only 'reward' within a successful continuous improvement programme is a high level of recognition when the improvement team present their idea and results to upper management. Benefit to the employee need not be spectacular, but should note Taylor's stipulation that, to be effective, a reward must come soon after the work has been done [12].

4. Conclusion

Today the general view of Taylorism is 'how can we overcome it?' By concentrating on the works of Taylor, and how they have been abbreviated, we can understand why this misconception has arisen. The task facing us now is to take advantage of this knowledge in order to enthuse employees with the newer and better 'lean' principles.

It has been said that when you hire a pair of hands you also get a brain, completely free [13]; With so many new, 'leaner', techniques requiring employee enthusiasm at the centre, we cannot afford to be without their co-operation. If we take lean thinking purely for its productivity gains then, as with Taylor's work, we will once again forsake the real benefits.

[1] Waterman, R.H. (1990) Adhocracy. New York, W.W. Norton, pp22-3.

[2] Rose, M. (1988) Industrial Behaviour - Research and Control. Harmondsworth: Penguin Books, p338.

[3] Taylor, F.W. (1911) Principles of Scientific management. New York, Harper & Brothers, p9.

[4] Taylor, *op.cit.* p33.

[5] Taylor, (1911) *op.cit.* p130.

[6] Taylor, (1911) *op.cit.* p133-4.

[7] Shingo, S. (1987) The Sayings of Shigeo Shingo. Massachusetts: Productivity Press, ppxv-xvii.

[8] Womack, J.P., Jones, D.T., Roos, D. (1990) The Machine that Changed the World. New York: Rawson Associates, pp49-50.

[9] Pascale, R.T. (1990) Managing on the Edge. London: Viking, p27.

[10] Taylor, (1911) *op.cit.* p128.

[11] Taylor, (1911) *op.cit.* p33-4.

[12] Taylor, (1911) *op.cit.* p38.

[13] Ainger, A. (1993) Human Centred Systems conference, Nov'93, The CIM Institute.

Advances in Agile Manufacturing
P.T. Kidd and W. Karwowski (Eds.)
IOS Press, 1994

Problems of Post Tayloristic Rationalization Strategies - Work in Globalized Production

Hartmut Hirsch-Kreinsen

ISF, Jakob-Klar-Str. 9, 80796 München, Germany

Abstract. The thesis of this paper is that the current changes in industrial work do not at all lead to a new "one best way". Indeed, many different development paths of industrial work are identifiable. An important reason for this are the contradictory rationalization requirements which result from the ongoing globalization of production. The following arguments are part of an ongoing international research project at the ISF, Munich on globalization strategies in the capital goods industry.

1. Development paths for industrial labor

At present, the development of industrial production processes is characterized by a departure from Tayloristic rationalization principles in almost all industrial sectors. As the current debate on the "lean production" model shows, increasingly those management concepts are being pursued that strive toward such goals as the flattening of hierarchies, a company wide decentralization process, a more systematic use of skills and group work structures, and a continuous improvement and optimization of production processes. Contrary to the claims and hopes of the "lean production" debate, however, these new management concepts do not get implemented wholesale in company reality. As an entire series of new empirical investigations demonstrate, a "plurality of new rationalization forms" can be found in nearly all western industrialized nations at present, indicating an irregularity and nonconcurrence of the current situation of crisis.[1] The result is that the development of industrial work is characterized by a broad spectrum of different development paths with varying consequences for work organization, tasks, and skills. At least three such paths of development can be identified: first, a "neo-Tayloristic" path; second, a development path of "skilled and cooperative" industrial work; and third, the path of "polarized" industrial work.[2]

The reasons for this situation lie in the numerous contradictory influences that companies are being confronted with at this time. As a result, diverging orientations and reactions present themselves to the companies which are more or less mutually exclusive. For example, opposing requirements result out of the contradiction between the turbulent demands of the sales markets which exert enduring pressure for changes in company structures on the one hand, and the persistance of deeply rooted Tayloristic organizational structures in the company on the other. Another example is the unpredictable development of the labor market, especially the very uncertain supply of skilled production workers in the future. Finally, very few companies have either the necessary personnel or financial resources, nor do they possess the know-how which would be required for a fundamental reorganization or their traditional structures and the realization of new management concepts.

2. Contradictions of globalized production

An important cause for the uncertainty in development trends of rationalization strategies and industrial work lies in problems that result from a continuing internationalization of production. On the one side, saturated domestic markets, growing competition, and increasing costs, particularly in research and development, push companies in many industrial sectors to a global expansion and standarization of their sales and production strategies; this calls for a resolute and worldwide orientation towards "economies of scale". On the other side, the globalization of sales and production creates pressure for intensified flexibility, an expansion of "economies of scope", and the acceptance of increasingly specific market conditions. This pressure comes from the increasing segmentation and regionalization of the world market, which, among other reasons, is caused by the creation of large trade blocks such as the EU or NAFTA and even more by a number of national policy measures designed to protect indigenous industries. This can be seen in the increase in the number of so-called "non-tariff barriers" in the last few years.[3]

Companies often try to come to terms with these contradictory demands of globalized production by changing their existing internationalization strategies, and what emerges is a broad range of differing strategies. At the one extreme lies a so-called "global strategy" which is oriented towards a world-wide homogeneous market and strives for company integration to accomplish standardization of production and products as well as the most far-reaching centralization of decision-making and key functions as possible. In this scenario, competition advantages are to be achieved on the basis of "economies of scale".

At the other extreme lies a strategy that can be conceived of as "transnational"[4]. This strategy only makes partial use of the advantages of standardization in certain company functions such as research and development, purchasing, and component manufacturing. It is characterized much more by strong regional ties, and the differentiated product and production strategies that go along with this. Company integration is thus not generated through centralization as with global strategy, but rather through a cooperative network-type coordination of regional and decentralized company units. In this scenario, competition advantages are to be achieved through a well-formed "economy of scope" and differentiated market orientations.

3. Consequences for the development of industrial work

The extremely diverse trends in the development of industrial work are mainly linked to the change in internationalization strategies. The paths of development do not only differ between company, industrial branch, and nation, as outlined at the beginning of this paper, but also within an internationally active company. Which forms of work get implemented in particular company units is dependent upon a whole range of nationally specific socio-economic factors to which the internationalization strategies of the company have to adjust. Particularly important in this regard are:

o   the regional labor market conditions and the availability of specific qualifications, which, among others, are affected by the nationally specific system of vocational training;

o   and related to this, the regionally specific "technological infrastructure" which comprises company external innovation resources such as know-how, service, and relations with advisors and producers.

Beyond these direct labor and skill related conditions, existing labor policy constellations, wage agreements, and co-determination rights of unions and works councils play an important role in the design of industrial work. They, in turn, are based on the particular national system of industrial relations and its long running practices and traditions.

4. Areas of action

The factors outlined here have direct significance for the development of industrial work in individual company units in various nations and regions. The long term competitive position of companies depends upon the extent to which they are able to correctly weigh the importance of the various influences on decisions regarding production strategies.

Frequently the transnational strategy is recognized as the most promising path. Due to turbulent changes on the world market, this strategy is most often found in the capital goods industry.[5]

The transnational strategy is oriented to the use of qualifications and skills specific to a respective region or nation and actively tries to realize forms of industrial work that are appropriate to the respective conditions. However, in which form this occurs is very much dependent upon the internal relations of such a company: for example, the ongoing processes of control and coordination between the company headquarters and the decentralized company units, the structure and influence of inter-company management systems and their instruments of control, and not to be forgotten, the particular "company culture", and thus the extent to which the need for a regionally or nationally specific design of industrial work will be taken into account.

These aspects characterize the relevant areas of action and design that make up a transnational company strategy.[6] The way in which such a strategy is concretely put into practice and the forms of industrial work which thereby result, will decide its future success or failure.

References

[1]   Altmann, N.; Köhler, Ch.; Meil, P.: No End in Sight - Current debates on the Future of Industrial Production Work. In: N., Altmann et al. (eds.): Technology and Work in German Industry, Routledge, London 1992, pp. 1 -11.

[2]   Hirsch-Kreinsen, H., Schultz-Wild, R.: Chances for Skilled Production Work in Computerized Manufacturing Systems. In: P. Brödner, W. Karwowski, W. (eds.): Ergonomics of Hybrid Automated Systems III, Elsevier, Amsterdam 1992, pp.147 - 154.

[3]   Porter, M.E.: Changing Patterns of International Competition. In: California Management Review, Vol XXVIII, No 2, 1986, pp. 9- 40

[4]   Bartlett, C.A.: Building and Managing the Transnational - The New Organizational Challenge.In: Porter, M.E. (ed.): Competition in Global Industries, Boston 1986, pp. 367 - 401.

[5]   Taylor, W.: The Logic of Global Business: An Interview with ABB's Percy Barnevik. In: Harvard Business Review, March-April 1991, pp.91 - 105.

[6]   Handy, Ch.: Balancing Corporate Power. In: Harvard Business Review, November-December 1992, pp. 59 - 72 .

Advances in Agile Manufacturing
P.T. Kidd and W. Karwowski (Eds.)
IOS Press, 1994

Overcoming Taylorism: Training and the development of work organization

Bruno CLEMATIDE
Danish Technological Institute, P.O. Box 141, DK-2630 Taastrup, Denmark
Jörg KLUGER
Berufsförderungszentrum Essen, Altenessener Straße 80/84, D-45326 Essen, Germany
Brian DILLON
Nexus Research, 9 Nth Frederik St., Dublin 1, Ireland
Jean-Yves MARTIN
Groupe Forège, Treize Septières, BP 129 F, 85603 Montaigu Cedex, France

Abstract. The need to develop new 'anthropocentric' forms of work organization to overcome the 'Tayloristic' paradigm is recently a main issue in companies as well as in institutions. The implementation of group based structures of work organization (e.g. the implementation of production islands, quality circles etc.) are becoming a big challenge for continuing vocational training (CVT), too. In the context of a common FORCE project, concepts and experiences from four national projects in the manufacturing area are outlined.

1. Introduction

It has almost become a commonplace to state that continuing vocational training has to play an important part to maintain the competitiveness of European industry. The need to develop new 'anthropocentric' forms of work organization to overcome the 'Tayloristic' paradigm is recently a main issue in companies as well as in institutions. Actually, concepts like 'lean production' and 'lean management' are discussed broadly, but cannot be taken over one by one.

The implementation of group based structures of work organization (e.g. the implementation of production islands, quality circles etc.) are becoming a big challenge for continuing vocational training (CVT), too. Following the recent problems during the implementation process of concepts like 'total quality management' and 'quality consciousness' throughout the hierarchies, there has to be both - a 'top-down' and a 'bottom-up' - strategy: Of course, it is a pre-requisite that the top management agrees to the guidelines and formulates the framework of the company's total quality management (TQM) 'philosophy' (top-down), but there has to be a parallel process-orientated implementation of an integrated technical, work organizational and social approach starting up from the shop floor level (bottom-up) as well.

In the context of a common FORCE project ('Better connection between in-house training and public continuing vocational training'), concepts and experiences from four national projects in the manufacturing area are outlined.

2. The need to re-design the workplace

Under the influence of modern electronic systems and new forms of work organization - overcoming the 'Tayloristic' paradigma - there is a big change from an industrial to an informational and communication based society. Thus, work is based on those elements of human work which cannot be formalised and automated. Occupational work is tied towards organizational forms which rely on the base of active collaboration, personal involvement and responsible and trust-based cooperation on the part of all employees throughout the hierarchy. The traditional relationship between specialist and personal qualifications are tending to reverse [1].

Information networking and strong forms of cooperation among previously - more or less - isolated departments, devisions and systems have already become practical reality. The integration of various activities frequently brings about radical changes into the formal and operational organization structure of the companies. Qualification requirements 'automatically' change as well with the various forms of work organization and with the nature and substance of work and activities [2].

3. Demands on learning and training concepts

Production and corporate devisions which were previously independent of each other are now merging together. Besides this, the concentration on traditional distributions of work are increasingly needing employees with interdisciplinary specialist capabilities, including in particular...
- sound technical knowledge and experience
- flexibility
- an ability to adapt to changing situations
- the activation of knowledge and experience for creative problem solutions
- social capabilities on a cooperative and communicative level (communication and co-operation skills, team skills, willingness to challenge, etc.) and
- an ability to act independently with
- personal responsibility within a broad scope of action [3].

Future-orientated vocational training must, as a contribution to personality development, consider the individual person as a whole. According to this, implementation and learning processes in the field of findind new forms of work organization must be shaped jointly by shop floor workers, process supervisors and the middle management as well. The focus is on 'shared visions' to overcome the traditional hierarchies and 'Tayloristic' forms of work organization.

4. Experiences from four national projects

The workshop will report on the experiences during the implementation of group organized work and quality circles in four national projects in different branches of manufacturing industry:

- Denmark: Beck & Jorgensen A/S, Hempel A/S - paint industry,
- France: Groupe Forège - wood industry,
- Germany: Vorwerk Elektrowerke KG - manufacturing of floor-care equipment,
- Ireland: Donnelly Mirrors Ltd. - manufacturing of car mirrors.

Very often the sphere of enterprise (in-house training) and the sphere of off-house continuing vocational training are too far apart - even in cases where both the public and other off-house training providers and the enterprises deliberately want to contribute to the development of anthropocentric production systems [4].

In our common FORCE project ('Better connection between in-house training and public vocational training') - with four different national contexts for continuing vocational training - we bring these two separate worlds together, and demonstrate how systematic preparation in the enterprises before courses, and last but not least, participative follow-up actions in the enterprises contribute to an understanding of a process of continuous learning, and to the emergence of new types of work organization.

References

[1] M. Brater: Between entitlement and reality. The effects of changed framework conditions on social positions, functions and qualification concepts for trainers in industry. In: P. Dehnbostel, *inter alia* (Ed.): New Technologies and Occupational Training ('Neue Technologien und berufliche Bildung'), Berlin and Bonn, 1992, pages 210 - 227.

[2] N. Meyer and L. Klaßen-Kluger. Furtherance of Competence to Act as the Objective of Occupational Training. Discussion paper, Berufsförderungszentrum Essen, 1993.

[3] M. Herpich, D. Krüger and A. Nagel: Technology input, organizational structure and qualification. Results of industrial case studies. In: P. Dehnbostel; *inter alia* (Ed.); New Technologies and Occupational Training ('Neue Technologien und berufliche Bildung'), Berlin and Bonn, 1992, pages 47 - 85.

[4] J. Kluger: Regional cooperation networks for qualifying towards operating production technology. In: P. Brödner and W. Karwowski (Ed.): Ergonomics of Hybrid Automated Systems III, Elsevier Science Publishing, Amsterdam, New York, London, Tokyo, 1992, pages 461-467.

Advances in Agile Manufacturing
P.T. Kidd and W. Karwowski (Eds.)
IOS Press, 1994

Human-Centred Systems:
The 21st Century Paradigm

Richard ENNALS

Rukesh KAURA

Kingston Business School, Kingston University

Andrew AINGER

Engineering Director, Human Centred Systems Ltd

Abstract. Human centred systems stretch from plant organisation to all elements of an organisation. Principles of skill based work, collaboration, participation and decentralised decision making are aided by a balanced management approach.

1. Introduction

World markets are now international, dynamic and customer driven: companies and their managers are not in control. Opportunities for product growth will no longer be derived from economies of scale but from economies of scope. The key question for industry is how seemingly conflicting goals can be achieved, and what is the appropriate manufacturing response. The answer determines corporate strategy across all industrial sectors, which in turn sets the scene for government policies in industry, education and training.

2. Two Ways Forward

Although there is no single best option to suit all market situations, two fundamentally different responses can be identified, with wide-reaching implications.

The orthodox technological response is in the tradition of Taylorist scientific management, seeking technological solutions which minimise dependence on the unreliable human element. Costs, including research, development and training, should be cut, and workforce headcount reduced. Financial engineering is given precedence over production engineering.

The alternative balanced response is based on organisation, people and technology, now being pioneered by companies such as BICC Group and Lucas Industries - the Human Centred Approach [Cooley 1989; Rosenbrock 1989; Brodner 1990,1994]. Human centred systems, designed and developed with people taking precedence over technology, extend from plant layout and management to all elements of an organisation, and derive from a long research and industrial tradition [Gill 1993]. At all levels, principles of skill based work, collaboration, participation and decentralised decision making are aided by an appropriately balanced approach [Goranzon and Josefson 1988; Goranzon and Florin 1990, 1991, 1992].

3. Human Centred Manufacturing in Europe

Action research case study experience has been provided by a series of European collaborative projects. Work supported by the ESPRIT Human-Centred CIM Project brought together different British, German and Danish traditions, linking researchers with industrial support in the development and pilot implementation of new specialist tools to support manufacturing [Corbett 1991].

The British DTI MOPS Programme BESTMAN Project (Best Practice in Manufacturing) moved the emphasis to quality and continuous process improvement in cell based manufacturing [Ainger 1994], again with cross-company collaboration to facilitate dissemination of good practice [Kaura and Ennals 1993]. New technology has empowered shopfloor groups of workers to take control of their level of the production process. The ACiT software system enables cell groups to plan and schedule production, managing bottom-up rather than the more conventional top-down approach. We see the emergence of "Executive Information Systems with a productive human face".

4. Towards the 21st Century

Commercial work continues with developments at Human Centred Systems Ltd in the participative design of emancipatory software tools for production planning and scheduling. This must be accompanied by changing approaches to management and management development, reflecting the realisation that the key asset of a company is the collective skill of the workforce.

At Kingston Business School we are working with Shakespeare Speechwriters Ltd to investigate the use of speech-driven systems in manufacturing: can the hand-held mouse be augmented by voice commands to provide more natural design tools? Can we talk to databases and word processors? What are the implications for skills and working practices?

As the technologies of computing, communication and broadcasting converge, there are new challenges in manufacturing and management. Snell and Wilcox, leaders in broadcast electronics standards conversion, have focused on factory design and corporate development to foster skill in high technology, increasing investment in research and development through the recession. Building on expertise developed in the BBC Research Department, the NASA of the broadcasting industry, they have sought to provide a creative environment for innovation in the tranquillity of a Hampshire converted Water Mill, and are seeking to broaden the management understanding of their key technical staff.

General principles have emerged from experience to date. Human centredness is primarily a process, rather than a term to be applied to physical products, but the design of a product along human centred lines has been shown to facilitate use in a human centred way [Corbett 1991]. Successful tools are designed in association with the user and with knowledge and experience of the context and culture of use. The designer can also be a builder and user, with the conventional professional division of labour partly overcome through the appropriate skill-enhancing use of new technology [Cooley 1989].

5. Education and Training

Human centredness has major implications for education and training, challenging current policies and orthodoxies [Ennals 1991, 1994]. Recent British government policies purport to have been based on competence, with a system of National Vocational Qualifications based on a Taylorist decomposition of complex skilled tasks into simple observable sub-tasks, each with performance criteria. Knowledge, understanding, the culture of working life, and learning by experience have all been downgraded or discarded, resulting in a new system which as yet lacks the respect of industry and education professionals.

There are vital lessons to be learned for industrial policy at company and government level. Industrial euthanasia is at the expense of the future of the workforce and of society. There is an alternative.

6. A Paradigm for the 21st Century

The pace of technological change is such that workers need flexible attitudes and underpinning transferable skills, must expect to participate in continuing education, and are likely to experience frequent career change. Technology is achieving commodity status, with competitive advantage provided through intelligent use. Superficial competence in the performance of simple tasks will not equip workers to compete in a changing technological environment.

Without investment in people as the core of manufacturing strategy and policy there will be no ongoing industrial base. Education must bring together technical knowledge and cultural understanding, bridging traditional gaps between disciplines and with industry. We must invest now for the 21st Century.

References

Ainger A.   Tools for Implementing Best Practice Cell Based Manufacturing, Proc 4th Int. Conf on Human Aspects of Advanced Manufacturing and Hybrid Automation, 1994

Brodner P.   The Shape of Future Technology: The Anthropocentric Alternative, London: Springer Verlag 1990

Brodner P.   The Two Cultures in Engineering in eds Goranzon B. and Cook J.   Skill, Technology and Enlightenment, London: Springer Verlag 1994 (in press)

Cooley M.   European Competitiveness in the 21st Century: Integration of Work, Culture and Technology, Brussels: EEC FAST Programme 1989

Corbett J.M., Rasmussen L.B. and Rauner F.   Crossing the Border: The Social and Engineering Design of Computer Integrated Manufacturing Systems, London: Springer Verlag 1991

Ennals R.   Artificial Intelligence and Human Institutions, London: Springer Verlag 1991

Ennals R.   Engineering, Culture and Competence in eds Goranzon B. and Cook J., 1994 (in press)

Gill K.   Human Centred Systems: Foundational Concepts and Traditions   in eds Ennals R. and Molyneux P. Managing with Information Technology, London: Springer Verlag 1993

Goranzon B. and Florin M. (eds)   Artificial Intelligence, Culture and Knowledge: On Education and Work, London: Springer Verlag 1990

Goranzon B. and Florin M. (eds)   Dialogue and Technology, London: Springer Verlag 1991

Goranzon B. and Florin M. (eds)   Skill and Education: Reflection and Experience, London: Springer Verlag 1992

Goranzon B. and Josefson I. (eds) Knowledge, Skill and Artificial Intelligence, London: Springer Verlag 1988

Kaura R. and Ennals R.   Human-Centred Systems: The 21st Century Paradigm Working Paper, Kingston Business School 1993

Rosenbrock H. (ed)   Designing Human-Centred Technology: A Cross-Disciplinary Project in Computer-Aided Manufacturing, London: Springer Verlag 1989

Advances in Agile Manufacturing
P.T. Kidd and W. Karwowski (Eds.)
IOS Press, 1994

"A Lack of Fit?" The Adoption of Japanese Style Manufacturing Techniques in Britain

Richard Mitton and Ian McLoughlin
Department of Management Studies, Brunel University,
Uxbridge, UB8 3PH, UK

Abstract. This paper reports the findings of a survey on the use of new production techniques in a sample of British establishments. It suggests that managerial perceptions of success are related to the degree of fit between technical and social relations.

1. INTRODUCTION

Considerable interest has been shown by British firms in new production techniques, in large part inspired by a perceived need to emulate Japanese competitors who have been a major source of inward investment in Britain since the early 1980s. It has been argued [1] that Japanese manufacturing methods involve far greater interdependency between the constituent elements of the production process. In particular, employees and work groups assume a position of potential power since this dependency bestows upon employees the capacity to create disruption, either intended or unintended, whose effects are likely to be extremely pervasive and have ramifications both up and downstream. The success of Japanese firms rests, not only on the technical solutions they develop to production problems, but also upon the development of appropriate human resource and industrial relations arrangements. These act to ensure that new 'pressure points' in the production system are not exploited by employees. This is accomplished by creating a complementary set of social relations which render the employee equally dependent on the employer. In short, 'there has to be a move towards goal homogeneity or ... mutually assured destruction' [2].

In Britain Japanese firms have been able to generate the appropriate set of social relations through personnel and industrial relations arrangements designed to ensure employee loyalty and commitment to the enterprise and at the same term ensure continuity of production and management discretion over the deployment of labour [3]. The corollary of this argument is that for British-owned firms to successfully assimilate Japanese-style production techniques they also need *inter alia* to transform their personnel and industrial relations arrangements along lines akin to that promoted in models of 'human resource management' [4]. If this is not accomplished, traditional adversarial - 'them and us' - relationships between British employers and trade unions may well result in sub-optimum results.

2. THE SURVEY

The purpose of this paper is to test this thesis through a sample survey of British and foreign-owned establishments operating in the electrical/electronics and vehicular sectors. The survey focused on two localities with strong concentrations of firms operating in these two sectors - the West Midlands and the 'M4 corridor' (Thames Valley to South Wales). The objective was to establish the extent of use and planned use of new production techniques and the degree to which they were accompanied by personnel and industrial relations arrangements of a 'human resource management' type which might be expected to generate appropriate levels of employee commitment and loyalty.

Three hundred works managers were polled in the summer of 1993. A total of 109 (36 per cent) responses were obtained of which 95 (32 per cent) were useable. Over three quarters of the establishments were British-owned. Of the foreign-owned establishments ten percent were American, six percent European and seven percent Japanese.

3. THE USE OF NEW PRODUCTION TECHNIQUES

Respondents were asked about the use of a variety of new production techniques (e.g. Quality Circles, Statistical Process Control, Total Quality Control, Just-in-Time, Operator Responsibility for Quality, Continual Improvement or Kaizen, Reductions in Set up Time and Cellular Manufacturing) at their establishment. Of all the establishments surveyed, 52 per cent used four or more of the techniques, 46 per cent between one and three, and only two per cent none. Of the British-owned establishments, 44 per cent used four or more, 55 per cent between one and three, and only one per cent none. Of the foreign-owned establishments, 80 per cent used four or more, 15 per cent between one and three, and five per cent none. Noticeably, all of the Japanese-owned establishments utilised at least one new production technique, 71 per cent using four or more.

Respondents were also asked about their plans for the introduction of new production techniques. In fact, 60 per cent said that they had no such plans and given the findings concerning extent of use, this clearly suggests that many establishments regarded innovation in this area as having come to a halt for the foreseeable future. Of those who had plans, the vast majority intended to introduce only one or two more of the production techniques listed. Whether an establishment was foreign-owned or not made no real difference to the existence of such plans. However, when the Japanese establishments were singled out within the foreign-owned category, over half of them did in fact plan to introduce further change (compared to under a half of the other foreign owned firms and just over 40 per cent of British-owned firms).

4. PERSONNEL AND INDUSTRIAL RELATIONS PRACTICES

Questions concerning the degree of 'strategic integration' in personnel and industrial relations policies and practices at the establishments were then raised. This has been identified as a defining element of the 'human resource management' approach [4]. The items were:

a policy commitment to job security, a policy of no compulsory redundancies, flexible job descriptions, team working, single status, performance related pay, appraisal schemes and formal assessment of workers, rationalisation of job categories, personnel specialism involvement in technical change and the absence of union recognition.

Overall, only 22 per cent of establishments reported using six or more of the 11 listed personnel policies and practices. However, 62 per cent said that between three and five of the policies and practices were present. The remainder had at least one of the listed policies and practices in operation. When looked at in terms of ownership, foreign-owned firms were more likely to have a broad range of the policies and practices listed than their British counterparts. For example, whereas 16 per cent of British-owned establishments had six or more of the policies/practices, this was the case for just under a half of the foreign-owned establishments, and 57 per cent of Japanese-owned establishments.

5. DEGREES OF FIT

Having established the extent of use and plans to use new production techniques, and the nature and range of personnel policies and practices in operation at the establishments, we were then in a position to judge the degree of fit between technical and social relations at each location. That is, where a wide range of production techniques were in use or were planned to be used (four or more), and a majority (six or more) of the listed personnel policies and practices were already in operation, we took this to be indicative of a high degree of 'fit'. In other instances, where a wide range of production techniques were in use or planned but the majority of listed personnel polices and practices were absent, this was taken to be indicative of a low level of 'fit'. In between these two extremes were instances where a minority of production techniques were in use or planned and only a minority of the listed personnel policies and practices were in operation.

In terms of a high degree of fit, this was only the case in 18 per cent of establishments. In the vast majority of cases (78 per cent) there was a lack of fit between the use and planned use of new production methods and personnel policy and practice. In the remaining four per cent there was a broad range of personnel practices/policies present, but few (three or less) production techniques. This suggests that these firms had the necessary social relations to support new production techniques, although they were yet to be introduced. In contrast, 12 per cent of British-owned establishments, compared to 42 per cent of foreign-owned firms and 43 per cent of Japanese establishments, had high degrees of 'fit'.

6. PERCEPTIONS OF SUCCESS

Overall 48 per cent of respondents in establishments which had adopted new production techniques claimed that this had been 'very successful' and the remainder 'moderately successful'. Of the establishments which exhibited a high degree of fit between the use of new production techniques and the personnel and industrial relations arrangements, 59 per cent of respondents viewed the production techniques as 'very successful,' with the remainder claiming the techniques had been 'moderately successful'. It would seem that the perceptions of success were more likely if supporting social relations existed.

7. SUMMARY AND CONCLUDING COMMENTS.

These findings echo those of earlier research conducted in Britain [1]. First, it was the case that new production techniques were widely used and/or planned in the surveyed establishments and that this was especially so in foreign-owned cases. Second, when it came to personnel and industrial relations policies and practices which might be deemed appropriate to the actual or planned use of these techniques, these were significantly less in evidence. Third, it was in the British-owned establishments that appropriate personnel and industrial relations arrangements were most likely to be absent and in the Japanese establishments (and to a lesser extent the foreign-owned) that they were the most likely to be present. Fourth, only a minority of establishments exhibited a high degree of 'fit' between technical and social relations. The absence of such 'fit' was particularly marked in British-owned establishments. Finally, managerial perceptions of success when using the new production techniques were more likely where a high degree of 'fit' was apparent.

However, we would argue some caution in the interpretation of these results. First, the 'fit' thesis is based on a rather crude contingency model that views commercial success as determined by effective organisational adaptation to changing technical circumstances. Second, and following from this, key questions concerning the choices managers have when seeking to adapt organisational arrangements, and the politics of change itself, are played down. For example, one argument presented in a study of Nissan's manufacturing site in Britain is that the congruence between technical and social arrangements generate employee compliance rather than commitment and that it is this which supports managerial definitions of their success [5]. Finally, the 'technical' imperatives generated by new production techniques are treated largely as a given. However, there is a strong argument for a more comprehensive understanding of the manner in which notions of 'TQM', 'JIT' or teamworking, for example, are themselves socially produced in adopting organisations.

8. REFERENCES

[1] Oliver, N. and Wilkinson, B. *The Japanization of British Industry*, Oxford, Blackwells, 1992.

[2] Pfeffer, J. (Cited in Ibid), Page 83.

[3] Wickens, P. *The Road to Nissan*, London, MacMillan, 1987.

[4] Guest, D. Human Resource Management and Industrial Relations, *Journal of Management Studies*, 24(5), 1987, 503-21; Storey, J. *New Developments in the Management of Human Resources*, Oxford, Blackwells, 1992.

[5] Garrahan and Stewart, *The Nissan Enigma, Flexibility at Work in the Local Economy*, London, Mansell, 1992.

Advances in Agile Manufacturing
P.T. Kidd and W. Karwowski (Eds.)
IOS Press, 1994

TECHNOLOGY AND PEOPLE AT WORK: TOWARDS BEST PRACTICE MANUFACTURING

Patrick Dawson and Verna Blewett

Department of Commerce, University of Adelaide, South Australia 5005

Abstract: The main objective of the paper is to examine the process of organisational change towards new manufacturing methods. A case study approach is used to illustrate the political process of negotiation and adaptation to managerial strategies of innovation and change. It is shown how successful organisational transitions do not occur overnight but take considerable time and commitment during ongoing processes of decision-making and consultation. The paper claims that employees are often the key to the development of more efficient manufacturing methods and that participative decision-making should not be restricted to solving operational process problems but, rather, should be extended to involve employees in management decision-making over the speed, direction and scale of shopfloor change, and on the design of human-centred systems.

1. Introduction

Manufacturing industry in the 1990s face the growing pressure of tight international and domestic markets, the need to achieve recognised quality standards in order to remain competitive, and the problem of achieving high levels of flexibility to adapt to increasingly volatile trading conditions[1]. In the search for more efficient manufacturing methods many companies are seeking to make more effective use of human resources in their pursuit of flexibility, quality and productivity[2]. Within manufacturing industries these changes have signalled a movement towards more flexible patterns of work and 'modularised' production arrangements[3], with an emphasis on quality and teamwork on the shopfloor[4]. In this paper, we seek to document two innovatory phases in the development of a skill-based lean manufacturing system in a case study of a South Australian (SA) automotive components manufacturer. The phases comprise: managing recovery and developing a climate of trust, and moving towards best practice manufacture.

2. Hendersons Automotive South Australia

Hendersons Automotive (SA) is the South Australian manufacturing plant of the Hendersons Automotive Group and is a major Australian supplier of seating components. The South Australian plant employs approximately 200 people over a two shift system and operates as a closed shop. The company operates in the metals industry, supplying seating components to each of Australia's automotive assemblers. There has been significant investment in state of the art equipment (coordinate measuring machinery, robotics and a transfer press line) that shares the factory space with some quite dated equipment.

3. Managing Recovery and Developing a Climate of Trust

In 1985 the company was facing a declining automotive market and some hard decisions. In early 1986, a new management team was established with a focus on quality, customer service and productivity, through a concerted push to improve occupational health and safety (OHS). At this stage, OHS issues were viewed as a means for attaining credibility amongst shopfloor people, who at that time were largely cynical and uncertain about the future of the company. Some of the physical changes embarked upon included making the flow of work more logical through improving the layout of the machinery and low cost measures such as painting and housekeeping.

Improvements in OHS was coupled with an increased willingness by management to consult with unions and shopfloor workers. This undoubtedly advanced the pace of workplace reform by contributing to the improvement in industrial relations. For example, formal consultative processes were developed, including a Safety Committee and a Works Committee. The safety committee was established before there was any requirement in the law for such a body and was therefore regarded as innovative by workers in the company. It began with a management appointed employee representative but was soon expanded to involve about six elected representatives. Following the introduction of the OHS legislation, the composition was altered to include all elected health and safety representatives. This is above the requirements of the legislation and has helped to maintain the high standard of OHS achieved by the company.

Steady improvement in efficiency, productivity and quality continued throughout this phase. By mid-1990 the company still appeared to be improving its health and safety performance but general morale in the plant seemed to have diminished. Formal consultative processes, outside of OHS, were not considered effective by the employees on the factory floor. As one interviewee recounted: 'The Works Committee was just a whitewash. It didn't do anything; just a lot of hot air. All management did was tell us what they were going to do. That's not consultation'[5].

Employees believed that their needs and ideas were not heard or considered. Although Kaizen Groups (continuous improvement groups) had been established in early 1990 as a forum for encouraging employee participation, many employees claimed that: 'The meetings were held too infrequently....and in any case those meetings didn't have any great affect on the company's operation - they were entirely related to product'[5].

By Mid-1991 the company had been featured in the media as an outstanding company, had won awards and had developed an enviable reputation in industry. However, people on the factory floor did not identify with this recognition and remained cynical: 'It's not like what you read in the papers. The managers get the limelight at the expense of the workers and don't give them sufficient recognition'[5].

At about this time the Human Resource Manager introduced the concept of a Consultative Committee, which was to consist of elected shopfloor representatives as well as shop stewards and management. To begin with it was to be in addition to the old Works Committee but the need to have one committee only was apparent very early in its life. Subsequently, the Consultative Committee has become integral to participatory change processes.

4. Moving Towards Best Practice Manufacture

During 1991, senior management attended a weekend course on World Competitive Manufacturing (WCM) conducted by the South Australian Centre for Manufacturing. A mission statement was written and a series of policy statements formulated setting the objectives for achieving sustainable competitive advantage. The company applied for Commonwealth Government funding and in December 1991, proved successful in attracting a grant of $420,000 from the Australian Best Practice Demonstration Program. Initially the award of the Best Practice funding was regarded with suspicion by the workforce and members of the Consultative Committee. For example, a visit by the Divisional Manager to a number of US automotive components manufacturers did nothing to foster goodwill within the factory (it was regarded as 'another perk for management' by many people on the shopfloor), even though it did provide the means for establishing contacts with benchmark companies in the US.

Between December 1991 and November 1992 the company: commenced international benchmarking; developed a sister relationship with a firm in the USA; purchased the consultancy services of Venture Industries to introduce a lean manufacturing program; put all 200 employees at the plant through a three-day training course in lean manufacturing techniques during September and October 1992; and encouraged the development of teams working with continuous improvement at the forefront of their thinking.

By the end of October 1992 the whole plant seemed excited. In mid-November 1992 a corporate restructuring resulted in the retrenchment of nine non-unionised, office personnel. There followed a dramatic change of mood. For example, in the wake of the restructuring eight key people chose to take a separation package and leave the company. Some of these people had been offered changed roles in the organisation, others expressed disillusionment at the restructuring by leaving for other jobs. The mood in the factory remained grim until the Summer shutdown in December 1992. It showed little sign of improvement in January 1993 but slowly recovered throughout the following six months. Further changes in management personnel, both at corporate and divisional level, and the introduction of incentive payments for team-based suggestions have assisted this recovery. By September 1993 the company was entering a new phase of development. All employees were participating in teams and in continuous improvement programs and there was improvement in the capacity to balance this with production and quality demands.

5. Conclusion

Hendersons Automotive (SA) is undergoing a process of continual change and innovation in the search for new efficient manufacturing methods which are able to combine technology and people in the development of more human-centred systems of operation. These ongoing transitions span a number of years and highlight the time, effort, and commitment required to successfully co-ordinate and manage these types of changes. Kaizen groups, world class manufacturing, lean production, quality circles, best practice, total quality management, and just-in-time manufacturing, are just some of the innovations which have been introduced at Hendersons over the last seven years.

The development of a Consultative Committee has been a central vehicle for promoting employee interest in participatory activities and team working. For example, one of the outcomes of the committee's work has been the introduction of incentive payments for team-based suggestions. Finally, we would argue for even greater involvement of employees in the design and development of collaborative skill-based work environments. Participative decision-making should not be restricted to solving operational process problems but, rather, should be extended to involve employees in management decision-making over the speed, direction and scale of shopfloor change, and on the design of human-centred systems. Employees are a key resource in the development of more efficient manufacturing methods and are critical to the success or failure of systems which seek to secure human involvement.

References

[1] Dawson, P. (1994) *Organizational Change: A Processual Approach* London: Paul Chapman Publishing.

[2] Wood, S. (1991:397). 'How do you manage a flexible firm? The total quality model', *Work, Employment & Society*, 5(3), 397-415.

[3] Buchanan, D. and Preston, D. (1992) 'Life in the Cell: Supervision and Teamwork in a "Manufacturing Systems Engineering" environment', *Human Resource Management Journal*, 2(4): 55-76.

[4] Dawson, P. (1991) 'Machine-centred to human-centred manufacture', *International Journal of Human Factors in Manufacturing*, 1(4): 372-38.

[5] Henderson's Employee Interviews conducted by Verna Blewett

Advances in Agile Manufacturing
P.T. Kidd and W. Karwowski (Eds.)
IOS Press, 1994

The Formation of Structures, Roles and Interactions within Agile Manufacturing Systems

Louis Brennan

Department of Industrial Engineering and Information Systems,
Northeastern University, 360 Huntington Avenue, Boston, Ma 02115, USA

Abstract. This paper addresses the emerging paradigm of agile manufacturing. The focus of this paper is on the formation of structures, roles and interactions to secure the successful implementation of agility. Attainment of agility by the enterprise requires the redefinition of the relationship between the customer and the enterprise components. This implies new structures, roles and interactions on the part of the enterprise and its human resources. This paper analyzes these issues in the context of the shop floor.

1. Introduction

Agile manufacturing is aimed at enabling the production of more highly customized products, when and where the customer wants them. Thus, economies of scope, involving the servicing of ever smaller niche markets, even to the level of single customer orders without the high costs traditionally associated with customized production, represents one of the key themes of the agile manufacturing paradigm. In addition, it embraces an enterprise view and alliances with other companies in the introduction of new products to the marketplace.

The focus of this paper is on the formation of structures, roles and interactions to secure the successful implementation of agility. Changes have been occurring in a largely piecemeal fashion in these areas within enterprises for more than a decade now. For example, flexible production technologies have enabled the emergence of flexible manufacturing systems and cells. The just-in-time philosophy has impacted the organization of production, while total quality management has changed employee roles within organizations. The emergence of concurrent engineering has redefined the relationships and interactions that exist between organizational functions. Whereas lessons can be learnt from these various experiences and applied to agility, attainment of agility by the enterprise requires the redefinition of the relationship between the customer and the enterprise components. This implies new structures, roles and interactions on the part of the enterprise and its human resources.

This paper analyzes these issues in the context of the shop floor. The structures that will likely emerge for the shop floor and the changing personnel roles and interactions are considered. The lessons that have emerged from the key changes that have been implemented within manufacturing enterprises over the past decade can be related to agility. Finally, the human resource requirements of shop floor and related personnel need to be evaluated in the light of these changes.

2. The Shop Floor

According to the APICS dictionary [1] shop floor control (SFC) is " a system for utilizing data from the shop floor to maintain and communicate status information on shop orders (manufacturing orders) and on work centers. Shop floor control can use order control or flow control to monitor material movement through the facility." The basic activities of a shop floor control system can be categorized into the four activities of short-term planning, execution, monitor and control.

The thrust in manufacturing has been towards the integration of all activities within the same system. This is implied by the word "total" : Total Quality Management (TQM), Total Productive Maintenance (TPM) and so on. Just-in-time and Manufacturing Resource Planning embrace all the manufacturing and other business activities. They emphasize the importance of consistent execution of the plans. A plan may be perfect and realistic, but unless it is executed consistently, it is useless. Triggered by this observation, this paper focuses on the sole executioner of the manufacturing plans : the Shop Floor Control function. A flexible and consistently performing shop floor, under rigorous control, is essential to success.

3. Existing Manufacturing Philosophies and Systems

Since SFC is the only executioner of the plans, its role in manufacturing is of the utmost importance. Using the basic activities of SFC (planning (short-term), execution, monitor and control), the implications of the leading manufacturing philosophies and systems are considered for SFC.

3.1 Just-in-time (JIT)

The basic idea in a JIT production system is to produce the kind of units needed, at the time needed, in the quantities needed [2]. Even though some of the JIT techniques can be applied to any kind of shop floor (in order to reduce setup times, inventories, defects, lead times and so on), the tendency is initially to achieve good quality control on the units being produced, followed by smoothed production and repetitive manufacturing (the latter can be realized through cellular manufacturing). Therefore, SFC is restructured in order to provide a fertile ground for the success of JIT. Planning is greatly simplified and Execution is very effective and triggered by demand. The Monitor and Control activity is accomplished with every worker responsible for the monitoring and control of his/her job assignment.

3.2 Manufacturing Resources Planning (MRP II)

According to the APICS dictionary [1], Manufacturing Resources Planning or MRP II is defined as follows: " A method for the effective planning of all resources of a manufacturing company. Ideally, it addresses operational planning in units, financial planning in dollars and has a simulation capability to answer what if questions. " The greatest effect of MRP II on SFC is that it offers immediate information from the shop floor to the manufacturing planning systems and other systems (e.g. marketing, accounting) and back to the floor. Thus, the four basic activities of SFC are integrated in the whole manufacturing system. SFC becomes more flexible, effective, visible and noticeable.

3.3 Hybrid JIT/MRP II

This is a combination of the two preceding philosophies to boost manufacturing competitiveness. Just-in-time is applied to all the business activities of a company with the support of the common data base provided by MRP II. The elements of Just-in-time are applied throughout the company which becomes a "shop floor". The beneficial effects on the shop floor of JIT can be spread throughout the company by means of MRP II.

3.4 Total Quality Management (TQM)

The APICS dictionary [1] defines TQM as " an inter-functional approach to quality management involving marketing, engineering, manufacturing, purchasing, etc.". Although TQM is included in the JIT philosophy, it is referred to separately here, for it can be implemented alone. On the other hand, TQM is essential for the success of JIT. According to TQM, the activities of the shop floor never stop being improved (kaizen).

3.5 Optimized production technique (OPT)

Within the APICS dictionary [1], OPT or Theory of Constraints (TOC) is defined as " a management philosophy developed by E.M. Goldratt which is useful in identifying core problems of an organization, finding effective second order (win-win) solutions and developing detailed implementation plans". The SFC focuses on the bottlenecks and is driven by the goal of increasing sales. The planning is done according to the capacity of the constraints (bottlenecks). The execution of jobs on bottlenecks is of the utmost importance. Monitor and control concentrate again on the constraints.

3.6 Computer Integrated Manufacturing (CIM)

CIM is defined as " the integration of the total manufacturing organization through the use of computer systems and managerial philosophies that improve the organization's effectiveness; the application of a computer to bridge various computerized systems and connect them into a coherent, integrated whole. " [1]. Planning is much faster and realistic, for the planning system is continuously provided with feedback from the shop floor. Execution is very much enhanced through sophisticated technologies such as CAD/CAM, CAE (Computer Aided Engineering) etc. Monitor and control data are accurate and automatically analyzed and converted to information. In addition, the information can be screened and only reported if certain management-by-exception conditions occur. Consequently, production managers can utilize their time more efficiently [3].

3.7 Concurrent Engineering

According to the US. Department of Defense, Concurrent Engineering is a systematic approach to the integrated, concurrent design of products and their related processes, including manufacture and support. This approach is intended to cause the developers, from the outset, to consider all elements of the product life cycle from conception through disposal, including quality, cost, schedule and user requirements. Concurrent Engineering is implemented by a multidisciplinary team which includes traditional project team members, a project manager, a supplier, a customer, and at least one representative from every business function : finance, accounting, marketing and sales, manufacturing and design, shop floor control etc. The effect on SFC of Concurrent Engineering is similar to that of MRP II.

4. Human Considerations

The newer manufacturing philosophies emphasize the importance of the human factor. To be successful all technologies (no matter how advanced) and philosophies depend on adequate consideration of the human factor. The heightened competition and the new trends in the manufacturing world require: i) multiple skilled employees and cross-functional training, ii) multidisciplinary teams, iii) work which demands greater initiative, creativity and flexibility, iv) higher levels of education , v) a holistic understanding of each employee's responsibilities, the interfaces with the equipment and colleagues, the location in the department and in the overall system, vi) continuous improvement (kaizen) and training, vii) employee participation (JIT), viii) goal-oriented rather than process-oriented labor, ix) product and quality responsibility by each and every employee (blue- or white-collar)

5. Instantaneous Engineering

Increasing demand for shorter lead times, lower cost and perfect quality coupled with shorter product life cycles and greater selectivity on the part of the customer inspires the concept of Instantaneous Engineering. This is the means of enabling customer driven manufacturing and can be considered as the realization of agile manufacturing from a shop floor control perspective.

Within an Instantaneous Engineering environment, the customer can switch from the current situation of choosing a product among the existing ones in the market, to the luxury of actively participating in many phases of the product life cycle, through a manufacturer-customer computer network. The term "instantaneous" relates to the process of designing (or capturing the need of the customer), fabricating and delivering the product in an immediate manner. Moreover, this offers advantages over the existing philosophies and approaches to both the customer and the enterprise.

Through Instantaneous Engineering, the Shop Floor Control function embraces all the manufacturing activities and becomes the nucleus of manufacturing. The customer obtains access to the shop floor through a computer network and can interact with the shop floor manager.

6. Conclusion

The transition from existing systems to Instantaneous Engineering is a difficult one and involves the surmounting of many challenges . Nonetheless, this concept heralds the manufacturing "way of life" as we approach the third millennium. This paper should encourage further research into the development and implementation requirements of Instantaneous Engineering.

References

[1] APICS Dictionary, 7th Edition, 1992.
[2] Y. Monden, Just-in-Time Production System. In: G. Salvendy (Ed.), Handbook of Industrial Engineering, 2nd Edition, Wiley, 1992, pp.2116-2130.
[3] J. Gaylord, Factory Information Systems. M.Dekker, 1987.

Part III
Concurrent Engineering and New Product Development

Advances in Agile Manufacturing
P.T. Kidd and W. Karwowski (Eds.)
IOS Press, 1994

Overcoming the Cultural Barriers to Implementing Concurrent Engineering

DE TUCKER
Edbro Plc, Bolton, BL3 6DJ, UK

&

R LEONARD
Total Technology, UMIST, Manchester, M60 1QD, UK

Abstract. This paper presents a case study describing how a typical medium-sized British manufacturing company has sought to adopt Concurrent Engineering (CE) in practice. It highlights the significant cultural differences which were found to exist between functional departments, and which previously served to inhibit the successful introduction of Concurrent Engineering. It then shows how the case-study company are working towards overcoming these barriers by adopting a more flexible organisational structure. The paper concludes by describing the positive changes which have been achieved so far, and this is contrasted by highlighting some of the negative aspects of adopting Concurrent Engineering.

1. Introduction

It is widely accepted by manufacturing companies that the adoption of Concurrent Engineering practices can significantly reduce the time taken to introduce new products into the marketplace. The philosophy underlying Concurrent Engineering is the consideration, at the earliest practical stage, of all possible constraints on a design, whilst also ensuring customer requirements are satisfied. Despite the simplicity of this concept, many companies experience great difficulty in implementing Concurrent Engineering. The reasons can be found by examining the inherent nature of organisational structure in the majority of manufacturing companies.

Since the factory became the common mode of production, management has needed to overcome the organisational difficulties associated with span of control. The most practical solution has been to group those people performing the same job function together, and to adopt hierarchical reporting structures. In this situation, individual members of a department tend to adopt the same values and goals as each other. These goals often conflict with the objectives of other departments, and the resolution of differences becomes a difficult process. Ultimately, this can lead to a misalignment with the overall strategic direction of the company, and result in sub-optimum competitive position.

The most appropriate way to engender a Concurrent Engineering culture is by forming cross-functional teams. However, simply placing people from different departments together in the same room will not overcome these deeply ingrained cultural barriers. It the benefits of Concurrent Engineering are to be maximised, management must review the entire product introduction process and adopt radical changes to long established procedures.

The rest of this paper describes how the case study company sought to address these critical issues.

2. Case Study

2.1. Background

The company are long established in the field of hydraulic engineering and sheet metal fabrication. In 1992 they merged with a French company who operate in similar markets. New product introduction now requires significant collaboration between the two companies.

The Technical Director highlighted several weaknesses with the traditional sequential engineering approach to new product introduction. The fundamental cause was identified as being a lack of communication and co-ordination between departments. This had led to unacceptably high development costs and the loss of potential sales revenue due to late product launch. It was proposed that the company should adopt a Concurrent Engineering approach to the next phase of product introduction.

2.2. The New Approach to Product Introduction

The Sales Department had identified a potentially lucrative market for a new form of multipurpose hydraulic lifting gear. The specification was evaluated by the Product Committee, who agreed there was significant potential, but that achieving maximum strategic benefit would require significant design innovation, and that profit maximisation depended on rapid product introduction. A multi-functional team approach to the project was proposed.

The first step was to find a suitable Team Leader who would co-ordinate the activities of all departments and arbitrate in any conflicts that were bound to arise. Thus it was more important for the leader to have strong interpersonal and diplomatic skills, rather than a high degree of technical ability. The success of the project would require circumvention of the bureaucratic channels which had characterised previous projects, so the leader would need a significant degree of authority and autonomy. This led to the choice of a senior manager of long standing, who was highly regarded at all levels of the company.

The Team Leader[1] was given full control over who should constitute the team members. This was because of his "..prior knowledge of people within the company. Because of the very aggressive timescales for the project, people were chosen on the basis of who had the best access to the information and technical resources needed to make the project work. The possession of the necessary technical expertise was taken as a foregone conclusion."

The initial design work was to be undertaken by the French engineers, who had the greater experience in the particular field. This added a new cultural dimension. A successful co-ordination between the French and the British teams, particularly in the early design stages, would have a critical impact on the outcome of the project. It was decided that the most effective way to ensure co-ordination was to appoint a single point of contact in each country to act as liaison. There were initial misgivings on the British side: "I realised if the project was going to work then I would be working closely with my French counterpart, but that we would rarely meet face to face. I knew that personalities could clash, especially as I couldn't speak a word of French! So a misunderstanding of technical detail was a distinct possibility."

Initial meetings were arranged both in France and England, and both leaders were given the option to opt out at that stage without prejudice. This was not necessary. Next the British team members were briefed as to what would be required of them and, more importantly, why this new approach was necessary. Again, all were given the option to

---

[1]   Unless otherwise stated all quotes are attributable to the British Team Leader

decline participation. No one did. The Technical Director commented, "I think everybody understood why this approach was so important to us. Tacitly, everybody new that we were not as effective in the market as we should be, even though our products have an excellent reputation. This policy was seen as vital to our long term survival, and it took precedence over departmental objectives. If successful everybody would benefit "

At project launch time, team meetings were held at least once a week, until a design concept had been approved in principle. Now a rapid prototype was required. Rather than the previous system whereby the project was regarded as a whole, it was decided to decompose the task into a series of logical sub-projects, with individual team members being responsible for particular objectives, but closely co-ordinated by the Team Leader. The smaller sub-projects were much simpler to manage, and the concurrent but co-ordinated activity of each team member resulted in a working prototype being developed in record time. This was a significant departure from the usual way of working. One Director commented, "..people were genuinely marching to the same drum. Everybody knew their responsibility and, critically, knew how their particular sub-project interfaced with the others."

The team meetings were an ideal vehicle to ensure co-ordination, but occasionally problems arose which needed urgent attention. In such an instance, the team members approached the Team Leader, who took remedial steps as he saw fit. If other team members would be affected, the leader would call them for a special meeting, otherwise everybody would be updated at the next progress meeting.

As the project progressed, the need for formal meetings diminished. Everybody could clearly understand what was happening, and the formal channels of communication were substantially replaced by informal ones. The progress meetings remained as the platform for official discussions. A core team of five people remained dedicated to the project. Everybody else became part of a transient team, called to meetings as required. These people were free to become involved in other projects.

Another important feature was the early involvement of shop floor personnel in the project. Because new manufacturing techniques were involved, it was important that personnel learn the new methods and become at ease with them before live production begins. The Technical Director stated; " Previously, the shop floor felt they were not given any chance to participate in decisions, and the management received little or no feedback from then. This created an 'us and them' culture and made the value engineering process rather ineffective." The prototype gave the workers an excellent opportunity to learn new techniques, and they were encouraged to make suggestions. This engendered a strongly co-operative atmosphere and provided a significant boosts to the morale of the workforce. They felt personally responsible for the success of the project. It was a matter of pride.

2.3 The Results

From product conception to market launch was scheduled to take eight months, a timescale previously unheard of. The project is currently running ahead of time. The Technical Director commented, "It was because we realised that Concurrent Engineering is fundamentally a people driven concept, not technological, that influenced our thinking. The real strategic benefits have occurred by remoulding our structure so that people can communicate effectively from the early concept stage, and by providing an effective feedback loop from those who undertake the manufacturing. It seems everybody wins!"

The company now have two further major projects underway, using the same 'people approach' to Concurrent Engineering.

2.4 Findings From the Case Study

The team approach has proved remarkably effective within the case study company. This success is attributable to several factors. Firstly, the choice of team leader is critical. His interpersonal skills serve as a key motivator for other team members, and the co-ordination of all parties is achieved mainly through his communication skills.

Secondly, the leader must have considerable autonomy and real authority to make crucial decisions, otherwise the bureaucracy of the functional organisation will continue to stifle progress. The Team Leader needs the full and visible support of the Project Committee. Except for reviewing periodic progress reports from the Team Leader, the Project Committee must adopt a hands-off approach to the day to day running of the project. Their role is to act as final arbitrator for any issues that the Team Leader feels cannot be satisfactorily resolved within the team itself. They may also be called upon to resolve conflicts which can emerge due to the overlapping authority of Departmental Heads and the Team Leader. If these occasions are rare it is a good indicator that the team is working well.

Thirdly, the size of the team affects its efficiency. A core team of four or five engineers proved to be ideal. The transitory team members tended to be the Heads of support departments such as purchasing and sales. This facilitated rapid access to the specific resources under their control.

Finally, it was noted that the protect duration was comparatively short when compared to the development of other larger and more complex products. When projects are scheduled to last for several years, there is a real danger of core team members loosing touch with technological developments in their specialist disciplines.

3. Conclusions

This paper has described the process of overcoming cultural barriers, rather than their removal. The long held differences between traditional functional departments do not disappear simply as a result of the adoption of a team approach to Concurrent Engineering. Indeed, the traditional grouping together of functional specialists helps to ensure alignment with new technological developments in the field. For companies competing in a commercially volatile and technically turbulent environment, a total removal of departmental differences could ultimately prove counterproductive. The most suitable approach to Concurrent Engineering is a hybrid system which simultaneously encourages departmental integration as well as promoting specialist technological development. The team approach adopted at the case study company indicates that a traditional organisational structure can be successfully adapted to bring about significant strategic benefits.

Advances in Agile Manufacturing
P.T. Kidd and W. Karwowski (Eds.)
IOS Press, 1994

The Key to Concurrent Engineering

William J ION
DMEM, University of Strathclyde, Glasgow, Scotland, UK

Abstract. This paper explores the approach taken towards implementation of concurrent engineering in a limited number of companies from Scottish manufacturing industry. It summarises and discusses the findings of company investigations and identifies the key factors that have influenced its introduction.

1. Introduction

In today's competitive business environment, a variety of key product factors need to be optimised to achieve competitive edge - innovation, value for money, costs, quality, reliability and time to market. The 'traditional' functionally organised product introduction process has proved to be incapable of meeting these demands. This has led many companies to introduce concurrent engineering working practices based around multidisciplinary design teams.

The benefits that concurrent engineering (CE) can bring to a company are well documented [1]. Typically a 30% lead time reduction and associated cost savings being claimed. Many household names like Nissan, Lucas and Ford have adopted CE and have reorganised their businesses accordingly.

The approach taken to the adoption of a CE philosophy tends to vary from company to company depending on a number of factors such as company size, market sector and company culture. The aim of this paper is to explore the approach taken towards implementation of CE in a number of companies from Scottish manufacturing industry.

2. Concurrent Engineering

The original definition of concurrent engineering coined in 1986 by the Institute for Defence Analysis (IDA) [2] is as follows;

'Concurrent engineering is a systematic approach to the integrated concurrent design of products and their related processes, including manufacture and support. This approach is intended to cause the developers, from the outset, to consider all elements of the product life cycle from concept through disposal, including quality, cost, schedule and user requirements'

This definition has since been interpreted or redefined in a number of different ways.

Many organisations will claim to be involved in CE through a number of approaches such as the formation of multidisciplinary product development teams or the integration of product design and production engineering functions. On their own these changes are unlikely to bring the full benefits that CE can potentially achieve.

The full benefits can be achieved only when a significant number of changes at all levels within an organisation have been made. These changes can be summarised as follows [3]:

1. The introduction of multidisciplinary teamworking involving personnel from all stages of the new product development process such as finance, marketing, design, manufacturing and purchasing (including subcontractors). Collocation of team members is highly desirable.

2. Simultaneous design of the product and manufacturing process.

3. The use of concurrent engineering tools such as Quality Function Deployment (QFD), Controlled Convergence Matrix (CCM) and Design for Manufacture and Assembly (DFMA).

4. The use of appropriate project management tools against clearly defined and agreed cost, quality and delivery targets specified to achieve complete customer satisfaction and business profitability. A fast and efficient communication structure is essential.

As the implications of CE reach into all aspects of a company's business it is essential that implementation has total commitment from the Chief Executive downwards.

3. Company Investigation

Full implementation of CE, as described above, is not always carried out and indeed it may not always be desirable or appropriate to do so. To gauge the level of implementation within Scottish manufacturing industry a study of a limited number of major Scottish businesses was carried out. It was hoped that this study would also give an insight into the key factors that influence the implementation of CE.

Five companies from different market sectors were studied. The basic characteristics of these companies are given in Table 1.

Table 1 - Company Characteristics

	Company Size	Industry Sector	Product / Market Characteristics
Company A	Medium	Power Generation	Systems Based; Low volume; High level of licensing and sub contractor involvement; Client driven.
Company B	Large Multinational	Computer hardware and software	High rate of new product introduction; Short development cycle; High volume.
Company C	Large Multinational	Banking equipment	Highly competitive market; Medium volume.
Company D	Medium	Fluid Handling	Low level of new product development; Evolutionary design; Medium volume.
Company E	Medium	Military equipment	High level of innovative, high technology, new product development; Low/medium volume; Client driven.

Information relating to the implementation of CE was gathered from a variety of different sources including published data, interviews and informal discussions with company representatives. The information gathered from the surveys is summarised in Table 2. For ease of interpretation this has been categorised under the four principal elements of CE, previously discussed in section 2.

Table 2 Summary of Company Investigations

	Teamwork	Simultaneous Design of Product and Process	Project Management	Use of Concurrent Engineering Tools and Techniques
Company A	Project teams - engineer based	Partial	Regular technical and project reviews	No
Company B	Multidisciplinary core teams, collocated	Yes	Corporate phase review process. High level of top down commitment	Yes
Company C	Multidisciplinary core teams, collocated	Yes	Structured product development process including phase reviews High level of top down commitment	Being implemented
Company D	Multidisciplinary teams, collocated	Partial	Regular Technical and project reviews	No
Company E	Multidisciplinary teams, collocated	Yes	Structured product development process. Strong staff commitment	Yes - actively developing new tools

Personnel from all five companies claimed that they were undertaking CE, even though only three of them (B, C and E) had made the major organisational changes necessary for full implementation. This is clearly demonstrated in Table 2. The other two companies (A and D) had made moves towards the formation of design teams by bringing manufacturing engineers into closer contact with design engineers - very few other changes were apparent. The utilisation of varied forms of multidisciplinary design teams formed the one common element among all five companies.

The contrast between the attitude and culture of these two groups of companies was quite marked. In general, the staff in companies B, C and E demonstrated a greater knowledge of and interest in company activities and structure. This had been enhanced by strong senior management commitment to the CE philosophy backed up by training and information programmes for all staff. A conscious decision to move towards CE within these companies had clearly been made at company management level. The changes in companies A and D appeared to have been made at a relatively local level and did not permeate throughout all functional areas within the companies. In fact, the use of the term concurrent engineering to represent the changes made seemed somewhat retrospective - a convenient name for a change that was previously underway, perhaps.

It is interesting to consider the possible motivation for companies B, C and E to undertake such wide ranging organisational changes. Two of these businesses (B and C) are part of large multinational organisations that operate in rapidly changing and fiercely competitive markets. Similarly, company E had been affected by a step change in competition within the military market, accelerated by the end of the cold war. Although competitive pressures have increased for the remaining two companies the changes have not been so severe.

It is clear that the businesses that have been subjected to the greatest degree of change have responded with the greatest level of reorganisation. In other words the characteristics of the markets in which they operate have forced them to reorganise in order to survive and/or remain successful.

4. Conclusion

The limited study undertaken cannot claim to be representative of Scottish manufacturing industry as a whole. It may, however, be considered to give a rough indication of the level of understanding and implementation of CE in Scotland.

The five companies investigated can be divided into two distinct groups. The first of these is characterised by a high level of organisational change and top level commitment to the concept of CE. Significantly the three companies in this group all produce innovative products and operate in highly competitive and volatile markets. Two of the three companies are part of American owned multinational corporations. The second group is characterised by a limited level of implementation of CE, consisting essentially of the formation of integrated design and manufacturing teams. The two companies in this group have their roots based in the Scottish medium to heavy engineering industry, both producing products of an evolutionary nature.

The contrast between these two groups indicates a number of key factors which have affected the introduction of CE into Scottish industry;

- company culture and ownership
- the nature of the product i.e. innovative or evolutionary
- the characteristics of the market sector

References

[1] J Hartley and J Mortimer, Simultaneous Engineering, Industrial Newsletters Ltd, 1990.
[2] Report R-338, Institute for Defence Analysis (IDA), 1986.
[3] B Miles and K Swift, development in computer aided concurrent engineering tools, Proceedings of Design for Competitive Advantage Conference, IMechE, 1994.

Advances in Agile Manufacturing
P.T. Kidd and W. Karwowski (Eds.)
IOS Press, 1994

New Product Development Strategies for Hong Kong Manufacturing Industries

Richard C M YAM   K S CHIN
Department of Manufacturing Engineering
City Polytechnic of Hong Kong
Tat Chee Avenue
Hong Kong
Tel: (852) 788-8420
Fax: (852) 788-8423

Esther P Y TANG
Department of Business Studies
Hong Kong Polytechnic
Hung Hom
Hong Kong
Tel: (852) 766-7129
Fax: (852) 765-0611

Abstract. The authors have carried out several studies through surveys and interviews to review the current product development strategies used in Hong Kong with emphasis on the applicability of adopting the proactive, quality and Time-to-Market approaches for the Hong Kong manufacturers. The study results infer that the majority of the Hong Kong manufacturers are undergoing substantial strategic changes. The corporate focus is moving from the low-cost manufacturing bias to the high- quality, customer-based and marketing/technology integrative emphasis. This paper presents the authors' findings.

1. Development of Hong Kong Manufacturing Industries

In recent years, the Hong Kong manufacturing industries have encountered keen competition from developing countries for their low cost manufacturing. And at the same time, the major customers of the industries have continuously demanded for high quality products at competitive prices. Many other unfavourable factors such as escalating land and production costs, shortage of labour, lack of government support, etc. are all forcing manufacturers of low added-value and labour-intensive products to move their production to developing countries [1]. With the 'Open Door' policy of China in the late 1970s, Hong Kong manufacturers have extensively shifted their manufacturing bases across the border to China. Hong Kong now is undergoing a critical transformation from a low cost manufacturing base to a high value-added, design and service orientated manufacturing centre. In approaching 1997, when the sovereignty of Hong Kong returning from Britain to China, linkages between China and Hong Kong will be more extensive and intensive. The authors' previous studies [2][3] indicated that Hong Kong manufacturers should go for product innovation, research and development, advanced manufacturing / hybrid automation and marketing information service and retain the low technology production in China. Through surveys and interviews [4][5], the authors have critically reviewed the new product development strategies being adopted by the Hong Kong manufacturing industries and their future trend.

2. The Trend of New Product Development Strategies in Hong Kong

The studies have identified the following three basic approaches being adopted widely by Hong Kong manufacturers in establishing their new product development strategies, namely; the proactive approach, the quality-focus approach and the Time-To-Market (TTM) approach.

2.1 Proactive Product Development Strategy in Hong Kong

The proactive product development strategy used by Hong Kong manufacturers implies concentrating on customer (marketing) and advanced technology (R&D). There is a general consensus among the respondents in our studies that in the 1990s, all the successful new products must be customer-oriented.

On the technology side, the Hong Kong manufacturing industries are traditionally reactive in terms of product development. This is probably due to their low-technology and labour-intensive nature. Hong Kong manufacturers are conservative to develop technological products because of high risk and lack of expertise/ government supports. The investment in R&D is the lowest as compared with foreign competitors . An imitative strategy based on quickly copying a new product from the competitors have been widely adopted for years. Moreover, a certain percentage of the Hong Kong manufacturers, particularly the small and medium companies, are OEMs (Original Equipment Manufacturers). They do not have product design function. All the designs come from their customers. This situation, however, has been changing in recent years.

Facing the keen global competition and increasing trade barriers in major markets, Hong Kong manufacturers are facing the problem that their OEM businesses are eroding due to the low-cost competition from other developing countries. They are also aware that the imitative strategy is no longer an effective means with the fading low-cost advantage and the rapid market changes that there is no sufficient time to react. They realize that a proactive approach in product development will make their business more successful in the long run. In our findings, there are now nearly 75% of Hong Kong manufacturers having their own design functions and many of them are using proactive approach.

It is also noted that some other factors are facilitating the Hong Kong Manufacturers in adopting the proactive approach. Although the Hong Kong government still insists the 'laissez-faire' policy that it would not encourage or intervene the local companies in making business decisions, it has recently implemented various direct and indirect policies to support the local companies to pursuit high-technology development. These include : the expansion of tertiary education especially in scientific and technological areas; the establishments of the Industry and Technology Development Council, the Hong Kong Technology Centre, and the Industrial Estates; and the financial funds for a large number of applied research projects in new technology development. Approaching 1997, China will become an important talent source for Hong Kong's technological development. Hong Kong manufacturers could finance and commercialize the Chinese technological research results, and make use of China's strong research and development base to develop the high technology manufacturing in Hong Kong.

2.2 Quality-Focus Product Development Strategy in Hong Kong

Most of our respondents were generally aware of the importance of developing quality products towards customers' satisfaction. ISO9000 is now very popular in Hong Kong manufacturing industries. The ISO9000 movement in Hong Kong was initiated by the

European Community (EC) that the suppliers of the products exporting to EC must be ISO9000 certified. Later, the customers of the Hong Kong manufacturers in other markets also indicate buying preference to ISO9000 certified companies. In order to help Hong Kong manufacturers qualify for this global requirement, the Hong Kong government has launched the ISO9000 certification scheme in Hong Kong for 6 years under the administration of Hong Kong Quality Assurance Agency (HKQAA). There have been more than 150 companies successfully certified by HKQAA, and many others are in the different stages of the application process and preparing for their certification exercise [6]. According to a government survey of Hong Kong manufacturing environment [7], about 42% of the Hong Kong manufacturers surveyed had plans to implement ISO9000. It reveals that the concept of quality assurance has well been accepted by the Hong Kong manufacturing industries.

Most of the studied companies were either had obtained the ISO9000 certification or were in the application process. They relied on a good quality system to reduce both time and cost in new product development. Nevertheless, the majority said that trade-offs decisions often appeared among quality, cost and time. How effective is ISO9000 on quality improvement is yet to be assessed. However, the commitment of most of the Hong Kong manufacturers on quality is evidenced. This will form a good platform for successful high-technology based product development.

2.3 Time-To-Market Product Development Strategy in Hong Kong

Many researchers advocated that the emphasis of manufacturing companies in the 1990s will be how to speed up the development of a new product. Or speaking in terms of time, the emerging way of thinking in the 1990s is how to reduce the 'Time- to-market'. The term 'Time-to-market' is generally defined as the elapsed time between product definition and product availability. TTM is a rather new concept for the Hong Kong manufacturing industries. As shown in table 1, 24.3% of the studied companies do not know the TTM concept. Among the 33 companies in the categories of implementing or planning to implement TTM, 23 companies (70%) are foreign companies. Based on the successful experience of their parent companies, they have quickly implemented the TTM strategies. Comparatively, local manufacturers are well behind in adopting the TTM approach.

Table 1 : Acceptance of TTM by Hong Kong Manufacturers

		No.	%
(1)	Realize the TTM benefits and implementing it in Companies.	21	28.4
(2)	Realize the TTM benefits and planning to implement it.	12	16.2
(3)	Know TTM concept but no special plan to implement it.	23	31.1
(4)	Do not know TTM concept.	18	24.3
	Total	74	100

Although most of the studied companies appreciated the benefits of being a market pioneer by quicker entry into the market, many companies opined that the strategies in determining appropriate market entry time and identifying the customer response to new products were more important than merely faster development cycle. For those studied companies which knew the TTM concept but had no special plan to implement it thought that with strong support by the R&D and Quality, the product development time would naturally be reduced. They also preferred their existing phase-by-phase development process to the overlapping process which has been widely accepted as a necessary change to reduce the TTM. Their major reasons for such a phase-by-phase approach are :

* *Their designers and engineers are accustomed to a phase-by-phase development process, confusion will appear in an overlapping process.*
* *The phase-by-phase approach provides a clear-cut accountability and responsibility of various departments concerned in the entire development process. It is better for management control.*

These conservative views on Time-To-Market may imply that Hong Kong manufacturers are still not be able to totally divorced from the low-cost manufacturing mentality.

3. Conclusion

To conclude, the key success factor for Hong Kong manufacturers in the 1990s is the corporate's total commitment towards proactive/quality/TTM approach. At present, the quality base established by the ISO9000 movement has formed a reasonably solid foundation for Time-to-market philosophy to be fully implemented. Through this continuous quality/time integrative approach, successful products will be fast to market with excellent quality and maximum customer satisfaction. In the future, with the China's technological support as well as the increasing emphasis on high-technology R&D in Hong Kong, Hong Kong manufacturers will soon be able to develop more technology-based and customer-oriented new products. On the other hand, the new emphasis on proactive/quality/TTM implies companies should go through a substantial philosophical, cultural and organizational change, in particular, the current conservative view on Time-To-Market. How well a company can handle these changes determines its future success.

Reference

[1]   Peat Marwick Management Consultants Ltd., The Restructuring of Hong Kong's Manufacturing Sector : A Critical Transformation, Executive Summary, 1989.
[2]   R C M Yam, K S Chin and J Yeo, The transformation of the Hong Kong Manufacturing Industries - A Strategic Review of the Hong Kong Clothing Industries, Transaction of ACME III/ICCM VI Joint International Conference in Technology Transfer and Comparative Management, Log Angles, USA, August 1993, pp.207-212.
[3]   R C M Yam, K S Chin and P Y Tang, Strategic Industrial Development / Technology Transfer Model for Hong Kong and China, Mathematical Modelling and Scientific Computing, Volume 4, February 1994.
[4]   R C M Yam and H Y H Lee, Success in Competitive New Product Development, Project Report, Manufacturing Engineering Department, City Polytechnic of Hong Kong, 1992.
[5]   K S Chin and T W H But, An Investigation on the Applicability of Concurrent Engineering in Hong Kong Electronic Products Industry, Project Report, Manufacturing Engineering Department, City Polytechnic of Hong Kong, 1993.
[6]   HKQAA, Register of Certified Companies According to ISO9000, Issue No.22, December 1993.
[7]   Hong Kong Industry Department, Hong Kong's Manufacturing Industries 1993, Hong Kong Government, December 1993, pp.290.

Advances in Agile Manufacturing
P.T. Kidd and W. Karwowski (Eds.)
IOS Press, 1994

Concurrent Engineering for Enhancing Worker Safety in Robotic Workcells

J. Graham, W. Karwowski, H. Parsaei, J. Zurada
University of Louisville, Louisville, KY, USA

Abstract. This paper describes new methodologies for improving the safety of human workers in robotic workcells through safety consideration early in the product/process design cycle. Specifically, cognitive engineering, sensory integration and quality function deployment approaches are investigated.

1. Introduction

Worldwide competitive pressures are forcing manufacturing organizations to utilize the best equipment and engineering techniques to design and bring to production new, high-quality, well-designed and competitively priced products within a very short timeframe. Concurrent engineering, based upon the concept that product design and process design should be integrated into a single step at the early stages of product development, offers a promising approach to responding to these competitive forces. Unfortunately, safety has often been relegated to a secondary phase of the process design cycle, where it is treated as an add-on or after-thought. There is thus an obvious need for incorporating worker safety criteria into the concurrent engineering process.

This paper will describe research into methodologies for incorporating safety considerations early in the product/process design cycle. Specifically, a new model for task allocation and hazard evaluation using the cognitive engineering approach will be presented, accompanied by new methods for sensory evaluation, selection and integration, and a new approach for safety system evaluation using quality function deployment. While this research is still in its initial phase, certain directions and conclusions can already be shared.

2. Cognitive Engineering Approach

Task allocation and hazard evaluation for robot sensory integration will be done using the cognitive engineering. This approach postulates three kinds of constructs in explaining the overall cognitive activity in the system: mechancism, representation and procedure. *Mechanism* refers to a basic set of information processing capabilities that are thought to underlie all more complex instances of cognition. *Representation* refers to the information about the external world that is stored inside the human-robot system, and is manipulated by the information processing mechanism. *Procedure* refers to the sequence of operations performed by the information processing mechanism on the representation to produce cognitive activity.

A model of cognitive processes suitable for application as an embedded user model in an intelligent human-robot interface must meet three sets of requirements: psychological, computational and operational. Psychologically, the model must deal both with the behavioral or observable aspects of human-robot interaction, and with the underlying cognitive processes that give rise to the observed behavior. The model must be computable in real-time. Finally, it must explicitly define the ways in which the robot operator will adjust his/her attention among tasks to compensate for the dynamics of the situation.

Zachary, et al. [1] proposed a cognitive modelling framework called COGNET, which stands for Cognitive Network of Tasks, whic conceptualizes the person as a network of cognitive tasks, each of which represents a partial strategy for solving some aspect of the overall problem. The flow of attention from one task to another is triggered by momentary changes in the problem environment, which may be the results of actions taken by the person or the results of actions of other agents and/or the environment.

An effective human-robot interface should perform four specific adaptive functions:
(1) Provide reminders or alerts to the user when the system believes the safety/hazard context is appropriate for initiation or return to a specific task in the COGNET network.
(2) Indicate the expected priority or order of precedence of tasks.
(3) Provide decision structuring assistance by identifying the internal organization of goals and subtasks within any given task identified as appropriate for initiation.
(4) Offer automated performance of any subtask, with the task instance adapted automatically to the current understanding of the problem.

3. Sensory Selection and Integration

A large number of sensory technologies have been suggested for use in robotic systems to monitor and enforce safety conditions for robot operations in industrial environments, with a somewhat smaller set of sensing systems actually developed and tested [2]. Current industrial robot systems operate primarily with perimeter guards and/or perimeter sensing devices [3]. One significant reason for the slow implementation of more sophisticated sensing devices is the problem integrating the sensory information into the control architecture of the robot so that time critical safety responses can be achieved.

Figure 1 shows an integrated detection and safety decision architecture using a neural network approach. The neural network detection unit combines several sensory inputs into a detection map, indicating the likelihood or belief that an unexpected obstacle (human or otherwise) is in the path of the robot. The neural network decision unit uses this combined detection map and other information, such as trajectory and velocity data, to formulate a strategy for avoiding a collision, and then transmits this strategy to the robot control. There are several advantages to this approach, primarily faster computation of collision avoidance decisions.

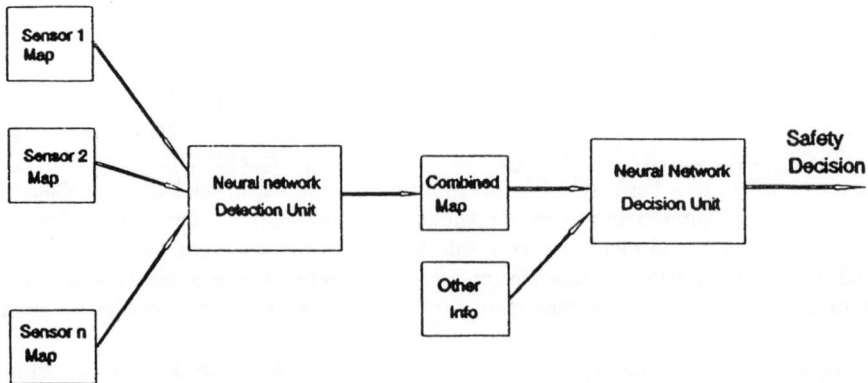

Figure 1 - Neural Network for Sensory Integration

The training of the neural network can be done off-line, using any of a number of training methods [4]. In this case it was decided to use a two layer neural network and error back-propagation training. The objective was to approximate the results which would be obtained from Dempster's rule of combination for a Dempster-Shafer representation of the sensory belief space [5]. The learning algorithm was simulated for several sensor configurations and input combinations. In each case 300 random training vectors were generated and used as input during the training process. The output of the neural net was compared to the exact Dempster-Shafer output to generate the error vectors. The network weighting factors converged after approximately 800 cycles of the training algorithm.

4. Evaluation through Quality Function Deployment

Quality function deployment (QFD) is a set of planning and evalution methodologies used for ensuring quality throughout each stage of the product development process. The 'House of Quality,' the basic tool of QFD, originated at Mitisubishi's Kobe Shipyard Site in the early 1970's. The foundation of the house of quality, is the belief that products should be designed to reflect the customer's desires, and that these desires should be translated into design targets and major quality assurance points to be used throughout production. The House of Quality is a conceptual map which identifies the relationship between customer requirements and engineering characteristics, using symbols to indicate the degree of positive or negative influence. This approach has gained wide acceptance in industry due to its ability to facilitate communications between the different constituencies in a manufacturing organization [6].

In a House of Quality chart, the customer attributes are listed in the extreme left column and their relative importance are specified in the next column. Along the top of the chart, those engineering characteristics that are likely to affect one or more of the customer attributes are listed. The relationship between the customer attributes and engineering characteristics are identified in the matrix. The objective measures, technical difficulty, imputed importance, estimated costs, and target values are associated with the engineering characteristics and shown at the bottom of the matrix. A preliminary House of Quality for safety design of a robotic workcell is shown in figure 2. The customer attributes in this case are safety attributes, and include cell area, safety zones, work environment, work related hazards, dependability, etc. The engineering characteristics, listed across the top of the matrix, are those factors under the control of the designers which can impact the safety attributes, and include controls, sensing systems, etc. The body of the matrix will be populated by either numbers or symbols indicating the interrelationship between engineering characteristics and safety attributes. The objective measures will be listed at the bottom of the diagram.

5. Conclusions

This paper has briefly presented three new methodologies that can impact the efficient and effective design of enhanced safety mechanisms in robotic workcells through their incorporation into concurrent engineering activities in product/process design. It is hoped that use of these methodologies can help give a new prominence to safety in the design phase of manufacturing systems, as opposed to being largely an afterthought.

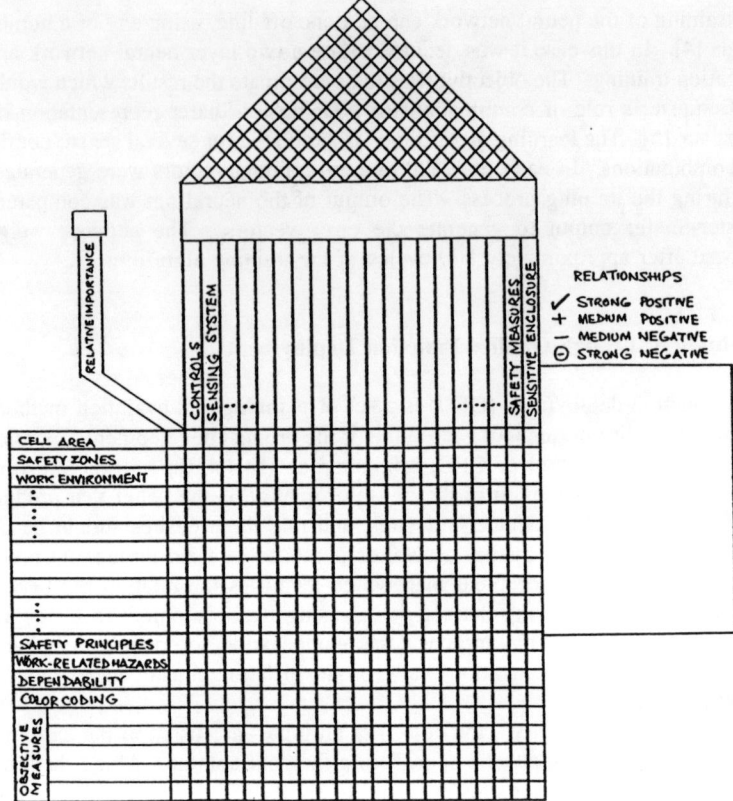

Figure 2 - Proposed House of Quality for Robot Safety

Acknowledgement

This work was supported under Award U60/CCU410085-01 from the United States National Institute for Occupational Safety and Health (NIOSH). Its contents are solely the responsibility of the authors and do not necessarily represent the official views of NIOSH.

References

[1] W. Zachary, J. Ryder, L. Ross, M. Weiland, Intelligent Computer-Human Interaction in Real-Time, Multi-tasking Process Control and Monitoring Systems. In: M. Helander, M. Nagamachi (Eds.), Design for Manufacturability - A Systems Approach to Concurrent Engineering and Ergonomics. Taylor and Francis, London, 1992.
[2] J. Graham, Safety, Reliability and Human Factors in Robotic System. Van Nostrand Reinhold, New York, 1991.
[3] American National Standard for Industrial Robots - Safety Requirements. Robotic Industries Association, Ann Arbor, Michigan, RIA/ANSI R15.06, 1992.
[4] J. Zurada, Introduction to Artificial Neural Systems. West Publishing, St. Paul, MN, 1992.
[5] G. Shafer, A Mathematical Theory of Evidence. Princeton Univ. Press, Princeton, NJ, 1976.
[6] J. Hauser, D. Clausing, The House of Quality, Harvard Business Review, 66(3), 1988, 63-73.

Advances in Agile Manufacturing
P.T. Kidd and W. Karwowski (Eds.)
IOS Press, 1994

A Design Environment for Concurrent Engineering

Owen Molloy[1], Therese-Lawlor Wright[2]

1 - CIM Research Unit, University College, Galway, Ireland
2 - Salford University, Salford, England

Abstract: In this paper, we look at design from the perspective of Concurrent Engineering (CE). It is necessary to develop a design environment to support CE, within which products may develop from conceptual to detailed design stages, with incorporation of life-cycle issues such as design for manufacturability, design for test, design for end-of-life, at all design stages. Through study of general design theories, we establish the fundamental activities of design which provide the basis for a design environment for CE.

1. Introduction

There are a number of major pressures currently making themselves felt in manufacturing industry [3], coming both from customers and the business environment, such as globalisation of the marketplace, environmentally benign production and recyclability and environmental costing. These pressures have raised the need to adopt the Concurrent Engineering (CE) approach to product development. The USA Defense Advanced Research Projects Agency (DARPA) Initiative in CE (DICE) states [4]: "CE is a systematic approach to the integrated, concurrent design of products and their related processes, including their manufacture and support". DARPA also concluded [7] that '..advanced computer software to assist a human team in considering all aspects of a product, including manufacture and logistical support, concurrently from the outset is essential for the development of high-quality products in the shortest possible time at affordable costs'. Therefore design methodologies to incorporate a true product life cycle view at the earliest stage in design are needed. Rather than attempt to develop a generic design methodology to satisfy a wide range of applications, we have researched the design process in relation to CE in order to produce an environment which supports the life-cycle view of design from the concept stage.

2. The Design Process

Research suggests that the following phases are common to most product design methods:

> Analysis - Problem (Customer Requirements) definition
> Functional requirements definition
> Conceptual design
> Detailed design

Other more detailed analyses of the design process from different perspectives have of course been performed. For example, Molloy et al. [6] and Tichem [11] suggest that initial process planning and detailed process planning be carried out in conceptual and detailed design, respectively.

3. General Design Theories

The axiomatic theory of design developed Suh [9], [14] describes the design process as the activity of mapping between domains (Figure 1).

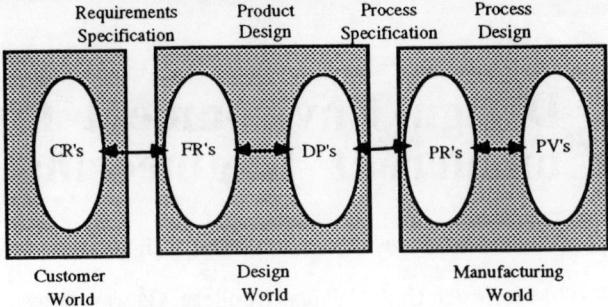

CR: Customer Requirement   FR: Functional Requirement
DP: Design Parameter   PR: Process Requirement
PV: Process Variable

Figure 1: Design as mapping between domains (after [8])

Yoshikawa [12] developed his general design theory as a methodology for the construction of CAD systems which tries to uncover the fundamentals of design in a scientific manner. In a similar manner to Suh, Yoshikawa argues that the design process is in fact a mapping process from the functional space to the attribute space, both of which are defined on an entity concept set (see also Bahrami and Dagli [2]). Therefore real design is possible as a convergent, evolutionary process, such that intermediate solutions are created and evaluated (metamodels) (Figure 2).

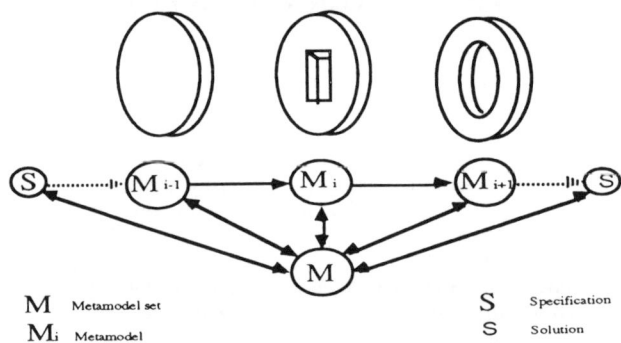

Figure 2: Evolution Model (after [13])

Our researches into the literature on general design theories and approaches to modelling the design process, have highlighted the basic underlying agreement between different views of the design process, as a mapping process between some requirement or specification expressed as the attributes of the desired solution, and the solution whose attributes best satisfy that specification. The value of the object oriented approach to this problem, in terms of modularity, intuitiveness, and hence ease of maintenance and expansion, is widely acknowledged. As stated by Warman [15], 'designers think in terms of features and objects, not in graphical entities', as seen in engineering drawings.

4. CE Approach to Design

To fully exploit, however, the advantages of CE mechanisms should exist to allow all phases of the product life to integrate and communicate effectively. One methodology which supports this aspect of CE and is well suited for team use is Quality Function Deployment (QFD) [1]. QFD is based on a matrix approach to design, mapping the requirements (starting with customer requirements) onto the means of achieving them. Therefore a series of charts may be developed which map the relationships between customer requirements, engineering characteristics, right through to production planning. However QFD offers the following advantages over other analysis methods [10]:

- It show mappings between requirements and components
- It shows potential requirements conflicts
- It shows the positive or negative impact of design elements on requirements
- It shows the positive or negative impact of design elements on other design elements

To allow concurrent development of the different product life aspects, a coherent, comprehensive product and process model is required. As stated by Krause et al. [5] *"product modelling is the key factor in determining the success of various product development strategies and industrial competitiveness in the future"*.

We can formulate a list of basic activities, involved in all stages of the CE design process.

- searching and matching to get entities / attributes (partially) satisfying requirements (e.g. customer requirements - functional requirements; functional requirements - design parameters)
- design structure manipulation
- creation of attribute values, targets, constraints and relationships
- constraint checking and analysis (reasoning & deduction, e.g. DFx)

5. A Design Environment for CE

Our work has shown the value of using a knowledge-based object-oriented approach to design systems and DFx and this is the approach adopted for building this environment. The design environment will assist the designer in building up the design starting from a specification or given customer requirements. Starting from given requirements, the designer should be assisted in finding objects which satisfy, or partially satisfy those requirements, by matching parameters. The QFD process can assist in finding those engineering characteristics necessary for the product, which can be used as inputs to the search for existing components and products. The product model must support the evaluation of constraints imposed by maufacturing and environmental requirements. Using an object oriented approach we can capture constraints information between objects as other objects. We can have many views of the product model, by selectively displaying the necessary information, while the actual data structure itself closely reflects the actual product structure. Rules can be applied to the product model to check all constraints, or constraints of a particular type (e.g. requirement satisfaction, or engineering conflict) to check whether they are violated. The designer may decide to trigger these rules or they may be triggered by specific design actions, such as the modification of a component parameter. Other types of constraints, which do not vary from product to product, such as environmental and DFx constraints, are represented as rules. These rules in turn will reference objects and their parameters. For example, DFA rules may contain references to specific assembly equipment parameters modelled as objects. Such rules are linked to the product model by the inference engine of the design environment, which searches the product model for objects whose parameters violate those rules.

An overview of the design environment architecture is shown in Figure 3. The following are descriptions of the main components of the architecture.

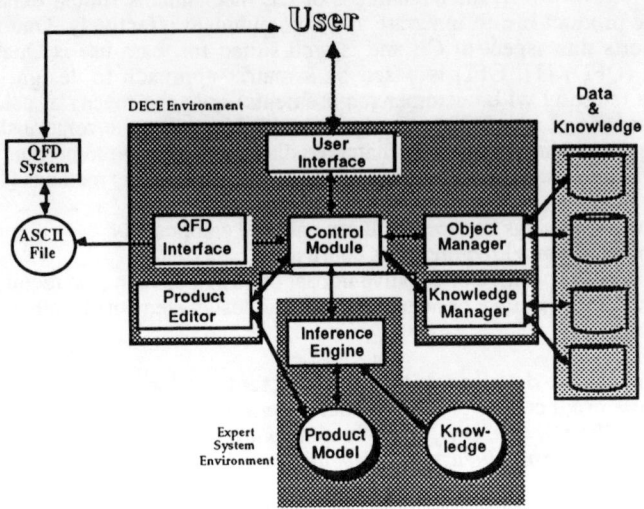

Figure 3: Architecture of the CE Design Environment

References

[1] Akao, Y. "Quality Function Deployment." 1990 Productivity Press. Cambridge, MA.

[2] Bahrami, A. and C. H. Dagli. "Models of design processes." Concurrent Engineering Contemporary Issues and Modern Design Tools. Parsaei and Sullivan ed. 1993 Chapman & Hall. London.

[3] Browne, J., P. J. Sackett and J. C. Wortmann. "The System of Manufacturing: A prospective study". 1992. DG XII CEC. Commissioned report.

[4] DARPA. "DARPA Initiative in Concurrent Engineering (DICE), Mission Statements". 1990. Defense Advanced Research Projects Agency (DARPA), USA.

[5]Krause, F.-L., F. Kimura, T. Kjellberg and S. C.-Y. Lu. "Product Modelling". CIRP Annals 1993: Manufacturing Technology. 42/2/1993. 1993. 695-706.

[6] Molloy, E., H. Yang and J. Browne. "Design for Assembly Within Concurrent Engineering". Annals of CIRP. 35. 1. 1991. 107.

[7] Reddy, Y. V., R. T. Wood and K. J. Cleetus. "The DARPA Initiative in Concurrent Engineering". Concurrent Engineering Research in Review. 1. Winter 1991/1992. 1991. 2-10.

[8] Sohlenius, G. "Concurrent Engineering". Annals of the CIRP. 41/2/1992. 1992. 1-11.

[9] Suh, N. P. "The Principles of Design." 1990 Oxford University Press.

[10] Thackeray, R. and G. V. Treeck. "Applying Quality Function Deployment for Software Product Development". Journal of Engineering Design. 1. 4. 1990. 389-410.

[11] Tichem, M. "Design For Manufacturing and Assembly: A Closed Loop Approach". International Conference on Engineering Design. ICED '93. August 17-19 1993. The Hague. 1993.

[12] Tomiyama, T., Y. Umeda and H. Yoshikawa. "A CAD for Functional Design". Annals of the CIRP. 42. 1. 1993. 143-146.

[13] Tomiyama, T. and H. Yoshikawa. "Extended General Design Theory." "Design Theory For CAD " Proceedings of the IFIP WG 5.2, Working Conference on Design Theory for CAD, Tokyo, Japan, 1-3 October 1985. Yoshikawa and Warman ed. 1987 North-Holland.

[14] Wallace, D. R. and N. P. Suh. "Information-Based Design for Environmental Problem Solving". Annals of the CIRP. 42. 1. 1993. 175-180.

[15] Warman, D. E. A. "Towards a System Architecture for Design For Manufacture". 1992. K.Four Ltd., Nene House, Town Bridge, Peterborough, U.K. Project Report.

Advances in Agile Manufacturing
P.T. Kidd and W. Karwowski (Eds.)
IOS Press, 1994

ECONOMIC MANAGEMENT OF THE PRODUCT DEVELOPMENT FUNCTION IN THE ERA OF CONCURRENT ENGINEERING

Torben Leinsdorff
Associate professor, The Technical University of Denmark

I. Introductory remarks

The present theory for economic management of the product development function concentrates on the economic efficiency of this function as an isolated part of the firm.

The theory of concurrent engineering has, however, led to an increasing understanding that the development function cannot be looked upon as an isolated part of the firm.

As a consequence of this, interest has arisen about the dispositional mechanism of the product development function. By the dispositional mechanism is here understood that dispositions (decisions) of the product development function decide the kind, the contents, the process and the efficiency of activities in other functions.

This recognition must be taken into account when discussing the economic management of the product development function. This management must not concentrate on the traditional economic efficiency of the function only.

It must also aim at ensuring that the dispositional mechanisms of the product development function work optimally for the manufacturing company as a whole. It will here be called integrated decisions.

On this background we have developed a model for the determinants to be taken into account in order to ensure the realization of this aim.

2. A model for the economic management of the product development function

2.1. The model in outline

The model operates with three maingroups of key factors for the economic management of the product development function:
- key factors about the formulation of goals to be met
- key factors about the measuring and following up phase
- key factors about communications

2.2. Key factors about the formulations of goals

The key factors about the formulation of goals to be met are:
- identification of the type of project to be done
- identification of the most important stakeholders of the project
- criterions of success for the stakeholders
- defining operational and quantitative areas of measurement

The key factors are to be seen on the background that - when the product development function is to make integrated decisions - a definition of goals to be met has to been done.

Main inspirators are here: Ehrlenspiegel, Noori and Olesen.

2.3. Key factors in the measuring and following up phase

The key factors in the measuring and following up phase are:
- measuring the areas of measurement
- following up the most important results.

The two key factors mentioned here are to be seen on the background that any control process must be based upon measurements and following up the results found.

Main inspirators are: Brown & Svensson.

2.4. Key factors about communications

The key factors about communications are:
- small project groups
- open and immediate problem solving
- open door policy
- a steering group staffed by top people
- early agreement
- staff of gatekeepers.

Communications are in general a main factor when integrating various functions of a firm.

In order to obtain a maximum of communications in the product development process it is important to have small project groups, an attitude to problems arising that they should be se solved openly and immediately, that there is an open door policy about the development process, a steering group staffed by top people, an early agreement on the details of the product to be developed and a staff of gatekeepers following the market development and informing the company about their findings.

3. Test of the model

The model was tested in one manufacturing company.

The firm analyzed uses a rather stiff system for cconomic management of the product development function. This system pays no attention to the type of development to take place. Together with an non-existing understanding of the need for integrated decisions it makes the economic management quite unsatisfactory. When speaking about communications, the firm does attend to some of the key factors of our model. However especially the policy of the open door and the use of gate keepers are very little developed.

On the other hand the key factors that the firm does not use now can easily be introduced into the firm. The main condition is that the attitude to product development is changed to a real understanding of the need for integrated decisions and that the system for economic management is changed accordingly along the lines outlined in our model.

On this background we found our model to be operational.

4. Final remarks

Product development and the successful economic management of this will be more and more important in the manufacturing company of the future.

Therefore it is important to develop methods for making this economic management function well. We hope to have made a little contribution to this.

Advances in Agile Manufacturing
P.T. Kidd and W. Karwowski (Eds.)
IOS Press, 1994

Concurrent Engineering - Key Implementation Issues

Dr Stephen Evans, Fiona Lettice and Palminder Smart
The CIM Institute, Cranfield University, Cranfield, Beds, MK43 0AL, UK

Abstract. A review of the literature, and the opinion of mature Concurrent Engineering practitioners, reveals the variety of issues faced by companies wishing to implement Concurrent Engineering. Key issues are shown to be primarily human or organisational in nature.

FAST CE, a sponsored project, aims to develop a workbook style implementation methodology that companies can use to increase the benefits generated by Concurrent Engineering while reducing implementation costs, risks and time. It does this by tackling the key human and organisational issues and by guiding the implementing company along routes shown to work in other implementations.

The key issues, the benefits associated with tackling them, and the implications of not addressing them effectively at an early stage in implementation are described in this paper. Mechanisms for dealing with the issues are explained and their inclusion in a prototype implementation methodology for Concurrent Engineering described. The implementation methodology has been developed through live application in collaborating companies.

1. Introduction

Concurrent Engineering (CE) is concerned with integrating all of the functional inputs involved in the whole product development lifecycle. The most effective way to achieve this integration is to form multi-disciplinary teams and task them with developing the product and its related manufacturing and support processes.

This paper will consider the human and organisational issues surrounding the implementation of Concurrent Engineering product development teams. The role of technology in CE is de-emphasised. Key issues are considered to be those that, if addressed early in CE implementation, will accelerate change and help companies to realise business benefits more quickly.

Some of the key issues are presented in Sections 3 to 5 of this paper, under the three stages of Preparation, Pilot Implementation and Expansion. These map on to the three-stage FAST CE Implementation Methodology.

2. Process for Arriving at Key CE Implementation Issues

An extensive literature survey of Concurrent Engineering and similar change management implementations was carried out. Regular forums have been held where mature CE practitioners share their experiences of CE implementation and discuss specific problems and solutions. Consulting to companies implementing CE provides the third source of issues facing those wishing to accelerate new product development performance improvement.

3. The Key Issues - Preparation for Implementation

Before a CE implementation is begun, it is vital that senior management are committed to the change. CE will evoke some issues that can only be resolved by those in senior positions, with an overall view of the whole business. Senior managers must make CE implementation a priority and be seen to be supporting it wholeheartedly by the entire organisation. A clear demonstration of commitment is to set up a Steering Committee and appoint a CE champion who are responsible for driving the change through the organisation and providing direction. Lack of senior management commitment is the single most given reason for CE implementation failure.

CE is a long term change programme. It often involves investing heavily in existing human resources at the beginning of a project to reap rewards later. If the costs and benefits are not understood, CE implementation may be abandoned before results materialise. Gaining a prior understanding of the benefits will help to improve commitment to change. This can be achieved by attending conferences, talking to people involved in CE implementation and visiting Best Practice companies.

Middle management may feel that they will lose control over decisions and budget, as well as losing people from their department. They must be assured that they will continue to have a role in the future, but that this will be different. If middle managers new role is not clearly and explicitly stated, they may remove resources from the CE team and slow product development progress. Middle managers should be actively involved in CE planning and implementation.

Senior Management must produce a CE implementation plan which shows the goals and targets for the CE implementation. The plan should show the resources needed, budget, and timescales. It will include detail about the first pilot project and outline for subsequent projects. Thorough up-front planning will help to prevent unnecessary confusion later. Potential risks should be identified and contingencies planned.

4. The Key Issues - Pilot Implementation

The next stage is to actually begin product development activities and launch the CE implementation with a pilot CE project.

Creating a supportive team environment is critical to the success of CE. The multi-disciplinary team should have a clear contract, negotiated up-front. This should detail the expectations of all parties involved, namely the Steering Committee, Team Leader, Functional Managers and Team Members. Attention should be paid to defining a clear product specification, detailing exact customer requirements. The benefits of having an explicit contract and mode of working are that everyone has clear expectations of the project and each other. The contract also provides a reference point for when conflict arises, allowing it to be quickly resolved. Another major advantage is that the CE team knows which decisions they are empowered and authorised to make, so speeding up problem resolution and task execution. The contract allows responsibility for day to day activities to be devolved to the team.

The team should be given ownership for the project, by holding the responsibility to develop their own project plan within guidelines stated by Senior Management. This helps to build commitment to the project and gives the team common purpose, direction and goals. The team are the most qualified to decide how to eliminate non-value adding activities from the product development process and can most easily identify where breakthrough savings can be made. It will give them a clearer understanding of the whole process and how all of the product development activities link together and affect one another.

If the team are not empowered or given a supportive working environment, they will be no more than a group and the benefits of team product development will not be realised.

The FAST CE Implementation Methodology creates the conditions necessary for high performance product development teams by defining Roles and Responsibilities for all participants. To illustrate, some typical roles and responsibilities for a Team Leader may include:

> To manage the team decision making process without making team decisions
> To ensure the project plan reflects reality and urgency
> If milestones are compromised will discuss with Steering Committee
> To listen, involve and communicate internally and externally
> To give guidance and support to team members

These form the working contract which must be developed and agreed by all parties. They are not imposed on the team, as the team play an active role in their development. These 'Relationship Statements' may be re-negotiated as the project progresses to reflect the changing requirements of all parties and as mutual trust grows.

FAST CE ensures that the Steering Committee sets a clear performance target for the team, which is aggressive yet achievable. This indicates to the team WHAT they must achieve. The team must then decide HOW it is to be achieved through the development of a detailed project plan.

Effective and efficient communication is central to speeding up the product development process. FAST CE recommends that the team be collocated, and should when possible comprise full-time team members. All the relevant functions should be involved in product development decisions as early as possible These mechanisms greatly improve communication, reduce external distractions and speed up decision-making processes.

The implementation methodology, which is undergoing a continuous development process of use-refine-use-refine, sets out to create specific management actions, early in the CE implementation process. These actions aim to give clarity and comfort to CE teams experiencing large changes in their daily work.

The methodology ensures that key issues are addressed using simple but proven techniques. It provides a 'common-sense' action-orientated approach to CE implementation. The underlying ethos emphasises the virtue of learning to manage a CE implementation by actually doing it. It provides the basic know-how and wisdom accumulated from best practitioners. Learning through doing exposes companies to new and valuable experiences which continuously improves understanding of how to maximise the benefits from CE.

The methodology is particularly designed to cater for organisational and contextual uniqueness as CE implementations will vary from company to company. Using key actions which improve the CE implementation process, individual companies can develop their own 'best practice' for product development.

Comfort is an important by-product of the working methodology which accelerates change in a reassuring environment.

5. The Key Issues - Expansion

Once the pilot project has finished, it should be reviewed and lessons carried forward to subsequent projects. CE will begin to spread throughout the organisation. This will mean that existing systems and structures will need to be realigned to accommodate a new way of working.

Existing performance measures and appraisal systems, which reward individual behaviours will need to be reviewed. They will be supplemented by team appraisal systems which reward individuals for team effort. If team behaviour is not recognised and rewarded, teams will lose their motivation to perform. Different career paths will need to be identified in a much flatter organisation structure.

CE implementation will cause a shift in emphasis from vertical functions to horizontal processes. Teams and management will be responsible for maintaining a process focus for product development. The teams will look for ways to continually improve and streamline the whole product development process. Management will change the organisational infrastructure to better support the process and institutionalise change.

6. Conclusions

Initial research has indicated that some issues are more significant than others and so need to be tackled by companies early in Concurrent Engineering implementation.

These issues relate to the creation of an environment where individuals can come together as a team and so out perform the sum of their previously individual contributions. Some examples of how such changes can be brought about, using a methodology focussed on early mangement actions, have been given. These illustrate the common-sense nature of the FAST CE Implementation Methodology.

References

[1] Clark K B and Fujimoto T (1991): <u>Product Development Performance: Strategy, Organisation and Management in the World Auto Industry</u>, Boston, Massachussetts: Harvard Business School Press

[2] McGrath M E, Anthony M T and Shapiro A R (1992): <u>Product Development: Success Through Product and Cycle-Time Excellence</u>, Boston, Massachussetts: Butterworth-Heinemann

[3] Johne A and Snelson P (1990): <u>Successful Product Development: Lessons from American and British Firms</u>, Oxford: Basil Blackwell

[4] Katzenbach J R and Smith D K (1993): <u>The Wisdom of Teams: Creating the High-Performance Organisation</u>, Boston, Massachussetts: Harvard Business School Press

[5] Reinertsen D and Smith P G (1991): <u>Developing Products in Half the Time</u> New York: Van Nostrand Reinhold

Advances in Agile Manufacturing
P.T. Kidd and W. Karwowski (Eds.)
IOS Press, 1994

Control, Contradiction and Complexity in a Pharmaceutical Research Company

Keith Randle and Al Rainnie

University of Hertfordshire Business School, Mangrove Rd, Hertford, Herts SG13 8QF,
UK. E-Mail:A.F.Rainnie@Herts.Ac.UK

Abstract. This paper addresses questions surrounding the control of the scientific labour process in a pharmaceutical research and development company. Set against emerging changes in both the structure of the industry and scientific method it focuses upon performance management and the management of working time as managerial control devices. The paper considers the tensions between the need perceived by managers for scientists to have a degree of freedom to act creatively and the tight controls actually operated by the company. It concludes that failure to resolve these tensions could lead to manifestations of discontent which could impact negatively on the continued commercial success of the company.

1. Restructuring Pharmaceuticals

A series of linked forces are having important effects on the pharmaceutical industry; the rise in the cost of R&D as public authorities require more thorough testing of drugs, the depreciating value of patents as patent life is eroded by testing procedures, pressure from public authorities to cut health care costs and the rise of generic drugs.

Competition, it is argued is based on marketing and development - the ability to bring drugs to the market quickly and sell them. della Valle & Gambardella [1] maintain that the new biotechnology base is encouraging a division of labour in innovation. New drugs increasingly result from complex relationships between agents with complementary assets. Drug *development* is still based on costly clinical trials, but research no longer requires lengthy screening necessitating large laboratories. Research into theoretical compounds provides opportunities for flexible and informal small and creative research-based organisations.

2. The Management of Research and Development

Glaxo R&D has a matrix management structure. Research is divided into a series of divisions based on scientific disciplines each led by a head of division. Project teams are multi disciplinary, with members drawn from across divisions. Each project is led by a leader who is responsible for its scientific direction, whilst the head of division retains responsibility for managerial concerns.

della Valle and Gambardella argue that small teams of talented scientists subject to loose control and with substantial autonomy in organising their work should be established in research companies. Companies will have to employ people who have strong academic backgrounds and who have internalised the values of the academic community. Clinical trials are required on any potential drug before it can be brought to market and these still require substantial financial resources, making them more effectively managed by larger hierarchical organisations. della Valle and Gambardella's approach to the managerial implications of the changing nature of research is somewhat simplistic and Shenhar [2] outlines an alternative approach in which projects are differentiated by their "technological uncertainty" leading to a fourfold classification: from low to super-high tech. Shenhar describes a set of tools essential to the management of these projects.

Both della Valle & Gambardella and Shenhar have adopted forms of technological determinism where management is a one dimensional activity, the most appropriate form of which can be read off from an examination of the product, process and personnel involved.

Friedman's [3] responsible autonomy-direct control dichotomy assumes a fundamental tension between the need to gain cooperation or consent from employees and the need to force them to do things that they may not wish to do in order that the goals of those in control of the labour process may be achieved. Research scientists would tend towards responsible autonomy. Others (Lowe and Oliver [4] and Hyman [5]) suggest that the relationship between management control and employee commitment is characterised by more complex "layers of control" which are not mutually exclusive and which may contain contradictions.

3. The Management of Research Workers

There was general agreement amongst the managers we interviewed that freedom of investigation, through relatively loose control structures is vital in a research environment. Set against this is the drive to produce profitable ethical drugs. This demands careful monitoring of projects to make sure that progress is being made towards developing a marketable molecule. Projects are closed on the basis of commercial decisions. Success in pharmaceuticals, it is argued, is based on an ability to combine capabilities such as running clinical trials and marketing with research. Management have to balance the perceived necessity of freedom of investigation with the demands of market driven monitoring, reviewing and closure of projects on a monthly basis.

Regular meetings review project progress. Managers are scientists, and so can keep track on developments, and project leaders have a day-to-day familiarity with both individuals and projects. However, there is a belief that at the level of the individual scientist there is a necessary degree of indeterminacy. To some extent management cannot know what scientists are doing, because by trying to control and codify the work, the risk is run of destroying creativity and commitment. The problem of controlling freedom, however, is manifested in the operation of both time management and reward systems. In addition project teams view themselves as being over-managed within a bureaucratic system that stifles innovation. Tight control over resources means that it is impossible to build on new ideas without approval.

4. "Trust Time"

From the mid 1980s, through to the early 1990s, the system of monitoring attendance at work within GGR evolved from flextime through to what is known as 'trust-time'. The original flextime system allowed employees to bank hours which could then be converted into days off to a maximum of twelve per year. This is now believed by management to be inappropriate, measuring attendance rather than activity or commitment. Under trust, core time consists of 10-12am and 2-4pm and employees should fulfil their commitment of a 37.5 hour week. It is hinted that those hours need not necessarily be spent on site. One clear gain for managers, under trust, is in the discretion they now have in determining employees right to take days off. But there is also the possibility that the system may lead to people being driven to exhibit 'good' behaviour, rather than actually working.

It may appear that a move from flex to trust time represents a move from direct control towards a system based more on responsible autonomy. This is not the case. Flextime gave workers a right to time off measured by a clock. Trust puts control of time-off in the hands of management and securing it is dependent on exhibiting performance and commitment in the eyes of line management. The result is two-fold; a reluctance to ask for days off, as this may be taken as indicating the wrong attitude and the necessity to demonstrate in a highly visible way that work above and beyond that expected is being performed.

The new system seems to have two benefits for management; in the name of greater freedom they are achieving higher levels of attendance (in days per year) and, further, the system of allocating days off is in the subjective control of line management, which has the added effect of pushing people to actively demonstrate high levels of performance. It also has the advantage of bringing the time management system into line with PRP.

5. Performance Related Pay

At GR&D individual performance is rewarded through a system of performance related pay. A pay review is the sole source of annual pay rises. There appears to be little guidance given to managers regarding the criteria that are to be adopted in assessment for PRP. This contributes to the subjective nature of the assessment process, but the explanation for this vagueness appears to lie in two areas: Firstly, the vast majority of projects initiated will not be successful in the sense of producing marketable drugs. Therefore performance cannot be measured in this way. Secondly, management perceive commitment to the team and to the project as essential to a successful outcome. Attitude, contribution to the team and motivation are the central elements defining outcome on prp.

Normative control relies on commitment and researchers at GR&D appear to demonstrate this. But we must differentiate between commitment to the job, work, or science and commitment to the employing organisation. Whilst employers seek to gain normative control through the internalisation of corporate values, scientists commitment may be to the work itself. The implication of this is that should the company decide to reallocate resources, resulting in the closure of a given project, employees might move to another organisation where personal goals can be met.

Normative controls arising from the internalisation of organisational values might be expected to negate the need for explicit forms of control represented by a system of PRP. Equally there may be a contradiction between the use of individual incentives and the use of project management techniques as a method of organising pharmaceutical research. Cannell and Wood[6] conclude that there is "real concern" about the effect PRP has on team working by overemphasising individual incentives. There is, in fact, little evidence that performance pay has any positive effect on employee task performance. There is then a tension beginning to emerge between matrix management structures, prp, appraisal and the trust time system. The system is highly and visibly subjective, giving great power to managers to act on their own perceptions of a persons contribution. This has led to a perceived need for people to be more active in demonstrating commitment and achievement. However, this is not straightforward. Individual employees may be better at communicating their successes, or even making their contributions appear more significant, than others. What we may find is that managers are actually appraising the ability to *communicate performance*, whether real or apparent, rather than the performance itself.

We have suggested that whatever problems arise from the interaction of management structures, prp and trust time etc may be largely suppressed until such time as the pressure of recession and restructuring on perceptions of the labour market is eased. However, it may also be the case that the effects of the reward systems will operate independently of any change in economic climate. The possibility therefore exists that employees could respond by being motivated, demotivated, unaffected or as suggested earlier, drawn into the "construction of a performance reality".

References

[1] della Valle F. and Gambardella A., (1993) " 'Biological' revolution and strategies for innovation in pharmaceutical companies", *R&D Management* 23 , 4,

[2] Shenhar AJ., (1993) "From low - to high-tech project management", *R&D Management* 23, 3,

[3] Friedman A., (1990) "Managerial Strategies, Activity, Techniques and Technology" in D Knights and H Willmott, *Labour Process Theory*, Macmillan

[4] Lowe J and Oliver N., (1991) "The High Commitment Workplace: Two Cases from a Hi-tech Industry", *Work, Employment and Society*, Vol. 5, No.3, pp 437-450 Sept.

[5] Hyman R., (1987) "Strategy or Structure? Capital, Labour and Control", *Work, Employment and Society*, Vol 1 No. 1 pp25-55 March

[6] Cannell M., and Wood S., (1993) *Incentive Pay - Impact and Evolution*, IPM and NEDC

Advances in Agile Manufacturing
P.T. Kidd and W. Karwowski (Eds.)
IOS Press, 1994

Resolving Conflict in
New Product Development

Helen MILL

University of Strathclyde, DMEM, 75 Montrose Street, Glasgow, G1 1XJ, UK

Abstract. This paper identifies factors which should be considered when deciding to introduce or continue a new product development program. By examining the different perspectives of mutlidisciplinary disciplinary teams and the tools currently used to assess project viability it is clear that conflict is probable. The use of the Project Review is discussed with reference to a pilot survey and its role in resolving discord is discussed.

1. Introduction

The decision to introduce a new product into development carries a high degree of risk and with the ever increasing use of multi-disciplinary teams associated with concurrent engineering the number of people involved in this decision making is growing. Each of these diverse disciplines will have different perspectives of priorities and acceptable levels of risk. As the project progresses there must be a constant review of these levels of risk. There are many decision making models for project selection and monitoring but most fail either because they do not fully consider the complex interactions of factors that can trigger failure or the models become so unwieldy that they are not implemented.

2. Factors Influencing the Success of New Products.

It is hard to identify factors that consistently trigger success as every new product development is unique in some sense. It is much easier to examine products in the market place and deduce the reasons for success or failure. This indicates only the tip of the iceberg as many new products are abandoned long before they reach the market place, but not before they have accumulated large development costs. It is estimated that in the pharmaceutical industry only 5% of new products ever reach the market. This is borne out by the results of the pilot survey which show that only one of the ten respondents had not had the experience of a project being cancelled, two of these cancellations occurred at the end of the development process and one after manufacture had begun. It is therefore important to deduce the factors that could lead to failure. These *risk initiators* must be monitored throughout the life of the project to ensure that the level of risk is acceptable for the stage of development and that risk reduction activities are implemented. The degree of risk that is considered to be acceptable will vary from industry to industry and from company to company dependent on both the aims of the company - market penetration, profit, diversification etc. and the company's financial position. The level of acceptable risk should be falling as the project progresses and certain factors relating to the customer needs should be clarified at a very early stage.

2.1 Identifying Risk Initiators

The starting point in establishing a list of risk initiators was a series of empirical studies looking at the performance of specific products in the market place. The first of these [1] conducted a correlation study of 177 Canadian firms to find the determinants of product success and identified 77 variables which could be grouped into 18 key areas and was followed by a further study of 252 new product introductions [2],[3] correlating their performance against organisation features of their development. Another study [4]examined 97 projects deemed to be failures and identified "managerial controlled" factors associated with failure. A further study [5] summarised the results of these previous analysis and drew up a list of key determinants which were matched against a survey of 52 firms to identify major variables relating to these determinants at the time of launch. These studies were used to create a list of potential risk initiators as shown in Table 1. This list was supplemented by information deduced from case studies.

Table 1 - Deduced Risk Initiators

Product Based Factors		Project Based Factors (Organisational)
Inaccurate Competition Analysis	Wrong Parts Tested	Lack Of Product Strategy
Inaccurate Market Analysis	Wrong Parts On The Product	Poorly Defined Design Process
Technology Loss	Parts Not Available	Lack Of Market Push
Volatile Market	Product Release Not Known	Technology Push
Inability To Manufacture	No Tolerance To Environment	New Market For Company
Inability To Sell	Rogue Products	Dominant Competitor
Wrong Target Customer	Retail Cost Too High For Market	Investment Source
Wrong Customer Needs Identified	Product Liable To Be Unsafe Use	Management Style
Inappropriate Targets Set.	Product Liable To Infringe Patents	Product Champion
Miss The Market Window	Lack Of Profit For The Company	Ignored Facts
In Adequate Development Skill	Wrong Product Image	Goals Unclear
Lack Of Integration	Costly Product / Process Disposal	Control / Feedback
Inadequate Technology	Degree Of User Abuse	Detailed Plan
Fails Legislation / Standards	Life Of Product Too Short	Communications
Unknown Technology (Company)	Maintenance Not Done	Contingency Plan
Wrong Concept Chosen	Interference From Outside Sources	Misalignment: Project To Company Goals
Wrong Development Personnel	Lack Of After Sales Service	
Errors In Analysis	Lack Of Product Distinction	Inconsistent Focus
Product Not Understood By Target Customer		
Concept Not Performing As Expected		

3. Conducting Project Reviews

The conclusions of the study by Cooper revealed some startling facts - in 75% of cases there was no market study, 77% had no trial market, 34% had no customer testing and 37% failing to conduct a financial analysis prior to developing the product and yet the evidence also showed that all these factors improve the chances of success. In contrast 85% of companies undertook a strong technical assessment of their product. It is clear therefore that project reviews can play a major part in the decision making process but with the wide range of personnel likely they are likely to be a potential arena for conflict. It is therefore important to investigate how these reviews can be conducted in a positive manner. A pilot survey of 10 companies was conducted to review current practise, the results are shown in Table 2. It is clear that multidiscipline teams are being employed and that marketing play a key role in these reviews. Although most respondents do conduct a review at the design specification stage most do not conduct such a review at an earlier stage and only 6 companies saw these reviews as a decision making forum others regarded it simply as a means of information dissemination. The topics of major importance shows a broad ranging discussion but, as with the earlier survey, shows a tendency for the technical issues to dominate.

Table 2 - Results of Pilot Survey

Questions	Choices		Results
Who? Who is invited to attend the project review meetings.	Company Directors	3	===
	Quality Assurance	6	======
	Manufacturing Personnel	9	=========
	Marketing Personnel	8	========
	Sales Personnel	4	====
	Purchasing Personnel	5	=====
	Selected Customers	2	==
	External Consultants	1	=
	Assembly Line Personnel	3	===
	Service Personnel	2	==
	Design Team	9	=========
When? At what point(s) in the new product development process are reviews conducted.	Never	1	=
	Before Project Specified	2	==
	After Project Specified	6	======
	Before Selecting a Concept	2	==
	After selecting a Concept	4	====
	At start-up Manufacture	5	=====
	During First Build	2	==
	At first customer release	2	==
	Regular timed intervals	3	===
Topics of Major Importance Measured by emphasis placed during the review (rated 1 -10)	Customer Preferences	7	=======
	Target Market Share	2	==
	Target Production Cost	8	========
	Market Window	6	======
	Configurations or Options	8	========
	Functions Available	10	==========
	Quality	8	========
	Reliability	7	=======
	Ergonomics	8	========
	Aesthetics	4	====
	Maintenance	4	====
	Installation	5	=====
	Performance Testing	10	==========
	Packing / Shipping	3	===
	Other	2	==

4. Decision Making Tools

As the review procedures are not being utilised to make the decision whether to initiate a new project it is important to identify the decision models that are available. These tend to fall into a number of distinct categories. Firstly there are economic models, these traditional accounting models which utilise expected costs and sales revenue and apply techniques such as Net Present Value to consider the time factor of investing money. There is evidence [6] that some effort is being made to include a wider range of criteria within such models but in general they fail to consider many of the deduced risk initiators.

Recently there have been a number of more complex models that have been created in an attempt to cover the wider picture. Such Multicriteria Models [7] are generally designed to suit a particular company and therefore could be tailored to include all of the risk initiators previously identified. Compensatory models [8] attempt to reduce a multicriteria problem down to a single aggregate value so providing both a means of ordering potential projects and a measure of the "distance" between them. However these models tend to assume that all the decision makers will hold a common view of the relative importance of one risk initiator to another. This is rarely true. In attempt to take account of this fact other models have adopted a further weighting system to reflect these differing views [9].

The main problem associated with using these advanced decision making models is their complexity and the amount of input information required. In a recent survey [10] it was found that the most commonly used means of selecting projects was a Financial Index method (65% of responders in both UK and Spain) which is based upon the traditional accounting methods. Other popular models were Economic Indices, Ratios (also based on accounting measures), Consensus Models and Checklists (10-20%). The use of checklists is also common within project reviews and quality audits. Although 10-15 % of the respondents were aware of "Dynamic Programming" or "Multigoal Programming" none of them actually utilised them in practise.

5. Conclusions

It is clear that the most favoured methods of project evaluation are simple economic models based on figures developed by marketing for projected sales and by engineering for projected costs. However these fail to consider many of the risk initiators such as changes in legislation, maintenance not performed correctly or a lack of after sales service. It is also clear that engineering , marketing and sales staff will consider different areas to be of prime importance.

It would seem therefore that the project review could be used to greater effect, particularly in the early stages of determining the potential viability of a project by utilising the risk initiators to provide a checklist or consensus model for project evaluation. This would provide the additional advantages that the decision making process would be consistent throughout the development project giving a clearer understanding to all participants of it's progress. The review would thus provide an opportunity for a structured debate and a common goal - to reduce project risk to acceptable levels.

References

[1] R. G. Cooper, The Dimensions Of Industrial New Product Success And Failure, Jnl Of Marketing, Summer 1979,P93-103

[2] R. G. Cooper, The New Product Process: A Decision Guide For Management, Jnl Of Marketing, Vol. 3 No 3, 1988, P238-255

[3] R G. Cooper & E. Kleinschmidt, An Investigation Into The New Product Process: Steps, Deficiencies And Impact, Jnl Of Production Innovation Management Vol. 3 Part 2 1986, P71-85

[4] J. K. Pinto & S.J. Mantel, The Causes Of Project Failure, IEEE Trans. On Eng. Management, Vol. 37 No 4; November 1990

[5] G. Lilien & E. Yoon, Determinants Of New Industrial Product Performance: A Strategic Re-Examination Of The Empirical Literature, IEEE Trans. On Eng. Management, Vol. 36 No 1; February 1989

[6] J.I. Ringuest & S.B. Graves, The Linear Multi-Objective R-And-D Project Selection Problem, IEEE Trans. On Eng. Management; 1989; Vol. 36; No 1;

[7] C. J. Bacon, The Use Of Decision Criteria In Selecting Information-Systems Technology Investments, MIS Quarterly; 1992; Vol. 16; No- 3; P335-353

[8] Baker & Freeland, "Recent Advantages In R&D Benefit Measurement And Project Selection Methods. Management Science Vol. 21 No. 10, 1975, P1164

[9] W.E. Saaty, Axiomatic Foundations Of The Analytical Hierarchy Process, Management Science, Vol. 32, 1986, P841

[10] A. M. Sanchez, R&D Project Selection Strategy., R&D Management Vol. 19, No 1, 1989

Advances in Agile Manufacturing
P.T. Kidd and W. Karwowski (Eds.)
IOS Press, 1994

Cooperation in Rapid Prototyping Environments

Hans-Jörg Bullinger and Joachim Warschat
Fraunhofer-Institut für Arbeitswirtschaft und Organisation (IAO), Nobelstr. 12, 70569 Stuttgart, Germany

Abstract: This paper gives an outline of the Rapid Prototyping Concept for Engineering Processes. A short introduction is followed by a description of the complexity of Product Development and of the resulting problems. New forms of team organization are proposed, together with CSCW-systems and a form of active knowledge representation of the experts, to make this complexity manageable.

1. Introduction

One of the crucial factors of competitiveness for companies today is a short time to market.

This imposes a very challenging problem due to the increasing complexity of new products, development and production processes as well as their organization, expert knowledge and technology.

Good examples are the new rapid protoyping technologies like stereo-lithography, selective laser sintering etc., which lead to short feedback-loops on the way from the idea via several prototypes to the final product. With these technologies more alternatives of the product can be studied and therefore the quality of the product can be improved. In organizing the product development according to a rapid prototyping process we get shorter learning cycles, better product quality and more flexibility [1].

2. Complexity of Product Development

Problems, today frequently mentioned, like too little orientation on customers, development times that are too long, product costs that are too high, and quality that is too low, are caused by an increasing complexity of the four main systems which make up the product develompent:

2.1 Product

The increasing complexity of products is expressed in their complex functionality, and a high number if varieties, construction groups and details. With only very few changes in the product, this leads to extensiv engineering activities with a number of improvement cycles, since the consequences of changes in the product model ar not forseeable and prototypes cannot be produced fast.

2.2. Technology

Conventional production processes, which often are very extensive, slow down the production of prototypes. The employment of generative production processes, and fast creation of virtual prototypes are going to be the main time-saving factors in this field [2].

2.3. *Organization*

Deeply structured organizations, and organizations with centralized management are often too unflexible to create a new product corresponding to the demdands of customers in only a short period of time.

Decentralized team organizations within the product development represents a faster method, but will demand new communication and information technologies [3].

2.4. *Knowledge*

The fast integration into a team is rendered more difficult by the fast increase of knowledge in connection with a high number of experts involved in the product development.
New forms of knowledge presentation and of communication will have to be developed for this process.

3. Cooperation of Experts

Cooperation of several experts within a team is one of the distinct features of product development, that distinguishes this process from the production work, where people of similar qulification work together.

Typical development teams, as,. e.g., for the development of a car seat, consist of designing, research and crash-computing engineers, specialists responsible for the calculation of costs, designers, bio mechanics, jurisconsults and experts on the materials used.

Another feature important in this field is the cooperation along vertical hierarchical lines.

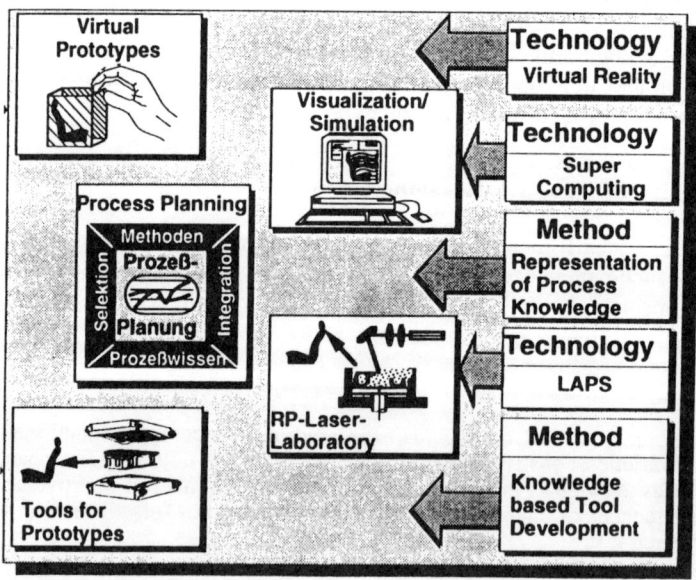

figure 1:   Cooperating experts and workplaces within a CSCE-network

This leads to new, decentralized organization and planning systems, which are includ computer-based tools. These systems are based on a highly complex set of rules for the coordination of activities of the individual team members. Temporal Reasoning as a field of default logics will be used to develop an apprapriate system, applying Allen-Relations to describe activities in time [4]. Thus the disadvantages of rigid planning methods, like network based planning, can be avoided as there are: no possibiltiy to display learning cycles, hierarchical structure of tasks, or unflexibility in comparing the actual state with the state required.

4. Computer Supported Cooperative Engineering (CSCE)

Since expert teams usually work on workplaces spread over a larger area, there is a need for development of specific communication technologies. Thus the creation of a multi-media network for the connection of work places, as it has been prototypical realized at the IAO, in connection with large capacity networks (ATM) is important, to make live video transmission of prototype processes or of virtual reality pictures onto the different workplaces possible. In particular, video conferencing has to be adapted to the requirements if it is to be used in product development as it is shown for different workplaces in ATM-networks (figure 1).

Thus design systems using CSCW are developed on the basis of active semantic networks, which define one workspace per staff member, e. g., one part of the Euclidean space in the case of geometrical design problems or of abstract workspaces for the functional design of the product. Between these workspaces date, or impulse for conflict solutions are transmitted in a regulated way.

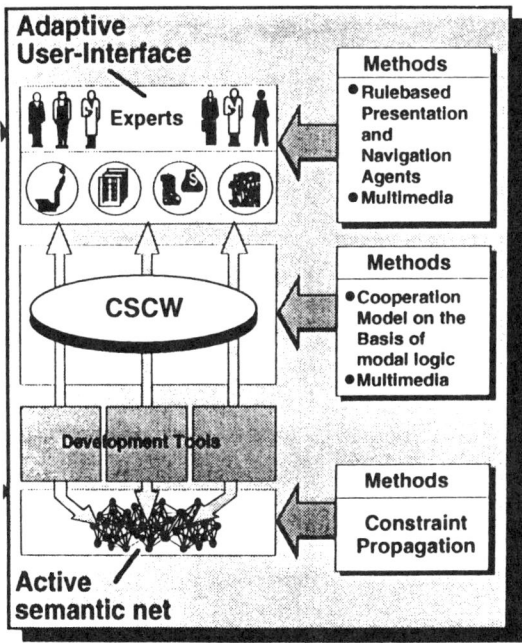

figure 2:   Expert Cooperation Supported by Active Semantic Net, Multimedia and Navigation Agents

5. Active Semantic Network and Knowledge Navigator

As basis for the cooperation serves an active semantic network in which, along with cause-effect sequences, knowledge on costs, production processes and quality as well as on forms of cooperation, organization and planning are filed. The active component is represented through methods that allow a constraint propagation within the network. Navigators, in the form of autonomous agents, are being developed to help the expert in detecting and extracting the knowledge he requires (figure 2).

6. The Rapid Product Development Center of Excellence in Stuttgart

A Center of Excellence has been set up in Stuttgart to further develop the organizational forms, information and communication techniques, and rapid prototyping technologies, mentioned above. This center involves 13 research institutes and the Daimler Benz Company with altogether more than 90 scientists.

References

[1]   Bullinger, H.-J.; Frech, J.; Warschat, J.: Development of Innovative Products. The Rapid
        Prototyping Approach. In: Orpana, V.; Lukka, A. (ed.): Production Research 1993. Proceedings of
        the 12th International Conference on Production Research, Lappeeranta, Finland, 16. - 20. August,
        1993, Amsterdam, London, New York, Tokyo: Elsevier 1993, p. 95-103.
[2]   IMS Intellingent Manufacturing Systems. International Conference on Rapid Product
        Development, 31. Jan. - 2. Feb. 1994, Stuttgart. FpF - Verein zur Förderung
        produktionstechnischer Forschung e. V., 1994, Nobelstr. 12, 70569 Stuttgart.
[3]   Baecker, R. (ed.): Readings in Groupware and Supported Cooperative Work: assisting Human-
        Human Collaboration. San Mateo: Morgan Kaufmann 1993.
[4]   Allen, J. F.: Planning as Temporal Reasoning. In: KR'91 Principles of Knowledge Representation
        and Reasoning. Morgan Kaufmann Publishers, Inc.: San Mateo, Calif. 1991, p. 3-15.

Advances in Agile Manufacturing
P.T. Kidd and W. Karwowski (Eds.)
IOS Press, 1994

Integration Mechanisms
Including Organizational and Technological Aspects

Hans H. K. ANDERSEN
Cognitive Systems Group, Risø National Laboratory, DK-4000 Roskilde, Denmark

Poul H. K. HANSEN
Department for Production, University of Aalborg, DK-9220 Aalborg Ø, Denmark

ABSTRACT Integration in manufacturing companies has become an important management area. However, the realization of integration is impeded by research fragmentation and lack of operational methods. This paper presents elements of an operational theory of integration in industrial enterprises. It builds on existing organizational and technological theory of integration and is furthermore embedded in a corporate strategic context by dealing with the questions of what, where and how to integrate. By combining a structure and a process oriented description of a company a new modelling technique is introduced. This modelling technique provides an identification and a visualization of the need for integration and a framework for designing integration mechanisms including operational goals for their performance.

1. INTRODUCTION

In order to support and facilitate distributed and dispersed work activities, modern organizations need assistance in the form of advanced computer systems. This is illustrated by the efforts in the area of Computer Integrated Manufacturing (CIM) to integrate formerly separated functions such as industrial design and process planning, marketing and production planning, etc., and by the efforts in the area of Office Information Systems (OIS) to facilitate and enhance the exchange of information across geographical distance and organizational and professional boundaries. Clearly the keywords in these efforts are *integration* and *information*.

Research is presently characterized by a great deal of interdisciplinarity seeking to give a broader understanding of integration. However, there has been no major breakthrough in terms of the introduction of frameworks for the study and design of integrated systems. One of the major obstacles seems to be the missing ability to visualize integration.

This problem is one of the major challenges for a current Danish research project "Integrated Production Systems (IPS)" [1]. Within the research area of Computer Supported Cooperative Work (CSCW) a current ESPRIT basic research project [2] focuses a corresponding problem related to the development of computer based supporting systems. The two authors are involved in each of the two mentioned research projects, and apart from presenting a temporary result the paper therefore also mark the first attempt to coordinate these research projects.

The specific objective of this paper is to contribute to the ongoing research process by presenting a modelling technique which, related to a specific enterprise, provides an identification and a visualisation of the need for integration and a framework for designing integration mechanisms.

2. THE OBJECTS FOR INTEGRATION

The practical relevance of integration is closely connected to the question of what to integrate, i.e. the objects for integration. In an industrial enterprise there are many potential objects for integration: departments, strategies, machines, computers, products, plans, goals, software applications, decisions, people, information, databases, knowledge, etc.

The issue of defining objects for integration must be viewed in parallel with the question of how a company is differentiated or dis-integrated. An enterprise may be viewed as a total system with input and output; but for practical reasons we divide the overall organization into departments and sections. There are several good reasons to divide a whole system into subsystems, e.g. to come up with manageable units and parts.

Lawrence & Lorsch [3] have contributed significantly to the discussion about how to integrate organizational units as opposed to differentiating them. We shall continue this tradition by introducing the organizational structure of departments and sections as one of the elements of our model for describing objects for integration.

Following the classical paradigm of organizational theory, we shall introduce an activity chain to represent the dimension of organizational processes [4]. An activity chain describes a series of physical and mental activities associated with the manufacturing of a part or a product, or associated with the handling of a customer's order.

Furthermore, we shall introduce the concept of a task which defines what has to be done, e.g. in terms of external requirements and conditions, internal constraints and specified objectives. Starting with the overall strategic task of the enterprise, i.e. its mission, we may develop a structure of tasks and subtasks.

A task will be carried out in one department or section; or it may be the common responsibility of two or more departments. Furthermore, in order to realize a given task, a series of activities is required. Hence, tasks relate activities to departments.

In the following section we shall show how these three objects for integration are interrelated and how integration mechanisms may be tied to a set of these objects.

3. THE MODELLING TECHNIQUE

Identification of the basic objects for integration has led us to suggest a modelling technique which combines a structure and a process oriented description of the company with tasks.

In figure 1 we have selected an activity chain related to a major issue, e.g. development of a new product, and a series of different activities have been identified. Above the activity chain we have shown the main tasks related to carrying out the activity chain. The tasks may be located either in one department or as the responsibility of two departments.

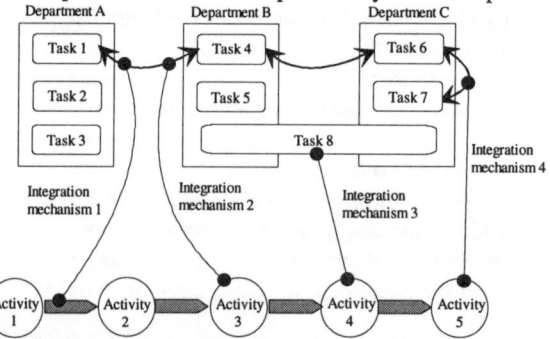

Figure 1   The DTA-Map depicting four integration mechanisms.

Figure 1 may help to identify areas where integration is important. For example, an integration mechanism is needed for the smooth transition between Activity 1 and Activity 2, requiring the coordination of Task 1 and Task 4. This coordination is also important for Activity 3; but may be of a different nature and thus calling for different integration mechanisms. In this way, the model which we shall call an DTA-map may depict the need for integrating *Departments* and *Tasks* in connection with carrying out *Activities* related to an activity chain. It may also initiate a discussion about which integration mechanisms to select.

In the following section we shall demonstrate the use of the DTA-Map in a case study of the production of technical documentation in an international manufacturing company.

4. CASE EXAMPLE ON ORGANIZING TECHNICAL DOCUMENTATION

The case study focuses on the identification and analysis of the characteristics of cooperative and coordinative work activities in the production of technical documentation related to new product development.

The technical documentation activities can be placed in the critical activity chain related to the new product development process as illustrated by figure 2.

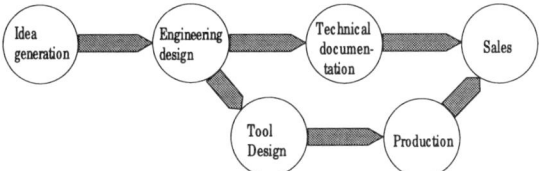

Figure 2   The critical activity chain related to the new product development process at ALFA.

As illustrated by figure 2 the technical documentation activities are running in parallel with the tool design and the production start. Before the products can be launched the technical documentation must be finished according to ALFA's status as an ISO 9000 certificate company.

A summary of the character of the identified nine critical integration mechanisms is shown in figure 3.

	Short Description
I-mech. 1	Coordination meetings (formal) and direct or mediated contact (informal)
I-mech. 2	Exchange of CAD drawings, draft technical documentation and various forms of product data
I-mech. 3	Two documentation scrutiny meetings (informal and formal)
I-mech. 4	Direct or mediate contact
I-mech. 5	Documentation review (informal)
I-mech. 6	Direct contact (informal) and monthly rounds (formal)
I-mech. 7	One or more documentation scrutiny meetings (formal or informal)
I-mech. 8	Direct contact (informal)
I-mech. 9	Documentation reviews

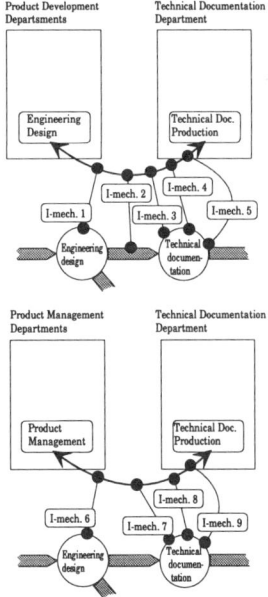

Figure 3   Summarize of the nine critical integration mechanisms identified.

5. CONCLUSIONS

In this paper we have presented elements of an operational theory of integration in industrial enterprises. It builds on the existing organizational and technological theory of integration, and furthermore it is embedded in a corporate strategic context by dealing with the questions of what, where and how to integrate. This has resulted in a modelling technique, the DTA-Map, which captures integration aspects related to activities along an important activity chain, the departments involved, and the relevant tasks. The DTA-Map has been tested in several case studies [5], and additionally an independent test is shortly reported in this paper [6].

Based on these tests we may conclude that the DTA-Map allow us to:

1. Define the need of integration, expressed in terms of activities, departments and tasks. This may provide a basis for defining operational goals for the performance of integration mechanisms.

2. Identify existing integration mechanisms in an industrial enterprise (technological and organizational mechanisms in combination).

3. Propose new or adjusted integration mechanisms which may improve the overall performance of the industrial enterprise.

The paper also mark a first attempt to coordinate two research projects on means for integration: The IPS-project [1] and the ESPRIT basic research project [2]. Our experience so far lead us to conclude that the results obtained in the two projects to a large extent are complementary and that the results in combination provide a beneficial framework for analysis and design of integrated manufacturing systems. We intent to continue our efforts to combine the terminologies and to include the two research viewpoints in our ongoing case studies.

REFERENCES

[1]   J.O. Riis (1994): "Integrated Prodcution Systems - Presentation of a Research Program", Technical University of Denmark & University of Aalborg

[2]   K. Schmidt (1993). Modes and Mechanisms of Interaction in Cooperative Work; In Computational Mechanisms of Interaction for CSCW; Schmidt, K. & Simone, C.; (eds.); COMIC Deliverable; Lancaster University.

[3]   P.R. Lawrence & J.W. Lorsch (1967): "Organization and Environment: Managing Differentiation and Integration", Harvard Business University Press, Boston

[4]   J. Frick & J.O. Riis (1991): "Activity Chain as a Tool for Integrating Industrial Enterprises", in Advances in Production Management Systems, E. Eloranta (Ed.), North-Holland, p. 331-319.

[5]   P.H.K. Hansen (1993): "Managing Integration in Manufacturing Systems - A Model of Objects and Mechanisms for Integration", Ph.D. dissertation, Department of Production, University of Aalborg

[6]   H.H.K. Andersen (1993): "Production of Technical Documentation in a Large Scale International Company", working paper, Cognitive Systems Group, Risoe National Laboratorie

Advances in Agile Manufacturing
P.T. Kidd and W. Karwowski (Eds.)
IOS Press, 1994

Computer Based Support for Cooperative Work in Engineering Design and Manufacturing

Eswaran Subrahmanian, Ira Monarch, Eric Gardner, Sean Levy and the n-dim group[1]
Engineering Design Research Center, Carnegie Mellon University, Pittsburgh, Pa, USA.

Abstract. Concurrent Engineering and its implications have been subject to discussion both in engineering research community and industry. In this paper, we use the theoretical framework for cooperative work developed by the Comic Project to explicate and further articulate the concepts of mechanism of interactions in the context of engineering design and manufacturing. We present a retrospective analysis of a case study involving the design of power trasnformers through the identification and composition of mechanisms of interaction. We are currently developing and implementing a computational environment that is a composition of several mechanisms of interaction.

1 . Introduction

Concurrent Engineering and its implications have been subject to discussion both in engineering research community and industry. These communities have adopted different meanings for the term concurrent engineering. For the engineering research community concurrent engineering has meant, for the most part, the use of techniques from artificial intelligence and other fields of computation to build cooperating set of tools from different areas of design and manufacturing using specialized representations and co-ordination mechanisms [1]. For industry, concurrent engineering has been interpreted as the creation of cross functional teams that include personnel responsible for all aspects of product life cycle [2]. Neither of these approaches to concurrent engineering sufficiently appreciates the importance of establishing and restructuring cooperating groups responsive to contingent situations nor do they acknowledge that this cannot be completely formalized. We agree with Schmidt (1993) that cooperative work is pervasive and that it occurs in forms that not explicitly recognized in existing organizational structures or in the development of design support tools[3].

2 . Theory into Practice -- Practice into Theory

On the basis of case studies and our ongoing effort on development of support systems for cooperative engineering work, we sketch here how the theory for computer supported

---

1. The n-dim group, other than the authors, consists of Robert Coyne, Allen Dutoit, Mark Thomas, Suresh Konda, Yoram Reich and Arthur Westerberg.

cooperative work developed by the COMIC project coheres with our development practice [4]. This theory is based on a number of case studies which identify mechanisms of interaction in cooperative fields of work. These mechanisms of interaction use stipulation and mediation by symbolic artifacts to reduce complexity. These symbolic artifacts are exemplary and indicative of the kind of things to look for when introducing computational symbolic artifacts into cooperative fields of work. There are even some clear cases of computational symbolic artifacts supporting mechanisms of interaction in cooperative work. However, in actual design situations these exemplars and precursors need to be significantly supplemented in order for appropriate mechanisms of interaction to be identified and computational symbolic artifacts to be created.

It is clear from our research and experience that designing such artifacts for actual engineering work environments involves close interaction between developers of computational support systems on the one hand and customers and users on the other. The developers bring their knowledge and experience producing software, their design and development tools, as well as their shells and prototypes for computer supported design work to negotiations with the customers and users who bring their knowledge and experience of a local work environment as well as their interest in improving productivity and cost effectiveness. To get these two design protagonists with their very different interests and perspectives to identify and agree on mechanisms of interaction is very difficult, though the COMIC theoretic framework does provide a common viewpoint, an abstract target into which these very different perspectives can be translated. This translation crucially depends on the specifics of the tools, shells, and prototypes that the developer brings to the design situation as well as the specifics of the customer's target work environment, organization and knowledge.

While we are very sympathetic to the theoretical framework developed by the COMIC research group, we believe it is important that this theory be further clarified and articulated in actual design practice. For one thing, there is a lack of tradition and experience in producing computer based symbolic artifacts. There are real qualitative differences between paper-based and computer-based symbolic artifacts in terms of the dynamics of interaction in producing and using them. People are already used to and comfortable with generating paper-based symbolic artifacts. This is not yet true of computer-based symbolic artifacts. Computers are relatively new on the work scene, and their use is second nature, if at all, only in a few contexts. By identifying cooperative work, mechanisms of interaction, and symbolic artifacts, the COMIC theoretical framework does set in relief the nature of some activities, tasks and artifacts in work settings for which computational support is relevant as well. However, the important differences between computer-based and paper-based artifacts as well as the lack of experience and practice with the former strongly suggest that this theoretical framework must be supplemented with practices, procedures, methods, techniques, tools, prototypes and shells which are elaborated and articulated in interaction with it. Moreover, experiences and data that are products and by-products of this interaction must also be fed back into a mutual articulation and elaboration.

3 . Cooperative work in product design and manufacturing

Product design and manufacturing involve shifting cooperative ensembles. While, mechanisms of interaction in these ensembles can be viewed as discrete and independent, they are also inter-related and constrain each other. A given mechanism of interaction may operate discretely and independently in timeframe, while the interdependencies between the mechanisms may operate in a different and longer time frame. Moreover, it is not possible to determine beforehand a sufficient set of mechanisms of interaction. Computational symbolic artifacts in these contexts may be crucially needed to support the co-ordination of these multiple mechanisms of interaction. Mechanisms of interaction can be expected, in any case, to operate very differently in a paper based environment than in a computer based environment.

While any mechanism of interaction is visible in its local setting, symbolic artifacts allow these mechanisms to function across locales. Computer based symbolic artifacts not only extend this functioning across locales but also address the *interrelation* among these mechanisms. This enables the composition of mechanisms of interaction for high levels of complexity.

The need to interrelate and coordinate mechanisms of interaction is borne out in our case study of transformer design [5]. Interpreting our study in light of the COMIC theoretical framework, it can be said that several mechanisms of interaction were encountered. One involved project schedule and status charts appearing on physical bulletin boards. More important are the symbolic artifacts serving as mechanisms of interaction contained in a design folder. A design folder is exchanged among many design participants. It contains multiple templates and other related documents which at any given time are only partially complete and sent to different participants for completion. When a discrepancy is discovered by one of the participants, a further template in the form of engineering change notices is brought into play. These templates represent a collection of mechanisms of interaction that need to be coordinated by the engineers responsible for the design. While the material existence of the design folder is *necessary* for co-ordination of these mechanisms of interaction, it is not sufficient. Currently, coordination is accomplished through interactions similar to conversation for action models by phone conversations and memos.

The case study also illustrates the existence of a classification mechanism that organizes prior designs. A hugh amount of information was classified. However, the classification uncovered in the case study was maintained by a single individual. Any access to this classification for collective use had to be through this individual. To make matters worse, every individual designer maintained his own classification of partial collections of prior designs. While a mechanism of interaction mediated by the classification artifact existed, it was fragmented and partially hidden, thus reducing its effectiveness in terms of access to information. Thus, this mechanism of interaction is only recognized informally and not integrated into the formal organization structure and procedures.

In summary there are two sets of mechanisms, one set operating within a project such as project schedules, drawings, engineering change orders, etc.; and another, the classification of designs over a long time period. These mechanisms are interrelated in that they affect refinement of design process over time. Based on this evaluation of the transformer design project, we are in the process of developing a computational symbolic artifact whose objective is to co-ordinate some of these above mechanisms of interaction.

The computational symbolic artifact we are developing addresses another problem in an engineering environment. In current design practice, computational design analysis tools have been introduced. Their introduction has changed practice while introducing another dimension of knowledge in engineering work -- the use and co-ordination of collections of such tools. This knowledge is for the most part tacit. The capture and representation of design tool use in the context of a particular design project is an important mechanism of interaction that is more easily realized in a computational medium. By creating this artifact, we are able to bring into the work process a mechanism that enables the replay and prototyping of designs using prior process paths. Further, it allows different engineers with different backgrounds and perspectives to critique and negotiate design changes. One of the results will be a reduction of engineering change orders. In effect, facilitation of concurrent engineering can be achieved through the identification and realization of mechanisms of interaction.

Acknowledgments: This work is partially supported by the NSF engineering design research center and a research grant from ABB.

References

[1] H. Park, M. R. Cutkosky, A. B. Conru, and S. H. Lee, An Agent Based Approach to Concurrent Cable Harness Design, *Artificial Intelligence for Engineering Design, Analysis and Manufacturing*, Vol 8 (1) (1994) 45-63.

[2] S. L. Konda, I. Monarch, P. Sargent and E. Subrahmanian, Shared Memory in Design: a Unifying Theme for Research and Practice, *Research in Engineering Design*, Issue 4, Springer Verlag, 1992, pp. 23-42.

[3] K. Schmidt, Modes and Mechanisms of Interaction in Cooperative work, in Computational Mechanisms of Interaction for CSCW, ed. C. Simone and K. Schmidt, Computing Department, Lancaster University, Lancaster, U.K., pp 21-104. - [COMIC Deliverable 3.1]

[4] COMIC Project Deliverable 1.1 and 3.1, Computing Department, Lancaster University, Lancaster, U.K.

[5] S. Finger, E. Gardner, and E. Subrahmanian, Design Support Systems for Concurrent Engineering: A Case Study in Large Power Transformers Design, In the Proceedings of the International Conference on Engineering Design, Heuritsa, Zurich, 1993, pp. 1433 - 1440.

Advances in Agile Manufacturing
P.T. Kidd and W. Karwowski (Eds.)
IOS Press, 1994

CAD MODELS ARE NOT MECHANISMS OF INTERACTION
Cooperative Aspects of Design For Manufacture

Henrik Borstrøm[1]*, Carsten Sørensen[2]
1. Institute of Manufacturing Engineering, Technical University of Denmark,
   DK-2800 Lungby, Denmark, Pho: (+45) 45931222, 4815, Net: proihb@uts.uni-c.dk
2. Cognitive Systems Group, Risø National Laboratory, DK-4000 Roskilde, Denmark,
   Pho: (+45) 46775148, Net: carsten@risoe.dk

Manufacturing organizations of today are facing a market where high flexibility and short leadtime is the name of the game. One of the ways to obtain higher flexibility and shorten leadtime is to implement a concurrent engineering process [1, 2]. Concurrent engineering increases the mutual interdependencies between engineering functions throughout the development process. Strengthening interdependencies leads to an increasing need in the engineering design process for the participants to engage in articulation work, e.g., allocation, coordination, negotiation, etc. of their activities [3, 5]. Most manufacturing projects are far too complex to rely on one omniscient and omnipotent actor maintaining a complete overview of the state of affairs at any given point in time [3, 6]. Design For Manufacturing (DFM) is in most cases a highly distributed activity, and can to a large extent be viewed as the process of articulating constraints and affordances between engineering designers and process planners [7]. Field studies in two Danish manufacturing companies documented the importance of ad hoc modes of interaction such as scheduled and un-scheduled project meetings, telephone conversations, faxes etc. as means of strengthening the mutual interdependencies between engineering design and process planning [3, 4, 6].

Computer Aided Design (CAD) work-stations play a major role in engineering design as flexible means of documenting decisions made throughout the process. CAD models are representations of the field of work, i.e., the product which is going to be manufactured. CAD models printed as blue-prints, represented on computer screens, faxed long distance etc. are important systems of reference in cooperative activities involving manufacturing engineers and process planners. The numerous detailed decisions made throughout the design process are reflected in the changes made to the model. The information in CAD models is, however, insufficient for carrying out a DFM process in a highly distributed manner, i.e., many participants with different areas of competence cooperating distributed in time and space. As argued by [3], articulation work mediated by changes made to the state of the field of work can only in a meaningfull way take place in very small and specialized settings. We do not view CAD models as mechanisms of interaction. A mechanism of interaction can be defined as "a device for reducing the complexity of articulating distributed activities of large ensembles by stipulating and mediating the articulation of the distributed activities." [8]. Except for the header information, the CAD model is a representation of the field of work. The only information relevant to who is or has been doing what in the header is typically: references to the location of the file containing the model, the signature of the draught-person and/or engineering designer, and the date it was produced.

---

* Author names are in alphabetical order only.

We will not discuss the content of CAD models. We are more interested in the types of information not stored in CAD models. Information which is of great importance for articulating decisions on interrelationships between engineering design and process planning decisions. The primary question asked in this paper is: How can CAD models be extended in order to improve the mediation and stipulation of articulation work in Design For Manufacture?

The field studies showed that although ad hoc modes of interaction might be one of the most important means of conducting concurrent engineering, they alone are not sufficient for handling the huge amount detailed decisions in a manufacturing project. When confronted with an abundance of detailed decisions that need to be articulated, organizations invent and adopt artifacts which can be interpreted as stipulating and mediating articulation work in order to reduce the complexity of this work. The fieldstudies identified several examples on such artifacts, e.g.: (1) The CEDAC board (Cause and Effect Diagram with the Addition of Cards) supporting articulation work between people from engineering design, process planning, electronic design, assembly, quality assurance, and quality control; (2) the product classification schema supporting distributed classification and retrieval of CAD models among engineering designers and draught-persons; and (3) the augmented bill of materials (ABOM) supporting integration between mechanical design, process

Augmented bill of materials: Experiment/functional model/prototype			
Instrument: *Name* Unit: *Unit name* Designer: *Name*	Type ID: *ID number* Product batch size *Integer* Draught person: *Name*	Page of pages Date:	
Categories	*1. component*	*2. component ...... 19. component*	*Responsible*
Component ID	*ID + version number*		Draught person
Description	*Text*		Draught person
Model name	*Model database ID*		Draught person
# Pr. instrument	*Integer*		Draught person
Model shop	*Check*		Draught person
Subcontractor	*Check*		Production planner
Production	*Check*		Process planner
Surface processing	*Check*		Project purchaser
New input materials	*Check*		Draught person
Machine ID	*ID number*		Foreman
CAM program	*Type of CAM prog.*		Foreman
Measure program	*Check*		Process planner
Foreman	*Initials of foreman*		Foreman
Delivery week	*Week number*		Production planner
Alt. delivery week	*Week number*		Production planner
Production time	*Estimated time*		Production planner

Controlled by process planner: *Signature*

Figure 1: The augmented bill of materials (ABOM). Each ABOM form holds information of up to 19 components for the same unit in the same instrument. All italics are our additions. The original is inverted, i.e., the rows in the figure are columns in the original ABOM. The original has 19 rows, corresponding to 19 different components. These are illustrated in three columns. This inversion is made in order to depict the whole ABOM and not only part of it.

planning, and production in relation to specifying and manufacturing experimental models, functional models, and prototypes [4]. In this paper, we briefly present the augmented bill of materials (ABOM) as an example of a paper-based artifact mediating and stipulating articulation work, and based on results from research within the Computer Supported Cooperative Work (CSCW) field, discuss which objects of articulation work can be considered for computer-based artifacts mediating and stipulating articulation work in Design For Manufacture.

The Augmented Bill Of Materials Supporting Design For Manufacture

One of the companies we studied is Foss Electric A/S. They manufacture instruments for measuring the compositional quality of milk. The personnel participating in projects include people from each of the areas of hardware design, electronical design, software design, chemical design, draught-persons, process planners, production specialists, as well as representatives from the model shop, marketing, quality assurance, quality control, and top management. The development from idea to final product involves a number of intermediate products: (1) A product concept defining the overall architecture and interaction between the involved technologies experimental set-up; (2) a few functional models (mock ups); (3) five to ten prototypes of the instrument used for verifying detailed ideas and designs; and (4) a test series of five to ten instruments in order to test manufacturability of the product.

One of the projects studied af Foss Electric was characterized by a relatively large number of both mechanical components and project participants. These circumstances led to the invention of an augmented bill of materials (ABOM) (see Figure 1). Earlier projects had used a more simple bill of materials, and in addition weekly meetings were held. The paper-based ABOM, together with a written organizational procedure, reduced the complexity of handing over CAD models for components from engineering design to process planning and production. It also served as input for scheduling activities in production. The organizational procedure stipulated who was responsible for which tasks when producing components for the experimental set-up, functional models, and for prototypes.

The engineering designer and draught-person fills in the information in the header. When the draught-person has finished polishing a CAD model, the model is sent to process planning via a server. He or she then fills in the first five columns (rows on the figure) containing information for each component, i.e., component ID, version number, textual description, database ID, batch size, whether the component are to be produced by the model shop, and whether new input materials are to be used. The ABOM is handed over, or sent by internal mail, to the responsible process planner. In process planning and production further information is added to the ABOM by the production planner, process planner, and foreman. They, amongst others, schedule machines, foreman, type of CAM program, measure program for quality control, delivery, and production time.

The ABOM supports distributed dispersion of tasks and responsibilities among actors and functions. The technical resources needed in connection with relevant tasks are also contained in the ABOM, i.e., the type of NC program and the designated machine. If new input materials for the component are to be used, there is a check-mark in a box, hence, there is a reference to material resources. The ABOM also contains references to field of work through the header, e.g., instrument and unit name, and batch size. References to time are made through the deadlines and the estimated times of delivery. As stated earlier, the ABOM represent an extension of a conventional bill of materials (BOM). When the BOM was used they also had the need for weekly meetings in order to negotiate, coordinate, and manage the distribution of tasks and responsibilities. The ABOM made these meetings redundant. It is reducing the complexity of articulation work among actors with different responsibilities regarding design, process planning, and production of components in a standardized way, structuring the route and relevant options a certain component or unit is going through.

Objects of Articulation Work in Design For Manufacture

Exploring how computational mechanisms of interaction can reduce the complexity of articulation work, Schmidt et al. [8] promotes a set of dimensions along which articulation work can be incorporated in mechanisms of interaction. We list these and show examples of attributes which can contribute to reducing the complexity of a distributed DFM process when incorporated in mechanisms of interaction:
1. *Actors*: Names of engineering designers and process planners. Who are are available and potentially relevant.
2. *Responsibilities*: Annotations of who is the owner of the CAD model or parts of the model. Who has filled in a given information. In case of a need for a meeting, who to contact.
3. *Tasks*: Goals, obligations and commitments, such as scheduling analysis of the properties of a new piece of machinery.
4. *Activities*: Jointly annotating CAD models. Negotiating the appropriate dimensions of, for example, a stamped hole in a plate.
5. *Conceptual structures*: Different types of classification schemes for determining importance of, for example; tolerances (which ones that are especially crucial), surface quality, level of competence of the participants, production prize of components.
6. *Informational ressources*: Pointers to written organizational procedures, general or local encyclopedias on engineering design and on process selection.
7. *Material ressources*: References to and annotation of characteristics of references to raw-materials, blanks, components, sub-assemblies etc.
8. *Technical ressources*: Descriptions of tools, fixtures, machinery, software applications. the operational characteristics such as machining tolerance, suitability for different types of materials, processing time and costs. Information on alternative processes.
9. *Infrastructural ressources*: Representations of rooms and buildings and operational characteristics such as capacity and location, turnaround time etc.
10. Demands and constraints posed by the *work environment*: directives and legislation.
11. The state of the *field of work*: The CAD model itself. Annotation of functionality.
12. The wider *organizational setting*: Corporate strategies. Linking to other projects
13. References to abstract systems of reference such as *time* and *space*.

The ABOM is a paper-based artifact, amongst others, reducing the complexity for mutual actors to cooperate around CAD models which are intrinsically computer-based representations of the field of work. It is, hence, not a far fetched thought to provide computer-based support of, in a distributed manner, to articulate work on CAD models with the purpose of DFM. By providing computer-based instead of paper-based mechanisms interaction, a number of opportunities and pitfalls arise. The are more comprehensively discussed by Schmidt *et al.* [8], and we here only provide a sample of important issues to discuss further. The paper-based artifact is in itself static. A computer-based mechanism or interaction can provide computer-support of dynamic aspects of articulation work, e.g., organizational procedures stipulating, for example, routing of information. No matter how well a paper- or computer-based mechanism of interaction is designed, there will eventually be situations when the standardized format will either have to be locally overruled, or perhaps permanently changed [3, 8]. In a paper-based form it is not a big problem to change the format by overwriting it. making permanent changes might prove more difficult. A computational mechanism of interaction must provide the functionality of overwriting the format if needed. Computational mechanisms also provides the possibility of having different participants choosing different views dependent on their needs in particular situations. The paper-based ABOM is restricted to the A4 format. All users see everything, although they might only need to see part of it most of the time.

Acknowledgments

Thanks to all the helpful people at Foss Electric A/S and Stelton A/S. Thanks to Peter Carstensen for participating in conducting and documenting the field study at Foss Electric, and to Kjeld Schmidt, Carla Simone, Peter Carstensen and Betty Hewitt for theoretical input. The research documented in this paper is partially funded by the Esprit BRA 6225 COMIC project, and by Fisker og Nielsens Fond, Ib Henriksens Fond, and the Danish Technical Research Council. All errors in this paper naturally remain the responsibility of the authors.

References

[1] Harrington, J., Understanding the Manufacturing Process. Key to Successful CAD/CAM Implementation. 1984, New York: Marcel Dekker.
[2] Lenau, T. Manufacturing Constraints in Concurrent Product Development. in Integrated manufacture and Design: The IT Framework. 1992. Heathrow, London: UNICOM.
[3] Schmidt, K., Modes and Mechanisms of Interaction in Cooperative Work, in Computational Mechanisms of Interaction for CSCW, C. Simone and K. Schmidt, Editor. 1993, University of Lancaster: Lancaster, England. p. 21-104.
[4] Borstrøm, H., P. Carstensen, and C. Sørensen, Two is Fine, Four is a Mess _ Reducing Complexity in Manufacturing, in Submitted for publication. 1994,
[5] Strauss, A., Work and the Division of Labor. The Sociological Quarterly, 1985. 26(1): p. 1-19.
[6] Sørensen, C. and H. Borstrøm, Small is Easy _ The Intricate Problems of Manufacturing an Ashtray, in Issues of Supporting Organizational Context in CSCW Systems, L. Bannon and K. Schmidt, Editor. 1993, Lancaster University.
[7] Sinclair, M.A., Human factors, design for manufacturability and the computer-integrated manufacturing enterprise, in Design for Manufacturability _ A Systems Approach to Concurrent Engineering and Ergonomics, M. Helander and M. Nagamachi, Editor. 1992, Taylor & Francis: London. p. 127-146.
[8] Schmidt, K., et al., Computational Mechanisms of Interaction: Notations and Facilities, in Computational Mechanisms of Interaction for CSCW, C. Simone and K. Schmidt, Editor. 1993, Esprit BRA 6225 COMIC: Lancaster, England.

Advances in Agile Manufacturing
P.T. Kidd and W. Karwowski (Eds.)
IOS Press, 1994

Computational Mechanisms of Interaction for Supporting Just-in-time Production Control

Betty Hewitt and Kjeld Schmidt
Risø National Laboratory, P.O Box 49, DK 4000, Roskilde, Denmark
betty.hewitt@risoe.dk, kschmidt@risoe.dk

Abstract. In modern industrial organizations, multiple actors are engaged in a myriad of complexly interdependent activities. The fluid meshing and control of these activities is achieved by the articulation[1] of the actors involved. To reduce the complexity and provide support for some of the functionality of articulation work, 'mechanisms of interaction' can be incorporated into computer based systems. A mechanism of interaction is an abstract device for stipulating and mediating articulation work, for example: timetables, scheduling schemes, kanban systems etc. A kanban system is a means of achieving just-in-time production control. This paper describes mechanisms of interaction, a field study of a kanban system and discusses the properties of a computational mechanism of interaction for just-in-time production control.

1. Introduction

In the design of conventional computer based systems, the core issues have been to develop effective computational models of pertinent structures and processes in the field of work (data flows, conceptual schemes, knowledge representations). While these systems are often used by many users in cooperative work settings, (e.g. database systems), the issue of supporting the articulation of cooperative work by means of such systems has not been addressed directly and systematically. If the underlying model of the structures and processes was 'valid', it was assumed that the articulation of the distributed activities was managed 'somehow'. It was certainly not a problem for the designer or the analyst.

In modern industrial and administrative organizations, the problems of articulating distributed activities are highly complex and the everyday social and communication skills are not sufficient in articulating the cooperative efforts of hundreds or thousands of actors engaged in a myriad of complexly interdependent activities, perhaps concurrently, intermittently, or indefinitely. Therefore to efficiently handle the complexity of articulation work requires artifacts that we have dubbed 'mechanisms of interaction'. Examples would be, scheduling systems, timetables, kanban systems, library catalogues etc. The first part of this paper discusses and describes mechanisms of interaction.

Field studies have been used with increasing frequency as a means of understanding the nature and complexity of cooperative work within organizations [1, 2]. Ethnographic techniques

---

1 By articulation we mean coordination, integration and management.

can be used to uncover the mechanisms of interaction used within organizations and in the second part of this paper we describe a field study conducted by Bjarne Kaavé [3] (Analyzed in Modes and Mechanisms of Interaction [4]) on a manufacturing company which uses the kanban system as a means of just-in-time production control. A kanban system can be viewed as a mechanism of interaction and is discussed from this perspective. This paper looks at how the kanban system is used in the cooperative work setting, how it supports the articulation of this cooperative work, and the problems encountered using it. From the description of mechanisms of interaction and the analysis of the kanban system in use, we discuss the properties a computational mechanism of interaction must therefore have for supporting articulation work for just-in-time production control.

2. Mechanisms of Interaction

A mechanism of interaction [4] is defined as a device for reducing the complexity of articulating distributed activities of large cooperative ensembles by stipulating and mediating the articulation work of these activities. In order to serve this function, such a device must have the following characteristics:

- It must be publicly available in that it does not solely reside 'in the head' of human actors and it must be persistent (i.e. independent of any particular situation). It must exist as an *artifact*.
- It must be loosely coupled to the state of the field of work so as to manipulate it independently of the field of work without any unwanted side effects. That is, it must be a *symbolic artifact*.
- It must provide affordances to and impose constraints on work articulation and it must make the state of work articulation publicly perceptible. It must provide a *standardized format*.

As a device for the articulation of cooperative work, a mechanism of interaction should be distinguished from other symbolic artifacts (e.g. traffic lights) and documents in cooperative work settings such as: letters, drawings, reports etc. Written documents are certainly symbolic artifacts but they merely provide a medium of communication for conveying information. However, managing the complex flow of information objects may require mechanisms of interaction. For example, prescribed forms and routing instructions and classification schemes such as library catalogues and thesauri.

Because the allocation of function between human and artifact is dynamic and evolving so must be mechanisms of interaction. Using a mechanism of interaction always requires certain social conventions and skills, but in highly complex cooperative work settings these conventions and skills must be supplemented and supported by a mechanism in the form of an artifact.

From this perspective a kanban system can be seen as a mechanism of interaction. A kanban system for production control consists of a set of cards acting as the coordination mechanism, both as a carrier of information about the state of affairs and as a production order conveying an instruction to initiate certain activities [5]. Aoki [6] conceives of the kanban as a "semi-horizontal operational coordination mechanism". The observes that this mechanism "crucially depends on the skills, judgment, and cooperation of [a] versatile and autonomous work force on the shop floor".

3. A Field Study of a Kanban System in Use

The Alpha company manufactures specialized optical appliances, and it has about 50% of the world market for this category of equipment [3]. The company currently produces about 6,000 units a year in 15 different models, each in 7 different variants. The product consists of a cabinet housing a complex array of electrical and electronic equipment, and optical instruments. It is a very rich and complex environment and production control has to be very carefully planned.

In Alpha, a kanban system is used in the shaping department which manufactures cabinets from sheet metal. Altogether seven different processes such as cutting, bending, welding etc. are involved. Setup times for the different machinery range from about 15 minutes to less than an hour, and the lot size is determined by the overall flow and setup time of specific machines. Lot sizes vary from up to 1,000 in the cutting processes to less than 50 in the welding and the sub assembly processes. These processes are de coupled by buffer stocks. Most metal parts go through more than five processes before assembly.

The basic idea is that loosely coupled but interdependent production processes can be coordinated by means of exchanging cards between processes. A particular card is attached to a container used for the transportation of a lot of parts or sub assemblies between work stations. When the operator has processed a given lot of parts and thus has emptied the container, the card is sent back to the operator who produces these parts. Having received the card he has now been issued a production order. The basic set of rules of a kanban system are [7]:

- No part may be made unless there is a kanban authorizing it.

- There is precisely one card for each container.

- The number of containers per part number in the system is carefully calculated.

- Only standard containers may be used.

- Containers are always filled with the prescribed quantity — no more, no less.

As documented in the field study, the kanban system suffers from several problems. 1) It is not adequate for coordinating manufacturing operations which have large fluctuations in the amount and type of products in demand. [5, 7]. 2) In a kanban system, information only propagates 'upstream' as parts are used down the line and the information ultimately conveyed has been filtered and distorted by successive translations along the line upstream. The kanban system therefore, does not provide facilities to anticipate disturbances or to obtain an overview of the situation.

Since Alpha is faced with extreme variations in demand for different models, operators constantly find that the configuration of the kanban system (the number of containers per part number and the quantity per container) is inadequate. They are therefore continuously changing the configuration by, for example, pocketing a card for a time, leaving a card on the fork-lift truck, ordering new lots before a container has been emptied, handing cards over directly, changing lot sizes, etc. This is possible because an informal network of clerks, planners, operators, fork-lift drivers, and foremen cooperate directly in controlling the flow of parts. A member of this network will for example explore the state of affairs 'upstream' so as to be able to anticipate contingencies and, in case of disturbances that might have repercussions 'downstream,' issue warnings. That is, the indirect, dumb, and formal kanban mechanism is subsumed under a very direct, intelligent, and informal cooperative coordination. The cooperative ensemble has 'appropriated' the kanban system in order to increase its flexibility. They have thus taking over control of the system, and are controlling the production far more closely and effectively than warranted by the kanban system. This is possibly because of their deep knowledge about lead times and inventories in the shaping processes, and the flow of information through the network of what Assembly needs.

4. Conclusion

In conclusion, the question is what will a computational mechanism of interaction for just-in-time production control give us that the manual kanban system does not. We need to overcome the problem of large fluctuations in demand, be able to anticipate disturbances and get an overview of the current state of the field of work.

One of the problems with the manual kanban system is that is tightly coupled to the field of work. This means that the speed and pattern of propagation of information is restricted by the nodes along the line upstream. Having a computational mechanism of interaction would de couple the state of the field of work from the production control system, thus allowing the users of the system to see at a glance what is happening both up and down stream and allow an overview of the system. The operators of each process could see the state of the field of work from the point of view of their own process, thus incorporating some of the flexibility currently achieved by direct cooperative coordination into a mechanism of interaction. As a computational mechanism of interaction is not closely coupled to the state of the field of work, it could be altered quickly to more effectively cope with the demands of the market (the configuration could be changed), and if necessary, circumvented. Any changes to it could be documented and saved, thus providing support for major changes to the mechanism of interaction. It would also be possible to simulate different configurations, such as: changing lot sizes, altering set up times etc. using the mechanism of interaction, thereby finding optimal configurations.

The concept of mechanisms of interaction is being further developed[2] to determine what other features they must have for supporting the articulation of many different types of cooperative work and how they could be represented formally so as to incorporate them in a systematic way into the design of CSCW systems.

References

1. Hughes, J.A., *et al.*, *The Automation of Air Traffic Control.* 1988, Lancaster Sociotechnics Group, Department of Sociology, Lancaster University:

2. Heath, C. and P. Luff, *Collaborative Activity and Technological Design: Task Coordination in London Underground Control Rooms*, in *ECSCW '91. Proceedings of the Second European Conference on Computer-Supported Cooperative Work*, L. Bannon, M. Robinson, andK. Schmidt, Editor. 1991, Kluwer Academic Publishers: Amsterdam. p. 65-80.

3. Kaavé, B., *Undersøgelse af brugersamspil i system til produktionsstyring*. 1990, Technical University of Denmark:

4. Schmidt, K., *Modes and Mechanisms of Interaction in Cooperative Work: Outline of a Conceptual Framework*. 1993, Risø National Laboratory:

5. Monden, Y., *Toyota Production System. Practical Approach to Production Management*. 1983, Norcross, Georgia: Industrial Engineering and Management Press, Institute of Industrial Engineers.

6. Aoki, M., *A New Paradigm of Work Organization: The Japanese Experience*. WIDER Working Papers, Vol. 36. 1988, Helsinki, Finland: World Institute for Development Economics Research.

7. Schonberger, R.J., *Japanese Manufacturing Techniques. Nine Hidden Lessons in Simplicity*. 1982, New York: The Free Press.

---

[2] This work is part of an ongoing project; COMIC, supported by ESPRIT basic research.

Advances in Agile Manufacturing
P.T. Kidd and W. Karwowski (Eds.)
IOS Press, 1994

Development Tools as a Catalyst for Teamworking

Helen MILL and William J ION
DMEM, University of Strathclyde, Glasgow, Scotland, UK

Abstract. This paper discusses the role that development tools play in assisting in teamworking in the new product development process. It describes three development tools, Controlled Convergence Matrix, Quality Function Deployment and Failure Mode and Effect Analysis, and discusses how they can assist in team building and effective communication throughout a project. The paper concludes by suggesting that development tools are an essential team catalyst and means of communication in a design project.

1. Introduction

Effective teamworking is the key to the success of concurrent engineering, but sustaining a high level of group performance with a group of professionals from a variety of disciplines can be difficult to achieve.

One of the key issues is communication - both in terms of ensuring that all parties have access to a common core of information and in having a common understanding of what that data conveys. This problem has been approached in many different ways, ranging from an integrated computer network [1] to monitoring all communications, including telephone conversations, and confirming all decisions in writing [2]. While technology has a large part to play in ensuring the correctness and timeliness of data being distributed it is clear that jargon is often used which can cause lack of comprehension and misunderstanding and in many cases suppresses the inevitable conflict between parties. Research has shown that it is necessary to expose such conflicts at a very early stage [3] without endangering the team cohesiveness.

It is clear, therefore, that the answer lies in negotiated settlements using a common language framework. Achieving such a common framework is not easy, however, one possible route is through the use of a number of development tools which can provide such a forum in addition to their prime function.

2. New Product Development

2.1 The New Product Development Process

The new product development process consists of a series of activities starting with the identification of customer need and ending with the sale, use and eventual disposal of a

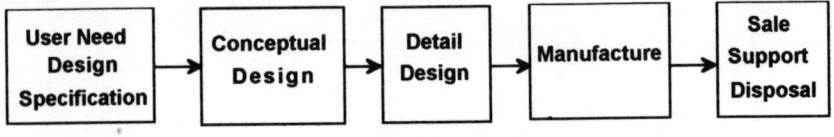

Figure 1

product. This process may be split into a number of different phases such as those shown in Figure 1. The way that a company organises its resources to undertake this activity can have a significant effect on its eventual success.

2.2 Design Organisation

In companies organised along 'traditional' lines different functional departments may be given responsibility for different aspects or phases of the process. This type of organisation has been shown to lead to protracted product development lead times, resources wasted on interdepartmental communication and poor design for manufacture - to name but a few [4]. This 'over the wall' type organisation reinforces and emphasises the differences between different functions within an organisation and does little to assist communication throughout the development process, let alone engender a team spirit and common ownership.

Recognising the deficiencies of the functionally organised project many organisations have embraced concurrent engineering and have introduced new working practices based around multidisciplinary design teams with an emphasis on reducing leadtime. Such working practices rely on the effective operation of multidisciplinary design teams with which the ownership of the project rests. As many members of such teams previously operated in functional departments there is a need to cultivate a common understanding and language between all team members. It is important to recognise that multidisciplinary teams can be organised in a number of different ways and that this common understanding must be imparted to both core team members who stay with the project throughout its life and seconded team members who may join the team for a part of the process.

In addition to the formation of multidisciplinary teams the implementation of concurrent engineering [5] is often accompanied by the introduction of a number of product independent tools and techniques which have been developed to reduce risk and to improve the effectiveness of the new product development process.

Three of the most important of these tools are Quality Function Deployment (QFD), Controlled Convergence matrix (CCM) and Failure Mode Effect and Analysis (FMEA). All three involve teamworking and play a crucial role in developing common understanding and communication at various stages in the development process. These aspects can be just as crucial to the success of a project as the output from the tools.

3. Development Tools

3.1 Quality Function Deployment

Quality Function Deployment (QFD) is a systematic process for translating the voice of the customer into product and process design, and then onto the operating systems for assurance that those designs are faithfully reproduced in manufacture. The 'customers voice' is satisfied by transforming the various needs into defined design requirements, which are in turn deployed into critical features in component part design. These critical features are then deployed through further QFD stages of operation - part design, process planning, and production control - to ensure that the critical customer-satisfying features are indeed produced. This whole integrating procedure uses a series of matrix charts as the tool for carrying the information.

The first QFD matrix identifies the relative importance of product attributes to the customer, maps the interrelationship between design features and functions and sets targets for the company's own products. The process relies heavily on the effective use of multidisciplinary teams who are involved in 'completing' the various QFD matrices.

A significant feature of the QFD process is that it spans the product development cycle from the voice of the customer through to manufacturing. This ensures that core team personnel are involved in the decision making process at the earliest possible stage and that they acquire an understanding of the customer requirements first hand. The physical process of completing the matrices is a powerful aid to resolving conflict and improving communication between disciplines. Team members meet face to face in a relaxed atmosphere and must assist each other by contributing their own specialised knowledge to the development of the QFD matrix. The matrices therefore act as a focus, concentrating effort and attention. Barriers between disciplines are broken down and misunderstandings due to use of jargon or inappropriate terminology are readily resolved. Additionally, new project terminology and jargon are often created in the early stages and consequently used throughout the project thereby imparting ownership to the whole team and helping in the process of team building.

The early identification of conflicting requirements and customers' priorities, which the matrices provide, bring possible areas of team conflict to the fore at the earliest possible opportunity. The early resolution of such conflict prevents divisions within the team later in the project when they are potentially more damaging. The common and explicit understanding which the matrices provide can assist in bringing such conflicts to an early and satisfactory conclusion. The process therefore provides a means of preventing long running and damaging disputes by bringing areas of conflict to the fore early in the new product development process thereby allowing early resolution through negotiation. The benefits of QFD to teamworking can be seen through the following quote [5].

"Everyone involved knows what the ultimate aim of the product is in terms of the customer's voice. The production engineer is not wondering whether he needs to maintain the specified tolerance on crankshaft balance; he can see what level of vibration is acceptable and together with the design engineer can organise and find how that level can be obtained."

3.2 The Controlled Convergence Matrix

The evaluation of design concepts has traditionally been the role of the design engineer alone, however, it is clear that other members of the new product development team also have a valuable role to fulfil by providing a different perspective. There is a real danger that poor communication and a lack of understanding between technical and non technical members will inhibit all the team participating in this most important phase. The controlled convergence matrix (CCM) [6] provides a vehicle for teamwork.

The CCM uses a matrix to compare each possible concept against a number of criteria chosen by the team, making use of the design requirements and the output of the first QFD matrix. By explicitly defining these criteria both in terms of how they can be "measured" and what constitutes a competitive advantage the team can gain a common understanding of what will contribute to the success or failure of the product under development. A common problem then emerges as it is often difficult for a team of such diverse background to agree on an exact criteria ranking order. When using the CCM such arguments become unnecessary as the technique seeks to optimise all criteria and the only ordering required is to identify the top third and the bottom third which is generally much easier to recognise. Once the criteria have been defined the best of the competition is chosen as a datum and the evaluation progresses by making a series of small decisions - does Concept A perform better, worse, or the same as the datum for Criteria 1 etc. so the expertise of each team member can be utilised where most appropriate.

Having completed the matrix the whole team can examine the overall pattern of results and suggest ways in which concepts can be changed to improve them. Such an informed debate can often produce new ideas but most importantly will produce a common understanding of the final decision, a better understanding of each other's disciplines and a consensus opinion which is vital to ensuring a team's cohesiveness.

3.3 Failure Mode and Effect Analysis

Once a concept has been chosen and development proceeds to a detail design phase it is imperative that it is engineered to a high standard. One method that is used to analyse a design for weakness is Failure Mode and Effect Analysis (FMEA). During this process each individual component is interrogated to find all possible modes of failure and ratings applied as to the severity, means of indication and probability of occurrence. This can be undertaken by the individual designer but it is more often conducted by teams including designers, manufacturing engineers, quality assurance and field service staff. This not only gives a wider field of experience to identify possible failures but gives visibility of future requirements for production, servicing and maintenance.

Such a thorough critique has obvious advantages in producing a better product but a team leader must be carefully picked not only for their expertise but also for their team building ability. Destructive criticism, whether justified or not, can cause disharmony and distrust throughout the team as individual's may be sensitive about their creation. It is important that the FMEA is used not only to identify faults but also as a forum to discuss ways to circumvent potential problems at all stages of the product development process. This may include the redesign of some components, new manufacturing procedures to ensure parts are correctly made and fitted or preventive maintenance action once installed.

4. Conclusions

Development tools, such as those described previously, when used at the right time in the new product development process provide:-
*   a common sense of purpose for the design team
*   a common language framework within the tools
*   a means of creating a common language framework for general interaction
*   increased involvement of all team members
*   a greater team consensus by better understanding of the decision making process
*   greater future visibility for the team

For these tools to succeed the following are required:-
*   a good facilitator or leader
*   a relaxed informal atmosphere where individuals are free to express their opinions
*   free access to up to date information
*   concurrent engineering organisation

This can be summed up by a quote from Brian Miles, Lucas Engineering
   "Tools can convert an average team into a high performance team."

References
[1] L.D. Burrow, Integrated Information Systems for Design and Manufacture, 1st Int. Design Manuf. Conference, 1990
[2] C. Hughes, Integrating Design in a Global Context, 1st Int. Design Manufacturing Conference, 1990
[3] Whitney, Designing the Design Process, Research in Engineering Design, Vol. 2 No. 1, 1991
[4] J. Hartley and J. Mortimer, Simultaneous Engineering, Industrial Newsletters, 1990
[5] D. J. Law, Integrated Tools and Techniques to Support Concurrent Engineering, Effecive Technologies for Engineering Success - Making CADCAM pay.
[6] S. Pugh, Concept Selection And Design Vulnerability, Design Prod Inst 4-5 Feb 1991

Part IV
Design and Implementation of Advanced Manufacturing Systems

Advances in Agile Manufacturing
P.T. Kidd and W. Karwowski (Eds.)
IOS Press, 1994

EUREKA Project - HITOP Development

Paul T. KIDD
Cheshire Henbury
PO Box 103, Macclesfield, SK11 8UP,
United Kingdom

Simone BLATTI
CIMCCSO
Route du Mont-Carmel 1, 1762 Givisiez,
Switzerland

Abstract. This paper outlines the background to the growth of interest in a technology implementation tool known as HITOP which was developed in the USA in the early 1990s. The nature of HITOP is described and the aims of an EUREKA project entitled *HITOP Development* are discussed.

1. Introduction

There is a growing interest in manufacturing industry, and in other industrial and commercial sectors, in methods and tools to support the design and implementation of advanced technologies, as well as in methods and tools for change management and to support the implementation of concepts such as concurrent engineering, total quality management, etc.

HITOP, which stands for the High Integration of Technology, Organization and People [1], is a tool to support the implementation of advanced manufacturing technology. HITOP was developed in the USA in the early 1990s, primarily by Ann Majchrzak (University of Southern California) and Mitchell Fleischer (Industrial Technology Institute).

Since the publication of the HITOP Reference Manual [1] there have been a number of HITOP applications in the USA (all very successful). However, the level of interest in the tool was, initially, quite small. In the spring of 1993, the Industrial Technology Institute (ITI) in Ann Arbor, Michigan, were approached by Cheshire Henbury Research & Consultancy (CHRC) with a view to selling HITOP in Europe and elsewhere. This resulted in an agreement for CHRC to sell HITOP. Shortly afterwards, ITI were also approached (independently) by the CIM Centre for Western Switzerland (CIMCCSO), with a proposal for translating HITOP into French and German in order to make HITOP more accessible to Swiss industry. This has led to further successful applications, for example, in GEC ALSTHOM T&D in Switzerland.

Subsequently, ITI, CHRC and CIMCCSO began to jointly consider further development of HITOP, forming an international consortium to undertake research and development work. The research project (entitled *HITOP Development*) is part of the EUREKA INTO (Integration of Technology and Organization for Quality Production) initiative [2].

2. HITOP As A Technology Implementation Tool

HITOP is a tool that helps with the implementation of manufacturing and information technologies. The HITOP analysis process enables the identification of organizational and job design implications of the critical features of the proposed technologies. This then helps with the identification of key task and skill requirements so that proper planning and development of human resources can be undertaken to meet operational needs.

HITOP comes in the form of a reference manual that provides step-by-step guidance, rationales for analysis, blank analysis forms, and worked examples. It covers a wide range of issues and is based on a six stage methodology.

The first stage of the methodology is concerned with assessing organizational readiness for change, which is followed in the second stage, by an assessment of the proposed technology, in order to identify its critical features. The third step is an analysis of the essential task requirements, leading to job designs which then leads to the fourth step, an assessment of skill requirements and how they will be met. The fifth step is concerned with determining how people should be rewarded. The final step is concerned with designing the organizational changes which need to be achieved given the technology and people requirements, which leads to the generation of a specific implementation plan.

The HITOP design tool therefore leads the analysts through: (1) An assessment of organizational readiness for change; (2) A definition of the critical technical features of advanced technologies; (3) The determination of essential job requirements, job design options, skills, training and selection requirements; (4) The determination of requirements and options for pay, promotion and organizational structure.

The analysis thus provides a direct and ordered consideration of critical technology, organization and people factors, and helps to identify those factors which require in-depth attention. The analysis also gives an expanded insight into the total organizational and people impacts of specific technologies, going well beyond skills and training.

HITOP allows the analysts to specify alternative organizations and different ways for managing people given specific technology plans. HITOP also provides guidance in determining the appropriate time for implementing technology plans, and helps to identify those equipment and system choices that are likely to create the greatest number of people and organizational problems, so that the people with responsibility for design and implementation may be better prepared to deal with these problems.

By performing HITOP analysis, the analysis team is guided by an iterative, systems based process in which all critical features of the organization, people and technology environment are systematically assessed and all implementable options are identified. This enables the analysis team to define the consequences of major decisions before those decisions are implemented. As a result, surprises downstream will be reduced and needed changes to the technology, the organization, or the people involved, can be identified.

3. Other Dimensions of HITOP

What joint technical and organizational design tools, such as HITOP, give us, is a means of dealing with organization, people and technology aspects of manufacturing enterprises, but for a specific purpose. HITOP is a technology implementation tool and its starting point is therefore technology.

While the starting point for HITOP analysis is technology, HITOP analysis provides an entry point into strategy development and change management processes. An important step in HITOP analysis is the assessment of organizational readiness change. During the process of assessment the existence or otherwise of a well defined manufacturing strategy, and the capability to manage change should become evident. In the event that HITOP analysis reveals deficiencies in one or both of these areas, opportunities exist to use the results of this stage of the HITOP analysis to focus attention on these deficiencies and to undertake the necessary intervention to remedy the problems identified.

Furthermore, HITOP is intended to show when technology options are not feasible (because of the organization and people implications), so that the technology can be redesigned to lead to different organization and people implications. In principle, therefore, HITOP can be used during technology development projects to highlight organization and people implication, before the design specification become fixed and hence difficult to change.

Another important point about HITOP, is that it is not necessarily a consultant's tool. Ideally, HITOP should be used by the people directly involved in design and implementation, and the role of an external consultant, if needed, should be to act as a facilitator and expert advisor. Whilst it is feasible for an external consultant to undertake the HITOP analysis, there are some disadvantages to this. Firstly, HITOP is designed to be a relatively low cost tool, and the problem with using an external consultant to undertake the analysis is that a considerable amount of the consultants time must be devoted to understanding the proposed technology plans. A far better approach is to use the consultant as a facilitator and advisor which involves much less cost. Secondly, a large amount of company specific knowledge could potentially be developed from a HITOP analysis. If the analysis work is undertaken by an external consultant, then no matter how much documentation is produced, a lot of this knowledge may well leave the company with the consultant. When the HITOP analysis is undertaken internally, then there is a much better chance of retaining this company specific knowledge.

An important feature of HITOP is that it is intended to be used by a team of people and not by one individual or professional group. To help ensure a successful outcome, HITOP analysis teams should consist of those people directly responsible for design and implementation of the technology, human resource experts, and the people affected by the technology. And in fact, the HITOP analysis process itself, can be used, with care, as a means of initiating team work and cultural change. This can happen because HITOP analysis using cross-functional teams, as described above, provides an opportunity for improving communications between functions, sharing problems and developing common values and experiences.

4. The Industrial Benefits of HITOP

HITOP has been used by companies such as Hewlett Packard, Digital Equipment Corporation, Solar Turbines (a subsidiary of Caterpillar), Douglas Aircraft, and General Motors in the US, and GEC ALSTHOM T&D in Switzerland, to achieve a high level of integration between technology, organization and people. HITOP has contributed towards improvements in both technology implementation and operational practices.

The experiences at Hewlett Packard in the US indicate the potential of HITOP. Using HITOP analysis and other team based concepts, Hewlett Packard achieved a three week ramp-up time to full production, at desired quality, for the fourth production line manufacturing formatter boards for the Hewlett Packard LaserJet IIIP printer. This ramp-up time was a significant improvement over the first production line, which had a ramp-up time of 18 months. The ramp-up times of the second and third lines were much shorter (owing to learning curve effect) but it still took five months to bring the third line to full production at desired quality. The massive reduction in ramp-up time was not the only benefit arising from the application of HITOP. The whole culture of the line was quite different from anything previously experienced. Communications between management and front line workers improved significantly and motivation and commitment soared. HITOP also facilitated empowerment, and so on.

5. EUREKA Project - HITOP Development

Although in the early days of HITOP there were only a limited number of applications, HITOP has proved itself to be a valuable tool - one worth supporting and developing further. EUREKA Project *HITOP Development* was therefore established with the aim of further developing HITOP by: (1) Making the tool more user friendly and developing a range of analysis aids; (2) Strengthening certain aspects of HITOP (e.g. assessment of readiness for change); (3) Extending the applicability of HITOP; (4) Integrating HITOP with other tools; (5) Investigating the generalisation of the underlying methodology to the implementation of concepts such as continuous improvement, concurrent engineering etc.

6. Acknowledgements

This paper is based on information provided by, and discussion held with Ann Majchrzak from the University of Southern California; Mitchell Fleischer, Mike Wood and Dick Shackson at the Industrial Technology Institute, Ann Arbor; and Scott Burton from Levi Strauss in San Francisco (formally leader of the HITOP team at Hewlett Packard's Boise plant). HITOP is a Trademark on the Industrial Technology Institute. Further details about HITOP and HITOP support services can be obtained by contacting the authors.

References

[1]   A. Majchrzak, M. Fleischer, D. Roithman and J. Mokray, Reference Manual for Performing the HITOP Analysis. Industrial Technology Institute, Ann Arbor, MI, 1991.

[2]   J.D. Gillis, Human Factors Collaborative Research in Manufacturing. These Proceedings, 1994.

Advances in Agile Manufacturing
P.T. Kidd and W. Karwowski (Eds.)
IOS Press, 1994

ACTION Integrates Manufacturing Strategy, Design, and Planning

Les GASSER and Ann MAJCHRZAK

Computational Organization Design Lab, Institute of Safety and Systems Management
University of Southern California, Los Angeles,. CA. 90089-0021 USA
+1.213.740.{4046/4023} {gasser/majchrza}@usc.edu

Abstract. This paper addresses several questions of how to strategically-plan, design, and implement highly-integrated, advanced-technology manufacturing systems, and how to assist planning/design processes with an automated, theory-based tool called "ACTION." ACTION addresses several critical issues of strategic planning (e.g. planning responsiveness, planning specificity, and generation of planning alternatives) by supporting the search and evaluation of alternative Technology, Organization, and People (TOP) integration strategies, and by linking high-level objectives and situational variables to detailed organization design features in a theoretically-motivated way.

1. Introduction

Proponents of strategic planning have stressed that it provides important structuring context for more tactical design and implementation decisions (e.g., [Steiner 79; Aaker 92]). In prescriptive accounts, strategic planning activities are carefully staged, sometimes quite explicitly, into phases that develop 1) analyses of external opportunities/threats and internal strengths/weaknesses, 2) analyses of gaps and matches between these internal and external arrangements, 3) strategies to exploit opportunities and counter threats, and 4) implementation mechanisms for realizing the strategies generated. A main attribute of a strategic-level planning approach is that it is focused on coordination among key decisions that affect an organization's ability to recognize, capitalize upon, or mitigate potential successes and failures, rather than on the particular content of such decisions.

Recently there have been several strong critiques of the value, purpose, and techniques of strategic planning (e.g., [Mintzberg, 94]). Some of the main content of these critiques is:

1) Planning as typically practiced is essentially a backwards-looking activity, because it relies on assessments of known states and known potential actions, projecting these into the future.

2) Typical planning cycles are long and hence they are not reactive and adaptive: planning decisions can easily be out of sync with current realities. For example, Glazier and Weiss [Glazier and Weiss, 91] studied (simulated) planning in conditions of increasing product change and reduced product life cycles. They found that under turbulence, planning organizations actually did more poorly than non-planners. This is consistent with well-known results in theoretical and simulated distributed activity systems (e.g. computer networks and feedback systems) in which shortened decisionmaking horizons (relative to the ability to make decisions) can lead to overall system performance that is chaotic in a technical sense (e.g. [Ishida et al., 92]).

3) The tremendous effort required to do planning tends to reinforce planning decisions and

remove flexibility. Paradoxically, under this critique, the more effort that has gone into a plan, the less likely that plan is to change over time, since more people and resources become committed to the plan and its underlying process as time goes on. This situation is problematic when the planning assumptions underlying planning decisions prove faulty or out-of-date (cf. 2 above).

4) Goal/strategy measures, even quantified ones, can be stripped of meaning and decontextualized when put into practice and locally interpreted, rendering the goals/strategies themselves meaningless, or, worse, useless.

5) In the end, a planning approach to strategy development, design, and implementation contributes little, because it turns out to be a restrictive, overly-complex, slow, embedded, unreactive process that actually impedes creativity and responsiveness. (Again, similar arguments have been made in a wide range of domains, including simulations of collective problem solving, e.g., [Durfee and Lesser, 88].)

Overall, these critiques reflect a view of planning as a stabilizing and long-term process relative to environmental change, and indeed this is often a main selling point of planning: to render a potentially unstable organization-environment relationship more predictable by thinking through and planning for future contingencies, and then taking proactive corrective action. Since planning means placing the future into the present (e.g. by modeling the future and analyzing it in the present), it also requires that the organization adhere to the plan in order to construct the planned-for future outcome.

The problems with this vision of strategic planning are twofold. First, it reflects the assumption that it is primarily the planned activities that transform an organization's current state into its future desired state. Second, is reflects the assumption that both an organization's internal and external relations are stable enough so that deductions based on present information will hold into the future. (This is a generalization, to planning, of a basic learning assumption: a learning system cannot learn about a random environment. Similarly, an organization can't plan unless some features of the world it is planning for are predictable.)

Alternatives to strategic planning thus come down to either 1) jettisoning planning altogether or 2) reconstructing planning and implementation activities to make them into a kind of continuous analysis and reaction process that will lead to a continuous realignment of organization-environment relations. Said another way, a rethinking of strategic planning would begin to link planning cycles specifically to the timing of the internal and external cycles they are meant to confront, in order to enact a closer coupling of organizational decisionmaking with both internal and environmental change.

2. Addressing These Problems in a Manufacturing Context: The Goals of ACTION

Our approach to manufacturing strategic planning in the light of these critiques is to begin to develop new planning tools, theory and methodology (in particular the ACTION project we describe below) to improve planning decisions as product and environmental changes increase, and as organization and product lifecycles decrease. Our hypothesis is that such tools can help refine planning content, planning effort, and planning time cycles to increase the precision and depth of knowledge that can be brought to strategic planning over less effort and a shorter reaction cycle. To the extent we can do this, we begin to move manufacturing strategic planning more toward the following ideals:

1) More rapid generation of strategic and tactical alternatives.

2) More comprehensive analysis of the impacts of alternative strategic choices.

3) Less effort expended to develop deeper knowledge of impacts and alternatives.

4) More rapid replanning cycles more directly tied to external and internal arrangements.

The benefit of improved, theory-based planning tools comes both from helping substantively with the ability to consolidate planning information, and by shortening the time needed to consolidate information. The aim is to have these joint effects multiply, keeping enterprises ahead of chaotic reaction points by improving both the quality of decisions and their timeliness relative to temporal thresholds for organization-environment instability.

3. ACTION's Support for Strategic Analysis

The aims of the ACTION project as related to strategic planning are to: 1) to capture both strategy and its planned articulation, using a well-defined and theoretically-defensible model, 2) to analyze tradeoffs among strategic and environmental elements for general internal consistency using Sociotechnical Systems (STS) theory, 3) to predict idealized operationalizations or articulations of strategy using STS theory, 4) to capture actual or planned-actual articulations of strategy, 5) to identify and prioritize gaps between idealized versus planned operationalizations, again using STS theory, and 6) to allow relatively rapid exploration of a space of alternative strategies, idealized articulations, and actual operationalizations.

In this way, ACTION is a tool which can help to link an organization's strategies to their consequences explicitly. Since ACTION's operationalizations of strategy involve many relationships 1) in great detail and 2) simultaneously, ACTION renders the meaning of strategies more definite and "web-like". The meaning of any individual strategic component is tied to many interacting organizational choices, rather than simply being abstracted away from context or reduced to a few subgoals. The idea is that strategies whose meaning is articulated through precise links to many organizational dimensions are harder to misinterpret. (Latour makes many similar arguments about building the strength and robustness scientific knowledge with his actor-network theory in [Latour, 87]).

4. Specifics of the ACTION Theory and Knowledgebase

ACTION's theory, methodology, and software efforts incorporate knowledge of technological and organizational features including: organizational objectives; specific technical system characteristics; variances (e.g., turnover, materials quality) job design; organizational unit structure; coordination among jobs and units; information, tool, and technology resources; skills and training needs and opportunities; decision-making discretion; customer involvement; performance management systems; values held by employees.

The ACTION TOP-integration model has been specialized to represent production areas of discrete-parts manufacturing organizations, and within this category, to four specific types of production context: short-life-cycle and general group technology cells, functional shops, and transfer lines.

The core knowledge developed in the ACTION Project---a comprehensive model of positive and negative relationships among these key organizational features---is called the ACTION TOP-integration Theory. An ACTION analysis reveals how a particular TOP design may be misaligned vis a vis this theory. The ACTION decision-support system helps users to capture and model specific organizational situations, apply theory and methods to them, and to perform "what-if" analyses. This tool supports four key functionalities, including both evaluation and design activities using summary and detailed organizational models. ACTION's summary-level analyses cover a few to several hundred relationships among two to several dozen features. ACTION's most detailed analyses cover thousands of relationships among several hundred TOP-integration features.

ACTION represents the TOP-integration features in a design or evaluation problem as a set of business objectives (goals) to be optimized within an environmental context that includes activity workflows and process variances (e.g. materials quality). ACTION's business objectives include several dimensions of throughput, quality, flexibility, manufacturability, etc. ACTION's summary analyses generate idealized organization profiles that are designed using STS principles and that take into account the organizational dimensions mentioned above. ACTION contrasts these idealized profiles against the actual projected organization designs envisioned by users, and performs a gap analysis to identify and prioritize differences. For design at the detailed level, a set of hierarchical constraints (called the probability models) that describe how to optimize for each business objective, is activated using many detailed attributes of organizational activities and features (e.g. inter-unit conflicts over skills and information, reliability of specific technologies, coordination patterns, etc.)

The ACTION theory and tool are being tested via thorough experimentation, use, and scientific validation procedures, and it is being incorporated into a structured framework for business re-engineering and organizational change, called the ACTION methodology. ACTION is a functioning system, now in use in 8 pilot sites in three manufacturing organizations.

5. Conclusion

Conventional approaches to strategic planning have been criticized as being too slow, inflexible, lacking in creativity, mismatched to envieonmental business cycles or rapid change, and difficult to operationalize. ACTION is addressing these problems by using a general, comprehensive, integrative, STS-based organizational theory and a set of automated decision support tools, specialized to discrete-parts manufacturing organizations. ACTION can improve planning processes by shortening planning cycles, improving the ability to generate and explore strategic alternatives using STS theory and knowedge, and operationalizing strategic concepts in ways that reduce interpretive ambiguity.

6. References

[Aaker, 92] David A. Aaker, Developing Business Strategies, John Wiley and Sons, Inc, 1992.

[Durfee and Lesser, 88] Edmund H. Durfee and Victor R. Lesser, "Predictiveness versus Responsiveness: Coordinating Problem Solvers in Dynamic Domains," in Proceedings of the 1988 National Conference on Artificial Intelligence, pp 66-71, 1988.

[Glazier and Weiss, 91] Rashi Glazier and Alan Weiss, Planning in a Turbulent Environment, Working Papers, U.C. Berkeley, April 1991. cited in [Aaker, 1992], pp. 19.

[Ishida et al., 92] Toru Ishida, Les Gasser, and Makoto Yokoo, "Organization Self-Design of Distributed Production Systems," IEEE Transactions on Data and Knowledge Engineering, 4:2, pp 123--134, April, 1992.

[Latour, 87] Bruno Latour, Science in Action, Harvard University Press, 1987.

[Mintzberg, 94] H. Mintzberg, The Rise and Fall of Strategic Planning, Free Press, 1994.

[Quinn, 80] J.B. Quinn, Strategies for Change: Logical Incrementalism, Irwin, Homewood, IL, 1980.

[Sarrazin, 77] J. Sarrazin, "Decentralized Planning in a Large French Company: An Interpretive Study," International Studies of Management and Organization, Fall/Winter 1977/78.

[Steiner, 79] George Steiner, Strategic Planning: What Every Manager Must Know. The Free Press, New York, 1979.

Advances in Agile Manufacturing
P.T. Kidd and W. Karwowski (Eds.)
IOS Press, 1994

Empirical Factors Interacting with the Development of CIM Strategies in Organizational Systems

Julia K. Kuark
Work and Organizational Psychology Unit,
Swiss Federal Institute of Technology (ETH), CH-8092 Zürich

Abstract: The complexity of computer aided integration projects makes the advantages of a conceptual overview incorporating the entire enterprise apparent, but precisely that is in fact lacking in practice. Only a few firms are in a position to demonstrate that they have methodical concepts. For the majority, their computer aided systems are the result of an evolution that more or less takes its own course. Results of a three phase empirical project disclose uncertain planning paths over a two year time frame. The presence of an explicit CIM strategy was not found to be associated with work orientation. In general, the companies with concepts appeared to have more experience with computer aided systems, a higher degree of computer permeation and a broader network span.

Introduction

Computer Integrated Manufacturing plays a key role in the further development of European industry. Because it is expected to enormously increase the potential flexibility and productivity of manufacturing companies, it is considered by many to be crucial to survival in the increasingly competitive markets of the future. In Switzerland, a major government funded 'CIM Action Program' has been launched in an effort to promote the development of CIM in the Swiss industrial community, one which is characterized by small and medium sized enterprises. Thus, many Swiss firms are finding themselves in a critical phase in which major strategic decisions are to be made. Most are, however, overwhelmed by the staggering amount of computer aided systems on the market today. While system vendors promise that their systems 'are just right for your needs', the sophistication of modern systems makes the way they function anything but transparent for the layperson.

Nearly all pertinent literature strongly recommends the making of an encompassing plan as the first step before implementing CIM components. The complexity of computer aided integration projects alone makes the advantages of a conceptual overview that incorporates the entire enterprise obvious. However, one of the key problems for decision makers and project engineers is that they do not, indeed cannot, know all the long term consequences of the decisions that they are currently making. And yet, decisions that are made today set parameters for the production systems that will be planned in the future. Thus, these sociotechnical systems begin to take an evolutionary course with a dynamic of its own. The danger that the resulting systems are inefficient or are unsatisfactory in other ways looms accordingly.

Project Design

At the Work and Organizational Unit of the Swiss Federal Institute of Technology, we contend that the successful design and implementation of CIM require an expanded approach that addresses issues of technology, people and organizations in an integrated manner [1]. Thus, a current research issue involves analyzing and evaluating the use of CIM components and networking in their respective contexts. A description of the main project as well as selected results are available in [2], [3] and [4].

An on-going investigation focuses on analyzing decision paths in the development of explicit and implicit CIM strategies as part of the aforementioned project [5]. In this framework, three levels of data in the capital goods industries of Switzerland have been acquired through a questionnaire survey (n=679), case studies (n=60), and open interviews (n=9). The two larger samples provide a basis for statistical evaluation of various planning and implementation factors. They will be enhanced with qualitative study of case specific development paths. These processes are being analyzed retrospectively so that both short and long-term effects in the organizational system become visible.

The survey phase provides an overview of parameters such as general approaches taken in implementing CIM. The case studies illustrate more detailed information on, e.g., the initial dates of the introduction of computer aided systems and subsequent generation changes. With the qualitative data obtained in the open interviews, the tracking of decision paths will give insight on interdependencies between the human, organizational and technical factors.

Selected results from [5] have been enhanced for presentation here with emphasis on factors interacting with the development of an official CIM concept.

Selected Results

The *survey* sample of 1990 consists of the machine building, electronic, and metalworking industries, where a total of 679 companies participated. Although the sample cannot be considered representative, it does provide a broad data base.

The most direct way of unveiling the aims of the company concerning the introduction of CIM technologies is to state the strategy explicitly and further, to make it public to employees. The importance of developing such a concept as a first step is emphasized in practically all relevant literature. Nevertheless, only 50 of 515 (10%) of the participants in the questionnaire survey (1990) replied that they do have a binding, overall CIM concept.

CIM projects, especially those with networking, are intentions that require several months or even years to realize. One indispensable resource in planning and realizing a CIM project is the personnel initiating, guiding and following it up. An additional advantage is offered by the continuity of project support. However, a total of only 125 firms (26%) stated that they have permanent task forces dedicated to the implementation of CIM or CAM/AMT. Twelve percent of the total sample responded that they have a CIM task force but no CAM/AMT task force, 5% have exclusively a CAM/AMT task force, and 8% have both.

The presence of an encompassing CIM concept, of course, does not guarantee the absence of problems during an implementation project. Unfortunately however, it cannot be illustrated here that having an encompassing concept reduces the number of problems encountered either; those cases with a concept mention problems more often that those without (n=515, χ^2=9.55, d$_f$=1, p<0.01). This is probably an indirect effect, as larger companies with a higher number of CIM components and networking, i.e. those that named more problems, are those that are likely to have concepts. Companies that are farther along with implementation are more likely to have a CIM task force (n=451, χ^2=61.28, d$_f$=4, p<0.001) or a CAM task force (n=436, χ^2=22.89, d$_f$=4, p<0.001), and those that have a task forces are also more likely to have binding CIM concepts (n=501, χ^2=41.94, d$_f$=1, p<0.001). It is possible that a task force is a step that leads to developing an explicit CIM concept.

The subsample selected for the *case studies* comprises 60 of the enterprises already examined. They were chosen on the basis of their willingness to participate, size and production demands. In comparison with the entire sample of 1990, the subgroup selected for the 1992 study shows a much higher degree of implementation of CIM components; differences of more than 20% are visible for the frequencies of the components (realized and partially realized) CAPPC, CAD, CAP, CAM and CAODA. The frequencies of the network lines of CAM-CAPPC, CAP-CAPPC, CAD-CAPPC and CAM-CAP are also more than 10% higher than those of 1990.

As an indicator of development trends within the 60 selected companies, the implementation of CIM components has been compared over the time frame of two years. The data was statistically handled as two dependent samples. The categories 'realized' and 'partially realized' were combined for this analysis. A general positive trend from 1990 to 1992 is observable for all components surveyed, but is only statistically significant in the case of CAQ (n=17, T=32, p<0.025).

The detailed analysis of whether or not planned components were actually implemented or not gives insight on the stability of the planning process. The data was analyzed to determine how many of the firms that reported in 1990 that they planned on implementing components within the next two years were actually able to do so, and how many that had not planned on doing so actually did. The directions of changes within the two year period in the respective subpopulations of those that had not yet realized each component are presented in Table 1. Statistical testing of significance was not always applicable due to the small size of the subpopulations. Significant trends in the direction of further implementation were found for the components CAPPC, CAD, CAODA and CAQ.

The logical development sequence would be movement in the direction from the category of 'no interest', to 'planning', through 'partial implementation' and on toward full 'implementation'. However, empirical examination of the movement between the categories reveals discrepancies in this sequence. For example, of the 6 cases in which CAPPC became at least partially realized by 1992, only half are accounted for by cases that claimed to plan to do so. The other 3 had, in fact, said that they had not been planning on implementing CAPPC within this two year period. Similarly, only 2 out of the 6 cases in which CAD was implemented between 1990 and 1992 occurred in companies which were planning it two years in advance. Thus, at least half of the significant trends towards increased realization of CAPPC and CAD are accounted for by cases that claimed not to foresee this development step in 1990. Moreover, of the 12 cases in which CAQ was newly implemented by 1992, 10 reported that they had no corresponding plans two years previously. Only for the component CAODA had the majority of the cases that introduced it planned to do so in 1990. Even so, only half of those that planned to realize CAODA in 1990 actually carried out this step by 1992.

Table 1:   **The Development of Systems between 1990 and 1992**
               ( ) = (already included) cases with an official CIM concept
               n.a. = not applicable   n.s. = not significant

Subgroups with resp. Components Not Implemented in 1990	1992						Wilcoxon Test Results		
	Still Not Planned	Still Planned	Planned and (Partially) Realized	Not Planned but (Partially) Realized	Newly Planned	no Longer Planned	n (ex. Ties)	T	p
CAPPC	1	4	3 (1)	3 (1)	0	0	6	0	<0.025
CAD	5	4	2	4 (2)	1	1	8	2.5	<0.025
CAP	0	3	2	2 (1)	1	0	5	n. a.	
CAM	5 (1)	0	3	0	0	0	3	n. a.	
CAA	32 (9)	4 (2)	1	4	0	3 (1)	44	7.5	n.s.
CAODA	3	5 (1)	5 (3)	1	3 (1)	0	9	0	<0.005
CAQ	19 (2)	6 (2)	2 (1)	10 (1)	0	2	14	5	<0.005

The numbers in parentheses in Table 1 signify those cases in which the presence of an official CIM strategy was confirmed either in 1990 or 1992. One would expect that a concept contributes to the reliability of the planning process, resulting in more cases with concepts in the column 'planned and (partially) realized'. Notably, they appear to be distributed among all the categories in Table 1. The recent recession undoubtedly accounts for a number of cases in which plans were postponed, but it does not explain the cases that said they had not planned on investing in these new technologies two years ago.

A significant trend towards the development of CIM concepts has been found; six companies that did not have concepts in 1990 stated that they did have one by 1992 (n=6, T=0, p<0.025). Of 15 firms that claimed to have a concept in 1992, 13 stated that it existed in the form of a written document.

While the results of the CIM concepts show a clear trend, the changes in the presence of CIM task forces do not. Thirteen of the 19 companies that already had a task force confirmed this in 1992, but 6 did not. On the other hand, 5 companies that hadn't organized one before 1990 did so before the second set of data was acquired. Although the question specifically stated permanent task forces, it is possible that project groups with limited lifetimes account for part of the positive answers.

Other factors associated with the presence of an official CIM concept and of a CIM task force were also examined. It was postulated that the presence of an explicit CIM strategy correlates with work orientation [1] based on a better awareness of the difficulties as well as the advantages of CIM. Moreover, a conscious, explicit strategy may be necessary for making structural changes in the direction of work orientation, because of the difficulty in overcoming the structural inertia of prevailing tayloristic systems. The evaluation of the data available from the first (n=679) and second (n=60) project phases could not support this hypothesis. No statistically significant associations between work and technical orientation and the presence of a CIM concept or a CIM task force were found. However, considering the hurdles of the ambiguousness in the definitions as well as the operationalization of the variables, the appropriateness of quantitative methods can be questioned. Early results of the qualitative study do speak for the effectiveness of the presence of an official strategy.

On the other hand, as work orientation increases, the relative number of people that work (in a task force or individually) as CIM planners also becomes higher. It may be an indication that work oriented companies practice broader participation in the planning of their systems. The absolute number of CIM planners was set relative to size class so that the influence of company size would be diminished. Although the number of CIM planners is, as expected, associated with computer permeation, no significant correlation between computer permeation and organization al orientation could be found. Hence, both the degree of computer permeation and organizational orientation apparently have interdependencies with the number of CIM planners. Due to the small sample size, the estimation of the strength of the influence of each of these variables is infeasible.

In general, the companies with CIM concepts appeared to have more experience with computer aided systems, a higher degree of computer permeation and a broader network span. With respect to the dates of first implementations, the medians of the groups with concepts are all lower, signifying earlier implementation, that those without. Additionally, approximately 75% of the companies with concepts and with CAPPC have already made at least one generation change, while a disproportionately high number of cases without concepts are still utilizing their original CAPPC systems.

Furthermore, both the indicators for total computer permeation and network span are associated with the presence of a CIM concept. The difference in computer permeation between the subsamples with and without concepts is highly significant. Computer permeation is a variable that essentially combines the size and number of computer hardware system components set into relation with three indicators of size of the company. Thus, the influence of company size should already be filtered out. The network span relates the reach of direct connections between the existing computer systems, where systems between different organizational units are considered to have a wider reach than those within a single organizational unit. The median of the group that reports to have concepts has a higher network span that without concepts.

Conclusions

Considering the emphasis in the literature on the making of an encompassing CIM strategy as an early, if not first, implementation step, very few of the companies in the surveyed sample stated that they had such concepts in practice. Only 10% confirmed the presence of official concepts in 1990. However, a statistically significant trend towards the development of such concepts was found between 1990 and 1992.

The analysis of the state of plans and implementation of CIM components before and after the two year period discloses fairly erratic planning paths. In three of the four groups in which statistically significant trends towards further implementation were found (CAPPC, CAD and CAQ), cases that had said they had no plans for implementing the respective components accounted for half or an even larger proportion of those that actually implemented between 1990 and 1992. Only for CAODA systems had the majority of those that implemented within the two year period claimed in 1990 to plan to do so. The presence of an official CIM strategy was not necessarily found to aid the stability of the planning process. A planning horizon even as short as two years seems to be elusive for the realization of CIM components.

Although the evaluation of the data available from the first (n=679) and second (n=60) project phases could not quantitatively support the hypothesis that the presence of an explicit CIM strategy is associated with work orientation, a positive correlation was found between work orientation and the number of CIM planners. The question of the interdependencies between the presence of a concept and organizational orientation will be examined further in the third, qualitative phase of the study.

In general, the companies with CIM concepts appeared to have more experience with computer aided systems, a higher degree of computer permeation and a broader network span.

References

[1] E. Ulich, Arbeitspsychologie. ISBN 3-7281-1731-5. Verlag der Fachvereine, Zürich, 1991.
[2] J.K. Kuark, T. Moll, A. Schilling, H. Schüpbach, O. Strohm, & E. Ulich, CIM in Switzerland: The Use of Computer Aided Integrated Production Systems. In H.J. Bullinger (Ed.), Human Aspects in Computing: Design and Use of Interactive Systems. ISBN: 0444 88775 X. Elsevier, Amsterdam, 1991, 1196-1200.
[3] O. Strohm, Ch. Kirsch, J.K. Kuark, L. Leder, E. Louis, O. Pardo, & E. Ulich, Computer Aided Manufacturing Systems: Work Psychological Aspects. In A. Bürgi-Schmelz et al. (Eds.), Computer Science, Communications and Society: A Technical and Cultural Challenge. Conference Proceedings, 22-24 Sept. 1993, Neuchatel, Switzerland, 221-238.
[4] O. Strohm, Ch. Kirsch, L. Leder, O. Pardo, P. Troxler, & E. Ulich, Work Oriented versus Technically Oriented Manufacturing Systems: Methods and Results of Selected Case Studies. In this volume.
[5] J.K. Kuark, Decision Paths in Planning and Implementing Explicit and Implicit CIM Strategies in Organizational Systems. Dissertation in preparation, Work and Organizational Psychology Unit, ETH Zürich.

Advances in Agile Manufacturing
P.T. Kidd and W. Karwowski (Eds.)
IOS Press, 1994

Participating in CIM Systems

Reinhard BACHMANN and Gerd MÖLL
Faculty of Social and Political Sciences, University of Cambridge, UK
Lehrstuhl Technik und Gesellschaft, University of Dortmund, Germany

Abstract. This paper discusses some theoretical problems of the design and implementation of advanced computer based information technology. With reference to Giddens' "Theory of Structuration" it will be asserted that traditional concepts of participation are insufficient to achieve successful systems implementations, because these concepts favour rather conservative organisational solutions. It will be argued that the full utilization of the potential of CIM technology needs both a radical change of the institutional properties and views of the managers and workers of the enterprise.

1.

It is a widely shared experience that the design and the implementation of highly integrated information systems in industrial organisations are related to problems which are much harder to cope with than it seemed to be in the light of the 1980s' CIM-euphoria. Too many projects failed and left software-ruins or were of very limited use for their investors. Thinking about the reasons for this discouraging outcome it was pointed out that many managers started their projects naively assuming that they could implement CIM without fundamental changes in the organisational structures of their firms (structural conservativism) and - despite the warnings of trade unionists and some politically minded scientists - pursued a rather plain "top-down strategy", which ignored the interests and the cultural knowledge of the employees. In the light of experience, many managers have changed their opinions recently and joined the "campaign for participation", a conversion which surprised traditional "participationists". The result in any case is that today there are very few people left who object to the call for participation.

No doubt, participation is a step in the right direction but in order to get a realistic understanding of the potential which lies in this approach and how this can profitably be used in regard to successful implementation of advanced information systems we need to gain some basic insights into the interaction of technology and the social reality of organisations. Although there is a growing body of literature on issues, which are situated in the "no man's land" between engineering and sociology, it is still hard to see solid ground in this area. From our point of view the missing link between both areas is a conceptual one, which can be developed on the basis of empirical research but is a theoretical one in its very nature. Without a theoretical concept which links technology and social reality the empirical findings which are offered by engineers or sociologists are vague and hardly any help for the crucial question: how can we control the processes of socio-technical innovation to gain efficient and socially rational outcomes? From our point of view, participation is one of the most important issues on the agenda for developing the innovative potential of advanced information technology within and across the boundaries of industrial organisations, but we have to understand the relationship between technology and social reality theoretically before we can make full use of advanced CIM systems. Otherwise arguments for participation are hardly anything else than political wishes, on which one can agree or disagree, just as one likes. The present situation in which the managerial side has given up its containment policy and paradoxically as it seems for many trade unionists, embraces participation as a precondition of efficiency introduces a situation which might provide a chance of a reasonable discourse on this subject going beyond the over-simplified political arguments of the 1970s and 1980s.

2.

In order to understand the role of CIM systems in industrial organisations it is most important to reconstruct the "constructivist character" of advanced computer technology and explore the flexibility of specific technologies on the basis of a general theoretical conceptualisation, which rejects the view of "autopoietic" technological developments. Technology should rather be regarded as designed and applied by social actors according to their shared cultural beliefs and their embeddedness in social institutional structures. That is to say, social actors inevitably transfer social reality into their products, implement cultural and social knowledge into technological artifacts, as well as interpreting and using them against the background of the social reality in which they live. Technology and social reality are intrinsically linked with each other because they incorporate knowledge, social norms and cultural meanings which are exchanged between both spheres, irrespective of whether the social actors are aware of this or not. Thorough empirical investigations can show that, and how social power relations and cultural beliefs are transformed and sedimented in technological artifacts and vice versa [1].

Giddens' "Theory of Structuration" [2], one of the most prominent approaches of contemporary general sociological theory, seems to be a rather fruitful conceptional framework for the analysis of technological artifacts within the dynamic process of shaping and reshaping the social reality of industrial organisations. Although Giddens himself has - as far as we know - neither made an attempt to apply his theory of the transformation of modern societies on the level of an organisational system nor to think systematically about a genuinely sociological understanding of technological artifacts, innovative organisation theorists have shown the transferability of "structuration theory" to specific settings of organisational action. If we follow this approach in reconstructing social reality of firms we observe genuinely social structures, such as norms of behaviour, distribution of power, exchange of knowledge between cooperating departments etc., and technological artifacts, such as CIM systems, as products of collectively acting human beings. These structures, whether technological or non-technological, are permanently produced and reproduced by social actors, who - on the other hand - have inevitably to orient their ongoing actions to the structurally crystalized forms of their knowledge and consciousness.

The concept of the "duality of structure" (Giddens) intrinsically links two ideas: that structures are the result of human beings' action and, at the same time, are the preconditions which channel, i.e. restrict and enable, the options available to social actors. The "duality of technology" (Orlikowski) [3] repeats the same view equating genuinely social structures and technological artifacts, which thereby are reconceptualised as part of the social reality of organisations. The structurational view of technology thus studies technological artifacts as potentially modifiable throughout their entire lifecycle. In the mode of design and development certain interpretive schemes (rules reflecting knowledge of the work being automated), facilities (resources to accomplish that work) and norms (rules that define the organisationally sanctioned way of executing that work) are built into the technology by human actors. In the mode of use the social practices of the human agents are conditioned by these in-built properties of the technology. At the same time the users act upon the institutional properties of their organisation, i.e. the use of technology is conditioned by the stocks of knowledge, resources, and norms existing in the user organisation. Accordingly, the systems technology applied in a certain enterprise contributes to the specific organisational structural framework of domination, signification and legitimation. It opens up and at the same time limits the scope of meaningfully referring action in the mode of use. On the other hand organisationally "contextuated" information systems are inevitably and permanently reshaped and reinterpreted by their users. In consequence, the implications of new technologies not only depend on the actions and motives of designers and managers but also on the institutional context in which the technology is implemented and on the autonomy and capabilities of the users.

3.

Advanced CIM systems are not primarily designed to maximize the efficiency of specific organisational departments or production stages. According to their system character they are geared to "systemic" rationalization of the total process of production, even across the boundaries of cooperating organisations. While the interpretive schemes, facilities and norms which were built into the early CIM systems provided comprehensive, centralised concepts of production control and thus reinforced a rather rigidifying and formalising mode of production, today's generation of CIM technology rests upon the notions of flexibility, decentralisation and work integration, since the limits of traditional modes of mass production and taylorism became quite obvious. That is to say, modern CIM technology embodies a systemic logic which essentially runs against the logic of fragmentation of work and rigid separation between planning and execution. In that respect it is not an exaggeration to state that the new generation of this technology will reach its full potential only when used in a manner which constitutes a radical break with the traditional institutional properties of the firms. In other words, the inherent properties of these CIM systems widen the scope of organisational choice and give way to non-tayloristic modes of production. They tend to incorporate the opportunities of work autonomy and power-balanced work cooperation rather than hierarchy and determination. But the institutional properties of most of the enterprises, characterised by fixed functional boundaries, hierarchical communication patterns and accompanying background assumptions that shape an organisation's approach to the use of technology, restrict the use of this possibility. The structures of domination, signification and legitimation which are associated witn tayloristic principles of work organisation and which determine the behaviour, the readings of reality and perceptions of interests of workers and management will not automatically change just by the implementation of modern CIM technology. Rather they have to be adjusted in a process of collective learning.

Today the institutional structures of most enterprises are the mirror image of the insular strategy of rationalization, which is orientated to increase the task perfomance in specific departments of the industrial organisation as well as towards cutting personnel expenditures. This mode of rationalization does not only strongly support the division of labor and hierarchical structures, but also the way workers and managers shape and reshape their knowledge, interests and preferences. Under these conditions it is not surprising that the implementation and application of the most advanced CIM systems lead to serious incompatibility problems.

To make use of the productive potential of advanced system technologies it will not be sufficient to give the potential users a chance to express their interests in convenient work places. From our research experiences we know that this traditional understanding of participation often diminishes the likelihood of radical organisational solutions. In the case of new system technologies, however, a radical change of the existing work routines, know how, interpretive schemes and norms seems to be necessary. In order to achieve that a kind of "second order learning" [4] must be obtained, which brings to light the deep seated forms of reading the reality, rules of orientation and habits of thought, which underly the forms of work organisation. While the traditional understanding of participation only refers to well known interests of the social actors, second order learning touches the basic structures of organisational reality. It takes place in critical discourses in which the participants become aware of the structural preconditions of their actions. Only if this can be achieved is there a chance to take control of the ongoing development of organisational reality by the social actors involved. They inevitably create and recreate the social world they live in but often do this rather sub-conciously.

Successful design and implementation of CIM technology obviously means more than just the willingness of management and workers to cooperate. It means balancing the interpretive schemes, facilities and norms, which are built into the system technologies with the institutional context of the firms and the routines, norms and perceptions of the users. For that reason it will be necessary to query the definitions of organisational problems and solutions which are often taken for granted and take steps towards a more radical view which is based on insights into the fabric of industrial organisations' social reality.

References

[1] Th. Malsch *et.al.*, Expertensysteme in der Abseitsfalle? Fallstudien aus der industriellen Praxis, Edition sigma, Berlin, 1993.

[2] A. Giddens, The Constitution of Society: Outline of the Theory of Structuration, Polity Press, Cambridge, 1984.

[3] W. Orlikowski, The Duality of Technology: Rethinking the Concept of Technology in Organizations, *ORGANIZATION SCIENCE* **3**, (August 1992) 398-427.

[4] C. Argyris and D.A. Schon, Organizational Learning. A Theory of Action Perspective, Addison-Wesley, Reading MA, 1978.

Advances in Agile Manufacturing
P.T. Kidd and W. Karwowski (Eds.)
IOS Press, 1994

Organizational Structure Creating Mechanisms for Implementing Automation in Manufacturing

Stefan TRZCIELINSKI
Technical University of Poznan, Management Engineering Department
ul.Strzelecka 11, 60-965 Poznan, Poland

Abstract. In this paper an attempt is taken to explain the mechanisms of creating of organizational units which deal in the management process. The mechanisms are shown using as an example the area of management of automation in manufacturing. The key point is the assumption that the changes of organizational structure follow by the changes of compatibility of functions which are performed in the structure. The compatibility is defined as a result of information flow, similarity of procedures and similarity of competencies which are needed to perform the functions. Depending on the dominated factor of compatibility different forms of organizational structure take place.

1. Introduction

Management process is an information-decision process and usually is performed by many individuals. Therefor it must be co-ordinated and co-ordination is a kind of meta-management process. The most intensive co-ordination covers the areas of activities (functions) which are assigned to organizational units. Organizational unit is both the social and technical system which is set up to perform these functions. Function is the basic substantial factor for existence of an organizational unit. The unit is created according to the following conditions [1, p. 225]:
- when the function which is needed is defined, and
- when the resources which are necessary to perform the function (people, equipment,...) are utilized on satisfactory level.

Usually a single function which describes the information-decision process detailed enough to be understood by the individuals is not enough resource-consumption to create for the function an organizational unit. Therefor functions must be grouped in complexes to reach the second condition mentioned above.

The main problem of a designer of the organizational management structure is:
- to define the functions which are needed
- to aggregate the functions in complexes to create for them the organizational units.

The method of solving the first problem depends on decomposition of functions beginning from the mission of company or the main function of its subsystem. To define these functions which are really needed the contingency approach have to be applied. The issue of this process takes form of tree of functions [5, pp. 102-127]. An example of such a tree is shown in chapter 2.

In next chapters the second problem is discussed using as an example the area of management of automation.

2. Functions of Management of Automation in Manufacturing

The functions which should be performed in the organizational system are predicted by some features of internal and external environment of the organization [2, pp. 43-44]. All the features which stay in the relationship to the functions (and organizational structure) are called *contextual conditions*. In case of functions of automation management, the examples of the predictors are:
- diversification of products
- type and variety of technology
- scale of production
- variety of offers of machines and equipment which can be delivered.

The general functions concerning management of automation may be the following:
1. Preparation of decision predictors for introduction automated manufacturing
    1.1. Product design
    1.2. Standardization of technological process
    1.3. Quality control and identification of reasons for production defects
    1.4. Balancing of production capacity
2. Economic analysis of automation in manufacturing
    2.1. Identification of product types and technologies which can be included into automated manufacturing
    2.2. Deciding the scope and the level of automation
    2.3. Establishing the needed level of investment and operating cost
    2.4. Estimating the rate of automated production
    2.5. Cost and benefits analysis of the automated production
3. Purchasing of equipment for automated manufacturing
    3.1. Collecting data about suppliers such as name, location and prices of equipment
    3.2. Analysis of suppliers bids
    3.3. Negotiation of technical and financial conditions with delivers
    3.4. Quality check of delivery
4. Assembly of automated machines into production system
    4.1. Description of type and number of supporting equipment
    4.2. Purchase of supporting equipment
    4.3. Installation of primary equipment
    4.4. Accepting testing
    4.5. Designing of software systems and programming machines for automated manufacturing
5. Personnel
    5.1. Hiring of new workers
    5.2. Selection of appropriate workers to serve automated processes
    5.3. Training of workers to cope with automated systems

3. Grouping function in complexes

The automation management functions stay in different relationships to many other functions of management of various areas of company activities, particular to:

- Research and development
- Production planning
- Capacity planning
- Managerial accounting
- Purchasing
- Maintenance
- Human resource management.

In different contextual conditions both the functions of automation management and the above functions are grouped in different way and allocated to different organizational units. For example when the rate of products is big, the products are diversified and new products are introduced often, the cost and benefits analysis of introducing automated manufacturing (function 2.5.) may be done by engineering department. In opposite circumstances the function is assigned to accounting department.

What causes that depending on the features of contextual conditions the same function may be allocated to different unites? There are three bases of grouping [6]:

1. Information flow among functions.

Functions cover different sequences of information-decision process. Because of this they are connected by flow of information streams. These functions which are connected by the most intensive flow of information should be grouped together and performed in the same organizational unit (section, department, division). In this way the minimal need of co-ordination is ensured.

This base (criteria) of grouping leads to the same organizational structure as grouping on the base of product, geography or market [3, pp.347-349]. However the organizational theory does not explain in which contextual conditions these criteria may be used and to which area of activities of company may be applied.

Grouping on the base of flow of information streams should be used in every circumstances but of course the flow must be identified and its intensity must be assessed according to the contextual conditions [5, pp.180-195]. For instance usually balancing of production capacity (function 1.4.) is performed in manufacturing department because of the information connections with production control. But if the scale of production grows and the company introduces new products frequently, the function is more intensively connected with process planning and therefor it is performed in engineering department.

2. Similarity of procedures.

According to this criteria, functions which are performed due to the same or similar procedures (algorithms) should be grouped together because this enables to employ and utilize an especial staff and equipment supporting performance of this functions. For instance if there are used only single NC or CNC machines, programming of the machines (function 4.5.) may be done by MIS department because of both the concentration of hardware and the needed qualifications of staff. Otherwise the function is performed in engineering department.

Concerning the area of production process this criteria is described in literature as grouping on the base of process and equipment [3, pp.349-350].

3. Similarity of competencies.

According to this criteria functions which required people with the same or similar competencies to perform them should be allocated to the same unit. For example because of the needed competencies to decide on the standardization of technological process (function 1.2.) and on the scope and level of automation (function 2.2.) both functions are performed in engineering department. In literature this criteria is known as grouping on the base of similarity of functions [3, p.347].

As a main criteria similarity of competencies is used in case of grouping functions for large units like department or division which have their internal structure. But in fact it is a limitation which must be respected when grouping functions. Therefore this criteria is very often combined with information flow and similarity of procedures.

4. Casual - Effect Mechanism of Structural Changes

Contextual conditions are the primary predictors of changing the organizational structure. But the changes take place when some secondary predictors appear with intensity exceeding the ability of existing structure to tolerate them. The secondary predictors are the following:
- setting up a new or eliminating of an existing function
- the change of importance of existing function
- the change of method or algorithm of performing function
- the change of labour consumption of performing function.

They cause the change of compatibility of functions which is a result of information flow among functions, similarity of procedures and similarity of competencies. Using the symbolic notation the compatibility is defined as:

$$C = I_f + P_S + C_S$$

where:
I_f - intensity of information flow between functions
P_S - similarity of procedures of performing functions
C_S - similarity of competencies needed to perform functions.
Let's describe the compatibility as it follows:
C_I - when the dominated factor is I_f
C_P - when the dominated factor is P_S
C_C- when the dominated factor is C_S

If F_1, F_2, F_3, F_4 mean different functions (or their complexes), the different forms of departmentation can be identified as it follows:
$C_C(F_1,F_2)$ - differentiation on the base of similarity of functions
$C_I(F_1,F_3)$ - differentiation on the base of product, market or geography
$C_P(F_3,F_4)$ - differentiation on the base of process or equipment
$C_C(F_1,F_2) = C_I(F_1,F_3) = C_I(F_2,F_4)$ - matrix structure.

References

[1] J. Boszko, Uogolniony model tworzenia efektywnych struktur podmiotowych w przedsiebiorstwach budowy maszyn, *Zeszyty Naukowe Politechniki Poznanskiej, Organizacja i Zarzadzanie*, 14 (1988) 223-242.

[2] R.B. Duncan, Characteristics of Organizational Environments and Perceived Environmental Uncertainty. In: M. Zey-Ferrell (Ed.), Readings on Dimensions of Organizations. ISBN: 0 87620 771 9. Goodyear-Publishing Company, Inc., Santa Monica, California 1979, pp. 42-58.

[3] B.J. Hodge and W.P Anthony, Organization Theory. ISBN: 0 205 11325 7. Allyn and Bacon, Inc., Needham Heights, Massachusetts, 1988.

[4] H. Koontz *et al*, Management. ISBN: 0 07 035377 8. McGraw-Hill, Inc., USA, 1980.

[5] E. Pawlowski and S. Trzcielinski, Projektowanie struktury organizacyjnej przedsiebiorstwa. Podstawy rozwijania i kojarzenia funkcji zarzadzania, vol.1 and 2, TNOiK, Poznan, 1987

[6] S. Trzcielinski, Mechanizmy przeksztalcen struktury organizacyjnej zarzadzania podsystemu zaopatrzenia materialowo-technicznego w przedsiebioestwie budowy maszyn, *Organizacja i Kierowanie* 1-4 (1988) 223-237.

Advances in Agile Manufacturing
P.T. Kidd and W. Karwowski (Eds.)
IOS Press, 1994

The Holistic Perspective for the Management of Technology

Hongyi SUN and Frank GERTSEN
Department of Production, University of Aalborg, Fibigerstraede 16, 9220 Aalborg,
Denmark

Abstract. It has been widely discussed that the organisational and strategic issues should be considered for technological innovations. However, most of the previous research dealt with either organisation and technology, or strategy and technology. This paper will report a research aiming to establish a holistic model linking organisation, technology, strategy as well as market factors. The managerial implications for practice and future research will also be discussed.

1. Introduction

According to literature and our experiences, we would like to argue that the research with a dualistic perspective involving either organisation and technology, or strategy and technology, is not sufficient, and that the organisation, technology and strategy should be considered from a holistic perspective.

The basic ideas of the holistic perspective include that (1) all these factors are necessary, (2) all the factors are interrelated, therefore (3) changes in one of the factors should be accompanied by changes in other factors. To provide empirical evidences for the argument, a survey was conducted [1]. A questionnaire was designed and twenty eight observations of manufacturing companies were collected in Denmark and Norway. The data were collected through on-site interview in order to secure the quality of the data. The sample companies are small and medium sized (SME) manufacturers in mechanical and electrical-mechanical industries. Their sizes range from 50 to 1000 employees.

Quantitative data about organisation (O), technology (T), strategy (S), market (M) and performance (P) were formulated into aggregated indexes by average. The dimensions of organisation, technology, strategy, market and performance indexes are as follows. (1) Market index measures the dynamics of market in terms of geographical distribution, frequency of new production and uncertainties etc. Additionally, the tendency of market terms of stability, rise and fall of the demand; (2) Organisation index measures the education, skills, use of matrix and teams, re-organising activities, employee participation, job rotation and communication of goals etc. (3) Strategy index measures the existence, formulation and spread of strategy, including market, product and manufacturing aspects; (4) Technology index measures the utilisation of AMT in production, design and management; (5) Performance index is a comprehensive measure of the managers' subjective feeling of goal achievement in the current year, and the objective long-term performance in terms of the increases of profit and market share etc.

In next section, propositions will be proposed and demonstrated by empirical evidences.

2. Propositions and Empirical Evidences

2.1. The Correlation between O, T, S, M and Performance Indexes

The relations between the factors and performance are described by proposition 1 as follows:

Organisation, technology, strategy and market are all related to performance. Those companies that pay attention to organisation, technology, strategy and market tend to have higher performance than others.

This proposition implies that organisation, technology, strategy as well as market factors are all related to performance. This can be demonstrated by the correlation between these factors and the performance. The statistic correlation and the significant levels (in parentheses) between M, O, S, T and P are as following: M-P: 0.62 (0.00), S-P: 0.49 (0.01), O-P: 0.33 (0.09); T-P: 0.39 (0.05). Organisation, technology and strategy are all positively correlated to the performance at the significant level of from 0 to 0.09. Although market dynamics are not correlated to performance, the market tendency is positively correlated to performance (0.62 at the significant level of 0.001). This implies that the four factors are indeed related to performance. The measure by the indexes reflect the extent of attentions companies pay to each factor. So, according to the correlation between these factors and performance, it can be concluded that those companies which pay more attention to organisation, technology, strategy as well as market tend to have higher performance than others.

2.2. The Correlation among M, O, S, and T

This section will try to look at the interrelations among all the factors, and the relation to the performance. The idea will be described by proposition 2 as follows.

The interrelation among organisation, technology, strategy and market are related to performance. Those companies that can properly link its organisation, technology, strategy and market tend to have higher performance than others.

This proposition can be demonstrated by the statistic correlation among these factors. The statistic correlation between two factors reflect their link or alignment. These factor are correlated rather significantly. However, the final intention is not only to look at their correlation, but the relation between their correlation and the performance. In order to verify whether the correlation is relevant to the performance, the correlation test was repeated for two groups of the sample. The sample was divided by the average of the performance (ranging from 21 to 49 with an average of 35) into two groups as figure 1 (a) and (b) show.

Group (a) is the high performance group (high-P, n=14), and group (b) is the low performance group (low-P, n=13). The results in figure 1 revealed that the higher performance group is correspondent to the higher correlation among all the factors. At the significant level is 0.07, only one of the correlation, between O and T, is significant in the low performance group, while 5 are significant in the high performance group. This implies that organisation, technology, strategy as well as market contributed to the performance in a collectively way. According to the above analyses, the second proposition can be accepted.

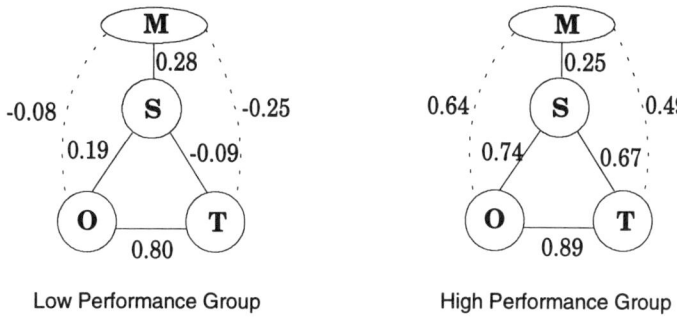

Low Performance Group   High Performance Group

Figure 1 The high performance is correspondent to high correlation

3. Establishment of the MOST model

Based on the above two propositions, a conceptual model which links M, O, S and T is proposed. It is called the MOST model representing the management and linkages of Market, Organisation, Strategy and Technology. The framework is illustrated in figure 2. Compared with similar previous models, the MOST model has the following new features. (1) The MOST model was established with an evaluation factor, i.e., the performance. (2) This model is empirically based. (3) This MOST model has an explicitly separated management factor. (4) This model involves also an environment factor, i.e., the market. (5) The MOST model has involved both the multi-factors and the multi-interrelationships. (6) This model is built from a holistic perspective.

Figure 2 The MOST model linking Market, Organisation, Strategy and Technology

4. Discussions

The MOST model has triggered the following discussions.

(1) Any of the three variables, organisation, technology and strategy, should not be designed or planned without knowing or considering the other two.

(2) When discussing the relationship between two variables, the third has to be taken into account. For example, the discussion about the relationship between the organisation and the technology is not sufficient without considering the strategy as well as market.

(3) If one of the three variable changes, the other two may change as well. However, the change will not happen automatically. It needs time for a company to change. And it also needs management intervention.

5. Summary and Future Research

In this paper, the authors have argued the necessity of a holistic perspective for the management of technology. Propositions are proposed and a conceptual model, MOST model was built. The model suggests that future research and practical should be conducted from a holistic perspective. The establishment of the MOST model was only a beginning. Research under this model can be continued in many directions, for example, the tools and methods for the linkage between organisation and technology, between technology and strategy and so on [2,3].

References

[1] Sun, Hongyi (1993) *Patterns of Organisation and Technology Development with Strategic Considerations, Managerial implications for Advanced Manufacturing technologies,* Ph.D. dissertation, University of Aalborg, ISBN 87-89867-24-6.

[2] Sun, Hongyi and Jens O. Riis (1994) "Organisational, Technological, Strategic and Managerial Issues along the Implementation Process of Advanced Manufacturing Technology: A General Framework of implementation guide", *International Journal of Human Factors in Manufacturing,* Vol 4 (1).

[3] Riis, Jens O. and Hongyi Sun (1994) "Organisational Changes and Technological Innovations under the Guidance of Manufacturing Strategy", In *the proceedings of the 4th International Conference on the Management of Technology,* Feb. 28 to Mar. 4, 1994, Miami USA.

Advances in Agile Manufacturing
P.T. Kidd and W. Karwowski (Eds.)
IOS Press, 1994

153

Interaction of Advanced Shop Floor Management Systems with Production System Organisation - An Exploratory Study

A. Lucas Soares[1], Nuno Romão[2], J.M. Mendonça[1]
[1] INESC/FEUP-DEEC, [2] Growela Portuguesa

Abstract. This paper presents a study about the influence of the innovating features of advanced shop floor information management systems on the production system organisation. An interaction model highlighting the influence on the organisation structure, information network and decision process is presented. This model is used as a base to the analysis of a case study in the portuguese automotive supplier industry. An application of the interaction model is devised in the joint specification of advanced information technology for shop floor management and the production system organisation.

1. Introduction

The application of information technology (IT) in the support of shop floor management is evolving from the individual control of traditional functions - scheduling, dispatching, monitoring - to more comprehensive systems, featuring the integration of these shop floor activities with maintenance and quality assurance. The potential improvement obtainable with Advanced Shop Floor Management Systems (ASFMS) is directly dependent on their organisational fitness features as well on the implementation strategies adopted. Recent discussions and practices on new organisation forms for production systems address the necessity of a holistic approach in the development and implementation of new manufacturing information management systems [1][4]. This paper presents some conclusions about the interaction between ASFMS and the production systems organisation, based in a field study realised in a Portuguese manufacturing company.

2. Innovating Features in ASFMS and their Organisational Impact

The main trends in the innovating characteristics of ASFMS which resulted from recent R&D projects are integration frameworks for the coexistence of heterogeneous shop floor applications and powerful decision support tools. The motivation to an organisational interaction study of the ASFMS is twofold: to assess the organisational impact of the new features of ASFMS and to foresee the support of ASFMS to new shop floor organisational forms. Two innovating features can cause major influences in the production system organisation: integration of heterogeneous SF related applications and modelling and simulation based decision support. Whatever the organisational structure, the primary visible effect of the integration of heterogeneous SF related applications is an efficiency enhancement, provided by the real-time interaction between those applications. However, it is dificult to attain major improvements due to remains of work practices and old coordination mechanisms embedded in a functionally divided structure. Modelling and simulation tools provide the SF management with the possibility of exploring "what-if" scenarios, prior to decision making on order sequencing, resource maintenance, etc. Object oriented user interfaces, being intuitive in use, improve users perception of the SF status. Although this feature can be used to support centralised SF control, it is also possible to implement the concept of Electronic Control Stations (ECS) [1].

If decentralised control and autonomy are required, better overall coordination and compliance with the company goals can be achieved.

2.1 New Organisational Forms in Production Systems - ASFMS Support

The relation of ASFMS with production systems organisation may follow two streamlines: matching an "as is" organisation or triggering an organisational change. The first approach leads most of the times only to marginal improvements or even to complete failure. The second approach encompasses a continuum of possibilities, ranging from reorganisation due to IT enforcement to radical and innovative changes supported by IT, this calling for a joint development of the IT system and the organisation. The integration of heterogeneous software applications is a basis to the integration of functions, usually vertically divided by marked hierarchies or horizontally by departments/sections. Graphic simulation and common modelling user interfaces can be seen as a perceptual linkage intra and inter working groups, aiming to atain an overall manufacturing coordination. However, this feature can also be used to reenforce autocratic management centralizing SF control, with the known consequences of lack of responsiveness and people demotivation.

2.2 A Model for the Analysis of the Organisational Impact of ASFMS

The model used in our study aims at highlighting the influence of the integrating features of ASFMS in the production system organisation in terms of structure, information network and decision process. The organisational model used is based in the work of Harrington [3], where the "perceptual organisation" paradigm is introduced. The introduction of IT systems with the integrating characteristics of ASFMS is likely to create imbalances between structure, information network and the decision process that must be assessed and corrected. The main critical technical factors of ASFMS influencing organisation are :

Data Integration and Distribution. This influences directly the information network by means of information flow automation, creation or deletion. The resulting information integration and distribution, influences structure and decision process. Another important effect is the possibility of disabling informal information flows.

Function Integration and Distribution. Applications integration and distribution directly influences the organisation structure in terms of job content, group formation and hierarchy relations.

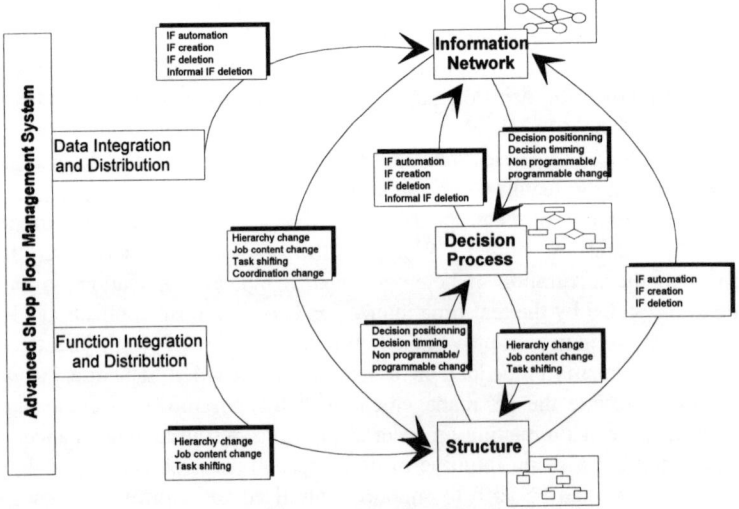

Figure 1 - Interaction Model

Interaction Model

This is a descriptive model of the interaction that can arise between ASFMS and the production system organisation. The model is used as an influence network enabling the cross influences of the critical technical features in the organisational structure, information network and decision process to be considered (Fig. 1)

3. Case Study - Introduction of an ASFMS in an Automotive Supplier

The objective of this study was to assess the organisational impact of the new features of ASFMS. An exploratory method based on observation and interviewing, not interfering with the actual technology introduction process was followed. The company was involved in a modernization process sponsored by the Esprit AICIME special action, whose goals were to install and assess CIME solutions in suppliers for automotive industry in Portugal.

3.1 Assessing the Readiness for Change

Automotive supplier companies are facing increasing market demands leading to the adoption of lean production and "Just-in-Time" concepts, resulting in a strong impact on the production system in terms of flexibility and quality. ASFMS can be used to improve a conventional "mechanical" organisation, but this is not enough to cope with the demands referred to above. Thus, the company's commitment to change must lead to a more "organic" organisation. ASFMS can provide a powerful aid in pursueing this objective.

3.2 PROFIT - Main Features and Implementation Strategy

PROFIT is a commercial ASFMS, resulting from the Esprit Project 5478 Shop-Control [2]. Based on an innovative IT infrastructure, it covers production scheduling and control, quality and maintenance management, monitoring, supervision and alarm facilities. Advanced modelling and simulation tools provide the user with the physical, functional and informational aspects of the SF, while granting an integrated homogeneous access to the different applications through context sensitive entry points. Animated graphic monitoring of SF status, as well as simulation of "what-if" scenarios, are major features in decision support. Shop floor applications share static database information where security, redundancy, consistency and integrity are ensured. Dynamic information exchange between applications is dealt with through messaging services enabling control interactions to occur in real-time. The IT infrastructure is based on widely spread standards allowing the distribution of applications as required by the SF organisation. This ASFMS was implemented following a phased introduction aproach. The stamping area was chosen as the pilot for the PROFIT implementation, with subsequent extension to other plant areas.

3.3 Innovating Features Organisational Influence

The modelling and simulation tool in conjunction with the monitoring and supervision applications, caused a marked influence in the information network. The critical technical feature here - data integration and distribution - deleted information flows and created new ones. The information flows deleted were: actual production status, logged by SF workers on paper, and later transmitted to the foreman; daily production report compiled by the stamping area foreman at the end of the day, and later transmitted to the planning department; and relevant SF events reported by the foreman to the planning department. These information flows were substituted by a direct link between SF and planning department using a data collection module updating the SF database. The interaction model shows that the change in the information network influences both the coordination mechanisms and the job structure of the planning department and the stamping area.

With the former information network, the area foreman was a gatekeeper of important information to the planning department. Now he is bypassed and his job content looses the task of the daily production report elaboration, with consequent impact on power relations.

Regarding the other aspect, application integration, the critical technical feature involved is function integration and distribution. Formerly, scheduling was performed by the planning department with dispatching by the area foreman. Short term plans   were developed by the planning department each two weeks in conjunction with the foreman, and later implemented by him. Problems ocurring during this period (lack of tools and materials, transporters problems) were, to a certain extent, managed by the foreman without the intervention of the planning department, because of his direct SF status knowledge. With the new application, this task was transfered to the planning department with the consequence of job empoorment to the foreman. According to the interaction model, this change in the structure influences the information network by automating the information flow downloading the short-term production schedules from the planning department to the foreman. The influence in the decision process comes essencially from the change in the information network. The direct link from the SF to the planning department created by PROFIT, caused the dispatching decision making to be partially shifted from the SF.

3.4 Results and Comments

The PROFIT implementation at this company is not a finished process, and some degrees of freedom exist with respect to organisation design. One major conclusion that can be drawn is that there is not a "technological determinism" in the production system organisation caused by this IT system. A range of organisational design choices can be made to best fit the company strategy. The company has chosen to adapt IT to the existing organisation. However, since this inovation process is intended to increase production system efficiency, atempts are made to follow a centralizing path, taking advantage of the powerful information management and decision suport provided. This is indeed an organisational choice, not a technical one. If it is a conscious choice, care must be taken to assure that changes triggered by ASFMS in any of the organisational elements, information network, decision process or structure, are tracked with consequent changes in the others.

4. Conclusion and Further Work

In this paper an overview of the possible organisational influences of the innovating features of ASFMS was presented. A case study in the automotive industry was introduced from which some conclusions from the ASFMS introduction process were derived. With this study practical knowledge was collected that enabled the creation and validation of interaction models for further assessement of the influence of ASFMS on the organisation. These models can only be improved by the analysis of further results regarding the experience of companies in introducing advanced IT systems in the SF management. It is possible to combine the influence network model with a knowledge base in a computer model, and use it as a basis of a computer aided tool for the joint specification of the SF information management system and the production system organisation. This is work presently being undertaken within the Esprit Project 8865 Real-I-CIM.

References

[1] Hirsch-Kreinsen, H. et al. (1991) - Technological Preconditions for Skilled Production Work in CIM - APS Research Paper - FAST Programme
[2] Mendonça, J., Schulte, J., McCarthy, S. - Real-Time Shop Floor Control: Modular System and Integration Tools - CIME Europe 93 - Amsterdam.
[3] Harrington, J. (1991) - *Organizational Structure and Information Technology* - Prentice Hall
[4] Campbell, A., Warner, M. (1990) - Managing Advanced Manufacturing Technology in *New Technology and Manufacturing Management*, Warner et al. (ed.) - James Wiley & Sons

Advances in Agile Manufacturing
P.T. Kidd and W. Karwowski (Eds.)
IOS Press, 1994

Facilitating new shopfloor roles within modern manufacturing

Sharon K. Parker and Paul R. Jackson

MRC/ESRC Social and Applied Psychology Unit, University of Sheffield, S10 2TN, UK

Abstract. Forming autonomous, team-based shopfloor jobs appears to be a critical element when introducing modern manufacturing practices. Evidence suggests these jobs enhance shopfloor performance and promote employee well-being. However, little research attention has been given to the types of shopfloor roles required with such work reorganisation, or to the factors which affect the transition to these new roles within brownfield sites. In this paper we explore these issues, drawing on the experiences of a company which successfully adopted a high-involvement approach.

1. Introduction

The need for flexible responses to changing market and technological conditions has seen the rise of new manufacturing practices, which to many commentators amount to a 'new manufacturing paradigm'. Various initiatives such as Just-in-time (JIT), Total quality management (TQM), and the introduction of new technology (collectively called Integrated Manufacturing; IM, [1]) are being widely introduced to enhance competitiveness. Evidence suggests these practices, by comparison with Taylorised factories, demand a higher level of shopfloor performance in which operators add value to both processes and products (e.g. [1], [2]). Individual competencies and learning "become a key competitive asset... that are at the heart of an organisation's adaptability and ability to learn" ([2] p. 5).

However, such 'high-performance' does not simply result from introducing IM practices. Rather, the changes required in people need to be facilitated and supported by changes to the organisational structure and other HR systems. More specifically, there is strong evidence that devolving control to teams of workers promotes better operator performance (e.g. allows them to use and develop their local expertise) whilst improving their well-being (see, for example, [1], [2]). It is further advocated that this work design be supported by 'high-involvement' HR practices that ensure employees have sufficient power, information, knowledge and ability to influence the company business [2].

Despite the value of such high-involvement working, there is a lack of research attention given to its introduction in brownfield sites. Rather, the focus has been greenfield sites such as Nissan where new forms of working seem to be more successful. Here we examine the experiences of a medium-sized electronics company (Company F) which successfully[1] introduced product-lines (i.e. semi-autonomous work groups organised around cells of products), TQM and JIT. We look firstly at the changes in shopfloor roles that were required, followed by factors which facilitated the transition to new roles.

---

[1] i.e. there were substantial improvements in productivity (lead time, quality, etc) with no detriment in existing high-levels of job satisfaction and psychological health.

2. Changing shopfloor roles

Operators: Many commentators have suggested that shopfloor employees need to taken on a new and more demanding role in the transition to high-involvement working. An empirical study of the performance requirements of operators in Company F supported this view [4]. Rather than reliably performing prescribed tasks, operators were expected to develop their role proactively in order to meet shared goals, such as customer satisfaction. This required that operators have: a wider, detailed knowledge of the production process; multiple technical skills; and a range of 'meta-competences' that can be applied across many tasks (such as communication, problem-solving skills). A fundamental - yet often neglected - requirement for high performance we identified was the need for broader role orientations [4], [5]. Operators needed to develop a strategic view of their role beyond narrow responsibility for individual tasks where they felt ownership for a range of production problems (e.g. an operator in Company F stated: "My actual goal is customer satisfaction... I'll do anything to make sure the job goes out on the day its meant to").

Supervisors: Although their role varies considerably across organisations, supervisors have traditionally had responsibility for the day-to-day operation of their area (e.g. allocating work, monitoring performance). However, their accountability tends to outweigh the control their decision-making authority and this, in combination with other historical factors, has meant supervisors can feel like 'lost managers' whose role is ambiguous, stressful and low in status and rewards [6]. Given the potential for further confusion with high-involvement working, it is critical that supervisors' roles are clearly defined. In complex production environments, we argue that the best performance results when supervisors adopt a 'first-line manager' role (see also [2], [6]). Minor disturbance-handling and daily operational concerns are devolved to the work team, while the supervisor's responsibility and authority over the major parameters of the section is increased. Most important, the supervisor is responsible not only for technical performance but for facilitating employee development so that operators are willing and able to function autonomously. The role therefore includes: fostering commitment; developing team members' skills, knowledge, orientations and confidence; providing goals and direction; competing for resources; managing boundaries; administering personnel policies. A more facilitating, motivating style is important as the types of self-directed behaviours required (such as using initiative) cannot be easily coerced, and direct control structures (e.g. bonus systems) are likely to have been removed. Person-management skills, such as conflict resolution, communication and consensual decision-making, are critical.

At Company F, supervisors initially found this developmental role difficult. They described how hard it was to get operators to make decisions and think for themselves (e.g. who said 'you're the boss, you do it'). It was tempting for supervisors either to continue doing the tasks themselves or to always devolve responsibility to the most able team members. Despite these early difficulties, the transition was managed successfully and teams learnt to function autonomously. At this stage, operators described how it was important that supervisors "give you space to work things out yourself". Supervisors then became more involved in management decisions about manufacturing strategy, planning for the introduction of new product-designs and technology, and developing human resource procedures (e.g. they were actively involved in the forming a new appraisal scheme). Thus, whilst the initial role of the supervisor may focus mostly on facilitating team development and performance, it is important that supervisors' experience is used positively beyond this phase, for example by acting as a technical adviser or consultant to groups working on projects for future changes.

Production support personnel: A defining feature of IM is the integration of previously distinct functions of manufacturing. This obviously has implications for the roles of support personnel (e.g. quality inspectors, production planners, maintenance staff). One approach, used for quality inspectors at Company F, gives production support staff more responsibility for developmental tasks while direct operators take on the lower level aspects of the work. Quality personnel were initially responsible for training operators to do their own inspection, and then became involved in auditing processes to ensure continuous improvements in quality. Another approach used for skilled test engineers was to incorporate support personnel directly in product-line teams. They were expected to transfer their knowledge of testing as well as learn new (mostly lower level) skills. For many test engineers this loss of status and reduced opportunity to use their fault-finding skills meant they left the company. Others, although initially resistant, have now carved a niche for themselves by taking on either more managerial duties (e.g. functioning as team leaders) or higher level technical jobs (e.g. continuous improvement of testing procedures).

3. Managing successful transitions

Participation and involvement in the change process: Involvement of supervisors and support personnel is critical given the uncertainty they are likely to feel over their future jobs, in combination with the critical role they play in the transition process. Although supervisors participated in the planning of change, the failure to include test engineers in this process may have contributed to their difficulties with the transition. Operators should also have the opportunity to influence the decision-making process.

Communication and information-dispersion: Acceptance of new roles requires an understanding of the wider picture and why change is occurring (e.g. knowing about company goals and strategies, competitors, markets). More information is also required on a daily basis for operators and supervisors to perform effectively in high-involvement settings. The dissemination of such information at Company F was facilitated by daily team meetings, regular company-wide communication sessions, a site-wide newsletter, and information systems that operators as well as managers could access.

Systematic training for operators: A lack of training for a pilot team, combined with high expectations about the new work system, meant operators became quickly frustrated with their inability to learn new skills. For the site-wide changes, systematic programmes based on training needs analysis were thus introduced. As well as fostering skill development, and allowing for it to be readily monitored, this commitment to training showed people that the company was serious about individual development. It is also important to have a wide conception of training - not just in operational tasks but in meta-competences (e.g. problem-solving) and team-working (e.g. how to run meetings, resolve conflicts).

Training and education for supervisors: Training in person-management skills will certainly be necessary for most supervisors moving into managerial roles. Perhaps less obvious is the need for higher levels of technical proficiency. This adds 'expertise' to the authority status, enables more specialist functions to be integrated within the supervisor's area, and facilitates their involvement in continuous improvement projects. Training and education may also enhance supervisors' traditionally-low status, increase access to informal management networks, and promote strategic understanding. At Company F, those supervisors who best made the transition to front-line manager had technical qualifications and were supported by the company to do Masters-level training. Given that supervisors in the UK tend to lack educational and technical qualifications [6], company promotion of further training is likely to be vital.

Consideration of long-term careers: For supervisors and support staff, it is necessary to think beyond the transition phase to consider and prepare for the role that will be taken on and how it will change. Part of this involves reconceptualising people's views of careers as upward progression to seeing sideways movement as a career advance. Similar consideration needs to be given to operators' careers. Some operators who are already highly skilled may 'top out' (i.e. have nowhere to go after learning all the skills) while others may not be able to cope with the demands of the new role.

Aligning rewards with company goals: High-involvement working requires that payment and promotion systems reward people for behaviours that match the company's strategic goals. Individual bonus schemes for operators, for example, foster an orientation that the amount produced is to be achieved at all costs, regardless of other issues (e.g. quality, customer demand). Group bonus schemes do not reward individual contributions and, if targets are constantly raised, operators can feel 'squeezed'. In Company F, operators were paid with a grading scheme with pay increments linked to appraisals.

Appropriate and fair appraisals: The criteria on which people are assessed need to be clear and appropriate. Uncertainty over what was expected at Company F was resolved by incorporating more up-to-date performance dimensions (e.g. ownership, social skills) into the appraisal scheme (see [4]). A continued challenge the company faces is to ensure appraisals are carried out fairly, particularly as supervisors' are less involved in the day-to-day performance of the team. Peer appraisal might need to be considered.

Structural changes: To facilitate the development of production-support roles, changes are needed to the organisational structure to align indirect functions more directly with the product groups they support. In Company F, quality personnel often experienced conflicting priorities because, although working as an integral part of production, they were not responsible to the production manager.

4. Summary

Major changes in roles are required in the transition to high-involvement working- not just for operators but also for supervisors and support personnel. More attention needs to be given to what these roles entail, and how the transition to them is facilitated. The importance of integrated and coherent human resource practices cannot be underestimated.

References

[1] Dean, J. W. & Snell, S. A. Integrated manufacturing and job design: Moderating effects of organisational inertia. *Academy of Management Journal* **34** 1991 774-804.

[2] Lawler, E. E. From job-based to competency-based organisations. *Journal of Organisational Behaviour* **15** 1994 13-15.

[3] Lawler, E. E. *The Ultimate Advantage*. San Francisco, Jossey-Bass, 1992.

[4] Parker, S. K., Mullarkey, S., & Jackson, P. R. Dimensions of performance effectiveness in high-involvement work organisations. *Human Resource Management Journal* (in press).

[5] Parker, S. K., Jackson, P. R., & Wall, T. D. Autonomous group working within IM: A longitudinal investigation of employee role orientations. In G. Salvendy and M. J. Smith (Eds). *Human-Computer Interaction: Application and Case Studies*. Amsterdam: Elsevier Publishers, B. V., 1993 45-49.

[6] Child, J. & Partridge, B. *Lost managers*. London: Cambridge University Press, 1982.

Advances in Agile Manufacturing
P.T. Kidd and W. Karwowski (Eds.)
IOS Press, 1994

Technological Transplants in Japanese Management Techniques

Shigenobu NOMURA
Aichi Institute of Technology, Yagusa–Cho,Toyota–City, Aichi 470–03, Japan
Kazuho YOSHIMOTO
Waseda University, Okubo3–4–1, shinjuku–ku, Tokyo 169, Japan
Aya HIROSE
Sanno University, Isehara–City, Kanagawa 259–11, Japan

Abstract. This paper deals with the problems involved in the application of traditional Japanese Management Techniques to Japanese–affiliated enterprises outside of Japan. We assume that Japanese management Techniques will be affected by management culture factors formed from management environment, and propose the idea of three types of technological transfer, application type, adaptation type, compound type. The possibility of technological transfer is verified by our proposal and investigation. Finally, the main factor encouraging Japanese Management Techniques are proposed and topics for a new management approach are presented.

1. Introduction

Japanese companies has been attracting much attention overseas for their high productivity and the excellent quality of their products. We think that Japanese Management Techniques have much to do with this high quality and great productivity. Against the background of Japanese culture, Japanese Management Techniques consist of a management culture of company loyalty, strong group consciousness, small group activity, a good education system, a respect for human dignity, collective decision making, and cooperation between capital and labor, among other factors(1).

Japanese Management Techniques incorporate a special method that has long been cultivated in the society and culture of Japan. If one attempts to introduce this management to another country directly, much cultural resistance is encountered(2). If it is in contradiction to a different culture that has been cultivated are greater. However, if the Japanese Management System is not in conflict with the basic culture of a certain country, it can be adapted to that country. The main theme of this paper is to discuss the possibility of adaptation of Japanese management Techniques according to our theory and investigation.

We investigated the results obtained in other countries in adapting Japanese Management Techniques in adapting Japanese Management Techniques in terms of TQC,JIT,Employee training, Ranking by Seniority, Labor Management Relations, small group Activities, morale improvement, Lifetime Employment System, etc. The investigation was performed by direct feedback through listening and by questionnaires sent to Japanese enterprises abroad. The target countries were the U.S. and Asian countries. We discuss how Japanese Management Techniques have been applied and how management has adapted to the culture of other countries.

First of all, various factors of management culture in Japan and other countries are clarified, and their characteristics are classified in terms of the various factors. Then we propose three types of technological transfer, application type, adaptation type and compound type. We assume that management culture, including each form of Japanese Management Techniques, has factors that will be affected or not affected by a particular race, culture or tradition. Each of has specific cultural aspects, which can be divided into changing, unchanging and intermediate. These aspects are the points in technological transfer. The possibility of technological transfer of Japanese Management Techniques is verified by our personal and investigation. Finally main factors which encourage a production system are proposed and future topics for new management approach are described.

2. Japanese Management

Japanese Management is a special form of activity that has been developing in Japanese society and culture. We assume that the relation between cause and effect in production activity and culture is shown in Figure 1. According to the society situation of the time , management environment creates a management culture forms Japanese–type production activities. Both types of production activities have materials and equipment as their means of production.

But in human–related matters, there are huge differences in terms of production activities, due to the peculiar culture of each. Figure2 presents more detailed factors in Japanese culture and Management culture. Characteristics of Japanese culture include a Buddhist attitude, groupism, and racial homogeneity. Management cultural contents were mainly classified into 5 factors(1),(3). The management environment enveloping an enterprise may be divided into the common environment and the environment specific to each and every enterprise. Factors in the common environment are policy, society and culture. Factors in the specific corporate environment include products, production, labor and market. The specific corporate environment has been created as result of the decision–making by each enterprise and differ with the individual enterprise. Therefore the management environment is closely connected with a management culture; then management culture will be are affected by that change, and new Japanese production activity will be created through the cooperation of labor–management.

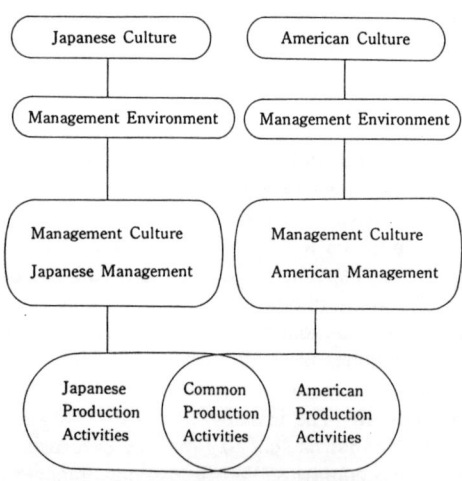

Figure 1 Culture and Production Activities

Japanese Culture		
Buddhist Mentality Groupism Racial Hamogeneity		

⇓

Management Environment		
Management Culture		

| Personal Factor Human Relation Factor Social Factor Job Factor Organization Factor etc | | |

Figure 2 Culture and Management Culture

3. Production Theory of Symbiosis

In order to develop Japanese Production Techniques with in various countries, we propose typical two approaches. One is to introduce a company system in response to the chracterristics of each country; the other to adapt the production system of that a given country to ones own company in response to production system of each country.

The former may be called an application approach and the latter an adaptation approach (4). Figure 3 indicates the relation between application and adaptation. Victor $\vec{a}$ expresses the direction of the production system of symbiosis and is defined as follows:

$$\text{vector } \vec{a} = \frac{Ax}{Ay} \qquad Ax : x \text{ element of } \vec{a} \\ Ay : y \text{ element of } \vec{a}$$

Vector $\vec{a}$ has length and direction, where length indicates the strength of introduction, and direction indicates strength of management axis and the adaptation axis.

In thinking about of the application and adaptation in terms of small group activity, we encounter a number of management culture factors: the value of work, the type of job, human relations, the ability to organize, and society. These factors can be classified into directly affecting factors and indirectly affecting factors, when introducing small group activity is indicated as Figure 4. The length and direction of vector $\vec{a}$ depends on the managerial strength of the enterprise. We must find the direction of a desirable vector $\vec{a}$ so as to maintain production activities of symbiosis.

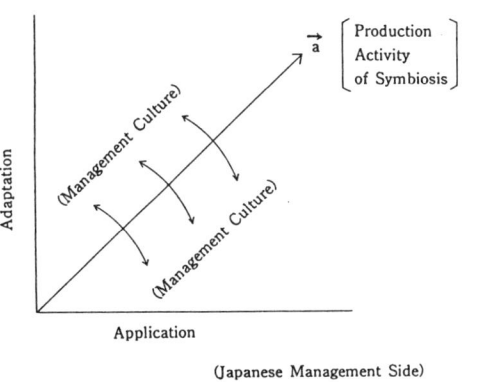

(Japanese Management Side)

Figure 3 Production Theory of Symbiosis

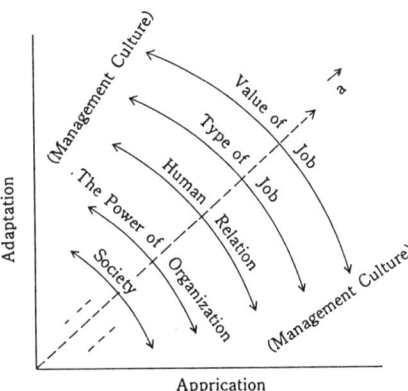

Apprication

Figure 4 A Small Group Activity

The introduction of Japanese management can be classified basically into three shown below:
1) application type
2) adaptation type
3) compound type
The interrelation of the three types is shown in Figure 5. In the application type, production is by the same system used in Japanese production. In the adaptation type, production is achieved by the same management techniques as used in the given country. The compound type is a mixture of 1) and 2). Most enterprise use the compound type approach. These processes are shown in Figure 6.

4. Management culture factors

Management techniques depend on the management environment at a given time. We consider that the management environment creates the management culture, and propose management culture
factors of the following three types.
1) changing factors: factors that can change in an early stage.
2) possibly changing factors: factors that can change slowly.
3) unchanging factors: factors that basically cannot change.
We discuss from the application standpoint technological transfer and these three factors.

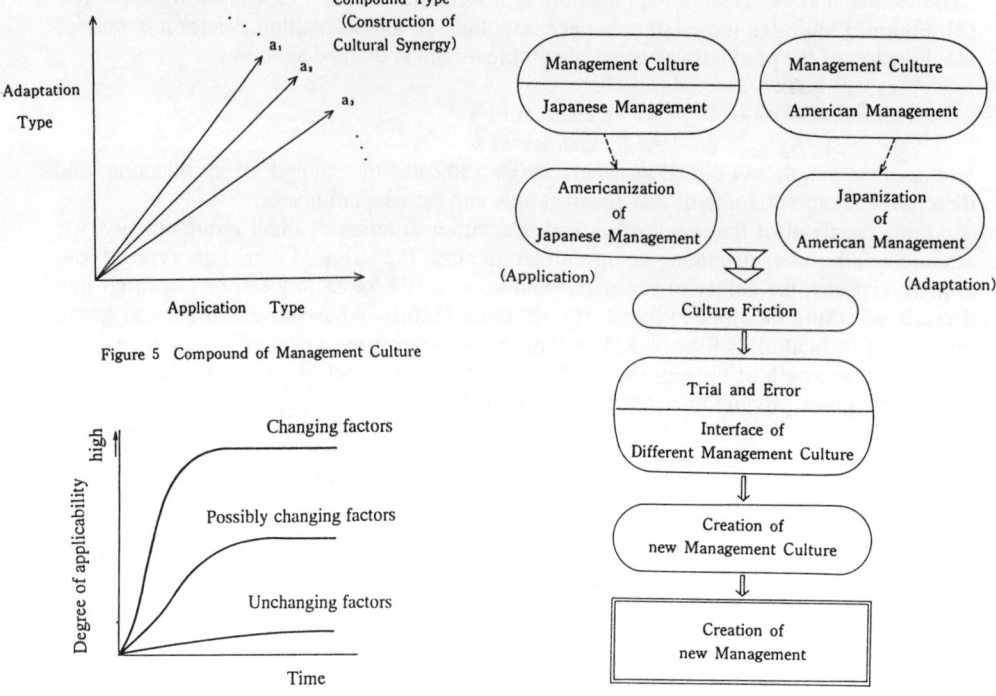

Figure 5 Compound of Management Culture

Figure 7 Relationship of Time and Applicability

Figure 6 Creation of New Management

Figure 7 indicates the relationship of time and the degree of applicability , and each of their factors is affected by the length of time. These factors relate mainly to technological transfer of compound type, and indicate that the higher the ratio of unchanging factors is, the more difficult the technological transfer becomes. The next problem to discuss is the structural elements of the management culture factors. Japanese management technique attaches importance to the human aspects developed in Japanese climate and culture. The foundation of this activity is groupism, a group consciousness based on human dignity and workers participation from the bottom up. In other countries, however, most people tend to be more independent, responsible and creative than Japanese. Then we propose five factors based on the awareness of groupism, and verified their effectiveness: human relations factor, personal factor, organization factor and work values factor.

5. Conclusion

The technological transfer problem of Japanese management technique was treated in terms of the relation between cause and effect of production activity and culture. The characteristics of the three types of models were classified and the existence of management culture factors was proposed. Also, the effectiveness of the five factors was clarified.

References
(1) T. Shishido: Japanese–affiliated enterprise in USA, Tokyo Economy Shinposha, 1980
(2) M. Murayama: Theory of Management Overseas Transfer, Soseisha, 1989
(3) S. Nomura: International production strategy of small and medium business, Annual
    Report of Society of Business Diagnosis (Japan), 1993

Advances in Agile Manufacturing
P.T. Kidd and W. Karwowski (Eds.)
IOS Press, 1994

CULTURAL ASPECTS OF THE DESIGN OF ADVANCED MANUFACTURING TECHNOLOGIES

J Martin CORBETT
Warwick Business School, University of Warwick,
Coventry CV4 7AL, UK

Abstract. This paper presents findings from a cross-cultural survey of CNC producer companies. Results indicate that despite serving different customers, no significant differences in technological design are apparent. Implications for the development of human-centred CNC are discussed.

1. Introduction

In recent years there has been a growing interest in designing human-centred AMT. Researchers and writers within this emergent tradition view current AMT design and implementation practices as essentially technocentric in nature. What remains unclear is the relative importance of customer-demand and technology-push in shaping such AMT designs. Within the context of the development of human-centred AMT, such a disaggregation is important as it helps inform where and how proponents of human-centredness need to focus their efforts in order to redirect current technocentric design practices.

2. The CAPIRN Survey

In an attempt to open up both the process and context of AMT design, a large survey of CNC producers in the UK, Japan and Germany was carried out by a team of researchers under the auspices of CAPIRN (International Research Network on Culture and Production), as the first phase in a cross-national study of AMT production chains. A total of 91 randomly selected CNC producers (17 from the UK, 29 from Germany, 45 from Japan) participated in the study.

2.1 Cross-cultural Differences

The CAPIRN questionnaire contained a question relating to the origins of the chief CNC design and development impulses in the producer companies. As the results in figure 1 reveal, respondents generally regarded customer demand as the major factor shaping their product designs. Marketing was also cited as a major factor. Additional questioning revealed that producers' perceptions of customer demand were surprisingly uniform - reducing lead times, higher quality machining and increasing production flexibility being cited by almost all respondents. Hence, at first sight at least, CNC design would seem to be rather more customer orientated than technology-led.

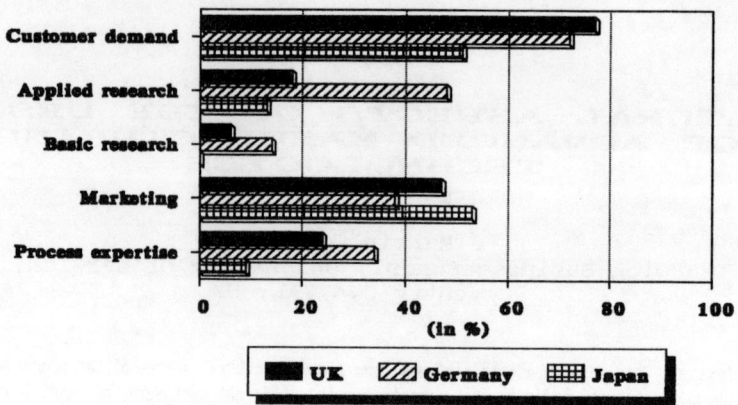

Figure 1: Origins of CNC Design Impulses

In an effort to open up the design and development process a little more, the CAPIRN questionnaire asked respondents: *how would you assess the relative importance of the following criteria with respect to optimal CNC machine tool design?* (see figure 2). The percentages in figure 2 relate to the percentage of respondents who regarded each criteria as 'extremely important'.

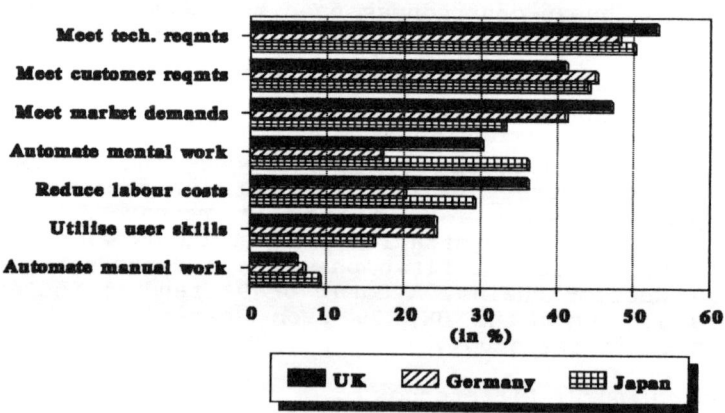

Figure 2: Optimal CNC Design Criteria

Again results here suggest that producers give a high priority to customer-driven CNC designs. Note also that meeting technical requirements is cited most often as a key design criterion. The extent to which these requirements are fully compatible with customer requirements is unclear. Indeed, additioning questioning indicate that only two options were generally available to customers - either 'off the shelf' or individually configured CNC machine modules (e.g. a choice of CNC controllers, motor drive units, and tool or work handling capability). This choice was offered by 60% of the Japanese and German producers, and 70% of producers in the UK. A second option (offered by 46%, 68% and 53% of German, Japanese and UK producers, respectively) offers customers the opportunity to submit a technical requirements profile. However, these requirements tend to be restricted to minor technical features (e.g. precision of offset measurements and the provision of live tooling). Thus, customer requirements would seem to be met only within producer-defined limits.

Figure 3 reveals some cross-national differences in perceptions of target CNC users, with 37% of Japanese producers targetting specialist office programmers as users (compared to 23% of UK and 15% of German producers). Note also that three times as many German producers gear their CNC designs toward operation by skilled workers than do their Japanese counterparts. However, an interesting anomaly is thrown up when one compares figures 2 and 3. Whilst 54% of German producers target their products towards skilled shopfloor operation, less than half this number (24%) see the utilisation of user skill as an important CNC machine tool design criteria. This again raises the suspicion that user targetting may represent more of a marketing ploy than an engineering design strategy.

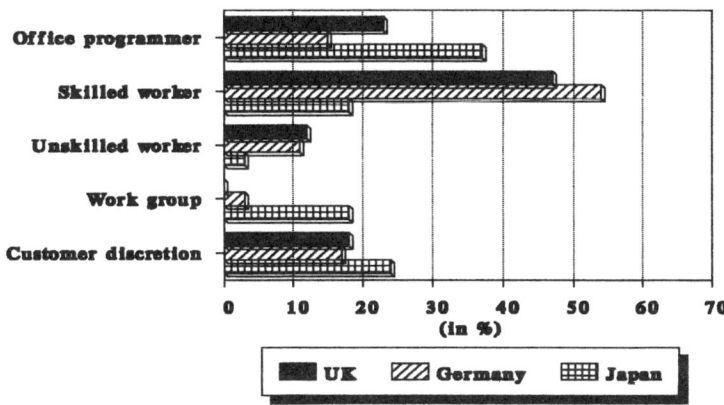

Figure 3: Projected Users of CNC

The cross-national differences revealed in figure 3 may also reflect more on marketing than design. Follow-up interviews with survey respondents revealed that producers in the three countries perceive themselves as operating within different market segments - i.e. they attempt to fulfil the needs of different customers. This, in turn, leads to the development of different types of design and marketing expertise within different production chains.

2.2 Cross-cultural Commonalities

Behind these apparent differences, however, a number of common factors emerge. First, marketing rhetoric relating to the meeting of customer requirements does not influence overall CNC design in any fundamental way (remember also that figure 2 shows that less than half the respondents regarded such an aim as 'extremely important'). Thus the influence of producers' in-house process expertise (i.e. knowledge of materials and AMT systems functioning) and applied research as chief CNC development impulses may remain substantial (see figure 1). Second, 26% of producers regarded the reduction of user's direct labour costs as an 'extremely important' CNC design criterion and 27% saw the automation of mental work (i.e. the embedding of planning and operational decisions into CNC software) in a similar light. This compares with only 18% of producers who saw the utilisation of user skills as 'extremely important' (see figure 2).

A third common factor, stemming directly from the previous two, is that many user companies need to take full advantage of training services offered by producers because their own skill base is not fully commensurable with the skill requirements of the purchased machines. Questionnaire responses reveal that 69% of producers engage regularly in after-sales training courses for both the office and shopfloor users of their products - evidence for both a cognitive disjuncture between AMT designers and AMT users, and an increasing user dependence on producers who have to assume training, financing and other services in addition to their usual technological services. Indeed, this dependency enables AMT producers to force the user companies into adopting systems that correspond best to the producers' own production, development and financing conditions. As a result the problems which user companies experience in attempting to master these suboptimal solutions increase their dependency on AMT producers.

Taken overall, results from the CAPIRN research suggest the key influence of marketing expertise within CNC producer organisations. The interests and organising principles of CNC producers centre on the fulfilment of customer requirements (within predefined technical limits) and the maximisation of market share and profitability. What is of particular interest in the case of CNC production in the UK, Germany and Japan is that although the companies sell their products in different markets, there are marked similarities in their perceptions of optimal CNC design (figure 2). Again this suggests marketing rhetoric has little influence over the CNC design process itself.

3. Conclusion: Towards Human-Centred AMT Design

The brief overview here suggests that human-centred researchers need to focus their efforts on actual CNC design practices within producer companies rather than on user companies. Unfortunately, at the present time, most effort is concentrated on implementation and use. Results presented here imply this may be too little too late.

Advances in Agile Manufacturing
P.T. Kidd and W. Karwowski (Eds.)
IOS Press, 1994

Human Factors In QR and EDI Implementation

Marcia PERRY

Victoria University of Technology, Melbourne, Australia

Amrik SOHAL

Syme Management, Monash University, Melbourne, Australia

Abstract This paper describes the introduction of Quick Response (QR) product replenishment and Electronic Data Interchange (EDI) amongst a group of Australian companies. The methodology being applied to introduce the new philosophy and concepts to the organisations involved in the program has itself resulted in significant improvements for all parties involved. This has involved organising supply chain companies into clusters and discussions at Roundtable meetings facilitated by an independent person. The paper discusses the outcomes of this program to date.

1. Introduction

In order to compete successfully manufacturers are developing strategies based on speed, agility and reliability. In the apparel industry this concept is frequently referred to as Quick Response (QR). Like Just-In-Time, QR embodies a generic philosophy as well as a loose set of practices. Much of the existing literature in this area points to the importance of the development of supply-chain partnerships for successful outcomes [1,2,3]. This paper reports on a government funded program aimed at speeding up the successful introduction of QR product replenishment and Electronic Data Interchange (EDI) in the Australian Textile, Clothing and Footwear (TCF) industry.

It is suggested that a there are a number of underlying human relations issues related to the successful introduction of a QR\EDI program. This paper discusses these issues, in particular group cohesiveness, establishment of good inter-company working relationships and the building of a climate where problems are openly shared.

2. The QR\EDI Program

Fifty firms are involved in this program and are clustered in groups of three supply chain companies already committed to QR. The objectives of the program are being achieved by set procedures embodying an initial company evaluation, the formation of "clusters", a QR audit of the company, Roundtable meetings, workshops and seminars, grant dissemination and before and after measurement of QR implementation success factors.

Our observations of the program so far suggest that the opening up of communication channels between the firms and the development of partnerships for mutual benefit has been a recipe for success. Pugh [3] and Parker [4] concur on the validity of the supply-chain partnership approach to QR implementation.

Outcomes of the program to date have been: the arrival at mutually agreed-to supply chain solutions that are beneficial to all parties; the sharing of information in a climate of trust; and top level commitment to the implementation of the changes and the championing of the new ideas to company employees

The methodology being applied to the QR\EDI Program is significantly contributing to its success. Firstly, it encompasses a positive problem-solving environment, confirmed by Bailey et al. [5] as a desirable element in working groups; secondly, it relies on skilled, neutral facilitation with leadership that is not aligned with a particular company and is seen by all as being fair; and thirdly, the participants are all high level company managers who see involvement as being beneficial to their companies. The facilitator's style could be described as encouraging, positive, affable and organised. At introductory meetings he adopted a "selling" style whereas during workshops his style has been that of "participating" where he has taken more of a back seat and where decisions have been made by the group.

An important competitive element of QR today is fast supply-chain product replenishment information transfer via electronic means. Integrated EDI adoption to this end is being promoted through the QR Program. EDI is gradually replacing the majority of paper transactions and is also used for payment in a growing number of cases. As being demonstrated by the QR program, QR\EDI benefits all the parties concerned, nor just the retailer. A disadvantage of full-scale Quick Response implementation is that it can require costly and time-consuming infrastructure changes within a manufacturing company. Indeed, implementation of large scale changes can be risky without total supply chain involvement, support, risk sharing and commitment.

3. Success of the Supply Chain Workshops

The problems mentioned above are being addressed by concentrating on eliciting top-down problem solving initiatives within clusters and firms; initiatives that can be effected quickly, without requiring large scale infrastructure changes and costs in the first instance. This initial process, witnessed in the cluster meetings, could be seen to be an "obstacle removal process" resulting in manufacturers building a more streamlined supply-line foundation on which to base strategic infrastructure changes, with less risk than perhaps in the past.

A startling example of this "obstacle removal process", which occurred in one of the initial Roundtable meetings is that of a production manager from a shoe manufacturing company explaining to the group about the "necessity" for tying up production in making over two hundred shoe samples to show buyers for the coming season because he had to show all possible styles and colours. His two suppliers also talked about having to supply large numbers of samples of their products to the manufacturer. The three men were quite taken aback when the buyer present said, "We don't need so many colours, we are more interested in the styles." This interaction indicated a very poor level of supply line communication prior to the meeting on what was a crucial area - that of product development and the elimination of waste activities [6]. It also indicated that the parties were "doing what they had always done", perhaps out of fear of losing business.

The outcome of the interaction was a resolution by all the partners engaged in the clusters to joint product development which would be more precisely related to market requirements and further reduce the number of sample styles and colours produced. Reference was also made to using interactive Computer Aided Design systems in the future, to replace colour sampling. It can be said that the outcome of the interaction was most positive, with creative solutions to member's problems being explored by the group.

The initial Roundtable meetings proved to be of invaluable importance in increasing understanding between cluster members. The usual format was a general discussion about how the companies in each cluster could improve their performance, both individually and collectively. Prior to the Roundtable meetings, participants had been asked to develop selfish "wish lists", that is, things they would like to happen in the supply chain to improve their own performance and profitability. These "wish lists" were circulated to cluster members prior to the initial Roundtable meeting. Normally the "wishes listed" by participants were what they either saw as shortcomings in their cluster partners, or what they knew had been achieved by other companies they wished the group to emulate.

It can be speculated that because the "wishes" by definition were framed as challenges, they tended to be thought of in a positive light, rather than as problems with negative connotations. From observation of Roundtable meetings and workshops, cluster group acceptance of "wish list" items and desire to solve the inherent problems was high.

The methodology applied; of each cluster company providing the other cluster partners with a "wish list" which is discussed at a Roundtable meeting, and is followed up later with a mutually agreed-to "action plan" has worked well, to date. It is a "cards on the table approach" where supply chain problems are dealt with openly and effectively. The role of the QR Program facilitator, too, has been a crucial one, and points the way to the use of external, impartial facilitators at supply-chain meetings in the future, beyond the QR Program lifetime. It is this person's task to galvanise the group towards becoming a viable problem sharing and solving team. This is not always easy when in the past the communication lines between participants have either been non existent (as has been the case with a number of manufacturers' suppliers and retailers) or the communication lines have been limited.

4. Outcomes of the QR\EDI Program

The outcomes of the program to-date has been very positive in achieving QR advances. The program has produced a range of significant efficiency improvements and economic outcomes, gleaned from workshop minutes, which are being achieved quickly and at little direct cost to the participants. These include:
- much better relationships than have ever been achieved before between buyer, manufacturer and supplier.
- vastly improved information flow up and down the supply chain.
- quicker exchange of vital information.
- elimination of "padding" of schedules", that is, parties are no longer asking for unreasonable schedules; no longer asking for larger quantities than actually required; no longer setting just-in-case" delivery dates much earlier than actually required
- fewer order cancellations.
- greatly improved cycle times in the supply chain.
- increased sales due to improvements in cycle times and correct sales analysis data being available quickly.
- reduction in costs and inventory at all stages of the supply chain.
- increased material availability by standardisation of product, raw material and colour.

There is considerable anecdotal evidence in support of favourable QR outcomes to companies. Some examples are provided below:

"The positive attitude towards partnering between the two managements has given the strongest links ever, which reflects in much better working relations and trading. This enthusiasm has been carried through to the shop floor."

"Our ability to plan production has improved greatly. We have halved delivery times, vastly improved dissemination of sales, marketing and production information to all parties, increased sales and virtually eliminated dead stock."

"The program will give us an advantage over importers by improving new product development procedures and increasing our ability to react to industry trends and retailers' requirements. We will be able to bring our new range forward by two months as a result of the program."

"There is potential for increased sales as a result of involving the retailer in the product design and forecasting stages."

"There has been a vast improvement in our ability to schedule production and hence fill orders."

These comments indicate a positive attitude towards the QR program amongst participants and demonstrate some early benefits to participants. The conclusion is easy to draw that without the QR program, and without the formation of supply-chain partnerships and increased communication, the old problems and frustration would continue and the barriers to QR would still exist.

Arising out of the cluster-team discussion of wish lists and subsequent action plans, some commonly occurring guarantees were given by retail buyers and clothing and textile manufacturers. These guarantees are indicators of improved response to customers along the supply chain. Overall, the supply chains' ability to service end-customer demand has been rising; directly as a result of the workshop process. An interesting point to note is that almost all of the above guarantees involve further, on-going communication at both supply chain and company level.

The issue of EDI implementation, considered highly important for QR success is being mostly handled on a case-by-case basis by consultants attached to the program, basically because of the specific technical expertise required. The consultants attend relevant workshops and communicate regularly with the QR program team, informing them of EDI implementation developments. EDI implementation towards full integration across a range of functions has been occurring steadily within participating companies, following consultant involvement.

The overall success of the QR program to date has largely been based on its supply chain relationship focus, accompanied by its team problem solving approach and dedicated facilitator leadership. The program objective of improving supply-line response rate to customer orders is already being achieved in many cases as a result of both creative workshop input and consultant support with EDI implementation.

References

[1] C. Troyer and D. Denny, Quick Response Evolution, *Distribution Management*, May (1992) 104-107.
[2] D. Kincade, N.D., N. Cassill and N. Williamson, The Quick Response Management System: Structure and Components for the Apparel Industry, *Journal of the Textile Institute*, 84\2 (1993) 147-151.
[3] L. Pugh, Keynote address: Quick Response Trading Partnerships: The Future of Manufacturing, *Quick Response 91 Conference*, Nashville, 1991.
[4] D. Parker, *Quick Response: The Only Way Forward for UK Clothing*, Working Paper, Hollings Faculty Clothing, Design and Technology, Manchester Polytechnic, April, 1989.
[5] J. Bailey, J. Schermerhorn, J. Hunt and R. Osborn, Managing Organisational Behaviour, Second Edition, John Wiley and Sons, Sydney, 1991, pp. 259-262.
[6] J. Tidd, Flexible Manufacturing Technologies and International Competitiveness, Printer Publishers, London, 1991, p. 94.

Advances in Agile Manufacturing
P.T. Kidd and W. Karwowski (Eds.)
IOS Press, 1994

Mediating between users and designers user involvement in design of a flexible sewing machine

Thomas Binder & Palle Banke
Human Resource Development, Danish Technological Institute, DK-2630 Taastrup

Abstract. This paper describes in brief experience gained within a development project under the BRITE/EURAM programme, seeking to develope flexible sewing technology for group work. In the paper it is discussed how a cooperation between machine manufacturers and clothing companies can be organized and where conventional design methodology has to be changed to allow for a fruitfull dialogue between designers and users.

1. Work and technology in the sewing shop

The clothing industry was for a long time a prominent example of tayloristic management practice. In the sewing shop, work has been divided into individual tasks with a cycle time down to a few seconds, and machinists have been working on piece rates in repetitive, monotonous and at the same time highly specialized jobs.In later years, the tayloristic sewing shop has, however, ceased in importance in Western Europe. A large part of mass production has moved to low-wage countries, and changes on the market have, in general, reduced the lot sizes. The result has been that a growing number of clothing companies are seeking a new and more flexible work organization.[1]

Technologically the turn away from specialized high-volume production has led to a certain degree of confusion. The clothing sector has, traditionally, been looking to the machine suppliers for new trends in manufacturing. As part of the sector is now turning towards re-integration of tasks and is giving higher priority to the human side of manufacturing, the suppliers seem to have difficulties in finding the road to take.

2. The GSS-project, designing sewing machines for group work

In most European countries, efforts have been made to introduce group work in the sewing shops in a way which combines the companies search for flexibility with opportunities for the employees to have more skilled and less strained jobs. One of the experiences from these efforts has been that existing technology tends to complicate such an organizational restructuring because of the high specialization of each type of sewing machine.

One this background, a Danish-Italian development project 'Group Sewing Systems' was initiated with financial support from the BRITE/EURAM programme. The aim of the project is to develop multi-purpose sewing machines specifically suited for sewing long-cycled operations in production groups. The project consortium consists of an Italian producer of sewing machines, an Italian and two Danish clothing companies, the Danish Clothing and Textiles Institute and our department of Human Resources Development at the Danish Technological Institute. The project started in 1990 and is running for 4 years with a total budget of approximately 30 man-years.

3. Conceptualization, from organizational need to technical requirement

From the outset, the project partners were brought together by a common understanding of the need for change in machine design. In the clothing companies, a number of specialized machines such as pocket automats, shoulder seamers and two-needle stitchers were used. If the sewing shop should be reorganized for group production, this type of machines formed a very visible bottleneck for the recombination of operations. For the machine manufacturer, however, the description of technical barriers for the group organization in the sewing shop was not in itself enough to define the design objectives.

Our role as researchers and consultants in the field of work life research was defined as the 'interpreters' who had to bridge the gap between organizational need and technical requirement. We made our contribution in two steps. First we did a number of case studies of how different types of new technology were implemented in the clothing industry. It was shown that management in the clothing companies in deciding to invest in the sewing systems studied emphasized conceptual aspects of the new technology. In contradiction to this result, it was also found that the systems when implemented tended to be integrated in the existing manufacturing practice. The reason for this apparent mis-match between intentions and actual use could be traced in two directions. In one direction, there seemed to be little or no follow-up on the decision of investment taken on higher management levels. In another direction we also found that the technical support organization of the machine supplier tended to neglect the broader perspectives inherent in the technology. For technicians and engineers setting up and providing training for the new sewing systems, the technology was very much dealt with as a 'unit to be plugged in' to the existing line.

The results described above were reported to the project consortium and made way for fruitful discussions both among machine designers and clothing companies[2]. Both seemed to conclude from the results that implementation had to be given more thought than what was found in the case studies. From there, we took a second step towards conceptualizing the new technology to be developed. The draft specifications for the design drawn up in the early stages of the project were poorly structured and did not envisage any coherent idea about the road to take. We found that what was needed was a differentiation between different design levels, and a pinpointing of design choices at each level which had direct or subsequent consequences for the organizational adaptability of the technology to be designed. We suggested four different design levels and opted for a deductive specification process where choices were made on the most aggregated level first and secondly carried on to the lower levels.

The levels we suggested were based on a tool perspective on technology with each level corresponding to an organizational level in the organization of the work in the production group. In this way our first level was the group level where the new machinery could be regarded as a tool for the overall task of the groups, in our case to produce finished jackets. We argued that a tool at this level means the physical arrangement of the technology into one or more units to which the group can assign an operational set of tasks. With this terminology we described the physical appearance of the line, the production cell or the dedicated automat to be examples of design choices. On the basis of the preparatory work we had made to establish production groups in the participating clothing companies, we could argue that the production island formed an attractive choice for the production groups in questio[3]. If the production island was accepted as a design choice, we could immediately derive design specifications. Just as important, the production island would form a point of reference for subsequent considerations on design for the next levels: work-station level, machine-system level and sewing-head and devices level[4].

This line of thought turned out to be very hard to adopt particularly for the machine designers. On the formal level, our suggestions were accepted as part of the framework for the preparation of the requirement specifications. From the process, however, it was clear that the designers focused their attention on the machine system level and rather saw the levels above as a 'wrapping' which could be furnished later on. For the continuation of the development work it was decided to employ a dual strategy. Preliminary requirement specifications were agreed upon to get the technical design process started. As a parallel to this, a first prototype was prepared for the sewing shops on the basis of designs previously made by the machine manufacturer. The practical evaluation of the prototype could then supply new suggestions for the design from the clothing companies.

4. Testing prototypes, when machinists enter the design office

The objective of the prototype testing was to stimulate the specification process. In this sense the testing differed from conventional prototype testing which mainly aims at validating functionality. What was sought for was a process of learning and searching in the sewing shop rooted in experience from a realistic work situation. To communicate the need for a firm user involvement in the sewing shops, we did, however, organize the test activities very similar to the normal process of implementing new technology. This means that we organized training sessions, scheduled and monitored the use of the prototype in daily production and carried out evaluations based on interviews, job observations and inspections of finished products.

The training sessions for machinists in the clothing companies proved to be the most fruitful occasions for design considerations. We and consultants from the clothing and textile institute used approximately one week together with the machinsts who were to operate the prototype. During training, important problems in the design could be disclosed, and possibilities for a change in work practice could be considered and reflected upon.

The use of the prototype in daily production did not provide to the same extent new information about problems and possibilities. Even though the companies had dedicated resources for the test activities, the routines of the sewing shops made it difficult for the machinists to create a learning environment where technical difficulties were not conceived as obstacles but as important inputs to the design process. The long-time use of the prototype did, however, consolidate the findings in the more intense training sessions and improved the knowledge of specific aspects of the machine functionality.

The first prototype testing gave important input to the design process, and demonstrated both to clothing companies and machine manufacturer the potentials of involving the end user. An important experience was that intensive sessions with the users gave more results that the long term evaluation in the production environment. It was also clear that an on-going and direct dialogue between users and designers needed to be established. For the testing of the final prototype it was therefore decided to involve the machine manufacturer more directly in the evaluation, and to concentrate our contribution on the research side on organizing extended sessions with the users.

5. How to play games with designers and users?

The GSS project will be finished at the end of 1994, and it is still too early to evaluate the outcome. A second round of prototype testing will be made, and results obtained here will influence the final result. From what can now be seen, two main conclusions can be drawn.

First, the project demonstrates that machine design can be adapted to specific needs and requirements in an organizational setting emphasizing skill and competence. From the second prototype launched at the beginning of 1994, important results in terms of operator-controlled flexibility and multi-purpose functionality have been achieved reaching beyond the frontline of mainstream technology for the sector.

Second, experience from the first three years of the project illustrates some important problems in the methodology of conventional machine design. For the designers it is apparently very difficult to deviate from a linear design process starting from a well-defined requirement specification. As discussed above, there is a very obvious need for a conceptual renewal of the approach to sewing-shop technology. We were able to demonstrate this need, but could not fully transform it into a workable part of the design process. When viewed in retrospect, the design process is rather inductive than deductive and usually takes as its starting point 'designs already made'. Furthermore, to adapt the practice of prototype testing, usually employed to verify functionality, to a test phase emphasizing a more substantial input to the design process turned out to be problematic in the sense that the feedback from the users came late and could not be unfolded in a genuine and longer-lasting dialogue involving mutual learning and searching among users and designers.

User involvement in the design process has been suggested and tried out throughout other areas of technological development. Particularly in the area of system design, participation strategies for design have been widely applied and reported in literature[5]. Within this tradition, it has been suggested to see design as a language game between different actors producing innovation where different goals and orientations meet particularly among users and designers[6].

We see this description of the design process as very relevant also to the process machine design. Compared to areas such as system design, more thought must, however, be given to the conventions and models among the design community in an area like mechanical engineering. Design of machines is not a fully open-ended process. The resources needed and the technical obstacles to innovation must be acknowledged if a fruitful debate with machine manufacturers is to be achieved. The models dominating the thinking of machine designers is moulded by the tayloristic heritage. If new models are to gain ground, we will need to induce opportunities for an open and comprehensive dialogue in early as well as later stages of the design process[7]. The designers need to adopt to a design process of mutual learning and searching giving up the ideal of a linear and well-defined process of problem solving[8]. We as the mediators will, on the other hand, need to take on our role as 'game organizers' with a more elaborate understanding of why and how designers need to build on past experience.

references

[1]   Antti Kasvio (ed.)Industry without blue-collar workers - perspectives of European clothing
         industry in the 1990's, Work Research Centre, University of Tampere, 1992
[2]   Palle Banke & Thomas Binder, Will New Technology 'help' Taylorism overcome the present
         crisis, in Antti Kasvio (ed.), op.cit.
[3]   Peter Brödner, Fabrik 2000, Alternative Entwicklungspfade in die Zukunft der Fabrik,, Berlin,
         Edition Sigma, 1985
[4]   Palle Banke & Thomas Binder, Design of Human Centred Technology in the clothing Industry:
         TA-approach to the Sewing Machine Technology, proceedings from the 3rd European Congress on
         Technology Assesment, Copenhagen, 1992
[5]   Greenbaum & Kyng (eds.), Design at Work: Cooperative design of Comuter Systems, Lawrence
         Erlbaum, New York, 1990
[6]   Pelle Ehn, Work-Oriented Design of Computer Artifacts, Arbetslivscentrum, Stkhlm, 1988
[7]   Thomas Binder & Klaus T. Nielsen, Industrial Culture and design methodology, in Rauner,
         Social shaping of Innovation and Manufacturing (tentative title), Bremen, 1994, (forthcoming)
[8]   Rauner, Rasmussen & Corbett, Crossing the Border, Springer, 1990

Advances in Agile Manufacturing
P.T. Kidd and W. Karwowski (Eds.)
IOS Press, 1994

Transfer of New Skills and Technologies in Advanced Manufacturing: Transformation Models

Valery F. Venda*, Douglas R. Strong*, Ilona V. Venda* and Oleg V. Shevyakov**

Department of Mechanical and Industrial Engineering, University of Manitoba, Winnipeg, R3T 2N2, Canada

**Russia State Committee on Higher Education, 33 Shabolovka, Moscow, Russia*

Abstract. Principles of the ergodynamics in advanced manufacturing are proposed. The ergodynamics studies work and technology transformation dynamics. It combines and develops further research and design methods of ergonomics, industrial engineering, engineering and applied psychology, economics, medicine, sport and fitness on humans in dynamic environments, and mutual adaptation with machines and environment. Ergodynamics studies particularly an important trade off in the manufacturing: productivity vs universality. A higher degree of structural system specialization provides for its higher maximal manufacturing efficiency but it narrows the range of acceptable environmental changes where high efficiency could be reached.

1. Introduction: Transformation dynamics of the manufacturing work structures

The methods of transformation dynamics [3,4] could be successfully used for economic analysis and planning skills and technology restructuring in advanced manufacturing [1, 5]. They are very convenient for analysis of work efficiency and flexibility when the products and production are being changed. In a recession, and especially in a time of strong competition, this is of great importance. Using transformation dynamics models of the work efficiency Q as a function of work efficiency-complexity factor F and time T , we can compare the economic features of a manufacturer's different tactics in improvement of human resources, workers' skills needed for a new production.

We analyzed analogous task for transformation of manufacturing facilities . To improve the human resources the manufacturer has also two choices:

1. Change the workers. Build a new plant elsewhere, using new workers. When the new plant functions well, he or she would terminate the workers at the old plant and sell the plant and land.

2. Using quality circles, production and team meetings, special courses and seminars, intensify continuing education and retrainig the workers and managers for a new production, tasks, functions. Transform previous skills and work structure into new ones a careful change in one part of his plant, using volunteers to work in this changed area. If the change succeeds, he or she would introduce it in the areas that need the same type of improvement [5].

The second approach is more humane and attractive, but it can work only by using transformation dynamics and a mutual adaptation between all major facility components as parts of a human-machine-environment system. This approach leads to the prospective policy of the "recycling" of human resources, skills, knowledge in dynamic manufacturing.

2. The ergodynamics in manufacturing

The ergodynamics studies work transformation dynamics [2, 6]. It combines and develops further research and design methods of ergonomics, industrial engineering, engineering and applied psychology, economics, medicine, sport and fitness on humans in dynamic environments, and mutual adaptation with machines and environment.

These are the principles of the Ergodynamics.

1. The criteria of work efficiency and complexity represent the goals that humans should achieve at work. A functional work structure represents the dependence of the work efficiency and complexity on the human's factors of mutual adaptation with the machine and the environment. The functional structure is modeled with a bell-shaped characteristic curve for work efficiency and a U-shaped characteristic curve for work complexity as the functions of the factors of mutual adaptation (factors of efficiency-complexity).

2. The actual work efficiency (values of certain efficiency criteria) depends on the functional structure and work environment. Maximal work efficiency occurs when the functional structure and work environment are mutually adapted.

3. Different work environments can lead to maximal work efficiency values if different work functional structures are used.

4. Interaction between two systems is most effective when the environment or factor of mutual adaptation between systems F corresponds to the crossing point of the characteristic curves of the system's functional structures.

5. Decreasing system efficiency always accompanies transformations of the system's structures S_i and S_{i+1}. The efficiency losses are lowest if the transformations go through a

2

state common and equal to the previous S_i and new S_{i+1} structures. This point is modeled by the crossing point of the structures' characteristic curves with the coordinates $(Q_{i,\ i+1};\ F_{i,\ i+1})$. Figure 1 shows a process of development of the system structure Si and its transformation into more advanced and productive structure Si+1. The whole process of the system development includes nine different phases: 1. S_i structure convergence; 2. S_i synchronization; 3. S_i plateau; 4. S_i de-synchronization; 5. Diverged S_i; 6. Transformation plateau of qualitative transformation of S_i to S_{i+1}; 7. Convergence S_i + 1; 8. Synchronized S_{i+1}; 9. Plateau of S_{i+1}.

The *transformation period* denotes a time interval when a system sustains losses. Losses occur because of the temporary drop in its efficiency during the transition from one structure-strategy to another. This period covers a time from the start of the efficiency drop (divergence of S_i) to when the efficiency Q_{i+1} of the new structure S_{i+1} reaches the maximal efficiency level: $Q_{(i+1)} = Q_{i\ max}$ (Figure 1).

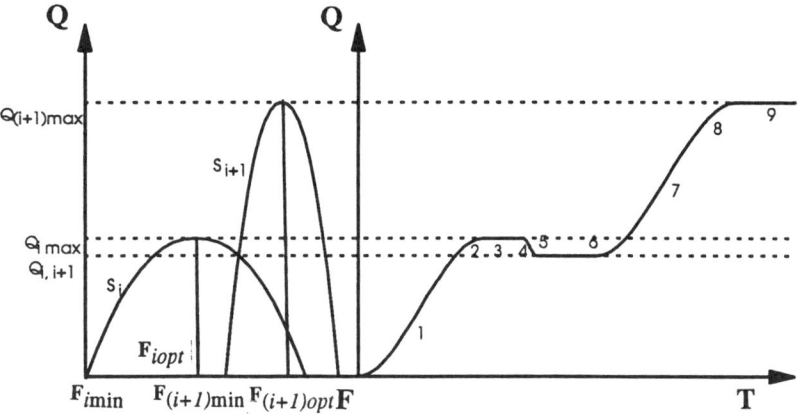

Figure 1.   Manufacturing system progress stages in development structure Si and its transformation into S_{i+1}[6].

3. Trade off in the manufacturing: productivity vs universality.

Any comparison, competition or cooperation will assume the presence of some factor, F_j, common to both systems; otherwise they will function in different intervals and even spaces, and not overlap. It is impossible to analyze a competition and compare the functional structure efficiencies of two companies one making aircrafts and another making toys.   A comparison of the profits of the companies gives a very limited information on human resources   and structural advantages. If two companies directly compit each other it is always possible to find a main factor influencing the efficiency (e.g. productivity).

At Figure 1 the $i+1$th structure has a higher maximal efficiency than that of the ith structure. This is due to the better degree of mutual adaptation of system components in the ith structure. That is, any change in some internal component will have a greater result for the other $i+1$th structure components than the same change for the ith structure. Hence, the slope of a characteristic curve is steeper for the $i+1$th structure than that of the ith structure.

Accordingly, any change in one component of the ith system affects the other components faster than it does in the $i+1$th system. Therefore, the characteristic curve slope is steeper in the former case. A range of the factor F values $\Delta F_i = F_{imax} - F_{imin}$ were the structure can perform with some efficiency Q>0 reflects a universality of the structure: the bigger ΔF the bigger universality of the structure (see Figure 1). The universality has a negative correlation with a maximal or average level of the system structure efficiency Q_{max}. Thus $\Delta F_i > \Delta F_{i+1}$ but $Q_{imax} < Q_{i+1max}$.

Specialized, convergent systems, (in which components are better adapted to each other) have a more efficient forecast ability than universal ones.

Specialized systems perform better in a narrow range of environmental changes (as the steeper slope of their characteristic curves shows). Such systems increase their efficiency faster for small changes to the environment-- they are more active in adapting the environment to their needs and interests but in very narrow condition range.

So, a higher degree of structural system specialization provides for its higher maximal efficiency. At the same time, it narrows the range of acceptable environmental changes in which a strategy may be realized with good efficiency. In living systems, the higher specialization and efficiency are chilled by a more standard and reflective (and thus more rapid) pattern of forecast, formation and actualization.

REFERENCES

[1] Hendrick, H.W. A macroergonomic approach to work organization for improved safety and productivity. In: S.Kumar (Ed.) Advances in Industrial Ergonomics and Safety, London: Taylor and Francis, 1992, 3-10.

[2] Venda, V. F. On the laws of mutual adaptation in human-machine and other systems. In: Karwowski, W. (ed.), Trends in Ergonomics. Amsterdam: Elsevier, 1986.

[3] Venda, V. F., and Venda, Yuri V. Transformation dynamics in complex systems, *Journal of Washington Academy of Science*, #4, December, (1991).

[4] Venda, Yuri V., and Venda, V.F. Introduction to the Transformation Dynamics, In: Advances in Industrial Ergonomics and Safety-IV, London: Francis and Taylor, 1992.

[5] Venda, V., Strong, D., Hawaleshka, O., and Rychlicki, B. Human factors and transformations of manufacturing technologies, In: Advances in Industrial Ergonomics and Safety-IV, 1992.

[6] Venda, V.F. and Venda, Y.V. Dynamics in ergonomics, psychology, and decisions. Norwood, N.J.: Ablex, 1994.

Advances in Agile Manufacturing
P.T. Kidd and W. Karwowski (Eds.)
IOS Press, 1994

The Implementation of FMS as an Innovation Process

Raimo HYÖTYLÄINEN
VTT Automation, Otakaari 7 B, Espoo, Finland
P.O.Box 1301, FIN-02044 VTT, Finland

Abstract. The adoption of new technological systems are considered as an innovation process. According innovation research, a new technological system has to be implemented in order to be called an innovation. Process innovation are defined from angle of the applying organization. The focus is on the planning, implementing and use of FM-systems and the related organizational innovation. A model for a new kind of implementation process and results on the emprical study are presented.

1. Introduction

Applications of flexible manufacturing systems (FMSs) have become a central factor in gaining both flexible and at the same time economic production. Many companies, however, seem to have difficulties in efficient organization of the implementation of new technologies [1]. The problem is that the companies do not have straightforward models and methods for defining the planning and organizational forms relating to new technological choices. A new promising starting point to analyze that problem is innovation research and its aspects to adoption of technical systems.

The FM-system can be seen as a technological process innovation the planning, implementation and use of which forms a complex techno-organizational innovation process. Many problems encounted in the introduction of FM-systems can be traced back to problems in the control of the planning and implementation process. Analyzing implementation processes from the angle of innovation research it is possible to develop new models for controlling that innovation process. Results reported are based on the case study concerning two FM-systems [2]. The FMS research begun in 1985 and completed in 1990. Some starting points behind and results of the study will be discussed in the following.

2. Shortcomings in innovation approach

Most often, when talking about technological process innovations a reference is made to the original innovation brought on by the manufacturer of the system. To the manufacturer, such an innovation appears as product innovation, which is then developed, marketed and delivered to the users. Lately, increasing attention is paid to applying the technological process innovation to the user organization, which is also looked at as an innovation process [3, 4, 5]. This is based mainly on two factors. First, one can talk about an innovation if it concerns the application of a technical system new to the user organization. Second, the new system has to be adapted to the existing practices and work methods of the company, which also calls for organizational change.

Systems based on flexible production automation can not be purchased like single machines. The entire operation practice of the plant is to change in connection with the investment and innovations undertaken in the application phase can, in many cases, seen to be as important as innovations in the development of basic machinery [4, 5]). Complex problems will often have to be solved in the implementation phase of technology. The suppliers of the system or machinery have only limited resources for solving these problems, which is essentially due to them not having sufficient knowledge and experience of the special circumstances of the applying target.The user organization of the system has operational knowledge based on experience and learning and can therefore solve problems and create innovations with which the various components of the system can be made to function as a whole [5]. Thus, the company specific technological know-how and its own innovations relating to the applications are important factors for company success.

The studies where the crucial role of the user organization in applying technical systems to their operation environment and in developing the systems further are found do not much specify the implementation process in any detail, this is to say in which phases of the implementation process and how innovations are made [4, 5, 6]. Less attention has, therefore, been paid to how and by whom new solutions and innovations are finally brought about and carried out inside the user organization. The studies have not touched upon the role of the system users in the innovation activity of the implementation process.

The role of the system users ("final users") has traditionally been forgotten in innovation research. According to Lundvall [7] this reflects to an extent the reality of a modern industrial system. The innovation process has become professional activity and the workers tend to end up as passive users or "victims" of new technology rather than subjects with an active role in the innovation process. Lundvall tries to question this work division. According to him the workers play a central role in the learning process taking place in production and resulting in many gradual innovations improving the system and its activity. Lundvall regards skilled and active system users as strengthening the entire " national innovation system". The organizational and institutional innovations contributing to user participation can shape into a social innovation improving the efficiency and productivity of the "national innovation system". A good example is the Japanese system of company management and organization, which supports the users in making gradual, incremental process innovations. The cumulative result is the opening of new possibilities for radical process and product innovations as well [8, 9, 10].

3. Planning concepts and implementation models

The changed and changing operational environment forces us to look at production as an ever adapting and developing activity process, as "a learning organization" [11]. The demands of continuous change and development make it difficult to draw a clear line between the planning and use of a manufacturing system as traditionally done [12]. The planning and innovation process of an FM-system have to be seen as an entire and continuous innovation process not ending in the implementation phase but reaching into operational activity. This means that planning cannot be considered to stop in the planning phase either, but to carry on into the implementation phase and operational activity as planning required by continuous improvement and development activity.

The companies seem to have a basic choice of three strategic "ideal models" in the adoption of production automation. Table 1 presents the comparison of these models.

Table 1. Comparison of "ideal models" of three implementation strategies.

	TECHNO-CENTRIC MODEL	USER-CENTERED MODEL	USE ORIENTED MODEL
Planning target:	Machine system	Operation system	Evolving operation system
- control system	Centralized computer control, standard solutions	Interactive and de-centralized control system	Solutions supporting and expanding users' role in system control
- human-machine interface	Man only supervises production	Man controls production and optimizes it	Man controls production, programs, optimizes and develops it
Organization of planning	Specialized planning and separate planning organization (no users included)	Cooperation between planning organization and users, "participating" planning	Integration of implementation into planning, cooperation of planners and users, joint training
Plan implementation	Strict task division between planning and use	User teaching and training by participation in planning	Implementation as the next phase succeeding planning
Work organization	Work rationalization and strict task division, "individual work"	Reorganization and expansion of tasks, group work	Work group organization, network relations
Use of professional skill	Strive for replacing human work by automation	Wide-scale use of the workers' professional skills	Users as problem solvers, planners and developers
Aim	"Unmanned factory"	"Skill based production"	"Learning" organization

The planning paradigm has influence on whatever is seen as the target of planning. The idea of the planning target, again, has a solid link to the way of organizing planning and plan implementation. The technocentric model focuses on planning a machine system, which takes place in a highly specialized and divided planning organization. It is also characteristic that the planning of manufacturing and control systems takes place separately and the division of work between planning and operation is strict. The user-centered model based mainly sociotechnical approach aims at planning of an operation system. The users cooperate with the planning organization in creating user oriented solutions. The use oriented model strives for evolving operational systems and active problem solving and development from the users (cf. lean production models) [13]. Cooperation of planners and users as well as joint training is central in creating this.

4. Results on company practices

As a process innovation an FM-system is evolving technology which is clearly demonstrated in the case study. One of main results in the study is that the implementation of FMS is a controversial and gradually proceeding process where problems encounted and their solving activity play a central role. In that process two mechnanisms are prevailing. The management and designers determine mainly the goals and realization of the change process, but the importance and influence of the systems users become emphasized in the implementation phase of a system. The systems users' inputs are badly needed in disturbance handling and activity to optimize the system operation.

Despite several benefits gained in two target FM-systems, the implementation of the systems run into many economic, technical and organizational problems which can be traced back to problems in the control of the planning and implementation process. The main reason for that is the contradiction between planning practice and work organization pursued. Although the companies strived for user-centred forms in work organization practices they held to the traditional planning practice according the technocentric model. After all, companies strive to renovate their planning and organizational practices. They do not, however, reach more than partly the new planning and organizational practices corresponding to the use oriented model which can be seen to quarantee the ability of an organization to adapt and develop in the long term.

It became clear that the switch to planning and implementation practices consciously striving for use of the innovation processes of the implementation and the supporting organizational solutions can not be carried out straight forward and at once but as a result of the aims of the various organizational levels of the company. Strive for new solutions is flavored with traditional operation practices of the company. The strive for new practices can be described as experimenting for "best practice" by trial and error.

References

[1] Boer, H. 1991. Organising innovative manufacturing systems. Aldershot - Brookfield USA - Hong Kong - Singapore - Sydney: Avebury.

[2] Hyötyläinen, R. 1993. The implementation of FM-system as an innovation process - the viewpoint of manufacturing. Helsinki University of Technology, Mechanical Engineering deparment, Industrial Economy, Ph.D thesis.

[3] Voss, C.A. 1988a. Implementation: A key issue in manufacturing technology: The need for a field of study. Research Policy, Vol. 2, No. 17, 55-63.

[4] Gerwin, D. 1988. A theory of innovation processes for Computer-Aided Manufacturing Technology. IEEE Transactions on Engineering Management, Vol. 35, No. 2, May, 90-100.

[5] Slaughter, S. 1993. Innovation and learning during imlemention: a comparison of user and manufacturer innovations. Research Policy 22, 81-95.

[6] von Hippel, E. 1988. The sources of innovation. New York - Oxford: Oxford University Press.

[7] Lundvall, B-Å. 1988. Innovation as an interactive process: from user-producer interaction to the national system of innovation. In: Dosi, G., Freeman, C., Nelson, R., Silverberg, G. & Soete, L. (eds.) Technical change and economic theory. London and New York: Pinter Publishers, 349-369.

[8] Urabe, K. 1988. Innovation and the Japanese management system. In: Urabe, K., Child, J. & Kagono, T. (eds.) Innovation and management: international comparisons. Berlin and New York: Walter de Gruyter, 3-25.

[9] Itami, H. 1988. The Japanese corporate system and technology accumulation. In: Urabe, K., Child, J. & Kagono, T. (eds.) Innovation and management: international comparisons. Berlin and New York: Walter de Gruyter, 27-46.

[10] Freeman, C. 1988. Japan: a new national system of innovation? In: Dosi, G., Freeman, C., Nelson, R., Silverberg, G. & Soete, L. (eds.) Technical change and economic theory. London and New York: Pinter Publishers, 330-348.

[11] Leonard-Barton, D. 1992. The factory as a learning laboratory. Sloan Management Review, Fall, 23-38.

[12] Nadler, G. & Robinson, G. 1987. Planning, designing, and implementing advanced manufacturing technology. In: Wall, T.B., Clegg, C.W. & Kemp, N.J. (eds.) The human side of advanced technology. Chichester - New York - Brisbane - Toronto -Singapore: John Wiley & Sons, 15-36.

[13] Womack, J.P., Jones, D.T. & Roos, D. 1990. The machine that changed the world. New York: Rawson Associates.

Advances in Agile Manufacturing
P.T. Kidd and W. Karwowski (Eds.)
IOS Press, 1994

IMPLEMENTING CELL BASED SYSTEMS IN A MANUFACTURING COMPANY

Professor N D Burns and Dr C J Backhouse
Department of Manufacturing Engineering
University of Technology
Loughborough
Leicestershire
LE11 3TU

Abstract. This paper describes a major organisation change programme taking place within a company involved in the manufacture of large engineered structures. As part of the improvement programme a number of contract, commercial and manufacturing cells were designed and implemented based upon "Lucas Natural Groups". In addition to the cell structure a new performance measurement display system that includes a skill assessment scheme is being introduced into the cells as a means of targetting local performance and suitable training programmes.   Finally, the cells are being linked together in an organisation structure based upon the "Stafford Beer", Viable System Model.

1. Introduction

The business based in the UK is a small-to-medium sized autonomous division of a much larger construction company. The performance of the business had been declining for several years and in 1990 during the early stages of the recession the company was making a loss, estimated at over £1M per annum. If it was to survive in an increasingly competitive market the company had to effect a turn-around in performance.

The main manufacturing processes in the business consisted of the fabrication of large structures, followed by the machining, assembly, and fitting of components to the completed product. It takes several months to a year to produce the product and the expenditure for each contract is measured in millions of pounds. It is important to be within the time and cost estimate defined in the contract to avoid penalty charges. Much use is made of critical path planning techniques.

2. Change Initiated

The poor profitability of the business had highlighted the need for change. To formulate and direct the change process, a managment task team was appointed led by a new managing director who had a clear vision of the future and who saw his challenge as improving the profitability of the business.

Initially, the new management carried out a SWOT analysis by asking personnel working inside the organisation, and also a selected range of customers, for their views about the performance of the company. From the strategic analysis of the business a number of decisions were made, some of which are listed below:-

(i)   In order to become a global organisation it was decided to reduce manufacturing at the factory and in the future to manufacture major fabrications close to the point of sale using a selected range of sub-contractors.

(ii)   To reorganise the commercial, manufacturing and design activities into an integrated cell based structure that facilitated overall project control and improved product quality.

(iii)   To raise the skill base of the company.

(iv)   To invest in modern equipment that would improve the ability of the business to meet customer needs in an effective manner and to better control quality, time and cost.

3. The Cell Structure

The cell based structure linked together the commercial, contract engineering, design and manufacturing activities. The structure of the system, particularly in terms of the information links, was influenced by the Viable System Model [1] derived by Stafford Beer.

Within this model five systems are defined, namely:-

SYSTEM 5 - the strategic level.
SYSTEM 4 - the external environment scanning sub-system.
SYSTEM 3 - the operational management, control sub-system.
SYSTEM 2 - the operational coordination activity.
SYSTEM 1 - the operational activity.

A schematic diagram of the integrated business systems implemented in the company is shown in Figure 1.

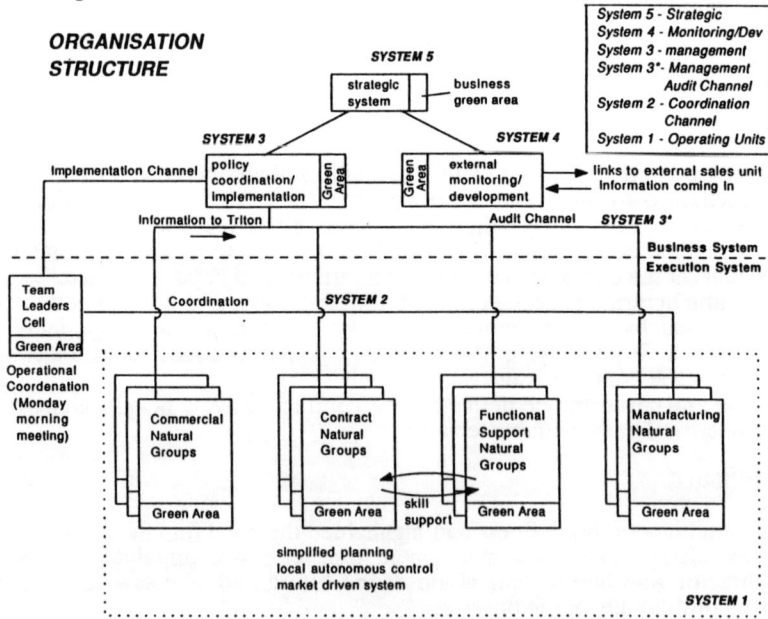

Figure 1
The integrated business cellular system.

To form the boundaries of the cells the shop floor and offices were reorganised using techniques such as Production Flow Analysis [2] and Lucas style natural groupings [3]. The primary aim was to define cells where there was as little backflow or crossflow of information or product as possible. An outline sketch of the reorganised shop floor is shown in Figure 2.

In the contract engineering departments the cells are organised around product families. Each cell has a small permanent core team consisting of a contract manager and senior design engineer and some support staff. Then, depending upon the work load, a team is formed by drawing people with the relevant skills from engineering support units. As the contract progresses the mix of skill changes and the membership of the teams also changes.

In addition to the contract teams there are also design support teams responsible for the engineering, costing and specification of common repeating modules in the total product range.

In the offices, at the present time, most of the cells are based around the functional activities, for example sales and accounts. However, these cells are being redesigned, by using a function/activity matrix which forms a core part of the Lucas Natural Group approach. Using the matrix enabled the identification of the main information flows in the business. The cells are being designed around these flows.

In order to enable local performance control a simple visual display system was established in each cell to show a number of local performance measures and additionally cross measures between cells. The cross measures showed how the cell was operating as a supplier and customer to other cells. In addition to monitoring local performance, the cell members hold a short meeting at the start of each working day.

Although the main measures for each cell were specified by management there was significant discretion concerning the design of the displays and any extra displays that a particular cell membership wanted. However, a guiding principle was that all the displays had to be simple and easily understood by visitors to the cell.

One of the more complex displays was the skills matrix. In shop floor cells this consists of a display chart that showed the machining and inspection skills of each individual within the group and their current status with training. This display greatly facilitated multi-skilling and cell planning and control.

In the contract engineering and design cells and some of the commercial areas of the business a more complex form of skills matrix was used. This system consisted of four levels as shown overpage.

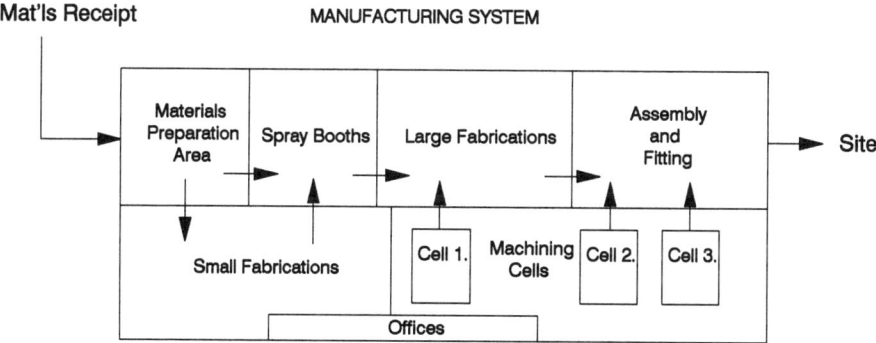

Figure 2
Reorganised Shop Floor

Individual skills	Skill level
- Professional Qualifications Institution Membership + Defined Experience	1
- Qualification + some Defined Experience	2
- Experience only or Qualified but no experience	3
- In training	4

In addition to each individual skill assessment an audit of business skill requirements was carried out by management. From the match of required skills and available skills training plans were devised for each individual.

4. Conclusions and the Benefits Achieved

The change to a cellular system was part of an overall integrated change programme within the company. It is hard to separate out the benefits associated with each component of the change programme, but the cellular concepts combined with the much better flow through the business directly contributed to much better contract control.

Previously, it was unusual for a contract to be completed within the cost and time estimates. Subsequent to the changes to cell based systems all the contracts were completed well within the cost and time estimates and the company was able to lower prices on many of its contracts.

The design of the cellular organisation was helped by the Viable System Model. The Model provided a framework, consisting of five sub-systems, that proved to be very beneficial in structuring the information links between the cells. The VSM also provided guidelines that helped with the design and positioning of performance measurements both inside the cells and as cross measures between cells.

The process of change is on-going and further work is continuing particularly to refine the engineering design process and the modularisation of the product structure. One of the major lessons learnt about change is that it is important that appropriate and, often simplified organisational design, is implemented before any automated or other expensive capital equipment is purchased. It is, therefore, important that considerable time is spent working with the people who are going to use the systems to evolve joint solutions where there is a strong sense of ownership.

References

[1] S. Beer, The Heart of the Enterprise, John Wiley and Sons, 1979.

[2] J.L. Burbidge, Production Flow Analysis, Oxford Science Publications, Clarendon Press, 1989.

[3] J. Parnaby, Lucas Manufacturing Systems Engineering Handbook, Lucas Engineering and Systems Ltd, 1991.

[4] D. Robey, Designing Organisations, A Macro Approach, Irwin Press, 1986.

Advances in Agile Manufacturing
P.T. Kidd and W. Karwowski (Eds.)
IOS Press, 1994

Introducing Lean Production
in a Shipyard

Anders Drejer and Frank Gertsen
University of Aalborg, Department of Production, Fibigerstraede 16, DK-9000, Denmark
Fax : +45 9815 3030, e-mail : I9AND@iprod.auc.dk and i9FG@iprod.auc.dk

Abstract. Since "The Machine that Changed the World" [1] was published in 1990, the concept of Lean Production has been a subject of great interest both from researchers and managers. Lean Production has proven capable of combining high performance in areas of productivity, delivery, quality, and flexibility with a high degree of commitment and motivation among the employees in the car-making industry. This paper presents the case-study of a Danish shipyard where attempts have been made to introduce Lean Production in order to increase performance in a business very different from the car-making industry. The case study was conducted as a longitudinal in-depth case study over a period of six months and demonstrates that the initial stages of introducing lean production in this case should focus on managerial and cultural issues. The objectives of this paper are to demonstrate how this has been done in the yard and to discuss the implications of further work on lean production.

1. Introduction

It cannot be questioned that the business environment of many industrial corporations has changed since the oil crisis in the early 1970's. Important trends are; technology, internationalisation, and lately the "greening" of legislation and consumer demands. On top of these changes, it is also important to consider the increased and fierce competition that many Western corporations face from the East. As a result of these trends, consumer demands have changed. While consumer demands before 1970 were directed primarily on price, today is demands are on price as well as quality, timeliness, and customised products.

1.1. The Concept of Lean Production

Ever since it became evident that Japanese corporations were becoming a serious threat to their Western counterparts there has been a great deal of interest for how Japanese corporations have achieved their competitiveness. Over the years different elements of Japanese production systems have been said to be the explanation of the success of Japanese corporations in industries such as car-making, ship-building, and steel-making. The conceptions of; automation, statistical process control (SPC), total quality management (TQM), just in time (JIT), and kaizen are examples of this trend. In 1990 the book "The machine that changed the world", [1], was published, demonstrating that the Japanese car makers have a large lead in different performance measures, most notably productivity, quality and development time.

In order to explain this Womack et al introduces the holistic concept of lean production consisting of the whole set of newly implemented Japanese techniques [1].

Lean production is seen as a unique way of producing goods - involving all functions in the value-chain from supplier, the company's internal functions (administrative functions, manufacturing, and the management functions), to the customers. In a recent world-wide survey, is was found that lean production has attracted a large amount of interest from managers, since lean production is rated as one of the most widely used programmes for yielding competitive advantage [2].

2. The case study at Shipyard Ltd.

2.1. Research aims of the Case study

In order to facilitate and understand the initial stages of the introduction of lean production and how the concept of lean production can be put into practice, the authors conducted a case study in a Danish shipyard (Shipyard Ltd.). Furthermore, the case study made it possible to gather experience from this process and compare this with other case studies and surveys from the literature. The case study dealt with two issues. 1) First, the top-management - and their actions - were followed during the initial stages of introducing of lean production and 2) Secondly, the authors conducted a number of analyses in connection with a pilot project to be undertaken in the yard.

2.2. Methods of the Case study

During the case study - which took place over a period of six months - more than 200 interviews and informal meetings were conducted and combined with other sources of information such as questionnaires and hard-data analysis. In total, more than four man-months were spent in the shipyard in connection with the case study.

2.3 The situation at Shipyard Ltd.

Shipyard Ltd. emerged in the mid-eighties as the result of a merger between three medium-sized shipyards. Following the world-wide crises within the shipyard industry two of the three yard's were later shut-down, and all the yard's activities were assembled in one location. Since that time and until 1993, Shipyard Ltd. has focused on the manufacturing-function and on centralising their organisation. As a result of the changes made, the yard's *organisation* was changed from having four to six managerial levels, which is a relatively large number compared to other Danish organisations. Furthermore, the organisation were functionally divided into a number of departments highly separated from each other. *The company culture* could be characterised by the word mistrust. This mistrust was apparent everywhere in the organisation - i.e. between the top-managers and the blue collar workers as well as the white collar workers, and between the blue- and white collar workers. Finally, the *management-style* could only be seen as centralised and non-democratic, since the employees had very limited knowledge about the goals and policies of the yard. Furthermore, the top-managers considered a "though" management-style as essential to regaining the yard's competitive position. In 1992 the top management of Shipyard Ltd. found that the yard was in essentially the same situation as the Western car makers. Their Japanese competitors offered the products at a price 30 % lower than that of Shipyard Ltd. - and were still able to deliver with shorter lead times and higher quality.

   At this point the CEO of Shipyard Ltd. read "The Machine that changed the World", [1], and found that lean production might be the answer to the yard's problems.

It was obvious to the CEO that the yard's competitiveness depended on reducing the lead-time, improving the productivity and quality, and at the same time improving the use of the company's human resources. Early in 1993 it was finally decided to introduce lean production in Shipyard Ltd, and the authors were given the opportunity to follow the initial stages of this process.

Table 1 : What was actually done during the initial stages of introducing lean production.
1. The idea of lean production was introduced to the top management group, who read relevant literature.
2. The vision of lean production was discussed and agreed upon by the management group.
3. The vision was spread to the middle managers and union representatives at the yard.
4. Negations regarding the vision along with information about lean production and the yard's competitive position was spread through the organisation.
5. Analyses and evaluation about the "lean-ness" of the yard was conducted.
6. A pilot project was, finally, agreed upon and initiated.
7. Furthermore, the top- and middle managers has planned to participate in an education programme aimed at changing the management-style at the yard. A few managers has left the yard in this process.
8. An implementation plan with a time horizon of 8-10 years, from February '94, has been launched.

2.4 Evaluation of Lean Production at Shipyard Ltd.

In the beginning of the process, the top management had a very limited view of lean production as confined to activities at the shop floor. Their plans for action were guided by this perception, as they suggested to change the job descriptions of the foremen and blue collar workers. However, this met very serious resistance from the union representatives in the shipyard, who made it impossible to start a planned pilot project for almost four months by protesting loudly. Negotiations continued during all this time. Analyses conducted by the authors demonstrated that low performance in the manufacturing function in most cases was caused by low quality of input from the design department (more than 80 % of the total number of engineering drawings per ship was changed at least once), which in turn was caused by low quality of input from the sales department (a substantial amount a changes in specifications for each ship) - all of which was caused by the compression of the total lead time from 18 to 12 months without significant changes in the organisation and information-flow of the work process. Therefore, it gradually became apparent to the top managers at Shipyard Ltd. that in order to achieve the benefits they wanted, the scope of the process had to be expanded to include white collar workers and administrative functions. Furthermore, additional analyses demonstrated that the company culture and mistrust between top management and the employees were perhaps the largest obstacle to the process - a message that took a little longer to gain momentum among the top managers. But after a long period of problems with the unions and negotiations, this changed gradually, and it became evident that the introduction of lean production had to take place in three stages; change of management and employee involvement, change of administrative work practices, and change of manufacturing work practices.

The results of the analyses can be illustrated by an evaluation of lean production at Shipyard Ltd. at the point of launching the process. Here the authors have made a diagnostic tool, where the contents of lean production, as described by references within this area, is divided into three categories. At this stage the "lean-ness" at each category was evaluated based on comprehensive analyses.

As for how to increase the "lean-ness", a number of "tools" deducted from the literature is included in the table. These tools should, in our view, only be taken as inspirations for possible actions and must be changed to fit the specific situation in which they are to be used, since the tools in the literature are based on the car making industry.

Table 2 : Diagnosis regarding the "lean-ness" of Shipyard Ltd.		
Category.	Judgement of "lean-ness"	Possible tools.
Management and employee involvement.	Mistrust was evident everywhere, and a "though" management-style.	Visionary management, fifth generation organisation, integrative and cross-functional goals, customer-oriented organisation.
Administrative and organisational practices.	Very sharp functional boundaries and bureaucratic practices.	Simple and effective communication systems, TQM, cross-functional teams.
Manufacturing practices.	Skilled and motivated workforce in a flexible environment.	Flow -oriented manufacturing, small batches, order initiated manufacturing.

In this case the analyses revealed that it was necessary to start with improving the management-employee relations, and Shipyard Ltd. is in the middle of a cultural change at this point in time - i. e. February 1994. This is a process that has a time frame of a couple of years, until it is believed that it will be possible to start working on the administrative practices and finally on the manufacturing practices. The agreement on a pilot project is used as an important first step for the introduction of lean production all over the shipyard, and additionally as an opportunity to gain experience and revise future actions.

3. Discussions and Implications

But can lean production be employed in other industries, and in other countries than Japan? Womack et al have no doubts about this question : "*In this process* [of studying the differences between mass-production and lean production]...*we've become convinced that the principles of lean production can be applied equally in every industry across the globe...*".

During the analyses, it became evident to the authors that the different tools might not be as appropriate to a shipyard as to a car manufacturer - and that they, even if they are useful, can not be employed in exactly the same manner. It was therefore chosen to view the elements of lean production as a "toolbox" from which one can choose the most appropriate tools in order to reach desired goals.

It also became evident that "lean-ness" in an organisation never will be measured to be either 0 or 100 % implemented, but rather somewhere in between - and that the need for changes will be different from company to company. We therefore suggest to evaluate where the need for improvements is largest, and starting the process towards lean production there. The results of the case study seems to indicate that the three categories of lean production should be treated in a hierarchic manner implying that management and employee involvement must be relatively "lean" before it is attempted to improve administrative practices, and finally manufacturing practices.

References

/1 Womack J. P., Jones, D. T., and Roos D, *The Machine that changed the World.*, Macmillan, 90.
/2/ Gertsen, F. & Riis, J. O, *Manufacturing Response to Strategic Challenges* Presentation at "the Fourth International Conference on Management of Technology IV", Miami, 1994.
/3/ Ohno, T, *Just in time - for today and tomorrow,* Cambridge, 1988.
/4/ Steudel, H. J. & Desruelle, P.,*Manufacturing in the Nineties,* Van Nostrand Reinhold, 1991.

Advances in Agile Manufacturing
P.T. Kidd and W. Karwowski (Eds.)
IOS Press, 1994

Integrated Human Factors Support of Advanced Manufacturing

H. A. Romero and J. C. Byers

Idaho National Engineering Laboratory, EG&G Idaho, Idaho Falls, ID, USA 83415-3855

Abstract. Traditionally, human factors (HF) has directly supported specific sub-systems involved in manufacturing such as management; support; and, operation functions. Although HF specialists have been successful supporting individual sub-systems, the real benefit of HF is realized when the support integrates the entire system. This is especially important in advanced manufacturing environments (AMEs) where manufacturing happens more quickly and is more dependent on the other sub-systems. Integration is accomplished when HF personnel are aware of the needs and outputs of each sub-system and those inputs and outputs are translated and optimized for the other dependent sub-systems. The benefits of integration include: improved organizational effectiveness; reduced product lifecycle costs; improved responsiveness to internal/external customers; decreased workplace injuries and illnesses; improved quality of worklife; and reduced likelihood or consequence of human error during production or product usage. This paper discusses the benefits and methods of integrating AMEs using a HF approach.

1. Introduction

The creation of a product or service has several functional and managerial aspects as depicted in Figure 1. Companies are realizing that the coordination of these aspects is vital to the creation of a low-cost, quality product that will allow them to compete globally [1]. Also, the integration of these aspects, especially the sociotechnical aspects, is vitally important to realizing the benefits associated with the introduction of new technology within the workplace. This view of HF diverges from the traditional view of HF. Traditionally, HF professionals have focused on the microergonomic aspects of the organization and not the macroergonomic aspects. Microergonomics is the traditional focus of the current generation of ergonomics. It is mainly concerned with optimizing physical and mental aspects of the human-machine interface. Macroergonomics concerns optimizing work systems at the individual, workstation/task, organizational, and environmental levels to achieve the maximum benefit from the integration of the social and technical systems. The main difference between the two approaches is macroergonomics' focus on the organizational and environmental aspects of the work systems. The sociotechnical system is a combination of two systems: social and technical. The social system comprises the human element while the technical system comprises the hardware and the manufacturing processes. Sociotechnical systems theory states that any organizational system requires a social system that links organization members with technology and with each other [2]. The social system encompasses job design, organizational structure and goals, formal and informal communication networks,

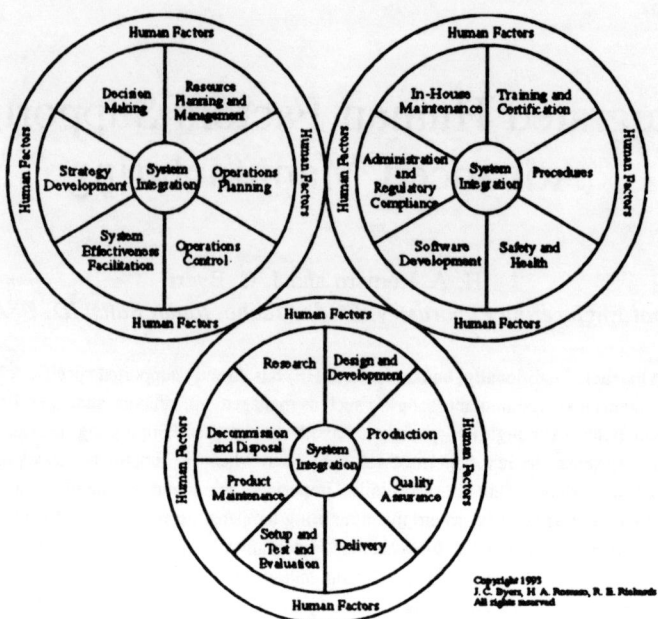

Figure 1. Human Factors Systems Integration

individual and group attitudes, assumptions, and beliefs. The technical system comprises the technology and the processes by which a product or service is created. Integration refers to matching the outputs from one system to the input needs of the joined systems. Integration can occur between the social and technical systems. Integration also can occur between processes and groups within the technical and social systems [12]. Realizing the full potential of any system or organization involved in creating a product or service requires the joint optimization of both the technological and social systems [13, 2]. Joint optimization will result in the greatest benefit for the entire organization.

Figure 1 depicts HF as external to each of three main functional areas or sub-systems of the organization. Additionally, it depicts HF as a source of contact between the functional areas suggesting a communications link. Viewing HF as an external consultant to the individual sub-systems, and a part of the organization, is extremely beneficial. HF personnel use various tools and methods to analyze tasks, environments, equipment interfaces, and workspaces. Acting as external consultants to the sub-systems helps the HF personnel to maintain the system perspective necessary for integration. This need for the system perspective is the reason HF personnel should be maintained positioned to provide and receive inputs from each system and functional area. HF professionals provide this support by optimizing the following functional areas: tasks, environment, equipment, personnel, work systems design, and documentation [9].

2. *Methods for optimizing the technological systems (microergonomics)*

HF personnel use the following tools to understand and enhance task effectiveness: function allocation, task loading, precision requirements, task feedback, error tolerance, and training [9]. Safety and industrial hygiene personnel are equipped to design working conditions that are not hazardous to workers. HF personnel are equipped to optimize the environment to increase productivity and reduce the number of errors that can occur. The human-machine interface should be designed to promote productivity. This design is accomplished by considering user operability, application, maintenance, accessibility, and standardized conventions and nomenclature. Personnel considerations that must be accounted for in system design include decision making, human adaptability, and user acceptance. HF personnel can also help ensure ease of transition during shift changes and appropriate responses during abnormal operating situations. The facility documentation including operating procedures, equipment manuals, and computer software should incorporate HF principles.

3. Methods of optimizing the social system (macroergonomics)

Several established techniques exist for optimizing the social systems, include participatory ergonomics, simultaneous engineering, quality circles, and participative management or empowerment. These methods have one thing in common: they involve people from all the levels of the organization in a combined effort to design and implement technology within an organization. This combined effort seems to work best when allowed to operate semi-autonomously. The reward system must be based on the contributions of the individual to the team effort, not to a particular product, must fairly account for the participation of the entire group. The appraisal system must recognize the individual and team effort [6].

Participatory ergonomics is an organizational design effort including structuring of tasks, authority, and work flow. The result is a framework for the organization. Participatory ergonomics requires ergonomic (or HF) experts to work with people from all levels of the organization to produce various alternatives for improving the functioning of the sub-systems and the integration of these sub-systems [11]. *Concurrent engineering* is a method of bringing people from all affected levels of the organization together to develop the process by which a product or service will be created or to design the product or service. The rationale for concurrent engineering is that impacts of the design or production process will be anticipated while the cost of making corrections is minimal compared to making changes after implementation. A *quality circle* is a small group of between three and twelve people working in the same or in dependent sub-systems, voluntarily meeting together regularly for about one hour per week in paid time, usually under the leadership of their supervisor, and trained to identify, analyze and solve some problems in their work, presenting solutions to management and, where possible, implementing solutions themselves [8]. *Participative management* is essentially an attempt to develop a horizontal or flat organization with the individual members empowered to make many decisions affecting their job. Yet, it must be remembered that there still must be a single head to the organization to provide direction, much like living organisms.

4. Joint optimization

Unless the focus is on jointly optimizing the social and technological systems, the benefits of integrating the organization will not be demonstrated. Joint optimization requires that the social optimization techniques be applied across the organization and focus on the impacts of changes to the technological system. In addition, the tools of systems analysis, functional analysis, task analysis, link analysis, and operational experience review are valuable diagnostic tools available to the HF professional and provide the information necessary to integrate the organization.

5. Barriers to systems integration

Workers who feel threatened by the introduction of new technology or lack the necessary skills to operate the new systems, will not accept the new system [4]. User acceptance is vitally important to achieving the benefits of new technology or even improving existing systems [10]. Training, appropriate job design, and appropriate implementation schemes are necessary to ensure worker satisfaction and acceptance. There are two type of barriers at this level: structural and personal. Structural barriers exist when the organizational structure is not prepared for integration. An example includes machine bureaucracies that are very structured and abhor the flexibility necessary for integration [7]. The personal factors include managers who fear loss of control that empowering employees assumes [5].

6. Benefits of integration

The benefits of jointly optimizing the social and technological systems include: increased productivity, quality, job satisfaction and workplace safety; and decreased turnover, absenteeism, and injury and illness rates [3]. These benefits are realized because people at all levels of the organization feel a sense of ownership and loyalty for the system that has been created. Additionally, each employee at every level realizes that change can be initiated to make improvements when required. Systems integration is the most appropriate method for deriving the benefits of implementing AMEs in industry. Optimizing the technological system through introducing new technology will not have the desired benefits due to impacts on the social system unless changes are made to both systems simultaneously.

References

[1]   Ainsworth, S, (1993) *Chemical Producers Stand By Worker-Management Teams*, Chemical & Engineering News, July, pp. 15-17.

[2]   Brown, O., Jr., (1990) *Macroergonomics: A Review*, Third International ODAM Symposium, Amsterdam: North-Holland.

[3]   Endsley, M. R., (1985) *Technological Change and Individual Adjustment*, Proceedings of the Human Factors Society, Vol. 29, p. 598-602.

[4]   Endsley, M. R., (1985) *The Toll of Technology on Job Commitment and an Implementation Process to Prevent It*, Proceedings of the Human Factors Society, Vol. 29, p. 870-874

[5]   Franklin, I., Pain, D., Green, E., Owen, J., (1992) *Job design within a human centred (system) design framework*, Behaviour & Information Technology, v. 11, no. 3, pp. 141-150.

[6]   Gupta, A., and Singhal, A., (1993) *Managing Human Resources for Innovation and Creativity*, Research Technology Management, May/June, pp. 41-48.

[7]   Hendrick, H. W., (1986) *Human Factors in Organizational Design and Management.*

[8]   Hutchins, D., (1985) Quality Circles: Handbook, Nichols, Co., New York.

[9]   IEEE (Institute of Electrical and Electronic Engineers), 1988, "Guide for the Application of Human Factors Engineering to Systems, Equipment, and Facilities of Nuclear Generating Stations," Standard 1023, October.

[10]   Noori, H., (1990) Managing the Dynamics of New Technology, Prentice-Hall, Englewood Cliffs.

[11]   Noro, K. and Imada, A., (1991) Participatory Ergonomics, Taylor and Francis.

[12]   Robinson, G. H., and Peterson, J. G., (1983) *Groups at Work: A Socio-technical View*, Proceedings of the Human Factors Society Annual Meeting, p. 566-570.

[13]   Trist, E., (1981) The Evolution of Socio-technical systems: a conceptual framework and an action research program, Issues in the Quality of Working Life, No. 2.

Advances in Agile Manufacturing
P.T. Kidd and W. Karwowski (Eds.)
IOS Press, 1994

Interrelationships between Strategies of Use and Development of Human Resources and the Design of Computer-aided Integrated Manufacturing Systems

L. Leder, O. Pardo & E. Ulich
Work and Organizational Psychology Unit, Swiss Federal Institute of
Technology ETH, CH-8092 Zurich

Abstract. Increasingly changing market demands are primarily met by the use of computer-aided manufacturing systems. Results in the GRIPS study, however, showed that the goals pursued with the implementation of technical systems were poorly attained. Apart from technical and organizational design problems many companies encountered qualification and training problems. Although qualification requirements related to computer-aided manufacturing systems were considered important by many companies, findings show that such skills were largely disregarded in the training offered. Furthermore, findings indicate that only few companies made use of their employees's skills and qualifications by means of user participation. The utilization and development of the employees's skills and qualifications in the participation process not only makes a substantial contribution to the joint optimization of technical, qualification and organizational design aspects, it also may prove economically advantageous in the long run.

1. Increasing Market Demands and Computer-aided Manufacturing Systems

Swiss manufacturing companies are faced with an increasingly competitive environment and rapidly changing market demands. The need for more complex, customized, high-quality products at competitive price calls for flexibility and is supposed to be reflected in the manufacturing conditions. The implementation and utilization of computer-aided manufacturing systems is considered an appropriate approach for coping with these requirements.

In the framework of a Swiss government funded CIM Action Program, a major research project GRIPS is currently being conducted at the Work and Organizational Psychology Unit of the Swiss Federal Institute of Technology (ETH). GRIPS focuses on the capital goods industry and comprises three consecutive project phases. In this paper, selected findings gained in the first phase, GRIPS I (written survey, N = 917) and second phase, GRIPS II (60 selected case studies) with regard to the utilization and development of skills and qualifications are presented and discussed.

2. Problems Encountered in Connection with the Implementation of Computer-aided Manufacturing Systems

Findings both in GRIPS I and GRIPS II indicate that goals related to economic efficiency and pursued with the help of computer-aided manufacturing systems are very poorly attained [1, 2]. Such goals include e.g. reduction of total cycle times, accuracy of keeping promised delivery dates, reduction of stocks, and increase in market flexibility. A clue for this poor goal attainment is supplied by the analysis of the problems encountered with the implementation and utilization of technical systems. In the GRIPS I study, 63% of the companies specified technical problems, 55% qualification and training problems, 49% problems related to organizational design, and 30% acceptance and motivation problems [1]. It can be inferred from these results that the implementation and utilization of computer-aided manufacturing systems per se does not automatically contribute to economic efficiency, and that a successfull implementation implies the joint consideration of technical, qualification and organizational design aspects. These findings point at a striking discrepancy between hard facts and visionary ideas connected with the implementation and utilization of computer-aided manufacturing systems.

3. Structure and Development of Skills and Qualifications

It is generally agreed that a successful implementation of computer-aided manufacturing systems implies the use of key skills such as self-reliance, capability for teamwork, interdisciplinary expertise, planning skills, and communicative skills [3]. In the GRIPS II study the importance of the following qualification requirements relevant to the utilization of computer-aided manufacturing systems were rated (see Table 1):

Table 1
Importance of Qualification Requirements in Connection with the Utilizaton of Computer-aided Manufacturing Systems as Rated in the GRIPS II-sample (N=55)

	very important	important	less important	negligible
accuracy in performing tasks	31%	21%	3%	0%
sense of responsibility	29%	25%	1%	0%
mental agility	25%	27%	3%	0%
cross-functional know-how and thinking	23%	25%	7%	0%
capability for team-work	21%	27%	6%	1%
conceptual skills	19%	33%	3%	0%
communication skills	17%	30%	6%	2%

It is striking that besides "accuracy in performing tasks" those qualification requirements are thought to be crucial which regard technical, methodological and social skills. Although many companies stress the importance of skills that allow employees to perform responsible, flexible and holistic jobs, only few offered training comprising social skills. In many cases, only after a technically-oriented implementation strategy had proven insufficient were skill requirements be reconsidered [4]. Furthermore, 30% of the companies had opted for the employment of additional manpower to compensate for qualification deficiencies. From a work psychological point of view it has to be noted that not making use of and developing the employees' skills and qualifications and tacit knowledge may prove disadvantageous with respect to human capital in the long run.

4. Use and Development of Skills and Qualifications by means of User Participation

Bridging the gap between required and available skills by means of "procurement of know-how" is one method for linking humane to technical aspects. This method implies a "reactive" fit, i.e. qualifications are adapted to requirements related to the use of computer-aided manufacturing systems. Another, "active" way of achieving a fit between humane and technical aspects is the participation path. User participation takes up the idea of technology as an option for design [5]. It offers the opportunity of incorporating and developing the users' knowledge and skills during the entire design and implementation process.

In the GRIPS II study user participation was examined in the framework of expert interviews with project managers on technical innovation projects recently accomplished. The following findings will concentrate on CAD, CAM, and CAPPC projects . Four aspects of participation were analyzed: representativeness, form, degree and subject of participation [6].

4.1 Representativeness of Participation

The representativeness of participation can be seen as a dimension varying direct participation, which refers to involvement of employees affected by the implementation, and indirect participation, which stands for involvement of representatives of the employees affected. The proportion of employees participating actively to the number of employees affected by the implementation on the whole was used as a measure for the degree of direct participation. The average rate of direct participation was 0.49. It varyied considerably with respect to the technical system implemented, ranging from 0.12 for CAPPC to 0.80 for CAD; for CAM the degree of direct participation was 0.56. Two explanations for these striking differences can be put forward. First, the number of employees affected was higher with CAPPC, the implementation of which affected 70 employees on an average (CAD: 9, CAM: 8). Secondly, participants were predominantly assigned to higher hierarchical levels. With regard to CAD as opposed to CAPPC, employees affected by the implementation fell under markedly higher hierarchical levels, e.g. constructing engineers.

4.2 Form of Participation

This aspect covers the relief of users from daily tasks and the extent to which participants were trained for the participation process. In CAM projects, only 25% of the companies conceded the participants part or full time project involvement, in CAD projects only 33%, in CAPPC projects, however, 57% the companies did so. Obviously, user participation was considered a full-time task by only a quarter up to half of the companies. Furthermore, only 45% of the companies offered the participants training for the participation process. Considering the significance of qualifying work design for efficiency and worker motivation [7], these results have to be termed unsatisfactory. In fact, empirical findings suggest that the opportunity to participate is connected to the motivation for high performance and further training [8].

4.3 Degree of Participation

The degree of participation reaches from the contribution of topic related information to the making of design suggestions. It is striking that only in 20% of the companies user participation encompassed design suggestions made by the participants. In 50 to 63% of the cases, participants only provided information. Allowing for the cases in which participants only obtained information, in 50 to 75% of the cases user participation was restricted to information. These findings suggest that most companies do not rely on their employees's expertise in their work when making decisions concerning the implementation of technical systems.

4.4 Subject of Participation

Decision subjects can be assigned to three levels, the operative, administrative and strategic level. In CAD projects participants were not involved in decisions related to the strategic level. In CAM projects 19% of the decisions were allocated to that level, in PPS projects 14%. Decisions concerning work tasks and processes were assigned to the administrative level. For all project types participation in work tasks related issues was higher than in questions regarding work processes. However, most subjects of participation were related to the operative level such as the selection of hardware and the design of user interface.

5. Effects of Participation and Conclusions

Results indicate that the utilization and development of the employees' skills and qualifications are of minor importance in most companies. Moreover, our data show that the underutilization of the users' know-how also applies to participation processes. The way most companies deal with participation can only be termed enhancement of acceptance and motivation. The work psychological point of view, however, is that a participation project designed in this way is likely to fall short of its expectations. This restricted view stands a good chance of leading to utilization problems. Moreover, it is not instrumental to the fit between technology and work organization. Indeed, the use of expert knowledge on all company levels in order to achieve a fit between technical resources and work processes requires a different course of action with regard to user participation.

The involvement of future users in the design and implementation process constitutes a substantial component of the concept of prospective work design: "Prospective work design is the deliberate anticipation of possibilities for personality development at the planning respectively designing stage of work systems through the creation of objective degrees of freedom for action which may be used by employees in different ways" [9]. Thus, user participation not only is connected with the utilization and development of skills and qualifications but also with organizational design, such as to optimize both the technical and the social subsystem in concordance with the concept of socio-technical systems design [10].

The implementation of computer-aided manufacturing systems calls for new skills which can be acquired in the course of the participation process. This form of "simultaneous qualifying" spares the companies the loss of specific know-how. Furthermore, the utilization and development of the users' knowledge in the participation process prevents companies from hiring and training additional manpower to deal with the implemented systems. In short, the participatory path may prove advantageous in the long run, even economically.

References

[1] Strohm, O., Kuark, J.K. & Schilling, A. (1993). Integrierte Produktion: Arbeitspsychologische Konzepte und empirische Befunde. In G. Cyranek & E. Ulich (eds.), CIM - Herausforderung an Mensch, Technik, Organisation, Schriftenreihe Mensch, Technik, Organisation (Hrsg. Eberhard Ulich), Band 1 (pp. 129-140). Zürich: Verlag der Fachvereine, Stuttgart: Teubner.

[2] Strohm, O., Kirsch, C., Kuark, J.K., Leder, L., Louis, E., Pardo, O., Schilling, A. & Ulich, E. (1993). Computer Aided Manufacturing Systems: Work Psychological Aspects. In A. Bürgi-Schmelz et al. (eds.), Computer Science, Communications and Society: A Technical and Cultural Challenge. Conference Prodeedings, Neuchâtel, Switzerland, 22-24 September 1993 (pp. 221-238). Lausanne: Beck.

[3] Köhl, E., Esser, U., Kemmner, A. & Förster, U. (1989). CIM zwischen Anspruch und Wirklichkeit - Erfahrungen, Trends, Perspektiven. Köln: TÜV Rheinland.

[4] Leder, L. & Louis, E. (1993). Zum Stellenwert von Qualifikation und Ökologie in Unternehmen mit rechnerunterstützten integrierten Produktionssystemen - Ergebnisse betrieblicher Fallstudien. In G. Cyranek & E. Ulich (eds.), CIM - Herausforderung an Mensch, Technik, Organisation, Schriftenreihe Mensch, Technik, Organisation (Hrsg. Eberhard Ulich), Band 1 (pp. 141-151). Zürich: Verlag der Fachvereine, Stuttgart: Teubner.

[5] Ulich, E. (1992), Arbeitspsychologie, 2. Auflage, Stuttgart, Poeschel.

[6] Spinas, P. (1989). User oriented software development and dialogue design. In M.J. Smith & G. Salvendy (Eds.), Work with computers: Organizational, management, stress and health aspects (pp. 200-207). Amsterdam: Elsevier.

[7] Duell, W. & Frei, F. (Hrsg.) (1986). Arbeit gestalten - Mitarbeiter beteiligen. Eine Heuristik qualifizierender Arbeitsgestaltung. Schriftenreihe "Humanisierung des Arbeitslebens", Bd. 77. Frankfurt: Campus.

[8] Baitsch, Ch. & Frei, F. (1989). Qualifizierung in der Arbeitstätigkeit. Eine theoretische und empirische Annäherung. Schriften zur Arbeitspsychologie (Hrsg. E. Ulich), Band 30. Bern: Huber.

[9] Ulich, E. (1989). Humanization of Work - Concepts and Cases. In J. Fallon, H.P. Pfister & J. Brebner (Eds.), Advances in Industrial and Organizational Psychology. North-Holland: Amsterdam.

[10] Emery, F.E. (1959). Characteristics of Sociotechnical Systems. London: Tavistock Institute of Human Relations. Document No. 527.

Advances in Agile Manufacturing
P.T. Kidd and W. Karwowski (Eds.)
IOS Press, 1994

Human Aspects of Obtaining Accurate Inventory Records

Roger A. LINDAU and Kenth R. LUMSDEN
Department of Transportation and Logistics, Chalmers University of Technology
S-412 96 Göteborg, Sweden

Abstract. This paper explains and describes how inventory record accuracy is impaired by the human factor. Manual involvement in data capture is common in companies today. It was, however, found that with automatic data capture techniques, inventory record accuracy is enhanced. Technical aid is not enough to obtain accurate inventory records, simplified routines and education being equally important. By combining new data capture techniques with simplified routines and education, inventory accuracy is improved.

1. Introduction

An inventory system will work efficiently only if records are accurate. All decisions on when or how much to order are based on the inventory balance of individual items. Keeping accurate inventory records is a necessity for any company and especially for those using MRP-logic [1]. File data used as inventory status data must be accurate, complete and up-to-date, if the MRP system is to prove successful or even useful [2]. An inventory record reporting inventories lower than the actual ones can trigger an unnecessary order, boosting inventory and wasting capacity. On the other hand, an inventory record reporting inventories as higher than they actually are can result in stockouts and perhaps also work stoppages [3]. Inaccurate inventory records trigger a chain reaction of problems, e.g. lost sales, shortages, missed schedules, low productivity, late deliveries and excessive expediting. To overcome some of these problems, organisations frequently order more than needed, creating excess inventory and high obsolescence [4].

The crucial activity when updating an inventory record is the data capture process. As data is often gathered first on paper sheets and then transcribed via keyboard, several opportunities for making data errors exist [5]. New techniques are today available, making the data capture process more accurate and enhancing the performance of activities dependent on high inventory record accuracy [6].

The aim of this paper is to describe how inventory record accuracy is affected by the human factor and how the accuracy of these records can be improved.

2. Method

2.1. Mail Survey

A mail survey was carried out among 320 Swedish manufacturing companies with 50 to 500 employees. The survey was aimed at store managers. The questionnaire was structured as follow:

- Brief description of the company
- Measurements regarding inventory (number of part numbers, number of transactions, turnover rate, inventory record accuracy, etc.)
- Routines for up-dating inventory records
- Technical aids used when up-dating inventory records
- Causes of faulty inventory records.

After two follow-ups the response rate was 54%.

2.2. Case Studies

The case studies were carried out as direct observations and as structured formal interviews with planning managers and logistics managers in four companies. The purpose of the case studies was to aquire a better knowledge of the data capture process and the strategies used by companies to improve their inventory record accuracy.

Table 1   Characteristics of the studied companies.

Company	Business	Turnover (M$)	No. of employees	Routine used for up-dating inventory records
A	Engineering	59	600	Manual
B	Engineering	23	230	Keyboard
C	Injection moulding	40	370	Bar-codes
D	Engineering	88	800	Bar-codes

3.   Results

The results show that humans are involved in data capture in most companies. The techniques used differ from a purely manual system (cardex system) to automatic data capture methods such as bar-codes. Manual systems were used by 5%, keyboard by 87% and bar-codes by 8% of the companies.

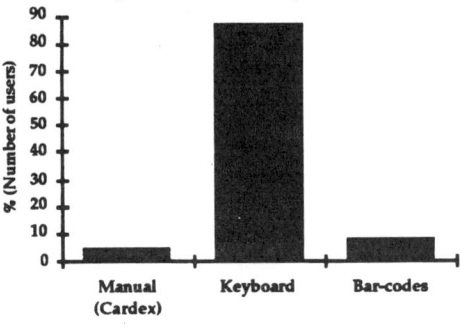

Figure 1   Data capture method used.

Humans are, however, involved to a certain extent in all data capture techniques, even when bar-codes are used. Inventory accuracy, which is measured as the absolute variance when a physical inventory or cycle count is performed, was shown to differ between the data capture methods used. The less human involvement, the higher the inventory accuracy.

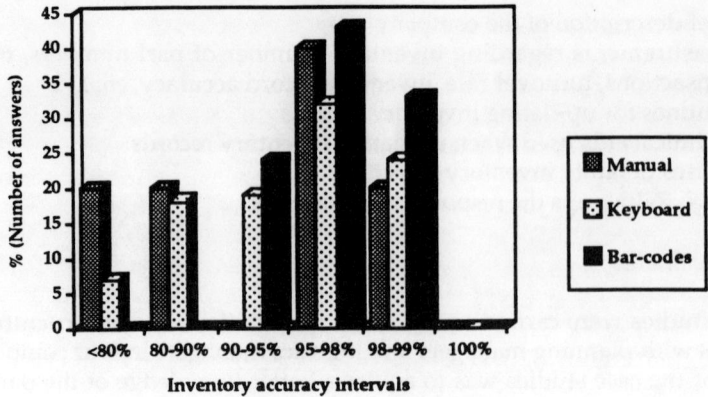

Figure 2   Inventory accuracy for different data capture methods.

Inaccurate inventory records were considered to be obtained for a number of reasons of which the human factor was considered the most important. The human factor includes weaknesses such as carelessness, poor discipline and absent-mindedness, and can therefore be seen as the overall cause of mistakes. However, when looking at more tangible causes poor routines, lack of education and lack of technical aids were identified as being the most predominat..

Table 2   Reasons for obtaining inaccurate inventory records.

Causes of inaccurate inventory records	Importance 6 ———▶ 1	
	(Low)	(High)
The human factor	1.2	
Poor routines	1.9	
Lack of education	2.1	
Lack of technical aids	2.2	
Software problems	3.5	
Personnel turnover	4	

Companies with a high inventory record accuracy also have a high transaction intensity.

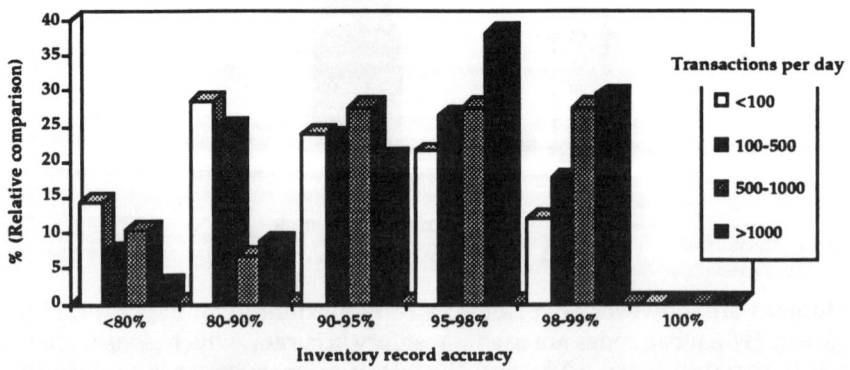

Figure 3   The connection between inventory record accuracy and transaction intensity.

The sum of each transaction intensity class is 100%, i.e. relative comparison.

These companies seem to be more aware of the importance of record accuracy and therefore spend more resources on minimising the influence of the human factor. This was verified in the case studies where companies with a high transaction intensity spend more resources on keeping inventory records accurate. One interesting finding was that there was a connection between the number of transactions and whether the company had defined inventory record accuracy. It was more common for companies with a high transaction intensity to have such a definition.

It is obvious from the results obtained, that bar-coding enhances inventory record accuracy. In the case studies it was, however, found that bar-coding alone did not improve the inventory record accuracy as much as initially anticipated. Only after simplification of data capture routines and education did the inventory record accuracy increase.

Being aware of the importance of inventory records, which companies with many daily transactions seem to be, also enhances inventory record accuracy, i.e. education is necessary in order to aquire an understanding of the importance of accurate inventory records.

4. Conclusions

Techniques such as bar coding have gained a high level of acceptance in industry. This paper shows, however, that there are more factors affecting the accuracy of inventory records than just a new data capture technique (NT). As humans are in most cases involved in data capture, automating this process enhances inventory record accuracy, fewer errors being fed into the item master file. However, as humans normally are the users of the automatic data capture equipment, a real understanding of the importance of accurate inventory records is not reflected in the use of new equipment. Errors can therefore still be fed into the item master file. Consequently, simplified routines (SR) and education (E) are needed too. Inventory record accuracy (IA) must be seen as a function of these three factors. This function can be expressed as $IA = f(SR, E, NT)$.

By being provided with these results, companies investing in automatic data capture techniques should be aware that simplification of routines and education of the individuals involved are as important as the new data capture technique.

References

[1]   G. Plossl, Production and Inventory Control. Prentice Hall, Englewoods Cliffs, 1985.
[2]   J. Orlicky, Material Requirements Planning. McGraw-Hill Book Company, New York, 1975.
[3]   E. Buffa and J. Miller, Production - Inventory Systems: Planning and Control. Irwin, Homewood, 1979.
[4]   R. Tersin, Principles of Inventory and Materials Management. North-Holland, New York, 1988.
[5]   R. Palmer, The Bar Code Book. Helmers Publishing, Peterborough, 1989.
[6]   R. Lindau, Automatic Shop Floor Data Capture and Its Impact on the Internal Material Flow (In Swedish). Report no. 12, Chalmers University of Technology, Göteborg, 1990.

Advances in Agile Manufacturing
P.T. Kidd and W. Karwowski (Eds.)
IOS Press, 1994

THE CONCEPT OF COMPANY SPECIFIC SOCIAL CONSTITUTION AS A TOOL FOR UNDERSTANDING THE INTRODUCTION OF PRODUCTION MANAGEMENT SYSTEMS

Christian Koch (Assistant Professor)
The Unit for Technology Assessment
Danish Technical University
DK 2800 Lyngby
Denmark

tel:45 31474715

ABSTRACT:This paper presents the concept of company specific social constitution. This is defined as "the total set of rules and norms that governs the attitudes to work and the working conditions of employeés".

The theoretical frame is used to analyze the introduction of Production Management Systems, PMS. This underlines the understanding of technological innovation as a ongoing social and political process taking place not only during design but also during implementation and use. Nevertheless the use of the theoretical frame in a danish context poses the tasks of identifying new key features, possibly new social groups and an analysis of different segments of danish industry.

1.Background

This paper describes some results of a project made for the Danish Union for Technicians (ref.[1]). It also represents some of the basic ideas of an upcoming research project to be made at "The Center for Interdisciplinary Studies of Technology Management" at The Danish Technical University and Copenhagen Business School. Most of the data was collected in 1993, covering two casestudies of enterprises utilizing Production Management Systems, PMS. And 6 portraits of suppliers of PMS, as well as portraits of their PMS-system.

The introduction of Production Management Systems often proves to be a cumbersome process. The visions of the supplier and his allies in the enterprise, meets reality and converts into practice. Instead of full realization, the system "degrades" into tools with more limited impact or the project strands and ends as failure. The reason for this can be a poor understanding of the enterprise as not only a welldescribeable organisation with structured tasks and clear goals to achieve. One has to include an understanding of the social network with complicated interaction, conflict and a long past. The introduction of PMS should be interpreted as a social, political process, where the result is highly dependent on the "collision" between the internal social network and the external promotors of PMS. There is a need for a deeper understanding of this "collision". This is offered by the concept of "-company specific social constitution" .

2.Theoretical Frame

The "company specific social constitution" as suggested by E.Hildebrandt (ref.[2]) is defined as

> *The total set of rules and norms that governs the attitudes to work and the working conditions of employeés.*

The central assumption is that social conflict is basic and that management needs to control it's employeés (ref[2] p.27 with reference to ref[3]). But in direct continuation of this, the argument is that the means of control today is sophisticated into "systemic control". Furthermore the social partners acting is a broad range of persons and groups.

The social constitution is a result of a long process. The internal social partners willingness to compromise and to abolish maximalistic point of views in a given period or topic leads to consensus rules governing future behavior. Conflicts where different opinions become central occur in other periods or in other topics and lead to a understanding of possible power. A series of policy-topics is by this process solved for a period. In some cases as a result of an internal process in other cases as a response to changing external conditions. It is difficult to change the compromises made on a given topic, but it is possible and necesserary. Some are under continious debate.

The central policy-topics is employment, pay, controlling   workperformance, work content and responsibility, career,   securing young and old employeés, education and training and participation.

E.Hildebrandt emphasizes that the social constitution is specific to a given enterprise or a branch. The specific properties of the social constitution are found in a specific context. E.Hildebrandt has investigated West German machine-manufacturing and underlines the following properties of the social constitution (ref [3]):

*Labour position
*Size of companies
*The relation between planning and execution
*The independence of departments
*Importance of cooperation between persons
*Coexistence of parallel structures and antagonism
*Position of shop steward
*Relation to local community

In brief this social constitution is characterized by a strong tradition among skilled bluecollar workers ("facharbeiter"), many small and mediumsized companies, short information paths, Purchase, Design and Assembly departments are relatively autonomous, many meetings between persons and negligible electronic communication, the employeés secures working conditions through direct bargaining in connection with tasks, the shop steward is recognized, but weak and the founder of the company and the workforce are local.

Transforming this theoretical frame into a tool for understanding of danish enterprises introducing PMS necessitates interpretation and restructuring. Thus it seems evident that danish industry consists of at least several segments of different social constitutions.

3.Results

In the project (ref.[1]) the investigation was focused on two cases, an electronics manufacturer and a wrapping-manufacturer. The method chosen was tentative; by interviewing key social actors (semistructured interviews), observing the planner and by dialogue on drafts of the report it was meant to approach the company specific social constitution. It is not possible to show the analysis as a whole in this short presentation. The result of the analysis in one of the cases is sketched below:

4.The electronics manufacturer

The diagram on the next page shows the main features of the company specific social constitution.

The company was founded in 1979 by the three owners.They are today managing director, production manager and manager of the software department. This "active ownership group" plays a central role in the company. The company has 42 employeés and produces a productprogramme of PLC-systems and modules.

The active ownership group quite early decided to follow a human ressource strategy towards the unskilled women workers in the assembly-department. This meant for example hourlybased (high) pay and jobrotation. In a period of 13 years the company expanded slowly but steadily. Therefore the group af women workers did not feel that they needed a club and a shopsteward. This need was overruled by the initiatives by management.

From 1988 the company organization was professionalized. The active ownership group couldn't manage the organization as a whole any longer. The production planner was hired to assist the production manager, but quickly took over most of the production management tasks and responsibilities. However the production manager maintained the material requirement planning and purchasing. From the early eighties to 1993 the main edp-assistance in the production area was a MRP-system that the production manager originally developed himself.

In 1993 the active ownership group decided to introduce a new PMS-system. This system was meant to carry out the MRP and the scheduling. It was decided to develop the system "inhouse" and the persons responsible are the production manager, the manager of the software department and a software developer.

Hence the production planner and the assembly workers wasn't "invited" to participate and the result was that the new systems schedulingmodule still haven't been used. This was anticipated by the planner. For a period he tried to convince the active ownershipgroup to buy a standard system. As a part of this effort the planner renewed his knowledge about PMS, by visiting suppliers and participating in various conferences. Nevertheless he wasn't able to convince management.

In this case the company constitution resulted in a reintroduction of a system that was already there. It wasn´t possible to utilize the potential of the planner and the workers. To do this the ownershipgroup would have to "violate" their own experience and capability. The resulting planning and scheduling process *does* work because of the planner's skills. So the partial failure of the system is not very important for the company. The future might bring attempts to reintroduce the scheduling module.

THE SOCIAL CONSTITUTION OF THE ELECTRONIC-MANUFACTURER

Property	Electronic Manufacturer
Labour position	Weak tradition among unskilled women workers highly dependent on management
Size of company	Small company
The relation between planning and execution	Short information paths room for improvising for planner, but not for workers
The independence of departments	Close links due to size Software design autonomous
Importance of cooperation between persons	Great, many contacts, few meetings No electronic communication
Coexistence of parallel structures and antagonism	Not evident some contradiction between plan and reality
Position of shop steward	The employeés secures working conditions mainly through confidence in management No shop steward, but a spokeswoman.
Relation to local community	Strong, local founders of company and local workforce But the local area is a major industrial community

6.Conclusion

The theoretical frame of "company specific social constitution" has proved to be useful in the analysis of the introduction of PMS. It underlines the understanding of technological innovation as an ongoing social and political process taking place not only during software-design but also during implementation and use.

Nevertheless the use of the theoretical frame in a danish context poses the tasks of identifying new key features, possibly new social groups and a analysis of different segments of danish industry.

REFERENCES

[1] C.Koch : The Technicians and Production Control. The Danish Union for Technicians, Copenhagen 1994 (in danish).
[2] E. Hildebrandt and R.Seltz:"Wandel betrieblicher sozialverfassung durch systemische controlle".Edition Sigma, Berlin 1989.
[3] H.Braverman: Labor and Monopoly Capital. Monthly Review Press, New York 1974.
[4] E. Hildebrandt:"Die betriebliche Sozialverfassung als Voraussetzung und Resultat systemischer Rationalisierung

Advances in Agile Manufacturing
P.T. Kidd and W. Karwowski (Eds.)
IOS Press, 1994

Implementation of hybrid MRPII/JIT system : a case study

J.A.A.SILLINCE and G.M.H.SYKES

Sheffield University Management School, 9 Mappin Street, Sheffield, S1 4DT UK.
Tel +44 +742 768555. Fax +44 +742 725103. E-Mail j.sillince@sheffield.ac.uk

Abstract. This case study investigates a partially completed process of MRP/JIT integration, including related cost accounting practices. One finding was that cost accounting tends to be done by operations managers rather than by accountants. Operations managers tend to use non-financial measures rather than financial ones, and they develop their own methods of getting them. A more formalised cost accounting system has been considered which would support further JIT advances, but so far shopfloor workers have opposed it.

1. Introduction

The primary method of data gathering for this case study was semi-structured interviews. The company studied made handtools and was located in Continental Northern Europe. It had an annual turnover of £4 million and employed 60 people on one site. It had in recent years restricted its product range (a rather precarious market position) in order to concentrate at the top quality end of the range on one particular type of handtool (called Tool X hereafter) in about 50 different versions.

2. Process

The average tool remained in the buffer store for approximately 6 or 7 weeks, before being drawn into the finishing stage. Here, the tools were etched, pointed, lacquered, fitted with wooden or plastic handles, and packed. Tools took 5 working days to pass through this finishing stage, compared with 4 to 6 weeks only three years previously. When a pack of Tool Xs was placed into the finished goods store, it was there from 1 to 2 weeks on average before despatch. In terms of lead times, then, the company had made considerable progress over the preceding two or three years. This had been effected, according to management, by the use of JIT-type and MRP-type practices.

3. MRP

The Material Requirements Planning program was written by the Managing Director. Past sales were used to create a forecast. Capacity planning was done globally on the basis of a day's production (one batch) of at least 2,750 items. The program calculated a production schedule in terms of each machine and in terms of actual stock in the system. Each day's batch had a bill of materials. However, the program did not calculate lead time, and so due dates for each machine were not calculated. This difficulty was circumvented because the lead time was relatively short: - 7 working days from Raw Material to buffer, and 5 working days from buffer to Finished Goods, so that daily WIP tracking, and spotting bottlenecks, was done on a visual and collaborative basis.

4. JIT

Shopfloor arrangements were made according to guesswork, collaboration, visual clues, and learned work-rhythm. The shopfloor machines were arranged in two horseshoe shapes, to reduce the available space for stacking materials near the production area, (a disincentive to forming long queues for machines) and to increase visual communication between work areas. The production route contained 12 to 14 separate but sequential processes beginning with cutting to length for the hot press. The batch was divided into clusters, each of which had a sheet showing what had to be done, together with the material required. These production control tickets also showed the date, product ID, date when material was issued, machine number, what the operator should do, and the time for each task. Operators were divided into groups, and planned their own sequencing and combining of items. Groups were planned in on sheets every day. If groups had achieved multi-skilling then the group was given complete control of its daily capacity plan. But multi-skilling was only partly achieved - workers were able to use different machines but most operators could not maintain them, so that special, higher paid maintenance operators were used. There were no terminals on the shopfloor. Also it was possible to keep track of WIP without entering it onto the computer all the time. There was not much paperwork on the shopfloor - no route cards for example. Thus some elements of JIT (partial multi-skilling, visual clues, MRP only for planning) were in place, meeting production control needs on a daily basis. At the time, the Production Planning Manager planned daily, but avoided the need to reschedule by having a 2 week planning list onto which new orders were placed. Rush orders required a new plan, and so he saw part of his function as protecting production from rush sales orders. It was obvious therefore, that the system was inflexible to the needs of customers, a business objective of increasing salience.

Continuous cost improvement was also an important element. Productivity gains were claimed to be 3% to 4% per year. Lead time and WIP had fallen considerably throughout the process. This was an important consideration since raw materials took 33% of costs. This improvement was to some extent due directly to the new Managing Director, who had arrived 4 years before. He changed the system from an exception-based to a standard-based one, eliminated large amounts of WIP and established a batch method, brought the lead time down, and wrote the partial MRP system. Much of the information had previously been in the head of the former Planning Manager. The order quantities were at last coming down, after initially large batch sizes. Dies in the hot press were being marked in order to keep track of the lifetimes. A die had an average lifetime of about 3,000 items, and each day's production was a batch of at least 2,750 items, so a new die was used each day, whether or not it was worn out. For some sizes of Tool X, the die life could be up to 10,000 items. Despite moves to JIT, the buffer still existed, and was 7 weeks' production i.e. 14% of annual turnover (about average in manufacturing industry). Finished goods stock was 3.5% of turnover (down from 5% a year before). Thus MRP/JIT had only reached an intermediate stage of integration. It was significant that, although the Financial Accountant made no estimate of stock holding costs in line with too much conventional practice, the operations managers were keen to see stock holding come down even further.

Most set up times were quite good (less than 10 minutes including cleaning) because there had been a drive 5 to 7 years before to bring them down. However, management was not completely satisfied and hoped eventually to bring them down further. The hot press provided the exception because its set up time was 0.5 to 1 hour (once per day). However, a set up/run time ratio of 10% was experienced on the bottleneck machine, i.e. Grinder No. 4 in Figure 1. This grinder was also a bottleneck because it was not duplicated, like some other machines, and the difficulty of the process led to many errors and scrapped work. One reason why set up times were not considered a problem was the fact that machines were only running 11 hours per day compared with 32 hours per day of available of available capacity (i.e. 2 machines for 16 hours per day). Machines were mechanical - not numerically controlled - and were on average about 10 years old. Because there were 2 of most machines, breakdowns were not a problem either. However, two years previously a breakdown log had been introduced for use in helping to decide when to buy replacement machines.

The company had a good relationship with its (single) steel supplier. They were able to keep a close eye on the analyses of quality, and they were able to plan together. The fact that there was only one supplier meant that any problems could be traced back to the supplier (in line with ISO 9000). They fixed details of amounts of different types of steel six months before delivery, which had to be made within a two week window. The average steel stock was 2 months' supply (it had been 6 months' supply 10 years before). The company looked at its stocks and steel needs every quarter and amended its estimates of need. Also most stock was held at the steel supplier, which the company did not have to pay for until delivered.

5. Quality

Quality management had had to take second place to production planning, on the argument that if production was chaotic then the quality would be poor. The company had begun to develop quality standard ISO 9000, although this had come as a result of pressure from customers (particularly local authorities and armed forces) rather than from pressure within the firm. For goods incoming, such as steel bars and plastic handles, it was intended to use a sampling plan compatible with ISO 9000. However, there were no inspectors. Warehouse staff did the sampling only if they had enough time, and there was no record kept of whether a sample had been missed. For goods outgoing, a mixture of sampling and 100% inspection (depending on customer pressure) was used. A stable, single-supplier relationship was important for controlling the quality of steel. Within the plant, statistical quality control was not used, although the very low level of scrap (1.5%) was constantly monitored by the Production Planning Manager. Batch sizes were also monitored, as were the number of Tool Xs produced from each 500 kg steel bar. This monitoring was required because count errors were about 5%. Sometimes people admitted they had made a mistake, but this was unusual. There were quality circles - an hour every fortnight in company time - but there were many problems with them. Supervisors did not like having to deal with the problems which were raised. After resistance by supervisors, the management had insisted that meetings should take place. The cultural awareness of quality as an issue was something that could have been improved, and also which could have suddenly got worse. Discipline was relaxed. For example, KANBAN would not have worked, because some operators would not have followed it completely, and these would have influenced others. Also, attention was not given to formalising, documenting, and inspecting for quality. The company admitted that quality was not the best it could have achieved, but that it was 'good enough' (a curious response given its 'top end of the market' strategy).

6. Accounting systems

Although operators worked in groups, and although there was some progress towards multi-skilling, cell-based costing was opposed by shopfloor workers. The cell would have been treated as a single work centre for cost accounting information such as running time, number of operations, parts processed, proportion of cell time devoted to each part, material costs and scrap. Similar non-conventional accounting methods have been used in JIT contexts (e.g. at Tektronix). The fear of operators was that it would expose errors or slow work, or impede pay rates linked to machine utilisation (irrespective of whether this causes queues downstream). Without such cell-based information it was difficult to reward groups for indirect labour effort. This prevented or made difficult further moves towards multi-skilling, preventative maintenance, stock control (of mini stocks to replace the buffer), and recording. In order to monitor accurately and control the trade-off between avoiding queues (not working a machine) and low machine utilisation rates (25% of costs are processing costs), such activity-based costs were needed. Also operators were not flexible about moving from one process to another. They spent days or even weeks on the same process, whereas management wanted movement on an hourly basis. There was no on the job training (apart from the maintenance section) which could have been used to motivate change.

The contrast with Japanese accounting systems is instructive. There companies use accounting systems more to motivate employees to act in accordance with long term manufacturing strategy (JIT, TQM, or whatever) than to provide management with precise data on costs, variances, and profits . For example, operations managers must decide priorities for cost reduction (e.g. improved flow versus quicker deliveries). In the case of operatives at the factory studied, the problem was to show them how wider skills and greater flexibility (multiskilling) and cellular working could increase autonomy without reducing wage levels. Yet only traditional accounting methods were used in reality.

The significant finding in the current case was that managers developed their own methods of keeping an overall feel for what was happening within the factory. They used more non-financial measures (20) than financial measures (12). Paradoxically, it may seem more beneficial for there to be no cost accountant (muddying the waters with irrelevant standard costings) so that operations managers are forced, as they were in this case, to develop measurements, samples, and informal information sources. This perception, if it existed, would be misjudged to the extent that activity-based costing has been shown to be another method by which operations managers can keep interest focused on the real domain of the factory operations, and to the extent that accounting can be used to motivate employees to meet strategic objectives.

Advances in Agile Manufacturing
P.T. Kidd and W. Karwowski (Eds.)
IOS Press, 1994

Design of Computer Aided Manufacturing Systems: work psychological concepts and empirical findings

C. Kirsch, O. Strohm, & E. Ulich
Work and Organizational Psychology Unit
Swiss Federal Institute of Technology (ETH), CH-8092 Zurich

The research project "GRIPS" is investigating on the design of computer aided integrated manufacturing systems from a work psychological perspective with the intention to develop and empirically support adequate design concepts. Evidence from a broad questionnaire survey indicates that most CIM implementations fail to meet the expectations associated therewith. Based on the assumption that only the joint optimization of social and technical system results in humane working conditions and economic efficiency, implementation and use of CIM systems has been investigated in 60 companies in Switzerland. The conceptual framework distinguishes technically-oriented and work-oriented design concepts. The later are characterized by the fact that, besides technology, work and organizational design and employee qualification are taken into account. The findings support the hypothesis that work-oriented design concepts are related to higher efficiency and better achievement of goals pursued with the use of new technologies.

Design Concepts for Computer Integrated Manufacturing

Manufacturing companies in Switzerland are currently confronted with increasing competition, complexity of products and processes and fluctuating market demands. In order to fulfill the demands for more flexibility, quality and competitive cost „Computer Integrated Manufacturing" (CIM) has been considered as the „general remedy". Yet the available systems originally were conceptualized for large companies and the small and medium-sized enterprises (SMEs) were overwhelmed by capital intensive new technologies. A government funded „CIM Action Program" has been established in 1989 in order to provide the needed support for SMEs, which had to deal with budgetarian limitations and lack of know-how in basic technologies. The goal of this program is to improve the economic competetiveness of the swiss enterprises. According to the „CIM Action Program" the emphasize in CIM is on „I", the integration of human ressources, technology and organization. This claim pays due regard to the work-psychological assumption that only the joint optimization of technology, qualification and organizational structures results in successful implementation of CIM. The holistic approach to implementation of new technologies has been referred to as „work-oriented" design concept for advanced manufacturing systems. Work-oriented concepts integrate aspects of technology, organization and people. Technology is not conceived as a constraint but as a tool that offers options for design. Technically-oriented concepts on the other hand, primarily address the implementation of technical systems and their components, while organizational issues, and human ressources aspects, such as e.g. employee selection and training, are neglected [1]. Work-oriented production systems differ from technically-oriented ones on four levels: enterprise, organizational unit, group and individual [2]. Work-oriented concepts are characterized by the structural features *decentralization, functional integration, work group autonomy,* and *complete task design.* On the contrary technically-oriented concepts are dominated by centralization, division of functions, centrally controled groups, and task fragmentation.

The Grips Project: Methods & Results

The goal of the GRIPS project is to investigate the implementation and use of computer aided integrated production systems in Switzerland, systematize the related design concepts and relate these to efficiency of computer aided production. The project consists of three consecutive phases. In the first phase GRIPS I a written survey has been carried out in the capital-goods industry and processing industry throughout Switzerland, inquiring into use and integration of computer aided manufacturing systems, aims associated therewith and problems during implementation of new technologies. In the second project phase GRIPS II detailed case studies have been conducted in 60 companies selected from the original sample. As research methods we used document analysis, company tours and expert interviews. Manufacturing conditions, work and organizational structure, implementation of new technologies, integration of computer aided production functions, employee qualification and training have been investigated. The goal was to identify work-oriented and technically-oriented production concepts regarding the four organizational levels. In the third phase of the project detailed sociotechnical system analysis is carried out in matched pairs of companies, that are comparable regarding company size and manufacturing conditions, but different regarding manufacturing structures (N=14). Document analyses, expert interviews, time based order-flow-analyses, observational interviews, task analysis (VERA, KABA), complete work shift analysis and questionnaires are used as methods for investigation.

The GRIPS I results (n=917) show that there is a high level of computerization in swiss companies. For example in the capital-goods industry more than 50% of the companies (n=679) already have implemented the components CAPPC, CAP and CAM. Almost half of the companies have implemented the components CAD and CAODA. None of the goals pursued by computer implementation has been achieved even by half of the companies rating them 'very important'. Technical shortcomings, problems with employee qualification, work organization and lack of acceptance or motivation of employees were frequently encountered problems. GRIPS II revealed that the ratio of computer aided production functions compared to conventional tools is highest in the indirectly productive departments, especially in the manufacturing scheduling department (73%), the purchasing and the production planning departments (68%). Computer aided systems are the least implemented in assembly (16%).

Concerning the manufacturing design the GRIPS II results indicate that 63% of the companies have rather centralized enterprise structures, and that most companies in our sample are functionally specialized on the level of the organizational unit. On the group level, work and organizational structures have been analysed in production and assembly. On the average a work group consists of 5 persons, with a range of 3 to 12 persons. Group autonomy was operationalized with reference to Gulowsen [3]. 53% of the companies claimed to have any form of group work. However only two companies have work groups with some influence on relevant decisions, 25 work groups are centrally controled. None of the groups can be categorized as „(semi-)autonomous work group".The percentage of work groups involved in decisions on specific work related questions is shown in Table 2. Regarding jobs on the individual level 86% of the tasks of workers in manufacturing can neither be described as challenging nor complete. All levels considered, only two companies in our sample can be categorized as work-oriented, two other companies show a slight tendency towards work orientation (Table 1).

Table 1. Classification of companies (n=60) according to their degree of work-orientation in production design

				- production design concept -			
	technology-orientation				work-orientation		
level	high	medium	low	low	medium	high
company		centralization			decentralization		
	2	14	22	14	4	4
organizational unit		functional division			functional integration		
	5	25	21	7	2	0
group		central control			local control		
	21	9	13	8	7	2
individual		fragmented tasks			complete tasks		
	3	14	29	11	3	0
overall work-orientation		technically-oriented design			work-oriented design		
	2	29	25	2	2	0

Table 2. Work group scope for excertion of influence on work related issues. Corr....significant correlations (p < 0.05) of group autonomy and attainment of the goals pursued with computer implementation (n=32).

object of decision	no influence	consul-tation	involve-ment	autonomous decision	Corr
distribution of tasks	13%	34%	34%	19%	*
subject of production / quality norms	44%	31%	22%	3%	*
choice of production method	23%	35%	35%	7%	
work shift scheduling	37%	28%	16%	19%	*
production volume	25%	44%	28%	3%	
choice of spokesperson	47%	28%	12%	13%	
group leadership	44%	25%	22%	9%	*
acceptance of additional work	72%	16%	9%	3%	*
recruitment of new group members	84%	10%	6%	0%	

Efficiency of Computer Implementation

Based on the hypothesis that work-oriented manufacturing structures can make a contribution to economic efficiency, goals pursued with the implementation of new technologies have been analyzed. None of the goals for computer implementation has been achieved even by half of the companies rating them 'very important' (Table 3).

Table 3. Percentage of companies rating the goal as „very important" and beeing „attained". Data from GRIPS I (1992; n=130-450) and GRIPS II (1993; n=59)

goal	percentage of companies rating the goal as			
	„very important"	„attained"	„very important"	„attained"
	GRIPS I		GRIPS II	
reduction of cycle time	65%	27%	68%	33%
increased accuracy of delivery time	68%	29%	64%	47%
reduction of material on stock	48%	14%	54%	44%
improved basis for calculation	40%	41%	52%	48%
increase in market flexibility	51%	27%	48%	46%
increase in internal flexibility	39%	28%	46%	48%

Statistical analysis of the relationship between goal attainment by means of computer-ization and work-oriented concepts of production design revealed that some goals are correlated to work-oriented design concepts on all levels except the level of centralization vs. decentralization (Table 4). On the level of the work group overall goal attainment by means of computer implementation was predominantly determined by the competences of the work group concerning specific work-related issues (Table 2).

Additional support for the profitability of work-oriented concepts was provided by dynamic balance-sheet analysis, that revealed a significantly higher increase in economic data from 1985 to 1991 for companies with work-oriented design concepts, as e.g. increase in cash-flow (T-test .20), turnover (T-test .14), capital assets (T-test .12) and joint-stock-profitableness (T-test .16).

Table 4. Significant correlations of goals attained by means of computer implementation and structural features of work-oriented production concepts (n=53, *p<0.05, **p<0.01)

goal	WO	D	FI	WG	CT
increased accuracy of delivery	.23	.17	.16	.29*	.06
correct manufacturing documents	.32*	.20	.12	.34*	.18
transparency of material flow	.29*	-.02	.11	.39*	.20
internal flexibility	.17	.26	.27*	-.07	.07
not missing the technical boat	.42**	.08	.23	.33*	.37**
motivation of employees	.28	.00	.11	.39**	.20

WO...Overall Work-Orientation, D...Decentralization, FI...Functional Integration, WG... Autonomous Work Groups, CT...Complete Tasks

Successful implementation of CIM systems requires the use of key skills such as self-reliance, capability of teamwork, interdisciplinary expertise, planning skills, and communicative skills. Training of technical and methodological, as well as social skills, is regarded as crucial for successful implementation of new technologies. The GRIPS II results demonstrate that in CIM systems the need for socially competent workers is increasingly recognized. However, the analysis of training programs for CIM systems shows that social skills generally are neglected. In many cases, only after a technically-oriented implementation strategy has proven insufficient, skill requirements have been reconsidered [4].

Conclusions for Design of CIM Systems

These findings suggest that the integrative approach towards man-technology-organization, (MTO approach) of work-oriented design concepts, provides an appropriate means for the implementation of computerized manufacturing systems. It furthers the attainment of goals pursued by implementation of new technologies and is related to economic efficiency of organizations [4].The technically-oriented approach of mere implementation of new technologies whereas still prevalent in many companies, generally is not rewarded with the intended improvements in flexibility and efficiency.

It has become evident that the key-factor for success is the joint optimization of man, technology and organization. Restructuring measures designed according to the following work psychological principles [2] are favorable:

- Organizational design prior to automation
- Education and training as a strategic investment
- Functional integration
- Local self-regulation.

References

[1] Ulich, E., *Arbeitspsychologie*. Verlag der Fachvereine, Zürich /Poeschel, Stuttgart, 1994, 3rd edn.
[2] Ulich, E., CIM - eine integrative Gestaltungsaufgabe im Spannungsfeld von Mensch, Technik, Organisation. In: G. Cyranek & E. Ulich (Eds.), *CIM - Herausforderung an Mensch, Technik, Organisation. Schriftenreihe Mensch, Technik, Organisation*. Verlag der Fachvereine, Zürich, 1993, pp. 29-43.
[3] Gulowsen, J., A Measure of Work Group Autonomy. In: L. E. Davis & J. C. Taylor (Eds.), *Design of Jobs* . Goodyear Publishing Company, Santa Monica, 1979, 2nd edn., pp. 206-218.
[4] Strohm, O., Kirsch, C., Kuark, J. K., Leder, L., Louis, E., Pardo, O., Schilling, A., Ulich, E., Computer aided manufacturing systems: work psychological aspects. In: A. Bürgi-Schmelz et al. (Eds.), *Computer Science, Communication and Society: A Technical and Cultural Challenge. Conference Proceedings, Neuchâtel, Switzerland, 22-24 September 1993*, pp. 221-238.

Advances in Agile Manufacturing
P.T. Kidd and W. Karwowski (Eds.)
IOS Press, 1994

A Human Factors Approach to the Selection and Implementation of MRPII

Brian McGARRIE
Newcastle Business School, University of Northumbria, Carlisle Campus

Abstract. This paper describes the selection and implementation of a Manufacturing Resource Planning (MRP II) system for a small manufacturing company. A model to aid the selection and implementation is highlighted. The emphasis throughout the model is on the human factors necessary for the successful implementation of MRPII. The differences between small/medium and large company implementations are shown. Further uses of the model are highlighted and future work indicated.

1. Introduction

Most research efforts in the Material Requirements Planning (MRP) implementation area are case studies from single organisations, offering little evidence of prescriptive ability and are little more than isolated observations of what occurred at a particular place and time [1].

Although this paper presents a single case study of a Manufacturing Resource Planning (MRPII) implementation, a model for selection and implementation of such systems is highlighted. This paper shall focus on the human factors present in the model.

1.1. Company Background

The company under study, manufacture and assemble specialist products for the road and rail industry. The company have approximately 55 employees and a turnover of £4 million, and are part of a large engineering group.

2. Selection Model

The model used for the selection and implementation of the new MRPII system for the company, was built from the identification of the requirements of manufacturing control systems [2].

The first step undertaken is the *Unique Need Statement*. This focuses on the unique needs of the company. The objective of this step, is to present a number of characteristics of the company's environment which establish the need for certain planning and control modules of an MRPII system. At the same time, the deficiencies of the current computer system become apparent.

After extensive analysis of over 50 characteristics which can effect the need for an individual manufacturing planning and control module, the number of characteristics was reduced to 12 in the case of the company under study.

The characteristics influencing the needs of the company under study were:

1) The demand pattern.
2) Regularity of production of individual products.
3) Rate of new product introduction.
4) The time taken to manufacture and assemble typical products versus the time the customers are normally prepared to wait.
5) The commonality of parts.
6) The number of bought-out components.
7) The number of purchasing and stock transactions necessary.
8) The number of products sold in a large number of variations.
9) Percentage of deliveries from suppliers arriving on time.
10) Quality of goods received from suppliers.
11) Yield and rework levels on the shopfloor.
12) The flexibility to cope with revisions of customer demand.

Following analysis of the characteristics, the main requirement was for a complete upgrade of the current purchasing system. This appeared mandatory. A move towards Just-In-Time (JIT) production and the pull technique also appeared beneficial, although there were a number of problems to overcome before such an approach could have been considered. The front end of the existing system appeared to be woefully inadequate, that is, the demand management and production planning, hence there was a strong need for Master Production Scheduling (MPS).

The second step of the selection model was to provide a checklist of the factors which must be overcome if successful implementation and operation of the individual control modules is to be realised.

The third step of the selection model involved developing a Functional Specification Document for the company. A 'best fit' analysis was carried out with the results of the software analysis. System vendors were short listed, and vendor client sites visited.

3. Implementation Model

This model is based on formulating an implementation strategy and is very much human factors based. A *'Way Forward'* document is presented to the company. This document outlines approximately 50 factors which have to be addressed before and during implementation. These issues were faced in conjunction with a traditional MRPII implementation plan [3]. Examples are now provided of some of the human factors which the company under study faced:

3.1. Encouraging participation of managers in the formation of the strategic review

Each manager prepared a brief statement, which is built up by the Managing Director to form part of the review document. This factor was important as regular involvement of top management in the planning process is absolutely essential. Otherwise, the business plan will be incomplete, and will contain unresolved conflicts.

These managers must have well-defined roles within the business strategy, otherwise, the strategic direction for the business cannot be reviewed properly. In addition, there will be problems with communication of any business plans, and no effective or active participation by employees. The end result will be poor business development or stagnation.

3.2. Evidence of top management having a positive attitude to change

At the group level, there was a very positive attitude to change. The Managing Director of the company had very much this same attitude. This is vital, as he had to be the initiator of change. However, the managers within the company seemed to have a reluctance to change. The Managing Director believed they were not 'hungry' enough. He didn't believe that they had the capacity for lateral thinking, and puts this down to the type of individuals they were. Managers were very much day to day control people. An example of this could be seen with regards to the proposed new computer system. If the Managing Director hadn't questioned the quality of the information coming out of the current system, both in terms of speed and effectiveness, he believes the managers would have been quite happy to keep it running until it finally broke down. Thus, the mangers were reactive rather than proactive. For successful implementation, mangers must be willing to change the way they run the business.

3.3. The existence of a managed process of making beneficial change

The company didn't have a structured approach to the management of change. However, the company did actively collect ideas although perhaps not encouraging people to the extent they ought to. The Managing Director believed that the people working for the organisation thought they were employed by the company, not participating in the company. He recognised that one way of gaining people's commitment was to get ideas from them. At the time of the study, the leading hands could recruit their own labour, subject to managerial scrutiny. Thus, they were building teams.

3.4. The company don't actively plan, manage and develop an educational policy

The company did encourage people to go to further education, if coinciding with the needs of the company. This factor is important, as resistance to change can be overcome by the dedication and commitment of management towards education and training. It was suggested that the education of the workforce should not only address the technical issues related to MRPII, but focus on alleviating fears related to changes in the interpersonal relationships as well.

3.5. Extent of inter-departmental co-operation

The Technical Department and Sales Department had a good relationship. However, the co-ordination between the Technical Department and Materials Department could best be described as fraught. This appeared to be down to the personality of the Materials Manager, who had been allowed to develop his role beyond his abilities, and assumed responsibility for other managers.

Interdepartmental co-ordination is an essential prerequisite for the implementation. The objective should be to achieve the integration of joint goals and responsibilities. Good co-ordination is characterised by departments helping out each other at will.

3.6. Accountability within the current system

Too many allowances were made because of the current system. In a small company, the system is going to impinge on all areas of the company. There should be an approach towards building accountability into peoples jobs.

4. Results

The company have successfully implemented an MRPII system by using the selection and implementation model. This model can be used for all discrete batch manufacturing companies.

The following advantages were realised when implementing within a small company:
Easier to obtain top management commitment.
Management more closely involved in the day to day running of the company.
More areas of shared responsibilities and teamwork, for example, the purchasing manager was also in charge of stores and manufacturing.
Less bureaucracy.

The disadvantages facing the company were:
Lack of personnel - unable to devote full-time members of steering committees and project groups
Few disciplines and procedures in place.

5. Conclusions/Future Developments

This model is now being used by a number of companies, large and small to select and implement new production planning and control systems.

Work is being undertaken to extend the model to supply chain management.

Initial work has been undertaken to develop this idea to the service operation environment.

6. References

[1] D.L. Turnipseed, O.M. Burns, and W.E. Riggs, An Implementation Analysis of MRP Systems: A Focus on the Human Variable, Production and Inventory Management, vol.33, no.1, (1992), 1-6.

[2] A. Kochhar and B. McGarrie, Identification of the Requirements of Manufacturing Control Systems: A Key Characteristics Approach, Integrated Manufacturing Systems, vol. 3, no.4, (1992), 4-15.

[3] J.P. McManus, Developing a Detailed MRP-II Implementation Plan, Production and Inventory Management, vol.30, no.2., (1989), 75-78.

Advances in Agile Manufacturing
P.T. Kidd and W. Karwowski (Eds.)
IOS Press, 1994

MOPS PROJECT BESTMAN

Mr Colin Brown
Human Centred Systems Limited
222 Maylands Avenue
Hemel Hempstead
Herts HP2 4FE

ABSTRACT

The MOPS (Manufacturing Organisation, People and Systems) initiative recognises that organisation and people are an important consideration in manufacturing systems implementation which hitherto has been largely ignored. Human Centered Systems Limited has been involved in developing usable information technology to help bridge the gap between manufacturing systems and their users who need to have the information separated from the data. Some of these developments have been within the framework of part-funded collaborations with companies in Europe. It was a natural extension to these developments that they should get involved in its practical implication and this recognition led to an invitation from the DTi to put forward a proposal for a place on their MOPS initiative. This they did and project BESTMAN was born. The following brief report attempts to explain the work and research done over the past two years and an indication of the tangible benefits of applying some of the methods in the five manufacturing companies who are partners in this project.

'BESTMAN' GENERIC MODEL

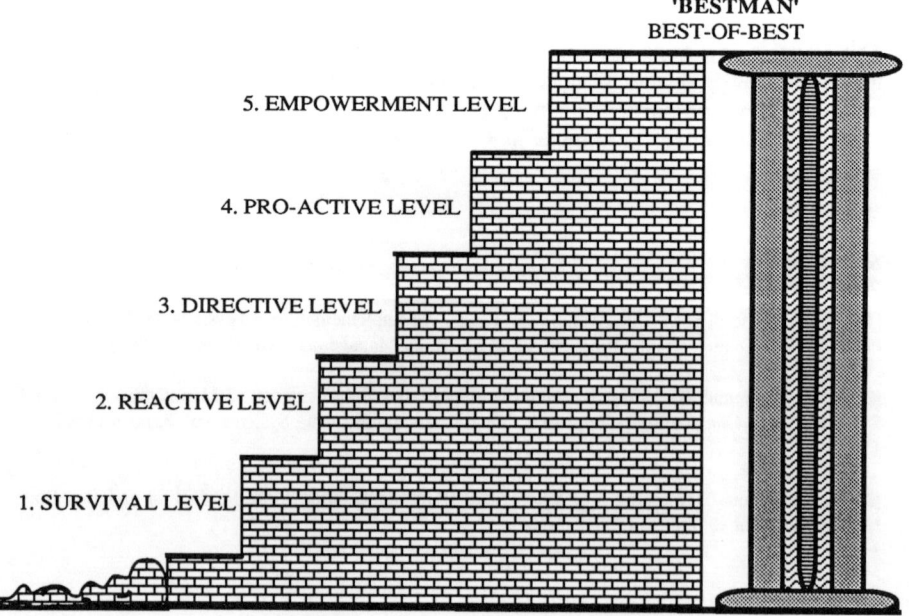

SYNOPSIS

The term BESTMAN was derived from BEST Practice Cellular MANufacturing and the original research was to be focussed on how manufacturing businesses can achieve excellence and, therefore, improved competitive edge, through the use of cellular manufacturing techniques and the proper use of appropriate supporting tools, methods and procedures. It was realised at a very early state of the Project that to concentrate on the manufacturing process only would not necessarily improve overall business performance. What was needed is an approach that addresses the requirements of the whole organisation and aligns all the processes to the overall objectives.

In this respect the term cellular is used in its broadest sense and cells should be defined in ways that specific businesses decide is best or most appropriate to their needs and circumstances. The concept is one of businesses within a business, of fostering internal and external customer/supplier relationships and the empowerment of employees to manage and control the resources of their business or cell. The overall business aims will be to improve customer service levels by reducing leadtimes and increasing throughput and at the same time minimise inventory and work-in-progress. In other words, to make each business (cell) as efficient and responsive as possible by providing the tools, methods and climate to achieve the overall goals. The research starts by establishing the current status of the company.

A Generic Model has been developed which describes various levels of business competence and with the aid of a benchmarking questionnaire and scorecard, the starting position can be identified. The questionnaire asks searching and provocative questions about key areas of manufacturing, business control such as; Customer Service, Organisation, Operations, Quality and Technology. From the responses received to the fifty or so questions it is possible, through a detailed analysis and attributing values to some the questions/answers, to determine a company's position on BESTMAN's Generic Model. This preliminary research is followed by a detailed Business Process Analysis which enables strengths and weaknesses in the company to be identified and become the focus of corrective action. At this juncture it is possible to introduce and apply the appropriate tools and methods which have been developed and refined through Project BESTMAN and effect a tangible improvement in the company's performance.

PARTNERS

As already described, the initial research is to chart the BESTMAN model, document the progress of give manufacturing companies as they go through the stages of the model, publish these in the form of case studies and identify the critical success factors for dissemination to British industry. The five manufacturing companies are:-

Drexel Equipment (UK) Limited

GEC Engineering (Accrington) Limited

ITW Finishing Systems and Products

Kysor Europe Limited

Temco Limited

The companies were chosen for their desire to move to Best Practice Manufacturing and also because of the diverse nature of products, manufacturing styles and locations so as to give as Generic a set of 'do's' and 'don't's' as possible. All the Partners were at various stages of re-organisation and manufacturing control systems implementation and all needed a different approach to determine the most appropriate solutions for their particular circumstances.

In all, ten case studies have been published covering the work done in the five Partners, describing the problems faced, detailing the approach taken and enumerating the tangible benefits gained. The major benefits have been in throughput improvement resulting in reduced WIP and inventory and increased customer service levels. The flattening of organisational hierarchies, reduced chains of command and elimination of non-value added activity as well as the simplification of processes, has brought about substantial overhead cost reductions. At least two of our Partners have moved form loss into profit over the last two years, in spite of the current economic climate and are now in a strong position to take advantage of the upturn.

SUMMARY

The BESTMAN collaborative group now has two representatives on the DTi's M90's Inside UK Enterprise "Best Practice" companies register and a third is almost at that status. The group headed by Human Centred Systems Limited through the practical application of the BESTMAN Generic Model principles and research, is developing usable solutions which will enable a business to:-

*Benchmark its position in terms of business competence.

*Analyse its key business processes.

*Determine an improvement path.

*Achieve BEST practice MANufacture.

The Project will also deliver a Toolbox of all the above mentioned techniques as well as a full write up of all the research material. This will help the business that recognises the need, to go about integrating these advanced manufacturing techniques with human skills and appropriate supporting information technology.

Advances in Agile Manufacturing
P.T. Kidd and W. Karwowski (Eds.)
IOS Press, 1994

Mobilising continuous improvement for strategic advantage

John Bessant, Sarah Caffyn, John Gilbert
Centre for Research in Innovation Management, University of Brighton, Falmer, Brighton, BN91PH, U.K.

Abstract

This paper reports on the results of a major research programme in the UK which has been looking at the experience of continuous improvement (CI). It provides an indication of the experience of CI implementation in the UK and the emerging challenges which this raises. Part of the research has concentrated on developing a framework methodology for implementing and maintaining CI together with a 'toolbox' of resources to support this; the paper presents an overview of this.

1. Introduction

Continuous improvement (CI) can be defined as an organisation wide process of focussed and sustained incremental innovation. It is a simple concept which builds upon the basic tendency of people to make improvements to what they do and the things which they produce; at its heart is the application of creative problem-solving to day-to-day work. The potential benefits of capturing and building upon this natural resource are well-known; it is, for example, clear that much of the Japanese quality and productivity miracle is founded on CI rather than heavy investments in new technology [1]. Nor is it a new idea; examples exist of formal and structured attempts to build upon workforce involvement and creativity dating back to the 19th century, and the quality 'guru' Joseph Juran referred to 'the gold in the mine' in his book published in the 1950s, when speaking of the innovative ideas which could be contributed by each individual worker [2]. In manufacturing CI is seen as one of the key elements in the emerging prescription for 'lean manufacturing' and it is acknowledged as central to the long-term effectiveness of more radical attempts at organisational restructuring such as business process re-engineering [3-5].

CI represents an important break with the hitherto dominant Ford/Taylor paradigm for business organisation since it assumes there is never a single 'best' way to carry out tasks but always a better. Thus the role of individual creativity moves from something to be excluded as a troublesome source of variation and interference with a well-designed system towards being central to a learning and developing organisation. The problem – and the challenge – is that its apparent simplicity and desirability hide a wealth of difficulties in successful implementation. These are, at their source, problems in organisational transition, requiring significant efforts in the re-design and re-shaping of organisations to enable participation and contribution on a large scale.

2. Related research

CI relates to at least two fields of research. The first is innovation research, where considerable activity has been seen over the past fifty years aimed at trying to understand the importance and dynamics of the process of technological change. It is possible to derive a number of guidelines for effective innovation management from this – see, for example, [6] – but most of this is concerned with essentially discontinuous, radical and occasional innovation. Such innovations are important but it is important to recognise the equally significant role played by incremental innovation, both in terms of reinforcing and extending the gains made by radical change and in its own right as a cumulative set of small changes [7]. In general we have considerable information and experience in managing major innovation, and this kind of activity is often the province of technical specialists; we have far less information about or experience of managing incremental change involving a much larger proportion of the workforce.

The second field is that of total quality management (TQM), where extensive discussion and experiment has followed the successful experience of Japanese manufacturing in the 1960s and 1970s [8, 9]. TQM represents both a philosophy and a set of practical activities which have a transforming effect on the ways in which organisations perceive and respond to problems in supplying their goods and services to others. At its heart is a change in the underlying value system towards one which places satisfying and even delighting the 'customer' as the prime business objective. Enabling this change involves a step change (often accompanied by some form of restructuring) followed by a sustained long-term process of continuous improvement. Although much has been written about TQM, it is clear that the hardest part of the prescription is the enabling and long-term maintenance of such CI activity.

3. Continuous improvement and organisational learning

A number of commentators have begun to identify CI as a key mechanism in organisational learning, representing one important way in which firm-specific capability is developed[10, 11]. As Sirkin and Stalk describe [12] the process is one of gradual evolution from crisis-driven problem solving towards gradual acquisition of competence and eventual mastery.

The first stage involves tackling basic 'firefighting' loops in which the immediate and most pressing problems are dealt with, as they crop up. This is less than satisfactory, not least because there is every chance that the same fires may flare up again unless steps are taken to prevent them and to get down to tackling the underlying root causes.

The second level is one in which the emphasis is on bringing the various processes which the organisation uses to perform its tasks under control. This means developing a clear understanding of those processes and what affects them, and measuring throughout to ensure they are coming under control. It is at this point that some form of standardisation is often sought, to try and 'fix' the gains made and ensure there is no slipping back. Standards like ISO 9000 are essentially frameworks for ensuring that processes are being monitored and controlled in this fashion – but achieving them can involve a lengthy earning process.

Once the organisation's processes are under control and problems are being prevented rather than simply treated, the next step is to begin to improve those processes through attacking the root causes of problems. This shift in emphasis is important; much of the early activity in continuous improvement is about error cause removal and problem elimination; by contrast this third stage is about looking for opportunities and experimenting with new possibilities.

Finally the organisation can begin to invent completely new processes, building on its by now deep understanding; at this point it will have developed some highly specific competencies which it can exploit for strategic benefit. The big advantages of such home-grown competencies is that they are hard for others to copy; a good example is the way in which Western car firms tried to emulate Toyota's successful production system. Although simple in concept it requires the organisation to unlearn many old ways of doing things and then to learn a new set, building up competence in the manner described above.

4. Diffusion of Continuous Improvement

CI is an attractive concept and has received considerable attention in recent years. This can be attributed to a combination of factors including:

- demonstrable success in Japanese, and latterly, Japanese transplant companies
- extensive promotion of the concept on the back of total quality management (TQM) programmes
- strong supply side drives for adoption of TQM (e.g. government requirements for certificated standards such as ISO 9000, EC promotion for consultancy support in TQM, etc.)
- low capital investment requirement
- low entry barriers in terms of skills and specialised knowledge
- recognition of under-utilised assets - 'with every pair of hands you get a free brain'

Not surprisingly, given this range of factors, the diffusion of CI concepts has been extensive. However, despite the considerable enthusiasm with which it has been embraced there is growing evidence that successful implementation of CI is not simple. Whilst many companies can point to significant strategic benefits, these only emerge after an extended period of time and as a result of considerable efforts at implementing and maintaining a supportive system for CI [13]. Several studies highlight the drop-out problem where the typical pattern is for short-term gains in the early stages of a CI programme, followed by a decline in enthusiasm and involvement [14].

Analysis of the problem of CI failure or fade-out suggests that there is no single cause but rather a collection of contributing factors. Recognition of the complex nature of this problem, and the widespread experience of difficulties in introducing CI led to the setting up of the CIRCA research project in 1992.

5. The CIRCA project

The CIRCA (Continuous Improvement Research for Competitive Advantage) project was set up as part of the wider initiative of the UK Department of Trade and Industry called MOPS - Manufacturing, Organisation, People and Systems. This initiative was an attempt to complement the R&D and technology transfer work which had been taking place in the field of advanced manufacturing technology, and was an important sign that government recognised the growing body of evidence suggesting that improved competitiveness derives not only from investment in hardware but also in organisational change [15].

The pattern of activity was essentially 'action research' around a set of major themes and problem issues in a core group of firms; a second strand of activity concerned the building and operation of an experience sharing and research network. This larger group of firms were linked by a common concern with CI, and through a series of workshops, visits, special interest seminars and other activities, was able to extend and share experiences in the area of CI. Nearly 100 firms have been involved in some way with the network, and there is now a research 'club' of 25 firms actively involved in CI research and a wider group who are users of the research results.

The project has been running for just over two years and is scheduled to continue for a further three. Although the networking is itself an important output as far as industrial users are concerned, there have been a number of other outputs. These break down into two groups – a diagnostic framework with which firms can position themselves and audit the health of their CI activity, and a suite of support resources to enable action to be taken to improve the process.

The diagnostic tool is based on a model of the CI process which we have been developing in the programme. This is essentially a 'benchmarking' type of framework, which we have been using to facilitate and enable learning about CI and how it can be improved. In essence our research suggests that there are five critical systems which need to be in place to enable successful CI. If any of them is missing, the chances are that the programme will not last into the long term. But even with all systems present, there is still considerable scope for development and improvement within each of them. Benchmarking of this kind is not so much about relative performance as about learning new tricks, new angles, new ways of dealing with particular CI issues by making use of the experience of others. Within the CIRCA programme this is enabled via a network , but the possibilities exist to extend to benchmarking activities.

The second element of the CIRCA work involves developing, collating, and making available tools and techniques which may help in dealing with particular issues identified in the diagnostic phase. These range from design of systems which are missing, through to various inputs of knowledge, through to specific tools and techniques targetted at particular problems. The emphasis is not on finding single 'best' solutions but on organising the variety available, so that firms have access to many different ways of dealing with typical CI problems. This wide range of resources is being managed in the form of a 'dynamic toolbox' – an evolving resource base which we are trying to develop in 'hypermedia' format, allowing users easy access to the elements most relevant to their needs.

6. References

1. Imai, K., *Kaizen*. 1987, New York: Random House.
2. Schroeder, D. and A. Robinson, *America's most successful export to Japan-continuous improvement programmes*. Sloan Management Review, 1991. **32**(3): p. 67-81.
3. Womack, J., D. Jones, and D. Roos, *The machine that changed the world*. 1991, New York: Rawson Associates.
4. Hammer, M., *Don't automate, obliterate*. Harvard Business Review, 1990. .
5. Johansson, H., et al., *Business process re-engineering*. 1993, Chichester: John Wiley.
6. Rothwell, R. *The fifth generation innovation process*. in *25th Anniversary Conference*. 1991. Science Policy Research Unit, Uiversity of Sussex:
7. Hollander, S., *The sources of increased efficiency: A study of Dupont rayon plants*. 1965, Cambridge, Mass.: MIT Press.
8. Atkinson, P., *Total quality management: Creating culture change*. 1990, Kempston: IFS Publications.
9. Garvin, D., *Managing quality*. 1988, New York: Free Press.
10. Garvin, D., *The learning organisation*. Harvard Business Review, 1993. .
11. Leonard-Barton, D., *The organisation as learning laboratory*. Sloan Management Review, 1992. **Fall**.
12. Sirkin, H. and G. Stalk, *Fix the process, not the problem*. Harvard Business Review, 1990. **July/August**: p. 26-33.
13. Bessant, J., et al., *Rediscovering continuous improvement*. Technovation, 1994. **14** (1).
14. Kearney, A.T., *Total quality - time to take off the rose-tinted spectacles*. TQM Magazine, 1992. .
15. Bessant, J., J. Burnell, and S. Webb, *Helping UK industry achieve competitive advantage through continuous improvement*. Industry and Higher Education, 1992. **September**: p. 185-189.

230   *Advances in Agile Manufacturing*
P.T. Kidd and W. Karwowski (Eds.)
IOS Press, 1994

Destiny and Organisational Issues

P D Pearce, A P Jagodzinski, M Dixon, K Wittamore,
D Mulhall, D V Clarke, D M Lewis, P M Shepherd,
& V A Lovitsky (United Kingdom)

Abstract. The key to industrial change involving new technologies is people - their organisation, their motivation and the opportunities they are given. Destiny includes a diagnostic PC software tool to assist manufacturing companies in the identification, analysis and improvement of:- present organisation structures and the way they really work, interpersonal working relationships and problems, and key organisational issues. It allows changes in organisational structures to be modelled and refined before being implemented. Thus, companies may develop organisational structures which actively assist the successful implementation of new technology.

1 Introduction

The Destiny project was undertaken in conjunction with the UK Department of Trade and Industry (DTI) and several manufacturing companies. The project was part of the DTI Manufacturing, Organisation, People and Systems (MOPS) Programme [1]. Destiny has been taken into Europe with AAMECO, another DTI funded project. Currently the project team is continuing work on the prototypes developed during the Destiny project, to produce refined software products and workbooks to deliver the techniques to the manufacturing industry.

In many companies people are constrained by organisational structures, dominated by traditional hierarchies and functional divisions, Peters (1989) [2]. If the maximum benefit is to be obtained from the implementation of new technologies, particularly in manufacturing, there is a need to move towards flatter organisational structures. Such structures need to have increased emphasis on peer group working and cooperation, making relationships and interpersonal skills of paramount importance.

An effective manager must be able to embrace and control change. People are the key to the effective management of change - how they are organised and motivated, and how they perceive relationships and opportunities. Destiny is an approach that uses a diagnostic PC tool to help managers understand how their companies really work, based on collecting the perceptions of significant employees and on modelling interpersonal/intergroup relationships. Destiny thus enhances the prospects for appropriate change, and decreases the chances of implementing costly changes that may not work well by offering the opportunity to experiment with organisational change on paper first, rather than with people.

Destiny assists in:
- <u>analysing</u> organisational structures and the way they work
- <u>identifying</u> interpersonal and peer group working relationships and problems
- <u>focusing</u> management attention on key organisational issues
- <u>modelling and refining</u> new organisational structures before implementation.

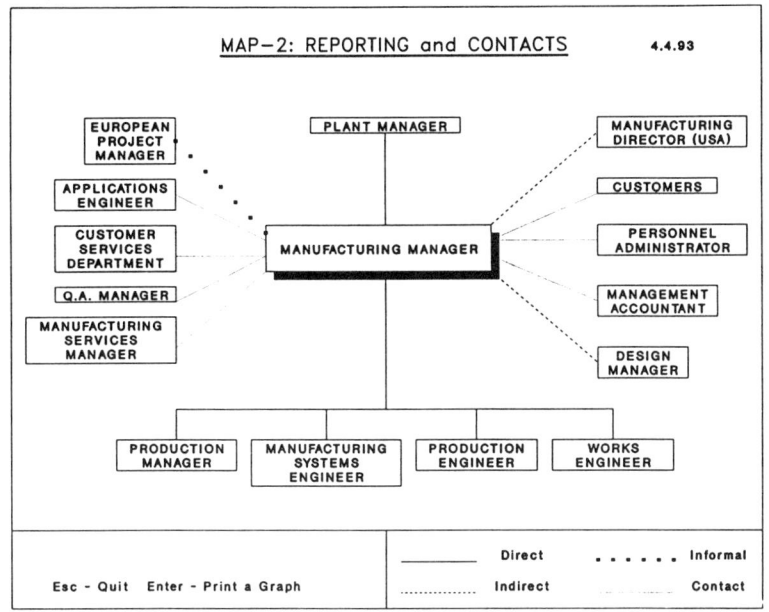

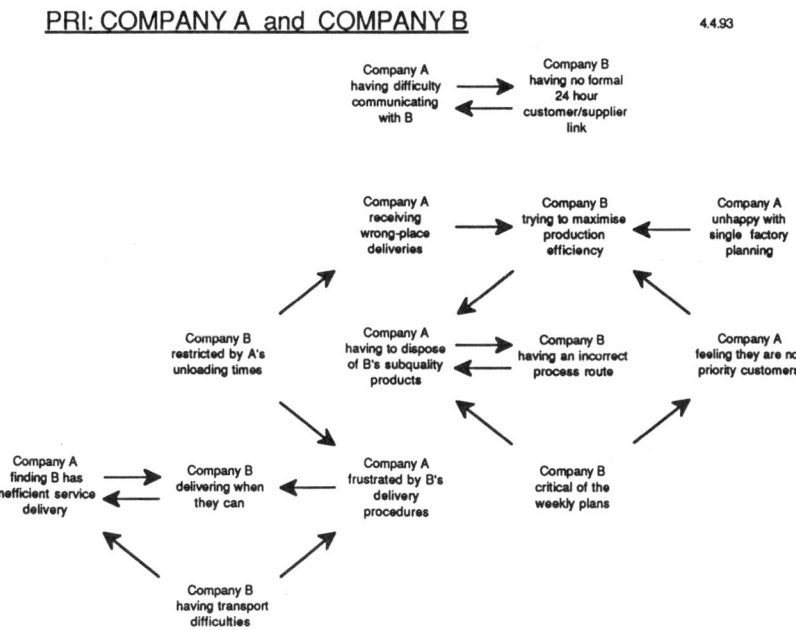

Figure 1

Examples of Destiny Charts and Graphs

2 Techniques

Destiny brings together modified versions of two existing and proven generic techniques, strengthens their application to manufacturing industry and converts them into an integrated PC tool. The techniques employed do not impose a doctrinaire solution but, instead, recognise the strengths and weaknesses of the individuals and groups already within the company. The sources of these two techniques are OPD Management Consultants Ltd and Dr David Mulhall, a clinical psychologist.

OPD Management Consultants Ltd have a proven paper technique for identifying and defining management roles and performing a preliminary analysis of the effectiveness of a company's organisational structure. This technique allows identification of general areas of potential difficulty and leads directly into Dr Mulhall's Professional Relationship Index (PRI).

Dr David Mulhall has developed a technique over a number of years (PRI) for providing a detailed analysis of specific problem areas. It produces a graphical representation of the dynamics of the complex inter-relationships involved, and highlights where they are, and are not working.

Destiny integrates these two techniques into a software tool for use by Human Resource Development (HRD) professionals, subsequently referred to as management consultants. The routine tasks are undertaken by a facilitator, who uses the Destiny software to collect information from interviews. This data is processed and given to the consultant as a series of annotated organisational charts, individual relationship charts, tables and natural language descriptions providing a global view of the section of concern. The consultant does further analysis, applies PRI to explore local issues which merit further investigation and highlights issues for consideration by the company prior to exploring potential solutions and modelling alternatives.

Details of the application of these techniques is given in [3] and [4].

3 Results

Examples of the charts and graphs produced by Destiny are given in figure 1.

Destiny has been used in several companies and the results clearly show the advantage provided by the embodiment of the techniques into software. These tests also highlighted areas for improvement and for further computer assisted analysis.

4 Conclusions and Further Work

The benefits of incorporating the OPD and PRI methodologies in software are as follows:
-       two complex techniques involving years of management consultancy experience and advanced computational psychometrics are disseminated to other management consultants including
    -       the business analysis contained in the OPD questions
    -       the representation of information through charts etc
    -       the psychometric analysis
-       the complexities of the methodologies are "hidden" by the software so that they can be adopted with minimal training

- the exhaustive analysis of data is performed to an extent well beyond the capability of the paper methodology; for example Destiny is able to compare all views (planned, actual and perceived) of organisational structures and identify classes of opinion
- human error is eliminated from the complicated statistical procedures required for the analysis of questionnaire data
- the considerable drudgery and time needed for the manual processing of statistics and drawing charts is eliminated.

The prototype software is being refined by redesigning sections to allow for greater flexibility in adding questions and for further analysis of answers. Intelligent processing will follow to provide more complex analysis. The possible extension of the techniques to cover a wider area of organisation issues is being investigated with the management consultants.

As a software tool Destiny can be used by any competent management consultant with a minimum of training for the modelling and refinement of organisational change to improve the chance of "getting it right first time", at a cost which is low enough to make it a realistic possibility. The relative ease and cheapness of applying these sophisticated techniques means that for the first time organisations can monitor their health on a regular basis. Destiny can support an information base about the organisation which can be built up over time and interrogated when necessary. The techniques employed do not impose a doctrinaire solution but, instead, illuminate the strengths and weaknesses of the individuals and groupings already within the company. The important decisions about how to solve the problems identified by these techniques remain the province of the managers and directors of the organisation. Destiny simply clarifies the problems.

The benefits of applying Destiny for companies contemplating change, or seeking a better understanding of their present organisation, are substantial and include:
- identifying areas which would benefit from change
- enhancing in-house management capability to effect change
- minimising reliance on external consultancy
- increasing employee loyalty and productivity
- improving overall management effectiveness and company performance
- exploring the best use of people to maximise returns from investment in advanced manufacturing technologies.

References

[1]   DTI, Advanced Manufacturing Research and Development:Summaries of DTI funded projects. Management of Advanced Manufacturing Branch, Department of Trade and Industry, 1993.

[2]   T Peters, Thriving on Chaos. Pan, London, 1989.

[3]   P D Pearce, A P Jagodzinski, K Wittamore, D Mulhall, D V Clarke, D M Lewis, P M Shepherd, & V A Lovitsky, Diagnosing Organisational Structures by Destiny. Proceedings of the Management of Technology Conference, Miami, February 1994.

[4]   P D Pearce, A P Jagodzinski, K Wittamore, D Mulhall, D V Clarke, D M Lewis, P M Shepherd, & V A Lovitsky, Destiny - Understanding How People Really Work in Manufacturing Organisations, IEE Colloquium/Workshop on "Management of Change", Digest No: 1994/056, 2 March 1994.

Advances in Agile Manufacturing
P.T. Kidd and W. Karwowski (Eds.)
IOS Press, 1994

Tools to help SMEs develop Skills in Strategy Formulation

G.Frizelle, M. Gregory *(Cambridge University)*

J. Harris *(Kewill Group Consultancy Services)*, G Ridland *(Chilton)*

Abstract. This paper describes the development of a methodology to help SMEs evolve appropriate manufacturing strategies and better operational control It records the results of a series of tests of the methodology involving senior management teams in manufacturing companies.

1. Objective

The objective of this project, called CIM2000, is to provide tools, based on existing research and leading practice which will help SMEs in the process of strategy development while building in-house management skills.

Small to medium sized companies (SMEs) tend to be unfamiliar with structured approaches to decision making and unaware of academic research that could be of value to them. This has been aggravated by lack of tools to make such work more accessible. However developments in computer technology have opened up new opportunities by making the collection, structuring and review of data much more affordable and user-friendly.

One area is the provision of advice on the development of strategy for improved operational control. Poor formulation can result in lengthy and expensive projects that are not germane to the company's needs. A consortium supported by DTI is involved in the development of new tools to meet this need.

2. Framework

2.1 A framework for developing the tools

The project involves the embodiment of existing research and leading practice in a commercial decision support system. The tools are developed within a collaborative framework to provide both the academic input and the means of developing a commercially robust product proven in the field. A consortium led by Kewill Group Consultancy Services (KGCS) includes Cambridge University, Strathclyde University and two industrial partners. One, Chilton is a private company engaged in textile production in the Marks and Spencer supply chain. The other, David Brown Special Products is a specialist gear manufacturer within the David Brown Group.

The Manufacturing Systems Research Group at Cambridge has a particular interest in the development of manufacturing strategy and its implications for various areas of manufacturing management, including the implementation of improved operational control. The lead partner, KCGS contributes its experience of leading joint academic/industrial consortia and as a leading supplier of operational control systems to make the deliverables available to the SME community. The involvement of industrial partners at every stage of the project ensures that the deliverables are both robust and usable.

The development phase included the preparation of a new systematic process, in the form of a paper-based Methodology. In its final version this incorporates some of the features of a Workbook, while offering a blueprint for the computer based tools that will subsequently be produced. It is based on an international review of relevant research.

This draft paper-based Methodology was tested and worked on by the other members of the consortium. The industrial collaborators, in particular, were able to validate the proposed routines using data from their own companies.

2.2 A framework for validation

Once the paper-based Methodology was felt to be relatively stable it was subjected to a programme of testing. Its dual function as Workbook and programme specification made it a rather cumbersome document. Thus field trials required a facilitator, who was familiar with the detail of the document.

Testing was done in two parts. The first was to conduct a formal exercise with the two industrial collaborators. To preserve impartiality, both trials included people with no prior exposure to the project. Then six additional companies were selected from the SME community, so that the tests could be extended to a variety of companies in distinct manufacturing sectors.

All of the tests followed the same format. A group of directors and senior managers met for a day. The collaborators supplied a team of three. One acted as facilitator to lead the Group through the paper methodology exercise. One helped with the administration of the exercise and did some of the supporting calculations. The third acted as observer and was given an observation sheet to complete.

3. Method

3.1 The paper-based Methodology

The immediate academic antecedents of the Paper Methodology were the work of Platts and Gregory [1], Richardson et al [2] and Frizelle [3]. However much of that research had its roots in the work of Skinner [4]. His emphasis on focus and consistency were made the basis upon which the methodology was constructed.

Its aim is to help management identify a number of Success Factors that are critical to the success of their particular business. It first focuses on a specific and homogeneous area of the company, called a Strategic Business Unit (SBU). It then asks management to specify their financial goal for the SBU and their preferred strategy for attaining it. These are then compared to templates drawn from the work of Richardson et al.. Consistency of purpose is sought between management's own goals and those suggested by the templates. Once a priority list has been agreed the Critical Success Factors (CSFs) are identified, and these provide a focus for the management team. The final stage is to suggest what is best practice in the agreed areas.

Two additional and important consistency tests are carried out to check for continuity of purpose within the management structure. The first queries if the future strategy take account of the way the company attained its present position. The second looks for congruence between company and manufacturing goals. This latter is a key link to helping operational management.

3.2 Structuring the tests

The methodology itself determined the test structure. It required the presence of the senior management team which reinforced consensus amongst team. Once the area of the business (the SBU) has been determined, the two major exercises are eliciting the success factors and comparing the SBU to the Richardson et al templates. Both require formalised question sets to be answered, which are quite lengthy. The earlier tests used an open forum approach, with someone chairing the session. He/she usually recorded the conclusions of the group. However it was found that peer pressure tended to make for conformity, particularly when preceded by extended discussion. Later exercises broke the attendees into smaller working groups, to minimise this effect.

The final stage of the exercise involved prioritising the list of CSFs. The observer then asked the participants to review the effectiveness of the exercise, and a questionnaire was left for them to complete.

4. Data Sources

The data required for the methodology is drawn from within the target enterprise. However, the two principal inputs to the structure of the methodology are existing literature and trials carried out in the eight companies.

Literature from within the consortium furnished two Workbooks covering strategy development and control system selection. However it was necessary to bridge a gap between them. The work of Richardson *et al* proposes six possible strategies a company can follow, called Corporate Missions by the authors. They also suggest four manufacturing types, called Manufacturing Tasks, which relate back to the Corporate Missions. Finally they list thirteen success factors whose priorities are determined by the Corporate Mission. This framework provides an essential link in the methodology.

The second data source were the tests themselves, both the information provided by the companies within the context of the Paper Methodology, and the observations from the observer's questionnaire.

5. Results

In all the tests the management team accepted the findings of the study and agreed the validity of the selected Critical Success Factors. Moreover the exercise made management more aware of the constraints that becoming more focused placed on their actions. In two cases the team felt they had been given a new perspective on their business.

However there were two other useful outcomes:

> -the methodology was seen as helpful in building cohesion amongst the team members based upon a common perception of the whole business
> -greatly improved prospects for implementation of new systems given a clearer view of objectives, approaches and interdependencies.

[1] K.W. Platts and M.J.Gregory (1990) Manufacturing Audit in the Process of Strategy Formulation. Int J.Ops and Prod Management Vol 10 No 9 pp 5 - 26.

[2] P. R. Richardson, A. J. Taylor and J. R. M. Gordon (1985). A Strategic Approach to Evaluating Performance. Interfaces. Vol 15 no.6 pp 15-27.

[3] G.D.M.Frizelle (1991) Deriving a Methodology for Implementing CAPM Systems Int J.Ops and Prod Management Vol 1 No 7 pp 5 - 26.

[4] W.Skinner.(1985) Manufacturing: the Formidable Competitive Weapon. Wiley New York.

Advances in Agile Manufacturing
P.T. Kidd and W. Karwowski (Eds.)
IOS Press, 1994

The Implementation of Process Innovation in Small Manufacturing Firms

Sotiris A. PAPANTONOPOULOS
Department of Engineering Management

Nicholas S. VONORTAS
Center for International Science and Technology Policy
and Department of Economics

and

Lan XUE
Department of Engineering Management
The George Washington University, Washington, DC 20052, USA

Abstract. Small manufacturing firms often lack the recourses to access, adopt, and successfully implement Computer-Integrated Manufacturing technologies. A study is being conducted to investigate the technical, economic, and organizational factors that affect the implementation of Computer-Integrated Manufacturing technologies in small and medium sized enterprises. The study combines the variable and process approaches in implementation studies by identifying the important variables and processes (social and temporal) affecting the success of implementation and examining how these variables and processes change over the process of implementation. The general framework of the study is outlined.

1. Process Innovation in Small Manufacturing Firms

Computer-Integrated Manufacturing (CIM) technologies are thought to be vital for U.S. manufacturing industries to regain their competitiveness in the 1990s and beyond. Yet, U.S. firms have been slower than their competitors in other industrialized countries in adopting CIM techniques. Moreover, even where such manufacturing techniques have been adopted, the results have not always been undoubtedly positive [1,2].

There are reasons to believe that the rate of adoption of new efficient manufacturing techniques may be even slower in the case of small and medium sized enterprises (SMEs) [3]. SMEs (355,000 of them in the U.S.) "...frequently lack expertise, time, money, and support to upgrade their current manufacturing operations, introduce new technologies and methods, implement better quality control, and improve workforce training" [4]. While

CIM technologies have the potential of providing significant productivity and quality improvements, they also require relatively steep capital investment, including significant start-up cost and learning costs, that smaller firms may not be able to afford.

On the other hand, it seems to be a well established fact that the primary need of most small manufacturers is for proven rather than very advanced automated technologies [5]. In addition to the aforementioned need for capital resources and well trained personnel, state-of-the-art technologies must generally be modified and developed further before they can be utilized productively to their full potential. That is, the risks and uncertainties involved may be too high for small manufacturers to handle on their own. Existing studies have shown that proven technologies have the advantage that they are already well-tested and they are readily procured, operated, and maintained. Such studies have also sounded small manufacturers out in that proven computer-based technologies such as computer numerically controlled machines and computer-aided design would be expected to solve these firms' priority problems [5].

Especially in mature industries with relatively poor prospects for sales growth, one would expect the small business manager to consider a significant capital expenditure in order to shift to a new manufacturing process technology to lie beyond the realm of medium term strategic choices [3]. The greater risks and uncertainties involved will probably be unacceptable in the short run, at least, and technology will tend to be considered more or less fixed. In its effort to increase productivity, management will thus be avert to significant fixed (sunk) costs; rather it will be likely to focus its attention on gaining greater control over variable production costs in order to make more efficient use of existing equipment and labor.

In light of the importance of SMEs for the manufacturing sectors of the U.S. and other industrialized countries and the difficulty of such firms in accessing, acquiring, and adapting advanced manufacturing technologies, a study is being conducted to investigate the technical, economic, and organizational factors that affect the implementation of CIM technologies in SMEs. The CIM technology at focus is numerically controlled (NC) machine tools and its advanced version, computer-numerically controlled (CNC) machine tools.

2. The Implementation Study

The study makes a clear differentiation between the decisions involved in adopting some new technology and in implementing the technology. The former relate to strategic and financial considerations. The latter are thought to start where the former end and relate to the factors that bring the new system (the acquisition of which has already been decided) up and running. The latter decisions, involved in implementing the technology, are the focus of this study.

2.1. A Classification of Implementation Studies

The implementation of process innovation has received increasing attention in recent literature. Implementation studies can be classified into two categories, the variance approach and the process approach [6].

The main objectives of the variance approach are to identify variables or factors affecting the success of implementation and to explore the causal relationships between these variables and measures of implementation success. These variables encompass technical/production, organizational/managerial, and economic/strategic aspects of the implementation process. If the variance approach is concerned with predicting levels of outcome from levels of predictor variables, the process approach is concerned with explaining how outcomes develop over time. A comparison of the variance and process approaches developed by Markus and Robey [7] (Table 1).

Table 1. A comparison of the variance and process approaches in implementation studies. Adapted from [7].

	VARIANCE APPROACH	PROCESS APPROACH
ROLE OF TIME	Static	Longitudinal
DEFINITION	The cause is necessary and sufficient for the outcome	Causation consists of necessary conditions in sequence; chance and random events play a role
ASSUMPTIONS	Outcome will invariably occur when necessary and sufficient conditions are present	Outcomes may not occur (even when conditions are present)
ELEMENTS	Variables	Discrete outcomes
LOGICAL FORM	If X, then Y; if more X, then more Y	If not X, then not Y; cannot be extended to "more X" or "more Y"

The study has combined the variable and process approaches by identifying what are the important issues/variables and important processes (social and temporal) and examining how these variables change over the process of implementation.

2.2. Outline of the Study

The general framework used to this end is provided in a theoretical paper by Dean et al. [8]. The premise of that work is that the adoption of advanced manufacturing technology does not ensure successful implementation. There is a real need to differentiate between the decisions leading to technology adoption and, after the funds have been committed, those ensuring its successful implementation.

According to Dean et al. [8], managers must consider three types of objectives. First, the *technical* objective - the sine qua non of effective implementation - requires that the system meets the expected technical performance requirements of the manufacturing process in question. Second, the *economic* objective would generally require that the financial position of the firm improves, even after considering the usually steep initial (fixed) investment cost. Third, the *political* objectives which can be both inter-organizational and intra-organizational. The latter require that the new system satisfies both its champions in the firm and its users in terms of achieving their respective goals and enhancing their organizational status. Successful implementation is defined as meeting all three objectives.

Implementation is looked at as a process, the relative importance of various implementation factors evolving as a company acquires more experience with the new technology. In addition, as the company's experience grows, the attention of the decision makers may be gradually shifting in the direction of factors in line with the true potential of the new technology (e.g., strive to achieve more flexibility, decrease the introduction time for new products, or acquire longer term strategic advantages) rather than the short-term tangible gains chased by beginners.

The study has borrowed the above general framework in order to investigate the implementation of CIM technologies in SMEs. Success of implementation was therefore defined as meeting all three objectives above though the third objective is interpreted in somewhat narrower terms (intra-organizational) to refer to human factors and managerial issues. Three sets of variables were chosen from the literature in order to describe the equivalent objectives.

Following the formulation of the process and the associated final objectives of the implementation of CIM technologies in SMEs, a thorough literature review has been conducted in the areas of industrial engineering, innovation and organizational change, and the economics of technological change. The literature review has resulted in a series of working hypotheses with respect to the factors that may influence the implementation of CIM technologies in SMEs.

These hypotheses form the basis for interview instruments used in case analyses. Case analyses are currently being conducted, involving data collection through plant visits in SMEs in the machine tool industry in the greater Washington metropolitan area. Participating firms were selected from the database of the National Tooling & Machining Association. The collected information will be analyzed, juxtaposed with the results in the literature, and tentative conclusions will be drawn.

References

[1] T.W. Garsombke and D.J. Garsombke, Strategic Implications Facing Small Manufacturers: The Linkage Between Robotization, Computerization, Automation and Performance, *Journal of Small Business Management* **27** (4) (1989) 34-44.
[2] D. M. Schroeder *et al.*, New Technology and the Small Manufacturer: Panacea or Plague? *Journal of Small Business Management* **27** (3) (1989) 1-10.
[3] M.R. Kelley and H. Brooks, External Learning Opportunities and the Diffusion of Process Innovations to Small Firms: The Case of Programmable Automation, *Technological Forecasting and Social Change* **39** (1/2) (1991) 103-125.
[4] P. Shapira *et al.*, Federal-State Collaboration in Industrial Modernization. School of Public Policy, Georgia Institute of Technology, Atlanta, Georgia, 1992.
[5] U.S. General Accounting Office, Technology Transfer: Federal Efforts to Enhance the Competitiveness of Small Manufacturers. GAO/RCED-92-30. Washington, DC, 1991.
[6] L.B. Mohr, Explaining Organizational Behavior. Jossey-Bass, San Francisco, California, 1982.
[7] M.L. Markus and D. Robey, Information Technology and Organizational Change: Causal Structure in Theory and Research, *Management Science* **34** (5) (1988) 583-598.
[8] J.W. Dean *et al.*, Technical, Economic, and Political Factors in Advanced Manufacturing Technology Implementation, *Journal of Engineering and Technology Management* **7** (1990) 129-144.

Advances in Agile Manufacturing
P.T. Kidd and W. Karwowski (Eds.)
IOS Press, 1994

Instruments of Psychological Work Analysis as an Attempt to Reduce Gaps Between Work Analysis, Evaluation and Design

Wolfgang G. Weber & Martina Zölch
Work and Organizational Psychology Unit,
Swiss Federal Institute of Technology (ETH), CH-8092 Zurich

Abstract. Instruments containing clear procedures for psychological work analysis and criteria for job design are pre-requisites for joint-optimization in advanced manufacturing systems. Unfortunately job analysis instruments often produce gaps between work analysis, evaluation and design. Reasons for this are discussed and an task analysis instrument which tries to bridge these gaps will be introduced. A guideline resulting from the application of this analysis instrument (and others) will be presented that supports the development of design proposals for work at CNC-machine tools.

1. Gaps between the Stages of Analysis, Evaluation and Design of Work

The development and realization of "work-oriented" design concepts (with emphasis on human skills and needs [1, 2]) requires psychological models and instruments which are not "technology oriented". Instruments containing clear procedures for psychological work analysis and criteria for job design therefore form an important component of joint-optimization in advanced manufacturing systems. However, such instruments must be able to meet specific demands to be effective in practice. Is this not the case, the instruments only suceed in producing so-called "gaps" between the analysis, evaluation and design stages of a project. This may be caused by their theoretical and methodical orientation and therefore only be of limited help in realizing changes in jobs. Thus, the first of these *gaps* yawns *between the theoretical models and their operationalization.*

Every method for task analysis whose purpose is to evaluate how humane work activities are, contains assumptions about the psychological processes of working people. But often these assumptions are not out-lined explicitly, let alone theoretically specified. Moreover, psychological demands which promote personal development (in the sense of required competencies and qualifications), as well as physical demands are often implicitly (mis)understood per se as work load [3]. Based on these design proposals, and the psychological models backing them decisions are made with far-reaching consequences for workers. Although there is general consensus regarding certain job characteristics such as complexity, autonomy and variety the call for more conceptual clarity is well founded [4, 5].

In addition to the theoretical aspects, one must also deal with the problem of the level at which task characteristics are defined and criteria of job design are to be applied. Assuming that there is consensus that tasks are characterized as a set of conditions by which the task performer is directly influenced, there is "the question of what tasks are and how they should be measured" [4, p. 211], even without mentioning discussions concerning the distinction between task and role, objective or subjective job characteristics and macro or micro level of analysis (cf. [5, 6]). There is often little or no guidance offered to the users of such instruments as to whether the application level of the criteria is the task, parts of the task, or the whole job [7].

Design research also often assumes that tasks are invariant across different types of jobs and job categories. This results in a universal claim on a broad field of application. Such analysis instruments (e.g. JDS, PAQ) use very general analysis and evaluation dimensions. They are useful for education research and sociological investigations, but they offer little support for design. In addition, the appropriateness of task analysis with general criteria instead of specific operationalizations for particular jobs should be questioned. Because of context specific demands, a criteria like 'job variety' should be defined for jobs in the office sector differently e.g. dealing with different types of information than in the production sector, where job variety can mean work with different materials and tooling methods. Global measures which are not specified for particular jobs make it inherently difficult to draw conclusions for concrete work activities and hence, concrete design proposals. For that reason Clegg legitimately asks [5, p. 139], "what predictions would one make comparing engine testers and storemen?" Therefore task analysis instruments must be oriented much stronger on task taxonomies [4].

The questions of what, why and how analyzed work tasks should be changed are often insufficiently answered; there may be only poor indications of what work task characteristics have been negatively evaluated and need improvement. Thus, a second *gap, between the chosen presentation of analysis results and its corresponding usefulness for the derivation of design proposals arises.*

Task analysis instruments often confront the task designer with an evaluation profile, which consists of a multitude of rather abstract dimensions or indices. These often redundant dimensions refer to very general activities like "speaking", "signal registration", and so on which are again usually rated in abstract evaluation categories. Thus, tasks elements which psychologically belong together for performing a task are isolated and removed from the concrete work activity and work units. Results of these tasks analyses cannot be fully related back to the concrete sequences of operations. This makes it considerably difficult to identify sub-tasks which can be servable to job enrichment.

Undoubtedly, inadequate presentation forms of analysis results are also due to the fact that work analysis questionnaires do not take the execution of the work task into consideration. The advantages of analysis in the setting of the work activity by means of interviewing and observing the worker carrying out the work task are not utilized and the sequences of operations not systematically recorded. Valuable information which would facilitate the design of work is lost when only the questionnaire method is employed.

2. Development of Design Proposals and Guidelines Supported by Theory-Based Task Analysis Instruments

To overcome the outlined problems, various work analysis instruments for the production and clerical sector have been developed and empirically approved in the German speaking work psychology community (e.g. [8, 9]). In the following, one work analysis instrument for tasks in the production sector [8, 10] and the derived guideline for the design of work at CNC-machine tools [11] will be presented. The way in which these instruments try to reduce the gaps between theoretical basis, work analysis and appropriate design steps will be discussed.

The VERA-instrument has been developed from the psychological theory of action regulation. It can therefore be called a *theory-based* analysis-instrument. By means of VERA the "regulation requirements" of industrial tasks can be identified. Meaningful regulation requirements consist of a high amount of planning and decision making components within a given work task. In such a context, existing qualifications can be expanded and new skills can be acquired. This instrument is essentially based on an elaborated 5-level-model of action regulation and a corresponding flow chart which enables the identification of regulation requirements of a work task. It is necessary to identify the parts of a task (sub-tasks) which comprise a high amount of planning and decision making requirements. The work task then may be assigned to one of five levels (in increments of half levels). The higher the level of a concrete work task, the stronger the personality-promoting potential is with respect to a particular job characteristic.

The unit of this type of analysis instrument is the *action related work task* and the working conditions which influence the operator directly during the working process. A work task is defined as *all operations performed that contribute toward the task goal* including utilization of tools, machines, information, co-operation partners, work pieces as well as cognitive processes that help to achieve the end result.

The analysis is usually carried out by trained examiners in the form of an *observation interview*. This means that the work analyst acquires the information for answering the questions of the instrument in an open dialogue with a proficient worker asking him or her in detail about his or her work while observing the work activity for several hours. The prerequisite for the later evaluation is the observation and recording of the working steps and conditions of the working activity in detail.

Because it is the work task that is the subject of analysis, the analysis procedure is considered to be *condition related*. This means that the regulation requirements are evaluated independent from a particular individual. It is not the working persons who are evaluated (e.g. their individual performance or working style), but the work tasks and the environment as conditions for their working activities. To identify the personality promoting potential (as well as the impairing potential of work-activities) independent of the individual workers is important because in this way *preventive* and *prospective work design* can be promoted [1, 2].

At this point it has become apparent that this type of analysis is not a short, abstract rating procedure. The work analyst has very carefully to apply the theoretical model to the concrete task and substantiate his/her decision for a certain regulation level. Lastly, the VERA-instrument supports the development of design proposals to increase the regulation requirements by providing redesign strategies for the specific analyzed working task and thereby reduce the gap between analysis, evaluation and design.

The VERA-instrument, in combination with other methods, was applied to *150 work tasks at CNC-machine tools* in Swiss and German enterprises. These tasks were analyzed in their organizational contexts, including the departments NC-programming and production planning and evaluated. The goal of this investigation was to identify the personality promoting as well as the impairing potential of various CNC-work tasks. Evaluating the VERA task protocols - obtained by the outlined method - a pool of sub-tasks could be extracted which are generally suited for job enrichment for CNC-operators. These are NC-programming, process planning for several machining techniques, several functions of job scheduling, preparation of measurement plans and programs and carrying out relative complex maintenance tasks. Jobs which primarily consisted of setting up a machine tool as well as feeding were not evaluated as personality promoting. The results show that CNC working structures have greatly differing potentials in terms of mental requirements, depending on the extent that the act of planning (e.g. NC-programming) is separated from the act of execution on the shop floor. In this way, *8 types of CNC-work structures* could be classified and *18 types of CNC work tasks* (e.g. universal skilled work, specialized skilled work, fixed production work) could be composed which differ with respect to the amount of planning and decision making requirements as well as their need for design. Based on this classification a guideline titled "Work at CNC-Machine Tools" (book and interactive software-version) has been developed for use in industrial settings [11].

By means of a question flow chart, in which characteristical functions of CNC work tasks are described, the user of the guideline first identifies the types of manufacturing structure and work task which correspond to the CNC-work task he/she intends to design. The types of work at CNC-machine-tools are distinguished systematically to what extent:

*   challenging sub-tasks such as shop floor programming or maintenance are required,
*   programming is required for one or more machining techniques,
*   the operator is responsible for production scheduling or shop floor control,
*   managing shop floor control is carried out by a semi-autonomous group. This means that the management of tools, for example, is in the hands of the operators, that the responsibilities for quality control are shared, and that there is a shared arrangement of job rotation.

Once the user of the guideline has determined the type of CNC work task he/she intends to design, he/she is referred to specific chapters containing a work psychological evaluation of the different types of CNC work tasks. In a further chapter, respective design proposals are given and illustrated by means of practical examples. Developed from the underlying classification, the design proposals contain suggestions for the conversion of limited into ambitious CNC work tasks. The scope of planning and decision making requirements of CNC work tasks could be enriched by integrating sub-tasks with a high VERA level into the original task definition. In this way, CNC-work-structures with low-skilled operator tasks can be changed step-by-step into work structures with highly skilled tasks. By means of such a theory based classification, the practical user can bridge the discussed gaps; the pool of personality promoting sub-tasks can be directly transferred into design proposals. Although extensive analyses were necessary to develop this useful guideline for work at CNC-machine tools, the user is spared the efforts of both familiarizing him/herself with the theoretical foundation, as well as for the extensive work analysis.

References

[1]   Ulich, E. (1994). Arbeitspsychologie (3. Auflage). Zürich: vdf / Stuttgart: Poeschel.

[2]   Ulich, E. (1989). Humanization of Work - Concepts and Cases. In J. Fallon, H.P. Pfister & J.B. Brebner (Eds.). *Advances in Industrial and Organizational Psychology* (S. 133-143). Amsterdam: North-Holland.

[3]   Weber, W.G. (1994). Psychologische Analyse und Bewertung computergestützter Facharbeit. München: Quintessenz.

[4]   Roberts, K.H. & Glick, W. (1981). The Job Characteristics Approach to Task Design: A Critical Review. *Journal of Applied Psychology, Vol. 66, No. 2*, 193-217.

[5]   Clegg, C.W. (1984). The derivation of job design. *Journal of Occupational Behaviour, Vol. 5*, 131-146.

[6]   Kidd, P.T. (1992). Interdisciplinary Design of Skill-Based Computer-Aided Technologies: Interfacing in Depth. *The International Journal of Human Factors in Manufacturing, Vol. 2 (3)*, 209-228.

[7]   Algera, J.A. (1990). The Job Characteristic Model of Work Motivation Revisted. In: U. Kleinbeck, H.-H. Quast, H. Thierry & H. Häcker (Eds.). *Work Motivation*. (S. 85-103). Hillsdale, New Jersey: Lawrence Erlbaum.

[8]   Volpert, W. & Oesterreich, R. (Hrsg.). (1991). VERA Version 2: Arbeitsanalyseverfahren zur Ermittlung von Planungs- und Denkanforderungen. Handbuch und Manual. (Forschungen zum Handeln in Arbeit und Alltag, Bd. 3, Hrsg. W. Volpert & R. Oesterreich). Berlin: Technische Universität.

[9]   Dunckel, H., Volpert, W., Zölch, M., Kreutner, U., Pleiss, C. & Hennes, H. (1993). Kontrastive Aufgabenanalyse im Büro. Schriftenreihe Mensch Technik Organisation (Hrsg. E. Ulich), Band 5 a & b. Zürich: vdf / Stuttgart: Teubner.

[10] Oesterreich, R. & Volpert, W. (1986). Task analysis for work design on the basis of action regulation theory. *Ergonomic and Industrial Democracy*, 7, 503 - 527.

[11] Weber, W.G., Oesterreich, R., Zölch, M. & Leder, L. (1994). Arbeit an CNC-Werkzeugmaschinen. Schriftenreihe Mensch Technik Organisation (Hrsg. E. Ulich), Band 6, Zürich: vdf/ Stuttgart: Teubner.

Advances in Agile Manufacturing
P.T. Kidd and W. Karwowski (Eds.)
IOS Press, 1994

Work-Oriented versus Technically-Oriented Manufacturing Systems: Methods and Results of a Case Study

O. Strohm, C. Kirsch, L. Leder, O. Pardo, P. Troxler & E. Ulich
Work and Organizational Psychology Unit, Swiss Federal Institute of Technology ETH,
CH-8092 Zurich

Abstract. The procedure and methods of the third phase of the research project GRIPS are presented. The goal of the investigation is the comparison of work-oriented and technically-oriented manufacturing systems. Selected results from one case study and a proposal for restructuring of this company are presented.

1. Work-Oriented versus Technically-Oriented Manufacturing Systems

Work-oriented design concepts - in accordance with the principles of sociotechnical system design - aim at the joint optimization of social and technical subsystems, of organization and technology, as well as the acquirement and utilization of workers´ skills and competence. Technically-oriented concepts, on the other hand, are those that primarily address the implementation of technical systems and their components, while organizational issues and human resources are neglected [1]. The implementation of computer aided integrated manufacturing systems affects four levels of the company: the enterprise, the organizational unit, the group and the individual. Based on this model, work-oriented manufacturing systems - as opposed to technically-oriented ones - are characterized by the features decentralization vs. centralization, functional integration vs. functional specialization, work in self-regulated groups vs. central control and qualified tasks vs. partialized tasks.

2. The Third Phase of the GRIPS Project (GRIPS III)

In the research project "GRIPS" (original "Gestaltung rechnerunterstützter integrierter Produktionssysteme") work psychological concepts for the design of computer aided integrated manufacturing systems are currently being developed and empirically examined. The GRIPS project is based on the assumption that work-oriented concepts - in contrast with technically-oriented concepts - contribute to humane working conditions as well as economic efficiency.The GRIPS project consists of three phases. The first phase (GRIPS I) was completed in the form of a written survey during 1990 in the machine building, electronic, metal-working and the processing industries throughout Switzerland (N=917). In the second project phase (GRIPS II), 60 companies were selected for detailed case studies with the aid of document analyses, company tours and interviews with experts of the companies.

In the on-going third phase (GRIPS III), matched pairs of work-oriented and technically-oriented companies on the basis of comparable manufacturing conditions and product range have been selected. They are studied in detail and juxtaposed with regard to features of e.g. sociotechnical design, job design, human-machine functional division and interaction, qualification structures, mental work load structures, employees' perceptions and redefinitions of the work situation as well as economic efficiency. A rigorous and approved analysis procedure is applied which incorporates e.g. document analyses, expert interviews, process-flow analysis, sociotechnical system analysis and task observations (see Table 1).

The investigation takes approximately 5-7 person weeks per case, depending on the size of the company and complexity of its production.

Table 1: The Steps of the Investigation in GRIPS III

1. Analyses of Manufacturing Structures at the Level of the Enterprise
Topics are the history of the organization, organizational goals, human resources, employee training and development, implementation and integration of computer aided manufacturing systems, quality management, etc.
Methods: Document analyses, expert interviews with general manager, production manager and department manager
2. Process-Flow-Analyses Along the Value-Add Chain
Analysis of the process-flow of 2-3 finished and typical orders
Methods: Document analyses, expert interviews with department managers and supervisors
3. Sociotechnical-Analyses in Selected Departements
The analyses concerns the input, transformation process, output, social and technical subsystem, technical-organizational design, obstructions, interruptions and fluctuations
Methods: Document analyses, expert interviews with department managers, supervisors
4. Objective Work-Analyses of Specified Key Positions
Objective analyses and evaluation of division of labor, content of tasks, human-machine functional division and interaction, etc.
Methods: Expert interviews with employees, observational interviews, task observations
5. Subjective Work-Analyses
Subjective data for the analysis and evaluation of the employees' perceptions and redefinitions of the work situation
Methods: Written survey of the employees
6. Strategies, Procedures and Milestones of the Sociotechnical History
Topics are aspired goals, project management, design of the process, employee participation, attained results, etc.
Methods: Document analyses, expert interviews with general manager, production manager, department managers, project managers

One key element of the investigations are the evaluations of sociotechnical systems and tasks. The evaluation of sociotechnical systems builds on ten criteria. Exeamples of essential criteria are:

1. The independence of the organizational unit
2. Polyvalence
3. Technical-organizational convergence
4. Self-regulation

The evaluation of regulation requirements of tasks is e.g. carried out with the instrument VERA (original „Verfahren zur Ermittlung von Regulationserfordernissen in der Arbeitstätigkeit") [2]. The VERA-Model distinguishes between five levels of regulation and 10 steps of evaluation. With this instrument, the qualification potential of tasks and the contribution of the taks to the development of the employees personality can be estimated.

2. Results of the Pilot Study

The GRIPS III procedure was applied in the context of a pilot study in a large company of the machine building industry. This was done as a company analysis and first step in a restructuring project. The goal of this project is the company-wide implementation of self-regulated work groups.

The analysis showed that the company is characterized by technically-oriented production structures with functionally specialized organizational units in the value-add chain. The sociotechnical evaluation of two exemplary work systems in manufacturing reflect this fact (see Figure 1).

Figure 1: Sociotechnical evaluations of two work systems in the manufacturing department

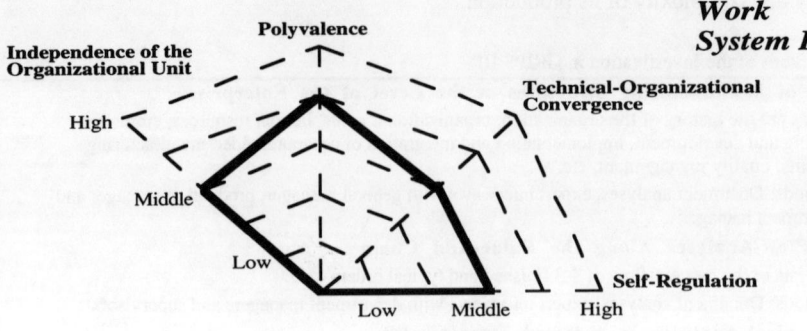

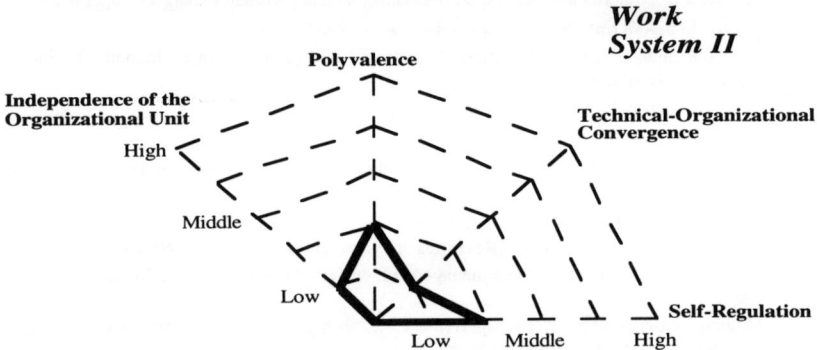

The independence of these work systems is reduced because e.g. the planning of the manufacturing, the production control and the NC-Programming is localized in central planning departments. The quality control is only partly integrated into the work systems. The independence of work system II is very low because the last manufacturing operations are carried out here before the assembly process starts and thus one of the main tasks of this department is the management of missing parts.

The regulation requirements, or VERA steps of tasks in the manufacturing department reflect the reduced independence, or partialized primary tasks of the different work systems (see Figure 2).

Figure 2: VERA steps of tasks in the manufacturing department (n=10)

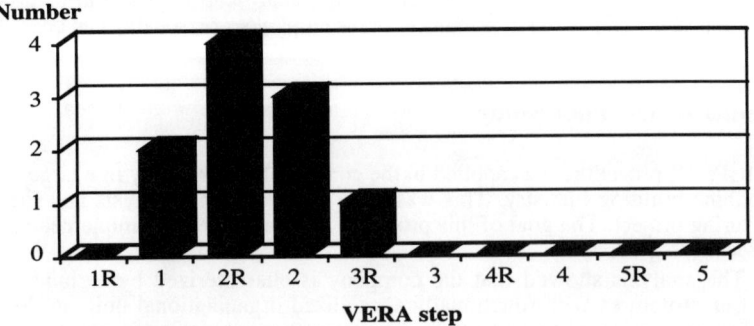

Most tasks contain elements of the planning of actions (step 2R and 2). However, the setting of goals (step 3R and 3), which is part of the operational planning or NC-

Programming is only required restrictively in one task. Thus, the tasks in the manufacturing department are characterized through incompleteness and reduced utilization of human resources.

3. Proposal for Restructuring

A central requirement for the design of self-regulated work groups are complete tasks with a task context within the organizational unit [1]. This requirement is not fullfilled in the existing functionally specialized production structures of the companay. The results of the whole companay analysis and the goal of introducing self-regulated work groups led to the proposal for restructuring in Figure 3.

Figure 3: Proposal for Restructuring [3]

In this structure overlapping production islands are suggested, with complete tasks and an internal task context. This production structure corresponds to functional integration on the level of the organizational unit as a requirement for the design of self-regulated work groups. The companys project team and the production manager has accepted this proposal to meet the goal of the project and as a requirement for human work conditions and economic efficiency. Currently, our proposal is being specified by the project team.

4. Conclusions

The experiences of this and a second pilot study, as well as the on-going investigations show that the procedure and methods used by GRIPS III is well-founded for the examination of work- vs. technically-oriented manufacturing systems and the research hypotheses. Restructuring processes, where organization before automation has validity, can also be started in a constructive way by this type of analysis.

References

[1]   Ulich, E., *Arbeitspsychologie.* Verlag der Fachvereine, Zürich /Poeschel, Stuttgart, 1994, 3rd edn.
[2]   Volpert, W., Oesterreich, R., Gablenz-Kolakovic, S., Krogoll, T., & Resch, M., *Verfahren zur Ermittlung von Regulationserfordernissen in der Arbeit (VERA).* Köln: TÜV Rheinland, 1983.
[3]   Strohm O., Troxler, P., & Ulich, E., *Vorschlag für die Restrukturierung eines Produktionsbetriebes.* Zürich: Institut für Arbeitspsychologie der ETH, 1994.

Advances in Agile Manufacturing
P.T. Kidd and W. Karwowski (Eds.)
IOS Press, 1994

KOMPASS
Complementary Analysis and Design of Production Tasks in Sociotechnical Systems

Steffen WEIK, Gudela GROTE & Martina ZÖLCH
Work and Organizational Psychology Unit, Swiss Federal Institute of Technology, (ETH) Zurich, Nelkenstr. 11, CH - 8092 ZURICH

Abstract. Based on the work psychological and especially the sociotechnical tradition a heuristic for the analysis and design of production tasks in highly automated systems will be outlined. By following a complementary approach, which explicitly pays attention to the question of function allocation between human and machine - as well as to the fact, that system design always has to mind and optimize the relation of human resources, technology, and organization - a set of 18 evaluation criteria will be proposed.

1. Introduction

Nowadays everybody - even orthodox engineers - seem to be more or less aware of the "ironies of automation", which do occur in all probability if only the classic aims of automation are followed, i.e. the replacement of human control, planning and problem solving by automatic devices and computers. The paradoxical conclusion by Bainbridge [1], "...that automated systems still are man-machine systems, for which both technical and human factors are important..." seems to be widely accepted. Strangely enough, according to Endsley [2], the practitioner who wants to improve the integration of human operator and technology seems to be left with "little but admonitions" to accomplish this task. What these practitioners are lacking in particular, is a systematic and criteria based support to handle the most basic design question in every human-machine system: Which functions (or tasks) have to be allocated to the human operator, to the machine or to both of them?

Allocation decisions are usually based on criteria, which are established and selected according to the *normative assumptions* of those who are in charge of the system design [3]. These normative assumptions concern in most cases the abilities of technology as well as the value and importance of human work in automated systems, which still is very often assessed either as cost factor, as risk factor, or in terms of quantitative efficiency compared to machine performance (economic, leftover or comparison allocation [4]). Already 30 years ago Jordan [5] pointed out, that none of these allocation strategies can really do justice to human features in a work system and that one should therefore turn to a *complementary* view of the relationship between human and machine. Such a view accommodates the fact that the capabilities of human and machine differ also in a *qualitative* manner. As a consequence they cannot replace but only *complement* each other.

In this paper we want to show how this concept of complementarity - which is cited quite often but still is rather poorly operationalized [6] - might be worked into a system design heuristic. By using the term heuristic we want stress the fact that we do not intend to develop a generally valid standard for the design of complementary human-machine systems, which we think is neither possible nor adequate, in particular not from a social scientist's point of view [7]. What we rather need are practical guidelines, heuristics or tools to support, moderate, and coordinate multidisciplinary design teams who try to develop different allocation scenarios in a participative manner [8] and give them the means to evaluate existing as well as planned allocation solutions by well-founded criteria (cf. section 3).

2. The objects and research design of KOMPASS

The design of work systems and (production-) tasks always takes place within the following three domains: *human resources, technology,* and *organization* [9,10]. In order to achieve a work oriented [11], holistic work design a number of approaches concerning especially the first two aspects have so far been proposed.

Design guidelines which cover the *organizational aspect* are nowadays well developed. This concerns mainly the discussion about the (semi-) autonomous group work and the sociotechnical systems approach [11,12,13] (cf. section 3, first level of analysis and design).

Criteria for the analysis, evaluation, and design of *humane, intrinsically motivating, and skill promoting working conditions* have been described in detail by the psychological action regulation and activity theories [11,14] (cf. section 3, second level).

As far as the *technological aspect* is concerned the design of human-machine systems is very often reduced to the design of optimal (i.e. user friendly) interfaces. Criteria for this kind of interface evaluation are well-known and well-tested. KOMPASS wants to surpass this rather ergonomic approach and concentrate more on the easily neglected question of how to allocate the actual *contents* of *complete* working tasks [15] to either human, machine or both. One of the main objectives of KOMPASS is to develop criteria which enable a design team to evaluate whether an already existing or planned human-machine system complies also in terms of its technical conditions with the demands of a complementary and jointly optimized [13] system design.

As already mentioned in the introduction the development of these criteria cover only one of the objects of the research project KOMPASS, whose aim is to provide industrial practitioners who are concerned with the introduction and adaptation of automated production processes, with a practicable and *participatory* tool, which can support *multidisciplinary design teams* in carrying out the following design steps:

(1) Identification of furthering and hindering preconditions for complementary design.
(2) Sociotechnical analysis and evaluation of a defined work system regarding work processes and tasks.
(3) Participatory development of different scenarios for the allocation of functions between human and machine but also between employees.
(4) Criteria guided evaluation of these scenarios.

This aim is basically followed in three phases, each involving different methods [8]:

(a) Development of an integrated set of criteria for function allocation, humane task design and sociotechnical design (see below).
(b) Framework for an organizational development approach to system design: Retrospective analyses of automation projects, looking at the automation philosophy and allocation strategies adopted, the degree of worker participation and the resulting organizational, task, and technical design. Interviews with system designers are carried out, in order to obtain information on currently used allocation strategies and the underlying rationale.
(c) Heuristic for the development and evaluation of design options: The results of the analyses in phase (a) and (b) are incorporated into a heuristic which will be tested for its usefulness in ongoing automation projects.

Currently, project activities concern mainly Phase (a). Based on a large body of literature covering all three aspects of complementary system design as well as on exploratory analyses of highly automated tasks, 18 criteria to support mainly the design steps (2) and (4) have so far been specified (cf. section 3). Newly developed criteria - especially those concerning the allocation of function - are now tested for their reliability and validity through *case studies* in different areas of automated production (metal and wood industries, manufacture of circuit boards). 30 double-analyses of different production tasks from a range of production types and levels of automation will be carried out.

3. Criteria for analysis and design of complementary productions tasks

First Level of Analysis and Design: *The Sociotechnical System* [adapted from 11,12,13]
1) *Completeness of realized work functions in the organizational unit* (Processing depth of manufactured products / feedback of goal conformity)
2) *Independence of the organizational unit* (Extent of dependence on other organizational units with respect to e.g. quality, supply, etc.)
3) *Task Interdependence* (Form of co-operation based on sequential, reciprocal, or pooled task interdependence)
4) *Polyvalence of the Operators* (Proportion of tasks the operators have competence and skills for, not necessarily responsibility)
5) *Autonomy of Work Groups* (Tasks which require joint decision-making and planning within the group)
6) *Boundary Regulation* (Supervisory tasks / Extent of required presence of the supervisor with the work group)

Second Level: *The Individual Work Tasks* [adapted from 11,14,16]
7) *Completeness* (Tasks with preparing, planning, executing, controlling and maintenance elements)
8) *Planning and Decision Making Requirements* (Tasks which require planning and decision making regarding content of work, equipment, work flow and results of work tasks)
9) *Variety* (Tasks whose execution encompasses dealing with different materials, tools, persons, types of products, procedures and interventions)
10) *Co-operation (Communication) Requirements* (Tasks whose execution implies or requires co-operation and communication)
11) *Autonomy* (Individual workers' possibilities of structuring their own work conditions, e.g. influence on the allocation of jobs, the time-management of jobs, the form of cooperation and organization)
12) *Opportunities for Learning and Personal Development* (Challenging tasks whose completion expands existing qualifications and requires new skills)
13) *Absence of hindrances* (Objective hindrances to the work which require additional effort, e.g. information which is not readily available or interruptions that are caused by malfunctions)

Third Level: *The Human-Machine System* [adapted from 6,17,18; see also 3]
14) *Coupling* (Closeness of coupling: Extent to which the technical system determines the activity within a human-machine system; extent to which the operator can influence the degree of coupling)
15) *Authority / Responsibility* (Relationship between operator and machine when it comes to questions of authority and responsibility / extent of decision competence over controlling the production process)
16) *Flexibility* (Flexible function allocation, i.e. either to the operator, the machine or both / possibility of changing these allocations in the course of time (dynamically) according to personal needs or abilities)
17) *Transparency / Proximity to the Process* (Possibility of comprehending the automated process, the interventions and the quality of the machining / possibility to get feedback through different sensory channels or organs)
18) *Technical Linkage (Integration / Dependency)* (Dependence of the observed technical system on other technical systems with respect to data transfer, e.g. requirements on immediate responses by the operator)

The operationalization of these 18 evaluation criteria has been worked into interview guidelines and are now tested for their practicability through exploratory case studies in several production companies. To get to know more about the usefulness of our concepts we are also planning workshops with representatives from industry and science (social scientist and engineers), to check whether our concepts are plausible to a greater audience.

4. Conclusions

The theoretical framework as well as some preliminary results concerning the development of evaluation criteria for the complementary analysis and design of automated manufacturing systems were presented.

Apart from the practicability, reliability and validity of the different criteria mentioned, we more and more get aware of the fact, that there are many unanswered questions concerning the relationships among these criteria. For instance the following questions might be looked at: Is flexible or even dynamic allocation of functions really a necessary and/or sufficient condition for complementary system design? How to deal with the discrepancy between coupling and proximity to the process? What are the advantages of tight coupling and the disadvantages of loose coupling? Further subjects of discussion might be: Are there any functions which never should be automated? And if there are such functions - how can they be determined? Are the need for control by the operator or the unpredictability and criticality of the function [6] good indicators? We hope that at least some of these questions will be answered by KOMPASS.

References

[1]   Bainbridge, L. (1983). Ironies of automation. In: Johannsen, G. & Rijnsdorp, J.E. (Eds.), Analysis, design and evaluation of man-machine systems. Proceedings of the IFAC / IFIP / IFORS Conference Baden-Baden 1982. Oxford: Pergamon Press, 129 - 135.

[2]   Endsley, M.R. (1993). The integration of humans and advanced manufacturing systems. *Journal of Design and Manufacturing, 3* (3), 177 - 187.

[3]   Weik, S. (in press). 'Complementary' instead of technological-centred system design - remarks on the problem of function allocation from a work psychological point of view. In: Wearn, Y. & Tauber, M. (Eds.), Task analysis for system design. Amsterdam: North Holland.

[4]   Bailey, R.W. (1989). Human performance engineering. London: Prentice-Hall International.

[5]   Jordan, N. (1963). Allocation of functions between man and machines in automated systems. *Journal of Applied Psychology, 47* (3), 161 - 165.

[6]   Clegg, C., Ravden, S., Corbett, M. & Johnson, G. (1989). Allocating functions in computer integrated manufacturing: a review and a new method. *Behaviour and Information Technology, 8* (3), 175 - 190.

[7]   Klein, L. (1993). On the collaboration between social scientists and engineers. In: Trist, E. & Murray, H. (Eds.), The social engagement of social science - a tavistock anthology. Volume II: The sociotechnical perspective. Philadelphia: University of Pennsilvania Press, 369 - 384.

[8]   Grote, G. (in press). A participatory approach to the complementary design of highly automated work systems. In: Bradley, G. & Hendrick, H.W. (Eds.), Human factors in organizational design and management - IV. Amsterdam: Elsevier.

[9]   Kidd, P.T. (1990). Organization, people and technology: Towards continuing improvement in manufacturing. In: Faria, L. & Van Puymbroeck, W. (Eds.), Computer integrated manufacturing. Proceedings of the sixth CIM-Europe annual conference. London: Springer, 387 - 398.

[10] Ulich, E. (1993). CIM - eine integrative Gestaltungsaufgabe im Spannungsfeld von Mensch, Technik und Organisation. In: Cyranek, G. & Ulich, E. (Hrsg.), CIM Herausforderung an Mensch, Technik und Organisation. Schriftenreihe Mensch Technik Organisation (Hrsg. E. Ulich), Band 1. Zürich: vdf / Stuttgart: Teubner, 29 - 43.

[11] Ulich, E. (1991). Arbeitspsychologie. Zürich: vdf / Stuttgart: Poeschel. 3. Auflage 1994.

[12] Emery, F.E. (1978). Characteristics of socio-technical systems. In: Emery, F. (Ed.), The emergence of a new paradigm of work. Canberra: Australian National University, 38 - 86.

[13] Susman, G.I. (1976). Autonomy at work. A sociotechnical analysis of participative management. New York: Praeger.

[14] Oesterreich, R. & Volpert, W. (1986). Task analysis for work design on the basis of action regulation theory. *Ergonomic and Industrial Democracy, 7,* 503 - 527.

[15] Ulich, E., Rauterberg, M., Moll, T., Greutmann, T. & Strohm, O. (1991). Task orientation and user-oriented dialog design. *International Journal of Human-Computer Interaction, 3* (2), 117 - 144.

[16] Dunckel, H., Volpert,W., Zölch, M., Kreutner, U., Pleiss C. & Hennes, H. (1993). Kontrastive Aufgabenanalyse im Büro. Schriftenreihe Mensch Technik Organisation (Hrsg. E. Ulich), Band 5 a & b. Zürich: vdf / Stuttgart: Teubner.

[17] Kraiss, K.F. (1989). Autoritäts- und Aufgabenverteilung Mensch-Rechner in Leitwarten. In: Gottlieb Daimler- & Karl Bunt-Stiftung (Hrsg.), 2. Internationales Kolloquium Leitwarten. Köln: TÜV Rheinland, 55 - 67.

[18] Corbett, J.M. (1985). Prospective work design of a human centered CNC-lathe. *Behaviour and Information Technology, 4* (3), 201 - 214.

Advances in Agile Manufacturing
P.T. Kidd and W. Karwowski (Eds.)
IOS Press, 1994

Communications Difficulties Within a Small Firm

Mr P R Barber, Researcher, University of Sunderland, Edinburgh Building Chester Rd, Sunderland SR1 3SD

Abstract. This paper is the product of a study within a small firm.
It identifies the communication problems within the small firm.
It proposes that the introduction of team work and a reduction in
scheduling are the solutions to these communication problems.

1. Introduction

This paper is the result of a study undertaken within a small to medium sized North East engineering firm with a turnover between £5 and £8 million a year. The purpose of the study was to investigate the administrative systems and human interfaces involved within the operation of a specific contract. The aim was to identify specific areas for improvement and to make recommendations on how these could be achieved.

The problems observed at this company are common to many small to medium sized companies. There is a commonly held view that small companies, because of their size, do not suffer from communication problems. Such problems were found to exist in the company under consideration.

It was considered that following a contract through the manufacturing process was the best way to determine the way in which the systems were operating. The method of investigation was by means of a series of weekly visits to the company over a period of 3 months.

2. Initial Research

The first task was to become acquainted with the contract. The contract was a typical one performed by this company for a foreign customer. As is common with these contracts the company's customer was only a subcontractor and the end user was two or three layers further down.

The second task was to identify the personnel and departments involved in the contract and specifically which department took overall responsibility for the contract.

The departments involved were:-

2.1 *The Contracts Department*, where a contracts engineer was given overall control of the contract.

2.2 *Production Services* where a draughtsman was placed in charge of preparing the work sheets, drawings, and the quality plan.

2.3 *Engineering Shop* where the supervisor / foreman was responsible for the manufacture of the parts for the contract.

2.4 *The Documentation Department* which was responsible for the collection and correlation of all the relevant paperwork.

2.5 *Quality Assurance* which was responsible for testing and quality control on the contract.

The day to day running of the contract was overseen by the Manufacturing Director. The majority of the functions outlined above were within his scope of control except the Contracts Department which was under the control of the Commercial Director.

After the initial investigation into what the contract was about and who was involved, the next stage was to follow the contract through the company. All contracts start with a "kick off" meeting which informs all the relevant people of the contract's content, start date, completion date etc. Following this the Contracts Department sort out the relevant contractual details and the drawings and specifications are sent to the drawing office.

The quality plan is then prepared by a draughtsman who examines the specific requirements of the contract. He then forwards it to the Contracts Department who forward it to the client, who in this case forwarded it to their client. This leads to the first problem. This is the long and convoluted lines of communication which exist, as there are often two or three layers of subcontractors between the company and the end user.

While the quality plan was being prepared and approved the drawings were prepared and work sheets drawn up. These were dispatched in lots to the engineering shop along with detailed schedules showing when each job is to be done. In this case while the drawings were being prepared a set of test components were manufactured and sent to an independent testing establishment for various tests specified by the end customer. The results from these tests were phoned through to the Drawing Office and confirmed in writing. This highlighted some of the internal communication problems within the company, as the Drawing Office and the Engineering Shop knew that the components had been approved but the Contracts Department was not aware of this.

When the test components were approved, work commenced on the contract according to the schedule produced, but this schedule was almost immediately out of date, as a quality problem appeared. This incident further highlighted the communication difficulties within the company as several different answers could be obtained as to what was being done about it and as to whether there was a problem at all. This situation also highlighted the climate of fear which exists within the firm as everybody tried to distance themselves from the problem. After the quality problem had been sorted out, production continued under new schedules, but these schedules in turn were broken as a machine break down created the need for more rescheduling.

When the work was completed and tested, the paperwork corresponding to the component was sent to the Documentation Department for correlation. Here the contract file increased in size considerably from the first days of the contract to its completion. The size of the contract file in the Documentation Department provides an indication of the progress of the contract.

3. Problems

From the above study the following problems were observed with the operational procedures and culture within the company.

3.1 A lack of communication between the Contracts Department and the rest of the departments involved within the contract.

3.2 Too much time devoted to detailed scheduling.

3.3 A lack of co-operation between people when things go wrong.

4. Proposed Solution

The proposed solution is to encourage team work within the company, so as to encourage people to pull together. The way to do this is to have progress meetings where everybody involved is present and information is shared. These meetings should be kept short, to a maximum length of half an hour.

The purpose of these meetings will be:-

4.1 To disseminate factual information so that all concerned are fully informed, and aware of contractual obligations and financial constraints.

4.2 To assess the current status of the contract.

4.3 To identify any potential problems early so that effective solutions can be formulated.

It is recommended that these meetings should be held once a week with individual ones for large contracts and an umbrella meeting for small contracts. There should be a representative(s) present from the following departments and for continuity these people should remain the same for each meeting on that contract.

4.4 Contracts.

4.5 Production.

4.6 Quality.

4.7 Testing / Inspection.

4.8 Drawing office.

4.9 Dispatch.   Only on a large contract.

4.10 Documentation. "   "   "   "   "   "   "

The meeting should be chaired by a representative of the Contracts Department as they have overall responsibility for a contract.

The chair will have a major role to play in ensuring that the meetings foster good relationships between team members. In particular the meetings should not be used as a mechanism for the allocation of blame and punishment of poor performance.

The people involved in these meetings should be encouraged to work together outside them on problems and not to try and find someone else to "pass the buck" onto. Recognition should be given to teams if they successfully work together on solving problems and achieving deadlines, budgets, quality targets etc.

The problem of scheduling can be tackled by providing the production supervisor with the start date and finish date of a particular contract and allowing him to work out his own detailed schedule, with the aid of the weekly contract meetings. For this to work he will probably have to be given a lot of support at the start of this scheme and it is probably best to try it out on some smaller less time critical contracts first.

The problem of communications with the client can be reduced by producing a schematic flow diagram showing who's who and how they all connect together. Then technical queries which inevitably crop up in such operations could be dealt with quickly and easily as the correct person to contact about such matters has been identified.

The solutions above if implemented properly will reduce the communication, administration and control problems within any similar small to medium sized company.

5. Acknowledgements

1. The engineering company involved. For allowing me to do my study and to publish this paper.

2. Mr B Attewell for his guidance.

Advances in Agile Manufacturing
P.T. Kidd and W. Karwowski (Eds.)
IOS Press, 1994

THE INFLUENCE OF HUMAN RESOURCES AND NEW TECHNOLOGIES ON SUCCESS IN SMALL AND MEDIUM ENTERPRISES

Ingrid Sattes, Ulrich Schärer and Simona Gilardi
Institute of Work Psychology, Federal Institute of Technology, Zürich

Abstract

This article reports findings from an interdisciplinary study of determinants of success in Small and Medium Enterprises (SMEs). The study was carried out in 50 firms with between 6 and 499 employees in three sectors: the metal-working industry, the machine-building industry and the electrical/electronics industry. The data was collected in structured interviews with chief executive officers, observation-interviews and document analyses. The results show that (1) there is no difference in economic success between firms of different sizes, but a significantly lower fluctuation in smaller enterprises (SEs). (2) Medium enter-prises (MEs) hire significantly more low or unskilled wor-kers and the average percentage of employees with appren-ticeship or higher education is over 55%.(3) In all firms there is a much higher chance for managers to benefit from paid further education than for the workers, and MEs offer internal formal training to a higher percentage than SEs. (4) Personal computers, workstations and CAM/CAD technolo-gies are almost as prevalent in SEs as in MEs, but SEs have a significantly lower proportion of computer-aided func-tions and a much lower degree of technical integration. Po-sitive correlations between the use of new technologies and economical success have been only found for MEs. (5) Comparing the two groups of enterprises that have a high de-gree of modern technology with or without a high degree of qualified staff and further education leads to the inter-pretation that new technologies alone are less successful than in combination with good human resources.

1. Introduction

The importance of SMEs in Switzerland can be illustrated by the number of employees working in SMEs in Switzerland, which is around 75% and is still growing against former assumptions [1]. To find out which factors are relevant for SMEs success in competition is topic of this research project.

2. The Project Design

The case studies presented comprise a part of a 3,5 year project of the Federal Institute of Technologies of Zurich, Switzerland, involving eco-nomists, engineers, computer scientists and work psychologists.The research topics cover strategies, technologies, structure and systems, entrepreneurs, human resources and the environment.

We undertook 50 intensive case studies in firms with between 6 and 499 employees in the three sectors of the metal-working industry, the machine-building-industry and the electronics. In a standardized questionnaire of all Swiss firms of these sectors the findings of the case studies are to be verified. 1677 questionnaires (return rate of 33%) are available for the analyses and the results will be published soon.

The results presented here are from case study data. An emphasis is put on the separate look on small (6 to 49 employees, SEs) and medium (50 to 499 employees, MEs) enterprises.

3. Economical and social success

During case studies a number of economical measures were collected. Here, only rentability is presented. There was no difference between the economical success measures of firms of different sizes or sectors. Table 1 shows the results which confirm the hypothesis, that SEs can strategically compensate for the disadvantages of their size.

Table 1: Measures of Economic Success per Size Class

Number of Employees	Rentability 1990
6-49	8.8%
50-499	8.2%

n.s.

From the point of view of work psychology and political economy, it is important not to concentrate on sheer economical success. We collected therefore data on fluctuation, absentism and accident rates. Table 2 shows that there is a significant difference in fluctuation caused by employees (for accidents and absentism as well) indicating higher social success for SEs.

Table 2: Fluctuation Rates per Size Class

Number of Employees	Fluctutation Caused by Employees	Fluctuation Caused by Enterprise
6-49	10.9%	2%
50-499	14.6%	1.7%
	$r=.2$, $p<.05$	n.s.

The result of lower social success of MEs contradicts to other empirical studies [2]. It can be explained by the observation that in SEs there is a tendency of higher social responsibility of the owner and higher personal identification of the employees. In our study, other possible explanations for a lower fluctuation were found in the conditions of work organization. Workers ratings of their possibilites for making decisions and the degree of variability in their jobs show a tendency to be lower, the bigger the enterprise. These are important factors of work which lead to an intrinsic work motivation, i.e. a personal need to solve a task as excellent as possible. We believe, that this characteristic of work in SEs is another important success factor, helping SEs to compete with bigger enterprises.

4. Qualification and further education

40 % employees with no or lower education and 53% with apprenticeship or higher edusation was found. There is a statistical significant lower degree of employees with no or lower education in SEs ($r=.25$, $p<0.5$). There are also differences between sectors. An overall index for evaluating human resources shows a tendency of higher values for SEs ($r=.28$, $p=.08$).

Additionally the handling of further education was evaluated. 50% of the firms offer internal training, but this often consists of rather unsystematic instruction for upcoming tasks. This assumption is supported by the fact that in firms with a lower qualification level there are sinificantly more internal cour-ses ($r=.30$, $p<.05$). There is a signficant lower degree of internal training in SEs ($r=.41$, $p<.05$).

In terms of paid, external training the results are alarming for both small
and medium enterprises. Table 3 shows, the managamers have a fairly good
chance of receiving training outside of the company, but for employees without management or leading functions the chances are alarmingly low. A look
at the topics of external formation proves, like know from literature [3]
that the emphasis lies on technological tasks. Leadership is important too.

Table 3: External Training in SMEs

	Percentage of SMES, which offer _Managers_ External Further Education	Percentage of SMES, which offer _Workers without Man. Functions_ Ext. Furth. Ed.
None	19%	43%
< 30% of Managers/Workers	29%	57%
31-50% of Managers/Workers	29%	-
51-100% of Managers/Workers	24%	-

Very weak associations between qualification level, further education and
economical success were found and only for SMs. But there is evidence, that
a strong further education policy is connected with less problems in innovation processes which are due to unmotivated personnel. This was the most
prevalent reason for innovation problems in our sample. The association
between human resources, further education and social success is stronger.

In summary, it appears that while MEs are able to function successfully with
weaker human resources connected with a strategy of high automation, SEs
tend to put emphasis on good human resources. In terms of training all firms
have a quite modest further education policy, but it does seem that only SEs
are able to utilize this investment to improve their economic success.

5. New Technologies

In our sample different patterns of computer use for different firm sizes
were found: MEs have more mainframes, SEs more work stations and personal
computers. Table 4 shows that in SEs significantly less computer-aided
CAODA, CAPPC, and CAP was found.

Table 4: Computer-Aided Functions per Size Class

Number of Employees	CAODA	CAPPC	CAD	CAP	CAM	CAQ
6-49	34%	35%	48%	21%	35%	17%
50-499	74%	70%	70%	63%	55%	22%
total	54%	52%	60%	50%	46%	20%
	p<.05	p<.05	n.s.	p<.05	n.s.	n.s.

If the degree of technical integration is defined as the number of
connections between the 8 most important components, again a significant
lower degree of technical integration in SEs is found, as well as for
computer-aided storage, assembly or transport.

The relationship between the use of new technologies and success measures
presents a very heterogenic picture. Positive associations between computer
use and economic success can generally only be found for MEs, correlations
for technological measures ranging between $r=.39$ and $r=.59$. For SEs there
are no or negative correlations. An overall estimation of the degree of the
use of new technologes comes to the same result.

One possible interpretation of this result could lie in the fact, that on an
average SEs implemented computers 5 years later than MEs. Also SEs claim to
offer significantly more customer-oriented products, indeed none of the SEs

does have a standard program of products. Their possibilities of achieving a higher productivity by the use of automation might be reduced.

A differenciation between two strategies is possible: (A) producing low scale with a high amount of customer-orientation which is not achieved with automation, but with strong human resources and (B) producing bigger scales with automation and a low level of qualification. We also want to look at a strategy (C), which emphazises technologies and human resources as well.

6. Use of Technologies, Human Resources and Economic success

In the sample we dichotomized SEs and MEs into groups (1) with low values in technologie and human resources, (2) with high values in human resources and low values in technologies (A), (3) with low values in human resources and high values in technologies (B) and (4) with both high values in technology and human resources (C). Tables 5 and 6 summarize the comparision of the 4 named groups for MEs and SEs. In both size groups strategy (C) is more successful than strategy (B). For MEs strategy (B), for SEs strategy (A) is most successful.

Table 5: Use of New Technolgies, Human Resources and Economic Success in MEs

	Average Rentability	Number of Firms
Technology (0) + Human Res. (0)	6.6%	5
Technology (0) + Human Res. (1)	3.5%	6
Technology (1) + Human Res. (0)	9.3%	11
Technology (1) + Human Res. (1)	13.7%	5
		F=15.8 df=2 p=.26

Table 6: Use of New Technolgies, Human Resources and Economic Success in SEs

	Average Rentability	Number of Firms
Technology (0) + Human Res. (0)	6.1%	8
Technology (0) + Human Res. (1)	22.3%	4
Technology (1) + Human Res. (0)	6.0%	4
Technology (1) + Human Res. (1)	7.9%	7
		F=4.8 df=2 p<.05

4. Conclusions

There is evidence supporting the assumption, that the efficiency of new technologies is higher when also human resources are good. An automation strategy which tends to lower the qualifications for a given task, e.g. prefers to hire more non-skilled workers, seems to be less efficient.

It appears that SEs have more difficulty in turning their efforts of implementing new technologies into higher economic success. In our sample SEs are more successful, which emphasize good human resources and further education instead of new technologies. Of course, this effect might be different for different industrial sectors and even for special products. This and other questions will be pursued with the large sample of the questionnaire.

References
[1] Leicht, R. & Stockmann, R. (1993). Die Kleinen ganz gross? Der Wandel der Betriebsgrössenstruktur im Branchenvergleich. *Soziale Welt, 44*, 242-274.

[2] Gaugler, E. & Martin, A. (1979). Personalunterschiede bei Klein, Mittel- und Grossbetrieben. *Personal, 31*, 22-24.

[3] Künzle, D. & Büchel, D. (1988). *Weiterbildung als Strategie für Region und Betrieb*. Bern: Haupt.

Advances in Agile Manufacturing
P.T. Kidd and W. Karwowski (Eds.)
IOS Press, 1994

Introduction of CAD in Small Danish Enterprises

Klaus T. Nielsen

Dept. of Environment, Technology and Social Studies, Roskilde University
Hus 11.2, P.O.Box 260, DK-4000 Roskilde, Denmark

Abstract. In a case-based study of CAD used in small Danish enterprises health &
safety, training and organizational issues are investigated. Some problems as
screens, chairs and upgrading courses are solved without delay, but more complex
issues are not even discussed. This pattern is ascribed to a conscious anti-bureaucratic organizational line in the companies. And that questions the widespread
belief that better planning is the way to get better working conditions.

1. Introduction

I will in this paper use the summary of results from a study of small CAD-using companies as a base for the discussion of how 'human factor'-issues should be handled in an
anti-bureaucratic organizational environment.

2. The study - why and how

The study was commissioned by the National Union of Technicians in Denmark. They
wanted an investigation into the status of CAD (and CAM) in small companies. The
union has for obvious reasons more contact with working places employing many technicians than with smaller places. The criteria for the sample of companies in this study
were that no more than 5 unionized technicians worked in them, they should be manufacturing companies, and preferably with less than 50 employees in total. All the companies were to be using PC-based CAD-systems and indeed all used AutoCAD.

The status the union sought concerned health and safety (H&S), education/trai-
ning, organizational changes and plans for the upgrading of the technology. The union
wanted to know the level of participation of their members in the CAD implementation
process, too, although this is not reported in this paper.

Four companies were covered by the study. The initial contacts were made
through the union members with whom 3-hour long interviews were made. Their immediate superiors and the managing directors were interviewed, too. A question guide were
used for the loose structured interviews.

3. Results

All 4 companies were in the business of producing either production equipment or acces-
sories to production companies. The number of employees in the companies were 22, 46,
52 and 66. All companies had 3 or 4 CAD-stations, and in one case 2 machines were
located in the same room, whereas in all the 3 other cases all machines were located in
one room.

All companies were inclined to upgrade to the latest version of AutoCAD quickly, but otherwise only one company had plans for investments in CAD-related technologies. This company were in the process of buying a CAM-system.

3.1 H&S Issues

Concerning the physical conditions most companies had good screens, good chairs, and 'modern' surroundings. The screens and chairs had received quite a lot of attention, and the general attitude at the time of the purchase of the systems had been: 'buy the best you can get'. The mice used had received much less attention, and generally were lacking behind the level of 'good practice'. The lay-out of the rooms with the CAD-systems didn't match 'good practice' either. Sunlight giving reflexes, machines making noise and high temperatures were not in one case only.

The worst case is in that sense quite illustrative. A small dark room with 3 machines placed with the back against a southbound window which were covered with a heavy dark curtain to avoid the sunlight. The light and the heat were bad both with the curtains drawn or not. When the room was turned into a CAD-room these issues had not been taken into account. For other reasons the technicians with their CAD-systems were in the process of being moved to another room. Despite the obvious problems in the old room, highlighted by the fact that labour inspection had been on the case, no considerations of the situation in the new room had been made, neither by the managing director nor by the internal (compulsory) work environment group.

I will return to strains due to work patterns later, but the general conclusion is that a few specific issues: screens, chairs and to a certain extend noise, heat and light reflexes, are taken into consideration especially at the time of system introduction, whereas lay-out, planning and work organizing (as we shall see later) appear to be to complex issues to be raised, if not to be dealt with.

3.2 Educational Issues

In one case the CAD salescompany used was chosen because the offered 'free education'. In reality this meant 1½ days proper training at the machines after the were installed plus what was termed 'lots of support'. After the 1½ days course the CADusers got a week to 'play' with the system to get accustomed with it. In none of the four case did 'start-up' training last longer than 3 days. Upgrading from one version of AutoCAD to the next were in all companies followed by a one day upgrading course.

In some of the cases other types of CAD-related training had been given, e.g. a LISP-programming course had been given to some of the technicians.

Although both union and employer organizations are fairly reluctant to give specific figures of duration when they recommend training and education, there are no doubt that the training given in these cases do not satisfy the recommendations. But on the other hand neither employers nor employees did mention any need for better education. After a couple of months everyone felt that the working speed and the quality of the drawings were satisfactory, that is at the same standard or better than before. (These judgments were nowhere sustained by any form of documentation.)

On the other hand all companies had experienced problems due to insufficient back-up procedures, and most companies after a period of time encountered problems with non-systematic storing of datafiles. Both types of problems can been seen as related to lack of 'good data behaviour or discipline', i.e. insufficient training.

3.3 Organizational Issues

In three of the companies the technicians worked project or product organized, whereas in the last company a functional division of labour where used.

Lets sketch the last case first. Here the sales function where in focus, sales functions and constructions functions were project organized leaving the drawing function out as comparably low skilled job, and giving the drawer a 'full working day at the screen'-job.

In the project organized companies construction and drawing where integrated. The general rule was one job (be it a drawing, a set of drawings to a specific machine or even a total project) one man. This gave more varied contacts with the workshop(s), costumers, etc. All these technicians had longer formal and/or practical education, and they reported broad use of their education. As an average they estimated 3/4 of their time in front of the screen, at the same time as reporting extensive use of the screen due to the constructive side of their work (compared with the intensive use associated with pure drawing). In one of the companies the project organization meant a quite uneven distribution of hours in front of the screen among the technicians.

Although the reported time in front of the screens might be a bit overestimated, both physical strains due to screen-monitoring and due to the use of mouse (in many cases quite unergonomic designed mice) call for attention. In most cases it is a matter of interpretation wether the time away from the screen do follow the rules newly implemented into the legislation due to the European Communities directive on work at screens (90/270/EEC). But the drawer in the case with functional division of labour most likely do not have a legal jobcomposition.

The introduction of CAD in only one of the case where followed by organizational changes. In this case - one of the project organized cases - the drawing (and construction) function had been centralized taking some of the more integrating features out of the technicians jobs. These jobs had been slightly deskilled, but this was not the general cases in the four studied companies. In the project organized companies the jobs must be described as broad and 'whole' with high decision latitude and good options for cooperation. With the exception of the just mentioned case CAD had only had unnoticeable effects on 'skill levels' in this sample of companies.

4. Perspectives - 'Human Factors'-planning in an anti-bureaucratic environment?

The number of technicians working in CAD-related jobs in the four companies were low; three to five. This of course is a part of the explanation why the above discussed issues have had so little attention in the companies as are the case. But I see it more as part of a strategic position held by the companies. 'We don't discuss problems before we have them' it is said, which is virtually them same as saying we don't bother to plan ahead, and although this attitude undoubtedly is more profound when it relates to H&S etc., it also applies to the general business line in the companies.

Is possible to survive on competitive market taking such a relaxed position to planning? Well; apparently yes. All the companies in the sample had high profiles on the Danish or even the international market for at least parts of their product range. These companies are not rational in minimizing input to output, but they are rational as being flexible and adaptive to their markets. These companies survive (although not necessarily all of them) and sustain jobs which (at least for technicians) are generally speaking more satisfying than jobs in larger companies and organizations.

First the level of 'anti-bureaucratism' can be overestimated. When it comes to the investments in CAD-systems all of the companies had used written documents as a part of their decision-making. Secondly not everybody in the companies was advancing the anti-bureaucratic line, e.g. in one of the companies the technicians immediate superior in general was inclined to a much more bureaucratic line than the rest organization; she was newly employed and came from a large organization, but she was employed partly to make internal procedures more formal.

Still it is fair to say that the 'anti-bureaucratism' were a conscious strategy for management and middle-management in all the companies studied. Not without exception and not unchallenged, but as the benchmark for other ways of dealing with problems.

It is this 'anti-bureaucratism' that makes simple publicly know H&S-issues as screens and chairs matters to be dealt with right away, whereas more complex H&S-issues as lay-out and organization simply are not on the agenda. It is this 'anti-bureaucratism' that makes training into a matter of hours rather than days.

In a series of booklets [1] addressed to companies introducing CAD/CAM the technicians union together with the employers association wrote: "..., and one must be prepared to go through a systematic process of planning. That is the key to CAD/CAM". Now this might work in larger organizations and it might even be successful in small companies as the studied ones if actually performed with wholehearted support, but it does not match the underlying strategy of 'anti-bureaucratism' active in these four companies. Comprehensive planning processes taking technic, organization, working conditions and education into account as the one put forward in the mentioned series of booklets in reality are inappropriate.

This is a problem for the union: 'how should they support their membership in getting better working conditions?', but it is a problem to 'human factor'/ergonomic-specialists and public agencies as the labour inspection, too. If planning is an inappropriate method to tackle problems of this sort what is then an appropriate method?

I don't pretend to have an answer to this question. Firstly my message is not to discard planning wholesale. There are advantages of proper planing which will benefit both competitiveness and H&S standards in many small companies. My point is just to confront the idea of planning with the reality of 'muddling through'. Secondly one should observe the fact that if issues is widely know in the public it affects the local decision making. This might be a solution to the apparently widespread use of nonergonomic mice as the cases of permanent injuries due to use of mice get known.

Thirdly I hope that giving up the idea of one rational way of making decisions and acknowledging the political [1] or even casual nature of the process might in time produce strategies for human factors more compatible with the strategic anti-bureaucratism of smaller companies.

[1]   'CAD/CAM ...', a series of 6 booklets published by Jernets Arbejdsgiverforening & Teknisk Landsforbund, (The employers association in metalworking industry and the technicians union), København 1988.

[2]   See International Journal of Human Factors in Manufacturing **3** (1) (January 1993); Special Issue: Systems, Networks and Configurations: Inside the implementation Process; Guest editor: Richard Badham.

Advances in Agile Manufacturing
P.T. Kidd and W. Karwowski (Eds.)
IOS Press, 1994

Human Factors In the Justification of an Advanced Manufacturing System

Gunnar S Bolmsjö and Per G Dahlén

Lund University Department of Production and Materials Engineering 211 00 Lund, Sweden.

Abstract. A central part of the design of a production system is to methodically weigh the production factors, labour and capital (machines, robots, etc.) and integrate them into a well-functioning unit. The purpose of this paper is to analyse the impact of human factors on the design of an Advanced Manufacturing System (AMS). The impact is illustrated in a case study from a Swedish engineering company. An investment in automated spot welding is justified in a Life-Cycle Cost-analysis. In this case, the company's economic incentives for automation were increased. Furthermore, the labour related additional costs motivated additional automation where all monotonous tasks were eliminated. The result was a highly automated manufacturing system, offering tasks with more variety and less static work load.

1. Introduction

Life-Cycle Costing (LCC) is a method of work usable when quantifying the costs related to an Advanced Manufacturing System during its life cycle. The typical LCC-graph for a production system is usually associated with a bathtub [1]. The costs are high in the beginning of the life-cycle because of purchase, installation, projecting and start-up costs. When the equipment is installed and working as intended, the costs decrease. In the final stage of the life cycle, the costs for repairs and disruptions increase, finally reaching a level that is no longer profitable. The life cycle can therefore be divided into three basic phases: the Acquisition phase, the Operation phase and the Disposal phase [2].

The production factor, labour, is usually regarded as a variable cost, rented by the hour. In this paper, a company is instead suggested to look upon the employment of a new employee as an investment made in training, education, instructor hours, reworks and scrap materials. The purpose of this paper is therefore to use the LCC-technique to carry through an analysis of how the costs for an employee over the employment cycle, influence the mix of the production factors, labour and capital, in a justification of an AMS.

2. Labour LCC analysis

It is possible to pinpoint a number of parallels to the traditional LCC-graph for production systems when estimating the costs for an employee during the employment cycle:
- The costs at the beginning of the cycle are high due to recruitment costs and introduction of the new employee [3].
- After a while these additional costs should decrease and the personnel costs come closer to the costs for wages and labour-related overheads.

- If the working environment is unhealthy, serious production disruptions may occur because of absenteeism or work injuries [4]. The costs for wages, vacations and fringe benefits, depending on the wage system, also may increase over time.

It is therefore natural to divide the employment cycle in a similar way to the life cycle for a machine in a traditional LCC graph. The basic categories that the personnel costs can be divided into are: *Employment costs, Operation costs* and *Disposal costs*.

A quantification of the LCC for an employee was carried through in a low automated assembly department, in a Swedish engineering company. The department was chosen because the work tasks carried out were highly repetitive and monotonous and therefore suspected to cause high disposal costs. The quantification is shown in detail in [5]. In this context it is enough to illustrate how a general LCC curve for an employee in the assembly department can be graphed, as in figure 1.

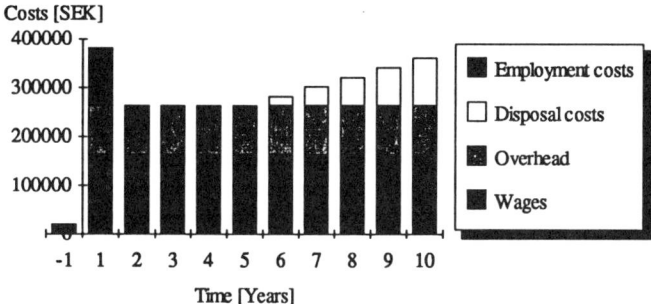

Figure 1. The LCC over the employment cycle for one employee in the assembly department (from [5]).

Since the labour turnover exceeded 10 percent, the average life cycle is suggested to last 10 years. The total LCC for one employee was found to be 3 041 500 SEK (1991 value of money).

3. Justification of AMS

The company concerned in the case study considers investing in robotics when planning for the production of a new generation of products. The choice is between recruiting personnel for manual assembly or investing in an AMS. In the project phase three different layouts were evaluated, all with different levels of automation.

Layout 1, is a low automated system with work tasks regarded as monotonous, similar to the original tasks in the department. The costs are therefore assumed to be in line with the quantified personnel costs above. Seven new employees have to be employed and trained and a small investment in welding equipment is needed.

Layout 2, is a semi-automatic system. An industrial robot does the actual spot welding and a fitter loads and demoulds the fixtures according to the speed of the robot. The time between loadings is one minute at the most, so the fitter is tied to the robot. The task is considered repetitive and with little content. It is difficult to give the fitter other work assignments than loading fixtures for the robot. Layout 2 does not improve the work situation for the remaining fitter. The labour turnover and absence are presumed to be proportionally unchanged, when compared with layout 1. Layout 2 requires additional investments in automation equipment, such as an industrial robot, fixtures, and a turn-table.

Layout 3, is a layout where the degree of automation has be further raised. A material handling robot performs some of the fitter's tasks from layout 2. If the production is automated according to layout 3, almost all of the monotonous tasks are eliminated. The fitter is then allocated other tasks, such as administration in the form of planning and material management. The physical stress decreases from the level causing illness to a decidedly lower level. Absenteeism is therefore presumed to be halved in comparison with layout 2. However, layout 3 involves surplus costs in the form of more expensive production equipment and production control:

• A material-handling robot is needed to replace the fitter in layout 2.

• There is greater demand on the magazines holding the material for the material handling robot than on the containers used to supply the fitter with material in layout 2.

• The more machines that need to be controlled, the more complex the control system of the robot cell. In layout 2, it is sufficient to program the robot for a certain series of actions and to control the spot welding with the help of signals from the control system. Indexing of the turn-table and loading from the containers are easily taken care of by the fitter. In layout 3, two robots must be aligned with the turn-table at the same time as the spot welds and the magazines are controlled. In this case, the synchronisation problems should be quite small and the ordinary I/O units of the robot should be able to deal with them. If more functions are necessary, these can be implemented in a PC, which is also used for production planning and similar tasks.

In figure 2 the estimated LCC for the labour-intensive assembly system in layout 1 is compared with the estimated LCC for layout 2 and 3.

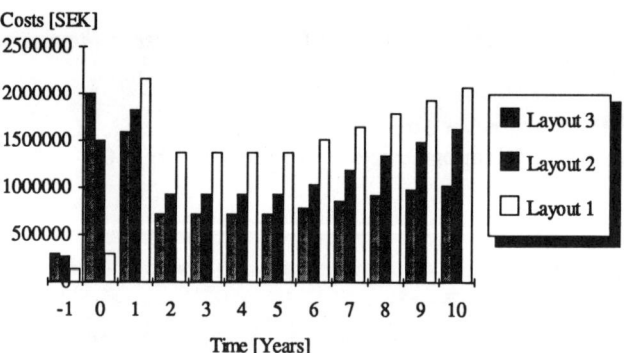

Figure 2. Estimation of the costs associated with manual, semiautomatic and automated spot welding.

The net present value (NPV), with an annual real discount rate of 3 percent, was calculated for the costs related to each layout. The NPV was thereafter divided with the number of products produced during the estimated life cycle, in this case 9 000 000 products during 10 years. The assembly cost per product for layout 1, 2 and 3 turned out to be approximately 1.6, 1.2 and 1.1 SEK.

Most parameters in the justification of an investment are nothing but rough estimations and prognoses for the future. When the uncertainty is severe a sensitivity analysis should be carried out. In this case, different labour related parameters influencing the LCC for the product are analysed. In layout 2, the second best alternative, some labour related parameters were varied ±50 percent. The NPV of the assembly costs was calculated after each variation. The cost per product can be read off on the Y-axis in figure 3. The X-axis represents the variation in the parameters analysed. Zero at the X-axis corresponds to the originally estimated costs at the Y-axis. 1,1 at the Y-axis equals the costs for the best alternative, layout 3. The sensitivity analysis indicates that with an increase of 50 percent of the length of the employment cycle, and a reduction in the disposal costs makes layout 2 competitive.

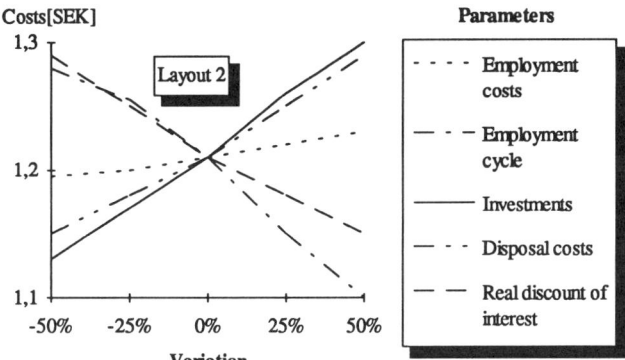

Figure 3. Sensitivity analysis of the assembly costs per product.

4. Discussion and some findings

The economic incentives for automation seem to increase when the total labour-LCC are included in the justification of an investment in robotics. In the example given, one important factor when justifying investments in robotised spot welding is the improvements in the working environment that an automated system offers. The result was a highly automated manufacturing system offering tasks with more variety and less static work load. The sensitivity analysis hints that without factors like absenteeism and labour turnover taken into account, the semiautomatic system would appear to be as cost effective as the AMS.

The analysis presented in this paper is, however, not sufficient to serve as a desision basis for an investment in AMS. The required length of the paper does not allow us to include multi-objective deterministic evaluation methods and the only probabilistic/stochastic evaluation was the sensitivity analysis of labour-related parameters. An adequate decision basis should include these evaluations and also focus life cycles of production equipment, revenues and flexibility.

Acknowledgements: This work is sponsored by the Swedish Work Environment Fund.

References

[1] Kapur, K., *Reliability and Maintainability*, In the Handbook of Industrial Engineering, Salvendy G., eds., John Wiley and Sons, New York, pp., 8.5.1-8.5.34, 1982.
[2] Nachi-Fujikoshi, Corp., eds., *Training for TPM - A Manufacturing Success Story*, Productivity Press Inc., 1990.
[3] Bekiroglu, H., Gonen, T., *Labor turnover: roots, costs and some potential solutions*, Personnel Administrator, July, pp., 67-72, 1981.
[4] Steers, R., Rhodes, S., *Major Influences on Employee Attendance: A Process Model*, Journal of Applied Psychology, Vol., 63, No., 4, pp., 391-407, 1978.
[5] Dahlén, P., Bolmsjö, G., *Human factors in a Life Cycle Cost Analysis*, Presented at the Eight International Working Seminar on Production Economics Innsbruck, Austria, 1994, (submitted for the preceedings).

Advances in Agile Manufacturing
P.T. Kidd and W. Karwowski (Eds.)
IOS Press, 1994

Beyond Implementation: Managerial Challenges in the Efficient Use of CIM Technologies

Lan Xue
Department of Engineering Management
The George Washington University, Washington, DC 20052, USA

Abstract. This paper argues that there is a need for research in CIM technologies to go beyond implementation stage. Recent trends in the diffusion of CIM technologies are used to support this argument. Some challenging issues in managing CIM technologies in the post-implementation stage are examined based on an empirical study.

1. Research on CIM technologies: the need to go beyond implementation

Computer integrated manufacturing (CIM) technologies have the potential to revolutionize the manufacturing industries by reducing manufacturing cost, increasing product quality, enhancing operational flexibility, and broadening marketing opportunities. CIM technologies include application of computers in various functional areas of manufacturing, such as computer-aided design (CAD); numerically controlled (NC) machine tools and flexible manufacturing systems (FMS); and computer-aided process planning (CAPP).

It has been widely recognized, however, that it is necessary to integrate technology, people and organization in order to realize the potentials of CIM technologies [3,5]. Studies in this respect have been proliferating. In particular, many recent studies on CIM technologies have focused on implementation issues [1, 2, 6, 7, 8]. Implementation in this context is often defined as covering the entire adoption and installation process, ending with the routine operations of the adopted technology.

However, little attention has been paid to the problems in managing CIM technology that may occur after the technology is successfully implemented. This paper argues that studies on CIM technology need to extend their focus beyond implementation stage and address these challenging issues. Recently, some researchers have begun to shift in this direction. One such example is [7]. It divided Implementation process into four stages: (1) initiation and justification stage; (2) preparation and design stage; (3) installation and training stage; and (4) routinization and learning stage. The post-routinization stage was included in the implementation process by this definition. The study stressed the need for continuous learning after a technology had been incorporated into the regular production.

While this study began to pay attention to issues in post-routinization stage, there are some weaknesses in its general approach of including post-routinization stage in the implementation process. One is that by mixing problems unique to post-routinization stage with other issues related to implementation, the former tends to be submerged by the latter. In order to fully understand the challenges ahead after a CIM technology is brought into a regular production setting, one must look beyond implementation, treating the post-routinization stage as a distinctive process deserving full attention.

Several new developments have made it imperative to shift the focus in CIM research beyond implementation. Due to length limitation, the only point to be elaborated here is the trend in the diffusion of CIM technologies.

Based on data from a survey of U.S. manufacturing establishments conducted by the U.S. Bureau of the Census [9], table 1 shows the percentage of operations dependent on CIM technologies in fabrication and machining in U.S. manufacturing establishments in 1991, and the anticipated changes in 3 years.

Table 1. Changes in the percentage of operations dependent on CIM technologies (1991-94)

% of operations dependent on CIM	not applicable	<10%	10% -24%	25% -49%	50%-74%	>=75%
1991*	9835 (23.4%)	11258 (26.8%)	6313 (15.0%)	5822 (13.8%)	4611 (11.0)%	4242 (10%)
1994* (anticipated)	9696 (23.1%)	8255 (19.7%)	6894 (16.4%)	6484 (15.5%)	5558 (13.2%)	5060 (12.2%)
difference between 1994 and 1991	-139** (-0.3%)	-3003** (-7.1%)	581 (1.4%)	662 (1.7%)	947 (2.2%)	818 (2.2%)

(* : each entry shows the number of manufacturing establishments which fall in the category and their percentage in the total population; **: the negative sign indicates that the number of establishments in this category would be reduced in 1994 compared with that in 1991.)

Since the "not applicable" category includes all plants which had not adopted CIM technologies at the time of survey, the small reduction anticipated in this category means that there would be very few new adopters between 1991 and 1994. Similar trends could also be observed for CIM technologies in other functional areas. On the other hand, existing CIM users would rely more and more on CIM technology for their operations. This shift to more intensive use is reflected in Table 1. While fewer CIM adopters would rely on CIM technologies for less than 10 percent of their operations, more CIM adopters would rely on CIM for higher portion of their operations, ranging from 10 percent to higher than 75 percent.

Further, implementing CIM technologies into regular production seems no longer to be a major concern for most current users. For example, when asked about the length of time from placement of order to full operation of CIM technologies used in fabrication and machining, 54.5 percent of the applicable respondents said it took them less than three months to do so. An additional 28.2 percent respondents needed three to six months. When asked about the most significant problems encountered in the acquisition or use of the CIM technologies, over 60 percent of the applicable respondents said that they encountered none. For those who did, two thirds of them cited overall cost as the major problem, rather than implementation related issues.

The implications of the above data are twofold. First of all, the diffusion of CIM technology has reached a plateau. This means that there will be very few new adopters in the next few years. Major activities will be in the area of intra-firm diffusion, where existing users will intensify their use of CIM technologies. Secondly, after many years of learning, both CIM suppliers and adopters seem to have gained enough experience to implement CIM technologies successfully. The challenge now is how to use these technologies better to achieve their full potentials.

2. Management of CIM technologies in post-implementation stage: challenging issues

While there are many challenging managerial issues during the post-implementation stage, the following discussion will focus on some of them that emerged from an empirical study the author conducted on CIM application in machining [10]. The general context of the study was manufacturing plants in which CIM technologies were used concurrently with conventional technologies in machining operations. The main data source for the study came froi.i a size-stratified national survey of over 2000 U.S. manufacturing establishments in 21 industries completed in 1987 [4].

At the time of the survey, the majority of CIM technology adopters (over 95%) also used conventional technology for their production. From Table 1, one can also see that by 1991, over 90% of the U.S. manufacturing plants still had to use conventional technology for at least 25% of their operations.

The concurrent use of both CIM technology and conventional technology for more or less similar metal cutting operations poses some managerial challenges to shop floor managers. One particular set of issues was the impacts of various managerial policies and practice on production efficiency. While there have been many studies indicating the need to change old managerial practice for better utilization of CIM technology, little has been said about impacts of such change on the efficiency of the conventional technology. For example, will the changes required by CIM technology produce the same benefits for operations using conventional technology? Should there be a differentiated management policies with regard to different technologies? Our current understanding on these questions is still quite limited.

In order to answer these questions, a regression model on production efficiency was constructed to test the efficiency impacts of various factors and to see weather such impacts would differ between CIM technology and conventional technology . Unit production time was used as the dependent variable while product attributes, different management policies and practice were used as independent variables. The analysis was conducted for 566 products machined by CIM technology and 527 products machined by conventional technology.

The results showed that impacts of various managerial variables on production efficiency were different when the technology used were different. For example, one variable tested was Talorist Management Practice, which was characterized by the existence of both specialists for quality control and production standard, and the requirement that written work orders be followed. The results showed that following a Talorist management practice severely retarded production efficiency under CIM technology, but such practice had little impact on production efficiency under conventional technology.

Another variable tested was Seniority System in promotion and firing. The existence of a seniority system were beneficial to both productions using CIM technology and conventional technology. But the magnitude of the impact for conventional technology was much greater than that for CIM technology, suggesting that retaining tacit and situation-specific skills through seniority system plays a more important role for production using conventional technology.

While these results are still preliminary, they have shown that using both CIM and conventional technologies in a regular production setting poses some new challenges for manufacturing managers. Such challenges demand a sophisticated understanding of the opportunities and difficulties presented by the new technology, and the cross-impacts between the new and the old. Researchers on CIM technology can contribute to such understanding by shifting their focus beyond implementation stage.

Reference

[1] J. Bessant and H. Haywood, Islands, Archipelagoes, and Continents: Progress on the Road to Computer-Integrated Manufacturing, *Research Policy*, No. 17, 1988.

[2] R. Jaikumar, Postindustrial Manufacturing, *Harvard Business Review*, Nov.-Dec. 1986.

[3] W. Karwowski et al, Integrating People, Organization, and Technology in Advanced Manufacturing: A Position Paper Based on the Joint View of Industrial Managers, Engineers, Consultants, and Researchers, *The International Journal of Human Factors in Manufacturing*, Vol. 4 (1) 1-19, 1994.

[4] M. R. Kelley and H. Brooks, The State of Computerized Automation in U. S. Manufacturing, Center for Business and Government, John F. Kennedy School of Government, Harvard University, 1988.

[5] P. T. Kidd, Organization, people, and technology, towards continuous improvements in manufacturing, in *Proceedings of the Sixth CIM-Europe Conference*, May 15-17, Lisbon, Portugal, 387-398.

[6] Dorothy Leonard-Barton, Implementation as mutual adaptation of technology and organization, *Research Policy*, 17 (1988), pp 251-267.

[7] Hongyi Sun and Jens Ove Riis, Organizational, Technical, Strategic, and Managerial Issues along the Implementation Process of Advanced Manufacturing Technology-A general Framework of Implementation Guide, *The International Journal of Human Factors in Manufacturing*, Vol. 4 (1) 23-36, 1994.

[8] Marcie Tyre, Task Characteristics and Organizational Problem Solving in Technological Process Change, Alfred Sloan School of Management, Working paper, WP # 3109-90-BPS, January, 1990.

[9] U.S. Bureau of the Census, The Survey of Manufacturing Technology: Factors affecting adoption, SMT/91-2, U.S. Government Printing Office, Washington, DC, 1993.

[10] L. Xue, An empirical analysis of manufacturing performance under computer integrated manufacturing technology and conventional technology, Working paper, George Washington University, 1993.

Advances in Agile Manufacturing
P.T. Kidd and W. Karwowski (Eds.)
IOS Press, 1994

The Human aspects of implementing advanced systems in a changing manufacturing environment; the case of international multi-site working group projects in Pirelli

Giorgio Basaglia[a], Marco Guida[b] & Louise C. Treanor[a]
[a] *European projects, Pirelli S.p.A.. Viale Sarca 222, 20126, Milan, Italy.*
[b] *Pirelli Informatica, Via dei Valtorta 48, 20127, Milan, Italy.*

Abstract. In the early 1990's Pirelli, like many of its counterparts, underwent a period of historic recession and consequent organisational change. This period of unrest has had a significant impact upon the way in which the company is managed in every dimension. Consequently the two way effect that the organisational and external environment has on the application of new technology and vice versa has become a central issue to Pirelli. This issue has lead to the evolution of a group-based, bottom-up, end-user driven approach for the development and application of advanced manufacturing systems. Such an approach has facilitated the coequal consideration of human and technical aspects in the industrial environment.

1. Introduction

Pirelli as an international group has traditionally given much importance to the human issues at work. This tradition has subsequently been compounded by the prevailing market conditions and has become an integral part of all areas including those of a more innovative and high risk nature such as the development of advanced manufacturing systems.

Today, the technology of advanced manufacturing systems is more than ever oriented towards the modelling of human knowledge and reasoning. Therefore, the development of such systems in an international, and hence multi-site context, such as Pirelli, has called for a group based, bottom-up, end-user approach where human issues play a key role.

Using the Pirelli Tyre Sector as a case example, the main areas of discussion for this paper will be provided by the following:

- An overview of the industrial domain in the past and present.
- The evolution of an approach for the development of complex manufacturing systems in the face of organisational dynamism.
- The effect that the approach and the application of advanced systems has had on the organisation and the outlook for the future.

The focus will in particular, be on recent developments and applications of advanced manufacturing systems in Pirelli through international multi-site teamwork with respect to the changing operational environment.

As the criterion for the development of such systems in Pirelli is very similar to that applied in CEC financed R&D programmes, it has been a logical step for the group to move further towards collaboration in such programmes as ESPRIT and BRITE/EURAM. In fact, much of the experience gained and reflected upon in this article can be attributed to Pirelli's involvement in certain CEC research projects.

2. An Overview of the Industrial Domain

2.1. New technology in the manufacturing environment - in the past

In the past the introduction of new technology of any type was primarily viewed to be a technical endeavour. New developments and their applications tended to be organised mainly from a technological point of view. These types of project were often headed by external 'technical experts' who were mainly visible in the installation phase only. The majority of considerations would be given to the potential impact that these new developments would have on the operating environment and more particularly in terms of:

- Technical integration into existing processes and system constraints.
- Technical performance in relation to productivity and quality.
- Reliability and maintainability

The technology of manufacturing systems themselves was based on traditional IT techniques which again strongly reflected the technological issues of the industrial application as opposed to the human issues. This technology push reflects the 'automation orientation' that has typified the direction of many manufacturing industries over the last few decades.

From an organisational point of view both the company's internal and external operating environments were relatively stable and characterised by the following factors:

- Continuity of personnel - 'a job for life'.
- Stability of structure - classic functional based organisational structure where specialisation categorised people and their roles.
- Autonomous single site production - where each factory tended to meet all the demand of its local markets.
- Security in the market - growing markets i.e. original equipment and replacement markets for tyres.

Therefore the technically based approach to the introduction of new manufacturing systems was conducive to such relative stability and continuity within the organisational and operating environment.

2.2. The Current Situation

Pirelli in recent years has undergone radical organisational change as a consequence of changing market conditions and internal corporate strategy. The market for tyres very much reflects the trends in vehicle sales and consequently suffered a decrease in growth which in some cases was actually negative.

The clients have become more involved in the operations of the factory and require ever improving products which can meet narrower tolerances and give better performance. The organisation has consequently tended towards a strategy of rationalisation resulting in less factories, more complex logistics and the distributed production of local demand. The reduced number of factories and the increased performance expected from each of them has highlighted the importance of 'multi-site' working for the improvement of common factors.

The vicious circle of stunted demand, increased customer requirements and reduced industrial resources, that necessitated rationalisation, led also to the realisation that survival was intrinsically linked to the exploitation of human resources. Although JIT, KANBAN and many other Japanese ideologies had for many years been practised in the industrial environment, this new period of crisis marked an even greater turn in management philosophy towards human based tools and techniques, (i.e. Total Production Maintenance, personnel mobility, multi-skilled/role work force, improved communication networks...)

In addition, to the changing environment which, pushing further emphasis on responsibility to the people involved, the orientation of new technological applications is also changing. In particular the aim of developing and introducing new manufacturing systems has significantly moved from the automation of the process and procedures to the support of the human based activities which in turn give the added value/quantity and flexibility to the manufacturing facility.

3. The evolution of an approach for the development of complex manufacturing systems in the face of organisational dynamism

The manufacturing environment like many others, is complex and contains many variable internal and external factors which together make for very dynamic working conditions. The initial evolution of Pirelli's approach to development of complex manufacturing systems was the systematic consideration and inclusion of the various internal and external influences.

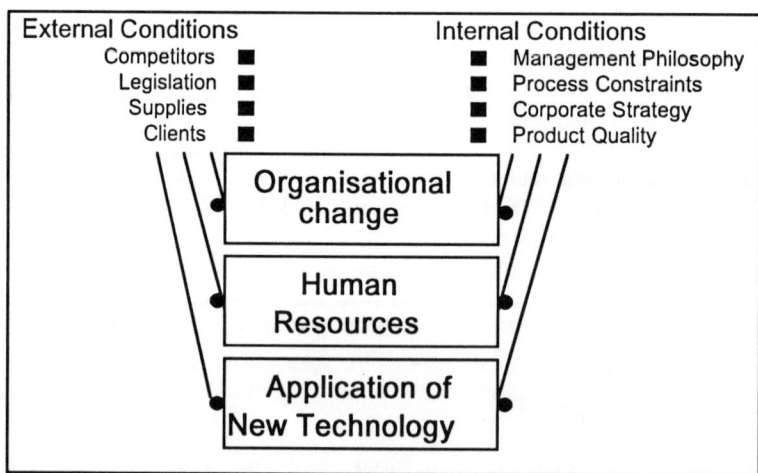

The increased emphasis on integrated logistics and distributed production led to the start of 'multi-site' group working as opposed to the more traditional practice of 'multi-functional' group working on common problems. By involving different factories in the development of a single system guarantees its transferability and applicability. Also the contributions of different sites to system design constitutes a richer source of knowledge and experience upon which to base the development.

The involvement of many sites was not enough to guarantee 'user' oriented developments. Therefore 'key' end users were identified and involved in the system development to ensure that all aspects of the system would be 'user driven' and hence bottom-up. The best and most feasible to bring all of these parties together was through international joint working group meetings which encouraged team bonding and effective information exchange. This approach in turn encouraged the following organised changes:

- An increased distribution of responsibility for remaining personnel. A rationalisation of the management layers caused the responsibilities per capita of remaining managers to increase.
- A change in decision making policies. The outcome of the organisational changes produced an environment where decision making needed to be made at lower organisational levels.
- A more efficient exploitation of available resources. The rationalisation during the organisational change, as well as reducing resources that were futile, enabled senior managers see where remaining resources could be exploited.

In addition to the organisational changes undergone by Pirelli, technological advances have also played their part in the evolution of an approach for systems development.

Technological advances have meant that:

- Human reasoning, experience and know how can be effectively modelled and exploited.
- Man machine interfaces have overcome many of the 'user friendly' barriers associated with more traditional manufacturing systems.
- The unification of computer standards and the evolution of integrated computer development tools allows the development of truly flexible and transferable manufacturing systems.

4. The effect that the approach and the application of advanced systems has had on the organisation and the outlook for the future

This approach helped emphasise the general awareness of human issues :

- real commitment from top management for innovative/new manufacturing.
- drive towards a higher skilled and multi-skilled work force.
- emphasis on communication and the importance of involving people at all levels of the problem under analysis - group working.

The benefits from a system point of view have been shown through

- New ideas and willingness from the shop floor to create new systems and improve existing ones.
- Ease of acceptance of the system into the manufacturing environment.
- Increase system suitability and efficiency for tasks involved.

Part V
Human Computer Interaction

Advances in Agile Manufacturing
P.T. Kidd and W. Karwowski (Eds.)
IOS Press, 1994

Keeping Mice
in the Control Room:
the pros and cons of mouse driven
interfaces for process control

C. Baber

Industrial Ergonomics Group, School of Man. & Mech. Eng.,
University of Birmingham, Birmingham B15 2TT

Abstract. Recent years have seen an increase in the application of computers to
process control operation. Computers have replaced wall-mounted mimic diagrams to
present screen-based plant overviews and operational details using advanced graphics
capabilities. In order to operate this technology, there has been a need to move away
from traditional 'knobs, buttons and levers' to computer-based control devices. In this
paper the relative merits of a particular class of computer-based control device, mice,
will be discussed with reference to process control operations.

1. Introduction

Computers were first introduced into control rooms in the 1960s. In traditional control
rooms the information received by operators was an analogue of plant functioning; dials
indicated changes in temperature or pressure. This means that the overview of plant status
could be obtained from a panel containing a variety of indicators. As panels are superseded
by computers, so the manner in which information is presented changes. Plant information
need not be presented according to plant topography; rather a variety of different formats
are possible. This has the potential for a far richer picture of plant activity than could be
obtained from dials. However, computers are not without their problems.

The operator can skip between pages of information on the visual display unit, which
can lead to any of the reported problems of 'navigation' in complex information structures.
Not only can there be problems in terms of searching information, but the use of computers
can effectively reduce the operators 'window' on plant activity. The changes in technology
also have implications for operators tasks. Baber (1991) notes that operators are required
to perform three types of activity relating to display technology: information search and
retrieval; data manipulation; control actions. While control actions are traditionally few,
computerisation has altered the way in which information is searched for and retrieved, and
offers a wider range of data manipulation activities than do traditional technologies.

There has been surprisingly little interest into what would constitute appropriate
devices to permit operators to control computers in the control room. Often the input device
used is the one which came with the computer which is likely to a keyboard, but which is

increasingly likely to include the mouse. While mice are popular in windows- based office systems, does this mean that they will be useful for process control operation ?

1.1   *The mouse*

The earliest version of the mouse was developed by Douglas Engelbart of SRI in the mid-1960s. In this version two wheels, positioned at right angles to each other, were used to convey the x-y movement of the mouse to the computer, in order to drive the cursor on the screen. In the mechanical mouse of today the wheels have been replaced by a ball mounted against a number of sensors. Mice are cheap, robust devices which can be used to perform a variety of display management tasks and operations.

1.2.   *Operating the mouse*

By recording the movement of the cursor across a screen, Barker et al. (1990) found that mouse movement time could be reliably decomposed into phases. A slight delay between stimulus presentation and movement onset was followed by a rapid ballistic phase, covering much of the distance to the target. This was followed by a number of additional movements around the target, with periods where the cursor was stationary, and a final pause before the end of the movement. It was suggested that the additional movements represented visual monitoring of mouse position due correction, and this was supported by a high corrrelation with test results from visual-spatial tasks. Thus, mouse use can be considered as a complex motor skill, utilising visual feedback in the target acquistion phase.

2.   Training Requirements

Apple Computer Inc. have proposed that it takes less than 15 minutes to become familiar with the rudiments of mouse operation. However, this optimistic figure hides some of the potential problems encountered by users of mice. Barker et al. (1990) found a significant difference between good and poor mouse users; with the poor users being not only less accurate, but also showing differences in speed and initial distance covered, with the poor group using more corrective movements. Good mouse users were able to employ visual information to monitor mouse position. It is probable that experienced users are also able to access reactive feedback from the mouse's movement, but we are not sure how this learning occurs.

There are several different mouse designs on the market, some with a single button and others with two or more buttons. For single button mice, it is sometimes necessary to 'double-click' the button and the inter-button timing seems problematic for novice users. For multibutton mice, there may well be problems in terms of remembering the definition of each button.

3.   Musculoskeletal Aspects of Mouse Use

The assumption that mice do not produce musculoskeletal problems seem to have passed into computer-lore with little or no questioning. However, in surveys of office staff using windows-based software, we have noted a high incidence of reports of musculoskeletal problems relating to the right wrist and shoulder; in many of these cases the mouse is positioned at a distance from the keyboard.

For large movements across the screen or to large targets, it is possible to move the mouse using full arm movement, but for small movements, either across small distance or to small targets, more precise control will be had from using wrist movements, often with the fourth and fifth fingers resting on the table surface. However, evidence is beginning to accumulate in the literature relating extensive mouse use to specific injuries, principally related to ulnar deviation resulting from holding the mouse and pressing buttons (Davie et al., 1991; Franco et al., 1992). As screens become both larger and more cluttered, with an increase in small targets, there may well be a mismatch between these different task requirements. As desk space is often at premium in the control room, this may become an important issue.

4.   Mice and Performance

4.1.   Task-fit

There are certain tasks for which mice are particularly well suited and other tasks for which mice are clearly unsuited, for instance,while mice are useful for pointing at and selecting objects on the screen, but that they may not be useful for tasks requiring a high degree of motor precision, such as drawing, nor are they useful for data entry. We examined the use of mice and touchscreens for controlling a simple process (based on Crossman's waterbath). We found a relationship between speed of object selection and device, i.e., performance was more faster with the touchscreen, and for changing object status, i.e., moving sliders etc. was easier with mouse. Thus, there is some variation between 'task-fit' and input device.

4.2.   Speed and Accuracy

The relative performance benefits of the mouse depend very much on the type of devices against which it is compared. If mice are compared against relatively crude positioning devices, such as cursor keys, a speed advantage can be obtained, but if direct pointing devices are used, such as light pens or touch screens, mice are outperformed. In terms of speed advantage then mouse performance must be considered in relation to a number of factors, not least of which will be devices against which it is being compared and the tasks being performed.

Some manufacturers offer variable gearing for control : display ratio, such that cursor movement over large distances can be accelerated. However, there is evidence to suggest that altering the c:d ratio is unlikely to have a beneficial effect on performance. Tränkle and Deutschmann (1991) have shown that there are several factors which play a far greater role in mouse use than that played by c:d ratio. These factors are target size, distance moved and, to a lesser extent, the size of the display.

This reinforces the notion of a relationship between movement and visual feedback and suggests that problems would arise if objects become closely grouped, as is often the case in process control diagrams. Sound ergonomic advice exists for placement of knobs on control boards and similar guidelines are required for objects placed on the computer screen.

4.3. Strategy

We have been investigating the interaction between device used for object selection and strategies people use for tasks such as fault diagnosis. For instance, a study comparing user performance on a simulated fault finding task with either a mouse or a touchscreen to select nodes for testing showed a significant interaction between device and strategy. The touchscreen led to users selecting more nodes to test, and working through all nodes in an particular area, while mouse, in contrast, led to users adopting a more efficient, flexible search within areas. It might be that the indirect devices require a different level of planning prior to operation, which encourages users to plan ahead rather than simply to act.

5.   Conclusions

One might assume that the selection of mice will be a matter of available space and user preference. However, we have proposed that there are considerations of 'task-fit' which suggest that the mouse is not always optimal, and that its use can influence performance and strategy. Secondly, there would appear to be a relationship between device use and risk from physical discomfort and injury. Thirdly, the effective use of mice relate to experience and strategy. We would conclude by observing that mice may well represent a convenient compromise as input devices in process control, provided that they are used in adequate space, that operators are given adequate training, that the objects on screens are sufficienty sized and spaced to permit efficient use and that consideration is given to the interaction between mouse use and operator activity.

6.   References

Baber, C. Speech Technology in Control Room Systems: a human factors perspective Ellis Horwood, Chichester (1991)

Davie, C., Katifi, H., Ridley, A. and Swash, M. 'Mouse'-trap or personal computer palsy Lancet 338 pp.832 (1991)

Franco, G., Castelli, C. and Gatti, C. Tenosinovite posturale da uso incongruo di un dispositivo di puntamento Medicina del Lavoro 83 pp. 352 - 355 (1992)

Tränkle, U. and Deutschmann, D. Factors influencing speed and precision of cursor positioning using a mouse Ergonomics 34 161 - 174 (1991)

Advances in Agile Manufacturing
P.T. Kidd and W. Karwowski (Eds.)
IOS Press, 1994

Effects of Tactile Feedback in Process Control, Exemplary in Mouse-Driven Interfaces

M. Göbel, J. Springer, H. Luzak

Institute of Industrial Engineering and Ergonomics, Aachen University of Technology, Bergdriesch 27, 52062 Aachen, GERMANY

Advanced process control systems are based on graphical user interfaces. Instead of keyboard manipulation, graphical input devices are primarily used for direct object manipulation and command control by menus, usually in form of a computer-mouse. The speed and precision of interaction plays an important role for the whole system performance. Using a conventional mouse, the user only can get visual information from the screen about the position of the mouse and the action that was initiated. In contrast, during the manipulation of real objects, the visual channel is only responsable for giving broader information about the action, while the motoric action itself is predominantly controlled by tactile information fed by interoceptive and exteroceptive sensory signals. Consequently, working with a standard computer-mouse requires concentration primarily on the visual system. In respect to this situation, it is proposed that a computer-mouse be enhanced with an additional tactile feedback to approximate more closely real object handling. To evaluate this hypothesis, a standard computer mouse was enhanced with a total of four tactile actuators, two lying under the fingers that controlled the mouse sideways and the other two under the mouse buttons. A comparative experiment with tracking, positioning and selectings tasks was carried out. While the overall performance decreased during tracking tasks, ballistic tasks were carried out 11-25 % faster. Movement characteristics were less conservative and the regulation circuit of man and machine with tactile feedback was less stable, but finally faster.

1. Working with a graphical input devices - Aspects of Human Workload

Advanced process control systems are based on graphical user interfaces. For direct object manipulation and command control by menus, a graphical input device, usually in form of a computer mouse, is primarily used. Commands no longer have to be kept in mind and the user has only to select the desired function by moving the mouse cursor to the corresponding position and pushing the mouse button. Consequently, cognitive load is reduced, but on the other hand motoric and sensoric load is increased due to the dynamic movement character. Especially for complex screen informations and during critical situations an accurate and fast movement reaction plays an important role for the whole system performance.

The comparison of information flow while using a computer-mouse and while handling real objects points out an essential difference: For the movement of a computer-mouse, subjects have two senses available, visual information and interoceptive information (muscle tension and joint angles). To handle a real object, one more important sense is available: tactile and kinesthetic perceptions give information about touching an object and about the exerted or backdriven forces [4, 6]. All these senses can be processed in parallel, since they make use of different human sensoric resources.

The absence of tactile and kinesthetic information leads to a concentration on visual infor-mation. Thereby, the construction of movement strategies is more difficult than in real environment, since the visual information processing requires more time than information processing of other senses [7]. Further, visual information processing is already required for other system operations.

It can be hypothesized, that an additional tactile or kinesthetic feedback, which transmits redundant information the about object handling on the screen will allow a more intuitive handling of screen objects and so increase task performances and decrease human strain reactions.

2.   Tactile Feedback Applications for Computer-Mice

With regard to the regular functions of cutaneous sensory information and their influences in a real environment, tactile information should be primarily used to enhance object usability [1, 8]. Even as a kinesthethic feedback seems to be more important for the handling of real objects, it is quite difficult to accomplish for a computer-mouse design. A vibrotactile feedback could be used as a virtual energy-field which would seem to be radiated by the objects on the screen and indicates the approximation of the mouse-cursor to an object [5].

For different applications adapted designs have been proposed: The horizontal positioning of the mouse-cursor to a specified object would be supported by vibrotactile stimuli on the side-planes of the mouse-body. As the cursor were to come closer to the object, vibrotactile intensity would increase. To allow the distinction between the different directions, the intensity of left and right side would vary during approximation.

For the vertical direction either the left or the right actuator would be placed more toward the front of the mouse than the other (and thus these actuators would be used both for vertical and horizontal direction) or the both mouse buttons are additionally used for tactile signal transmission. This kind of representation would generate a redundant feeling of the display while moving the mouse, just as the display-content would be additionally engraved into the table-plate. With a matrix of multiple actuators in each mouse-button, a higher spatial resolution may be useful for trained operators or for visually disabled persons. The size of the vibrotactile detection area would decide if either the approximation or the exact positioning were predominantly indicated.

3.   Experimental Investigation of Additional Tactile Feedback on Performance Parameters

A standard computer-mouse was enhanced with a total of four electro-magnetic actuators, two lying under the fingers that controlled the mouse laterally and two others under the mouse-buttons [2, 3]. To consider the different kinds of movements, three types of experiments were executed:

•   A Tracking Task to study controlled movements. A vertical line moving in a horizotal direction on the screen had to be followed by the mouse-cursor as exactly as possible. At random points the movement changed abruptly its direction.

•   A positioning task to study ballistic movements. The mouse-cursor had to be driven as fast as possible towards a vertical line which was suddenly appearing at a random position on the screen.

•   A selecting task to study menu selection tasks. The mouse-cursor had to be driven into a suddenly appearing field, and afterwards the left mouse-button had to be switched.

The used screen was a standard 14"-VGA-color display (CRT, 640•480 pixel), and mouse sensitivity was set to default-values (linear translation, maximum movement amplitude 81mm). All tasks were tested with 22 subjects (16 males, 6 females, 20 to 35 years old). Each task was tested with and without tactile feedback, and with narrow and broad detection area, each with 3 to 6 repetitions. All subjects were accustomed to working with conventional computer mice, but had not worked before with a tactile feedback. To become familiar with tactile feedback and the task requirements, subjects underwent a demonstration program and a test-phase with original tasks before starting the tests.

The overall performance was measured as well as sequentially separated movement parameters to enable a more precise interpretation (fig. 1 and 2).

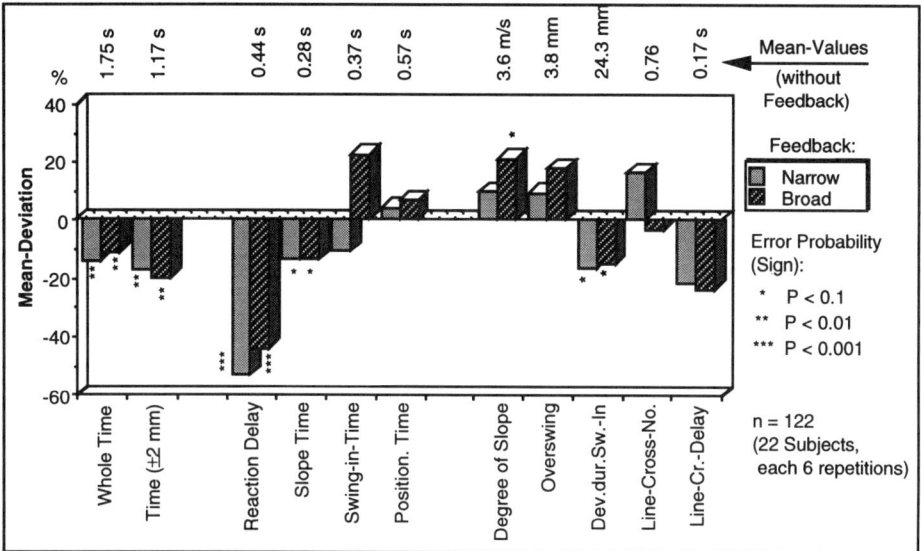

Fig. 1:   Changes in performance characteristics by an additional tactile feedback during the execution of positioning-tasks

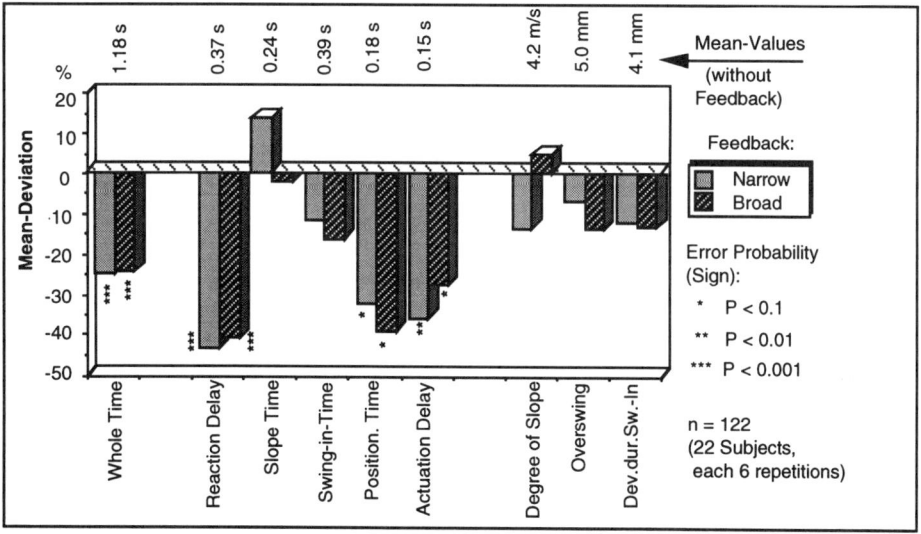

Fig. 2:   Changes in performance characteristics by an additional tactile feedback during the execution of selecting tasks

For the tracking task, the mean deviation, as a measure for the total performance, increased slightly with tactile feedback. With narrow feedback, subjects tended to move ahead, while with broad feedback they had a tendency to move more behind the line. The line-cross-frequency increased in both cases. The reaction delay after an abrupt change in movement direction was slightly reduced, but the overswing amplitude increased. Explanations to such a reaction may be, that either subjects were not familiar enough with tactile signals to incorporate them into motor control, or that the tactile signal is not well suited to have an intuitive influence on human motor control.

Regarding the positioning task (fig. 1), the time requirement for the whole task is reduced by 11 to 14% . The time requirement for the positioning into a range of ± 2 mm is more strongly reduced by 20%, while reaction delay was strongly shortened to about half the time. The overswing amplitude is also increased, but in spite of this, tactile feedback helps to reduce the deviations during swing-in by a higher line-cross-frequency. The use of tactile feedback leads to a less controlled movement with lower stability but better total performance.

For the selecting task (fig. 2), considerable faster actions during nearly all movement sequences can be measured, resulting in a total reduction of action duration of 25%. If it succeeds to optimize the slope phase and the swing-in phase, improvement potential can be estimated to about 35% on the performance side.

4 . Conclusion

Even a very simple tactile-feedback-device leads to significant changes in movement behavior. Especially in the case of selecting tasks, which represent the most frequent utilization of computer-mice in process control, a reduction of 25% in time required to actuate will bring a considerable increase in human performance. The changes in reaction strategies imply a more direct and intuitive reaction if tactile feedback is available. Therefore, aside from the measured increase in total performance, a considerable decrease in human strain may be assumed. It becomes visible, that the adaptation of the tactile feedback characteristic to the user is not yet optimized. Especially during the swing-in-phase of ballistic movements, subjects tended to overestimate their control-performances. For future applications not only a static characteristic, as was used for these experiments, but a dynamic one may lead to further improvements.

References

[1] BLUME, H.-J., BOELKE, R.: Mechanokutane Sprachvermittlung. Fortschritteberichte VDI, Reihe 10, Nr. 137. Düsseldorf: VDO-Verlag, 1990.

[2] CRAIG, J.C.: Difference Threshold for intensity of tactile stimuli. Perceptions and Psychophysics, Vol. 11 (1963), 150-152.

[3] GESCHEIDER, G.A., CAPARO, A.J., FRISINA, R.D., HAMER, R.D., VERILLO, R.T.: The effects of a surround on vibrotactile thresholds. Sensory Processes, 2 (1978), 99-115.

[4] HENSEL, H.: Somato-viszerale Sensibilität. In: KEIDEL, W.-D. (Hrsg.): Kurzgefaßtes Lehrbuch der Physiologie. 6. Aufl. Georg Thieme Verlag. Stuttgart, New York, 1985.

[5] HILL, J.W.: The perception of multiple tactile stimuli. Technical Report No. 4823-1, Stanford University, Electronic Laboratory. Palo Alto, CA, 1967.

[6] JUNG, R.: Einführung in die Sinnesphysiologie. In: GAUER, KRAMER, JUNG (Hrsg.): Physiologie des Menschen, Bd. 11: Somatische Sensibilität, Geruch und Geschmack. Urban & Schwarzenberg. München, Berlin, Wien, 1972.

[7] TAGHAVI, A., PENNING, J.: Über die menschliche Reaktionszeit nach periodischen und aperiodischen optischen und akustischen Reizen. Pflügers Arch. ges. Physiol. 319, 1970.

[8] TEKANO K., STUDENT, C.: Bewegungsorganisation und Reaktionszeit. In: Handbuch der Ergonomie, Bd.1. Carl Hanser Verlag.München, Wien, 1989.

Advances in Agile Manufacturing
P.T. Kidd and W. Karwowski (Eds.)
IOS Press, 1994

Psycho-physical Stresses and Strains Arising at Mouse-driven Interfaces used in Process Control

Kurt Landau
Gerhard Wendt
University of Hohenheim, Department of Ergonomics
Fruwirthstrasse 48, D-70593 Stuttgart,Germany

Abstract. This paper reports on a case study from the chemical industry in which a before/after comparison was made between a classic process computer control procedure and a mouse-driven work station with two VDU's using the windowing technique. A polygraphic concept was used to measure the comparative stresses and strains. The investigation showed that the hardware and software largely complied with the relevant EC standards. AET analyses show that the work station lies within the normal range of informatory-mental stresses as compared with other VDU work stations. The investigation of strains revealed no increases in strains which could be interpreted as resulting from fatigue or exhaustion of the operator.

1. Definition of the problem

The conventional process computer control rooms used in the chemical industry and in process engineering are currently being replaced by work stations using VDU's exclusively. It can be assumed that this will result in considerable shifts in the stresses and strains affecting the operators.

The fact that the work is performed exclusively on VDU's can result in an intensification of the effort required from the process control operator, as it is not uncommon for several control functions to be amalgamated and displayed on a single VDU system. It must also be expected that the job profile, which has previously involved identification of system failures and initiation of compensatory action, will change to one of pure supervisory control. Supervisory control may demand a higher degree of strategic thinking for the correct selection of menu windows on the various hierarchical levels. The operator must possess the necessary qualifications and it is essential for him to be adept in the execution of time-critical procedures. The nature of the operator's interaction with colleagues working in outlying parts of the chemical plant or in other control rooms will also change, in that it will, in part at least, take place by VDU.

2. The work system

This paper reports on a case study from the chemical industry (manufacture of chloromethane and chloroethylene) in which a before/after comparison was made between a classic process computer control procedure ("Cubicle Mod5") and a mouse-driven work station with two VDU's using the windowing technique ("Operator Station").

In the older system the operator works standing/ walking/ sitting in an airconditioned, artifically lit room in which the working environment is otherwise more or less neutral. His main functions are:

- setting target values
- eliminating system failures
- starting up and stopping the installation

- monitoring the running of the installation
- checking process data
- miscellaneous tasks.

Whereas in the older work system the man-machine interaction occurs mainly via function keyboards on the individual pieces of equipment and via set wheels or other regulatory devices, in the new situation it is mainly via mouse actions. In our investigation the number of operator actions per time unit varied between 0.24 and 0.70 actions per minute.

3. Method

A polygraphic concept was used to measure the comparative stresses and strains. This included:
- the AET technique for analysing stresses
- the use of checklists to assess the ergonomics of the hardware and software (90/270/EWG, DIN 66234, ISO 9241 ANSI/HFS 100-1988)
- the investigation of work sequences and work paths (Landau, Stübler, 1992)
- the measurement of pulse rate and cardiac arrhythmias, electromyograms and electro-
     oculograms (Rohmert, 1979)
- the use of a questionnaire to analyse subjective perception of strains and physical problems (Landau, Stübler, 1992).

The investigation was performed on three working days, each with an early and a late shift, and two different operators. Basic data on the test persons and detailed information on the trial can be referred to in Landau and Wendt (1994).

4. Analysis of stresses

An ergonomic analysis of the hardware using a check-list covering 95 items showed clearly that the character display on both VDU's was too small by international norms. Another criticism was that the mouse keys were not freely programmable so that regularly recurring commands could not be placed on the right key of the mouse. The hardware achieved a score of 86% of the maximum theoretically achievable points on the check-list. The software analysis covered 117 items relating to the criteria:
- suitability for the task   - conformity with expectations
- self-explanatory display   - in-built safeguards against breakdowns
- controllability   resulting from errors

It achieved the rating "good" with a score of 84% of the theoretical maximum. The weaknesses lay in the dialog design. One negative point was the mixture of English and German dialog. Details of the ratings obtained in the check-lists can be referred to in Landau and Wendt (1994).

A comparison of the stresses by the AET procedure shows that no dramatic shifts in stresses occurred as a result of the change from Cubicle Mod5 to the operator station (Fig. 1).

Computer-aided monitoring work was involved in both cases, albeit with different computer generations. The frequency of use of regulatory devices occurs in roughly the same order of magnitude as in conventional control rooms, but there are shifts from set wheels and switches to keyboard and mouse operations of a different nature which is not identifiable in this profile.

There were no special stress factors resulting from environmental influences. No new stress situation arose as a result of the social or organizational working conditions. As is to be expected, the increase in VDU work led to a rise in demands in the field of visual information reception and the same applies to the proprioceptive feedback from the mouse operation. There is an increase in the amount of knowledge required. This is attributable to the EDP skills demanded.

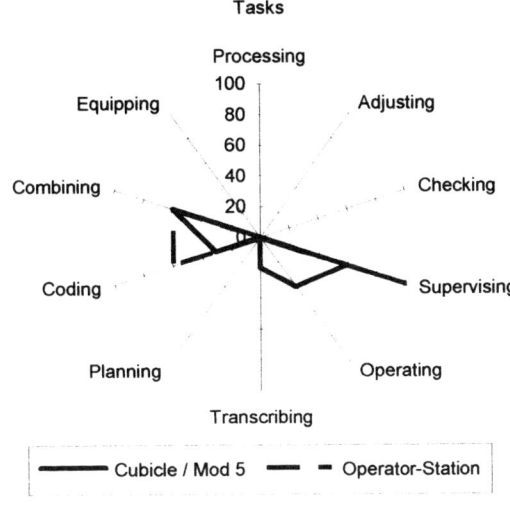

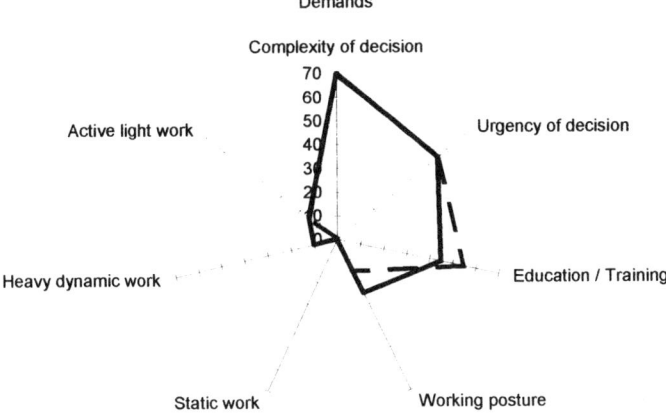

Fig. 1: Comparison of AET stress profile for Cubicle Mod5 and operator station

An analysis of the physical work involved shows that the work is now performed mainly in a sitting posture, and this means that stresses arising from standing postures are eliminated. The amount of heavy dynamic work of the legs has also diminished as a result of the virtual elimination of walking activity. The degree of unilateral dynamic work of the hand-arm system remains unchanged.

5. Analysis of strains

The following table summarises the differences in strains arising from operation of the conventional Cubicle Mod5 system and work at the operator station (Fig. 2). It must be remembered that the sample studied was small and that the results are therefore not generalizable.

			Conventional control panel work ("Cubicle Mod 5")		Operator Station	
			absolute	scaled	absolute	scaled
No. of operaror actions [1/min]		M	0,42	-	0,26	-
Heart rate		M	104,75	115,23	97,50	107,25
[1/min]		SD	5,29	5,82	4,66	5,12
Heart rate arrhythmia		M	2,60	2,69	4,21	3,61
[-]		SD	2,12	2,21	1,27	1,32
Myograms	M. Occipito Frontalis	M	38,55	40,75	73,99	77,31
[% of an MVC contaction]		SD	9,00	9,53	10,28	12,01
	M. Trapezius	M	25,78	52,19	43,46	46,47
		SD	9,54	18,98	12,31	11,97
Oculograms	horizontal	M	32,32	34,16	44,45	43,10
[% of maximum value]		SD	14,07	14,94	21,66	20,45
	vertical	M	41,57	39,94	42,63	45,52
		SD	16,85	16,36	19,78	20,52

Fig. 2: Number of operator actions as a measure of stress and comparison between conventional control panel work and the operator station using four physiological parameters to determine strains (Mean statistical values for two test persons in each of two shifts for each work system)

The mean heart rate registered during operation indicates that the level of physical stresses does not cause the limina for long-term energetic strain to be exceeded. The incidence of cardiac arrhythmias was highest during work at the operator station. This indicates that the psycho-mental strain is lower there than in work at the conventional control panel.

Neither the heart rate time sequences nor the other physiological parameters reveal any trend related to the time of shift and it is therefore possible to assume that the psycho-physical condition of all the test persons is in a steady state. Scaling of the parameters to the number of operator actions per time unit reveals only minor increases in strain during work at the operator station. Interviews with the operators before and after work failed to reveal any increases in strains arising during work at the operator station (see Landau and Wendt, 1994).

References

DIN 66234, Bildschirmarbeitsplätze, Parts 1 to 8
Guideline of the Council of the European Communities (EC) dated 29 May 1990 (90/270/EWG) on minimum safety and health requirements for work with VDU's. In the Official Gazette of the European Communities No. L 156/14 of 21 June 1990
Human Factors Society, American National Standard for Human Factors. Engineering of Display Terminal Workstations (ANSI/HFS 100-1988), Santa Monica, Ca, 1988
International Standards Organization (ISO), Visual Display Terminals (VDT's) Use for Office Tasks. International Standard 9241, Part 3, Ergonomic Visual Display Requirements, Geneva, Switzerland. ISO
Landau, K.,Stübler, E. (Hrsg.): Die Arbeit im Dienstleistungsbetrieb, Stuttgart, Ulmer Verlag 1992
Landau, K.,Wendt, G.: Stress and strain of operators in supervisory control systems of chloroethylene (vinyl chloride) and chloromethane (methyl chloride) plants. Report no. 947, Institute for Work Organization, Bad Urach, 1994
Rohmert, W.: Determination of Stress and Strain at Real Workplaces: Methods and Results of Field Studies with Air Traffic Control Officers. In: MORAY, N. (ED:): Workload, its theory and measurement, Plenum Press, New York and London, 1979, 423-443
Rohmert, W, Landau, K.: A New Technique for Job Analysis, London, Taylor & Francis, 1983

Advances in Agile Manufacturing
P.T. Kidd and W. Karwowski (Eds.)
IOS Press, 1994

The Human Interface with Virtual Reality and Its Impact on Advanced Manufacturing

Lon N. Haney and Henry A. Romero

Idaho National Engineering Laboratory, EG&G Idaho, Idaho Falls, ID, USA 83415-3855

Abstract. This paper introduces a vision of using virtual reality to increase human performance for advanced manufacturing activities. The benefits to advanced manufacturing include increased effectiveness and reduced costs for research, design, development, quality and production. Human factors research is needed to determine how to structure the virtual environment to optimize human performance and reliability for practical applications.

1. Background

Virtual reality (VR) is an emerging technology having the potential to significantly shape human endeavors. The use of virtual environments (VEs) have been postulated to increase efficiency, performance and safety in communication, scientific inquiry, education, medicine, engineering, design, industry, process control, commerce, space exploration, aviation, navigation and traffic control (land, sea, air, orbital, and planetary), robotics, teleoperation, telepresence, architecture, chemistry, genetics, travel, and recreation, etc. In order to fulfill this potential, numerous issues must be explored, including human factors.

Null and Jenkins provide a compilation of discussions on potential VR applications for space exploration, telerobotics, and aeronautics [1]. Sheridan and Zeltzer report on the development of a virtual cadaver for use by medical students, and of research for a VR system for air traffic control [2]. Moshell and Hughes describe a vision of how VEs will transform the educational process for the twenty-first century [3]. Haney presents a vision of how human performance and reliability in complex systems could be increased by using VR to reduce cognitive complexity and increase situation awareness [4]. Caudell and Mizell describe development efforts and applications at Boeing for the use of VR in production job aids (e.g., replace expensive, bulky, maintenance intensive wiring forms and mandrils) and design tools [5]. Ellis provides an overview of the nature and origins of virtual environments [6].

One potential benefit of the use of VR is increased efficiency and safety for all phases of manufacturing from design to product distribution. A mechanism for these benefits would be the use of VR technology for the enhancement of design and operational environments for interactions with, and control of, the manufacturing process.

2. Virtual reality

There are many emerging and evolving concepts and definitions of virtual reality. Any representation that emulates reality (i.e. a drawing, a photograph, a movie, an audio recording) is, in a sense, a virtual reality. The concept of virtual reality discussed in this paper involves a software structure which allows some representation of a physical or conceptual object, environment, world, data set, or other entity, and the requisite hardware to present the representation to the human senses. The representation could be dynamic and interactive. The virtual representation could be combined (e.g. overlaid) with the real world if desired (e.g., transparent VR).

Possible hardware could include head mounted devices (i.e. helmet or eyephones) capable of presenting high fidelity three dimensional (3D) visual and 3D auditory displays to an individual. Three dimensional visual displays with a wide field of vision (e.g. 150 degrees or greater) can facilitate an experience of "immersion". Immersion is the experience of a feeling, emotion, or sensation of being in the display rather than external to it. Special gloves, body suits, exoskeletons, or some combination of these could provide high fidelity haptic displays providing sensations of touch, vibration, and force feedback. Thermal displays could also be designed and provided.

Part or all of the users body or some more abstract representation of "self" may be represented in the display. Interaction with the virtual environment may be accomplished through gestures, voice, or using real or virtual (i.e. existing only in cyberspace) 3D widgets such as a flying mouse or a wand. Scale and movement in the VE would be under the control of the user.

Using a VR interface, interaction with a machine or process could be designed to support natural human capabilities and styles (i.e. 3D vision, pointing, touching, hearing, and talking). Human physical characteristics and individual experience in the real world result in particular types of interactions with the surroundings that are innate to humans and are characteristic of the human condition. Hypothetically, these interaction styles could be used to facilitate correct perception of important information with less cognitive demand than is required using more artificial interactions such as the abstract symbology inherent in typical mimic or text displays.   A plant or system could be represented in a virtual environment such that the operators could virtually see, hear, and feel the systems and components to determine status. Critical states, parameters, and safety functions could be presented in more natural and intuitive ways with reduced need for multiple alphanumeric displays requiring cognitive integration. Salient visual (e.g. shape, color, movement, etc.), auditory and haptic cues for abnormal conditions or failed components could be presented.

Relationships between multiple systems could be shown in a natural way. Control of the level of detail may be analogous to "looking closer". Control of scale and easy movement could facilitate appropriate interaction within the virtual environment according to operator needs. Natural voice and gesture commands could be used to manipulate systems, components, parameters and the VE.

Manipulation of components and parameters could occur directly rather than through controls and displays "separated" from the process. The manufacturing process could be experienced as well as manipulated in a first hand manner. Performance could be enhanced by reducing cognitive complexity and increasing situation awareness. Human reliability could be increased by reducing the opportunity and probability of critical human errors during normal operation and while responding to accident conditions.

3. Potential uses and benefits of VR for manufacturing

The potential uses and benefits associated with the use of VR and VEs in manufacturing environments involve reduced costs associated with manufacturing, research and development, production, and prototype development; decreased research and development time; improved designs; increased system reliability; improved customer responsiveness; and removing humans from potentially hazardous environments. Transparent VR can be used as a production and maintenance aid increasing reliability, decreasing maintenance and production time, improving training, and increasing safety. Through the use of VE and VR, the designers and developers can actually "see" the piece or system being designed and the manner in which the piece or system functions in operational environments. Haptic displays will allow designers and potential users to operate controls and evaluate instrumentation configurations in "real" environments receiving force-feedback responses from the designed piece or system. Plant and system representations in VE will allow more accurate simulations increasing reliability and ascertaining "real" reactions from potential operators. Finally, the use of VE and VR in prototype development will reduce all quality costs including the cost of materials, scraps, redesign efforts, and lead time to production. VE and VR in prototype development can allow designers to test exotic materials with relatively little cost and determine system parameters without investing in expensive hardware development. The use of VE and VR can facilitate timely transition from automation to manual control should automation fail by increasing situation awareness.

4. Research needs

Efforts by leaders in VR technology have focused on creating hardware and software systems and initial "virtual worlds". Less effort has been made in terms of scientific investigation of the characteristics of a VR interface which would best promote human performance. Significant diversity is arising concerning the nature of the VR interface. Extensive guidelines and standards have been developed for 2D and 2 1/2D CRT and analog human-machine interfaces. For example, Gilmore, et al, at the INEL developed extensive human factors guidelines for 2D interfaces [7]. Validated human factors guidelines for the development of the VR interface do not exist.

Human factors research is needed to determine how to structure the virtual environment to optimize human performance and reliability for practical applications. Development of rigorous "standards" for VE at this stage of the evolution of the technology may not be prudent in terms of potential limitations of creative development. However, development of human factors guidelines would facilitate practical applications and enhance the usefulness of the technology.

Research issues include: methods of visual, auditory, and haptic representation; methods of interaction; as well as simulation sickness and emotional response. For example research is needed to determine how best to represent hardware and parameters (e.g. system components, physical status, temperature, pressure, flow, etc.) to present intuitive representations to the operator while minimizing or eliminating the use of alphanumeric text. Color, shape, movement, auditory, thermal, and haptic coding are possible mediums for representation. Issues concerning combining VR and the real world to facilitate cognitive integration of information requires investigation. Research into representation of off normal events and alarm representation and handling is important. Also, how to present procedures to the operator in the VE requires careful investigation. Research is needed on methods of interaction for controlling hardware and parameters and for controlling scale and movement within the VE. For example, determining the appropriate gestures, voice commands, or 3D widgets that are best in terms of minimizing error and facilitating appropriate control. Also, because the applications may involve extended exposure of the operator to the VE, issues relative to simulation sickness and emotional response to the VE are important topics for research.

5. Conclusions

Practical implementation of VR without addressing these issues could result in less than optimal applications, even having human performance and reliability at levels lower than more traditional interfaces. Failure to address these research issues would tend to limit the practical applications of VR and would slow the expansion of the state-of-the-art for VR.

References

[1]   Null, C. H. and Jenkins, J. P., NASA Virtual Environment Research Applications, and Technology, <u>A National Aeronautics and Space Administration White Paper</u>, October 1993.

[2]   Sheridan, T. B. and Zeltzer, D., Virtual Reality Check, <u>Technology Review</u>, October 22, 1993.

[3]   Moshell, J. M. and Hughes, C. E., Shared Virtual Worlds for Education, Proceedings of the 4th Annual Virtual Reality Conference & Expo. (Mekler Publishing Co.), San Jose, CA., 19-21 May, 1993.

[4]   Haney, L. N., Virtual Reality for Human Performance and Reliability in Complex Systems, Proceedings of the 12th Int. System Safety Society Conference, New Orleans, LA., July 1994, In Press.

[5]   Caudell, T. P. and Mizell, D. W., Augmented Reality: An Appplication of Heads-Up Display Technology to Manual Manufacturing Processes, IEEE publication, 659-669, January, 1992.

[6]   Ellis, S. R., Nature and Origins of Virtual Environments: A Bibliographical Essay, <u>Computing Systems in Engineering</u>, Vol.2, No. 4., 321-347, 1991.

[7]   Gilmore, W. E., Gertman, D. I. and Blackman, H. S., <u>User-computer interface in process control: A human factors engineering handbook</u>, Academic Press, Inc., San Diego, CA, 1989.

Advances in Agile Manufacturing
P.T. Kidd and W. Karwowski (Eds.)
IOS Press, 1994

Abstraction Hierarchy Representation of Manufacturing: Towards Ecological Interfaces for Advanced Manufacturing Systems

Anne-Marie Kinsley and Joseph Sharit
Department of Industrial Engineering, State University of
New York at Buffalo, 342 Bell Hall, Buffalo, NY 14260-2050, U.S.A.

Kim J. Vicente
Department of Industrial Engineering, University of Toronto,
4 Taddle Creek Road, Toronto, Ontario, CANADA M58 1A4

Abstract. This paper develops the theory needed to apply the ecological approach to interface design to advanced manufacturing systems in terms of Rasmussen's abstraction hierarchy. Differences with continuous systems are noted and their effects on interface design are discussed.

1. Introduction: Ecological Interface Design

Rasmussen's abstraction hierarchy (AH) [1] and skills-rules-knowledge taxonomy [2] have been rich conceptual tools for studying human interaction with complex computerized systems. They have also been the basis for applied work, including the theory of ecological interface design (EID) [3], [4]. The goal of EID is to support the operator at all three levels (skills, rules, and knowledge) of performance, simultaneously exploiting the efficiency of perceptual processing and enhancing problem-solving, by making visible the invisible constraints in the system. Among other requirements, EID specifies that the information content of the interface must include representations of the work domain at all levels of the AH, so as to provide a normative visualization of system functioning.

EID was developed in the context of process control and to date has been applied only to continuous systems (e.g., [5]). Despite a previous unsuccessful effort to apply the AH to an FMS [6], there are good theoretical reasons for believing that the applicability of EID is not limited to process control [7]. However, the invisible constraints in discrete manufacturing are different in nature from those in process control. This paper is a systematic attempt to extend the theory of EID to advanced manufacturing systems (AMSs).

2. AH Representation of an AMS

Very briefly, the AH contains at each level a description of the entire AMS, including its topology, in terms of one set of objects and activities. Furthermore, means-end relations exist between successive levels, such that each level gives an implementation of (i.e., a means of carrying out) the description at the next higher level, and an explanation of (i.e., the end served by) the next lower level. (For a more detailed description, see [7].)

The level of physical form is the lowest and most concrete. Here the make, model, and location of every component in the system are specified. The physical connections among components create a topology of all the possible travel paths of the pallets that circulate in the system. For example, a description at this level could include a machining centre connected to a power-and-free conveyor by a turntable, which indicates that pallets can flow between the machining centre and the conveyor. It would also include the computer equipment (minicomputers, PCs, PLCS, and so on) in the system.

The level of physical function describes the system in terms of the functions performed by the components in the previous level's description. For instance, the machining centre might implement functions such as cutting and grinding, and a PC might implement the function of sequencing and AGV dispatching. These functions are linked together into not one but several topologies, corresponding to the process plans of the parts manufactured by the AMS. As an example, a given part might require milling, followed by drilling, followed in turn by tapping; this sequence would be contained in the topology for that part. The sequence is implemented, at the level of physical form, by a route that visits in turn machines that perform milling, drilling, and tapping. (Such a route is a subset of the physical form topology.) If, as is typical in AMSs, some physical functions are implemented by more than one machine at the level of physical form, there will be several such routes.

The third level, that of generalized function, specifies the general processes carried out by the system: loading, unloading, setup, machining, assembly, inspection, material handling, and scheduling. At this level it becomes clear that the system is partially self-organizing; that is, one element (the scheduling function) may change other elements without any input from an external controller. Again, the functions at the previous level are implementations of the functions at this level—physical functions such as sequencing and AGV control implement the general function of scheduling, physical functions such as drilling and tapping implement the general function of machining, and so on.

At the level of abstract function, the system is described by topologies of flow of conceptually distinct abstractions of the system. In a process control system, the abstractions would be mass and energy. For an AMS, they are mass, delay, value, and priority. These topologies are described in more detail below.

Finally, in the most abstract level of the hierarchy, the level of system purpose, the AMS is described in terms of the purposes for which it was designed: to produce parts correctly and efficiently. The abstractions found at the previous level instantiate and measure the fulfillment of these goals.

2.1 Topologies at the Level of Abstract Function

The mass topology describes the flow of material in the system, from the source (loading dock or other point at which pallets enter the system) to the sink (unload dock) through a series of inventories (all the other components). This flow is governed by a law of conservation of mass, which states in informal terms that a pallet can move only from an occupied space to an unoccupied one.

The delay topology describes the time each pallet has spent in the system. There is a delay inventory corresponding to each pallet. When a pallet visits a component for an operation (machining, travel, waiting, etc.), delay is added to the corresponding inventory. The delay sink corresponds to the pallet leaving the system upon completion. The flow of delay is described by a law stating that the predicted completion time is the sum of the initial time at entry, the delays due to all past operations, and the predicted delays due to all operations to come.

Similarly, the value topology sees the pallet as an inventory, and certain system components are sources of value. The initial value of a part (the value of the raw materials) comes from a source corresponding to the loading dock; a machining process adds value to the pallet, but waiting in a buffer does not. As in the delay topology, completion of work on the pallet corresponds to a value sink.

Finally, there is a topology of priority. Again, the pallets correspond to inventories, the loading dock provides a source of initial priority, and the completion of work corresponds to a sink. However, the priority topology includes other sinks and sources that do not correspond to system components but are instead the result of decisions made by the system controller. The most straightforward case is when, of two parts, the one of higher value or the more delayed one takes priority. At other times, the laws may be quite complicated. For instance, part A being manufactured for a one-time client will have a lower priority than an otherwise identical part B being manufactured for a client who may place a large repeat order on the basis of the current one. Part B's priority may be higher than that of a higher-value part or an overdue part.

3. Comparison of Continuous and Discrete Systems

In continuous domains such as process control, the mass and energy flow topologies contained in the level of abstract function correspond to physical laws (e.g., of thermodynamics) that govern system behavior. The lower levels of the AH are tightly coupled to the upper ones. Thus, given the control settings made by the operator, the behavior of the system—whether or not it will satisfy the goals—is determined by the laws. The laws themselves are not under the control of the operator.

In discrete manufacturing, by contrast, the physical coupling within the system is much looser. Many decisions must be made in the course of system operation: which routing through the AMS a pallet should take, in which order a machine should process jobs, how priorities should be assigned, and so on. These decisions amount to choosing which of several possible topologies will apply. The behavior of the system depends therefore not only on current settings of low-level parameters and on physical laws but also on future control decisions. Thus, some constraints in discrete manufacturing systems are "intentional" [7]; they are not hard-wired into the system, but must be chosen.

4. Using the AH for Ecological Interfaces

To fulfill the fundamental aims of EID, an interface for an AMS must reveal all levels of the AH and the means-end relations among them, thus making visible the invisible links between actions and goals. Thus, each of the topologies must be represented, as must the scope of possibilities for choosing among alternative topologies. This leads to an interface that shows whether and how well goals are being met (levels of abstract function and system purpose) as well as the action possibilities the operator can use to improve system performance (lower levels and alternative topologies). Thus, for example, an ecological interface would show a part's predicted finishing time, its due date, and all the sources of delay. It would also show all the sources of value (a subset of the sources of delay). The operator would be able to look at alternative topologies to find one with fewer sources of delay.

Individually, some of these elements seem simpler to represent than others. A time line can be used to show delay and a mimic diagram to show mass flow, but it is more difficult to imagine how value or priority should be represented. In combination, with the interrelationships that arise from the fact that the same physical components contribute to more than one abstract function, the task becomes even more challenging. Moreover, a number of decision support functions are likely to be needed in order to assist the operator or supervisor in evaluating alternative courses of action.

To be truly ecological, the interface must also present the information in a direct-manipulation format [3]. It is clear that the creation of ecological interfaces will require designers to exercise their imaginations.

5. Conclusion: Prospects and Benefits of Ecological Interfaces

While it is far from straightforward to create a design along the lines described in the previous section, the potential advantages of an ecological interface warrant a significant design and development effort. Such an interface would provide a normative "external mental model" [4] by showing the available alternatives for links that are chosen rather than hard-wired, and the effect of each alternative on functions such as timeliness and value. This gives AMS supervisors (or other users who may exercise local control over parts of the system) a rational basis for individual decision-making and provides a common representation of the system to support group decision-making.

Ecological interfaces are also intended to reduce users' memory load, because information available on the interface does not need to be remembered, and to favor the replacement of high-level cognition with less effortful, more efficient perceptual processing.

Given the importance of human behavior for AMS performance [8], the potential benefits of EID are well worth pursuing. It is hoped that this AH analysis will serve as a useful basis for the design of ecological interfaces in advanced manufacturing. As a first step, an experiment is being prepared to study control of a model AMS using an interface developed according to the framework reported here.

Acknowledgement

We are grateful to Jens Rasmussen for very helpful comments.

References

[1]   J. Rasmussen, Information Processing and Human-Machine Interaction: An Approach to Cognitive Engineering. North-Holland, New York, 1986.

[2]   J. Rasmussen, Skills, rules, knowledge: Signals, signs, and symbols, and other distinctions in human performance models, IEEE Transactions on Systems, Man, and Cybernetics SMC-13 (1983) 257-267.

[3]   K.J. Vicente and J. Rasmussen, The ecology of human-machine systems II: Mediating "direct perception" in complex work domains, Ecological Psychology 2 (1990) 207-250.

[4]   K.J. Vicente and J. Rasmussen, Ecological interface design: Theoretical foundations, IEEE Transactions on Systems, Man, and Cybernetics 22 (1992).

[5]   K.J. Vicente, Memory recall in a process control system: A measure of expertise and display effectiveness, Memory and Cognition 20 (1992) 356-373.

[6]   S.P. Krosner, C.M. Mitchell, and T. Govindaraj, Design of an FMS operator workstation using the Rasmussen abstraction hierarchy. In: Proceedings of the 1983 International Conference on Systems, Man, and Cybernetics. IEEE, New York, 1983, pp. 353-364.

[7]   J. Rasmussen and A.M. Pejtersen, Mahawc Taxonomy: Implications for Design and Evaluation (Report Risø-R-673(EN)). Risø Laboratory, Roskilde, Denmark, 1993.

[8]   Blumberg, M., and Alber, A. (1982). The human element: Its impact on the productivity of advanced batch manufacturing systems. Journal of Manufacturing Systems, 1, 43-52.

Advances in Agile Manufacturing
P.T. Kidd and W. Karwowski (Eds.)
IOS Press, 1994

MUSE : A Structured Human Factors Method for *US*ability *E*ngineering

Kee Yong LIM
School of Mechanical and Production Engineering, Nanyang Technological University,
Nanyang Avenue, Singapore 2263, SINGAPORE.

John LONG
Ergonomics Unit, University College London,
26 Bedford Way, London WC1H 0AP, UNITED KINGDOM.

Abstract. Existing problems of human factors input to system development are
highlighted. A solution is proposed comprising the development of a structured
human factors method that may be integrated with a similarly structured software
engineering method. Such a structured human factors method is described.

1. Introduction

Human factors input to system development is often constrained to late evaluation activities
following system implementation. At this stage of system development, design specifications
become 'frozen' and resistent to change. Consequently, late human factors involvement is
non-optimal since its recommendations would tend to become too costly, or be considered
impossible to implement. This unsatisfactory situation is referred to generally as the *'too-
little-too-late'* problem of human factors input to system development.

One solution to the problem is to integrate explicitly the design activities of human factors
and software engineering *throughout* the system development cycle. In this way, human
factors inputs (with respect to their scope, timing, granularity and format), may be set
appropriately against the system development context. To support integration, conceptions of
complete design processes of the individual disciplines need to be defined explicitly. Such
conceptions have been established in structured analysis and design methods, as exemplified
by their explicit stage-wise design products, processes, and notations. By exploiting these
methods, human factors and software engineering design concerns that intersect may be
identified more easily for integration. The exploitation of structured analysis and design
methods is also desirable, since the methods (and computer tools that derive from them) are
already well established in software engineering. Without the development of similar methods
(and hence tools) for human factors, its contributions to system development may become
less effective in the future. In particular, existing conceptions and techniques of human
factors design should be developed into structured analysis and design methods. Efforts in
this direction led to the development of a structured human factors method named *MUSE*
(*M*ethod for *US*ability *E*ngineering). An overview of the method is described in this paper.

2. MUSE: A Structured Human Factors **M**ethod for **US**ability **E**ngineering

Generally, MUSE comprises three phases that address the following human factors concerns:

(i) *Information Elicitation and Analysis Phase*. Background information to support design is derived and processed at this phase. In addition, basic human factors analyses are also undertaken in the two stages of the phase (see Figure 1);

(ii) *Design Synthesis Phase*. This phase is concerned with the definition of the initial system and its sub-systems. One or more conceptual designs of the new system is/are then synthesised. Three stages comprise the phase (see Figure 1);

(iii) *Design Specification Phase*. The three stages of this phase address the detailed specification of the new system (see Figure 1). In particular, human factors specifications of a user interface design, are derived.

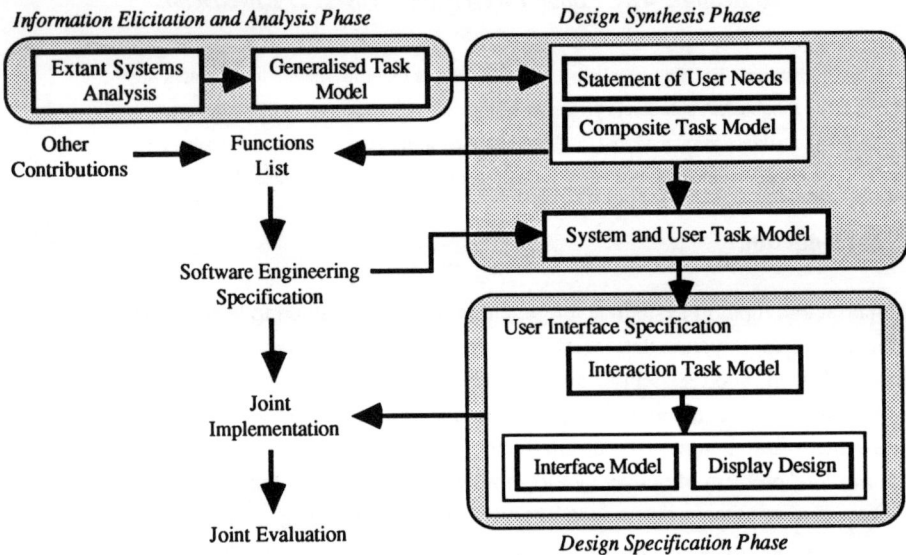

Figure 1 : A Simple Representation of MUSE, a Structured Human Factors Method

A brief description of the design stages of MUSE follows.

(1) *Extant Systems Analysis Stage*. As indicated earlier, background information to support system development is generated at this stage. In particular, information on the current system (and other extant systems if appropriate) is elicited from users and other sources. The information is analysed to uncover critical user needs and problems; salient characteristics of current tasks; features and rationale of the existing user interface design; etc. The objective of the analysis is two-fold. First, information is gathered to set new design requirements against the context of the current system. Second, extant system designs are assessed with respect to the new system. Together, these activities support a better understanding of the implications of the new design. On the basis of the information, the potential and consequential effects of

porting (or non-porting) of particular design features from the extant system to the new system, may be considered. For instance, assessments of possible transfer of learning effects (both positive and negative effects) would be supported.

(2) *Generalised Task Model Stage.* This design stage is concerned with processing the information gathered at the preceding stage. The objective is to generate (predominantly) device independent descriptions to support analytic mapping between specific extant design features and new system requirements. To this end, generalised task models are derived to characterise the conceptual design of *new* system and each *extant* system analysed. By comparing the models, appropriate extant design features may be identified and selected for porting to the new system. Similarly, new design features that conflict with the current system may be avoided later.

(3) *Statement of User Needs Stage.* At this stage, conclusions of extant system analysis are collated and summarised with respect to the user. The statements derived constitute a set of human factors requirements for the new system. By formulating user needs explicitly, their incorporation into an overall statement of requirements for the new system, is facilitated. A user-centered basis for developing the new system may thus be established. To this end, the information documented should include problems experienced by existing users; the design requirements, constraints, and rationale of existing and new design features; performance criteria; domain semantics; etc.

(4) *Composite Task Model Stage.* A conceptual design of the new system is generated at this stage, to support the allocation of functions between the human and computer. In particular, a composite task model is synthesised from compatible sub-sets of the generalised task models derived earlier. On-line and off-line tasks are then designated in the composite task model.

(5) *System and User Task Model Stage.* This stage addresses the decomposition of on-line and off-line tasks (see (4) above). Specifically, *on-line task* components of the composite task model are decomposed to generate a *system task model* to define the cycles of human and computer actions entailed by the interactive task. Similarly, *off-line task* components are decomposed to generate a *user task model* to describe tasks not supported by the computer. During decomposition, appropriate design features of extant systems that are consistent with the statement of user needs, may be considered for incorporation.

(6) *Interaction Task Model Stage.* At this stage, an *interaction task model* is derived by decomposing the *human action components* of the system task model. A device level description of user inputs required by the interactive task is derived. Points for triggering and removing computer displays may then be defined by locating the start- and end-points of coherent groups of user inputs. The resulting specification provides a *user-oriented perspective* of the interactive task (assumed to be error-free at this stage -- see (8)). Generally, two rules of thumb guide the derivation of an appropriate interaction task model. First, the model should be decomposed to a level understood easily by design team members and end-users. Second, the terms used its description should be consistent with the primitives of the adopted user interface environment (e.g. screen objects and actions) and hardware (e.g. basic keystrokes). On the basis of an appropriate interaction task model, design activities of the remaining stages of the method may then be undertaken. Note that the development of training programmes and user manuals would also be supported by such a task model.

(7) *Interface Model Stage.* A set of human factors descriptions (termed interface models) of the behaviour of user interface objects, is derived at this stage by decomposing *computer*

action components of the system task model. The models specify how the appearances and behaviours of user interface objects would change in response to user inputs, and to changes in attribute states of domain entities. The specification derived provides a *computer-oriented perspective* of the interactions required to perform the on-line task.

(8) *Display Design Stage.* This design stage addresses three aspects of human factors design. First, potential user errors are identified by examining each input prescribed by the interaction task model. To avoid such errors, modifications may be considered to 'design-out' the error occurrences. Problems that can not be resolved may then be accommodated by a more tolerant design and a greater emphasis on user training. To address these concerns, design iterations may be required with two stages that precede the present stage. Second, *static* characteristics of computer displays are specified; namely the composition and layout of screen displays of computer support functions, and error, feedback and help messages. Third, the *dynamic* characteristics of computer displays are specified; namely the context for actuating particular displays to support the human-computer dialogue and the interactive task. Thus, the description specifies how, when and what computer functions and messages are presented to support each stage of the user's task. Note that the specifications derived at this stage should be consistent with the interface and interaction task models derived earlier.

This account completes an overview of MUSE. For a more detailed description of the method, the reader is referred to [1] and [2].

3. Summary

Case-studies conducted during method development, have indicated MUSE to be a promising structured human factors method. For more effective incorporation of human factors in system development, MUSE has also been integrated with a particular structured software engineering method (see [1]). To this end, design inter-dependencies between the disciplines have been identified explicitly in the integrated method to facilitate design collaboration. Results of case-studies conducted during method development were generally positive.

Both projects to develop and test the methods have now been completed. Presently, a follow-up project is underway (funded by the European Community) to trial and disseminate MUSE at a number of industrial sites. Results of the project will be reported at a later date.

Acknowledgements

Part of the research associated with this paper was carried out for the Procurement Executive, Ministry of Defence (UK). Acknowledgements are extended to individual members of staff at CA2 (RARDE). However, views expressed in the paper are those of the authors and should not be attributed to the Ministry or its staff.

References

[1] Lim, K. Y., Long, J. B. and Silcock, N., 1992, Integrating Human Factors with the Jackson System Development Method : An Illustrated Overview. In : Barber, P. and Laws, J. (eds.), Ergonomics (Special Issue on Cognitive Ergonomics III), 1992, 33 (12), Taylor & Francis, London.
[2] Lim, K. Y. and Long, J. B., 1994, MUSE : A Structured Human Factors Method for USability Engineering. Book to be published by Cambridge University Press.

Advances in Agile Manufacturing
P.T. Kidd and W. Karwowski (Eds.)
IOS Press, 1994

Human Factors in the Design of a Mass Spectrometer Human-Computer Interface

Jennifer Winstanley, VECTRA Technologies (UK) Ltd
Europa House, 310 Europa Boulevard, Gemini Business Park,
Westbrook, Warrington, Cheshire WA5 5YQ

Abstract. The research concerned the computer interface of a scientific instrument used for chemical analysis; a mass spectrometer. Software had been developed by the company to allow operators to control the mass spectrometer via a computer interface replacing a system using analogue dials, but they had found user resistance to its implementation.

The objectives of the research were twofold. For the company, it was hoped that considering how operators perform the task and the expert knowledge used in the control of these instruments would allow a more usable and acceptable system to be developed and thereby improve its commercial viability. For the researcher it was a way to demonstrate the benefits of considering human factors in the design of a complex interactive system.

The work included the design of an ergonomic interface and an assessment of its usability and acceptability. This was done by developing a simulation of the current interface and comparing it with a simulation of an ergonomically designed interface. A combination of laboratory-based experimental work and the more ecologically valid approach, employing users of the system in a work setting, was adopted. An iterative approach was taken, revising the interface design as comments from analytical chemists were received before the final experimentation phase. The new design of interface was assessed using three measures; performance accuracy on one selected task, the time taken to perform the task, and user preferences assessed through verbal protocol analysis.

The results revealed that novice subjects show better performance scores for the revised design, but showed a non-significant time difference in performance. Analytical chemists carried out the task more quickly with the revised design, but showed no benefits in performance accuracy. Novice users are able to set the instrument more accurately when ergonomically designed. Once expert, users meet the required standard of accuracy in a shorter space of time and the interface is preferred.

1. Introduction

The desire to make the operation of their mass spectrometer cost-effective and more readily used by novice users resulted in the company transferring from a system which had been in use for many years. The original system combined analogue dials and oscilloscope presentation of information and the intention was to change to a computer-based system. Experienced operators felt that the new system was slower and less easily used than the original. Such was the concern that a compromise system was developed and preferred, with separate analogue dials used in conjunction with a computerised facility.

The researcher was asked to apply Human Factors knowledge with a view to improving the interface, by making it more usable and acceptable to those who currently set-up the mass spectrometer. This should, in the future improve its commercial viability.

The researcher considered the issues associated with the introduction of computer technology. The advantage of the flexibility provided by computer screen displays is countered by limitations, such as information presented to operators visually over a small surface area. There are differences between the information processing capacity of the human operator and that of the computer. The human operator can process information 'in parallel' but they have a lower capacity for serial processing. The computer, on the other hand, has a high serial processing capacity. This discrepancy results in limitations on interface design.

Problems can arise when operators try to change to the use of a new interface from a task where specific behaviours are well-learnt. For example, the computer generally uses one control device for all actions, where human operators often learn to use several. Driving a car would be a good example, where the car can be steered at the same time as changing gear and braking. This way, an individual can learn to operate a control without visual search, relying on kinaesthetic sense to direct their movements. When a computer interface is implemented, where one device is used to perform several tasks and the operator relies on only visual sense, the result is longer search time and more distraction from thinking about the task [1].

Research into this usability and acceptability problem had the potential to assist the company to move over to a completely computerised system, which meets the capabilities and addresses the limitations of those who will be using the instrument.

The methodology used was one where usability and acceptability would need to be measured. 'Usability' has been defined as "the extent to which users can exploit the potential of the system given that task needs are met". Acceptability "describes how willing users are to use a system in their own organisational context" [2]. Three factors were identified in attempting to operationalise the terms "acceptability" and "usability" for this application:

1. Accuracy of the tuning procedure which will ultimately result in a more accurate analysis of the substance, therefore achieving the primary goal of the operator at this task.
2. The time it takes to perform the task, which will give an indication of the ease by which the task can be achieved.
3. Do they prefer the new interface to the one they currently use? Is it more acceptable than the present system? An important consideration where customer sales are involved.

2. Methodology

Initially verbal protocol analysis from analytical chemists performing one task on the mass spectrometer, and demonstrations of the existing system, provided the information used to write the software to simulate the system. Successive versions of the simulation were assessed by analytical chemist for accuracy of representation of the real interface. Successive iterations allowed the simulation of one of the main tasks performed on the system to be realistically simulated; that of tuning or adjusting the electromagnetic plates within the instrument to achieve the most accurate analysis of the sample. Once the simulation was considered to be an accurate representation of the system the experimentation phase could begin.

The knowledge elicitation aspects of the task analysis provided an indication of the problems experienced by operators. Three main areas of concern emerged and the revised design attempted to address these areas:

1. The increased processing of information from one source (i.e. visually) required by the computer presentation of information.
   Attention can be 'zoomed in' on one aspect or can be 'zoomed out' to encompass more information. However, attentional capacity is limited when the field of view is enlarged in order to encompass more information. In this complex task the operators will focus-in on critical aspects of the screen and be unable to process the information which replaces the dials of the original analogue device.
2. Density of information on the screen.
   The screen showed an enormous amount of information which was not specifically required for the task. This information had the potential to produce an interference effect.
3. The way in which the control information was presented.
   The part of the display which replaced the original knobs and dials was presented in a way that was not compatible with operator's mouse-control abilities. The way in which these were used would be completely revised.

Novice subjects were used to represent the customers, although in reality many customers do have some experience in the use of a mass spectrometer, unlike the novice subjects used in this study. Analytical chemists were used from the company to represent experienced users.

The system's acceptability and usability was assessed using three measures, *usability* was assessed by (i) performance accuracy and (ii) the time taken to perform the task, and *acceptability* was assessed by (iii) verbal protocol analysis with experienced users. The first trial was utilised as a familiarisation exercise for novice users.

3. Results

Statistical analysis, using the T-test, showed that novice users demonstrate a significant improvement with the revised design of the computer interface compared to the simulation of the original design, but showed a non-significant difference in the time taken to perform the task (see Table 1).

Table 1: Between Subjects Design - Novice users

Usability dimension	Original Design	Revised design	t-value
Performance	23.19 σ 4.52	27.48 σ 4.02	**2.40 p<0.05**
Time	209sec. σ 83.54	239sec. σ 50.70	**1.06 p>0.05**

Experienced users showed a significant improvement in the time taken to carry out the task with the revised design of computer interface but did not show a difference in performance accuracy between the original and revised designs. The non-significant result was still in the direction expected, with subjects maintaining their high standard of performance (see Table 2).

Table 2: Within Subjects Design - Experienced users

Usability dimension	Original Design	Revised design	t-value	
Performance	31.54 σ 6.53	34.12 σ 7.51	**0.56**	**p>0.05**
Time	294sec. σ 116.82	173sec. σ 60.22	**2.80**	**p<0.05**

Perhaps of greater importance is that analytical chemists preferred the revised design to the original and found it easier to use, which augurs well for future customers.

The results provide good evidence that the ergonomically designed interface is more easily used than the original computer interface, and of the benefits to be gained by incorporating Human Factors in the design.

4. Discussion

Consideration of the task in terms of the cognitive as well as behavioural aspects of operation, through verbal protocol analysis, allowed the computer interface to more readily meet user requirements, capabilities and limitations.

The serial processing capability of a computer system cannot be matched by the human operators in this task. Operators who previously had feedback on task operation provided by a combination of knobs and dials giving tactual feedback as well as visual presentation of information, ensured that one sensory mode was not overloaded. The computerised system required that all information was processed visually by the operator.

The possible future advantages to be gained by this research is in the partial automation of some of the tasks i.e. produce an expert system, by the use of verbal protocol analysis. Although not statistically tested, there seemed to be a large difference in performance scores between novice and experienced users. The benefits of a hybrid system would be that the knowledge used by experts, if it could be incorporated into the system, would allow potential customers to benefit from this expertise to provide superior analysis of samples and also from ease of use of the system.

As with much research, the company and technology has moved on since this research was carried out. The design of interface and the operation of the mass spectrometer is no longer comparable to the one the researcher studied. Therefore, the success of such a design in "the real world" has not been assessed.

References

[1] Bainbridge L, VDT/VDU Interfaces for Process Control. In: Wilson J R, Corlett E N and Manenica I (Eds), New Methods in Applied Ergonomics. Taylor & Francis, London, 1987.
[2] Richardson S, Operationalising Usability and Acceptability: A Methodological Review. In: Wilson J R, Corlett E N and Manenica I (Eds), New Methods in Applied Ergonomics. Taylor & Francis, London, 1987.

Advances in Agile Manufacturing
P.T. Kidd and W. Karwowski (Eds.)
IOS Press, 1994

309

Using Visual and Auditory Components in Multimodal Interfaces for Control Environments

Dawn Michelle Roberts
F. Layne Wallace

Dept. of Computer & Information Sciences   4567 St. Johns Bluff Rd., S.
University of North Florida   Jacksonville, FL 32224 USA

Abstract. In the computer control environment, a delicate balance must be maintained between presenting too little information and "cognitive overload." The present study tested this balance by using two forms of information displays, graphic and textual, in a power plant control situation. Additionally, three different forms of auditory information were presented with each of the displays. The subjects were asked to monitor information related to power plant operations and perform an alternative task (completion of a spreadsheet) to provide a more realistic testing environment. Response times, completion times, and accuracy served as dependent measures. Both independent dimensions (visual and auditory) showed significant differences among groups on the dependent measures. Graphic information seemed to be superior to plain text information and auditory alarms appeared to be beneficial with a minimum of cognitive overhead.

1. Introduction

In the computer control environment, a delicate balance must be maintained between presenting too little information and avoiding "cognitive overload" [1]. Logical would tell us that having auditory information to supplement the visual information should lead to faster response times and more accurate performance. However, Polson [2] presents three forms of interference which may degrade performance when switching from one modality to another; attention switching, modality switching, and task switching. Also, Wickens [3] postulated that there are definite costs associated with cross-modal information presentation. He suggested that there are two categories of theories which may account for the effects of multiple modes of information presentation; Structural theories and capacity theories. The capacity theories seem to be most useful in a descriptive role. Cross-modal information presentation seems to be primarily related to task structure. Years earlier, Colquhoun [4] had performed a series of studies which examined the effects of adding auditory information to visual information. The auditory information seemed to be primarily attentional while the visual information was conceptual.

An alarm system interface should satisfy two requirements: A) that the alarm interface be effective; and B) that the interface not incur a cost in total human performance, either in time or accuracy [5]. Since most alarm systems are embedded within an overall control system, it is important that the alarm interface not degrade performance of other tasks. Work by Wickens [3] suggests that a cross-modal alarm system might be most effective if the cost of modality switching is kept to a minimum.

While many of these studies examined the aspects of cross-modal information presentation, few of them set the research in a control situation where multiple tasks are required of the subjects. The present study simulated a control situation based on an electrical power plant.

2. Methodology

The current study attempted to examine the effect of two types of visual presentation of information and three types of audio information presentation on user performance. The testing environment consisted of two computers, a control status computer and a general task computer. The subjects were asked to create a spreadsheet from detailed specifications and, simultaneously, monitor the simulated electrical power plant operations and correct any problems which arose. The primary dependent variables in this situation were accuracy and response times to crisis conditions.

2.1 Research Tasks

The research tasks required the subjects to monitor nine status indicators on a computer screen simulating a electric power plant status monitor. Another computer was used by the subjects to complete a spreadsheet task. This arrangement was based on an existing electrical power plant situation. The nine indicators on the status monitor represented three separate water tanks. Each tank had indicators for temperature, volume, and pressure. Each indicator displayed the current value and the safety range for that parameter. An alarm condition existed if the current value was outside the safety range for any of the indicators.

The second computer was used to simulate a typical situation where the operator is expected to accomplish daily tasks while monitoring the status displays. The task chosen for this study was to create a spreadsheet to manage a yearly budget. The level of complexity for the task was such that the subjects would have to do some planning as they were creating the spreadsheet but was not so complex as to prohibit the subjects from finishing the spreadsheet during the research session time period (40 minutes).

2.2 Research Design

Two information display formats were used for this study; text and graphics. Text information took the form of numbers to represent the current values of temperature, volume, and pressure. The graphic presentation used bar graphs to show the current levels of the tank characteristics. Safety ranges were denoted for the text presentations by displaying the safe high values over the safe low values with the current value displayed between them. The graphic information display was organized into bar charts with the safety ranges noted on the side of the scale for each chart. The current value was displayed by moving the bar up or down on the scale. The general screen layout was the same for both text and graphic information displays.

The three forms of auditory feedback were no sound, a beep for all alarm situations, and a different "realistic" sound for each tank characteristic. The "realistic" auditory signal for an error condition in temperature was a high-pitched whistle similar to that of a teapot. When the value for pressure set off an alarm, a "pop" was produced. The sound for irregular water volume was a high-pitched tone which decreased in pitch.

This study used a two-way (2x3) factorial design with the types of visual display (text and graphics) and the types of auditory feedback (no sound, beeps, and "realistic" sounds) as the independent variables. The dependent variables were minimum response time, maximum response time, average response time, minimum completion time, maximum completion time, average completion time, response errors, and spreadsheet error scores. The response errors were divided into tank errors, attribute errors, and change errors. Alarm situations occurred at random intervals between three and five minutes. The number of alarms given during a research session was used as a covariate. As soon as an alarm condition was initiated, the computer program started an internal timer and recorded the time when the subject made the first keystroke to correct the alarm as well as the time when the subject had corrected the error condition.

3. Results and Discussion

The data analysis was divided into 4 parts; demographic characteristics, response times, completion times, and accuracy. The demographic analysis showed no unusual findings. No significant differential effects for response accuracy were found with either information display type or auditory information presentation technique. No significant differences were found with spreadsheet accuracy or time taken to complete the spreadsheet.

3.1 Response Times

The response time measures were divided into minimum response time, average response time, and maximum response time. The minimum response time was examined to determine best-case response to a crisis. The average response time was examined because it is often not possible to depend on responding in a minimum amount of time. The maximum response time was used to determine a worst-case scenario.

Text information display was found to have a significantly higher minimum response time than graphic information display. The mean for the text information display group was 4.86 seconds while the mean for the graphics display group was 2.29. There were statistically significant differences among all three levels of auditory feedback groups. The mean minimum response time for the group which received "realistic" sounds was 2.23. The group which received no auditory feedback had a mean of 30.18. The group which got a beep had a mean of 6.53.

A significant difference in the maximum response time was noted between the groups which received any type of auditory feedback and the group which received no auditory feedback. The mean for the group receiving no auditory alarm was 101.12. The mean for the group getting a beep alarm was 16.86 and the mean for the realistic sound group was 12.31. This would seem to indicate that the worst case scenario was somewhat alleviated by any sort of additional information about an alarm condition.

The groups that were given information in textual form had a significantly slower average response time than did groups which were given graphic information displays. The mean for the text group was 16.6 while the graphic group's mean was 11.47. The group which received no auditory feedback was significantly slower to respond to an alarm than were either of the two groups which received active auditory signals. The mean for the group which received no auditory information was 30.2. The group receiving a beep had a mean of 6.6 while the realistic sound group had a mean of 5.4.

3.2 Completion Times

An ANCOVA was computed for all three crisis completion time measures with visual information presentation and auditory information presentation as the independent variables. The dependent variable was the completion time. The number of alarm conditions was used as the covariate.

The graphic interface proved to provide significantly faster completion times than did the text interface. Groups using the text interface had a mean minimum completion time of 7.5 seconds. The graphic display groups had a mean of 4.5. This difference seems to be primarily due to the amount of time needed to initiate a response.

The analysis of the minimum completion times for the auditory groups showed a significant differential main effect between the group which received no auditory information and the two groups which did receive auditory information. The mean for the group receiving no sound was 8.5 seconds. The beep group had a mean of 5.0 and the realistic sound group had a mean of 4.4.

There was a significant difference between the group receiving no auditory feedback and the two groups receiving auditory feedback. The mean for the no-sound group was

103.6 seconds. The beep group had a mean of 23.8 and the realistic sound group had a mean of 19.1 seconds.

The groups using the text interface took an average of 19.9 seconds to correct the situational error. The graphic display group took 14.4 seconds to complete the correction task. The group which got no auditory information had a mean of 32.7 seconds. In the groups which received auditory feedback, the beep group had a mean of 9.7 seconds and the realistic sound group had a mean of 8.9 seconds.

4. General discussion

The current study attempted to examine the effect of different types of information presentation techniques and different types of auxiliary auditory information methods within a practical control environment. The two types of information displays were pure text (with a graphic component of logical information grouping) and a graphic display consisting of bar graphs. Three types of auditory feedback were studies: no auditory feedback, a beep alarm, and an alarm of "realistic" sounds.

Graphic information display seemed to be generally superior to textual information display. The current study took extra care to ensure that the information to be displayed was analogous to that found in an actual control situation while, at the same time, not giving the subject so much information as to induce "cognitive overload." The findings of this study are agreement with many previous studies, in particular, that of Hanson, et al. [6] who examined digital displays versus analog displays in a control setting.

Supplemental auditory information also proved to be more effective than no auditory information. Additionally, dimensional metaphors, coupling a unique change in pitch with a specific alarm situation, resulted in significantly lower minimum response and error correction times. The reduction of alarm condition correction times through the use dimensional metaphors did not seem to degrade performance on other tasks, in this case a spreadsheet creation task. One word of caution, during the protocol analysis of the videotapes of the users, the subjects who were given a simple beep as an alarm signal seemed to confuse the alarm beep with a beep denoting an error in the spreadsheet. Based on this, the authors advise using an auditory alarm which is novel in the control environment.

References

[1] Sutcliffe, A., *Human-Computer Interface Design*, Macmillan Education, Ltd.: London, 1989.
[2] Polson, M. C., C. D. Wickens, and H. A. Colle, "Human Interactive Informational Processes," in *Intelligent Interfaces: Theory, Research and Design*, P. A. Hancock, and M. H. Chignell, ed., Elsevier Science Publishers, Amsterdam, 1989, pp. 129-164.
[3] Wickens, C. D., L. Fraser, and J. Webb, "Cross-modal interference and task integration: Resources or preemption switching." *Proceedings of the Human Factors Society 31st Annual Meeting*, October, 1987, pp. 679-683.
[4] Colquhoun, W., "Evaluation of Auditory, Visual, and Dual-Mode Displays for Prolonged Sonar Monitoring in Repeated Sessions," *Human Factors 17*, 1975, pp. 425-437.
[5] Sorkin, R., B. Kantowitz, and S. Kantowitz, "Likelihood alarm displays." *Human Factors, 30(4)*, 1988, pp. 445-459.
[6] Hanson, R., et al. "Process control simulation research in monitoring analog and digital displays." *Proceedings of the Human Factors Society 25th Annual Meeting*, 1981, pp. 154-158.

Advances in Agile Manufacturing
P.T. Kidd and W. Karwowski (Eds.)
IOS Press, 1994

Interface Agents for Effective Human-Computer Coordination in Hybrid Automation Systems

Wayne W. Zachary and Monica Weiland

CHI Systems Inc., 716 N. Bethlehem Pike, Lower Gwynedd, PA, USA 19002

Abstract: The COGNET framework provides a means of building intelligent interface agents around cognitive models of human expertise in a domain. An additional set of software tools were developed to support the process of embedding the models into intelligent interface agents. This paper describes the COGNET framework and supplementary agent-building tools, and discusses the potential uses of interface agents for hybrid automation systems.

1. Introduction

For tomorrow's manufacturing environment, manufacturing processes must be maximally efficient in use of resources, such as machine-time, storage space, and delivery schedules. This requires that manufacturing processes need to be scheduled tightly to take maximum advantage of available resources. However, delivery delays, shortages, and other unforeseen events require workers to make ad hoc changes to manufacturing processes, often without access to knowledge of the complex system-wide ramifications of their decisions. Even if operators have the information, the ability to use the information effectively in such a time-constrained environment is limited given human limits in cognitive processing. This problem has caused many experts to call for Flexible Manufacturing Systems (FMS) using high levels of automation to control the complexity. Despite the push for more automation, many researchers in the field contend that the human operator must remain in the loop to maintain system effectiveness [1].

The success of these emerging automation technologies, therefore, is dependent on hybrid automation -- the effective integration of man and machine. The need for hybrid approaches is most critical in those areas where human judgment, knowledge, and decision-making are critical for overall system effectiveness. The introduction of hybrid automation, however, is creating a new type or human 'supervisor', with unique problems. Advanced hybrid automation systems provide real-time control over often multiple tasks, allowing increased speed and precision of operations. However, this capability often out-strips the ability of the human system supervisor to make timely use of it. Moreover, as the computer systems grow more complex, their human users have increasing difficulty in understanding and efficiently applying the underlying system functionality now at their disposal. These problems suggest that cooperation between human and computer in hybrid automation needs to be mediated, in order to mesh the distinct capabilities of human and computer. Research has begun to focus on using the human-computer interface to solve such problems, by creating 'interface agents' whose sole purpose is to enhance the cooperation between human and computer in hybrid systems.

One obstacle to development of interface agents in manufacturing applications is the fact that these applications are, by their very nature, highly complex, and closely tied to specific task domains. Unlike generic or horizontal applications (e.g. database searching), interface agents in manufacturing applications therefore require substantial amounts of domain-specific and task-specific knowledge in order to be useful to the system and its human operators. This makes their development potentially lengthy and costly. In addition, manufacturing applications are typically limited in the number of end-users. This means there are fewer customers across which to amortize the development costs.

Realization of the economic and competitive potential of interface agents in work-specific applications is dependent on an ability to generate, maintain and evolve interface agents in specific application domains in a cost effective manner. To reduce costs, the authors and colleagues at CHI Systems are developing an advanced, flexible workbench for the engineering of interface agents. The workbench is based around a general methodology and an underlying model for development of intelligent interface agents. The methodology and model, are collectively called COGNET, for COGnition as a NEtwork of Tasks [2].

2. COGNET: A Framework for Intelligent Agents

COGNET was created to provide a means of building intelligent interface agents around cognitive models of human expertise in that agent's domain. COGNET focuses on the class of problems where the person needs to perform multiple, often competing tasks in real-time and share attention among them, and where the role of momentary context and problem history is critical in the overall cognitive process. These real-time, multi-tasking problems are typical of manufacturing applications.

COGNET provides both a framework for viewing human-computer interaction, and a set of techniques for building models of computer-based problem solving in specific domains. In the COGNET framework, the human computer-user is seen as interacting with an external environment through the medium of the computer system (and specifically, through that system's human-computer interface). The person is implicitly assumed to be in a work-setting, and therefore pursuing some high-level goal with regard to that external environment. Within this overall goal, the activities of the person appear as a network of tasks. Some of these tasks compete for the person's attention, others may be complementary, and still others may need to be performed essentially in parallel. Each task represents a specific 'local' goal which the person may pursue to achieve or maintain some aspect of the overall, high-level goal and contains a chunk of procedural knowledge. The way in which the procedure is instantiated, however, can be heavily dependent on the past evolution and/or current state of the problem, making the task context-dependent. As procedures, tasks will typically define a substantial span of activity, and will be performed over some period of time. However, tasks may interrupt one another, and a given task may be interrupted and resumed several times as the person copes with the on-going sequence of events in the problem environment (as viewed through the user-computer interface).

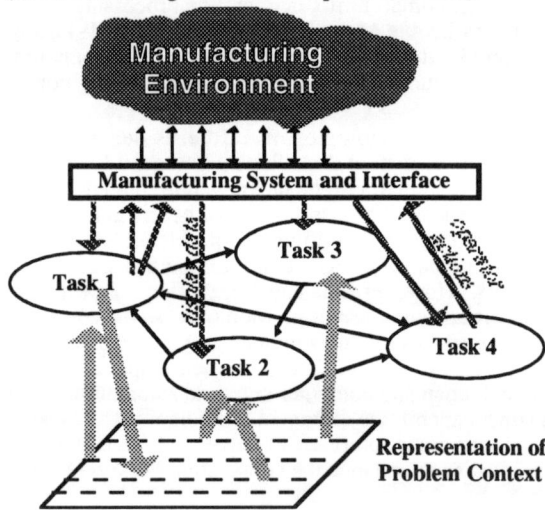

Figure 1   COGNET View of Real-time, Multi-Tasking Problem Solving

Although the tasks may compete for attention, they work together in the aggregate to solve the overall problem with which the user is faced. What unites these separate chunks of procedural knowledge into a global problem-solving strategy is their use of a common problem representation. This problem representation is declarative, and provides the problem-context information on which attention demands are mediated and task-performance is adapted. As a given task is performed, the person gains knowledge about and/or makes inferences about the situation, and incorporates this information back into the problem representation. However, as the problem representation evolves, it can change the relative priority among the tasks competing for attention, allowing a new task to capture the focus of attention. Much of the information about the problem is also gained from perceptual processes, for example by scanning and noting information from displays and controls at the workstation. This overall framework is pictured in Figure 1.

COGNET provides three notational elements to represent the various aspects of the real-time, multi-tasking problem-solving process:

- a notation to describe the representation of the current problem instance, including elements of its evolution and present inferred state. This notation is generalized, multi-panel blackboard structure of the kind reviewed in [3].
- a notation for describing the information processing and associated person-machine

interaction associated with each cognitive task. This notation is related to the GOMS notation [4], but includes extensions that allow for accessing and creating information in the blackboard that contains the current representation of the problem context. Other extensions allow for the interruption, suspension, and subrogation of the current task to other tasks.

- a production rule notation for describing the processes by which information, once registered, is perceived and introduced into the current representation of the problem (i.e., blackboard). These independent processes are called perceptual demons.

COGNET has been used to develop detailed models of several complex domains, including en-route air traffic control [5], telecommunications operations, command and control [6], and vehicle tracking [7]. In addition, the model for vehicle tracking was used to develop a fully functional intelligent agent for vehicle tracking [8]; an interface agent for a helicopter cockpit is currently under development [9].

3. COGNET-Based Intelligent Agents

The basic structure of a COGNET-based intelligent interface agent in an operational environment is pictured in Figure 2. The domain specific model is formalized and embedded into the interface, and is used to interpret the interactions between the user and the system (i.e., the HCI transactions) and to generate recommendations for user behavior. The interface agent can help the human user of an automation system to:

- manage attention among competing demands;
- designate and/or delegate tasks and subtasks for human vs. machine execution in a context-sensitive manner;
- interpret automation system activities and functionality; and
- manage computer-generated information by context and/or by individual human operator.

Interface agents can also help the automation system to:

- present information to the human in its proper context ;
- adjust activities and displays to individual users; and
- make inferences about the human user's actions, intentions, or goals.

For architectural details see [8].

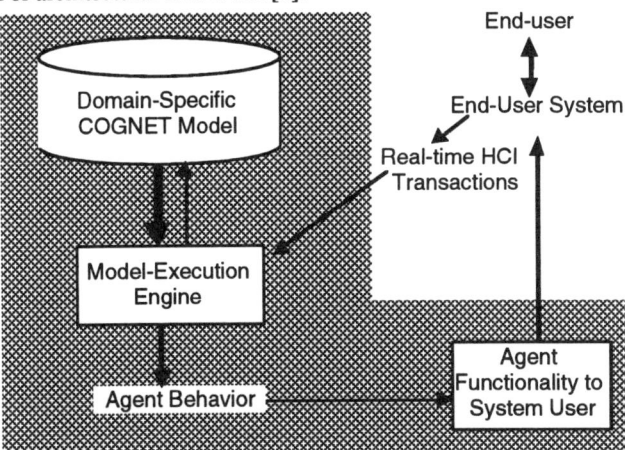

Figure 2.   Organization of a COGNET-Based Interface Agent

4. Agent-Building Tools

The initial attempts to embed COGNET models into intelligent interface agents proved successful, but extremely complex and time-consuming. As a result, a toolset was developed to support the embedding process. This toolset is called BATON, for Blackboard Architecture for Task Oriented Networks, [8]. BATON currently contains four software tools. Given a fully-specified blackboard in COGNET notation, one tool generates a C-language data structure with the same organization/slots. A second tool provides a set of C functions for manipulating (i.e., create, delete, read) messages on the blackboard data structure, and for evaluating the blackboard for the presence of specific patterns of information (i.e., specific contexts). Given a fully specified set of perceptual demons, the third tool generates a set of C functions that monitor the external environment and place information on the problem blackboard as it is encountered. Finally, given the set of conditions, expressed as blackboard patterns, under which each task will begin to vie for attention, the fourth tool will construct C functions to monitor the blackboard for those specific patterns.

Even with BATON, however, the COGNET-based interface agents produced to date have all been designed and implemented in a customized manner that is highly integrated with the end-user system in which they were embedded. As a result, the software to implement an agent was not sufficiently generalized to make it easily reusable. This makes the development of COGNET-based agents dependent on the expertise of those familiar with the details of COGNET and its underlying philosophy. The lack of an engine to support reusable modules for rapid prototyping of interface agents has been a barrier to wider use and broader transfer of this technology to broadbased usage in hybrid manufacturing environments.

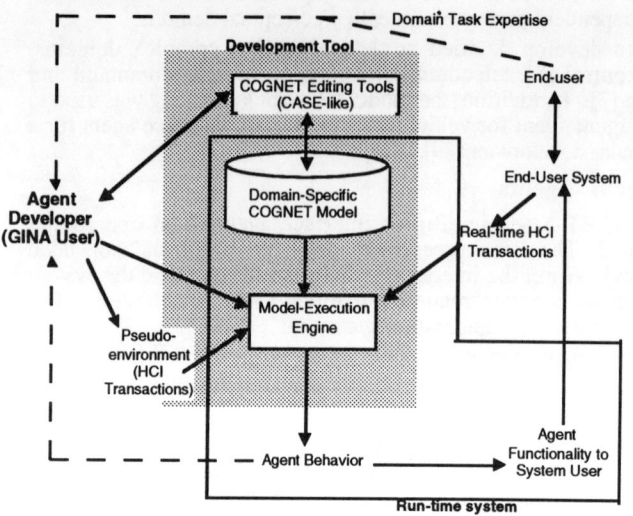

Figure 3   A Generator for Interface Agents (GINA)

Current research is underway to create a tool for rapid development of COGNET models and their translation into embedded interface agents of the type illustrated in Figure 2. This tool has two inter-related parts, and is called simply GINA -- the Generator for Interface Agents. The first is a development workbench, which supports the construction (i.e., design, coding, debugging, testing) of the interface agent (Figure 3). The second is a run-time environment, which supports the execution of the fully-developed interface agent once it is actually embedded in the end-user computer system.

References

[1] Ammons, J., & Govindaraj, T. (Sept./Oct. 1988). Decision models for aiding FMS scheduling and control, in *IEEE Transactions on Systems, Man, and Cybernetics*, New York: IEEE.

[2] Zachary, W. (1989). A context-based model of attention switching in computer-human interaction domains, in *Proceedings of the Human Factors Society 33rd Annual Meeting*.

[3] Nii, P. (1986). Blackboard systems: The blackboard model of problem solving and the evolution of blackboard architectures, *AI Magazine* , 7, 2, 38-53.

[4] Card, S., Moran, T., & Newell, A. (1983). *The psychology of computer-human interaction*. Hillsdale, NJ:Lawrence Erlbaum Press.

[5] Seamester, T., Redding, R., Cannon, J., J., & Purcell, J. (1993). Cognitive task analysis of expertise in air traffic control. *International Journal of Aviation Psychology* 3(4), 257-283.

[6] Zachary, W., Zaklad, A., Hicinbothom, J., Ryder, J., & Purcell, J. (1993) COGNET Representation of Tactical Decision-Making in Anti-Air Warfare, in *Proceedings of the Human Factors and Ergonomics Society 37th Annual Meeting* (pp 1112-1116), Santa Monica, CA: Human Factors Society.

[7] Zachary, W., Ryder, J., Ross, L., & Weiland, M. (1992) Intelligent computer-human interaction in real-time, multi-tasking process control and monitoring systems, in M. Helander & T. Nagamachi (Eds.). *Human Factors in Design for Manufacturability*. New York: Taylor and Francis.

[8] Zachary, W., Ross, L., & Weiland, M. (1991). COGNET and BATON: An Integrated approach for embedded user models in complex systems, in *Proceedings of 1991 International Conference on Systems, Man, and Cybernetics*, New York: IEEE.

[9] Zaklad , A., Weiland, M., Zachary, W., Voorhees, J., and Fry, C. (1993). Active man-machine interface for advanced rotorcraft, in *Proceedings of the American Helicopter Society 49th Annual Forum*. St. Louis, MO: AHS.

Advances in Agile Manufacturing
P.T. Kidd and W. Karwowski (Eds.)
IOS Press, 1994

A Graphical Display To Support Human-Computer Decision-Making In Production Scheduling

Peter. G. Higgins
School of Mechanical and Manufacturing Engineering, Swinburne University of Technology,
P. O. Box 218 Hawthorn 3122, Australia
higgins@mechman.mm.swin.edu.au

Abstract. This paper discusses the symbolic objects forming a graphical interface for a "hybrid" human-computer scheduling system. Visual features of these objects represent the job attributes that a human scheduler may use in deciding how to allocate jobs to machines and to arrange the order of processing.

1. Introduction

The processing of jobs within small-batch manufacturing environments requires assignment of finite resources. The creation of a schedule typically involves resource analysis and assignment, deciding job routes, dispatching jobs to work stations, and, monitoring the status of tasks and stations — all done with due regard to capacity constraints. Finite, specifiable set of options and a single, clear objective are uncommon. Usually there are no clearly stated boundaries and strict rules for manoeuvring within these boundaries Furthermore, the processing of jobs usually does not follow the schedule as originally planned. Schedules have to adapt to unanticipated events: for example, new jobs, changes to priorities, and machine breakdowns.

To aid personnel at the shop-floor level to plan under such circumstances, a "hybrid" human-computer scheduling system, bringing together intelligent human behaviour and "intelligent" computer behaviour, is being developed [1]. In this paper the author discusses the **form of a graphical interface** for this system.

2. Limitations of the Gantt Chart

For some time schedulers have used Gantt Charts as a tool. They show the sequence of operations and the expected utilisation times at each resource. Interactive scheduling systems are available that use Gantt charts as their primary interface. The computer applies heuristics to build a schedule. The user then modifies the resultant Gantt chart. A rebuild-algorithm then reschedules activities that temporally follow these manual changes. The interplay between human and computer continues until some satisfactory schedule evolves. Often the measure of satisfaction is but a single indicator of performance.

This approach may lead to inappropriate decisions. Systems that apply standard OR (Operations Research) heuristics consider only, at most, few of the job attributes that may affect a schedule's performance. By letting the computer lead the decision-making process, the scheduler may not deliberate upon the effects of significant attributes.

Human schedulers should intercede during the schedule's construction process, instead of merely modifying a Gantt Chart after the computer has built it. On considering **all** the job attributes, the scheduler can decide which heuristics to apply to a chosen set of jobs. A Gantt Chart is then the result of human-computer decision process. Such an interactive build requires a display of all the job attributes that may affect a schedule, so that the scheduler may consider all the dependencies and conceivable interactions.

3. Why Graphics?

The job attributes could be displayed as alphanumeric characters, so why use graphics? Cognitive studies support superiority of graphic representation. Perceptual inferences such as distance and size determinations, spatial coincidence judgments, and colour comparisons, allow users to obtain the same information as more demanding logical inferences such as mental arithmetic or numerical expressions. Various studies found that when comparing size of unseen objects, subjects use mental imagery. Therefore using graphical size for representing attributes that refer to spatial size provides an opportunity to present the user with an explicit representation supporting mental imagery [2]. Encoding techniques such as colour, shading, and spatial arrangements can help the scheduler search and group jobs. If the graphics support preattentive and parallel visual-search further benefit may follow.

Graphics allow for greater information density than alphanumeric displays. This is a distinct advantage where there are many attributes.

The form of representation affects functionality and utility of an artifact [4]. By displaying the attributes as graphical objects particular aspects of the domain are mapped such that the features suggest and remind the user of the set of possible operations. The design challenge is the setting up of the mapping so that the semantics used by persons knowledgeable of the particular manufacturing domain is directly visible by the observer. The scaling must be chosen such that the desired data displays patterns easily recognisable by the observer.

4. Attributes as Graphics

A scheduler's internal representation is not limited to job attributes alone. It also includes decision rules based on the interplay of the attributes of different jobs and the characteristics of different machines. This internal representation is transformed by the interface into a surface representation that the person who is making the decision can interpret [4]. Using this surface display, the scheduler goes beyond what is represented. Seeing information in the relationship between attributes, the scheduler can form sets and subsets of jobs. Different groupings can result in different schedules, meeting different, and possibly opposing, objectives. In choosing which grouping to use, the scheduler interplays the internal intrinsic information with everything else about the jobs, machines and environment that have not been encoded [3]. Grouping behaviour is affected by the surface representation as its form suggests and reminds the scheduler of the set of possible outcomes [4].

4.1.   Graphical form

The style and format of the interface affect the useability of the device. There are some rather ingenious means for depicting data. Graphical objects that are Kleiner-Hartigan trees, or derivatives, can be very expressive for clustering nominal data (see figure 1) [5]. Their shape exposes dominant clusters. The organising of a sequence of jobs requires pattern-matching across jobs. In Kleiner-Hartigan trees the branches of a stylised tree represent attributes. With the Star Symbol Plot, the chord lengths represent attributes. Another popular form is the Chernoff Face, in which different facial features represent different attributes.

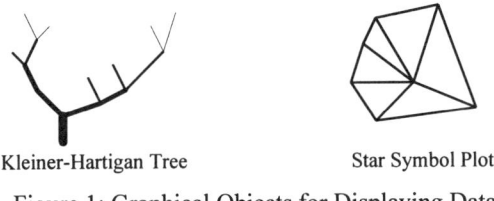

Kleiner-Hartigan Tree   Star Symbol Plot

Figure 1: Graphical Objects for Displaying Data

While these graphical forms have the advantages of high data-density and ease of recognising differences during searches for objects with like features, they are overly derived. The surface representation needs to conform to Norman's Appropriateness principle. Even if their design allows a scheduler to readily cluster jobs on specific features that produce schedules that are "good" performers, there is a large cognitive hurdle between tree branches, star shapes and facial features and the attributes of jobs. The question of Semantic Directness arises [4]. Does the representation support the user's conception of the task domain? 'Does it encode the concepts and distinctions in the domain the same way the user thinks about them?

4.2.   Case Study

Discussion of these issues needs to be in the context of an explicit production environment. The interface for the "hybrid" scheduling system was developed for a company that printed fan-fold paper for computer-printers. Instead of single geometric shape representing all the attributes, a graphical object for each object used a variety of means for displaying the different attributes. The representational form depends on the way the attribute that it represents affects scheduling decisions. Dimensional size, showing variable size, can be of use for displaying variables that are preferably monotonically decreasing or increasing. Spatial location is used for the representation in circumstances where an attribute is preferably kept constant between jobs. The form of the display conforms to Norman's additive and substitutive scales [4]. An additive scale makes it easy for the human scheduler to compare processing times by scanning vertically. While additive scales generally suit size comparison, for cases requiring an exact numerical value, Arabic notation is clearly superior (see Figure 2).

				Cylinders	Colours	Processing time	Width	Presses
1.	16504	Gillette	06/05		2		229	
2.	16667	TDGCar	25/05		1		345	
3.	16498	Triad	30/04		2		244	
4.	16748	Jarvis	27/05		2		305	
5.	16356	Triad	/		1		246	
6.	16537	Option	20/06		1		229	
7.	16556	RPHG	/		3		254	

Figure 2: A partial display of objects representing job attributes

Substitutive scales befit the display of depth of printed form and colour. For the printing cylinder, the horizontal location of the bar shows its size, and the bar's height expresses the depth as a fraction of the cylinder's size. This display of depth allows the scheduler to see in a sequence of jobs where there is a change in depth. A change in cylinder size requires a major set-up. The display of width is numeric. A symbol connects the widths when a job uses the same cylinder but a larger width than its predecessor. This shows a minor set-up for changing a perforating tool. For colours, the horizontal location of vertical bars shows commonly-occurring colours. To print the same colour on the front and back of a form requires two modules. Where this occurs, the bar has twice the height.

As the comparison of widths is not a pattern matching aspect, the use of a numeric display has the most clarity, as the scheduler does not have to decode the graphic to obtain the value of the measurement. A graphical prompt is used when comparing widths. It aids the search by showing where there is violation of the soft-constraint for sequential jobs to be in monotonically decreasing order when they are using the same cylinder size. The use of graphical symbolism for displaying process time helps to aid searches for jobs that can be fitted into a desired manufacturing period. If, for example, a job is sought to fill the time remaining until the end of the shift, say three hours, the scheduler may fix his/her eye on the three-hour interval on the processing time display and then scan vertically. Any job that is to the left of this datum can be processed in the required time.

The next stage of the research prorgramme is to compare under experimental conditions the performance between a solely alphahnumeric display and this display.

References

[1] P. G. Higgins, Human-Computer Production Scheduling: Contribution to the Hybrid Automation Paradigm. In: P. Brödner & W. Karwowski (eds.), Ergonomics Of Hybrid Automated Systems - III. Elsevier, 1992, pp. 211-216.

[2] R. E. Eberts et al. Four Approaches to Human Computer Interaction. In: P. A. Hancock & M. H. Chignell (Eds.) Intelligent Interfaces: Theory, Research and Design. Elsevier Science Publishers , 1989, pp. 69-127.

[3] J. Bertin, Graphics and Graphic Information-Processing. De Gruyter, 1981.

[4] D. A. Norman, Cognitive Artifacts. In: J. M. Carroll, Designing interaction: Psychology at the human-computer interface. Cambridge University Press, 1991, pp. 17- 38.

[5] J. M. Chambers et al., Graphical Methods for Data Analysis,.Wadsworth International Group, Duxbury Press, 1983.

Advances in Agile Manufacturing
P.T. Kidd and W. Karwowski (Eds.)
IOS Press, 1994

Graphical Intelligent Interface for Hybrid Decisions in Manufacturing Workshops
An Alternative to Automatic Workload Curves and Scheduling

Catherine THURIOT*; Marie-Françoise VALAX**
LAAS-CNRS; 7 Avenue du Colonel Roche 31077 Toulouse Cédex; France
**ER-CNRS 15; 5 Allée Antonio Machado 31058 Tolouse Cédex France.*

Abstract: In the workshops of discrete and diversified productions, traditional aid tools do not fulfil the decisional autonomy demands. We propose a new approach for the design of different aids, based on a more progressive and hybrid (Man-Machine) elaboration of associated workload and operations plans. Such an approach implies a decomposition of the decision system into distributed structures of decision centers, allowing both the insuring of local autonomy of the decisions, and their consistency according to the global objectives of production. This approach has been evaluated, from logical, technical and ergonomic points of view.

1. Introduction

Our previous studies on effective decisional processes in manufacturing systems in the case of discrete and diversified productions [1], revealed a real demands of autonomy for the workshop deciders (workshop supervisor, foreman, team leader., etc.). Consequently, the design of any tool aid for the work organization, has to consider this requirements, as well as the constraints associated with the given organization (consistency of the decisions related to the global production system and to the local workshop constraints).

The proposed approach of *workload and associated operations plans*, is based on a decomposition of the decision systems into distributed structures of decision centers, from which the decisions are transmitted in terms of constraints to be respected by their lower center(s). The local autonomy of a given center is then characterised by the *different types of margins* issued from the decisions of its upper center(s), the respect of which insures the global consistency of the decision system [2].

In this paper, we present the theoretical principles and concepts of our approach for the design of computerized aid tools for work organization in workshops and the first steps of its evaluation.

2.Problem

Traditionally, the decisions which are taken upper (or before) workshop by the planning center consist in determining a first operations plan. In such a plan, extreme dates, as earliest beginning and latest end ones, are fixed and considered as "boundaries" or constraints to be respected for each planned operation.

Those calculations realized with infinite capacities (i.e. : without considering the limited availability of the resources, named resource constraints), lead to *workload plans* for each aggregate resource (as operators team, or pools of identical machines), which are temporally distributed sums of work quantities required by the operations concerned by the given resource. The eventual taking into account of resource constraints is then insured by the elaboration of a *scheduling*, i.e. a second operations plan. According to detailed resource allocations, sequences and beginning dates are calculated for all operations, sometimes letting time margin for some of them.

Such workload plan and scheduling present some difficulties:

The workload plan allows middle term decisions of workload regulation (moving in time : delaying or advancing, adding or removing workload quantities). It makes a distinction between earliest curves (obtained when the planned operations are all distributed at their earliest beginning dates) and latest ones (when operations are distributed at their latest end dates). It helps to have a "hazy idea" of the temporal evolution of the work to be done, foreseeable risks of over or under-load and available global load margin. But this kind of information remains quite fixed, and one cannot use any direct way to exploit or handle the effective load margin : the workload regulation is done by moving some operations in time (delaying or advancing) or in space (sub-contracting, modifying the resource allocation, etc.).This traditional approach induces errors issued from too premature local scheduling [3]. Moreover, the workload plan does not give any information about earliest beginning and latest end dates of the operations which risk to be inconsistently moved out of their temporal boundaries.,

The scheduling allows allocation resource and/or time decisions. It consists in a diagram or an alphanumeric list showing a sequence of operations defined by a duration and associated with an arbitrary and partial time margin. One is then confronted with a unique and strict plan in which the information of all effective time margin, other admissible sequences between operations or resource allocations, are not presented, obstructing in such a way the use of all available autonomy. In addition, this uniqueness of plan does not allow the decider to appropriate himself a sufficient distance from it : he is generally confined to local and insufficient corrections without having the possibility to evaluate their impact on the global scheduling.

3. New approach: workload and operations associated plans.

In order to avoid upper errors and difficulties, we propose a more progressive approach implying successive steps of elaboration of workload and associated operations plans, which allows to directly handle and reason about aggregate workload margins or about time margins.

The decisions issued from the planning and concerning a given aggregate resource can be graphically represented as follows :

Figure 1. Graphical representation of an operations plan

This is the first planning or the *first operations plan*. Each pair of rectangles -white and black, respectively named *window* and *task*- associated with a planned operation, figures :

- its earliest beginning and latest end dates (extremities of white rectangle),

- its duration (length of the black rectangle which can be translated inside the corresponding white one, in the way that the considered decisions are done before any scheduling).

The workload plan relies on a decomposition of the time horizon into *adjacent time intervals* [4]. Each interval of the temporal structure is then associated with two levels of workload : maximal and minimal levels, obtained from automatic sums of work quantities required by the realization of the set of operations concerned by the given interval, and extreme positions of the tasks inside their windows [5]. Those workload levels can be compared with the capacity and profitability levels of the given aggregate resource, i.e. : the maximal work that can be done and the minimal work that must be done.

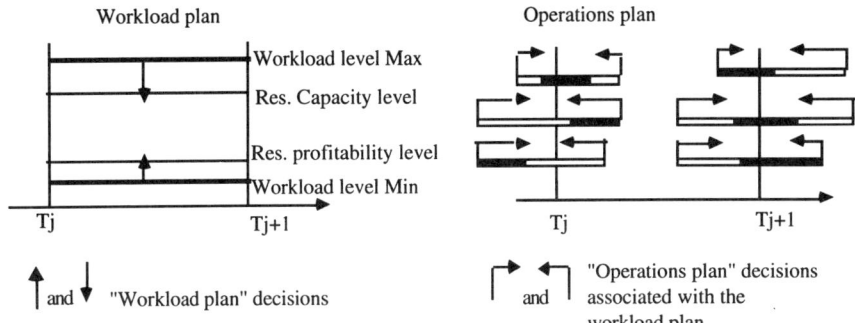

Figure 2 : One interval of a workload plan and associated operations plans

In figure 2, Tj and Tj+1 are temporal markers which determine the situation and length of an interval of the temporal structure. The surface between the two extreme workload levels figures the *load margin* available to the decider during the given interval. He can then modify those levels : for example, by pulling down the maximal level (or pulling up the minimal one) in order to avoid any risk of overload (or under-load). Some automatic processes of workload sliding or transfers which have been studied [1] can be graphically figured by "communicating vessels" mechanisms on the levels of adjacent intervals : they give an anticipation of local decisions on the global horizon. The new workload levels, so rectified on each interval during this first decisional step, lead to a new workload plan which constitutes the goal to be reached in the second step.

In the second step the decider have to translate the upper decrease in workload margins on the operations plan. In order to do so, he must reduce time margins (figure 2) by acting on the lengths of some windows which meat the temporal landmarks Tj (moving forward earliest beginning dates, or moving backward latest end ones). Such actions have to be done so that sums of work quantities, required for the realization of the set of operations concerned by the given interval and obtained from extreme positions of the tasks inside their new windows, really correspond to the new extreme workload levels

In other respects, in the operations plan, the decider can also directly "consume" the available time margins. Decision-making can take into account more detailed resource constraints, as its maximal instant availability (i.e., the number of present operators in the team or available machines in the pool at any time). Such considerations can use some of the automatic mechanisms of deductions and propagation of the *Constraints Based Analysis* [6]. In order to introduce this resource constraints for the time margins decrease, the decider can move tasks on different spaces corresponding to the temporal availability of resource items (operators, machines). This moves allow him to schedule and/or allocate resource to operations. In real time the computer insures the consistency of this actions by the dynamic calculation and representation of locks or boundaries.

In addition, successive hierarchical decomposition of the time horizon, will offer the possibility to reason and decide by following progressive steps of more and more detailed workload plans (elaborated on shorter and shorter intervals), associated with more and more precise operations plans (with more and more narrowed windows), leading to *the best precise level for the considered situation*, which would be, but not necessarily, a scheduling.

4. Approach validity

This approach has been evaluated, from logical, technical and ergonomic points of view.

From logical point of view, our approach seems to present important advantages in comparison with traditional approaches associated with existing aid tools :
- all available autonomy is known by the decider at any time, and is represented by the different remaining margins that he can handling or "consume";
- the consistency of his decisions according to the global production objectives (and particularly to the planning upper decisions), is insured by the graphical boundaries and

impossible actions (for example advancing an earliest beginning date, or increasing a maximal level of workload). Such locks could be removed in case of constraint relaxation, requiring some processes of constraints negotiations with upper decisions center(s) (2).

- the distinction between aggregate and detailed data allows specific and "separated in time" reasoning, with the possibility, at any time, to directly handling the workload or its associated operations plans.

- the risks of errors related to the corrections of the scheduling, and more particularly to the perceptual evidence of the chronology or sequence, decrease. The overlaps of some windows reveals the existence of different admissible sequences (and so scheduling). As well, those risks of errors decrease by more progressive reasoning issued from the successive hierarchical decomposition of the time horizon.

From technical point of view, our approach has been applied to the design and realization of the software MARGE [4] : Graphical and Ergonomic Aid Module for the decisions of workload Regulation. It presents the upper different plans and their associated graphical possibilities of actions. The distinction between "Man", "Hybrid Man/Machine" and "Machine" strategies is related to the different possible control levels : from "machine assisted actions" to "all machine calculated solutions". MARGE has been used for a part of the ergonomic evaluation of the approach.

The ergonomic validation has concerned until now the use of the workload plan. 12 unpracticed subjects had to solve consecutively three problems of workload regulation of increasing complexity. The success rate of the 36 analyzed protocols is equal to 69,7%. Learning is fast, from 16,7% for the first problem, to 91,7% for the second problem, to 100% for the third one.

But some numerous and persistent errors bring about reasoning on margins. For example : "increasing the maximal workload level" (52,8%), "global regulation on one level" (minimal or maximal) forgetting the other one (41,7%). Those errors have to be particularly analyzed regarding improvements in data presentation and users' training.

5. Conclusion

The proposed approach seems feasible from a conceptual point of view. Some graphical representations and mechanisms have been analyzed and computerized (in MARGE). The fast learning that has been observed on the problem-solving of workload regulation enable us to be confident of its compatibility with the cognitive processes of the "potential users" , even if some revealed errors invite ourselves to study the improvements of data presentations and the users' learning. More, the present results are issued from a laboratory experiment and will have to be confirmed on real situations.

References

[1] C.Thuriot and M.F.Valax, Computer aids to work organization in hybrid manufacturing,in *Ergonomics of Hybrid Automated Systems III* by P. Brödner and W Karwowski (eds) Elsevier Science Publishers B. V , 1992, pp. 117-123

[2] J. Erschler and C.Thuriot, Approche par contraintes pour l'aide aux décisions d'ordonnancement. Eds de Terssac G. and Dubois P. *"Les nouvelles rationalisations de la Production"* CEPADUES Editions, 1992.

[3] E.D. Sacerdoti , *A structure for plans and behavior..* American Elsevier Publishing company inc., 1977.

[4] C. Thuriot and F.Torres, Interface graphique pour l'aide à la décision de répartition de charges de travail.*ERGO'IA 92.* Biarritz,.1992.

[5] C. Thuriot, Decision Aid Approach for Workload Distribution. *Production Planning and Control.* Ed Taylor and Francis (sous presse)

[6] J.Erschler, L.Lopez and C.Thuriot, Raisonnement temporel sous contraintes de ressources et problèmes d'ordonnancement. *Revue d'intelligence artificielle,* 5(3), 7-32, 1993.

Advances in Agile Manufacturing
P.T. Kidd and W. Karwowski (Eds.)
IOS Press, 1994

The Development and Implementation of Advanced Control Systems at BNFL: The Application of Human Factors

J Reed

British Nuclear Fuels, Engineering Division, Risley, Warrington,
Cheshire, WA3 6AS, UK

W Harper

British Nuclear Fuels, Sellafield Technical Department, Seascale,
Cumbria, CA20 1PG, UK

This paper describes BNFLs approach to the integration of human factors in the research, development and implementation of high-technology control systems into existing operating plants and those in various stages of design. A project to assess an on-line expert system employed on an operating plant to aid early fault detection and diagnosis is described. This includes the development of assessment guidelines, task analysis, interviews and a detailed simulation study.

1. Introduction

The benefits of using advanced control and expert systems in process plants are acknowledged by many industries. This is a key development area in process control technology, and is likely to expand in the future. Increased productivity, the reduction of downtime and decreasing the likelihood of human error are key desirable results to be acquired. However, this presents a major challenge to effectively apply this new technology in large and complex plants, encode the experience and knowledge of the best operators and engineers and subsequently gain the real benefits that it has to offer in relation to enhanced plant safety and economic performance.

In this respect, a range of human factors considerations are well recognised as being of primary importance in potentially high-risk nuclear plants, BNFLs own early experience highlighted the importance of human factors issues from project inception through to delivering the system in the control room.

There is limited experience on which to draw information on operator interaction with such emerging systems, particularly in the context of process control. Therefore the integration of human factors in the exploration and application of advanced control systems is a necessary component of BNFLs R & D programme.

2. Background

BNFLs human factors work is superimposed onto the relevant technical and engineering groups throughout the company. This spans a diverse range of applications necessitating different human factors techniques, expertise and strategies to be adopted. The human factors involvement varies in size, duration and stage in the project lifecycle e.g. research into the utility of a new system to the company, design and development, evaluation and looking ahead to the type of control systems and plant operating requirements of the future.

3. Development of Human Factors Guidance

Guidance has been developed within BNFL on the human factors aspects associated with implementing on-line expert systems in the control room. This includes all aspects of human factors paying great attention to the operator interface design, including advanced VDU graphics facilities, operator input requirements and interaction with the system. Such aspects are of particular significance to expert systems which may be safety-related, used infrequently but required to provide effective operator support for difficult tasks in time limited high workload scenarios. The guidance is based on an extensive literature review, a review of experience gained within BNFL and discussions with other organisations possessing relevant applications.

The main aspects covered are as follows:-

* strategies to ensure the operator does not become over reliant on the system or de-skilled.
* the provision of a hands-off system to keep any potential increase to the operators workload to an absolute minimum.
* means to avoid the system falling into disuse.
* the involvement of operations personnel throughout the design and evaluation process.
* evaluation methods and analysis, including simulator trials.
* attention to operator needs in the detailed design of the interface (e.g. information content, format of presentation, the combination of diagnostic and plant status information, alarm/ message signals, control facilities, positioning in the control room).
* information requirements and facilities for the operator to assess the reliability of the system output.
* ways to encourage operator acceptance.
* operator training in how to use the system, the appreciation of the capabilities and limitations and ensuring that the appropriate level of trust is adopted.

The guidance had some limitations in its usefulness as it was not always possible to define clear principles. This can be attributed to several reasons, although largely due to the present lack of operational knowledge. Also, some judgement and interpretation of information in the context of process control was required, if not directly related. The process of developing the guidance generated a list of areas where further clarification was necessary in order to fully address certain human factors aspects of expert systems. Unsurprisingly, a number of the topics that arose related to the design of VDU systems in process control, regardless of their function as an expert system. It is also worth noting that it was found difficult to provide the detailed information required by the system developers, and further guidance would need to be less generic and much more specific to the different types of applications - if possible.

4. Evaluation Study

Expert system applications are currently in operation or at an advanced stage of development at a number of BNFL plants, plus feasibility studies are being undertaken for new plants.

Figure 1: BNFL Springfields B26 Control Room: The PROMASS interface is the VDU system on the right.

An ongoing programme of work is applying human factors to an on-line expert system called PROMASS (Process Operations Management System) installed in a fully operating control room in BNFL Fuel Division at Springfields (Figure 1). This system has been successfully providing early warning and fault diagnosis information since 1989 for the Integrated Dry Route (IDR) process for the conversion of Uranium Hexaflouride to Uranium Dioxide, which is a fundamental stage in the production of fuel for Advanced Gas-Cooled and Pressurised Water Reactors.

The windows-based VDU user interface of the PROMASS application contains many features that include not only diagnostic messages, but also extensive interactive time-plot, mimic and information facilities.

The objectives of the human factors project are:-

* improve the operability of the system via the identification and implementation of modifications to the operator interface, organisation and training system.
* investigate the utility of different techniques for data collection, representation and analysis in real applications using real operators as subjects.
* apply the expert systems guidelines.
* obtain a better understanding of operator behaviour in process control and expert systems usage.

4.1. Task Analysis

A comprehensive task analysis for the monitoring and control of the IDR process was conducted. This clearly identified the role of PROMASS in supporting the operators objectives. Detailed records were made of the operators information requirements and sources of process indications and feedback.

4.2 Operator Interviews

A total of 9 operators were interviewed, covering a range of experience. Feedback on usability, interface design, training and usefulness was obtained. The interviews also extracted information on the operators attitude and acceptance of the system.

4.3 Guideline Application

Based on the task analysis and interview information, PROMASS was assessed against the guidelines previously identified by BNFL for expert systems (Section 3) and other guidelines/criteria for advanced VDU-based systems in process control - for features such as alarm presentation, mimic design, trend displays, navigation.

4.4 Results and Lessons Learnt

The evaluation identified successful aspects and areas for improvement. The operator interviews and guideline evaluation generated recommendations regarding the interface design, training, instructions and the overall system functionality. The interface is being modified to incorporate display design changes to enhance the usability of the system.

The techniques produced a clear definition of the actual role of the expert system in plant operation and specified how, when and why it is used - there were a few interesting contrasts between this and the original design intent. The study also highlighted the differences in attitude towards the system, and identified the reasons for this.

A range of lessons were learnt, and all of high value in respect to altering this particular expert system and the development and implementation of new systems. Overall the importance of involving the operations personnel early on, as part of the team to identify the problem to be addressed was certainly apparent, as well as continual active participation during the development, evaluation, implementation and review process.

4.5 Simulation Study

A full off-line PROMASS system was employed to further evaluate the system under dynamic conditions. The experimental trials comprised a 1.5 hour scenario requiring the operator to diagnose the cause of 7 faults, some of which were related. It was designed such that the level of task complexity, information density and workload varied within this period. For completion, the other VDU-based control systems available in the control room were also simulated.

To test the feasibility and usefulness of various selected human factors data collection and analysis techniques, a small pilot study was run with 4 operators. Briefly, the experimental procedure comprised an initial familiarisation and training period. Verbal protocols, activity sampling and interviews (- ad hoc during the scenario, plus structured interviews at specific freeze points and on completion of the trial) were used to collect and characterise the data. All trials were videoed. The next step of data analysis is at an early stage. Techniques to identify and describe the cognitive tasks and to model operator behaviour are currently being explored.

5. Conclusions and Future Work

The results of this study have been valuable for a variety of reasons. For example, work is underway to increase the usability of the Springfields IDR expert system. The feasibility and usefulness of this type of study was proven within BNFL, and a great deal of information and operating experience was captured for current and future expert systems application - the features that worked and the ones that failed to achieve their objective.

BNFL is also active in a number of collaborative projects in the field and therefore maintains a detailed awareness of the current state of the technology. Major BNFL funded development projects are underway at a number of UK universities, both to support the current existing applications, and to develop other advanced process monitoring and control techniques. The technology can now be regarded as being mature with a number of high quality proprietary systems available. The content of future R & D work takes a company-wide business-led perspective. Through the development programme, which places particular emphasis on human factors, BNFL is in a very strong position to exploit this new technology.

Acknowledgement: The authors would like to acknowledge M Carey (RM Consultants) and OECD Halden Reactor Research Project in their work to develop the guidelines and undertake the evaluation study, respectively.

Part VI
Reliability, Safety and Health Issues

Advances in Agile Manufacturing
P.T. Kidd and W. Karwowski (Eds.)
IOS Press, 1994

Human-Machine System Reliability Using Fault Tree Analysis

Mahmoud A. ABU-ALI

Industrial Engineering Department, University of Jordan, Amman, Jordan

Jerry L. PURSWELL and Robert E. SCHLEGEL

School of Industrial Engineering, University of Oklahoma, Norman, Oklahoma 73019, USA

Abstract. A methodology and a software program are presented to estimate the minimal set of events and to calculate the overall probability of occurrence of a system failure. The methodology is based on the fault tree analysis technique and Boolean algebra rules and postulates. This methodology and software can be used for retrospective analysis of accidents and failures, and for prospective analysis for system design and failure analysis. The software can be utilized by people with little or no experience in fault tree analysis. The software develops a Boolean equation from the descriptive statements entered by the user. It also provides a means for sensitivity analysis for calculating the overall probability of occurrence of failures.

1. Introduction

During the 1960's there was a considerable increase in interest in human reliability within the context of complex military systems. This interest was due to reports providing evidence that human error accounted for a high percentage of system failures [1], [2], [3], [4]. Technological advances and automation have made machines more reliable, thereby increasing the proportion of system failures due to human error [5]. These advances have also made machines more complex, shifting human error from the operation phase to the design, manufacturing, testing, installation, and maintenance phases [5].

Machine reliability engineering is a well-established field, supported by many mathematical and empirical tools and methods. Fewer tools and methods exist to evaluate human reliability. In the context of human-machine systems, there is a need for methods and tools that can be used to analyze and estimate human error. However, there is a greater need for methods and tools that can incorporate machine reliability and human reliability to estimate the overall reliability of the system.

This paper presents a methodology and a computerized tool to estimate the overall reliability and the minimal set of basic events that may cause the failure of a human-machine system. The methodology is based on fault tree analysis (FTA), a technique that was developed by Bell Telephone Laboratories in 1962. FTA employs deduction to reason backward from the top event (failure) through intermediate causes to the basic causes of the failure. The current technique employs Boolean logic to represent the relationships between these causes.

Software was developed that enables FTA to be performed by users with little or no experience with the technique. The program can be used to conduct retrospective analyses such as accident investigation and to conduct prospective analyses such as evaluation of system performance or system design.

2. Methodology

FTA is based on Boolean algebra and requires three assumptions for its application. These assumptions are (1) events are combined using AND and OR operations, (2) basic events occur independently, and (3) the probability of occurrence of a basic event is a constant value. The FTA methodology presented here consists of five algorithms which include:

1. construction of the fault tree,
2. derivation of the Boolean equation,
3. conversion to an equivalent binary equation,
4. generation of the minimal events set, and
5. calculation of the overall probability of occurrence.

2.1. Construction of the Fault Tree

This algorithm is used to construct the fault tree and constitutes the following steps:

1. Identify the top event. This would be an accident or an incident in the case of retrospective analysis, or one of the most critical incidents or system failures in the case of prospective analysis.

2. Identify the causes of the top event and determine if all of the causes must occur (AND) or only one cause need occur (OR) to produce the top event.

3. Identify the causes that can be further analyzed (intermediate events) into their elementary causes and repeat Step (2) until causes cannot be further analyzed.

There are three types of basic events: normal, undeveloped, and independent. Normal events are expected to occur naturally in the system and need not be analyzed further. Undeveloped events cannot be fully developed because of a lack of information about their level of significance. Independent events do not depend on other events and cannot be further analyzed.

2.2. Derivation of the Boolean Equation

This algorithm generates a Boolean equation from the statements provided by the user. The procedure consists of the following steps:

1. Assign upper case letters (A ... Z) to the top and intermediate events, and lower case letters (a ... z) to the basic events.

2. Starting from the top event (A), substitute the causes for each intermediate event while respecting the relationship operation (AND or OR).

3. Repeat Step (2) until all variables in the equation are lower case letters (basic events).

4. Apply the laws and postulates of Boolean algebra to eliminate any repetition of terms and to express the equation in the sum-of-product format. The sum-of-product format provides sets of AND'ed variables (basic events) which are then OR'ed to form the equation.

2.3. Conversion to an Equivalent Binary Equation

This algorithm converts the Boolean equation into an equivalent binary equation. This procedure is necessary and is a preparation step for the minimal set generation and the system overall probability calculation. This algorithm comprises the following steps:

1. Determine the number of basic events involved in the current fault tree. This will determine how many binary digits should be used in each set of events (product term).

    e.g.:   Number of basic events = 3, labeled a, b, and c.

2. Arrange the product terms in ascending order according to the number of basic events in each term.

    e.g.:   A = ab + ca + acb.

3. Order the letters in each product term (e.g., a, b, ... z).

    e.g.:   A = ab + ac + abc.

4. For each term, insert a "1" in the location of the letter of the basic event if the letter exists in the term and insert a "0" if the letter does not exist in the term.

    e.g.:   Equivalent binary term of "ab" = 110 since "c" is not included.
    Equivalent binary equation of A = 110 + 101 + 111.

2.4. Generation of the Minimal Events Set

This algorithm minimizes the binary equation by removing terms that are already represented by other terms in the equation. The procedure utilizes Boolean algebra and constitutes the following steps:

1. Starting from left to right, compare the term with every following term. If for every "1" in the term there is a "1" in the same location in the following term, then eliminate the following term since it is represented in the equation by the term under consideration.

2. Repeat Step (1) for all terms except the last one. The resultant binary equation contains the minimal set of events.

    e.g.:   Minimal equation A = 110 + 101   (111 is represented by 110).

2.5. Calculation of the Overall Probability of Occurrence

This final algorithm calculates the overall probability of occurrence of the system failure (F) under consideration. System reliability (R) with respect to this specific failure can be calculated using the relationship R = (1 - F). This algorithm requires knowing the probability of occurrence of each basic event and assumes events to be independent in occurrence. It comprises the following steps:

1. Calculate the probability of each term using the product rule since terms are in product format, that is consisting of AND'ed basic events.

    e.g.:   assume P(a) = 0.5, P(b) = 0.9, and P(c) = 0.8,
    then P(110) = P(ab) = 0.5 $\times$ 0.9 = 0.45.

2. Calculate the overall probability of the system by considering that terms are OR'ed.

    e.g.:   assume A = ab + ac,
    then P(A) = P(ab + ac)
            = P(ab) + P(ac) - P(ab $\bullet$ ac)
            = 0.45 + 0.40 - 0.18
            = 0.67.

3. Software

The software was developed for IBM-compatible personal computers to run in the MS-DOS environment[1]. It was written in Turbo Basic which (1) provides a non-formal editing style, (2) minimizes debugging errors, and (3) generates an executable version of the source code (.EXE). The system consists of a set of modules where each module represents one of the algorithms discussed in the methodology section. These program modules can be executed from a menu module which provides a user-friendly interface.

The first module constructs a new tree (fresh start). It requests that the user input descriptive statements for the failure, the immediate events that lead to the failure, and the type of these events (normal, independent, undeveloped, or intermediate). It also asks whether all these events (AND relationship) must occur in order to cause the failure or if any one of the events (OR relationship) may cause the failure.

The module provides clear prompting messages, necessary error messages, and allows users to make mistakes and to correct them. Users are asked to input information for the intermediate events similar to the information required for the failure. The program module provides three windows on the monitor: (1) a dialogue line window where subjects are requested to enter the necessary information, (2) a dialogue record window which shows the information entered by the user for the current event, and (3) a stack window which shows all the events entered and requiring further analysis with respect to their dependency (hierarchical) order.

The other program module that requires information from the user is the module for calculating the overall probability of occurrence. Users are asked to input their estimation of the probability of occurrence for each basic event. An additional module provides the user with a sensitivity analysis tool by allowing modification of the basic event probabilities to examine the effect on the overall probability of occurrence of the failure. It also illustrates the effect of the modification on each term.

4. Conclusions

The presented methodology and software (1) provide a systematic approach for analyzing accidents, incidents, or system failures, (2) identify the important aspects in relation to the failure under consideration, (3) allow concentration on one failure or incident at a time, and (4) provide a tool that can be used to analyze system failure quantitatively.

References

[1] A. Shapero et al., Human Engineering and Malfunction Data Collection in Weapon System Programs, WADD Technical Report, 1960, pp. 36—60.
[2] D. Meister, Analysis of Human Initiated Equipment Failure During Category I Testing OSTF-1, Technical Report Ro54. San Diego, CA: General Dynamics/Astronautics, 1961.
[3] J. Cooper, Human Initiated Failures and Malfunction Reporting. In IRE Transactions in Human Factors in Electronics, HFE, 1961, pp. 104—109.
[4] H. Willis, The Human Error Problem, Technical Report 62-76. Denver, CO: Martin Marietta Corp, 1962.
[5] D. Embrey, Human Reliability in Industrial Systems: An Overview. In Proceedings of the Human Factors Society 20th Annual Meeting. Santa Monica, CA: Human Factors Society, 1976, pp. 12—16.

---

[1] The software is available from Dr. Mahmoud Abu-Ali at the Industrial Engineering Department, University of Jordan, Amman, Jordan. Phone: 962-6-843-555, FAX: 962-6-810-472.

Advances in Agile Manufacturing
P.T. Kidd and W. Karwowski (Eds.)
IOS Press, 1994

Modeling Cognitive Aspects of Human Error in Dynamic Tasks

Lon N. Haney, Wendy J. Reece, Cheryl J. Wilhelmsen, Henry A. Romero
Idaho National Engineering Laboratory, EG&G Idaho, Idaho Falls, ID, USA 83415-3855

Abstract. This paper discusses the current state of human reliability analysis (HRA) in terms of analysis of cognitive errors. The concept of incorporating cognitive science in HRA to facilitate identification and modeling of cognitive errors, and the need for identifying appropriate data for quantitative estimation is presented. Advanced manufacturing environments (AMEs) move the human into a more cognitive role requiring dynamic decisionmaking throughout the day. Cognitive errors can result in increased quality costs, decreased safety, and decreased effectiveness and efficiency of the AMEs. Fully characterizing, quantifying, and anticipating cognitive errors will allow system designers and users to reduce the likelihood or consequence of these errors.

1. Background

Techniques of Human Reliability Analysis (HRA) have been developed and used to model and estimate the probability of critical operator errors in complex systems. HRA has been widely used for a variety of applications. For example, the INEL has applied HRA to the analysis of nuclear power plant operation [1,2,3,4] and in the assessment of pilot error in advanced technology commercial aircraft [5]. The analysis of human error would also be applicable to the reliability of manufacturing operations and facilities. The analysis of human error to increase human reliability could benefit manufacturing by reducing lead time, downtime, and the costs of significant human error events and poor quality (e.g. scrap, rework, redesign, warranty costs). Results of a thorough HRA can serve as input to operations and production management, equipment configuration, and workforce development issues.

Several approaches have been developed and used for the modeling and quantitative estimation of human error. These approaches include: data based decomposition techniques [6], the use of simulator data to estimate error rates for categories of error types [7], development and use of time reliability correlations based on simulator data [8,9,10], simulation modeling [11,12], and formal expert estimation techniques [13,14]. One of the most widely used techniques, The Technique for Human Error Rate Prediction (THERP) [6] is most defensible for analysis of proceduralized tasks but provides little guidance for the estimation of human error probabilities (HEPs) for cognitive tasks.

Some of the existing techniques do account for cognitive tasks. However, these techniques use simplistic modeling for characterization of cognitive aspects of tasks or rely on expert estimation for the generation of HEP estimates. For example the Human Cognitive Reliability [8] method accounts for cognition only by categorizing cognitive behavior into three types, skill, rule or knowledge based, and providing a time reliability correlation for each to estimate HEPs. This taxonomy fails to capture potentially important aspects and influences of cognitive behavior (see Section 2). The Cognitive Environment Simulation [12] provides a computer simulation of human cognitive behavior resulting in a "stream of consciousness" type output, however estimation of HEPs using CES output relies on expert estimation. A need exists for the development of defensible HRA approaches to model the important aspects and influences of cognitive behavior in dynamic tasks that will support the estimation of HEPs for those tasks without significant reliance on less defensible expert estimation approaches.

2. Cognition

Contemporary research in the field of cognition and human-system interaction explores aspects of human reliability and decision making in dynamic systems [15]. The study of cognition has evolved from a traditional emphasis on modeling human decision making behavior [16,17]. Components of information processing include attention, memory, cognitive representations (mental models), feedback, and learning. Empirical analysis has demonstrated that people routinely employ heuristics, or rule-of-thumb strategies, when making decisions [18]. Cognitive errors are often due to the application of an erroneous strategy, logical deficiencies, or missing or inaccurate information. Motivation and stress can also contribute to error.

The quantification of human error due to cognition has not been fully accounted for in HRA methodology. Accurate estimation of errors in trouble-shooting, predictive judgments and other types of problem solving tasks would contribute a significant component to the modeling of dynamic tasks.

3. Cognitive aspects of advanced manufacturing environments

The task requirements for humans in advanced manufacturing environments (AMEs) are significantly different than standard manufacturing environments. AMEs require the human to make decisions concerning technical aspects of their jobs that may not be routine, proceduralized, or anticipated. AMEs are characterized by computer controls, automated machinery, remote operations, etc., that significantly alter the interface between the person and the process. These changes move the human into a more cognitive role making various decisions throughout the day. The non-routine, non-proceduralized nature of these events results in dynamic tasks with significant cognitive aspects. Incorrect decisions and actions occurring within the dynamic environment can result in decreased efficiency and effectiveness of the AME. Specifically, incorrect decisions will result in an increase in quality costs (scrap, rework, inspection, testing, etc.) and a decrease in the efficiency of normal operations. Additionally, incorrect decisions resulting in inappropriate actions during abnormal operations will significantly decrease the safety of the AMEs.

4. Research and Development Needs

The process for incorporating existing cognitive taxonomies and heuristics into quantitative HRA models requires investigation. Cognitive task analysis should be designed to collect required information about human actions in complex tasks. This information would serve as input to a more complete model of human actions, accounting for information processing and decisionmaking elements. Research is needed to identify existing quantitative data, and to assess how cognitive variables affect existing HEP estimates. New studies as well as existing laboratory results could be utilized to generate probability estimates and help identify factors for mathematically adjusting HEPs to account for cognitive variables. Existing sources of observational data should be assessed as well.

5. Conclusion

Modeling of human actions in complex systems would be greatly enhanced with the additional analysis of the cognitive elements of human performance. A dedicated analysis of cognition would yield insight into quantitative HRA evaluations of dynamic tasks which place additional cognitive demands on performance. By assessing the cognitive components of dynamic tasks, we can anticipate and circumvent potential costly errors. This knowledge can be used to increase reliability, safety and efficiency in the manufacturing environment.

References

[1]   Brownson, D.A., Haney, L.N. and Chien, N.D. (1993). Intentional Depressurization Accident Management Strategy for Pressurized Water Reactors, NUREG/CR-5937, EGG-2688.

[2]   Galyean, W.J. and Gertman, D.I. (1992). Assessment of ISLOCA Risk-Methodology and Application to a Babcock and Wilcox Nuclear Power Plant, NUREG/CR-5604, EGG-2608.

[3]   Kelly, D.L., Auflick, J.L. and Haney, L.N. (1992). Assessment of ISLOCA Risk-Methodology and Application to a Westinghouse Four-Loop Ice Condenser Plant, NUREG/CR-5744, EGG-2649.

[4]   Kelly, D.L., Auflick, J.L. and Haney, L.N. (1992). Assessment of ISLOCA Risk-Methodology and Application to a Combustion Engineering Plant, NUREG/CR-5745, EGG-2650.

[5]   Nelson, W.R., Byers, J.C., Haney, L.N., Ostrom, L.T., and Reece, W.J., (1992), Lessons Learned from Pilot Errors Using Automated Systems in Advanced Technology Aircraft. Proceedings of American Nuclear Society Topical Meeting on Nuclear Plant Instrumentation, Control, and Man-Machine Interface Technologies, Oak Ridge, Tennessee.

[6]   Swain, A.D. and Guttman, H.E. (1983). Handbook of Human Reliability Analysis with Emphasis on Nuclear Power Plant Applications, NUREG/CR-1278, SAND80-0200.

[7]   Beare, A.N., Dorris, R.E., Bovell, C.R., Crowe, D.S. and Kozinsky, E.J. (1983). A Simulator-Based Study of Human Errors in Nuclear Power Plant Control Room Tasks, NUREG/CR-3309, SAND83-7095.

[8]   Hannaman, G.W., Spurgin, A.J. and Lukic, Y.D. (1984). Human Cognitive Reliability Model for PRA Analysis, NUS-4531, Electric Power Research Institute.

[9]   Weston, L.M., Whitehead, D.W. and Graves, N.L. (1987). Recovery Actions in PRA for the Risk Methods Integration and Evaluation Program (RMIEP), Volume 1: Development of the Data-Based Method, NUREG/CR-4834/1 of 2, SAND87-0179.

[10]   Operator Reliability Experiments Using Power Plant Simulators Vols. 1-3, EPRI Report NP-6937, July 1992, Palo Alto, CA.

[11]   Siegel, A.I., Bartter, W.D., Wolf J.J., Knee, H.E. and Haas, P.M. (1984). Maintenance Personnel
        Performance Simulation (MAPPS) Model: Summary Description, Volume 1, NUREG/CR-3626/1 of
        2, ORNL/TM-9041/V2.

[12]   Woods, D.D., Pople, Jr., H.E. and Roth, E.M. (1990). The Cognitive Environment Simulation as a Tool
        for Modeling Human Performance and Reliability, Volume 1: Executive Summary, NUREG/CR-5213/1
        of 2.

[13]   Comer, M.K., Seaver, D.A., Stillwell, W.G. and Gaddy, C.D. (1984) General Human Reliability
        Estimates Using Expert Judgment, NUREG/CR-3688, SAND84/7115, Volumes 1 and 2.

[14]   Embrey, D.E., Humphreys, P.C., Rosa, E.A., Dirwan, B. and Rea, K. (1984). SLIM-MAUD: An
        Approach to Assessing Human Error Probabilities Using Structured Expert Judgment, Volume I:
        Overview of SLIM-MAUD, NUREG/CR-3518/1 of 2, BNL-NUREG-51716.

[15]   Bainbridge, L., Lenior, T.M.J., and van der Schaaf, T.W. (Eds.) (1993). Special Issue: Cognitive
        Processes in Complex Tasks, Ergonomics (Volume 36, 11) London: Taylor & Francis.

[16]   Tolman, E.C. (1949). Purposive behavior in animals and men. (Reprinted, University of California
        Press) New York: Appleton-Century-Crofts (1932).

[17]   Estes, W.K. (1959). The statistical approach to learning theory. In S. Koch (Ed.) Psychology: A Study
        of a Science (Volume 2) New York: McGraw Hill.

[18]   Kahneman, D., Slovic, P., and Tversky, A. (Eds.) (1982). Judgment Under Uncertainty: Heuristics and
        Biases New York: Cambridge.

Advances in Agile Manufacturing
P.T. Kidd and W. Karwowski (Eds.)
IOS Press, 1994

Risk Orientation, Complexity and Dynamic Function Allocation in Human-Machine Systems

E. Pascoe, N. Pidgeon and P. Barber
Birkbeck College, London University, London UK,

Abstract. An experiment using a simulation of a hazardous process control task examined the effect of individual risk orientation and task complexity on decisions underlying function allocation between manual and automatic controllers. The results indicated a difference in the pattern of function allocation between Risk-Averse and Risk Seeking subjects. An explanation for the diversity of allocation patterns is proposed in terms of 'safety-constrained learning' which limits the experience of Risk Averse operators to 'safe' process states and impairs their ability to cope with 'out-of-bounds' scenarios.

1. Introduction

Recent theoretical models of risk behaviour have identified that individual risk orientation (a tendency to Risk Seeking or Risk Aversion) is one of the factors influencing decision making under risk [1]. In process control tasks a Risk Averse strategy is one which prioritises safety over production gains, while a Risk Seeking strategy prioritises production gains over safety. Empirical studies of process control decisions [2] suggest large individual differences in the use of control strategies. One aim of this study was to investigate if some of these differences in the allocation of control functions between automatic and manual controllers could be explained by differences in individual operators' risk orientation. It was predicted that the Risk Averse operators would use more automatic control (as this strategy offers safety by minimising the probability of exceeding the safety bounds) as compared to the Risk Seeking group. The Risk Seeking group was expected to show a stronger preference for manual control in comparison to the Risk Seeking subjects, as this strategy (albeit less safe) maximises the potential for achieving good production figures. The second purpose of the study was to examine the effect of changes in task complexity on the operators' function allocation. Increased complexity was achieved by introducing an additional goal [3], that of having to deal with a fault and it's consequences in terms of disturbance effects on the system behaviour.

2. Method

22 subjects completed the process control task (12 males and 10 females, all aged between 26 and 31 years old). Subjects had no previous experience of a process control task. The task used was a dynamic, real-time computer-based simulation of a water-alcohol distillation system (PROCESS, see [4] for a detailed description). Control of such a distillation task is common in the process industry. The degree of automation of the simulation was in conformity with modern plant. In this simulation the operator's main goals are to maximise production within safety constraints. This can be achieved by varying the Feed Stock Pump and Temperature Settings by either manual commands or using automatic controllers. A warning is triggered when the process flow parameters violate certain safety bounds. Bonuses were offered for achieving performance above average but the safety goal required subjects to maintain the process within the set alarm safety bounds and penalties were deducted for each alarm caused.

A 2x2 mixed design was used. The first independent variable was Risk Orientation (between subjects) with two groups of subjects; Risk Seeking and Risk Averse. The second (within subjects) independent variable was Complexity, operationalised as the number of goals the subjects had to attend to simultaneously. In the Low Complexity condition the simulation was performing normally without disturbance. In the High Complexity condition a third goal was introduced, that of coping with a pre-programmed fault operationalised as a leak in the Feed Pump controller.

Operators were divided into Risk Seeking (RS) and Risk Averse (RA) on the basis of their choices in a lottery task (for a detailed description see [1]). They were trained to operate PROCESS, to manage the fault and instructed about the dual-goal (safety versus production) nature of the task. The characteristics of automated and manual control strategies were explained. Subjects had to control the process task using Feed Pump and Temperature controllers on 8 experimental trials, lasting 6 minutes each (4 of Low and 4 of High Complexity). Two sets of measures were used to describe the strategies and outcomes in the task. The first set of measures describe the strategies in terms of percentage of trial time spent using automatic settings for the Feed Pump and Temperature controllers, and the overall level of interaction with the system expressed as the average number of control actions per trial. The second set of measures related to outcomes and included Performance measured in total production achieved per trial and Safety measured by the number of alarms caused by exceeding the set alarm safety bounds.

3. Results

An analysis of variance indicates, that for both controllers (Feed Pump and Temperature) in both Low and High Complexity the Risk Averse operators spend more time on automatic controllers than the Risk Seeking group. This was demonstrated by the significant main effect of Risk Orientation for the Feed Pump ($p > 0.01$, $F(1,20) = 16.02$) and for the Temperature Controller ($p < 0.01$, $F(1,20) = 10.11$) .

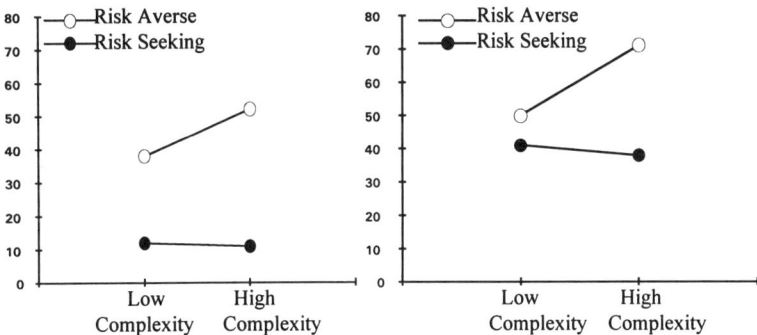

Figure 1. Percentage of time spent using   Figure 2. Percentage of time spent
          automatic feed pump   using automatic temperature controller

The response to increased complexity was different for both groups (see Fig.1 and 2). There was a significant interaction of Risk Orientation and Complexity for the Feed Pump (p<0.01, F(1,20)=10.52) and for the Temperature controller (p<0.01, F(1,20)=10.86). The Risk Averse operators increased their reliance on automated control strategy for the feed pump controller (from 37% in Low Complexity to 52% in High Complexity, significant at p<0.001, F(1,20)=18.57). The same pattern was observed in the Risk Averse group on the temperature controller (from 50% of trial time in Low Complexity to 71% trial time in High Complexity, as confirmed by simple effects significant at p<0.01, F(1,20)=16.17).

The Risk Seeking operators maintained the same control strategy (over 50% of trial time in manual mode) in both Low and High Complexity condition for both controllers (see Fig.1 and 2). Regarding control actions, the Risk Seeking operators made more, in both High and Low complexity conditions as compared to the Risk Averse group (p<0.001, F(1,20)=30.75). There was also a difference in Performance, with the Risk Seeking operators achieving more production than the Risk Averse group in both Low (RS=20.9 t/h and RA=16.2 t/h) and High Complexity (RS=14.5 t/h and RA=9.2 t/h) (p<0.01, F(1,20)=14.3). The Risk Seeking group had a worse safety record (1.7 alarms per trial) than the Risk Averse group (0.7 alarms per trial), causing more alarms in the Low Complexity condition. However, the Risk Seeking operators obtained a better Safety record in the High Complexity condition (3.2 alarms per trial) as compared to the Risk Averse group (4.9). The interaction of Risk Orientation and Complexity was significant at p<0.001, F(1,20)=19.13; for detailed analysis of these outcomes see [5].

4. Discussion and Conclusions

The results show that the Risk Seeking subjects selected manual control strategies for over 50% of trial time on both subsystems. Risk Averse operators, developed strategies based on the more extensive use of automatic controllers, using this strategy for over 50% of trial time except for the Feed Pump under Low Complexity condition (37%). This finding may be explained by the Risk Averse operators prioritising Safety over

Performance, and thereby selecting the automatic control strategy in the expectation of minimising bad outcomes (exceeding alarms safety bounds). The Risk Seeking group demonstrated preferences for the manual controllers, which may be explained by their prioritising Performance over Safety. The increase in complexity was expected to lead to a shift towards automatic function allocation in both groups in line with previous findings that a fault led to an increase in the use of automatic controllers [6]. This prediction was confirmed only for the Risk Averse operators. The Risk Seeking group maintained their manual strategies despite the increase of complexity. Analysis of the Risk Orientation effects on function allocation strategies combined with the analysis of Performance and Safety findings indicates, that the manual control strategy was coupled with good Performance in both High and Low Complexity, and a better Safety record in High Complexity as compared to the automatic control strategy. The strategy of relying mainly on automatic controllers selected by the Risk Averse group might have allowed them to avoid alarms in Low Complexity trials, but limited their experience with out-of-bounds process states. Hence ironically, using a safety-oriented control strategy based on automatic controllers may lead in the long term to an overall decline in safety.

The general picture emerging from our analysis in the context of the current task is the somewhat counterintuitive conclusion that Risk Seeking subjects cope better with faults under High Complexity. The explanation for this finding can be sought in terms of individual differences in risk orientation leading to 'safety-constrained learning' in Risk Averse subjects. Such type of learning appears to result in the development of control strategies that prevent the Risk Averse group from experiencing out-of-bounds process states, limiting their ability to cope with the occurrence of faults or a general increase in task complexity. The differing patterns of function allocation and safety record indicate that training sensitive to individual risk orientation should be considered to counterbalance the Risk Averse operators' safety constrained learning strategies.

5. References

[1] Lopes, L. (1987) Between hope and fear: the psychology of risk. *Advances in Experimental Social Psychology*, Vol 20, pp 255-295.
[2] Rasmussen, J. (1986) Information Processing and Human-Machine Interactive: An Approach to Cognitive Engineering. Amsterdam: Elsevier
[3] Brehmer, B. and Allard, R. (1991) Real-time dynamic decision making. Effects of task complexity and feedback delays. In J.Rasmussen, B. Brehmer and J. Leplat (Eds), *Distributed decision making: Cognitive models for co-operative work.* Chichester: Wiley.
[4] Jelsma, O. and Bijlstra, J.P. (1990) PROCESS: Program for Research on Operator Control in an Experimental Simulated Setting. *IEEE Transactions on Systems, Man and Cybernetics*, Vol 20, pp.111-132.
[5] Pascoe, E. and Pidgeon, N. (in press) Risk Orientation in Dynamic Decision Making. In Selected Proceedings of 14th Conference on Subjective Probability, Utility and Decision Making. Aix-en-Provence, France, 20-26th August 1993.
[6] Lee, J. and Moray, N. (1992) Trust, control strategies and allocation of function in human-machine systems. *Ergonomics*, Vol 35,(10), pp 1243-1271.

Advances in Agile Manufacturing
P.T. Kidd and W. Karwowski (Eds.)
IOS Press, 1994

343

Extensions of dynamic task allocation concepts for complex systems

Igor Crévits, Serge Debernard, Marie-Pierre Lemoine, Patrick Millot
Laboratoire d'Automatique et de Mécanique Industrielle et Humaine — URA CNRS 1775
Université de Valenciennes
Le Mont Houy — BP 311
59304 Valenciennes Cedex
Tél. : (33) 27 14 12 34 — Fax : (33) 27 14 12 94
e-mail : crevits@univ-valenciennes.fr

Abstract. Dynamic task allocation is a form of man-machine cooperation which consist in introducing in the control and supervisory loop of a process, a task allocator which shares tasks between a human operator and an automated system. Such a cooperation aims at optimizing process functionning and regulating operator workload. Several laboratory studies have shown feasability and interest of such approach but are confined to very simple process. This paper propose extensions of DTA concepts for complex processes with structures not taken into account and characteristics to be realized.

1. Introduction

Since several years it is now certain that the presence of man is needed in the monitoring and supervision of great process, and even vital. Numerous studies have insisted on this presence. We are interesting, in this paper, in the problems of man-machine cooperation and especially in the principles of dynamic task allocation (DTA). These principles have been stated in the 80's and gave birth to some laboratory studies. Nevertheless, they rely on particular and relatively simple characteristics of tasks and processes.

By widening the characteristics taken into account, this paper suggests some extensions of the DTA principles. First of all, we remind the principles and goals of the DTA. Then we introduce the state-of-the art in this field to determine two classes of deficiency based respectively on the characteristics of the tasks and on the characteristics of the processes. In the second section we suggest some extensions of basics principles meeting the first class of differences. At last, in the third part we propose other extensions for the second class of differences.

2. Dynamic task allocation

2.1. Principles

DTA consists in doubling the human operator (HO) with an automated system (AS) of type agent able to take in charge some tasks said shareable. This handling evolves with time to reach a double aim : (i) to obtain the optimum working of the system, (ii) to regularize the workload (WL) of the HO i.e. to avoid the overloads which damage its abilities in short term and to avoid underloads which damage its abilities in long term.

The idea of sharing implies, for the process, several characteristics. [Debernard93] expressed four ones : (i) the process has to be able of multitasking in order to constantly let the HO with tasks to do, (ii) an automated system able to perform some is available, (iii) in the system operation, work overload for the operator can appear, (iv) finally the whole system can not be automated, it means that the presence of an operator is essential.

The general structure of a system based on DTA is as following (figure 1). The role of the task allocator is to inform the two decision-makers of the tasks they have to handle and to avoid that the HO or the AS to assign to themselves a task allocated to the other. The control of task allocator allocates in affect the tasks i.e. it drives the task allocator. This function can be performed either by the HO, it's then called explicit DTA (EDTA) or by a dedicated automated system, in this case it's known as implicit DTA (IDTA) [2]. We will not go back on the advantages and disadvantages of one another of those modes presented in [1].

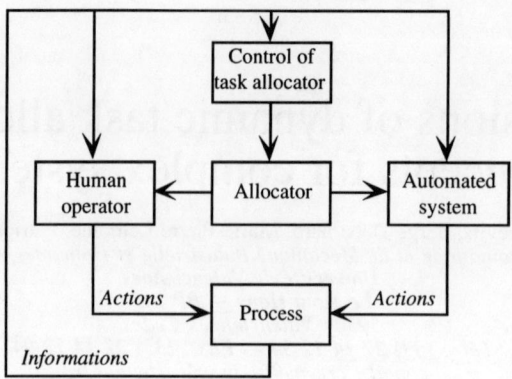

Figure 1 — Dynamic task allocation structure

Because the IDTA presenting a bigger scientific interest, studies related to it first tried to define the politic of the control of the allocator. [3] uses the queueing theory. [4] bases his work on a predictive model of the operator actions. [5] uses the system efficiency prediction. At last, [6] integrates a double goal of containment of the WL between two minimum and maximum thresholds, and optimization of the process efficiency. Those different approaches allowed the definition of the theorical foundations of the DTA and shown it's feasability. Nevertheless, they only interested in quite simple simulated process where the tasks are independant and always shareable. In the following of this paper, we try to extend these works to more complex situations and that as part of IDTA.

2.2. Insufficiencies

When the DTA is applied to a complex process, two classes of problems appear. The first one is due to the complexity of the tasks to do and to the consequences it drag to the control of the allocator. Indeed, in previous studies, the simplicity of tasks permited on one hand an easy recognition of these tasks by the control of the allocator and on the other hand to consider the tasks as always shareable, the two decision-makers having always the competences to fulfill them. But the thrust in the process operation requires to verify at every time the shareability of the tasks and therefore the allocation.

The second class of problems deals with the characteristics of the process. Functionally the basic principles of the DTA consider process or functions which are clearly identified, unitary and isolated; yet industrial processes are often made of several systems each of them achieving it's own function but interacting with the other to reach an overall goal. Then there are two kind of cooperation : (i) between functions ; (ii) man-machine for some functions. Meanwhile, in the principles stated previously, DTA does not take into account the second king of cooperation.

For every insufficiency class, we suggest extensions to DTA.

3. Dynamic complex task allocation

The first class of insufficiency comes from the tasks complexity and from their ability to be interconnected. To take into account, it is necessary to study the modifications to bring to take into account on the allocator and at last on the affectation politic integrated in the control of the allocator.

3.1. Control parameters

The aim of the control of the allocator is to distribute the shareable tasks among the decision-makers. Nevertheless, because the tasks are more complex, the notion of shareability can not be considered any more as static, for the AS is not always able to process some tasks and that in some context. In fact, we considerthe AS competences as unstatic. Generally speaking, we have to determine the competences of the two decision-makers, the area of these competences is the set of the shareable tasks.

The second parameter to take into account is the processing capacities of the two decision-makers. If, for the AS, this capacity is easily definable, it is not so simple with the HO for whom the processing capacity has limits reflecting his WL.

After all, the control of the allocator has to have informations issued form the process at it's disposal, to determine the set of tasks to do. This function can be particularly complex to implement because from the set of tasks to do and from the decision-maker competence it has to determine the set of shareable tasks. This is even more critical when the tasks are strongly dependant. The best solution is to build groups of shareable tasks which are independant from other tasks and to assign a group to only one decision-maker. Nevertheless it is not always workable if a task, in an independant group, is not shareable.

3.2. Task allocator

It plays a double role. It first has to forestall each decision-maker of the tasks that it has been affected. This is to prevent coordination conflicts harmfull to the efficiency of the man-machine system. Depending on the processes, it more or less easy to do.

The second role is to avoid decision conflicts among the decision-makers. Conflicts which appear when the shareable tasks, or group of shareable tasks, can not be made independant. It is necessary to establish a real cooperation among the decision-makers and that by informing each decision-maker of the decision of the other one. It is there too, not so simple notably from the HO to the AS.

3.3. Control algorithm

Two classes of algorithms are conceivable. The first one, the simpler one, consists in realizing task sharing in real-time in function of entry parameters which are shareable tasks, competences and capacities. Two subtypes of algorithms can be conceived according to the transferability of shareable tasks. This one can be defined as the ergonomical quality of task transfer from a decision-maker is going to do to the other decision-maker. Generally, simpler a task is, better the transferability is. When transferability is good, algorithm can be preemptive (PDTA) i.e. it can be transfer tasks form a decision-maker to the other, at every moment. This class of algorithms is very interesting because it allows better regulation of the WL of HO and performance optimization of the man-machine system. On the contrary, it can question the own work planning the HO have which can be ergonomically harmful. In such a situation and when the tasks transferability is bad, non-preemptive algorithm (nPDTA) could be used consisting in not questioning the task allocation when decided. Then the regulation of WL and performance optimization is more difficult, especially when not expected situations occur.

When these problems happen, it is better to use the second class of algorithm which consists in working in prediction. So it is necessary to predict tasks to be realized, which is a problem, and moreover to plan shreable task allocation. One of the major ergonomical advantage is that it is possible to forestall the HO who can plan his work. Nevertheless earlier the prediction is, more important the incertainty about allocation planning optimality is. Then this planning can evolve in time but can be more or less modifiable. So this modifiability depends on the incertainty of the tasks prediction and on the ergonomical criterion of non-modification of the planning of the HO.

Now we present the second class of insufficiencies linked to the processes structure.

4. Dynamic task allocation in complex systems

4.1. Structural characteristics of the processes

In order to face up to the increasing complexity of industrial processes, two principles are used. (i) The decomposition of the processes into sub-systems allows to decrease this complexity, sub-systems are solving reduced problem by cooperating in a global goal. (ii) The hierarchical organization allows to organize and structure sub-systems in function of the processes characteristics.

Some sub-systems or functions can be the object of a DTA. Then this one can benefit from the hierarchical organization and the cooperation between sub-systems. So it is necessary to differentiate two forms of DTA, the internal one and the external one.

4.2. Internal DTA

Definition. *A function forms the object of an internal DTA when the task allocator is only ensured by a decision-maker into the function.*

The study takes only into account : (i) an isolated function in a process ; (ii) the characteristics of the tasks to be realized in this function ; (iii) the competences and capacities of the HO which take in charge this function ; (iv) the competences and capacities of the AS associated to the HO.

4.3. External DTA

Définition. *A function forms the object of an external DTA when the task allocator is external to the function.*

4.4. Mixed DTA

Finally a mixed approach of the two types of internal and external DTA can be viewed i.e. a function realizes in internal its own DTA, but takes into account external informations about it. This approach can be considered as decision aid integrated into the internal DTA. Then the decision-makers in the function can validate, invalidate or modify the proposed allocation.

Such structures are possible in the case of parallel functions. The functions being the same can communicate between themselves the used allocations in the case of overloads. In the sequential processes, each function has a goal to be realized, based on the result of the previous function which depends on the task allocation. But this approach is very interesting in the hierachical processes where the functions have different temporal views of the process. Each function refines the results of the higher function. When DTA is used, function prepares the task allocation of the lower function and defines its own task allocation from this one of the higher function.

5. Conclusion

There is two ways to judge to the quality of a DTA. At first, cooperation between the two-decision-makers and actors must be real. The cooperation must not be a constraint for the HO in his decisions and actions. Secondly the global system performance must be higher, without increasing the WL of the HO.

The transferability and the modifiability of the allocation planning offer, when it is possible, more suppleness in the internal allocation and allow to adjust more accurately the WL and the performance. The external allocation offer other suppleness in supressing the allocation within the function

At the moment the new principles of mixed DTA are studied and applied to the context of air-traffic control in the framework of the SPECTRA project, study realized in collaboration with the CENA (Centre d'Etude de la Navigation Aerienne). On a first function, of tactical level, is implemented a DTA between radar controller and conflict solving system [1], and the second strategical level ensures the task allocation [7].

Bibliographie

[1]   S. Debernard, *Contribution à la répartition dynamique de tâches entre opérateur et système automatisé : application au contrôle de traffic aérien*, Thèse de doctorat, Université de Valenciennes, France, 1993

[2]   C.A. Rieger, J.S. Greenstein, *The allocation of tasks between the human and computer in automated systems*, Proceedings of IEEE International Conference on Cybernetics and Society, New-York, USA, 1982, pp 204 to 208

[3]   Y. Chu, W.B. Rouse, *Adaptative allocation of decision-making and task allocation in human-machine systems*, Proceedings of annual conference on manual control, vol 21, pp 9.1 to 9.11, 1985

[4]   J.S. Greenstein, M.E. Revesman, *Application of a Mathematical Model of Human Decision Making for a Human-Computer Communication*, IEEE SMC vol 16, no 1, pp 142 to 147, 1986

[5]   N.M. Morris, W.B. Rouse, S.L. Ward, *Studies of dynamics task allocation in aerial search environment*, IEEE transaction on systems, man and cybernetics, vol 18, no 3, 1988

[6]   P. Millot, *Supervision des procédés automatisés et ergonomie*, Edition HERMES, 1988

[7]   I. Crévits, Multi-level cooperation in air-traffic control, Fourth annual conference on Human-Machine Interface and Artificial Intelligence in AeroSpace, Toulouse, France, September 28-30, 1993

Advances in Agile Manufacturing
P.T. Kidd and W. Karwowski (Eds.)
IOS Press, 1994

Technological Uncertainty, Job Control, and Operator Strain

Sean Mullarkey, Paul R. Jackson & Toby D. Wall
MRC/ESRC Social and Applied Psychology Unit, University of Sheffield.
United Kingdom

Abstract. Using Karasek's theory of work stress we evaluate the combined impacts of AMT system errors, and three facets of job control on operator strain. Interaction effects between system errors and job control were observed for analyses involving Timing control only. We argue that adequate assessment of moderator effects for system errors requires additional job control measures that tap the opportunity operators have for the management of AMT system disturbances. Interpretation of results focused upon the ways in which AMT system disturbances can constitute positive or negative events, depending upon the level of work 'traction' associated with the operation of AMT.

1. Introduction

The purpose of this study is to evaluate the combined effects of AMT system characteristics and job control on worker mental health. The theoretical framework within which this study has been conducted consists of Karasek's Psychological Demands/Decision Latitude Model [1]. This theory asserts that mental-health and psychological work stress reactions arise out of the combination of two important factors in the workplace: psychological demands and decision latitude, or work autonomy. The basic proposition is that psychosocial stress reactions are likely to be more severe in jobs that are characterised by high psychological demands and low work autonomy, than in jobs where greater autonomy is experienced. This is because low levels of work autonomy constrain employee's ability to cope or manage with workplace demands by impeding appropriate preventative, ameliorative, or avoidance responses. Karasek's theory has been tested on large samples of Swedish and U.S. workers and has been shown to predict the prevalence of a variety of mental-health outcomes including depression, exhaustion, pill consumption, job dissatisfaction, and absenteeism.

The workplace demands that we chose to examine in the context of this study of advanced manufacturing technology consisted of two characteristics associated with AMT system performance: *technological uncertainty* and *technological abstractness*. Technological uncertainty reflects the extent to which an AMT application is vulnerable to disruption, leading to downtime; and technological abstractness refers to the extent to which these disruptions are ambiguous in nature, or difficult to conceptualise and comprehend. High levels of this latter aspect lead to error recovery that is hard to achieve, and analysis of causes of problems that are hard to define. We chose these characteristics for two main reasons. Firstly, recent research in human-computer interaction has identified system breakdowns as significant stressors in computer-controlled administrative work [2, 3]. Technical deficiencies represent a major aspect of office and AMT environments [4], yet the impacts of system errors on operators of AMT systems have yet to be examined. Secondly some authors have presented strong arguments that technological 'abstractness' or 'equivocality' represents a major stressor in high technology environments [5, 6], yet to our knowledge the effects of this aspect on operator strain has yet to be evaluated in any work setting.

In considering work autonomy as a potential moderator of these AMT demands we found it useful to refer to earlier work that has highlighted the multi-faceted nature of job control [7, 8]. We have chosen to focus upon three aspects of job control identified as being of particular relevance to alternative AMT applications [8]. These consist of: *boundary control*, which is concerned with the extent to which operators are involved in secondary production activities supporting primary operating tasks; *method control*, which reflects the discretion operators have over the methods used to carry out their work; and *timing control*, reflecting the extent to which operators can make decisions concerning the initiation, termination and pacing of operations, rather than these being determined by the requirements of the technology.

Setting our study broadly within Karasek's framework we predict that these different facets of job control will act as significant moderators of the impacts of uncertainty and abstractness on job-related strain. We therefore predict that job control will have a greater beneficial impact on job related strain for systems characterised by high abstractness and high uncertainty than for systems characterised by low uncertainty and abstractness.

2. Method

Seventy-two operators of seven distinct AMT applications used in the production of printed circuit boards and electromechanical control modules participated in this study. All were employed within an electronics company situated in the South of England. Data was collected in the form of an opinion survey that formed part of a broader study of organizational change. Multiple item measures were included for our two technology variables: Technological uncertainty (4 items, alpha .72), measured the extent to which machines required human support, adjustment, and attention because of a breakdown; and Technological Problem Complexity (5 items, alpha .75), our measure of technological abstractness. Items in this latter scale tapped the extent to which machine problems were hard to solve, whose causes were difficult to identify, and whether clear procedures existed for error correction. The three job control measures consisted of: Method Control (alpha .69), Timing Control (alpha .75), and Boundary Control (an alpha was inappropriate for this scale since it tapped a heterogeneous set of secondary task activities). The first two are described in detail elsewhere [9]. Four outcomes measures were used: the General Health Questionnaire (alpha .91), Job-related Anxiety (alpha .83), Job-related Depression (alpha .81) and an additional 7-item measure of Intrinsic Job Satisfaction (alpha .88).

Separate moderated hierarchical regression analyses [10] were carried out for each of the outcome variables across each aspect of job control. Demographic variables were entered first, to partial out potential confounding effects of age, sex and length of service, followed by technological variables (step 2), the relevant control variable (step 3), the three two-way combinations between the technology variables and control variable (step 4), and the final insertion of the full three-way product term.

3. Results

Results for the regression analyses across the three control variables show no moderating effects for both boundary control and method control. Strong and consistent interaction effects arise however with combinations of Timing control and technology characteristics. Regression results pertinent to these interaction effects are presented in Table 1.

Table 1. Beta weights and ΔR^2 values the for two and three-way interactions between Timing control and technology characteristics.

	GHQ	Anxiety	Depression	Job Sat.
Tcon * Unc	-0.25 *	-0.44 **	-0.29 *	0.10
Unc * PC	-0.10	-0.07	-0.09	0.19
Tcon * PC	-0.33 **	-0.20	-0.26 *	0.32 **
ΔR^2	.17 **	.22 **	.15 *	.12 *
Tcon * Unc * PC	0.33 *	0.30 *	0.26	-0.14
ΔR^2	.06 *	.05 *	.04	.01

PC; Problem Complexity; Unc; Technological Uncertainty; Tcon; Timing Control;
* p < .05; ** p < .01

Our results show that Timing control exhibits consistent two-way interaction effects with our technology characteristics. The two-way combination of timing control and technological uncertainty accounts for significant amounts of variance in GHQ, job-related anxiety, and job-related depression. Over and above these effects are the two-way effects between Technological problem complexity and Timing control. This latter interaction term explains significant amounts of variance in GHQ, job-related depression, and intrinsic job satisfaction. Our results also reveal an additional three-way interaction which account for significant further variance in GHQ and job-related anxiety, independent of both two-way effects.

When plotted, the two-way interactions involving Timing Control and Technological uncertainty show that at high levels of uncertainty, increases in Timing control are associated with decreases in job-related strain; and that at low levels of uncertainty increases in Timing Control are associated with the converse effect, i.e. increases in job related strain. The form of the three-way interaction shows that this pattern is much stronger at lower levels of problem complexity than at higher levels.

4. Discussion

The pattern of our results appear, at first glance, to lend partial support to our hypothesis that increases in job control at high levels of uncertainty will attenuate the effects of technology demands. This effect however arises for increases in Timing Control only. No effects were observed for Boundary and Method control. Additionally we found an unexpected increase in strain levels at low levels of uncertainty with increases in levels of Timing control. Two questions are appropriate here. Firstly, why did we find interaction effects for Timing Control, and not for Boundary and Method Control? Secondly, how can we explain the somewhat unexpected increases in job strain at low levels of uncertainty with increasing Timing Control?

It is clear that the nature of our findings are clearly inconsistent with what would be expected on the basis of Karasek's model, and what has been found in previous research. Increases in operator control can have substantial impacts on system uptime and the ability of operators to manage and prevent downtime occurrences [11]. Increasing the opportunity for active involvement in error rectification can provide important learning opportunities and greater challenge for operators. Such outcomes would be expected to be accompanied by increased motivation and mental well-being. These considerations raise serious doubts about whether the interactions present in our results reflect true attenuation of AMT system stressors.

In trying to understand why we did not find any moderating effects for Boundary and Method Control we found it useful to consider the content of, and variation in, these measures. We found that within both measures there appeared to be very little in the content of the items that tapped those aspects of job control concerned with the management and control of system errors. We also found that the absolute level and variation in the boundary control scale-scores for our sample were so low as to preclude the existence of moderator effects. Indeed we were informed in a later visit to the company that a company policy existed which allowed operators ten minutes to attempt to fix a system error, after which time an engineer had to be called. These observations highlight two important points. Firstly, for the analysis of moderator effects in job stress models, it is important for work autonomy measures to reflect aspects of the job that provide opportunity for the management of the particular stressors under consideration. For our study it appears that existing job control measures were too general for our purposes. This suggests that future evaluations of error demands and job control associated with AMT should incorporate measures that more directly tap different levels of operator involvement in fault management. Secondly, it is important to establish that there is sufficient variation in control to help account for outcome variance. Our findings, thus, do not run counter to the view that enhancements in boundary and method control for operators of AMT applications do not lead to beneficial effects on mental health.

An important issue arises concerning interpretation of the interaction effects between timing control and system uncertainty. Of the three job control aspects, Timing control would appear the least related to the management of system errors and error complexity. It would be a difficult task indeed to argue that operator discretion over initiation, Termination, and pacing of work activities are instrumental in reducing demands arising from AMT system errors. There appears nothing intrinsic to this control concept which establishes it as a resource for prevention, and active management of system uncertainty.

The only role in coping responses it may play involves the ability for operators to engage in avoidance responses, i.e. to 'put off' engaging in work activities for high uncertainty systems. This latter possibility does not appear to be a useful explanation since operators were not allowed to engage in such behaviours within this company setting, and it does not explain the increases in strain associated with increases in Timing control under low uncertainty conditions. Rather than having partially met our predictions, our results may reflect effects which are qualitatively different from the moderating effects expected from Karasek's stress model.

We have tentatively rested on one interpretation of these interactions between Timing control and system uncertainty. This is based on a reconceptualisation of theTiming control construct itself. We feel that, in the context of the operation of new technology, Timing control may tap into aspects of Baldamus' notion of 'Machine Traction' [12]. Machine Traction is a concept which reflects the degree to which machine operations "produce in the operator the feeling of being drawn along" [12: p. 63] by the standard operational requirements of the machinery. Baldamus describes this rhythm or pull of work as leading to a state of mind which is experienced as relatively pleasant, and one which provides a 'relief from tedium' [12: p. 59]. This experience is perhaps similar to what Csikszentmihalyi [13] has described as 'flow experience'. Viewed in this way high levels of timing control constitute low traction conditions. These may well arise as a result of high cycle times. It is easy to see why such conditions, in combination with low uncertainty, may be a source of strain, since there may be little for operators to do once a highly reliable machine has been set in operation and long gaps between task activities. Such conditions may prove monotonous for operators and machine breakdowns may provide a welcome break to this tedium. This interpretation is supported by one operators comment that "the only time anything interesting happens around here is when something goes wrong".

When high traction conditions (i.e. low timing control) are associated with low levels of uncertainty then these are the prime conditions for the development of positive reactions to workflow structure. Conversely, high system uncertainty may introduce conditions of chronic interruption to the development of this relative satisfaction. It is possible that frequent and chronic interruption effects may account for the high strain reactions we observed with this combination. Interruptions have been identified as a necessary and sufficient condition for the generation of arousal [14]. Further evidence for, and exploration of these processes will form the basis of future investigations.

5. References

[1] Karasek, R.A. (1979). Job demands, job decision latitude, and mental strain: Implications for job redesign. Administartive Science Quarterly, 24, 285-307.
[2] Johansson, G. & Aronsson, G. (1984). Stress reactions in computerized administrative work. Journal of Occupational Behaviour, 5, 159-181.
[3] Carayon-Sainfort, P. (1992). The use of computers in offices: Impact on task characteristics and worker stress. International Journal of Human-Computer Interaction, 4, 245-261.
[4] Hirschorn, L. (1986). Beyond Mechanisation: Work and Technology in a Post Industrial Age. MIT Press, Cambridge, Mass.
[5] Weick, K.E. (1990). Technology as equivoque: Sensemaking in new technologies. In P.S. Goodman, L.S. Sproull and Associates, (Eds.) Paul S. Goodman, Lee S. Sproull and Associates. San Francisco: Jossey-Bass.
[6] Frese, M. (1987). Human-computer interaction in the office. In C.L. Cooper & I.T. Robertson (Eds) International Review of Industrial and Organizational Psychology. John Wiley & Sons.
[7] Breaugh, J.A. (1985). The measurement of work autonomy. Human Relations, 38, 551-570.
[8] Wall T.D., Corbett, J.M., Clegg, C.W., Jackson, P.R. & Martin, R. (1990). Advanced manufacturing technology and work design: Towards a theoretical framework. Journal of Organizational Behaviour, 11, 201-219.
[9] Jackson, P.R., Wall, T.D., Martin, R. & Davids, K. (1993). New measures of job control, cognitive demand, and production responsibility. Journal of Applied Psychology, 78, 753-762.
[10] Cohen, J. & Cohen, P. (1983). Applied Multiple Regression/Correlation Analysis for the Behavioural Sciences (2nd Edn.). LEA, Hillsdale: NJ.
[11] Wall T.D., Jackson, P.R. & Davids, K. (1992). Operator work design and robotics system performance: A serendipitous field study. Journal of Applied Psychology, 77, 353-362.
[12] Baldamus W. (1961). Efficiency and Effort:An analysis of Industrial Administration. Tavistock: London.
[13] Csikszentmihalyi, M (1975). Beyond Boredom and Anxiety. San Francisco: Jossey Bass.
[14] Mandler, G. (1984). Mind and Body: Psychology of Emotion and Stress. WW Norton & Co: NY.

Advances in Agile Manufacturing
P.T. Kidd and W. Karwowski (Eds.)
IOS Press, 1994

Defining Degree of Automation

Z.G. Wei, A. Macwan, J.H.M. Andriessen and P.A. Wieringa
Laboratory for Measurement and Control, Faculty of Mechanical Engineering,
Delft University of Technology, Mekelweg 2, 2628 CD Delft, The Netherlands.

Abstract. Degree of automation (DOA) in supervisory control is defined as a function of the number and nature of the tasks performed by a human operator and realized by automatic control. A weighting scheme is employed to compute a numerical value of DOA. Two interpretations of weight are discussed. Objective weight is used to account for the effect of each task on system performance. Subjective weight is related to the mental effort associated with each task. Approaches to estimate the objective and subjective weights are presented.

1. Introduction

Automation in modern industry has been motivated by the desire to reduce operator workload and to improve system performance, and is prompted by developments in computer and information technology. Even though automation reduces the operators' workload, it may have adverse effects. Bainbridge [1] has discussed limitations of increasing automation. For example, the tendency to automate what is easiest and to leave the rest to the operator may lead to less coherent and more complex tasks and may result in an overall degradation of system performance. Higher levels of automation imply higher system complexity, which affects both operator and system performance [2]. The relationship between system performance and complexity could be a function of the level (or degree) of automation.

Degree of automation (DOA) is used as an indicator of the level of automation. The concept of DOA and a qualitative scale of DOA [3] for supervisory control have been suggested "to be used in discussing how automation and human actions can be balanced and in discussing safety control problem" [4]. In supervisory control, tasks are allocated to humans or automation by criteria and methodologies. From this viewpoint, tasks are categorized into three groups [5]:

(1) Tasks which must be allocated to human
(2) Tasks which can be performed by human or automation
(3) Tasks which must be automated.

Clearly, any discussion on degree of automation is limited to tasks in the second group.

Since benefits of automation are also accompanied by limitations, it is useful to investigate effects of automation on operator and system performance. For a quantitative assessment of the effect of automation on performance a numerical value is needed. This number may be useful for the development of a generic model on the relationship between system performance and complexity, or DOA and for the comparison of different systems. Thus, in this paper, we present an approach to define and estimate a numerical value for DOA. The next section presents principles and considerations in developing a definition for DOA.

In the third section, methods to assess relevant quantities are briefly discussed. The last section includes conclusions and possible applications in conducting laboratory experiments.

2. Definition of degree of automation

According to Sheridan [3], a definition of degree of automation in supervisory control is related to how many tasks are allocated to humans and automation. The discussion that follows is based on this principle.

2.1 General definition of DOA

As mentioned above, DOA is a function of the number of automated tasks. Thus, a simple definition of DOA could be the ratio of number of automated tasks and all tasks. However, most tasks are not the same by nature, therefore the definition of DOA should also account for the specific tasks that are automated. We propose a weighting scheme to account for this:

Degree of Automation is the ratio of a weighted sum of automated tasks and a weighted sum of all tasks.

Mathematically, this is expressed as:

$$DOA = \frac{\sum_{i=1}^{N} t_i w_i}{\sum_{i=1}^{N} w_i}$$

where t_i is the task allocation indicator, which is assigned a value of 1 if task i is automated and zero otherwise; w_i is the weighting factor for task i and N is the total number of tasks.

While defining the weighting factor w_i, it became clear that there are two distinct interpretations of w_i. These are shown in Figure 1.

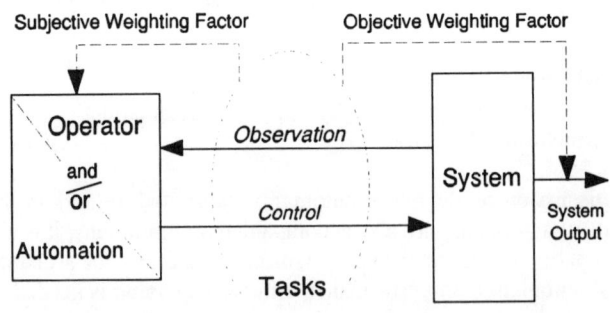

Figure 1 Schematic definition of weighting factors

As seen on the right side of Figure 1, w_i reflects the effect of task i on the output (performance) of the system. We refer to this as the objective definition of w_i. On the other hand, in order to be executed, each task requires an effort. Focussing on the mental effort required from the operator to execute a task, we arrive at the subjective definition of w_i as shown in the left side of Figure 1. These two definitions of the w_i will be explained in subsequent sections.

2.2 Objective weighting factor

Each task has a distinct effect on system performance. The effect of a task on system performance should be independent of the fact that the task may be performed by operator or by automation. Thus, a task could be weighted by considering its effect on system performance.

Using this definition of w_i, DOA is an indication of the fraction of system performance that is controlled by automation. Approaches to estimate objective w_i are presented in section 3.

2.3 Subjective weighting factor

From the perspective of the operator, each tasks has a unique level of mental effort. Mental effort is a subjective indicator and has more dynamic feature [6]. It is a function of a number of factors such as task complexity, task difficulty, etc. as they are perceived by the operator. Thus, allocation of some tasks to automation implies a reduced mental effort for the operator. We use mental effort as a subjective indicator of weight in the definition of degree of automation.

According to this definition of w_i, DOA is a fraction of mental effort that is transferred to automation. In the next section, a method to assess subjective weighting factors will be outlined.

3. Assessment of weighting factors

In this section, approaches to estimate objective and subjective weighting factors are addressed briefly.

3.1 Assessing objective weighting factors

By its very definition, objective weighting factors are independent of the human operator. Two approaches to their estimation are proposed. The first is an experimental approach. System performance is measured for a reference case where all tasks are executed correctly. Subsequently, each task is failed and system performance is measured. In each case, the change in performance with respect to a reference case is used to estimate w_i.

Alternatively, system analysis can be carried out to estimate w_i. This would involve judging the effect on system parameters as a function of task failure. Engineering analysis can then be used to calculate the effect of the parameters on system output.

3.2 Assessing subjective weighting factors

When subjective weighting factor is measured by mental effort, the approach to assess the latter can be employed to assess weighting factors. Considering a subjective rating scale approach, the so-called Rating Scale Mental Effort (RSME) developed by Zijlstra [6] could be used to assess the weighting factor. The RSME is a simple subjective rating scale that measures the costs of work behavior. For the detail of the RSME, the readers are referred to reference [6].

The steps to rate subjective weighting factors based on, for example, the RSME, are:

(1) Select a number of operators or subjects;
(2) Ask each operator to give an estimate (rating) of weighting factors;
(3) Obtain mean values [6];
(4) Perform sensitivity analysis.

4. CONCLUSIONS AND POSSIBLE LABORATORY APPLICATIONS

In this paper, a definition for the degree of automation (DOA) is proposed. The basic principle is to define DOA as a function of the number and the nature of automated tasks. Specifically, the ratio of a weighted sum of automated tasks and weighted sum of all tasks is used to estimate a numerical value of DOA. Two interpretations of the weight, an objective one and a subjective one, and methods to assess (estimate) each, are presented.

In the near future, laboratory experiments will be conducted to investigate operator and system performance as a function of DOA. Insights gained from the experiments will be used for studying task allocation between humans and automation in various industries.

References

[1] L. Bainbridge, The Ironies of Automation. In: J. Rasmussen, K. Duncan and J. Leplat (Eds.), New Technology and Human Error. John Wiley, 1987, pp. 271-283.
[2] H.G. Stassen, J.H.M.Andriessen and P.A. Wieringa, On the human Perception of Complex Industrial Process. A paper presented in the 12th IFAC World Control Congress, Sydney, Australia, July, 1993.
[3] T.B. Sheridan, Telerobotics, Automation and Human Supervisory Control. The MIT Press, Cambridge, Massachusetts, 1992.
[4] T. Inagaki and S. Hasegawa, Safety-control of Large-Complex Systems, In Proceedings of the 13th European Conference on Human Decision Making and Manual Control, Kassel, Germany, June 22-24, 1993.
[5] H.E. Price, The Allocation of Functions in Systems, *Human Factors* 27 (1985) 33-45.
[6] F.R.H. Zijlstra, Efficiency in Workload Behaviour. Ph.D. thesis, Delft University of Technology, Delft University Press, Delft, The Netherlands, 1993.

Advances in Agile Manufacturing
P.T. Kidd and W. Karwowski (Eds.)
IOS Press, 1994

355

Accidents in
Automated Manufacturing Systems

Vesa VANNAS, Markku MATTILA
Tampere University of Technology, P.O.Box 589, FIN-33101, Finland

This paper gives information about the accidents that have occurred in automated manufacturing. The data is based on 35 accident situations in Finnish workshops. The results showed that automated manufacturing related accidents do not occurred often, but the injuries that result are severe. Injuries are often connected to production disturbances and are frequently caused by human errors.

1.   Introduction

Production automation has eliminated many traditional risks in manufacturing, but along with the more complex systems and the increased forces, new types of risks have appeared that also affect people working outside the danger zone [1],[2]. Automated machines, the organization of work, human actions, the work environment, and lack of ergonomic design have been found to have an impact on safety [3],[4]. Previous research also shows, that most accidents in automated manufacturing take place during production disturbances [5],[6],[7]. Despite this, however, information about the accidents in advanced manufacturing systems has not been systematically collected in Finland.

2. Materials and methods

The data was collected by using a two-page questionnaire which requested information about one serious accident related to the automated manufacturing in the respondent's plant. The questionnaire was mailed to the safety manager of every Finnish workshop, that had more than 10 employees; 800 workshops met this requirement.

In the questionnaire information was requested concerning employee groups, work activities, type of injuries, contributing factors, the type of automated equipment, operating modes and safeguarding involved in the accident. The accident was defined as being automated manufacturing-related if it occurred in an automated manufacturing system or in part of a system, if the automated equipment had controlled the energy that caused the accident, and if equipment was automated so that it could start a movement, change direction, or alter its function without the intervention of an operator [5].

3. Results

The results are based on the responses of 87 companies, among which 52 reported that no automated manufacturing related accidents occurred in their plant. About 63 % of the analyzed accidents (N = 35) occurred in the context of stand-alone automated equipment (Table 1:). In almost half of the cases the type of individual equipment was a CNC- or NC-machine and 20 % of the accidents involved a machining center. Other equipment types involved in the accidents were industrial robots, other material handling equipment, and other auxiliary manufacturing equipment.

Table 1.   Number of automated manufacturing-related accidents by type of manufacturing system (N = 35)

Type of manufacturing system	Number of accidents	%
Stand-alone equipment	22	62,8
Transfer line	9	25,7
FMS / FMC	3	8,6
Other system	1	2,9
TOTAL	35	100

Most of the injured persons were operators (77 %). In three accidents the injured people were maintenance personnel and in three cases it was the foreman of the system. The activities performed by the injured person were related to the production disturbance in 65 % of the accidents. The activities performed at the time of the accident were mostly setting of the machine, clearing a blockage and general system operation (Table 2.).

Table 2.   The activities performed at a time of the accident (N = 35)

Activities	Number of accidents	%
Loading / unloading	4	11,4
Inspection	3	8,6
Other normal operation	5	14,3
Setting	12	34,3
Fault finding / retrification	2	5,7
Maintenance /repair	4	11,4
Clearing a blockage	5	14,3
TOTAL	35	100

In 39 % of the cases the system was in automated mode, and in 39 % of cases it was in manual mode. In seven accidents the system was stopped but not isolated and in one case it performed a slow speed movement. The movement was programmed in 51 % of the accidents. In one fourth of the accidents the movement was unexpected.

Safeguarding was inadequate in 42 % of the accidents. The most common deficiencies were that there was no guard or that the guard allowed access to the danger zone.

Table 3.   Main defects in safeguarding in accidents where safeguarding was inadequate (N = 17)

Main defect	Number of accidents	%
No guard	7	41,2
Guard allowed access to the danger zone	6	35,2
Guard removed	2	11,8
Interlocked guard failed to stop all relevant parts	2	11,8
TOTAL	17	100

About 10 % of the accidents led to severe injuries, so that the injured person could not return to his / her normal duties (Table 4.). There were no fatal injuries. Over half of the accidents were defined as major injuries, where the injured person lost 3 or more workdays. In these accidents injured persons lost an average of 26 workdays.

Table 4.   Number of accidents by severity of injury (N = 35)

Severity of injury	Number of accidents	%
No lost time	4	11,4
1-3 lost workdays	11	31,4
More than 3 lost workdays	18	51,4
Unable to return to normal duties	3	5,8
TOTAL	35	100

Main body parts injured were fingers and hands (Table 5.). Multiple injuries also resulted from a number of accidents, and head and trunk were involved in many injuries.

Table 5.   Number of accidents by injured part of body (N = 35)

Injured part of body	Number of accidents	%
Finger	13	37,1
Hand	6	17,1
Arm	1	2,9
Head	4	11,4
Back of the neck	1	2,9
Back	1	2,9
Trunk	3	8,5
Foot	1	2,9
Multiple injuries	5	14,3
TOTAL	35	100

The most common injury was a contusion or bruise (Table 6.). Another common type by nature of injury was a cut. Other injuries occurred only in occasional cases. In 43 % of the accidents the injury was caused by being caught in, under, or between moving machine parts or other movable equipment, and in 40 % of the accidents the injury was caused by being struck by moving equipment, parts, or flying objects.

Table 6.   Number of accidents by nature of injury (N = 35)

Nature of injury	Number of accidents	%
Contusion / bruise	14	40,0
Cut	8	22,9
Amputation	3	8,6
Fracture	1	2,8
Spratches / abrasions	2	5,7
Sprain / strain	1	2,8
Multiple	3	8,6
Other	3	8,6
TOTAL	35	100

The factors which contributed to the 35 accidents were divided into three categories: control system, workplace layout, and human factors. Human factors were contributed in every analyzed accident, and in many cases there were several human factors, which affected the same accident. The most common human factors were human error by the injured person, which contributed in 57 % of the accidents, and the following of incompatible procedures (29 %). Workplace layout contributed to 40 % of the accidents. The deficiencies found in the layout were improper placement of safeguarding, lack of material handling equipment, and dangerous workplace because of lack of safety barriers and other safeguarding. The control system contributed to 26 % of the accidents. In these cases control system was not suited to the task or it malfunctioned.

Discussion

The results show, that accidents connected to automated machines and manufacturing are rare, but the injuries are severe. They are mostly caused by moving equipment or machines, which often have unexpectedly extensive force. To reduce the amount and severity of injuries more attention should be paid to the preventive maintenance, and prevention of situations, like production disturbances or repairs, in which the system function differs from the normal operation. The personnel should be trained to handle disturbance situations with safe working methods. To reduce the amount human errors new methods should be developed in job training to make personnel more conscious of disturbances and risks.

Acknowledgement

This paper is based on research project financed by the Finnish Work Environment Fund.

References

[1]   W. Karwowski, M. Rahimi, and T. Mihaly. Effects on computerized automation and robotics on safety performance of a manufacturing plant, *Journal of Occupational Accidents*, 10, 1988. pp. 217-233.

[2]   W. Bauer and F. Kolhaas, Work safety in hybrid automated production systems illustrated on a flexible manufacturing systems. In: W. Karwowski and M. Rahimi (eds), Ergonomics of Hybrid Automated Systems II, Elsevier, Amsterdam, 1990. pp. 943-949.

[3]   J.R. Edwards, Accidents on Computer Controlled Manufacturing Plant and Automated systems in Great Britain 1987-1991. London: HSE. Unpublished Technical Report. 1993.

[4]   M. Mattila, T. Tallberg, V. Vannas and J. Kivistö-Rahnasto. Fatalities at advanced machines and dangerous incidents at FMS implementations. *International Journal of Human Factors*. 1994. (in press).

[5]   R. Kuivanen. The Impact on Safety on Disturbances in Flexible Manufacturing Systems. In: W. Karwowski and M. Rahimi (eds), ergonomics of Hybrid Automated Systems II, Elsevier, Amsterdam, 1990, pp. 951-956.

[6]   J. Järvinen and W. Karwowski. Accidents in advanced manufacturing systems: New study results. In: Proceedings of the Fifth Annual National Robot Safety Conference, October 12-14, Robotics Industries Association, Novi, Michigan, 1993.

[7]   M. Döös and T. Backström. Description of accidents in automated materials handling. In: S. Marras, W. Karwowski and J.L. Smith (eds.), The Ergonomics of Manual Work, Taylor and Francis, London, 1993. pp. 653-656.

Advances in Agile Manufacturing
P.T. Kidd and W. Karwowski (Eds.)
IOS Press, 1994

Technical defects behind accidents in automated production

Tomas BACKSTRÖM and Marianne DÖÖS
Division of Social and Organizational Psychology,
National Institute of Occupational Health, S-171 84 Solna, Sweden

Abstract. A study of accidents in automated production has been conducted at 21 workplaces in Swedish manufacturing industry. Accidents occurred at an estimated 4% of the automated installations each year. In two-thirds of cases the person sustained an injury from a part of the equipment for the handling or conveying of the product. Industrial robots have a low relative risk compared with other equipment. In three-quarters of cases a technical problem influenced the accident process. Frequently, the machine stopped because the product/work piece had become stuck or wrongly positioned. Technical defects causing accidents had often been in existence for some time, without counter-action having been taken.

1. Introduction

Accidents are a significant work-environment problem in automated production [1]. This paper describes the extent to which different kinds of automated equipment are involved in occupational accidents. The accidents are further analyzed with regard to kinds of defects and problems of a technical nature.

Estimates of the relative risk of accidents at different installations can offer opportunities for setting priorities for future preventive activities. However, it is unusual that such estimates are made. There is seldom access to data on how many pieces of equipment of different kinds there are at the companies under study.

Several studies suggest that a large proportion of accidents in automated production occur in the correction of disturbances [2, 3]. This indicates that technical problems are often a part of the chain of events behind automation accidents. A closer examination of the problems may generate suggestions for further research and development.

2. Material and method

Data were gathered from 21 work sites over a two-year period (May 1988 - May 1990). The companies concerned came from different industrial sectors, but mainly from the engineering industry. Denominator data on the equipment come from a questionnaire to which the companies responded. It was not possible to obtain all information from all companies, and certain figures are estimates made by the company in question.

Data on accidents at the companies come from the mandatory occupational injury reports. Just over 4,000 accidents were reported over the period. Strategic-sampling and manual-selection methods were employed to obtain a representative sample of automation accidents. The sample covers 127 cases, equivalent to an estimated 65% of such accidents at the companies in question. The ratios for relative risks used in this paper are calculated from just these 127 cases. Thus, as a result of the missing data the real relative risks are underestimated. The relative risk reported here is around 65% of the actual relative risk.

To obtain in-depth descriptions of technical defects, information from comprehensive investigations of 76 automation accidents was utilized. These investigations were carried out for a parallel study of the same companies over the same period [3].

Automatic equipment refers to an installation (or a part of an installation) which, without the direct intervention of a human being, *can either initiate a machine movement, or change its direction or function.*

An accident is defined as an unintended and branched sequence of events, where the final event occurs suddenly and results in an injury to a human being. By automation accidents are meant accidents occurring at an automated installation, where the equipment has controlled (or should have controlled) the harmful energy.

A technical problem is defined as the equipment not functioning as intended, which leads to an acute event. Either someone intervenes, or the equipment functions incorrectly, e.g. a deviant machine movement or the ejection of a work piece or tool.

3. Results

3.1. Equipment involved in automation accidents

At 18 of the companies under study (information from three companies is lacking) there were 2 056 automated installations. Over the two years, 106 automation accidents where identified at these companies. This means there was a relative frequency of 26 automation accidents per thousand automated machines per year. Adjusting upwards for the missing data generates the estimate that each year automation accidents occur at 4% of the automated installations.

The equipment has been classified both according to its principal function (table 1) and by the function of that part of the installation which injured the person (table 2). Only a fifth of the injuries occurred in the part of the equipment that serves its principal function. The variation in relative risk between machines with different functions has many causes; it can, for example, depend to which extent material handling equipment was used.

In a third of cases the principal function of the equipment that injured the person was that of machining. Approximately half of the companies' installations had a machining function and, as a result, the relative frequency was not excessively high. Equipment for joining has the highest relative frequency. See table 1.

Table 1. Principal function of the equipment at which the injury occurred. Proportions of automation accidents and relative frequencies of automation accidents per 1 000 installations per year for equipment with different functions.

Function of the equipment	% of automation accidents (n=127)	Relative risk. No. of auto. accs per 10^3 equip't per yr.[*]
Conveying	13	16
Deforming	12	33
Machining	32	17
Joining	20	70
Other (e.g. surface treating, packing)	22	41
Unclear	1	-
For all equipment	100	26

[*] Information on number of installations lacking from 3 companies.

Table 2. Function of that part of the equipment which injured the person in automation accidents.

Function of part of the equipment	% of automation accidents (n=127)
Principal machine function	20
Handling work piece	31
Conveying work piece	31
Other (e.g. tools store, shavings)	12
Unclear	6
Total	100

Certain installations are sufficiently well-defined for it to be possible to go down to a more detailed level and thereby relate the number of accidents to questionnaire data on how many installations there are of each sort. It can then be seen that welding equipment has the highest relative frequency of automation accidents, a relative frequency of 66 automation accidents per thousand automated welding installations. Welding accidents tend to lead to relatively short periods of sick-leave, a median of four days in comparison with eight days for machining accidents. The risk of an accident at a lathe lies approximately at the mean for all installations. Industrial robots are relatively safe, at just over half the mean risk.

In most automation accidents the person sustained an injury from a part of the equipment for the handling or conveying of the product. See table 2. By "handling equipment" is meant those parts of the installation which move work pieces in and out of the machine or hold them in position when its function is performed. By "conveying equipment" is meant equipment which moves work pieces between, for example, different installations or between installation and warehouse.

3.2. Technical defects and problems

In three-quarters of the automation accidents some type of technical problem was referred to in the occupational injury report. Usually, the problem led to trouble with the equipment so that the operator had to change task, e.g. to reposition a part that was sitting crookedly. The accident then occurred during the implementation of the task that had arisen as a result of the technical failure. In one tenth of cases a technical failure was the direct cause of injury; e.g. a machine movement took place despite the machine being in the stop position. See table 3.

From among the 76 comprehensively-investigated automation accidents, defects and problems of a technical nature were referred to in 86% of cases. This indicates that further investigation may lead to the discovery of technical problems behind more accidents, but it is unclear whether our study may have been affected by systematic features of the missing data.

A quarter of the automation accidents were preceded by disturbances in the materials flow, a part becoming stuck, or getting into a crooked or otherwise faulty position. The risks with production disturbances are described in Döös & Backström [3]. In every tenth case a machine movement occurred despite activation of the safety stop. See table 4.

Table 3. Involvement of technical problems in the accident process.

Involvement of technical problems	% of automation accidents (n=127)
Technical failure caused the injury	11
Technical problem changed task	65
No technical problem reported	24
Total	100

Table 4. Types of technical problems in automation accidents.

Types of technical problems	% of automation accidents (n=127)
Work piece became stuck, got into a crooked or faulty position	24
Machine part became stuck, got caught, jammed etc.	5
Machine stopped by another person or for unknown reason	7
Work piece became loose or was not properly secured	1
Machine movement despite activation of safety stop	9
Other deviant and unexpected machine movements	6
Other problems, remedied during operation: leaks, shavings, sensor problems, etc.	15
Other problems, stopping the machine: repairs, fault tracing, adjustment, etc.	9
No technical problem reported	24
Total	100

The technical failure has been investigated further in 28 of the 76 comprehensive investigations. The technical failures specified in the accident investigations were most frequently caused by jammed, defective or worn-out components. In two cases, a failure was caused by a computer-program error, and in one by electromagnetic interference. More than half (17 of 28) of the defects had been present for some time but not remedied. Only in five of the 28 cases where a technical failure/deviation was referred to, had the defect not manifested itself previously. Some faults had been repaired only to re-appear later. Certain defects had been present right from the time of installation, while others resulted from wear and the impact of the environment.

4. Discussion and conclusions

The results suggest that there is a need to raise the level of safety in those parts of automated installations which handle or convey parts/work pieces. This study, however, does not suggest any particular motive for further research on the safety of industrial robots.

Most automation accidents involve technical problems. This does not mean, however, that the accident problem can be solved only by means of technical research and development. The choice of manner of working when a problem arises is an important determinant of the risk level. The way in which a person works is influenced by many factors, e.g. the design of the interface between man and machine, and the safety culture at the workplace. Research on the factors influencing the operator's manner of working when technical problems arise can make a valuable contribution to safety in automated production systems.

References

[1] T. Backström and M. Döös, A comparative study of occupational accidents in industries with advanced manufacturing technology. *International Journal of Human Factors in Manufacyuring,* to be published.

[2] M. Döös and T. Backström, Disturbances in production - a safety risk or a chance for development in the human-computer interaction context. In M. J. Smith and G. Salvendy (Ed), Human-computer interaction: Application and case studies. Proceeding of the fifth intenational conference on human-computer interaction. Orlando, Florida, Aug, Vol1 (1993) 809-814.

[3] M. Döös and T. Backström, Production disturbances as an accident risk. Paper presented and published at Fourth International Conference on Human Aspects of Advanced Manufacturing & Hybrid Automation. Manchester, England, July 1994.

Advances in Agile Manufacturing
P.T. Kidd and W. Karwowski (Eds.)
IOS Press, 1994

Workplace Safety Analysis: Fuzzy Logic- Algorithmic Method

Alexander ROTSHTEIN
Vinnitsa Polytehnical Institute, Department of Computer - Based
Inmformation and Management Systems, Khmelnitsky Shosse 95, Vinnitsa,
286021, Ukraine

Abstract. This article represents some description of the main principles of a new method for man-machine systems risk assessment which can be called logic- algorithmic fault tree analysis. The suggested here method is based on the logic-algorithmic description of a man-machine system and in contrast to the conventional fault tree analysis it allows to conveniently take into consideration the ability of a man-operator to detect and to avoid harardous situations.

1.Introduction

While designing workplace and new equipment it is necessary to be sure they are providing the required level of man operator safety. Fault tree analysis (FTA) [1] is a very popular tool for risk assessing of complex large-scale systems. A fault tree provides a logical and hierarchical description of an accident (top event) in terms of sequences and combinations of malfunctions of individual components and adverse operating conditions (basic or fundamental events). By resorting to a fault tree, the reliability or safety of a complex system can be computed in terms of the propabilities of occurence of the basic events. In case of lack of proper data about the propabilities of occurence of harardous events we can use the FTA by fuzzy propability (or fuzzy FTA [2]), in which the propabilities of events changed for the possibilities of events, viz. a fuzzy set defined in propability space.

But using conventional and fuzzy FTA it is very difficult to take into consideration the ability of a man-operator for detection and for avaiding a hazard situation in the man-machine system process functioning. These specifics can be easily described based upon the algorithm of man- operator activity on the workplace [3]. A new method for risk assessment in man-machine systems entitled *fuzzy logic- algorithmic fault tree analysis* is proposed in this article. This new method is based upon the following principles:

*Application of logical functions for description of the events connected with appearence of hazards situations (logical part of the fault tree).

*Transfer from logic-algorithmic description to propabilistic description to make it possible to calculate the quantitative level of risk.

*Representation of the source data about propabilities of events leading to hazard situations in a fuzzy numbers form.

*Using fuzzy logical evidence and fuzzy knoweledge- based expert system for taking into consideration differnt factors influencing upon probabilities and possibilities of events connected with appearance, detection and avaiding of hazard situations.

2. Construction of the Logic-Algorithmic Fault Tree

The process of task execution on the workplace can be represented as a consequence of labour operations. An example of logic- algorithmic description for one operation is represented in Figure 1.

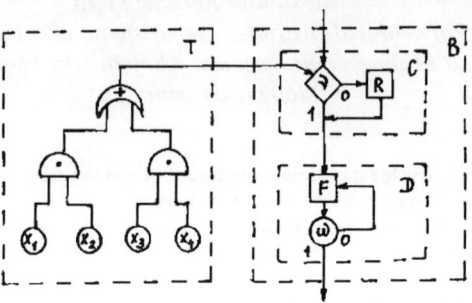

Figure 1. An example of a logic-algorithmic description

The left part of this figure is a conventional fault tree in terms of [1,2] for tools necesseary for operation execution, i.e. logic function

$$T = (x_1 \wedge x_2) \vee (x_3 \wedge x_4), \tag{1}$$

where T - top event which characterize the accidental condition with labour tools; x_i - fundamental events which characterize the malfunctions of system elements.

The right part of Figure 1 is an algorithm of man-operator activity during labour operation execution which, using algebraic language [3], can be described as

$$B = \underset{v}{(E \vee R)} \underset{\omega}{\{F\}}, \tag{2}$$

where: v - check of the logic condition "Are the labour tools in order?" ($v = 1$- yes; $v = 0$- no) and according to formula (1)
$$v = \overline{T} = (\overline{x}_1 \vee \overline{x}_2) \wedge (\overline{x}_3 \vee \overline{x}_4);$$

R - repairing of the labour tools; E - fixation of the results of control ; F - execution the labour operation on the workplace; ω - check of the logic condition: "Has labour operation F been execufed correctly? ($\omega = 1$-yes; $\omega = 0$-no).

So, Figure (1) or formulas (1) and (2) give the united logical- algorithmic description of the man-machine system, which allows to take into consideration:

by using logical part

*malfunction and failures in the labour tools (work clothes, air- conditioning, lighting, etc.), which can lead to an accident on the workplace;

by using algorithmic part

*detection and avoiding by the man-operator of malfunctions and failures in the labour tools;

*noncorrect execution of the labour operations which can lead to the hazardous situation.

*detection and avoiding by man-operator of errors during t labour process execution.

3.Evalution of the Logic-Algorithmic Fault Tree

For transfering from logic-algorithmic description of man-machine system functioning process to propabilitic description, which allows to estimate the level of risk it is necessary to use formal ruls stated in Table 1, where:

P_{x_i} - propability of event x_i occurence;

P_C^1 , P_D^1 and P_R^1 - probabilities of correct execution of operators C,D and R, correspondingly;

P_ν - probability of event: $\nu = 1$; P_T - probability of event: $\nu = 0$;

k_ν^{00} and k_ω^{00} - probabilities of malfunctions and errors detection during execution of ν- and ω- inspections, correspondingly.

Table 1. Rules for transfering from logical-algorithmic structures to the models for calculation

Logic and algorithmic operations	Symbols on the fault tree	Models for calculation
Logical AND: $x_1 \wedge x_2 = A_1$		$P_{A_1} = P_{x_1} \cdot P_{x_2}$
Logical OR: $x_1 \vee x_2 = A_2$		$P_{A_2} = 1 - (1 - P_{x_1}) \times \times (1 - P_{x_2})$
Linear structure: $C\,D = B$		$P_B^1 = P_C^1 P_D^1$
Branching structure: $\underset{\nu}{(E \vee R)} = C$		$P_C^1 = P_\nu + (1 - P_\nu) \times \times k_\nu^{00} P_R^1 = (1 - P_T) \times \times P_T k_\nu^{00} P_R^1$
Iterative structure: $\underset{\omega}{\{F\}} = D$		$P_D = 1 - (1 - P_F^1) \times \times (1 - k_\omega^{00})$

Formulas in the right part of Table 1 were received in the assumption that during ν- and ω- inspections there are no errors of the type "false alarm " [4].

The sequence of these formulas use is defined by the logic- algorithmic fault tree structure which is shown in Figure 2 for the structure represented in Figure 1.

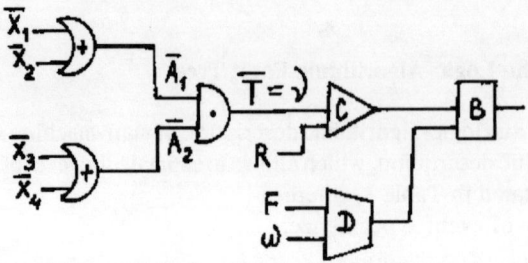

Figure 2. An example of logic-algorithmic fault tree

The methodology of undefined source data representation using fuzzy numbers and technique of taking into consideration different man-machine factors influencing upon these source data using fuzzy logical evidence are described in [5].

4.Conclusions

The above description principles and designed mathematical models are implemented in a dialog expert-simulation system for risk assessment.

This system allows: (1) to input the logic- algorithmic structure of events into the computer for risk assessment; (2) to form fuzzy knoweledge base about man-machine and enviroment factors influencing upon propabilities and possibilities of different events connected with each hazard situation; (3) to calculate propability of risk for the given variant to man-machine system; (4) to change the man-machine systems logic- algorithmic structure and parametrs of hazard events for safety improvement.

Further development of the suggested here method is connected with creation of some formal theory of logic- algorithmic fault tree design with required level of safety with involved expenses for its provision.

References

[1] R.E.Barlow and P. Chatterjee, Information to Fault Tree Analysis, Rep. OCR 73-70, Operations Research Center Universety of California, Berkely, 1973.

[2] H.Tanaka, L.T.Fan et al. Fault Tree Analysis by Fuzzy Probability, IEEE Trans. Reliability, R-32, (5), 1983, pp.453-457.

[3] A.Rotshtein, Design of Labour Systems: Formalized Theory. In: Willian S. Marras et al. (Ed.), The Ergonomics of Manual Work. ISBN: 0748400605. Taylor and francis, 1993, pp.457-460.

[4] A.K.Gramopadhyc, C.G.Drury et al., A Framework for Function Allocation in Inspection. In: P.Brodner and W.Karwowski (Ed.), Ergonomics of Hybrid Automated System III. ISBN: 0444895205, Elsevier, 1992, pp.249-257.

[5] A.Rotshtein, Fuzzy Model of Labour Systems Functioning Reliability and Quality. In: P.Bronder and W.Karwowski (Ed.), Ergonomics of Hybrid Automated System III. ISBN: 044489520 5, Elsevier , 1992, pp.135-145.

Advances in Agile Manufacturing
P.T. Kidd and W. Karwowski (Eds.)
IOS Press, 1994

The Design of a Laser Safety Advisory Tool

B. Soufi[1], A. Clarke[1], L. Vassie[2] and J. Tyrer[2]

[1] Department of Computer Studies,   [2] Department of Mechanical Engineering
Loughborough University of Technology
Loughborough, Leics. LE11 3TU

Abstract. This paper presents research which aims to enhance the awareness of, and access to, laser safety information. This is achieved by producing a conceptual model of the task of laser safety audit which forms the basis of an advisory system that provides a reliable, quick, and consistent means for the analysis of laser safety problems. The approach followed can be applied to similar problems requiring systems of advisory nature.

1. Introduction

Lasers are used for a variety of applications ranging from laser processing of materials to holography and displays. There are different hazards associated with using lasers depending on the application and the environment of use. Such hazards include optical hazards affecting the eyes, high voltage electrical hazards, and fume hazards i.e. lethal emissions from laser processing of materials.

Appropriate safety controls need to be in place to contain hazards. The determination of appropriate controls requires a comprehensive assessment of hazards that considers the complex interaction of relevant parameters. Recent surveys established that manufacturers and users of laser systems had difficulty in performing such assessments (Vassie et al. [1] found that about 50% of laser manufacturers had difficulties in performing an assessment of hazards). Difficulties in dealing with the hazard assessment of lasers are due to two main factors:

(1) Laser safety information, available in British Standard BS:EN:60825, is generally too technical to be of practical use.

(2) Lack of information in some specific areas such as fume emissions and enclosure strategy.

2. Overall Approach

A case for a different approach to laser safety was clear. Relevant practical information needed to be presented to a range of laser users that would easily understand and use it. To address the problem, a disciplined approach was required that would start by studying the users and establishing their requirements. It was also necessary to elicit knowledge from experts, since the expertise in the practical assessment of hazards is vested in a number of laser safety experts.

The knowledge thus acquired was used to construct a laser safety assessment model which then formed the basis of a computer-based decision support system.

3. User Requirements

The project's approach to the identification of user requirements drew on various methods, but mainly the HUFIT PAS (Taylor [2]), and HTA (Annett et al. [3]) for task analysis. Interviews were carried out with potential users. The data acquired, and the results of the task analysis formed the basis of specifying and categorising user requirements, described below.

3.1 The interview programme

A programme of structured interviews was carried out with a representative sample of the laser user community. Interviewees were selected so that the spectrum of laser types and applications was adequately covered. The sample included Health and Safety inspectors, manufacturers of lasers, and users of lasers with a diverse range of applications.

Interviews were held at the locations of participating companies/organisations, and were recorded for subsequent transcription and analysis. Various information was sought in the interviews but mainly to support the analysis of user requirements and characteristics. Information was obtained e.g. about interviewees' backgrounds and involvement in laser safety and also the data used when carrying out laser safety tasks.

The characteristics and requirements of these users were subsequently described in the form of profiles of five user groups (Factory Inspectors, Specialist/Principal Inspectors, Product Suppliers, Laser Systems Users, and Manufacturers). The classification of users into these groups resulted from consideration of the variability of the tasks, characteristics and requirements across the user population.

3.2 Task analysis

To obtain a description of user requirements, user task analysis was necessary. Some of the requirements would result from studying the users but it was important to understand the tasks they performed and the environment in which they were carried out. It was also important to know the capabilities of available system technology as users often do not know what they need and the ways in which a computer system could support their work.

Based on hierarchical task decomposition, tasks were described in terms of sub-tasks, goals and possible outcomes. The sub-tasks that are common between the user groups, and the tasks to be supported by the computer system were identified.

3.3 Significant findings

It became clear that there were different user groups that could benefit from a computer-based laser safety advisor. The nature of such a system depends, however, on the specific requirements of the selected user group(s). Furthermore, the characteristics of the users would give rise to additional requirements for the design of the system and its interface. It was therefore decided to focus the design of the system on targeted user groups. These were Factory Inspectors, and Laser System Users: these two groups had broadly similar requirements and characteristics.

4. Design and development of a laser safety advisory system

4.1 System characteristics

The user requirement study, and analysis of available system technology, established that the design of the system should meet certain requirements, viz. effective means of managing information in both textual and pictorial forms. It should also allow the user to easily and rapidly navigate through the system. The system should support two different modes of retrieving information: overview (to enable rapid access to general safety information), and detailed analysis (to provide decision support in carrying out a comprehensive assessment of hazards given a specific laser problem).

4.2 Approach to system design and knowledge representation

The approach to the design of the system was conditioned by the requirements of the users and the characteristics of the task. To meet requirements for the system, the knowledge acquired from experts, and task descriptions obtained by task analysis needed to be represented in the system in a manner that supported the two modes of retrieving information described above. As a result of the analysis of laser safety assessment tasks, a laser safety assessment model was constructed (see Fig 1).

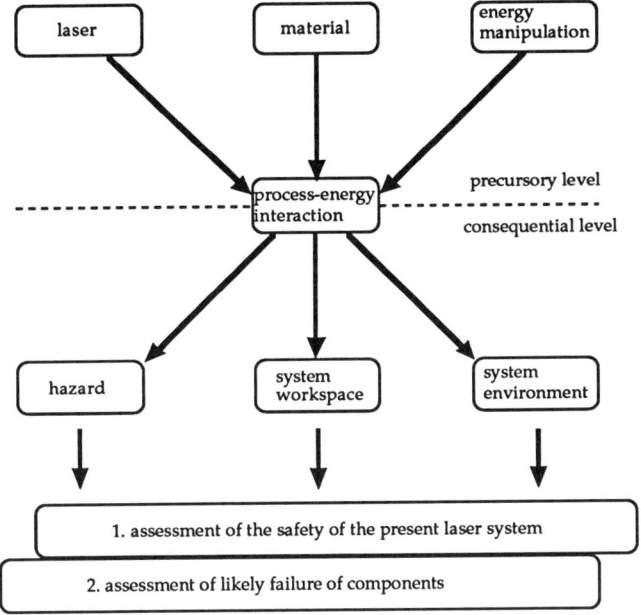

Fig. 1 A model of laser safety assessment tasks

Briefly described the model has the following structure: there is a 'Precursory' level, showing the three main domains of knowledge which together form the knowledge prerequisite for making safety assessment. The three domains are laser, material, and energy manipulation (the methods and techniques of beam delivery). There is 'Process-Energy Interaction' which represents the specific instance of interaction of laser energy, in a specific application, for which assessment has to be made. Completion of the process-energy interaction leads to a 'Consequential' level. Analysis of safety at this level will deal with types of hazard, the system workspace, and the system environment.

This model represents a general model of laser safety assessment, and can be applied to different application areas such as material processing, entertainment and display, and R&D.

4.3 System interface and its use

The material processing application area was selected for the prototype decision support system. The laser safety assessment model described above was applied to the material processing domain and used as the basis for developing the "detailed analysis" mode of decision support provided by the system. The system interface was designed to be transparent so that it reflected the structure of the model. This would assist in communicating the task structure to the user.

Elements of the model were used to derive the interface structure. In use, interaction between the user and the system is a two-stage process. The first stage is the specification of problem parameters by the user. The second is the display of recommendations by the system to the user. The structure of the interface and dialogue for both stages reflected the elements of the model.

The system also supports another mode of providing information. General safety information can be obtained under the headings of Laser, Application, and Class. For example, in material processing, information relating to the lasers CO_2, Excimer, and Nd:Yag can be obtained. Similarly, process-oriented information can be obtained by selecting a particular 'Application'. For a 'Laser' the typical processes (Applications) that it can be used in are displayed and vice versa. The classification information is based on the

British Standard (BS:EN 60825). Information about a particular 'Class' contains the definition and examples of the class, the manufacturer's requirements, and user precautions for that class.

4.4 Evaluation of the system

Since the approach to the design of the system was driven by user needs, early evaluation was a feature of the development work.

User feedback was obtained regarding methods of presenting information, the match between system functionality and task needs, and interface structure and characteristics.

The evaluation data obtained included performance data, and attitude data. Performance and attitude criteria are discussed by Shackel [4]. In our case, performance data was obtained from a recording of the interaction between the user and the system. Attitude data was obtained by studying user responses to interview questions and a questionnaire.

Overall, the evaluation confirmed the approach to meeting user requirements, and provided information that was used to improve the system's functionality and interface.

5. Applicability to other advisory or safety domains:

Following the confirmation of the success of the present research in producing a viable computer-based advisory system to improve the safety aspects of laser usage in advanced manufacturing, the potential applicability of the approach to other subject areas was studied.

A modelling tool was designed that will allow generic models similar to the laser safety assessment model (Fig 1) to be defined. Application of this modelling approach to the hazards associated with ionising radiation showed great potential and the tool is currently under development.

6. Conclusions

The problem of providing laser safety information to potential users engaged in the tasks of hazard assessment has been successfully tackled. A prototype decision support system has been developed, and can be used to support the tasks of targeted user groups. The approach to knowledge modelling and system development has been found to be of potential applicability to similar advisory subject areas.

References

[1] Vassie L H, Tyrer J R, Soufi B and Clarke A A, "Lasers and Laser Applications in the 1990's: A Survey of Laser Safety Schemes", Optics and Lasers in Engineering 18, 339-347, (1993).

[2] Taylor B. The HUFIT Planning Analysis and Specification Toolset. Proc. INTERACT '90, Diaper, D et al.(Eds). Elsevier Science Publishers pp 371-376.

[3] Annett J, Duncan K D, Stammers R B, and Gray M J. Task Analysis (London: HMSO), Training Information Paper No. 6, (1971).

[4] Shackel B. Usability - Context, Framework, Definition, Design and Evaluation in Shackel B and Richardson S (Eds) Human Factors for Informatics Usability. Cambridge University Press, (1991).

Advances in Agile Manufacturing
P.T. Kidd and W. Karwowski (Eds.)
IOS Press, 1994

Safety and automation in high-risk production systems - as perceived by system operators

Gudela Grote and Cuno Künzler
Work and Organizational Psychology Unit, Swiss Federal Institute of Technology
ETH-Zentrum, 8092 Zürich, Switzerland

Abstract. The relationship between automation and production safety is analyzed on the basis of questionnaire responses and oberservations at different workplaces in a chemical and a transportation company. Perceived work conditions as well as perceptions of production safety and organizational safety measures are compared between operators in highly automated production systems and in "conventional" production. Differences are dicussed in terms of possibilities for the enhancement of safety through - or despite - automation.

1. Introduction

Automation is usually aimed at increasing productivity and allowing higher levels of complexity in production processes. Especially in production systems with high risk potential, another important goal of automation is the exclusion of the operator as a "risk factor". However, this strategy produces the "ironies of automation" described by Bainbridge [1], e.g. using the operator as a backup for a technical system that was designed to perform more safely than the operator. Another problem is the gap between the knowledge used under "normal" conditions and the knowledge needed to cope with critical incidents. Even for highly qualified operators the difficulty arises that they can exercise their knowledge less and that with increasing automation less practical production skills and more theoretical system knowledge are needed [2].

Mainly in high risk systems like nuclear power plants and military aircraft, where it is most obvious that humans have to be kept "in the loop" in order to be able to fulfill their final responsibility for production and safety, these problems have been addressed by developing new approaches to function allocation that allow a complementary design [3]. In less "risky" work environments like industrial production, there are few empirical studies addressing the relationship between safety and automation. In a study by Ruppert and Hoyos [4], ratings on a safety diagnosis instrument were compared for production systems with different degrees of automation. No consistent correlations between risks and level of automation were found. Another study by Wehner et al. [5] compared correlations between subjective assessments of danger, health risks, risks for materials, and potential for human errors by maintenance workers in two production lines with different degrees of automation. As error potential was unrelated to the various risk indicators in the less automated production line while there was a high correlation between perceived risks and error potential in the highly automated line, the authors argue that automation leads to a detrivialization of error. In combination with the results of the first study, these findings indicate that risks per se do not differ, but the correlation between risks and human behavior changes as a function of automation.

The present study again looked at perceptions of workers in more and less automated systems, however with a different focus. Risks and error potential were not of interest, but rather differences in the perception of organizational safety indicators including technical support as well as in job perception in general. The study was conducted as an exploratory analysis, there were no specific hypotheses formulated.

2.   Method

The study was carried out as part of a research project concerned with the development of concepts and methods for the improvement of safety in high-risk production systems by means of comparative studies in four chemical companies and one transportation company. The methods used include the analysis of documents like organizational charts, safety regulations, descriptions of work procedures, and accident reports, observations of produc- tion workers during their regular work in different production areas with different degrees of automation, questionnaires for production workers dealing with their work situation and their views on safety in their company, and interviews with safety officers and managers in production and technical support concerning safety policies, automation philosophy, general principles of work organization etc.

For the purpose of the study the questionnaire responses by production workers in several plants of one chemical company and by traffic controllers in the transportation company were focused on. The questionnaire contained four parts, (1) an already existing measure for subjective work analysis [6], (2) a number of newly developed items concerning job conditions (e.g. time pressure, insufficient information, social conflicts) and safety-related aspects of the job (e.g. potential danger for own person and for others, decision-making in critical situations, safety-related supervision of others), (3) several newly developed items concerning various safety indicators that have been identified as discriminating between organizations with many vs. few accidents by previous research [7, 8], and (4) several open questions concerning the general understanding of safety and safe behavior, potential risk factors, and conflicts between production requirements and safety in day-to-day operations. The questionnaire was handed out by supervisors of the respondents. Fifty chemical workers and 51 traffic controllers filled out the questionnaire individually during their official work hours and sent them back directly to the researchers.

Among the chemical workers and the traffic controllers, two subgroups were formed based on data provided by company experts concerning the level of automation in the plants and train stations. The sample of chemical workers was drawn from a total of seven plants, two of which were classified as highly automated. The plants covered the range of production and product types in the company, the risk potential was classified as comparatively low in the two automated plants and in three of the other plants. The sample of traffic controllers was drawn from a total of 29 train stations, ten of which were classified as highly automated. The stations ranged from small stations in villages to stations in metropolitan areas. The employees working with highly automated systems in both companies tended to be younger, having spent fewer years in the company and in their current position. These differences were not statistically significant, however.

3.   Results

Table 1 shows the responses of the four subsamples on selected parts of the question- naire. While there were only minor differences on job perception for the high versus low automation groups, work conditions were perceived as more stressful by the respondents in highly automated work systems. Chemical workers in automated plants perceived generally fewer risks for themselves, other people, and equipment as workers in less automated plants, which corresponds to the objectively lower risk in these plants. For the traffic con- trollers this was only the case for risks for oneself, perception of risks for others and equipment was actually higher in the high automation subsample.

Table 1.   Questionnaire responses by chemical workers and traffic controllers

| | Chemical Workers | | | | Traffic Controllers | | | |
| | High Automation (n=8) | | Low Automation (n=29) | | High Automation (n=19) | | Low Automation (n=32) | |
Variable	Mean	SD	Mean	SD	Mean	SD	Mean	SD
Job perception								
Autonomy	2.93	.37	2.92	.63	2.74	.62	2.89	.61
Variability	3.96	.90	4.00	.70	4.57	.42	4.60	.33
Task transparence	3.85	.67	3.98	.58	4.26	.46	4.33	.43
Demands	4.16	.74	4.09	.58	4.84	.20	4.89	.22
Career opportunities	3.63	.45	3.48	.62	3.98	.55	3.88	.59
Social support	4.00	.59	4.02	.75	4.32	.46	4.15	.52
Cooperation	3.71	.82	3.98	.72	4.04	.67	3.99	.65
Adequate work load	3.31	.53	3.56	.81	3.38	.64	3.33	.68
Adequate difficulty	3.33	.73	3.72	.88	4.04	.57	3.95	.63
Work conditions								
Unforeseen events	3.00	1.20	2.25	1.04	3.79	0.92	3.32	1.05
Time pressure	3.00	1.07	2.75	1.40	4.26	0.81	4.06	0.85
Poor information	3.00	1.07	2.36	1.19	3.79*	0.79	3.10	1.11
Poor instructions	2.75	0.71	2.29	1.05	3.00	0.88	2.48	0.96
Concentration	3.88	1.25	3.57	1.20	4.89*	0.32	4.58	0.72
Parallel tasks	4.00	0.76	3.79	1.15	4.72	0.46	4.68	0.56
Isolated work	2.00	1.31	1.31	0.97	1.53	1.02	2.03	1.47
Monotonous tasks	2.38	1.30	1.83	1.00	1.53	0.84	1.65	0.88
Safety								
Risks for own person	3.25	1.16	3.36	1.57	1.89	1.10	3.81	1.60
Risks for other people	2.75	1.04	3.25	1.55	5.21	1.55	5.00	1.39
Risks for equipment	2.25	0.89	3.11	1.52	5.00	1.70	4.39	1.80
Humans as risk	36.0*	19.6	49.2	15.7	50.1	25.4	58.2	19.6
Technology as risk	40.6	20.4	28.0	8.9	34.5	24.3	26.0	18.0
Organization as risk	23.3	11.2	23.7	12.9	15.4	10.2	15.5	9.8
Ergonomic design	2.25	0.71	2.39	0.83	2.68	1.45	2.87	1.21
Technical support for safe operation	2.00	0.76	2.17	1.07	1.42**	0.69	2.34	1.23
Technical support for decision making	1.38**	0.52	2.50	0.92	2.02**	0.91	3.18	1.12
Safety First	2.88**	0.64	1.86	1.06	1.10*	0.31	1.46	0.95
Global satisfaction with safety measures	5.63	0.52	5.34	1.11	5.00	1.25	5.00	0.91

Note. * $p \leq .05$ (two-tailed t-test), ** $p \leq .01$ (two-tailed t-test); job perception and work conditions: scales ranging from 1 'no agreement' to 5 'complete agreement', and 1 'never' to 5 'daily', respectively; risks for people and equipment: scales ranging from 1 'never' to 6 'more than 30% of weekly work time'; human, technological, and organizational risks: distribution of 100 points for the three risk factors; aspects of technical system and Safety-First principle: scales ranging from 1 'fully correct' to 5 'not correct at all'; global satisfaction with safety meaures: 7-point "smilies"-scale with higher values expressing higher satisfaction.

The assessment of humans and technology as risk factors differed between the high versus low automation groups in both samples. Humans were seen as contributing less to accidents, while technology was seen as contributing more by respondents in highly automated work systems. There were no differences with respect to the perception of organizational risk factors. Respondents in less automated work systems were less satisfied with the available technical support, and in the chemical company they were also less satisfied with safety measures in general. There was a major difference between the companies regarding the perception of the Safety-First principle: traffic controllers in more automated stations perceived more adherence to this principle compared to their colleagues in less automated stations, while the reverse was true for the chemical workers. On most other safety indicators, e.g. safety training of personnel, definition of safety responsibility, and quality of regulations, there were no differences between the two subsamples in each company.

4.   Conclusion

Space limitations do not allow comment on a number of interesting results, e.g. the differences in job perceptions between chemical workers and traffic controllers in light of their objective work tasks, the generally more critical responses of the traffic controllers regarding their work conditions, and the differences between the companies regarding conflicts between production requirements and safety [9].

As regards the relationship between automation and safety, the results suggest that operators in high-risk work systems perceive technology as support for safe production and decision making and at the same time attribute more responsibility for accidents to technology. This result differs from previously reported findings, which indicate a stronger link between risk and human error in automated production [5], and raises the question of a possible diffusion of responsibility through automation. Another critical finding concerns the perceived higher stress in automated work systems, e.g. through more unforeseen events, poorer information and instructions, higher demands on concentration, and - in the chemical plants - more isolated and monotonous work. This is a common finding in studies on the effects of automation, also pointing to less favorable conditions for the enhancement of safety.

Automation neither "automatically" enhances nor decreases safety in production. A particularily interesting aspect of the relationship between automation and safety that should be studied further is the question of a diffusion of responsibility between human operator and technical system.

References

[1]   Bainbridge, L. (1982). Ironies of automation. In G. Johannsen & J.E. Rijnsdorp (Eds.), Analysis, design and evaluation of man-machine-systems (pp. 129-135). Oxford: Pergamon.

[2]   Böhle, F.& Rose, H. (1992). Technik und Erfahrung. Arbeit in hochautomatisierten Systemen. Frankfurt / Main: Campus.

[3]   Bailey, R.W. (1989). Human performance engineering (2nd ed.). London: Prentice-Hall International.

[4]   Ruppert, F. & Hoyos, C. Graf (1993). Arbeitssicherheit und "Neue Technologien". Zeitschrift für Arbeitswissenschaft, 47, 1-10.

[5]   Wehner, T. , Nowack, J. & Mehl, K. (1991). Über die Enttrivialisierung von Fehlern: Automation und ihre Auswirkungen als Gefährdungspotentiale. In T. Wehner (Ed.), Sicherheit als Fehlerfreundlichkeit - Arbeits- und sozialpsychologische Befunde für eine kritische Technikbewertung. Opladen: Westdeutscher Verlag.

[6]   Udris, I. & Alioth, A. (1980). Fragebogen zur subjektiven Arbeitsanalyse (SAA). In Martin, E., Udris, I., Ackermann, U. & Oegerli, K., Monotonie in der Industrie (S. 61-68 und 204-207). Schriften zur Arbeitspsychologie, Band 31. Bern: Huber.

[7]   Cohen, A. (1977). Factors in successful occupational safety programs. Journal of Safety Research, 9, 168-178.

[8]   Zohar, D. (1980). Safety climate in industrial organizations: Theoretical and applied implications. Journal of Applied Psychology, 65, 96-102.

[9]   Grote, G. & Künzler, C. (1994). Safety culture and its reflections in job and organizational design: Total Safety Management. In G. Apostolakis (Ed.), Proceedings of the Conference PSAM II, San Diego, March 1994. New York: Plenum Press.

Advances in Agile Manufacturing
P.T. Kidd and W. Karwowski (Eds.)
IOS Press, 1994

Production disturbances
as an accident risk

Marianne DÖÖS and Tomas BACKSTRÖM
Division of Social and Organizational Psychology,
National Institute of Occupational Health, S-171 84 Solna, Sweden

Abstract. Disturbances in automated production give rise to hazardous tasks. Results from 76 comprehensively investigated accidents showed that the handling of disturbances to production was the most common task being undertaken when an accident occurred. This was followed by regular materials handling and maintenance. The majority of the injured were operators, who also had a higher relative accident frequency than maintenance personnel and setters. Around 40% of the injured persons were in the area where the accident took place several times a day or more frequently. The belief of injured persons was that they personally or the technology had the risk under full control at the time of the accident.

1. Introduction

Using automated machinery does not eliminate accident risks. Certain chains of events occurring within an installation or along a line of machines give rise to a type of accident risk which is peculiar to advanced manufacturing technology (AMT). The consequences of a human error at an automatically controlled machine are often more difficult to foresee and can be seriously injuring. A technical failure may have consequences which are not predicted, and there may be latent defects in the equipment. Machine movements can be started unexpectedly, without being directly initiated by a human being [1, 2, 3].

The extent and severity of the accident phenomenon in automated production has been summarized in an article by Backström and Döös [4] where it was concluded that AMT accidents are a significant problem and that special risk is attached to the operators' handling of production disturbances.

Production disturbances have close links with accidents since operators often have to enter areas where dangerous machine movements can occur when the machine is not functioning correctly and the risk for unexpected machine movements is unusually high. From the operator's point of view entering these areas was a completely normal procedure, not perceived as dangerous.

2. Aims

The current study was conducted for the purpose of obtaining more in-depth knowledge of accidents at automated machines than it was possible to obtain from mandatory occupational injury reports. Its aim was to describe how and when automation accidents occur, and to identify the important and interacting causes of accidents at computer-controlled installations. In this particular paper, the relation between occupational accidents and production disturbances is analyzed.

3. Material and method

Over a two-year period (May 1988 - May 1990) comprehensive investigations were conducted of 76 accidents at 15 companies using automated equipment. The emphasis in the study lay on the engineering industry, but companies in e.g. the food, rubber and wood processing industries were also represented. The investigation method permitted the identification of a multiplicity of factors – technical, individual and organizational - related to the occurrence of accidents. The method [2] was based on the theoretical framework of deviation and energy analyses and SMORT [5]. The investigations were carried out by safety engineers at the companies. On average an investigation took about four hours, and included a structured interview of about an hour with the injured person. No claim is made that the study group constitutes a representative sample. An analysis of missing data, for example, showed that relatively mild accidents and accidents sustained by production personnel were under-represented in the sample, while serious accidents and accidents among maintenance staff were over-represented.

Automatic equipment refers to an installation (or a part of an installation) which, without the direct intervention of a human being, *can either initiate a machine movement, or change its direction or function.*

An accident is defined as an unintended and branched sequence of events, where the final event occurs suddenly and results in an injury to a human being. By automation accidents are meant accidents occurring at an automated installation, where the equipment has controlled (or should have controlled) the harmful energy.

4. Results

Behind each accident there are normally a number of interacting circumstances of an organizational, human and technical nature. In this paper attention is paid to how technical problems and defects lead to hazardous occupational tasks. The depth of the investigations conducted enabled us to become acquainted with early phases in the course of events resulting in the accident. The full body of information showed that there were technical causes behind as many as 86% of the accidents. In a further few cases a technical design deficiency led to the equipment being wrongly set up. Only in 9% of the cases no technical defect of any kind was reported as being involved in the accident. Technical failures and defects behind automation accidents are also described in Backström & Döös [6].

The technical problems usually led to some form of disturbance to production which operators and other personnel were in the course of remedying when the accident occurred. In 61% of the 76 accidents the injured person was occupied with the handling of some kind of production disturbance. This was by far the most common class of occupational tasks involved, followed by regular materials handling (e.g. removing finished parts). See table 1.

Table 1. Occupational tasks at the time of accident (n=76).

Occupational task	Percentage share
Handling of disturbance	61
Regular materials handling	14
Maintenance	9
Re-setting the equipment	5
Other (e.g. transportation, instruction)	7
No task	4
Total	100

Table 2. Type of disturbance handling at the time of accident (n=46).

Type of disturbance handling	Percentage share
Freeing a product that had become stuck	33
Correcting position of product/work piece	24
Prompting sensor impulse	11
Freeing jammed machine part	7
Simple kinds of fault tracing	7
Arranging for re-start	4
Cleaning	4
Setting the equipment	4
Work normally conducted by robot	2
Other	4
Total	100

Disturbance handling usually involved freeing or correcting the position of a work piece that had become stuck or was wrongly positioned. Other disturbance handling involved prompting a sensor impulse, freeing a jammed machine part, carrying out simple kinds of fault tracing, and arranging for re-start. See table 2.

As stated above, these technical problems often led to the undertaking of hazardous tasks. Our classification of their consequences shows that the most common problems of a technical nature concern disturbances to the materials flow in the form of stuck/jammed or wrongly positioned work pieces or parts. This type of problem initiated one of two rather similar sequences of events: 1) the part was freed and came into position, the machine received an automatic signal to start, and the person was injured by the machine movement initiated; 2) there was not time for the part to be freed/repositioned before the person was injured by a machine movement that came unexpectedly, more quickly or was of greater force than he/she expected. These sequences of events are illustrated in figure 1. Corresponding sequences can be described for other kinds of accidents.

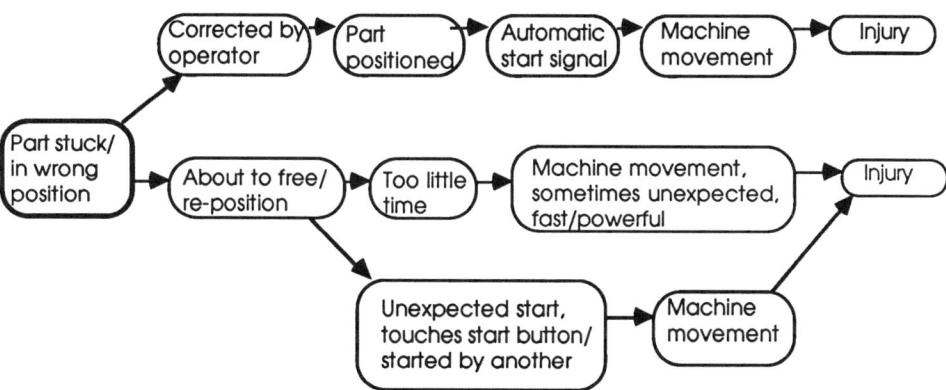

Figure 1. Alternative sequences of events after a part has become stuck or wrongly positioned.

Around two-thirds of those who sustained injuries from automated machines were production workers or operators. The tasks to which tasks they devoted most time were monitoring/surveillance and regular materials handling. Maintenance staff made up 10% of the injured. For them, maintenance work itself and the handling of production disturbances accounted for most of their working time. A parallel study of automation accidents [6] has shown that operators also had a higher relative accident frequency (15 automation accidents per 1,000 operators per year) than setters (11 per 10^3) and maintenance workers (6 per10^3).

The handling of production disturbances was a hazardous task for all occupational groups. It was mainly in the course of handling a production disturbance that members of all groups (except maintenance workers) were injured. This applied irrespective of whether it was a task to which they devoted a lot or only a little of their time. For maintenance staff, injuries primarily occurred in maintenance work itself.

An idea of the experience that the injured persons had had of similar hazardous situations can be obtained from investigating how often they had been in the area where their injury was sustained. Around 40% of the injured persons were in the hazardous area as frequently as several times a day; in some cases, they spent large parts of the working day there.

The great majority of those who were injured when working with automated equipment reported that they had deliberately entered the area where the injury took place. The risk zone was regarded more or less as a normal workplace, and the risk was considered as being under the full control of either the person or the technology.

5. Discussion and conclusions

The results of the study have been applied in a wide variety of ways. As well as being used at the investigated companies themselves (and also other companies), they have become a cornerstone of the recently developed Riv-method [7] for risk analysis of accidents and production disturbances executed on the shop floor. Moreover, the results have provided a basis for further research on and the in-depth problematization of the relation of production disturbances to both accident risks and experiential learning at the workplace. Seeing the production disturbances as opportunities for learning could also be of importance in expanding the work content of machine operators.

References

[1] T. Backström and L. Harms-Ringdahl, 1984, A Statistical Study of Control Systems and Accidents at Work. *Journal of Occupational Accidents* 6 (1984) 201-210.
[2] M. Döös and T. Backström, Learning from mistakes: an investigation model with examples of results, Research report 1990:1, National Institute of Occupational Health, Solna, 1990. (In Swedish).
[3] J. Reason, Human error. Cambridge University Press, Cambridge, 1990
[4] T. Backström and M. Döös, A comparative study of occupational accidents in industries with advanced manufacturing technology. *International Journal of Human Factors in Manufacturing*, to be published.
[5] L. Harms-Ringdahl, Safety Analysis. Principles and practice in occupational safety. Elsevier Applied Science, New York, 1993.
[6] T. Backström and M. Döös, Technical defects behind accidents in automated production. Paper presented and published at Fourth International Conference on Human Aspects of Advanced Manufacturing & Hybrid Automation, July 1994, Manchester, England.
[7] M. Döös and T. Backström, The Riv-method. Joint Safety Council, Stockholm, 1994. (In Swedish).

Part VII
Skill and Knowledge Enhancing Technologies

Part VII

Advances in Agile Manufacturing
P.T. Kidd and W. Karwowski (Eds.)
IOS Press, 1994

Workorganization and Skill Formation for Shopfloor Oriented Technologies

Heinz-H. Erbe, Jürgen Petereit

Institut für berufliche Bildung und Weiterbildungsforschung,Technische Universität Berlin
Franklinstr. 28/29, D-10587 Berlin, Germany

Abstract. SME`s with small batch production use in most cases numerically con-
trolled processing of materials and computer based shopfloor control of tasks to
meet the market demands of high quality to suitable costs and guaranteed times of
delivery.
The concept of shopfloor oriented technologies embedded in a workorganization,
which requires the development of human skills and experiences seems to be the
best way to proceed in the direction of more flexibility for these enterprises.
But using those technologies effectively the companies have to shift large parts of
responsibility for the production process to the workers at the shopfloor overcoming thereby the tayloristic division of work.
On the other hand the workers should be able to take this responsibility. That
means they need "in process" training additionally to their basic qualification in
handling numerically controlled machines.
This contribution reports experiences, made in the EC-programm COMETT with
SME´s of Portugal, Spain and Germany, in skill formation within the manufacturing process of SME`s. A concept of continuing formation to develop human re-
sources will be given to discussion. It has been developed a step by step procedure
in respect of the particular difficulties of these enterprises to free workforce for
qualifying processes.

1. Shopfloor Oriented Technologies

Shopfloor oriented technologies are preferably
used in SME´s with small batch or single
production. These technologies should support the
skill of the workers but not replace it.
In so far we can define criterias:
A. man-machine interfaces like
  - flexible machine tools with easy access to
    the working area,
  - manual <u>and</u> numerically controllable ma
    chine tools(figure 1)
  or at least
  - graphical- interactive programmable CNC-
    controls
  - help screens to support the diagnosis
    with respect to maintenance

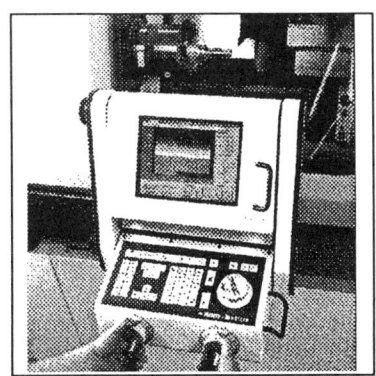

CNC-Plus/Keller/NUM/Realmeca
Figure 1

B. machine-machine interfaces like
    - at the shopfloor usable PC-supported NC-programming packages for different CNC-controls but with identical user surfaces like these and with NC-programme transfer in both directions
C. planing devices like
    - conventional planing boards or shopfloor control systems usable by skilled work ers for decision making with respect to scheduling tasks (no PPS)
D. quality evaluation support like
    - measuring devices connected to Statistical Process Control
    - coordinate measuring machine usable by all skilled workers at the shopfloor

2. Needed Qualifications

Studies have revealed that the main part of the throughput time (about 85%) is allotted to the idle time [1]. The classic rationalization reserves, for example the technical improvement of single machines, only lead to a small share of the throughput time's reduction. To achieve another reduction of the throughput time new reduction possibilities have to be found.

The main target of to-days rationalization efforts clearly aims at work organization, planing and control. It has been proposed to decentralize tasks of the central field to the shop-floor into the responsibility of skilled workers [2].

The trend in the development of PC's in recent years opens new markets for the deployment of low cost shopfloor control. It allows decentralized computer structures with defined task areas to be created within a company. These task areas are detailed scheduling and job supervision, which were previously reserved to central manufacturing

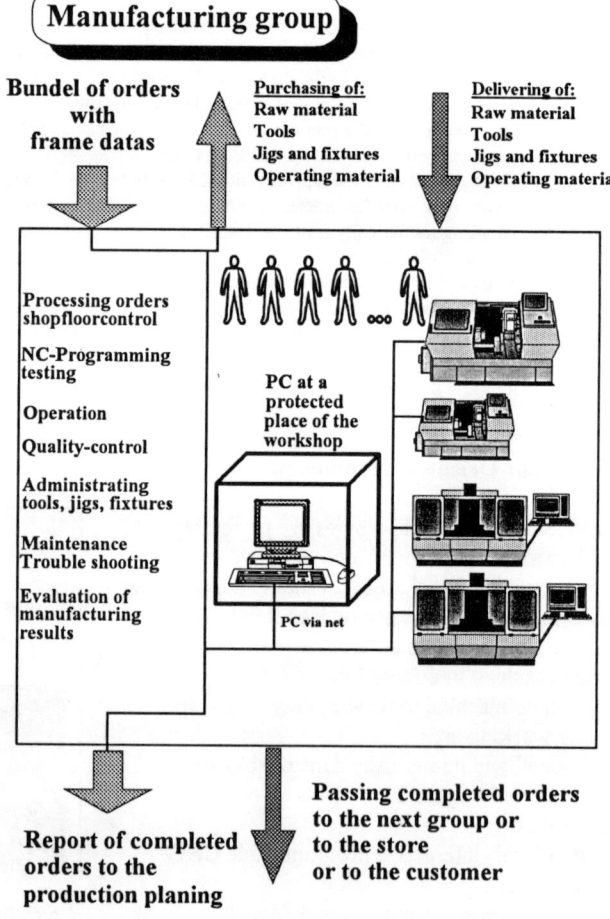

Figure 2

control with access to main frames.

This decentrally available intelligence allows access to be gained from production areas to data bases. With appropriate distributing the areas of responsibility a shopfloor control has been created which is capable of reacting rapidly to new factors arising within the companies or externally.

The man - computer interface is particularly important, because the software should be handled by workers whose main tasks are the material processing.

Because these tasks are new the workers have to get experiences in handling CNC-machine tools and should get as well qualifications in

- shopfloor control i.e. fineplaning of tasks when rough planing datas of complete orders are given,
- cooperation with other groups or departements when problems in scheduling occur
- planing of maintenance to increase the availability of the machine tools and other devices
- acertain required quality and conclusions to necessary alterations in the manufacturing process.

The qualifying process is directed to handle machine-/process-data and its evaluation and the observing of the quality of the manufactured parts straight at the machine tool with respect to surface requirements, tolerances of measures etc. and for immediate protocolling by feeding the datas to processing with computers.

On the other hand the workers should be capable of processing datas from roughplanning of customer orders to tasks at machine tools with respect to minimize throughput time or to meet the scheduled datas of delivery or shipment.

To support decisions and to learn from decisions done before, the evaluation or at least the results of the evaluation of machine- and process-datas should be in the hands of the skilled workers. They should be able to interpret statistical datas for observing quality e. g.

It is the intention of the qualifying concepts to foster a permanent continuing formation process at the shopfloor with a minimal input from outside. This depends strongly on the workorganization at the shopfloor. SME`s have to recognize the needs of giving more way for decisions to be done at the shopfloor i.e. less dividing of labour to get the required flexibility and quality of products.

3. Qualifying Concepts

SME´s are suffering from an always pressing time problem with respect of scheduled orders, which burdens the shopfloor. The qualifying concept has to consider these circumstances [3].

The concept follows a pick up - method as indicated in figure 3. Small groups of workers are learning at tasks belonging to orders the company has to execute anyway but the scheduled time for these tasks are not pressing. The workers get support of an external trainer or later of a meanwhile experienced skilled worker of the companies shopfloor.

The process runs in two phases in its turn divided in several steps:

- orientation about the training tasks
- training phase of the first group of workers
- transfer of the training tasks to the next group, which will be supported by meanwhile trained workers as well and so on.

The second phase starts when all workers within the qualifying process have run the first phase and are now able to handle the new tasks mentioned above. Within this phase the experiences will be deepened and enhanced. While within the first phase the workers should have 2 hours daily for training, within the second phase weekly training of 2 hours with all workers are sufficient to discuss problems of learning and actual difficulties with the new tasks.

This in short describes a method of continuing formation of workers at the shopfloor. It should not be a May-fly but a continuos process in SME's to foster its competence at the market.

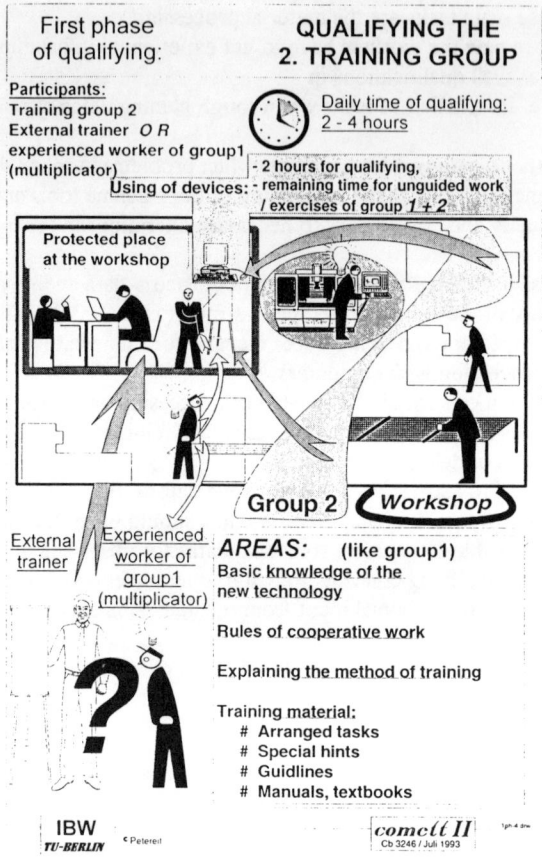

Figure 3

References

[1] T. Martin, E. Ulich and H.J. Warnecke, Appropriate Automation for Flexible Manufacturing, *Automatica* **26** (1990) 611 - 616.
[2] H.-H. Erbe, Skilled Work and Computer Based Manufacturing at the shopfloor. In: P. Bröder and W. Karwowski (Eds.) Ergonomics of Hybrid Automated Systems III. ISBN : 0444 8952 05. Elsevier, Amsterdam, 1992, pp. 435 -440.
[3] H.-H. Erbe, A. Potthast and W. Siebel (Eds.). Leitfaden für Innovationsentscheidungen. ISBN: 3 - 929796 - 03 - 1. RKW Verlag, Eschborn, 1993.

Advances in Agile Manufacturing
P.T. Kidd and W. Karwowski (Eds.)
IOS Press, 1994

The Development of a Theoretical Model for Predicting Skills Requirements in Advanced Manufacturing Settings

Richard Koubek, Gavriel Salvendy and Kuo-Hao Tang
Purdue University

Abstract. The object of the ongoing research is to find a more efficient way for prediction of skill acquisition and personnel requirements in the advanced manufacturing systems to cope with the emerging time-based manufacturing environment. A hybrid model is presented which integrates a number of diverse approaches into a single framework. The interactions among the various components are expected to provide explanatory power beyond that available considering each component separately.

1. Introduction

As computer-based manufacturing systems become more complex, the performance of humans in such systems becomes more critical for success due to the distribution of cognitive and decision making tasks to the shop floor. In order to maintain a manufacturing competitive advantage, the proper selection and training of personnel capable of performing in these complex environments becomes critical. The objective of this research is to develop a theoretically-driven engineering-science-based systematic model of human skill requirements associated with the operation of Advanced Manufacturing Technologies (AMT).

A variety of theoretical approaches are reviewed as basis of skill and knowledge assessment. These model components include ability taxonomy, knowledge structure, learning hierarchy, cognitive resources, and automatization. In order to integrate these components into a coherent framework, the relationships between the components are analyzed and a theoretical framework is presented. The theoretical framework provides a complete view of the learning processes as well as the dynamic interactions between ability requirements, knowledge structures, resources requirements, and other model components. It is expected that these interactions can provide explanatory power beyond that available considering each component separately.

2. Review of Approaches

Learning Hierarchy: One of the well established learning theories based on information-processing theories is Gagné's learning hierarchy. Gagné [1] was one of the first people to differentiate mental processes into higher and lower levels of mental processes. There are six levels in the learning hierarchy proposed in his later work [2]: *association, chaining, discrimination, concept, rule,* and *higher-order rule,* or, *problem solving.* Throughout the learning hierarchy, a distinction is made between features which are external and internal to the learner.

Learning in Knowledge-Rich Domain: Among many theoretical paradigms, Anderson's [3] ACT* suggests four basic learning mechanisms for acquiring skills in a knowledge-rich domain: *declarative recording, strengthening, knowledge compilation,* and, *generalization and discrimination.* Knowledge compilation is the way how declarative knowledge is transformed to procedural knowledge. Two mechanisms, proceduralization and composition, are used to achieve the transformation. Proceduralization transforms a domain-general production to a domain-specific production by using constants to replace variables in the domain-general production. Composition is the process of combining multiple productions into a single production. Generalization and discrimination mainly concern inductive learning. Generalization creates a new production rule that captures what similar individual production rules have in common. On the other hand, discrimination deals with the fact that such rules may be too general and need to be restricted.

Dual Processing Code Theory: Dual processing code theory is better known as automaticity or automatization. The foundation of this theory is the distinction of two types of qualitatively different cognitive processes: controlled and automatic processes [4],[5]. A controlled process requires conscious control and therefore costs cognitive resources. In contrast, an automatic process requires little or no attention and therefore needs very few cognitive resources.

Ability Taxonomies: Abilities are relatively stable attributes of an individual's performance. A specific task is said to require certain abilities in order to perform to an established criterion [6]. The objective is to identify and define the fewest independent ability categories which can describe performance for various tasks in a meaningful way.

Knowledge Structures: This approach suggests that the way humans structure their knowledge is a significant determinant of performance. Adelson [7] found expert programmers have a more abstract knowledge structure and focus on what a program does while novices have a more concrete representation and focus on how the program operates. In more recent research, Koubek and Salvendy [8] have proposed three levels of knowledge structures. In addition to Adelson's novice and expert level, a third level of super-expert is identified.

Cognitive Resources Theories: Kahneman [9] proposed that there is a single undifferentiated, sharable pool of resources available to all tasks and mental activities, which is termed as single-resource theory. The limitation of single-resource theory is that it cannot account for several aspects of the data from dual-task studies [10]. Based on these results, a multiple-resource theory with three dichotomous dimensions is proposed by Wickens, which argues that people have several pools of different capacities of resources.

3. Construction of Conceptual Model for Skill Requirements

Although each of the above cited approaches by themselves is insufficient for modeling skill requirements in AMT settings, they do provide different views from different aspects. Gagné's learning theory provides the view of hierarchical learning. Wicken's multiple-resource theory and Fleishman's ability taxonomy fit into the internal learning conditions of Gagné's learning theory and can be considered as sources as well as constraints during learning. While learning requires resources, Schneider and Shiffrin's dual code theory, can be integrated as a dynamic view of returning resources after automatization of a task. When learning moves up the hierarchy, it is reasonable to assume that operators will develop a higher level of knowledge structure to efficiently use domain knowledge. Koubek and Salvendy's three levels of knowledge structures therefore can be mapped to the learning hierarchy to facilitate learning processes. Finally, Anderson's knowledge compilation can be incorporated into the learning hierarchy to elaborate the details of knowledge transformation at the higher levels of learning. With all these considerations, a conceptual learning model is outlined in Figure 1. With this conceptual framework, a more complete, multi-faceted view of skill requirements for AMT operators is expected.

According to the conceptual model, the skill acquisition processes require dynamic interactions between ability requirements, knowledge structures, resources requirements, and other model components. These interactions between model components are the most important information provided by the integrated model if it is validated. The main hypothesized interactive behaviors, which are also the rationale behind the conceptual framework, are discussed below.

Learning Hierarchy and Ability Taxonomy: Three pieces of information are needed here for developing the interactions between learning hierarchy and ability taxonomy.

(1) Two phases can be determined by Gagné's learning hierarchy. One is the processes of attaining new skills and moving up the hierarchy. This is represented by thick arrow lines on Figure 1. The other is the process of applying skills at the obtained learning levels. This is represented by the text on the learning hierarchy in Figure 1. The first can be termed as learning phase and the second as performing phase.

(2) From Gagné's learning hierarchy, moving up the learning hierarchy requires that the internal conditions are suitable. These internal conditions include all previous stages of learning as well as the abilities required by the next level.

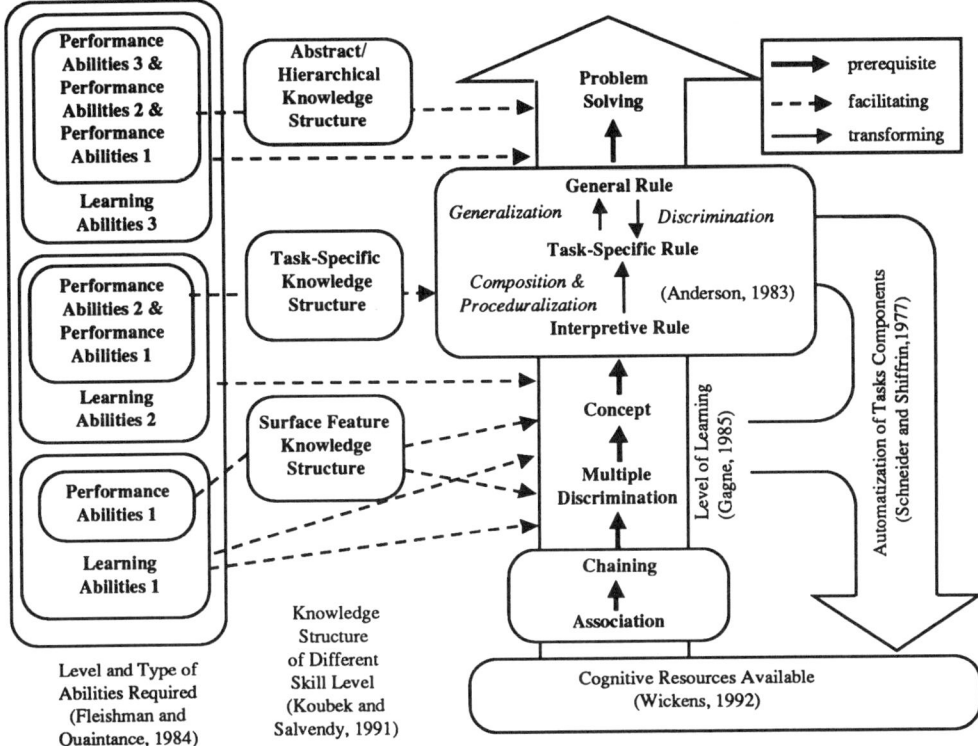

Figure 1. Hypothesized conceptual framework for learning processes in AMT environment

(3) On the other hand, research [11] has shown that abilities important for initial stages of learning are often different from those required for skilled task performance. Combining all three pieces of information, it can be reasonably assumed that the abilities required to move up to the next targeted level are different from the abilities required to perform the task either at the original level or next targeted level. Here, abilities required for learning can be termed as learning abilities and abilities required for performing can be termed as performance abilities.

Learning Hierarchy, Automatization, and Cognitive Resources: Learning hierarchy, automatization, and cognitive resources are all interrelated. In order to perform a task or learn a new skill, a certain amount of resources is required. As learning accrues, more cognitive resources are required for moving up the levels of learning. While this is true, repeatedly performing a consistent task gradually releases cognitive resources through automatization. Thus, operators learning a new skill may spend most of their resources at a certain level. Due to lack of resources, they are not able to move up to the next level at this time. After the operators become familiar with the skill at this level, a portion of their cognitive resources will be released through automatization of the consistent components. The operator now may have enough cognitive resources for learning at the next level. This process will continue until either ability limits or cognitive resource limits are met. In this process, the cognitive resources of shop floor workers are not wasted, but rather are continuously re-invested to acquire new states of learning, thereby increasing the efficiency of AMT systems.

Knowledge Structure, Learning Hierarchy and Ability Taxonomy: Koubek and Salvendy's three levels of knowledge structures can be mapped to learning hierarchy to improve learning processes. As learning moves up the hierarchy and the tasks become increasingly cognition intensive, more deep knowledge and reasoning activities will be involved in rule-based related tasks. The original surface feature knowledge structure becomes insufficient to perform these tasks. Operators need to develop a more abstract, higher level of knowledge structure to cope with the increasingly complex domain knowledge. Thus, the task-specific knowledge structure will be developed gradually. If the learning processes still go on, the task-specific knowledge structure will not be appropriate for the highest level of learning because problem solving related tasks require both deep and broad views of domain knowledge. The abstract/hierarchical knowledge structure then may be developed for efficient use of domain knowledge. Therefore, Figure 1 shows that the knowledge structures can be mapped to appropriate levels of learning in order to obtain better performance.

Knowledge Compilation and Learning Hierarchy: As previously stated, there is no learning mechanism in Gagné's learning hierarchy. This insufficiency causes difficulty if the conceptual model is to be transformed into an operational model. Since the focus of this research is at higher levels of learning, Anderson's knowledge compilation fits into this category well. By viewing Gagné's learning hierarchy, the defined concept is actually a form of declarative knowledge and therefore can be replaced by interpretive rules in knowledge compilation. On the other hand, the rule level in Gagné's learning hierarchy accounts for the procedural knowledge part in the knowledge compilation process and therefore the rule level can be used by both task-specific rules and general rules. Thus, knowledge compilation can be incorporated into Gagné's learning hierarchy at both defined concept and rule levels. From Figure 1, it can be seen that the knowledge required for problem solving may come from any of the three types of rule.

4. Conclusion

In the proposed theoretical framework for predicting skills requirements in AMT, various theoretical approaches have been integrated into a coherent framework. From above discussion, it can be concluded that in order to move up the learning hierarchy, several requirements should be satisfied: 1. All knowledge learned from previous levels of learning is prerequisite. 2. The training abilities and performance abilities with appropriate ability levels corresponding to the skill to be learned must exist in the learner. 3. An adequate amount of cognitive resources for learning new skills is required. 4. If the appropriate knowledge structure exists, it will facilitate the learning process. Also, the overall view of this theoretical framework suggests that the abilities and cognitive resources are the key inputs contributing to learning processes. From Figure 1, these inputs from the ability taxonomy and cognitive resources are the primary drivers pushing up learning processes. Therefore, the relationships between learning and these two model components are critical in this framework.

At the completion of the ongoing research, a new and effective methodology will have been developed based on the model for predicting, in a systematic and structured way, the knowledge, skills and abilities required to operate, maintain and manage AMT in the workplace. This tool will also enable corporations to concurrently select appropriate personnel in accordance with skills and knowledge requirements derived from the model while developing the new AMT hardware. With personnel who have abilities matching the task requirements and a targeted training program, the implementation of new manufacturing technology will have a smoother start-up and faster learning curve. As such, significant economic benefits will be achieved.

References

[1] Gagné, R. M. (1965). *The Conditions of Learning.* New York: Holt, Rinehart and Winston.
[2] Gagné, R. M. (1985). *The Conditions of Learning and Theory of Instruction* (4th Edition). New York: Holt, Rinehart and Winston.
[3] Anderson, J. R. (1983). *The Architecture of Cognition.* Harvard University Press, Cambridge, MA.
[4] Schneider, W. and Shiffrin, R. M. (1977). Controlled and Automatic Human Information Processing I. Detection, Search, and Attention. *Psychological Review,* 84, 1-66.
[5] Shiffrin, R. M and Schneider, W. (1977). Controlled and Automatic Human Information Processing II. Perceptual Learning, Automatic Attending, and a General Theory. *Psychological Review,* 84, 127-190.
[6] Fleishman, E. A. and Quaintance, M. K. (1984). *Taxonomy of Human Performance.* Academic Press, Inc.
[7] Adelson, B. (1984). When Novices Surpass Experts: the Difficulty of a Task may Increase with Expertise. *Journal of Experimental Psychology,* 10(3), 483-495.
[8] Koubek, R. J. and Salvendy, G. (1991). Cognitive Performance of Super-Expert on Computer Program Modification Tasks. *Ergonomics,* 34(8), 1095-1112.
[9] Kahneman, D. (1973). *Attention and Effort.* Englewood Cliffs, NJ: Prentice Hall.
[10] Wickens, C. D. (1984). Processing Resources in Attention. In *Varieties of Attention* (63-101). Edited by R. Parasuraman and R. Davies, New York: Academic Press.
[11] Salvendy, G. (1969). Learning Fundamental Skills - A Promise for the Future. *AIIE Transactions,* 1(4), 300-305.

Advances in Agile Manufacturing
P.T. Kidd and W. Karwowski (Eds.)
IOS Press, 1994

Development and Validation of an Operational Model for Predicting Skills Requirements in Agile Manufacturing Systems

Kuo-Hao Tang, Richard Koubek and Gavriel Salvendy
Purdue University

Abstract. In the proposed theoretical framework for predicting skills requirements in AMT, a variety of theoretical approaches are integrated into a coherent framework. Among all the components in the framework, ability taxonomy and cognitive resources are identified as the most important factors associated with skills acquisition and learning processes. The operational model is therefore based on these two factors. The concept, ability margin, is introduced as the difference between the ability level of an operator and the ability level required by the task. The model suggests that the amount of resources required by an ability to perform a task is a decreasing exponential function of the ability margin. To validate the model, a system of hypotheses are developed. It is expected that the cognitive resources required for performing a task can be predicted in part from ability margins.

1. Introduction

From a review of the literature [1],[2],[3], it appears that researchers use abilities and cognitive resources as two independent factors to predict or measure performance. While both abilities and cognitive resources are correlated with task performance, from the conceptual framework given by Koubek et al. (1994), both of them are sources as well as potential constraints for moving up along the learning hierarchy. However, it is obvious that an operator with high abilities is unlikely to spend the same amount of cognitive resources to learn or perform the same task as someone with low abilities. This observation suggests that there is some relationship between ability levels and cognitive resources. In the next sections of this paper, the hypothesized relationship between abilities and required cognitive resources is depicted and mathematically modeled. In order to validate the model, a system of hypotheses is developed. If the model is supported, it can serve as the foundation for an operational method to predict skill requirements and learning processes.

2. Construction of an Operational Model

Following the conceptual framework, the multiple-cognitive resource theory is assumed appropriate here. From Wickens' [3] multiple-cognitive resource model, both of the encoding (perceptual) and central processing (cognitive) stages in each modality and code combinations (i.e., visual-spatial, visual-verbal, auditory-spatial, auditory-verbal) are identified as a distinct resource pool. The resource pools associated with the response stage will not be considered here since these resource pools basically correspond to psychomotor and physical abilities. For the same reason, only perceptual and cognitive abilities from Fleishman's [1] ability taxonomy, which have strong connections with cognitive tasks, will be considered in this particular study. These abilities are termed as mental abilities in later discussion.

Suppose there is a mental ability set of total N possible abilities within the human task domain. Let S denote the ability set and s_i, i=1..N denote each ability in S. For each mental ability in S, a scale is defined so that the level of each ability can be determined for a specific operator as well as for a task. Let a_i denote the ability level possessed by a specific operator for ability i and $\max(s_i)$ the maximum ability level on the scale. Thus,

$$0 \le a_i \le \max(s_i), \quad \forall \, a_i, \, i = 1..N \tag{1}$$

For each task, a subset of N' mental abilities with different ability levels from S can be determined as the minimum requirements for performing the task. Let S' denote the ability set required by the task, b_i the minimum ability level requirement for ability i.

$$0 < b_i \leq \max(s_i), \quad \forall\, b_i \in S' \qquad (2)$$
$$b_i = 0, \qquad \text{otherwise}$$

In order to perform the task, the following conditions must be satisfied:

$$b_i \leq a_i \qquad \forall\, a_i, b_i\ i = 1..N \qquad (3)$$

According to Kahneman [4], cognitive resources will be allocated to different mental activities during the process of performing a task according to some allocation policy. On the other hand, it is obvious that a mental activity requires some mental abilities in order to carry it out. While there are numerous mental activities, there is only a limited set of mental abilities. Therefore, it is reasonable to further decompose the resource allocation policy down to the ability level. Since the purpose here is to identify the relationship between cognitive resources and abilities, the potential interference between different cognitive resource pools should be avoided. As such, it is further assumed that, throughout the research and experiments, the abilities required for the task only load on a single cognitive resource pool, e.g., visual-spatial pool. Let R denote the total capacity of the cognitive resource pool under consideration possessed by a specific operator. Let r_i denote the resources allocated to ability i for performing a task. Since the task requires the ability set S', we have:

$$0 < r_i \leq R \qquad \text{If } b_i \in S',\ i = 1..N \qquad (4)$$
$$r_i = 0, \qquad \text{otherwise}$$
and
$$\sum_{i=1}^{N} r_i \leq R \qquad (5)$$

To find out how cognitive resources are allocated to the mental abilities, the factors affecting resource allocation policy need to be identified. Two factors are considered as the most important factors here: different types of abilities, and the differences between the ability levels possessed by an operator and the ability levels required by the task, i.e., $(a_i - b_i)$, where i=1..N. The differences are defined as ability margins, m_i.

For the first factor, different types of abilities, it is obvious that the amount of resources allocated to abilities may differ from each other due to the nature of the abilities regardless of other factors. For example, in general, the amount of resources required for perceptual speed ability (e.g., perceiving an object) may be quite different from the amount of resources required for visualization ability (e.g., rotating the geometric object in the operator's mind) even though they both load on the visual-spatial pool.

For the second factor, ability margins, it is reasonable to assume that as the ability margin increases, the quantity of resources available for other tasks of learning will also be increased. The larger the margin of an ability, the less the resources are required. This assumption means that if people have superior abilities, they need relatively few resources to perform a task compared with one whose ability levels are barely sufficient to perform the task.

To identify the ability-resource curve, we can further assume that the ratio between amount of decreased resources due to increased ability margin, Δr_i, and initial resource level, r_i, is proportional to the increased ability margin, $\Delta(a_i - b_i)$. As the ability margin keeps increasing, eventually the resources level will decrease to a certain level, β_i, and not change anymore. This value can be considered as the cost for concurrence. Let α_i denote the decreasing rate so this description can be mathematically modeled.

$$\frac{\Delta r_i}{r_i} = -\alpha_i \Delta(a_i - b_i) \qquad (6)$$

as $\Delta \to 0$, we have:

$$\frac{dr_i}{r_i} = -\alpha_i d(a_i - b_i) \qquad (7)$$

Now, consider the characteristics of an ability-resource curve: The total resource level is equal to R and cost for concurrence is β_i. Thus, we have the conditions as follows:

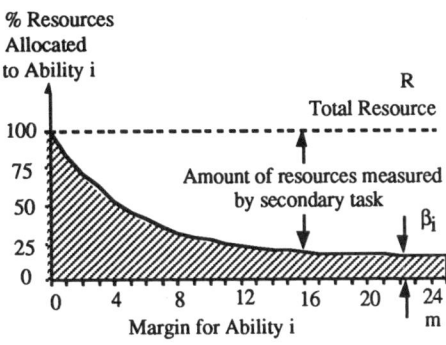

Figure 1. Hypothesized Ability-Resource curve.

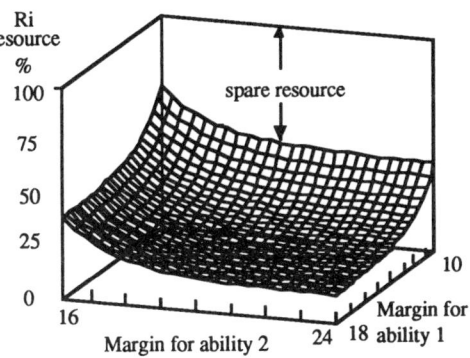

Figure 2. Ability-resource hyperplane
for testing additivity.

$r_i = R$ if $a_i - b_i = 0$

$r_i = \beta_i$ if $a_i - b_i = \infty$

By integrating Equation (7) and using these conditions, we have:

$$r_i = [R - \beta_i] \, e^{-\alpha_i m_i} + \beta_i \qquad \text{where } m_i = a_i - b_i \qquad (8)$$

Figure 1 shows how a typical ability-resource curve looks. Now, given a regular task with multiple abilities, an operator can perform a task and achieve a satisfactory level only when the following condition is satisfied:

$$R' \equiv \sum_{i=1}^{N} \{[R - \beta_i] \, e^{-\alpha_i m_i} + \beta_i\} < R \qquad \text{where } m_i > 0, i = 1..N \qquad (9)$$

In Equation (9), both abilities and cognitive resources are used as constraints. The ability constraint is given by $m_i > 0$, $i = 1..N$, and the resource constraint is given by $R' < R$. Through the proposed ability-resource curve, these two constraints can be considered together.

3. Validation of the Model

In order to validate the operational model, or, Equation (8) and (9), a series of hypotheses is developed.

Hypothesis One: The resources required by an ability which loads on that specific pool of resources is a decreasing exponential function of the ability margin.

This hypothesis is to validate the exponential relationship between resources required and different levels of ability margins, which was proposed in the previous section. If Hypothesis One is validated, the shape of the ability-resource curve will be identified.

Hypothesis Two: There is no interaction between abilities which load on the same resource pool in terms of the resource allocation policy, which means that additivity exists among different abilities.

Fleishman's ability taxonomy is identified from a series of interlocking experimental factor-analytic studies to ensure dependable predictions of human performance from one task to another (Fleishman, 1984). Thus, when used for predicting performance, there are few or no interactions between abilities. Since the ability taxonomy is designed in this manner, it can be reasonably assumed that the independent nature of these abilities extends to resource allocation policy. If the additivity exists, which means there is no interaction between abilities with regard to resource allocation, then the ability-resource response surface plotted by the series of secondary tests or other MWL measures should look like Figure 2.

If Hypothesis Two is supported, then the relationship between resources and abilities can extend to a task with multiple abilities. Thus, the information obtained from several single ability-resource curves can be combined to predict the total resource requirements.

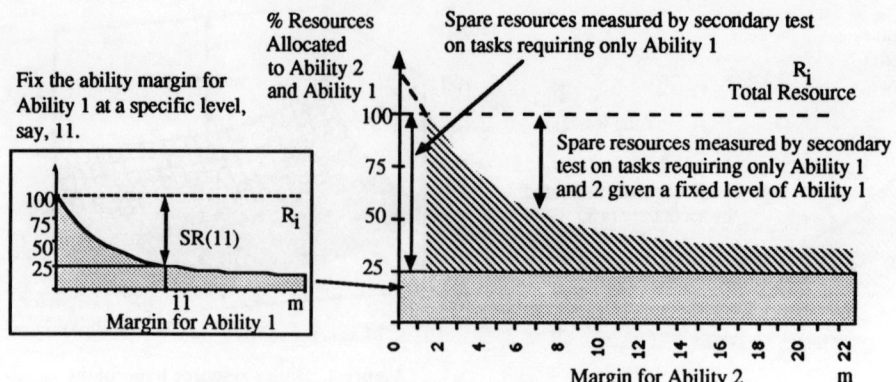

Figure 3. Measure ability-resource curve and total resource
capacity by using a task requireing only Ability 1 and 2.

Hypothesis Three: The ability-resource curve and total resource capacity can be measured by two identified "shapes" (decreasing rate) of ability-resource curves loading on the same resource pool.

Assume that the shapes, or, decreasing rate, of two abilities loading on the same cognitive resource pool have been identified. In order to find out all the parameters for ability-resource curves and total resource capacity, a task which requires only two abilities can be used for this purpose.

If the ability margin for Ability 1 is fixed at a certain level, say, 11, from Figure 3, the SR(11) can be measured and SR(11) will be the amount of resources that can be used for other abilities. Now given the task which only requires Ability 1 and 2, if we hold the difficulty of Ability 1 at an ability margin equal to 11 and vary the ability margin for Ability 2, then from Figure 3, a series of secondary tests can be performed to find out the spare resources when increasing the ability margin for Ability 2. Comparing this series of spare resources with SR(11), the partial ability-resource curve for Ability 2 can be determined. It is called a partial ability-resource curve because a part of the curve can not be determined from this approach due to the fact that part of the resources are used by Ability 1. In Figure 3, this part is shown by a dashed line. The partial ability-resource curve can be completed by extrapolation. Thus, the total cognitive resource level can be determined. By exchanging Ability 1 and Ability 2 in Figure 3, the ability-resource curve for Ability 1 can also be determined. The total cognitive resource level obtained by extrapolation for both abilities should be the same even though the decreasing rates of both abilities may differ from each other.

4. Conclusion

After the operational model is validated, it will provide a linkage between abilities and resources which can have important theoretical implication. In terms of practical application, the resultant tool will enable corporations to concurrently select appropriate personnel in accordance with skills requirements derived from the model while developing or ordering new technologies. With personnel who have abilities matching the task requirements and an improved training program, the implementation of new technologies will have a smoother start-up and a faster learning curve. As such, the economic benefits will be achieved, for example, by reducing time to market for products using new hardware and software at the shop floor.

References

[1] Fleishman, E. A. and Quaintance, M. K. (1984). *Taxonomy of Human Performance*. Academic Press, Inc.
[2] Wickens, C. D. (1980). The Structure of Attentional Resources. In *Attention and Performance VIII* (239-257). Edited by R. Nickerson, Hillsdale, NJ: Erlbaum.
[3] Wickens, C. D. (1984). Processing Resources in Attention. In *Varieties of Attention* (63-101). Edited by R. Parasuraman and R. Davies, New York: Academic Press.
[4] Kahneman, D. (1973). *Attention and Effort*. Englewood Cliffs, NJ: Prentice Hall.

Advances in Agile Manufacturing
P.T. Kidd and W. Karwowski (Eds.)
IOS Press, 1994

A Decision Centered Design Approach to Case-Based Reasoning: Helping Engineers Prepare Bids and Solve Problems

David W. Klinger
Klein Associates Inc.
582 E. Dayton-Yellow Springs Road
Fairborn, OH 45324-3987

Abstract. The objective of a project sponsored by the U.S. Air Force, Wright Laboratory at Wright-Patterson AFB (Contract F33615-89-C-5702) was to determine if a case-based reasoning system could be applied to aid in the preparation of bids for manufactured parts. Klein Associates developed and installed the system, Bidder's Associate, in the business unit of a manufacturer of jet engine parts. Our approach was to study the decision processes of experienced engineers so that we could develop a system to support (not bypass) them. The interface was designed so that pertinent information, which would help them make accurate situation assessments and inferences, was made available at appropriate points in the estimating process. Bidder's Associate has opened up the corporate database and allowed the engineer to look through previous work experience to see if the company has either manufactured or bid a part that is similar to the part to be bid or built. Information can now be extracted from a prior job, then adjusted and incorporated into the current case. Early evaluation has been quite positive, several success stories have been noted. Soon after completion of the prototype the development site overhauled their software system. Bidder's Associate is being modified so that it will fit within the new framework.

1. Introduction

Bidder's Associate (BA) is a case-based reasoning system which was developed to aid in the cost estimation of manufactured jet engine parts. The system was installed in a metal fabrication and manufacturing company which specializes in high tolerance, difficult to make, jet engine parts. Due to the complexity of the parts, errors were commonplace and cost estimates were, very often, simply a shot in the dark. Different engineers perform the job differently and what was learned by one was seldom passed on to others. Corporate memory was being lost due to turnover and lack of means for obtaining feedback or sharing experience. In short, mistakes were being repeated and successes were not. BA became a useful system which provides the means for capturing and maintaining corporate history and makes this history accessible.

2. What is Case-Based Reasoning?

Case-based reasoning (CBR) is based on the notion that humans often use reminding as a primer for information retrieval. How many times have you seen someone that reminds you of an old friend? You may be reminded by the person's hair, mannerisms, or facial expressions. We attach these cues, or characteristics, to people and when we see them in someone else, we remember the old friend. This type of reminding is called visual-based reminding [1]. Other types of reminding (process-based, goal-based, etc.) happen in much the same fashion, certain cues spark retrieval.

As humans we do not remember everything, limitations on memory and recall make this impossible. This is where the strength of CBR surfaces. CBR allows the user to determine what cues are relevant for a particular situation, and then query the system to find any previous cases that match these important features. Also, the user who remembers bits and pieces of a particular case can inform the system of the remembered cues and the CBR system "fills-in" the picture.

3. Manufacturing, CBR, and Decision Centered Design

The manufacturing domain was selected for various reasons. First, like most domains, corporate memory was being lost. Second, there is an art to the estimating process. It cannot be reduced to rules and procedures. A process engineer must construct a part using various resources. These resources include numerous types of materials, various manufacturing processes (welding, forming, tweaking, etc.), and an unlimited number of services provided by vendors outside the particular fabrication unit. The order of utilization of these resources is extremely important. Particular metals, for instance, must be tweaked after being heat treated, others before. Third, many of the parts are similar to previously bid and manufactured parts. For example, a rear bypass duct and a forward bypass duct may be similar, yet their part-number designations can be very different.

Prior cases can be a powerful source of corporate memory and expertise. Allowing the engineer to ask the system "have we ever . . .?" provides the engineer the opportunity to learn from prior manufacturing successes and failures. A part is often bid, manufactured, and delivered without any feedback to the engineers about how well the job was performed or, more importantly, about how accurate the original estimate was. This loss of corporate memory can be devastating.

The simple retrieval of previous cases can be easily accomplished using established CBR algorithms. These "canned" systems retrieve cases rather quickly from a large database, but the usefulness of the retrieved cases cannot be counted on. This is not a trivial matter. Earlier generation CBR systems were developed to display the application of the technology but did little in the area of retrieving meaningful, useful cases [2]. In order to avoid this trap, we concentrated on the decisions the users made as they developed cost estimates. We watched them prepare estimates and conducted Critical Decision method (CDM) interviews. CDM is an interview strategy that uses a set of cognitive probes to elicit information about actual nonroutine incidents that require expert judgment or decision making [3]. We documented the steps the engineers went through and we noted the cues and factors which helped them determine the manufacturing process, the purchasing of materials, and the determination of scrap rates. Of central importance were the factors which caused an engineer to say "This part is much like a part I previously developed an estimate for." It was these factors that drove the development of the search algorithm and search features.

The decision analysis pointed to several critical areas. First, the engineers spend a fair amount of time looking at the drawings. This sounds rather obvious, but what we found was they were trying to develop a three-dimensional image of the object in their heads. Once they had developed the image, they would mentally rotate it to "see" what the finished part would look like from different perspectives. We were told by an experienced engineer that "if a process engineer can't develop a three-dimensional image of a part, that person shouldn't prepare the estimate." Second, as simple as it may seem, engineers often have trouble remembering the size of a part. The drawings of a part are drawn to scale and as the engineer goes through the process of developing a three-dimensional image, they often forget about scale. They imagine the part to be about the size it is on the drawing. This is seldom the case. Third, when developing the manufacturing process schedule, the engineers envision the part moving from station to station until it has been completed. They seldom envision errors, the part moves through the system flawlessly until a perfect part is produced in record time. Again, these parts are extremely difficult to produce, their specifications are quite exact. Therefore, errors on the shop floor are commonplace, yet the engineers fail to consider this when developing a cost estimate. This could be the most serious error in terms of cost accuracy that a process engineer can make.

4. What Does Bidder's Associate Do?

Case retrieval is the heart of BA. It is here that the engineer can "look" through previous experience. The engineer queries the system, based on some critical features, as to whether the company has ever bid or manufactured a part similar to the one currently being bid. Simply put, the engineer asks the system to look through the records for a part that meets certain criteria. For example, if the current part is made of AMS 5599 (an aluminum alloy) and must be heat treated and is square, the engineer can enter these search features, and BA would "look" through the case base for a similar part. We were able to pinpoint the critical cues, the causal factors, and the relationship between these cues and factors that would result in the retrieval of meaningful cases. This analysis resulted in 14 features that can be searched for any given case. These features are weighted based on the importance given to them by process engineers during our interviews. The engineer may alter these weights in order to tailor each search to each unique instance. It should be noted that the three most used search features directly relate to the findings from the decision analysis. That is, the engineer is queried early on in a session with BA regarding the part to bid's shape, size, and complexity. These features help to trim the retrieved cases down to a workable few.

After a search has been initiated, BA displays a list of the retrieved bids. These retrieved bids are listed in descending order based on a similarity score. The score reflects how well the cases matched in regard to the searched values and their respective weights. The engineer can quickly glance through the information to see how each of the cases matched the selected features. This allows the engineer to select useful, similar cases to aid in the formulation of the new bid.

Once a similar case has been selected by the engineer, the information from that case is displayed. All 14 search features, and their respective values, are displayed for both the retrieved case and the current case. Also, if the retrieved case has any manufacturing history, that history is displayed. This allows the engineer to quickly look through a case to determine what can be utilized in constructing a bid for the desired part. Manufacturing information includes the manufacturing centers, the order in which they were used, and any written instructions or problems which were present when the retrieved case was manufactured. This is often the most fruitful information for the engineer. For instance, if there were numerous problems encountered, thus increasing costs, a bidder can use that information to submit a more accurate bid.

Utilizing information from previous history has many benefits. The simple reminding function helps the engineer to repeat successes and side-step previous errors. It is also important for new engineers to quickly learn how things are done in a particular shop. Therefore, allowing the novice to ask the system "have we ever. . .?" or "how do we . . .?" offers the novice the chance to gain insight into the workings of the company. This cuts down on training time for the novice, and allows the experienced worker to spend more time doing, and less time teaching.

5. Summary

Case-based reasoning has long been used as a tool for reminding. Only recently have CBR systems been developed for real-world applications. The success of these systems hinges not on the speed of information retrieval, but on the ability to retrieve meaningful data. To accomplish this, the system must be centered around the user's decision process. That is, the system, including the interface, must "think" about the cases the same way the engineer's think about them. A mismatched system and user will fail to communicate, and frustrated users will return to their old way of doing business. Cases should be consistent and easy to find. The cues used for retrieval must be able to stand up over time. What is important today must also be important tomorrow.

Evaluation of BA has been quite positive, several success stories have been noted. The process engineers reported that BA was "intuitive . . . . easy to use . . . it feels like I developed it." Soon after completion of the prototype the development site overhauled their software system. Bidder's Associate is being modified to fit within the new framework.

References

[1] R. Schank, Dynamic Memory: A Theory of Reminding and Learning in Computers and People. Cambridge University Press, Cambridge, MA, 1982.
[2] Stottler, R., & Klinger, D. W., A Case-Based Reasoning System for Manufacturing Bid Preparation. Presented at the Florida Artificial Intelligence Research Symposium (FLAIRS), 1992.
[3] G. Klein *et al.*, Critical Decision Method for Eliciting Knowledge. *IEEE Transactions on Systems, Man, and Cybernetics*, **19**(3), 462-472, 1989.

Advances in Agile Manufacturing
P.T. Kidd and W. Karwowski (Eds.)
IOS Press, 1994

Developing a company wide estimating, design and installation system

Albert C.K. LEUNG
Wormald Ausul (UK) Ltd., Grimshaw Lane, Manchester M40 2WL, U.K.
Raymond LEONARD
Total Technology Department, UMIST, Manchester M60 1QD, U.K.

Abstract. The estimation of costs is a vital concern of every manufacturing firm. Cost and performance information, which reflect the actual nature of operations, are vital for effective management in high-tech environments. In addition, how design knowledge influences the cost estimation process, and how design engineers integrate manufacturing cost considerations into the engineering design process are also important issues today. In this paper, the authors present a method in which a company wide computer system, PROTECT, was developed to assist a fire protection company in cost estimating and design, and it describes a way to generate interfaces between design application programs and a commercial Computer-Aided Design System. The advantages of the new system in terms of cost, speed, accuracy and company competitiveness are also highlighted.

1. Introduction

In this paper, a detailed description of the existing cost estimation and design systems in a fire protection company, the potential areas for improvement, and possible approaches to improvements are presented and discussed. The selection of a paradigm for system analysis and design, and the actual development process of the new system are described. A commercial viability survey was conducted to show the potential benefits and costs of such a development. Issues such as how the estimators and designers should cope with the increasingly more powerful and "intelligent" computer software and CAD systems were reviewed.

2. Existing Operation System

2.1 Background

Wormald International are part of the worlds largest fire protection corporation, and they design, manufacture and install protection systems for complex situations such as steam driven turbo-generators. To protect such hazards, the need exists for accurate quotations, correct designs, and efficient component acquisition & assembly. Previously, these tasks were accomplished by a hybrid approach of stand-alone, manual and computer systems and CAD draughting packages, operating in separate areas of the company. This carried the possibility of a duplication of information, extended lines of communication and significant throughput times. Detailed procedures were thus needed to avoid the dangers of inaccuracy, inconsistency and inefficiency being generated in the process, especially when large variations always exists between contracts.

2.2 Problems with Cost Estimation

The traditional approach to estimating a contract has the following problems:
- If the original estimate is too high in price, the contract is not won even though actual costs are competitive.
- If the original estimate is too low in price, the order is won but the profit margin is reduced or eliminated.
- Extended lead times on projects, due to insufficient equipment specification or an under-estimation of man-hours.
- Requests for contract estimates come at random intervals, so by the time staff are assigned to the contract they have to quickly produce the quote to meet the deadline. This allows no time to strike up a relationship with the client, and to clarify/amend client requirements at this early stage.
- It is time consuming and expensive to train new staff.
- Accurate quotations take a long time to produce.
- A lot of paperwork is generated, which is then interpreted by another member of staff once the order is taken.
- Orders are passed on to the manufacturing division in various forms, so a lot of time is spent clarifying different issues with the salesman originally responsible.

2.3 Problems with Design

Many problems arise during the transition period from Cost Estimation to Design, generally they are:
- The protection details passed on were not directly usable, so each design has to restart.
- The protection configurations taken by designers and cost estimators are sometimes different, therefore the estimate may not reflect the true cost of a contract.
- Insufficiently defined material specifications may lead to increased material costs.

Other problems that arise during the design stage are:
- Variations in designs, of similar contracts, by different draughtsmen means that some designs may not be optimum.
- Lengthy design times may delay a projects deadline, and also affect the company's competitiveness.
- A wide difference between contracts requires designers to remember a large quantity of information about different design procedures. Thus they may have to review the relevant manuals every new project, especially for protection systems and standards which are not usually met.
- It is time consuming and expensive to train new staff.

3. Solution

3.1 Requirements

To confront these problems, the company proposed the development of a software package and extended CAD system to allow both sales estimating and design staff to deal with these variations more effectively. This involved university collaboration via the Total Technology research programme at UMIST. The objective was to produce a system which would incorporate information from design and estimation manuals, fire protection standards, and know-how from experienced staff. This would ultimately allow the distribution of knowledge concerning the corresponding protection specification and procedures to all company designers and estimators. The system was to make design decisions by producing optimum designs directly from inputs, and produce outputs for both estimators and designers.

3.2 Selection of a paradigm for approach

For users, the ideal characteristics of software are: easy to learn; convenience in use; well explained in printed manuals and onscreen help messages; fast and efficient at performing the tasks for which it was developed or purchased; flexible, malleable, easy to use in various ways; compatible with other software products likely to be used; and good value when comparing the features offered to cost [1]. For developers of such software, the major concerns are: the selection of paradigm; the software's performance; the five quality factors [2]; the selection of development tools; and development costs.

Many cost estimation and design software tools of today were developed by conventional approaches, such as Structured Analysis and Design [3], and Expert Systems [4]. However, in this application, the extent and complexity of the system made those approaches inappropriate. As exemplified in [5], Expert Systems do address very complex domains, however they do not have their own theoretical support for system analysis and design, and are built mainly on the knowledge and reasoning logic of domain experts rather than on a real world model. Hence they may not represent the real systems. In addition, there were no suitable expert system building languages or expert system shells on the market which could interface with the currently used Computer-Aided Design Package and Estimating/Accounting software. In the end, an object-oriented approach was selected, because the methodologies offered more regarding software correctness, maintainability, and reusability etc. [6].

3.3 Potential Benefits and Development Cost Estimation

The main benefits of building the proposed system include: the production throughput time of a contract could be significantly reduced. Hence the company could take in more orders, and thus be more competitive and profitable; the loss in experience upon the retirement of key personal is not so costly; and optimum results can be achieved.

Two years time was scheduled to build the system, during which period a close collaboration between system developers and domain experts was required. By using the various empirical software development cost estimation techniques [7], a development cost estimation was carried out. The results of the estimate were encouraging, and provided the development team and senior management with a momentum and budget for later development.

3.4 Development Process

Following the Object-Oriented Themes [8] and Methodologies [9],[10], a full range of system analysis, design, programming and test & validation were conducted [11]. The system built, namely PROTECT, effectively co-ordinates the Estimating/Accounting software, AutoCAD, a major Hydraulics Package, and Material & Price databases. PROTECT has one central knowledge base, which is shared by all the connecting modules, and the inputs to PROTECT only contain the configuration details of the protected domain. Interactive action is required concerning the clients special requirements, such as different standards, special type of materials etc. Outputs of the software are formulated in the formats required by estimators and designers.

The integration of mechanical design with the CAD systems to build an intelligent CAD system is a very important issue today [12]. PROTECT achieved this by creating DXF drawing files for AutoCAD, thereafter, it produces preliminary drawings of all major arrangements in 2-D and 3-D with essential entities and dimensions. Therefore designers avoid the need to start each design from zero base and make major design decisions.

3.5 The role of PROTECT in the company

Fire protection domains are very complex, and consist of a wide range of applications with various design procedures and principles. PROTECT can only accommodate a small part of them, for which, PROTECT provides all the essential information required. However, this information may still need further refinement by estimators and designers to cover unstandardized features, such as ladders and vents etc. Therefore, PROTECT works as an "intelligent" human partner, rather than a full replacement, which has never been the objective of anyone. In the authors opinion, in such complex design and manufacturing domains, computer software and human collaboration is the best way to increase productivity.

4. Conclusion

The principal aim of software engineering is to help produce quality software. This paper has described the application of the object-oriented methodologies in a fire protection environment, in which a system, designated PROTECT, has been successfully built. Most of the objectives proposed have been achieved, and the company's productivity has been greatly improved. Thus PROTECT is now being extended to other fire protection domains.

References

[1] Beiser,K., Selection Software for Libraries, *Databases*, Vol.:16, Issue:2, p.18-29, April 1993.
[2] Meyer, B., *Object-Oriented Software Construction*, Prentice Hall, Englewood Cliffs, N.J., 1988.
[3] Yourdon, E., and Constantine, L., *Structured Design*, Prentice Hall, Englewood Cliffs, N.J., 1979.
[4] Buchanan, B.G. and Shortliffe, E.H., *Rule-Based Expert Systems*, Addison-Wesley Publishing Company, Massachusetts 1984.
[5] Klahr, P. and Waterman, D.A., *Expert Systems: Techniques, tools and applications*, Addison-Wesley Publishing Company, Massachusetts, 1986.
[6] Cox, B., *Object-Oriented Programming: An Evolutionary Approach*, Addison-Wesley Publishing Company, Masschusetts, 1986.
[7] Arifoglu, A., A methodology for software cost estimation, *SIGSOFT Software Engineering Notes*, Vol.: 18, Issue:2, p.96-105, April 1993.
[8] Booch, G., *Object-Oriented Design With Applications*, Benjamin/Cummings, Menlo Park, California, 1991.
[9] Shlaer, S. and Mellor, S.J., *Object Lifecycles: Modeling the World in States*, Yourdon Press, Englewood Cliffs, N.J., 1992.
[10] Rumbaugh, J., Blaha, M., Premerlani, W., Eddy, F., and Lorensen, W., *Object-Oriented Modeling and Design*, Prentice Hall, Englewood Cliffs, N.J., 1991.
[11] Leung, C.K. and Leonard, R., The application of Object-Oriented Methodologies in Fire Protection Environment, *IEE Colloquium on the design, implementation and use of object-oriented system*, Digest No. 1994/003, London, 10 January 1994.
[12] Jayaram, S. and Myklebust, A., Device - Independent Programming Environments for CAD/CAM Software Creation, *Computer-Aided Design*, Vol.:25, No.2, p.94-104, February 1993.

Advances in Agile Manufacturing
P.T. Kidd and W. Karwowski (Eds.)
IOS Press, 1994

Intelligent Tutoring by
Knowledge Refinement with Version Space

Evgeny SMIRNOV
Scientific Research Institute of Informatics, AI Lab, PO Box 137, 1784 Sofia, Bulgaria

Nikolay I. NIKOLAEV
American University in Bulgaria, Blagoevgrad 2700, Bulgaria, nikolaev@aubg.bg

Abstract. This paper studies the possibilities for simulating intelligent tutoring employing a specific method for knowledge refinement which uses the version space technology. The process of refinement is a process of building a model of the student. For input an overgeneral and inconsistent teaching domain theory from the tutor's practice is supplied. This input and the training examples, provided also by the tutor, guide a process of learning a correct model of the student. The training examples are, first, preclassified as positive and negative by the student and second, processed by incremental learning algorithms for deriving student's concept definitions of the tutor concept under study.

1. Introduction

The Intelligent Tutoring Systems (ITS) architecture [1], [8] includes: (1) an explicit model of the domain theory and an expert program that can solve problems in the domain; (2) a model of the student and (3) a model of the tutor. While the first part of this architecture is well studied, the recent models of the student and of the tutor do not facilitate building of ITS which are tailored to student's learning needs and goals. Some decisions of these problems suggest viewing the student modeling as machine learning research. Others suggest designing of tutoring shells for use by teachers in different domains.

A new method for building ITS by knowledge refinement with version space, which considers deriving a student's model as knowledge refinement of an overgeneral inconsistent domain theory, is presented in this paper. The knowledge refinement is performed by theory-based concept specialization (TBCS) techniques [6]. Employing TBCS characteristics, the tutoring model generates new teaching data which are subsequently classified by the student. The resulted preclassified data form the student's input for learning. This input and the model of the tutor together guide the TBCS processes. The student model is merged with the correct model of the domain so as to explain the student behaviour comprehensively.

The presented method is problem independent and can be used in different teaching domains. It is taken as a basis for development of a teacher's shell by the authors.

2. Knowledge Refinement with Version Space

The knowledge refinement task in this paper aims at finding a consistent domain theory from an overgeneral inconsistent one. The theory can be overgeneral because of two types of errors: (1) incorrect rules are present in the theory and/or (2) there are rules which lack of constraints in theirs premises. That is why the knowledge refinement process is a sequence of TBCS problem decisions.

2.1. The Theory-Based Concept Specialization Problem

The definition of the theory-based concept specialization problem is as follows:

Given: * Goal concept (GC): A concept definition.
    * Domain theory (DT): A set of first order clauses to be used in explaining how the training examples
      are instances of the goal concept.
    * Training set (TS): A set of positive and negative examples of an unknown concept SC, which are
      positive examples of the goal concept (it is supposed that SC is a specialization of the goal concept).

Determine: * A necessary and sufficient description of the concept SC that is implied by the positive examples and
      not by the negative ones considered in the context of the domain theory.

This TBCS definition supposes, first, that the goal concept is operational in the domain theory; second, the set of examples is a correct set of examples of the goal concept and of the concept SC; and third, the domain theory is perfect for the goal concept and not perfect for the concept SC.

2.2. Representation and Ordering of the Concepts

The finite set of concepts CDL is defined as follows:

$$CDL = \{ \ c \ | \ E \in S(\ TS \ ) \ and \ DT \cup E \vdash c \ \}$$

where S(TS) is the power set of the training set TS and E is an arbitrary subset of TS.

Definition 1: (Generalized Concept Partial Ordering) A concept c_2 ($c_2 \in CDL$) is more general than or equally general to another concept c_1 ($c_1 \in CDL$): $c_1 \leq c_2$ in the context of a domain theory DT, iff $DT \cup c_1 \vdash c_2$.

The CDL together with the generalized concept partial ordering form a description space. Let denote by C an arbitrary subset of CDL, under a set TS' such that TS' $\in$ TS:

$$C = \{ \ c \ | \ B \in S(\ TS' \ ) \ and \ DT \cup B \vdash c \ \}$$

The set C can be considered as a set of candidate descriptions of the concept SC to be learned. The task is to prove that every such subset C can be presented by boundary sets.

Theorem 1. 1. If $DT \cup E' \vdash c'$ and $DT \cup E \vdash c$ and $c' \geq c$ then $DT \cup E \cup E' \vdash c'$
            2. If $DT \cup E' \ |/\text{-} \ c'$ and $DT \cup E \ |/\text{-} \ c$ and $c \geq c'$ then $DT \cup E \cup E' \ |/\text{-} \ c'$,
            where E and E' are arbitrary subsets of training set TS (proved in [6]).

Theorem 2. Every subset C of CDL is convex.
    Proof: It must be proved, that for all c_1 and c_3 that belong to C, and c_2 that belongs to the set CDL such that $c_1 \geq c_2 \geq c_3$ follows that c_2 also belongs to C (definition of convexity [2]). From the dependencies: $c_1 \geq c_2 \geq c_3$, $DT \cup TS_1 \vdash c_1$, $DT \cup TS_2 \vdash c_2$ and $DT \cup TS_3 \vdash c_3$ using theorem 1 is concluded that: (1) $DT \cup TS_1 \cup TS_2 \cup TS_3 \vdash c_1$ and (2) $DT \cup TS_2 \cup TS_3 \vdash c_2$. Therefore: $TS_1 \cup TS_2 \cup TS_3$ belongs to S(TS'). From this follows that $TS_2 \cup TS_3$ also belongs to S(TS'). Hence according to the definition of the set C: $c_2 \in C$. ¤

Theorem 3. If the CDL is finite then all subsets C of the CDL are definite [2].

Theorem 4. All subsets C of CDL can be presented by boundary sets.
    Proof: Theorems 2 and 3 help to conclude that every subset C of CDL is convex and definite. Therefore upon the theorem for representing all subsets C of CDL by boundary sets [2] follows that C can be presented by boundary sets. ¤

The above Theorem 4 makes it possible to maintain every subset C of CDL according to the version space (VS) representation formalism:

$$VS\langle \ G,S \ \rangle = \{ \ c \ | \ \exists s \in S \ and \ g \in G \ such \ that: \ s \leq c \leq g \ \}$$

where: E_p and E_n are sets of positive and negative examples of the concept SC,
$G = \{ \ g \ | \ DT \cup E_n \ |/\text{-} \ g$ and $DT \cup E_p \vdash g$ such that does not exist g': $g \leq g'$ and $DT \cup E_n \ |/\text{-} \ g' \ \}$, and
$S = \{ \ s \ | \ DT \cup E_p \vdash s$ such that does not exist s' : $s' \leq s$ and $DT \cup E_p \vdash s' \ \}$.

2.3. Guiding knowledge-intensive version space

Two kinds of incremental algorithms for guiding version space exist. The first one supposes learning conjunctive concept descriptions. Such algorithms are the Candidate Elimination Algorithm [3] and Incremental Non-Backtracking Focusing (INBF) [7]. The second kind of incremental algorithms supposes learning disjunctive concept descriptions. Recent algorithms of this type are Space Fragmenting (SF) I and III [4], [5]. The conjunctive algorithms have lower space and time complexity than the disjunctive ones. The disjunctive algorithms can learn both conjunctive and disjunctive concepts compared with the conjunctive algorithms which are capable of learning only conjunctive descriptions.

The general scheme for guiding knowledge-intensive version space, which integrates those kinds of incremental learning algorithms, is:

> for each example *i*
> > if *i* is a negative example of the concept SC then
> > > [-Retain in *S* only those elements which are not implied by *i*.]
> > > -Make descriptions in *G* { and *S* }, that are implied by *i*, more specific only to the extent required so they no longer are implied by *i* [, and only in such a way that each remains more general than some elements in *S*].
> > > -Remove from *G* any element that is more specific than some other element in *G*.
> > else / *i* is a positive example of the concept SC /
> > > [-Retain in *G* only those elements which are implied by *i*.]
> > > -Generalize with *i* elements of *S* that are not implied by *i*, and only in such ways that each remains more specific that some element in *G*.
> > > -Remove from *S* any element that is more general than some other element in *S*.

In this general scheme the parts of the algorithm which belong to the conjunctive case are enclosed with brackets [] and those which belong to the disjunctive case with braces { }.

The main problem with CEA and SF I is that they both fragment the set G exponentially in the number of the negative examples and require many positive examples for converging the sets S and G. The INBF and SF III overcome the first shortcoming by using only near miss negative examples. A negative example is near miss in respect to s when the example does not imply s and does imply some description which can be received by modus ponens of s and some rule from the domain theory. In this way fragmentation is avoided and both algorithms have polynomial space and time complexity when learning a concept.

3. Architecture of the Intelligent Tutoring System

The proposed architecture of Intelligent Tutoring System includes:
(1) explicit overgeneral and correct models of the domain theory provided by the teacher who design the system. The overgeneral theory is constructed upon the main types of student's mistakes in his teacher's practice and has both kinds of errors identified in section 2;
(2) a model of the student that identifies, at a detail level, what the student understands;
(3) a tutoring model that provide a graph of the concepts to be learned by the student, strategy for tutoring and presentation of new material. The strategy for tutoring includes knowledge for selecting concepts for learning and knowledge for guiding the inference of the model of the student when learning a concept;
(4) an explanation model that explains student mistakes in the process of tutoring using the correct domain theory.

The management of the presented ITS is accomplished by the following tutoring model. A teaching concept is identified by the student from a given preliminary concept graph. The choice of new unlearned concepts is made using knowledge for choosing successful and unsuccessful concepts already learned by the student. The derivation of the model of the student is as follows:

Given: * An overgeneral domain theory that defines a concept to be learned by the student;
         * Knowledge for guiding the inference of the model of the student.

Determine: * Student concept definitions of the current teaching concept.

The derivation of the model of the student is considered a TBCS problem and is solved by the techniques presented in section 2.3. The formation of concept examples is carried out with knowledge for guiding the inference of the model. It determines a medium concept description between S and G as a function of the student's behavior when learning this concept. After that, an example is generated that does not imply any element from S and does imply the medium concept in the context of the overgeneral domain theory. The student classifies the generated example as positive or negative for the concept from his point of view. The example is used for updating the sets S and/or G by some version space strategy. The process continues untill the sets S and G converge. The resulted descriptions from S and G represent the student definition of the teaching concept.

A part of the overgeneral domain theory, that can imply the resulted definitions in the presence of all their possible positive concept examples, and the concept definition itself form the student domain theory for this concept. This theory presents the model of the student for this concept.

After learning some or all concepts the explanation model explains the student his correct and incorrect decisions. This is made by merging the decision trees of the correct theory with the decision trees generated by the student.

4. Conclusion

This paper studies the possibilities for simulating intelligent tutoring employing a specific method for knowledge refinement which uses version space technologies. The process of refinement is considered as a process of building a model of the student in the teaching domain. The contraction of the version space is caused by examples, preclassified by the student, generated on the basis of the distances between the lower and upper boundaries and rules from the model of the tutor.

This paper on the new ITS architecture proposes student modeling to be considered machine learning research.

References

[1]   W. Clancey and E. Soloway, Artificial Intelligence and Learning Environments, *Artificial Intelligence* **42** (1990) 1-6.

[2]   H. Hirsh, Theoretical Underpinnings of Version Spaces. In: Proceedings Twelfth Int. Joint Conf. on Artificial Intelligence, IJCAI-91, Sydney, Australia, 1991, 665-670.

[3]   T.M. Mitchell, Generalization as Search, *Artificial Intelligence* **18** (1982) 203-226.

[4]   E.N. Smirnov, Space Fragmenting- A Method For Disjunctive Concept Acquisition. In: du Boulay,B. and Sgurev,V. /eds./, Artificial Intelligence V- Methodology, Systems, Applications, Elsevier Science Publ., North Holland, 1992, 97-104.

[5]   E.N. Smirnov, Space Fragmenting III- A Polynomially Method For Disjunctive Concept Acquisition, *Int. Journal of Information Technologies and Applications* **1**, (1993) 322-335.

[6]   N.I. Nikolaev and E.N. Smirnov, A General Model for Integrating Empirical and Analytical Learning with Version Space, Technical Report (1-3-94), Department of Computer Science, American University in Bulgaria, March 1994.

[7]   B. Smith and P. Rosenbloom, Incremental Non-Backtracking Focusing: A Polynomially Bounded Generalization Algorithm for Version Space. In: Proceedings Tenth National Conf. on Artificial Intelligence, AAAI-90, 1990, 848-853.

[8]   E. Wenger, Artificial Intelligence and Tutoring Systems, Morgan Kaufmann Publ., Los Altos, CA, 1987.

Advances in Agile Manufacturing
P.T. Kidd and W. Karwowski (Eds.)
IOS Press, 1994

Decision Support for Flexible Manufacturing

J. Stahre and A. Johansson
Dep. of Production Engineering,
Chalmers University of Technology
S-412 96 Göteborg, Sweden
email: jost@pe.chalmers.se and aj@pe.chalmers.se

Abstract. The paper presents a theoretical model for the interaction between human operators and advanced manufacturing systems. The model includes five possible roles for the operators, a general description of manufacturing processes, and the description of various ways of communication through the human/process interface. The model, and a general data collection method, are being used in case studies. The method produces specifications for necessary operator training and education, and for customised operator decision support.

Introduction

The efficiency and economical success of advanced manufacturing systems, such as FMSs, rely heavily on its operators to handle unforeseen events and system failures. However, due to the complexity introduced by the high demands on integration and flexibility, the work tasks of these operators are becoming increasingly difficult. It is important to provide efficient support, enabling experienced operators to use their manufacturing skills, without having to resort to computer specialists each time a problem occurs. Such decision support must be tailored to suit the requirements for a specific manufacturing system, thus the need for design methods and suitable theoretical models arise. The approach in this paper is a theoretical operator/process interaction model. The model is supported by a data collection method providing specifications of decision support, training, and education for operators in advanced manufacturing systems.

Theoretical background

For many years there has been advanced research regarding the operator situation in continuous process industry. Research has been justified by the high investments and risks involved in such industries. Considering the high complexity of modern manufacturing systems, accumulated results from process industry research should be in parts transferable. In a crossdisciplinary research effort, called The *Humanufacturing Project,* a supervisory control model previously proposed by Sheridan [1] has been adapted for use in discrete parts manufacturing. The original model describes five roles of human operators and their interaction with processes. The adapted model for human—process interaction in manufacturing is presented in figure 1. The new model integrates the three main parts of a complete manufacturing system, i.e. the manufacturing processes (including machines and products), the human operators, and the interface between humans and machines.

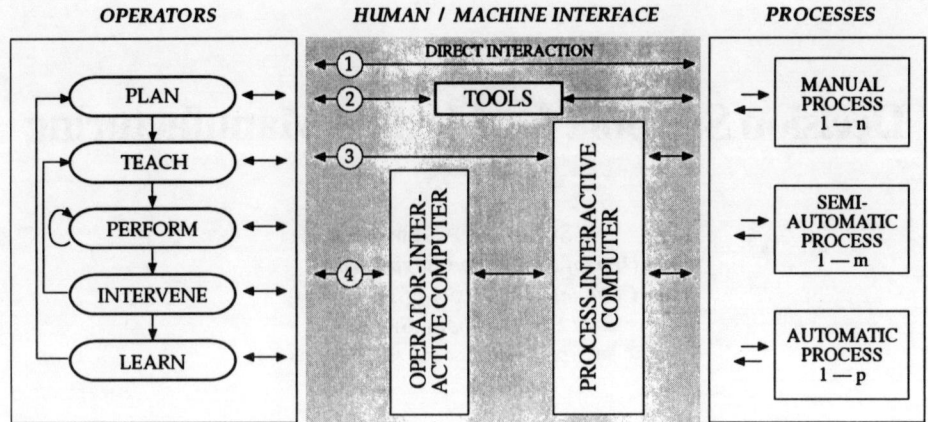

Figure 1. Operator/Process Interaction Model (OPIM)

Operator roles

The human operators in an advanced manufacturing system can be modelled using
Sheridan's roles of the human supervisor [1], in which the operator/supervisor may be seen
to act in five different capacities, as seen in figure 1, i.e.:

- *Planning* defines ⟶ ;source allocation strategy based on accumulated knowledge of the
  manufacturing system components, process demands, tool and material availability.
  The plan is dynamic and may be revised as the situation changes.
- *Teach* means transferring the plan to the system, i.e. programming of machines, robots,
  etc., or direct adjustment of tools, sensors, etc.
- *Perform* means that the operator observes execution of the programmed plan. He
  acquires and estimates process status, makes minor adjustments of automatic processes,
  and performs manual tasks, such as assembly. The operator constantly evaluates the
  system in relation to desired goals. The role is a closed control loop in itself. The
  original definition of this role, according to Sheridan, was *monitoring* and the inclusion
  of manual tasks such as assembly is an expansion, or generalisation, of the original
  model to adapt to discrete manufacturing tasks.
- *Intervening* is conditional and requires a failure or a system halt. If the cause is a
  previously known fault or the normal termination of a task, the operator takes action
  simply by starting the next process step. This generally means that the operator returns to
  the teaching/programming role in which a new process is started, thereby closing the
  intermediate control loop.
- *Learning* is entered every time an unknown error occurs, or if a normally terminated task
  is followed by an entirely new process. The operator should record immediate events
  surrounding the unknown error, thus accumulating experience not only for himself, but
  in computer records available to other operators. The operator's change in process
  understanding may lead to a change in resource allocation strategy. Subsequently, the
  operator will return to the planning role, thus closing the outer control loop.

Human/machine interface

The second part of the OPIM describes different modes of communication provided to the
operators. In process industry extensive parts of the total information flow passes through
several computer systems before reaching their destination (as in figure 1, index 4).

Manufacturing operators, on the other hand, will frequently interact *directly* with the manufacturing process (index 1), thus the demand for an expansion of the term *Human/ machine interface*. Operators may also communicate with the processes using mechanical or powered tools (index 2), or through computer systems closer to the process, i.e. robot controllers or PLCs (index 3). Information may travel any combination of routes depending on the task at hand and the equipment available.

Manufacturing processes

Any complex manufacturing system in operation may be divided into subsystems, or processes, e.g. the automatic machining of a part, the quality control of a product, or the manual assembly of two parts. The processes may be categorised into fully automatic, semiautomatic, and manual processes. The three categories, or combinations thereof, are all present in a majority of manufacturing systems. Research shows that manual processes alone, may occupy an operator as much as 50% of the time [2]. The manual, automatic, and semi-automatic processes may receive orders, or feed back sensor signals, through various means of communication described by communication arrows in figure 1.

The OPIM serves as a base for further research on how to efficiently support operators in various kinds of discrete parts manufacturing, e.g. Computer Integrated Manufacturing (CIM), Flexible Manufacturing Systems (FMS), and Flexible Automatic Assembly. Support may be added in several ways, varying from sophisticated technical systems to specialised on-the-job training.

Case studies

The OPIM has emerged during case studies in Swedish industries. A CIM system and an FMS have been thoroughly studied by a crossdisciplinary research team. While carrying out the cases, a general data collection method has been established [3]. The method aims to isolate work tasks that are considered complicated by the operators, and to roughly specify the requirements to support a specific task. Data collection employs a five-step method in order to systematically extract useful information.

To briefly describe the data collection method, step one identifies operator competence levels and attitudes, and general problems in the manufacturing system. In step two the operators' work tasks are identified in some detail. In step three the operators are asked to grade their work tasks by level of difficulty. The more difficult tasks are grouped in three categories: Tasks that are generally considered complicated, tasks that have a long learning curve (some operators consider it difficult, others easy), and tasks that are known only by a few individuals. Each of these categories can be considered a threat to productivity and reliability in the manufacturing system. In the fourth step all operators are interviewed individually about the selected tasks. In the fifth step the selected tasks are evaluated in relation to the operator roles in OPIM and in relation to the three levels of human behaviour (skill-, rule-, and knowledge-based behaviour) as defined by Jens Rasmussen [4]. Skill-based tasks should preferably be supported by on-the-job training, while knowledge-based tasks would need a more general education to support the operators. Rule-based tasks seem to be suitable for computerised decision support (e.g. expert systems) or detailed checklists.

Discussion

The operator/process interaction model in combination with the data collection method will pinpoint vulnerable factors and simplify the implementation of decision support in advanced manufacturing systems. The two also provide a framework for cooperation with software developers and management, by delivering rough specifications for decision support that will actually help the operators in a specific, complicated situation. There is otherwise a considerable risk that sophisticated that software is provided to the operators, either complicates the situation further, or does not give them support at an appropriate level to be of any help.

Experiences from the case studies are positive. Results from the FMS study are presently being implemented and the method is being used as a design tool in the development of a new automated flexible assembly system. To demonstrate the benefits of rule-based decision support a prototype has been implemented in an experimental CIM cell at Chalmers University. The prototype, Albert, is a rule-based expert system providing support for several of the operator roles described in OPIM [5] [6] and the experiences with Albert are so far very good.

Conclusion

A model for the interaction between operators in modern manufacturing systems and the manufacturing processes has been presented (OPIM). The model is based on research from continuous process industry, and case studies performed in Swedish industry seem to verify the model. A five step data collection method has been used in the case studies as an operational tool when defining training, education, and decision support needs for a specific manufacturing system.

Acknowledgements

The authors would like to thank Prof. N. Mårtensson and L. Mårtensson. Acknowledgement is also due to the National Swedish Board of Technical Development (NUTEK) and the Swedish Workhealth fund, who have generously sponsored the project.

References

[1] T. B. Sheridan, Supervisory Control. In: G. Salvendy (Ed.), Handbook of Human Factors. Wiley, New York, 1987, pp 1243-1263.
[2] B. Edgren, G. Johansson, L. Mårtensson, The Job Content in Flexible Manufacturing Systems, FMS. *Manufacturing Systems*, Vol.19:4 (1990) 303-308.
[3] J. Stahre, Evaluating Human/Machine Interaction Problems in Advanced Manufacturing. (Submitted for publication) In: *Computer Integrated Manufacturing Systems*, ISSN 0951-5240, Butterworth—Heinemann
[4] J. Rasmussen, Outlines of a hybrid Model of the Process Plant Operator. In: T.B. Sheridan, G. Johannsen (Ed.) Monitoring Behaviour and Supervisory Control. Plenum, New York, 1976, pp 371-383.
[5] J. Stahre, A. Johansson, Operator Decision Support in Manufacturing Systems — Error Detection and Restart using an Expert System. In: Proceedings of Robotics Workshop. ISBN: 91-7871-136-3, Linköping University, Linköping 1993.
[6] A. Johansson, J. Stahre, Albert — a Decision Support Tool for Operators in Manufacturing Systems. To be presented at: 4th Int. Conf. on Human Aspects of Advanced Manufacturing and Hybrid Automation, IOS Press, Manchester, 1994.

Advances in Agile Manufacturing
P.T. Kidd and W. Karwowski (Eds.)
IOS Press, 1994

Development of a Novel Method of Knowledge Engineering for the Creation of Hybrid Automation

Mica R. ENDSLEY
Thomas M. ENGLISH
Muralidharan SUNDARARAJAN
Texas Tech University, Lubbock, TX 79409 USA

Abstract. The problems of elicitation and acquisition of expert knowledge are reexamined in light of a theory of situation awareness. An approach to knowledge engineering is proposed in which a methodology for assessment of situation awareness is used to obtain information on experts' perceptions of the environment and on the decisions they make in various perceived situations. The perceived situations and corresponding decisions are used to define well-structured, constrained learning tasks for artificial neural networks or other learning architectures. A limited study illustrating some aspects of the proposed approach to knowledge engineering is described. In this study, a highly-constrained neural net was trained to emulate the decisions of expert fighter pilots as to which potential targets in a simulated radar display should be attacked. In testing, the net achieved test-set accuracy of 74%, which is identical to the average obtained when the experts were scored in the same manner.

1. Introduction

For the past decade, considerable effort has been directed toward the development of decision support systems as aids in a variety of settings from medicine to the operation of numerous complex systems. A continuing problem in this effort, however, has been the inability of researchers to develop robust decision models for these systems that duplicate the decision making capabilities of "experts". This has largely been due to problems in ascertaining just how experts go about performing highly complex cognitive functions.

Difficulties in the knowledge engineering process have been well documented [1-5]. A primary difficulty stems from the fact that expert knowledge is generally deeply embedded in complex mental models or schema, and, as such, is difficult for experts to verbalize [6, 7]. In reporting on mental events, people seem to be able to verbalize *what* they would do, but not necessarily *why* they do it. While some human cognition appears to be amenable to expression in the if-then rule format commonly used in expert system coding, much decision making, particularly that of experts, appears to rely upon the retrieval of set formulations of actions from memory based on a high level understanding of the situation. The processes used are below conscious awareness in these cases. Unfortunately, as experts are usually the subjects of interest in system development work, this problem poses a major stumbling block.

Current approaches to dealing with this problem in knowledge elicitation are also not without their difficulties. The use of verbal protocols and process tracing techniques during real-time decision behavior help to get at more of an expert's knowledge, but unfortunately carry the burden of analyzing large volumes of poorly articulated data. Another major approach has been to remove the knowledge engineer from the process by allowing the expert to interact directly with a computerized tool for eliciting and modeling knowledge. Unfortunately this technique merely sweeps the verbalization problem under the rug without dealing with the inherent difficulties faced by the expert. Hence, an alternate method of retrieving and modeling the knowledge of experts is greatly needed.

This paper explores the use of a novel technique for obtaining decision related information from experts and the application of computer learning techniques to model that knowledge to create decision support systems. The results are applicable to a wide range of problem areas, including intelligent control systems, advanced manufacturing systems, power plant control rooms and aircraft flight operations.

2. Approach

The approach explored focuses on effort in two major stages: *knowledge elicitation*, in which data is obtained on experts' assessments of domain situations, and *decision modeling*, in which machine learning techniques are applied to the acquired data to infer models of decision processes.

2.1. Knowledge Elicitation

The basic approach concentrates on modeling the "mental model" of the human expert by assessing the state of that model over time. Mental models have been defined as "mechanisms whereby humans are able to generate descriptions of system purpose and form, explanations of system functioning and observed system states, and predictions of future states" [8]. These models typically embody a person's understanding of some system, and are used to guide decision making associated with that system.

A situational model has also been posited [9, 10]. A situational model can be defined as "a schema depicting the current state of the mental model of the system" [11]. This situational model has been more widely called situation awareness (SA). SA is defined as a person's "perception of the elements in the environment within a volume of time and space, the comprehension of their meaning, and the projection of their status in the near future" [12]. Situation awareness is the main precursor to decision making in any complex and dynamic environment. It encompasses not only an awareness of specific key elements in the situation, but also a gestalt comprehension and integration of that information, along with an ability to project future states of the system. These higher levels of SA are critical to effective functioning in complex environments.

The situational model captures critical features of the mental model in various states. Recently, a technique has been developed and validated for measuring the state of a person's situational model [13, 14]. The Situation Awareness Global Assessment Technique (SAGAT) assesses operator SA during dynamic simulations of the task environment. Using SAGAT, the simulation is stopped at random times and operators are asked questions to determine their SA at that particular point in time. Their responses are normally compared to what is actually happening in the simulation to determine the accuracy of operator perceptions, thus providing an objective measure of SA.

A new application of this technique is to use it to reconstruct the operator's mental model. The use of the random sampling technique insures that a wide variety of situations, or states of the mental model are examined. The questions asked of the subject with SAGAT merely require the reporting of perceptions of the state of system elements and the resultant goal-directed conclusions drawn from these perceptions. This type of data is generally readily reportable by subjects [11, 15]. Subjects are not required to report *why* certain conclusions were drawn or the processes used in arriving at them which is the type of knowledge that subjects have historically been very poor at reporting [7]. SAGAT merely requires subjects to state what they know (decision inputs) and what they would do (decision outputs). As SAGAT records the inputs and outputs of these processes at numerous states, it should be possible to reconstruct the decision processes themselves from that data.

In addition to providing a method of getting at embedded knowledge, this approach provides further useful insight into the expert's cognitive processes. By measuring the subjects' perception of the situation it may be possible to more accurately model the link between those perceptions and resultant decisions than is possible from the actual situations themselves. In particular, these perceptions include information about how experts go from many pieces of independent data to an integrated comprehension of the situation. As much of expert decision making will be based on this higher level understanding, it is a key step in the modeling process that would otherwise be missed.

A second benefit of this approach derives from the fact that when situation awareness is poor (or partial), regularities in the relationship between *perceived* situations and decisions should be much stronger than regularities in the relationship between *actual* situations and decisions. Learning how an expert handles situations is simpler if one perceives the same situation as the expert. This issue is particularly relevant in highly complex and dynamic systems, where operators must switch their attention between parts of the situation. Switching of attention can lead to misperceptions in parts of the situation other than that currently focused upon. The gap between perceived and actual will increase as the complexity of the situation increases.

2.2. Decision Modeling

The use of SAGAT for collecting this type of data in dynamic systems is well developed and tested. The development of a working computer model from this data depends on an effective machine learning tool. A wide range of techniques for learning decision rules from input-output pairs exists. In the last five years, artificial neural networks and associated learning procedures have been applied effectively to a wide range of problems. Sensational improvements over traditional techniques have been obtained in processing of sensory data, in tasks such as speech recognition and handwritten letter classification [16, 17]. Neural networks have also been used as substitutes for major components of conventional expert systems. A common application of neural nets in these hybrid systems is to determine which of a set of rules should be applied in a given situation [18].

3. Empirical Results

Research partially investigating the efficacy of this approach was recently conducted. Ten expert fighter pilots were presented with simulated radar displays indicating locations of enemy aircraft. As part of a SAGAT data collection process, each pilot indicated for each display his perception of the situation and which of the aircraft he would attack. The objective was then to train a neural net to make the same decisions. The networking approach transforms enemy range-and-azimuth data into values called "threat factors." Threat factors are presented first to a neural net which selects large regions of the display where enemies will be attacked. For each large region of attack, refined threat factors are presented to a second neural net, which selects small regions of attack. This system has been evaluated by presenting it with displays it has not previously "seen," and determining for each display whether the decision matches that of at least one of seven experts. The system's accuracy of 74% was identical to an average of 74% obtained when the experts' decisions were scored in the same way.

The study also illustrates that a limited set of training data supports the learning of only a limited number of model parameters. By decomposing the decision into two steps, we were able to train two small nets instead of one large net, and the total number of learned parameters was small relative to the size of the training set. In general, data collection requirements can be minimized by designing a learning system with *inductive bias* — a built-in propensity to learn decision strategies that are reasonable in the particular problem domain. Useful inductive bias can be supplied by the system designer through a moderate degree of familiarity with decision-making in the problem domain. For instance, the designer might determine from the experts that certain situation and decision variables are important and can construct the learning model to pay particular attention to these characteristics in the presented data. This is knowledge of the *form* of the situation-to-decision map, rather than knowledge of the map itself, and is data which is fairly easily obtainable from the experts.

Overall, this effort confirmed the feasibility of a machine learning approach for creating effective models of human decision making from SAGAT data. In addition, several principles for engineering effective learning systems were established that should be applicable to the development of future systems.

4. Conclusion

The general idea of using machine learning to obtain knowledge that experts cannot articulate was stated more than ten years ago [19]. The novelty (and strength) of the approach we propose to investigate is that it exploits the theory of situation awareness in assessing human knowledge representation within a rapidly changing and dynamic system. SAGAT furnishes a method for determining the inventory of situation variables in a particular domain and for subsequently collecting the data to be supplied to a learning system. This provides data which is not otherwise available and takes advantage of the benefits of machine learning within the overall enterprise of knowledge acquisition.

5. References

[1] A. Barr and E.A. Feigenbaum, The handbook of artificial intelligence. Pitman Books, London, 1982.

[2] M.H. Chignell and J.G. Peterson, Strategic Issues in Knowledge Engineering, *Human Factors* **30**(4) (1988) 381-394.

[3] A.L. Kidd and M.B. Cooper, Man machine interface issues in the construction and use of an expert system, *International Journal of Man Machine Studies* **22** (1985) 91-102.

[4] A. Kidd, Knowledge acquisition for expert systems: A practical handbook,Plenum Press, New York, 1988.

[5] D.A. Waterman, A guide to expert systems. Addison-Wesley, Reading, MA, 1986.

[6] S.E. Dreyfus, Formal models vs. human situational understanding: Inherent limitations on the modeling of business expertise. Operations Research Center, University of California, Berkeley, 1981.

[7] R.E. Nisbett and T.D. Wilson, Telling more than we can know: Verbal reports on mental processes, *Psychological Review* **84**(3) (1977) 231-259.

[8] W.B. Rouse and N.M. Morris, On looking into the black box: Prospects and limits in the search for mental models. Center for Man-Machine Systems Research, Georgia Institute of Technology, Atlanta, GA, 1985.

[9] J. Roschelle and J.G. Greeno, Mental models in expert physics reasoning. University of California, Berkeley, CA, 1987.

[10] T.A. VanDijk and W. Kintsch, Strategies of discourse comprehension. Academic Press, New York, 1983.

[11] M.R. Endsley, Situation awareness in dynamic human decision making: Theory, in *First International Conference on Situational Awareness in Complex Systems.* Orlando, FL, 1993.

[12] M.R. Endsley, Design and evaluation for situation awareness enhancement. In: Proceedings of the Human Factors Society 32nd Annual Meeting, Human Factors Society, Santa Monica, CA, 1988, pp.97-101.

[13] M.R. Endsley, A methodology for the objective measurement of situation awareness, In: (ed.) Situational Awareness in Aerospace Operations (AGARD-CP-478), NATO - AGARD, Neuilly Sur Seine, France, 1989, pp. 1/1 - 1/9.

[14] M.R. Endsley, Predictive utility of an objective measure of situation awareness. In: Proceedings of the Human Factors Society 34th Annual Meeting, Human Factors Society, Santa Monica, CA, 1990, pp.41-45.

[15] M.R. Endsley, Situation awareness in dynamic human decision making: Measurement, in *First International Conference on Situational Awareness in Complex Systems.* Orlando, FL, 1993.

[16] T.M. English and L.C. Boggess, Back-propagation training of a neural network for word spotting. In: Proceedings of the ICASSP-92: 1992 IEEE International Conference on Acoustics, Speech and Signal Processing, IEEE, New York, 1992, pp.357-360.

[17] T.M. English, M. Gomez-Gil, and W.J.B. Oldham, A comparison of neural network and nearest-neighbor classifiers of handwritten lower-case letters. In: Proceedings of the 1993 IEEE International Conference on Neural Networks, New York, 1993, pp.1618-1621.

[18] S.I. Gallant, Connectionist expert systems, *Communications of the ACM* **31** (1988) 152-169.

[19] D. Michie, The state of the art in machine learning, In: D. Michie (ed.) Introductory readings in expert systems, Gordon and Breach, London, 1982, pp. 208-229.

Advances in Agile Manufacturing
P.T. Kidd and W. Karwowski (Eds.)
IOS Press, 1994

Investigating CNC Lathe Usability Issues using Verbal Protocol Analysis

Sanjay Batra[1], Ram R. Bishu[1], and James D. McManis[2]

[1]Center for Ergonomics and Safety Research
[2]College of Engineering Machine Shop

University of Nebraska-Lincoln
175 Nebraska Hall
Lincoln, Nebraska, 68588, USA
(402) 472-3495
Fax: (402) 472-2410

Abstract: In this investigation, we examined issues of CNC usability by a verbal protocol analysis observational study. Five experienced machinists performed machining tasks during two experimental sessions with retrospective and concurrent protocol analysis. The important findings include: (1). programming tasks are complex and have a very high cognitive load, (2). the CRT display is vital to process monitoring because machinists have poor perceptual access to the machining compartment, (3). machinists have incomplete mental models of the machine tool controller, and (4). poor interface design constrains machinists from engaging in exploratory learning behavior.

INTRODUCTION

A consequence of automation is that systems are more sophisticated and complex. The nature of worker's jobs on the shop floor have drastically been altered by technology. It is well known that Computer Numerical Control (CNC) has dramatically affected machinists' jobs. No longer do machinists need to control the operation sequence and timing, instead they must setup, program, adjust the machine tool and supervise its operation (Corbett, 1985). Machinists are required to perform trivial infrequent tasks, such as material handling and cleaning the machine, and in contrast they are required to respond to ill-structured novel situations. It is obvious that the cognitive load of CNC tasks is greater, while reliance on the traditional skills of machinists is lower. To answer present and future challenges that advanced technology present on the shop floor, we advocate a cognitive ergonomics approach to machine tool design. Such an approach should examine machinists' prior knowledge or mental models, existing manufacturing skills, and cognitive and physical capabilities. The key component of the machine tool is the operator-machine interface. We feel that a machine tool interface has potential to enable machinists to develop the abilities to use the machine and to facilitate effective error handling by filtering, organizing, and displaying pertinent information. This paper reports our initial efforts in examining machine tool usability issues by using verbal protocol analysis (VPA) to examine how machinists' behave with the interface.

BACKGROUND

Researchers examining operator-machine issues in discrete parts manufacturing have suggested general design principles or guidelines for designing operator interface of CNC machines (Bullinger, et al., 1984; Corbett, 1985; Carus, et al., 1993). They imply that: (1). operators are better suited for handling uncertain situations and performing problem-solving activities, (2). operators must have conceptual knowledge of the system for effective intervention, (3). information should be presented in a manner compatible with human memory; (4). humans often learn by trial and error and exhibit exploratory behavior; and (5). the machine tool should allow for multi-modal sensation.

Our approach to analyzing machine tool usability involves empirical evaluation. Instead of the more traditional performance based behavioral paradigm, we are employing verbal protocol analysis (VPA) which involves the analysis of data from subjects "thinking aloud". Protocol analysis is becoming a very popular method and has been applied in a number of fields including cognitive sciences (Ericson and Simon, 1984), process control (Sanderson, 1990), and software usability (Bowers and Snyder, 1990).

METHODS

Five experienced machinists (A through E) from four different local companies participated in this experiment. All machinists had a least 2 years of experience with CNC. All experimentation was performed on a Mazak CNC lathe in the College of Engineering and Technology Machine Shops at the University of Nebraska-Lincoln. Two video cameras were used to record machinists' behavior and verbal protocols. The task set for each session had one "warm up" part to machine from an existing program already in the machine and a part that the machinists had to write, debug, and run. Each machinist was given the task instruction packet and was instructed to perform the task as independently as possible. In session one, the machinists verbalized while reviewing the videotape (retrospective VPA). Machinists came back for session two within 3 days of session one. We video taped them bringing up the warm-up part and running it while performing concurrent verbal protocols. We encoded verbal protocols into categorizes that include: (1) procedural, (2) problem-solving, (3) prior conceptual knowledge, (4) strategic, and (5) help verbal protocols.

RESULTS AND DISCUSSION

The results in Table 1 indicate that these processes are highly task dependent. Figure 1 shows the mean frequency for all procedural protocol sub-categorizes by task. We observed significantly more monitoring or checking against a goal state protocols (PL2) when machinists were running a part and debugging a program graphically. There were three distinct graphics debugging tools: (1). figure check, (2). tool path check, and (3). simulation. Although all machinists used the figure check, only two of them verbalized in detail about it. Surprisingly only machinist B made intermediate checks of the figure shape during programming while the others only checked the figure shape of the program when they had completed the entire program. Perhaps this is because the graphic debugging tools are a relatively new development in CNC programming; machinists may be clinging to old practices and not using these tools optimally. The results show that machinists progressively look for more subtle parameters in programming as they progress through these three graphic debugging tools. There was a great deal of individual differences in the interface display information that they did monitor. Each machinist seemed to have a unique monitoring model.

When machinists were programming they verbalized more single decision questions (PH1) and decision answers (PH2). This was due to the dialog structure of the Mazatrol control language of CNC lathe. They are prompted to provide pertinent information with conversational dialog.

There was relatively little analysis of the programming problem before machinists actually started programming. Machinist A was the only one to identify all the part features during the concurrent session. This individual also programmed the part the quickest. Interestingly, every other machinist forgot to program the end facing, perhaps this is due to less complete print analysis and program planning. Machinist A had the most programming experience on the Mazak CNC lathe. Research in expertise across many domains have consistently shown that experts tend to spend more time and be more thorough in analyzing problems (Glaser and Chi, 1989).

Machinists wanted to look at another program as a memory aide. Machinist C even tried to go back to look at another program. Machinist B wanted to start with one of the existing program that was similar to the current program. This implies that machinists would find it valuable to be able to examine general programming segments to help them as a memory aide. The interface of the Mazak CNC lathe made it difficult to browse other program segments. A machinist calls up another program by remembering the program file number and the display does not allow both programs to be displayed at once. A programming code browser may prove to enhance programming performance of machinists. Another memory problems observed during programming was that machinists often forgot tool numbers which corresponded to the tool's location in the tool turret. All of the machinists experienced problems recalling the location of the tools they needed. To find this information, instead of going to the tool file that had the tool definitions, they would try to index the turret in order to see the location of the tool they needed. This approach proved to be problematic because the machine mode had to be changed in order to go from programming to indexing the turret.   Machinists seem to be less comfortable looking up information with the machine software than a mechanical means. Perhaps this is due to a lack of overall mental models of the software information structure relative to the structure of the hardware components.

Some machinists expressed problems with terminology used on the programming interface such as "Bar" for turning operations and "Edge" for end facing. The abbreviations used on the column headings of the display also caused a little confusion. In addition some of the icons caused some confusion on the part of machinists. We surveyed a number of CNC machine tools and discovered that there does not seem to be standard terminology and icons. This may present problems with training and transfer between machines.

Machinists were having problems with the status display when they were single blocking through the program. Single block mode requires machinists to manually advance through the program steps by pushing the single block button. For precautions and safety, the first part of a new program is usually done in this mode because the machinist has more direct control of the advancing tool. Four of the machinists felt extremely uncomfortable with the status display as they were running the part especially in single block mode. Often when they pressed the single step button the machine was performing a calculation machine code and it would not translate to a tool motion. There was no feedback on the CRT that the machine was processing. The screen did show the general process line that the machine was on but there was no real indication of where the tool would go if a machinist would press the single step button. This problem is compounded because of visibility problems in the machining compartment due to a small viewing window, low lighting, and dripping coolant. Machinists B and D expressed that for this activity they prefer to display more detailed G-codes so they could predict the tool motion from the display. The conversational language seems to be more abstract and does not map to detailed machine behavior in the minds of the machinists. The status display needs to have some type of predictive capabilities.

CONCLUSIONS

In this study, we identified a number of usability issues that pertain to CNC. These issues include: (1). programming tasks are complex and have a very high cognitive load, (2). the CRT display is vital to process monitoring because machinists have poor perceptual access to the machining compartment, (3). machinists have incomplete mental models of the machine tool controller, and (4). poor interface design constrains machinists from engaging in exploratory learning behavior. Our findings are based on research with a particular machine tool controller and may be limited. However, findings 2 and 4 listed above have been echoed in field case studies (Carus, et. al., 1992).

Table 1 Significant machining task dependent main effects

Verbal Protocol Category	F(6, 24)	PR>
Total Number of Protocols	16.82	0.0062*
Procedural(PRO):	14.07	0.0001*
Single Procedure or Step (PL1)	6.10	0.0005*
Monitoring (PL2)	20.82	0.0001*
Declaring a Decision Question (PH1)	13.83	0.0001*
Decision Answer (PH2)	21.04	0.0001*
Problem-Solving (PS):	6.52	0.0004*
Prior Conceptual Knowledge (PKN)	6.64	0.0003*
Strategic (STR)	3.55	0.0117*
Requests for Help (HLP)	3.87	0.0077*

F

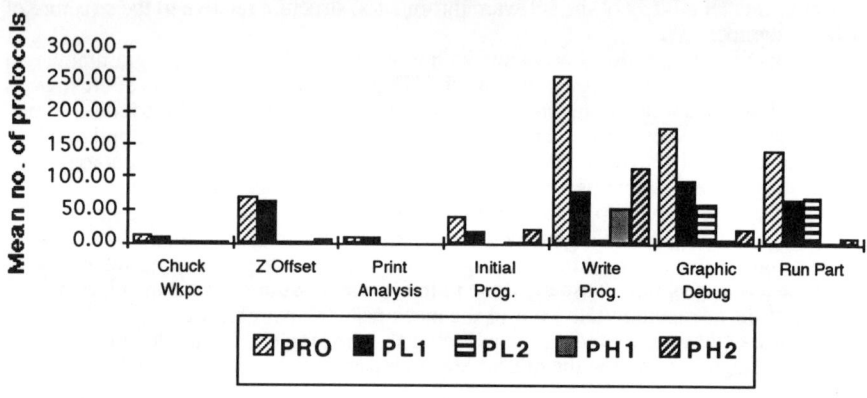

Figure 1

REFERENCES

Bowers, V. A. and Snyder, H. L., (1990). "Concurrent Versus Retrospective Verbal Protocol for Comparing Windowing Usability". Proceedings of the Human Factors Society 34th Annual Meeting, 1990, Denver, CO, pp. 1270-1274.

Bullinger, H. J., Fahnrich, K. P., and Sprenger, M., (1984), User-Oriented and Task-Consistent Programming Interfaces. In the Proceedings of the 1st Conference on Human Factors in Manufacturing, London, UK.

Carus, U., Nogala, D., Schulze, H.,)1992), "Designing CNC-Machine-Tools for Experience-Guided Working - an Interdisciplinary and Psychological Approach". In P. Brodner and W. Karwowski editors, Ergonomics of Hybrid Automated Systems III. Elsevier Science Publishers B. V., Amsterdam, pp. 237-242.

Corbett, J. M., (1985). "Prospective Work Design of a Human-Centred CNC Lathe". Behavior and Information Technology, 4(3), pp. 201-214.

Ericson, K. A., and Simon, H. A., (1984), Protocol Analysis: Verbal Reports as Data. MIT Press, Cambridge MA.

Glaser, Robert and Chi, Michelle, T. H., (1989). "Overview". In The Nature of Expertise, edited by Chi, Michelle, H. T., Glaser Robert, and Farr, Marshall, J., Lawrence Erbaum Associates, Hillsdale, NJ.

Sanderson, P. M., (1990), "Verbal Protocol Analysis in Three Experimental Domains". Proceedings of the Human Factors Society 34th Annual Meeting, Denver, CO, pp. 1290-1284.

Advances in Agile Manufacturing
P.T. Kidd and W. Karwowski (Eds.)
IOS Press, 1994

New Functionality for CNC Supporting Skilled Workers during Set-up and Automatic Cycle Phase in Manufacturing Freeform Surfaces

Prof.Dr.-Ing.habil. Dieter FICHTNER
Dresden University of Technology, Institute for Production Engineering, Chair for Production Automatization and Control Technology, Mommsenstrasse 13, 01062 Dresden, Germany

Abstract. NC programs for machining free form surfaces often contains inadequacies which are detected only on the shop floor. The programs must be modified, again in the NC office because of the lack of NC software tools on the shop floor. Worker's skill is not enough used. Time and money is lost. In order to change this situation the CNC functionality must be enhanced. The new NC functions are listed and partly illustrated in this paper.

1. Introduction

NC programs come to the shop floor in such a state that they have to be modified. This is caused by the fact that the people of the NC office do not know all the conditions existing on the shop floor, e.g. the actual blank dimensions or the reachable tools. But in the state of art in manufacturing free form surfaces one cannot change NC programs at the numerical control. For this purpose software tools do not exist. The NC programs return to the NC programming office. This process loop costs a big amount of time and money.

When there are software tools supporting the worker the above mentioned disadvantages can be reduced (see fig. 1). Besides that the attractiveness of production work will arise.

2. Investigations

Investigations were started for user support in NC machining free form surfaces.

In order to catch all the needs for a new type of CNC which will give the worker more opportunities to present his skill and to introduce it into a NC program an analysis of the current situation in enterprises was made. The enterprises enclosed in investigations belong to the industrial branches automotive industry, automotive supplier, energy machine building industry, injection mould industry, and forging industry.

The investigation was carried out by a team of researchers from some german institutions. Involved were Fraunhofer-Institut für Produktionsanlagen und Konstruktionstechnik Berlin, Gesamthochschule Kassel - Universität - Institut für Arbeitswissenschaft, Institut für Arbeitsingenieurwesen der TU Dresden, Institut für Fertigungsinformatik der TU Dresden,

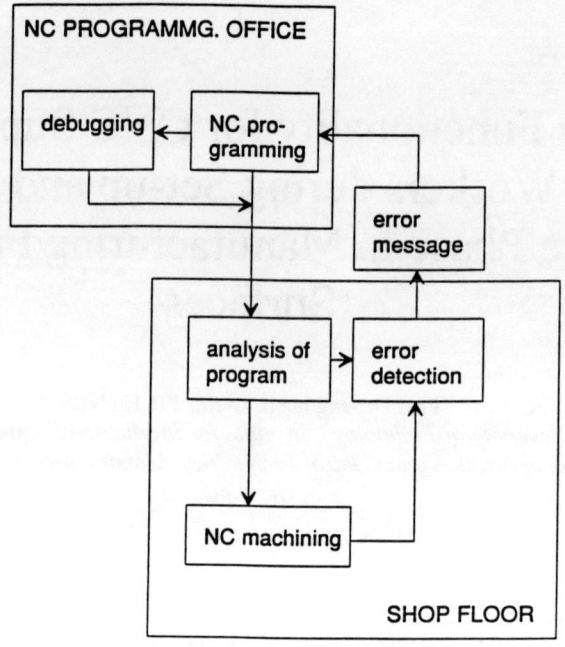

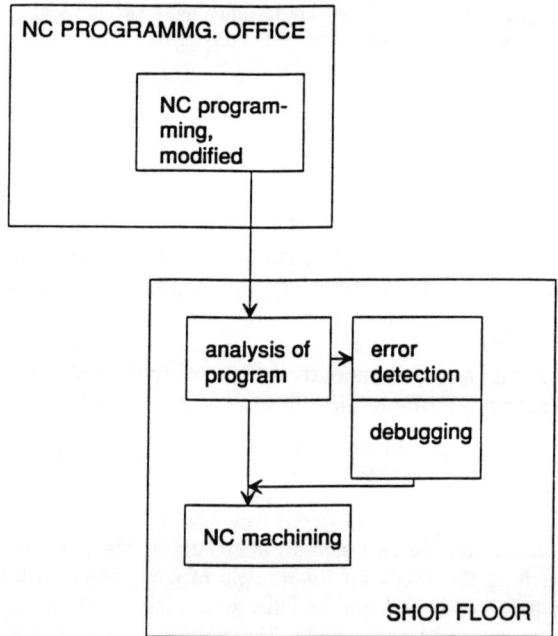

Fig. 1: NC-programming without and with CNC internal milling path generator

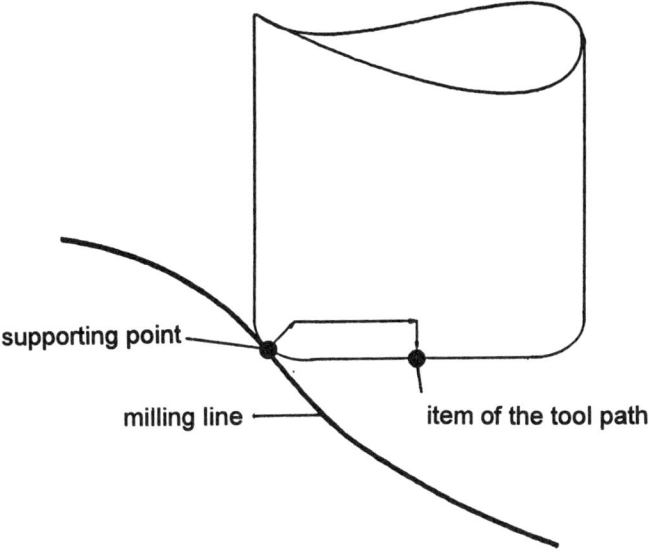

Fig. 2: Principles of tool path generation

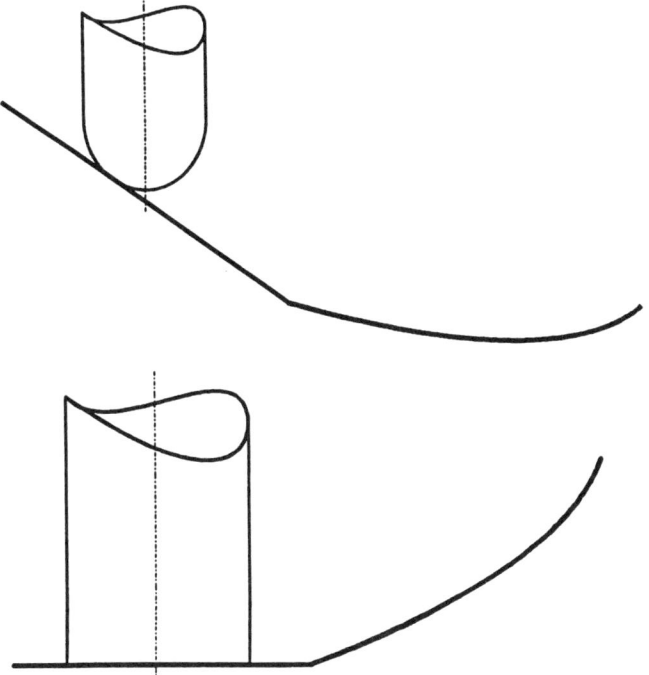

Fig. 3: Rotating the free form surface for better machining conditions

Institut für sozialwissenschaftliche Forschung München und Institut für Steuerungstechnik der Werkzeugmaschinen und Fertigungseinrichtungen der Universität Stuttgart.

The investigation was supported by the german authority Projektträger "Arbeit und Technik" and funded by the Bundesministerium für Forschung und Technologie.

The results of the analysis show that the shop floor has the opportunity to realize a higher efficiency of NC machining. 15 up to 30 per cent of the machining time can be reduced [1].

Basing on results of the investigation a range of new NC functions was implemented in the controller software.

3. New Elements of Functionality

Which are the new elements of functionality ? They are:
- generation of the milling path and coordinates for the 3 axis,
- reading the surface coefficients describing the free form shape in the standard format VDAFS,
- manipulation of surface positions (rotating, translating or mirroring) from the point of view of best machining conditions (see fig.2),
- selection of appropriate milling tools from a CNC internal catalogue,
- CNC internal generation of the milling path depending from shape and dimensions of the milling tool (see fig. 3),
- graphic simulation of machining processes, and
- pick up the same milling path after undisered interruption of machining process.

With this enhancement of CNC functionality the worker has a good mean in his hands to use his skill in preparing NC machining free form surfaces.

Using the new CNC software the amount of input data is very small in relation to a full scale NC programm. The detailed data which the control needs will be generated within the control unit using a so-called milling path generator [2][3]. In such a way the number of errors, especially of transmission errors, will be drastically reduced.

References

[1] D. Kochan (Ed.) Werkstattorientierte Nutzerunterstützung bei der Freiformflächenfertigung (Shop Floor Oriented User Support for Machining of Free Form Surfaces). VDI-Verlag GmbH, Düsseldorf, 1993. ISBN: 3-18-148502-0
[2] P. Wisniewski, CNC-interne Fräsbahngenerierung für doppelt gekrümmte Flächen (CNC Internal Milling Path Generation for Sculptured Surfaces). PhD thesis, Dresden University of Technology, Faculty for Mechanical Engineering, 1990.
[3] D. Kochan and D. Fichtner, User Support and Experiences for CAD/CAM Process Chaines for Five Axis Milling. In: Olling, G.J.; Kimura, F. (Eds.) Human Aspects in Computer Integrated Manufacturing. Proceedings of the IFIP TC5/WG 5.3 Eight International PROLAMAT Conference. Tokyo, 24-26 June 1992. ISBN: 0-444-89465-9. Elsevier, Amsterdam, 1992, pp. 639-653

Advances in Agile Manufacturing
P.T. Kidd and W. Karwowski (Eds.)
IOS Press, 1994

Override Logging - Development and Design of a New Function for CNC Machine Tools to Support Experience Guided Work

Sören STRIEPE

Institut für Arbeitswissenschaft der Universität Gesamthochschule Kassel
Heinrich-Plett-Str. 40, D-34109 Kassel, FRG

Abstract. 128 skilled workers were observed and questioned in detail about their work-action when planning and adjusting feed and cutting rates. Their work procedures were documented and the deficits of the available CNCs described. A new CNC function was developed which documents manual override changes, and which supports the worker in directly transferring these rates into the NC program. This logging function was designed and evaluated both in a laboratory and in a practical industrial setting.

1. Introduction

CNC machine tools are the most developed and widespread components in computer-integrated manufacturing. Their role as an instrument for realizing company production goals and the intentions of the individual worker in the production of work pieces, makes them the object of scientific interest.

A key problem area is lack of user-friendliness. Current systems display inadequacies with respect to their usability when applied to the actual machines on the floor. This results in additional workstrain for the operator, as well as in loss of production efficiency. To date, efforts to achieve user-friendlier systems have not been satisfactory. In particular, the support of work action in NC program compiling, testing and adaptation, and in the optimization of the cutting process, has not been achieved. Experts (manufacturers, users of the machines, ergonomic specialists) particularly criticize programming-dialogs which are too awkward, and the gap in feedback between complete pre-programming and the actual experienced cutting process. A major problem found is the lack of support available in setting cutting values (feed rate, cutting rate, depth of cutting) while compiling and running-in NC programs for CNC machine tools [1] [2] [3] [5] [7].

The goal of this study was to develop, design and test a CNC-component, which logs manual override cutting and feed rates, and which permits the direct transfer of the empirically determined technological rates to an NC program. This was to be carried out on the basis of an analysis of skilled workers' work-action when setting the rates, while compiling and running-in the NC programs.

2. Procedure and Methods

This study was caried out as a part of the CeA-Forschungsverbund (Assoc. for Computer Supported Experience-Guided Work) [6] [4]. The override logging function was developed in close cooperation with the Institut für Arbeitswissenschaft der Universität Gesamthochschule Kassel (IfA-GhK) and the Institut für Werkzeugmaschinen und Fertigungstechnik der Technischen Universität Berlin (IWF) [8].

Based on this concept of subjectifying job action [1], an analysis was made of characteristics of skilled operators' work action when setting cutting values. The focus was on small and medium sized industries and specific departments in larger industries involved in custom manufacturing and small- and medium-scale fabrication.

For the basic analysis of work action, a total of 128 skilled workers from 30 different metal working industries were interviewed. They were experienced in using CNC technology and NC programs and currently occupied in CNC turning and CNC milling. They worked with a total of 14 different CNC makes. The workers were observed operating CNC machine tools and were later questioned in partially-guided interviews conceived on the basis of their observed actions. This investigation was carried out in two steps. First, 26 of the workers were extensively questioned about their work with CNC machine tools. This was followed by a general survey of the other 102 skilled workers, who were questioned about the technical support available in setting feed and cutting rates during the compiling and running-in of NC programs.

The subsequently developed override logging component was designed and tested in two steps with the help of prototypes, both in laboratory and in practical industrial settings. These tests were carried out in collaboration with 5 skilled workers, who were observed and questioned about their work actions with the new override-logging function.

3. Observed Aspects of Work Action

Feed and cutting rates in current self- and pre-made and in past tested NC programs were tested and often adapted by the skilled workers. The testing procedure included adapting, checking and evaluating the programmed rates. Skilled workers see NC programs, and especially cutting values, as provisional, that is, subject to adjustment according to actual conditions. Cutting rates are often improved during set up with respect to aspired work piece qualities, tool wear and production times. In fact, adaptations are frequently necessary.

While running-in, the operators check the cutting values by reading the NC program on the display unit and by testing and varying the cutting conditions while machining. When checking and optimizing the cutting values during the machining process, operators orientate themselves mainly by means of directly perceptible process indicators such as noises, vibrations, chip properties and partially visual cutting sequences.

Override functions, which are used to regulate the feed and cutting rates, prove to be of central importance. Such functions allow direct manual intervention in the process and process changes become immediately ascertainable.

By varying the machining conditions, skilled workers arrive at better machining characteristics. Mainly by trying out different feed and cutting rates, by means of the override functions, workers arrive at a better coordination of these rates and the machining process, than by adopting the rates inherent in the NC program. The newly modified rates are important for later repeats and should normally be saved in the NC program. Replacing preprogrammed feed and cutting rates is difficult due to the complicated technical support available for acquiring, checking and modifying these rates.

4. Technical Concept

Promise for better technical support of work action shows a new CNC override logging function, which documents manual override changes and supports the worker in directly transferring the new feed and cutting rates into the NC program following the set up [8].

The basic idea is, that the manual override interventions be recorded and easily transferred to the NC program, following the running-in process. The technical realization requires installing two new modules in the CNC.

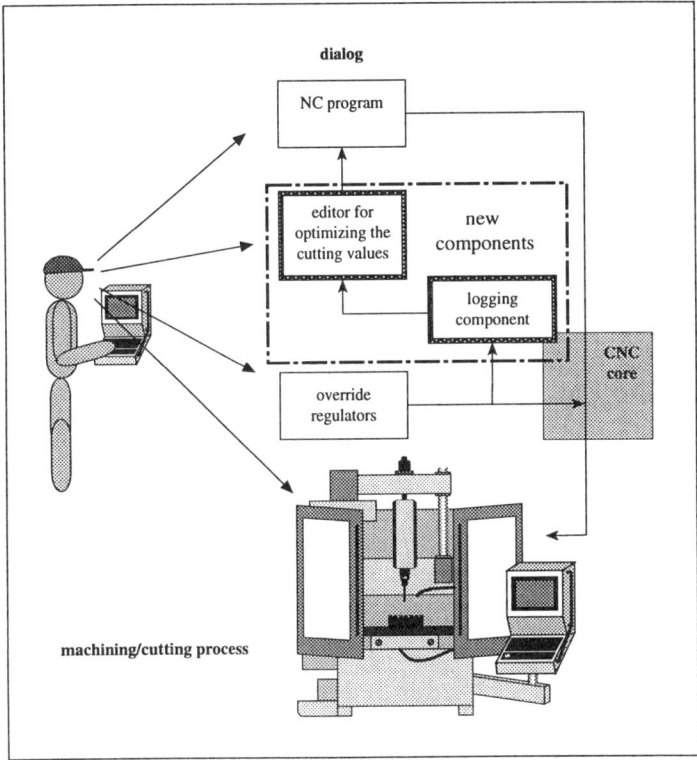

figure 1: New CNC components for the realisation of override logging

The first component is an override logbook which logs the feed and cutting rates, as well as the block number, G-functions and the actual position. Each time the override switch is activated, the component logs the rates.

The second component is an editor which facilitates coordination between the data which is to be recorded and the NC program. The NC program and the log appear side-by-side on the screen. If while running-in the NC program, the expert arrives at and regulates various optimal cutting rates (ie. a continual decrease in the feed rates while drilling), then these rates are all recorded under the respective "actual position" parameter. The block is broken down into several blocks and is shown line for line. The worker can then decide, if the run-in rates should be adopted into the NC program.

5. Testing the prototypes

The function of the override log basically suits several operations (ie. turning, milling, grinding). To begin with, a prototype was developed for turning on a Böhringer lathe with a Sinumerik 880T control and an external PC using Windows 3.1. Another prototype for milling was installed on a Hüller-Hille machining center with a Sinumerik 840M control.

The override logging prototype was first tested in a laboratory setting and later for three-months in a practical industrial setting with the collaboration of 5 skilled workers. Their work on the machines and with the new components was observed and later discussed in interviews. The problems concerning editing and the screen picture, that arose with the first prototype in the laboratory tests, were corrected.

The study showed that workers can well integrate the override log into their work action. They intensively use the override function for the feed rate while running-in the program. Afterwards, the workers review the program and log, picturing in retrospect the machining process and their interventions. In this way, they can decide on the adoption of the regulated rates. The certainty of easy and reliable logging and transfer will probably lead to less exact pre-set rates. Using the concept, machining courses and cycles only require rough cutting rates. The required degree of planning exactitude decreases as a result of improved optimizing points. The skilled workers only need to pre-write outlines or rough programs. These sketches then serve as a reference and a basis for fine tuning with the help of the override-, logging- and transfer-functions. When running-in with these components, skilled workers expect less interruptions due to complicated program changes in the editor, and less inconvenience due to memorizing or noting down.

6. Conclusion

This study on the work action involved in setting rates has revealed technical deficits of presently used CNCs. The work action in planning, testing and modifying feed and cutting rates should receive better support in future systems. Modifications to feed and cutting rates can be simplified by means of an override logging function.

References

[1] F. Böhle and B. Milkau, Vom Handrad zum Bildschirm. Campus, Frankfurt, 1988.
[2] F. Böhle and H. Rose, Erfahrungsgeleitete Arbeit bei Werkstattprogrammierung. In: H. Rose (Ed.), Programmieren in der Werkstatt. Campus, Frankfurt, 1990, pp. 11-96.
[3] A. Bolte, Planen durch Erfahrung. Institut für Arbeitswissenschaft, Kassel, 1993.
[4] A. Bolte and H. Martin (Ed.), Flexibilität durch Erfahrung. Computergestützte erfahrungsgeleitete Arbeit in der Produktion. Institut für Arbeitswissenschaft, Kassel, 1993.
[5] A. Bolte and S. Striepe, Fertigen nach Programm oder Programme nach Fertigung? In: A. Bolte and H. Martin (Ed.), Flexibilität durch Erfahrung. Institut für Arbeitswissenschaft, Kassel, 1993, pp. 29-43.
[6] Institut für Arbeitswissenschaft der Gesamthochschule Kassel (Ed.), Erfahrungsgeleitete Arbeit mit CNC-Werkzeugmaschinen und deren technische Unterstützung. Institut für Arbeitswissenschaft, Kassel, 1992.
[7] H. Martin and H. Rose (Ed.), CNC-Entwicklung und -Anwendung auf der Basis erfahrungsgeleiteter Arbeit. NW-Verlag, Bonn, 1992.
[8] U. Metzler and S. Striepe, Override-Protokollierung - eine neue CNC-Funktion, Zeitschrift für wirtschaftliche Fertigung und Automatisierung ZwF 9 (1993) 392-394.

Advances in Agile Manufacturing
P.T. Kidd and W. Karwowski (Eds.)
IOS Press, 1994

Report on Efforts to Standardize Terminology in Machine Tools

Azim Houshyar
Department of Industrial Engineering, Western Michigan University
Kalamazoo, Michigan 49008-5061, U.S.A.

Abstract. There is a need for a national database system that gives access to online information on all parts, subcomponents, and components of the machinery manufactured by machine tool builders. Existence of such a database requires the existence of a set of standard terminology for the parts being used in the machinery manufactured by the machine tool builders. This paper reports the efforts of several organizations to standardize the terminology used by the machine tool builders to identify their products and their components.

1. Introduction

Manufacturing industry has shown a great deal of interest in reducing the Life Cycle Cost (LCC) of any piece of equipment. One of the first steps in making the machinery and equipment more efficient is to institute a national database system that gives access to online information on the performance of parts, components, and subcomponents of the machinery manufactured by different machine tool builders. Development of such a database requires the use of standard terminology for different components, which in turn, demands full support of the users of machine tools to negotiate a set of standard terms for different parts, subcomponents, and components of these machinery to be used for storing and retrieving information at a national level.

A national organization was chartered to coordinate the efforts on paving the road to an improved generation of machinery and equipment manufactured by the U.S. machine tool builders. As a first step in developing a national database, the responsible organization conducted a series of workshops and brainstorming sessions to which all major machine tool builders were invited. The workshops were to acquaint the participants with the concept of reliability and maintainability and their relation with the life cycle cost and overall performance of a piece of equipment. The brainstorming sessions were to give the participants an opportunity to sort through different terminology and adopt a standard set of nomenclature for their common parts, subcomponents, and components.

As a logical continuation of the venture, the chartered organization pursued the concept of developing a database that can be used to keep track of the reliability performance of common parts, subcomponents, and components according to their source of manufacturing. The author was involved in every step of the venture, and was summoned to develop a software that would process and maintain the information on the performance of the machinery and equipment, and would generate a series of useful statistics on the effectiveness of those equipment. It is hoped that the results of this study will be a better understanding of the performances of different machinery, and a better estimate the life cycle cost of each machinery and equipment.

2. Background

The concept of up-front engineering and continuous improvements in the design and development of machinery and equipment is becoming more and more important in today's competitive world. Reliability and maintainability (R&M) are vital characteristics of machinery and equipment that enable U.S. manufacturers to be world class competitors. Predictable R&M of the manufacturing machinery and equipment is a key ingredient in maintaining production efficiency and the effective deployment of Just-In-Time principles. Improved R&M lead to lower total life cycle costs that are necessary to maintain the competitive edge. Highly reliable and maintainable production machinery offers the means for producing consistently high quality products at lower costs and at higher output levels.

Owing to an increasing demand for a means of communication between a company and its suppliers, each of the auto-industries, independently, pioneered efforts to develop reliability and maintainability guidelines for use by their purchasing agents in dealing with their manufacturing machinery and equipment suppliers, with Ford Motor Company's Guideline having been issued in January 1990, and General Motors' in October 1991.

Even though these publications along with many others were seemingly requiring similar collaborations between the company and their suppliers, they were perceived by the suppliers as another burden they had to comply with. In an effort to achieve a more harmonious and collaborative relationship, a third party played the role of assembling a committee of the representatives from all the parties involved and originated a common guideline that would be beneficial to all the organizations. In April 1992, National Center for Manufacturing Sciences sponsored a study, involving more than 50 organizations, that resulted in the publication of a *Reliability and Maintainability Guideline for Manufacturing Machinery and Equipment* [1]. The guideline embraces the concept of up-front engineering and continuous improvement in the design process for machinery and equipment. The Executive Summary of the Guideline has highlighted the direction for any future venture in improving the performance of the machinery via collaboration of builders and users:

a.   The requirements for R&M force a partnership between the supplier and user of manufacturing machinery and equipment, and the successful implementation of a R&M program requires a strong commitment from both the user and the supplier management team.

b.   Both members of this partnership must understand what equipment performance data is needed to ensure continued improvement in equipment operation and design, and must exchange this information on a regular basis.

c.   The efficient collection and feedback of equipment operation data is critical to a successful R&M program. Feedback of data provides a basis for improved R&M in new designs for future programs. And, no doubt, the users of the machinery can help the suppliers to improve their products by giving them access to their field data. The supplier also can inform the user of a predicted failure by analyzing the collected data from other users.

d.   Access to data is critical to predicting failure rates, trending, as a tool for manufacturing engineering to perform root cause analysis of repeating failures, and for providing focus to continuous improvement activities.

e.   For this reason the maintenance department and the machine operator should keep accurate and timely records of each failure, the symptoms and the corrective R&M data feedback to the supplier is important for improving the reliability and maintainability of existing designs of equipment.

3. Formation of a Common Vocabulary

As per the guideline successful implementation of R&M depends upon thorough communication between the user and the supplier. Communication must begin at project conception and continue through the entire life of the equipment to ensure that the problems are identified, root causes determined, and corrective action implemented. Attainment of reasonable levels of R&M requires planning, goal definition, a design philosophy, analysis, assessment, and feedback for continuous improvement.

Upon completion of the task of composing and publishing the guideline, many of the members of the advisory committee inquired about the means to realize some of the objectives of the guideline. The advisory committee, which consisted of a respectable representation of machine tool builders and users, was interested in striving to initiate a standard set of terminology for the machine parts, subcomponents, and components. They believed that existence of such a measure was a requisite to any data collection and exchange of information between different manufacturers and users of machine tools. This task, which subsequently evolved into a three-phase project, required us:

a.   To adopt a common set of terminology for the machine parts, subcomponents, and components to be used in storing and using data on the performance of machinery and equipment;

b.   To establish a computer software that could be used to store and maintain information and generate useful statistics on the performance of the machinery; and

c.   To initiate publication of a series of performance measures across a variety of operating environments for the leading machinery and equipment types used in automotive manufacturing.

The advisory committee postulated that the current exercise of using different terminology for the parts with similar function is obstructing any effort to accumulate and/or exchange information on the machinery parts, subcomponents, and components. And, as a logical ensuing step they voiced their desire to adopt a standard set of terminology for the common parts, subcomponents, and components. Using those terminology, the committee was interested in the development of a computer program that would store the information on different characteristics of the machinery and equipment manufactured by machine tool builders and used by auto-industry. Data include all the important characteristics of the equipment such as time between failures, duration of the downtime, scheduled maintenances, and the reasons behind failures. Hence the next envisioned phase of the study will be a compilation of the collected data into a group of reports that would be available on a restricted basis.

This paper reports the attempts that have been made to standardize the terminology used by the machine tool builders to identify their products and their components. The results, when published, will be conducive of a better understanding of the performances of different machinery, and presents a tool for the better estimation the life cycle cost of each machinery and equipment. The endorsed strategy was to choose the most common terminology to the builders. This approach resulted in several meetings with the interested machine tool builders who were notified of the intent of the project and were asked to submit a detailed list of basic structure, equipment type, devices, components, and parts manufactured in their site. Currently, the collected information is being sorted out. Based on the input from the builders, in the near future, a set of standard terminology will be selected for adoption and presented to the builders for their comments. Upon their final approval, the set of standard terms will be published and used in all future communications.

4. Construction of a Database Software

As an ensuing step in the process of developing a closer relationship between the machine tool builders and the users and to assist them improve the performance of their equipment, the author was summoned to develop a software program to be used to store and manipulate the information on performance of the equipment. The software will be used by the operators to enter the data, and by the design engineers and reliability engineers to study the system's performance.

Using dBase IV as the underlying programming language, a program is developed and tested that serves the above-mentioned objectives. The software has the following characteristics:

a.   The procedure is simple and does not require any programming knowledge. Moving from one menu to the next is by entering the required data points and following the instructions on the monitor;

b.   The procedure requires minimal data entry:

1)   Most entries are from the pop-up menus that reduces memorization of the names of parts, equipments, individuals, organizations;

2)   The operator has the opportunity to choose the current information, to look at, to edit an existing data and/or add new data points;

c.   The software generates useful statistics on the performance of the machinery, a system of equipments for a given shift and/or for a given period of time.

The software is in its final stages of pilot-test, and will be available for commercial use in a few months.

5. Discussion

This paper reports the results of an ongoing reliability and maintainability undertaking that has been proposed to improve the overall equipment performance of the machinery and equipment manufactured by the U.S. builders and used by the auto-industries. It is believed that publication and dissemination of a guideline on reliability and maintainability for the machine tool builders can be a notable step in informing the industry of the importance of R&M in reduction of overall life cycle cost of their products. Additional steps include standardization of the terminology used by different builders to identify their equipments and their components, creation of a database for storing data on the performance of those machinery and equipment and generating statistics on their effectiveness, and publication of the results of those performance characteristics (e.g. mean time between failures, etc.) for the benefit of future users of those machinery and equipments.

References

1.   Reliability and Maintainability Guideline for Manufacturing Machinery and Equipment, *Society of Automotive Engineers*, Inc., Warrendale, PA, 1993.
2.   A. Houshyar , Application of Reliability and Maintainability Methodology to Manufacturing Machinery and Equipment, *International Symposium on Robotics & Manufacturing (ISRAM'94)*, August 14-18, 1994, Hawaii.

Computerized Training of Electro-discharge Machining: Effects of Display Style

B. J. Donohue, R. R. Bishu, K. P. Rajurkar, S. Batra

Center for Ergonomics and Safety Research
and
Center for Nontraditional Manufacturing
University of Nebraska-Lincoln

University of Nebraska-Lincoln
175 Nebraska Hall
Lincoln, Nebraska, 68588, USA
(402) 472-3495
Fax: (402) 472-2410

Abstract: This study investigated the effectiveness associated with computer based training of electro-discharge machining processes. Six instructional units were developed for electro-discharge machining processes in SuperCard. Both alphanumeric and graphical versions of the interactive tutorial were created. Twenty four subjects (12 experienced, 12 novice) interacted with both tutorials during two sessions (one per session). Performance was evaluated based on questions after each unit and behavioral interaction data. Results indicated that computer based training has merit in training electro-discharge machining processes. Graphical display may help users to visualize processes that are difficult to perceive with direct observation.

INTRODUCTION

Recent advances in non-traditional manufacturing have left an educational gap. Training and skill enhancement are critical, due to the uniqueness of these processes. Transferring both skill and knowledge through personalized training has been a long sought goal. Advances in computer capabilities allow more of these processes to be carried out in a multi-media environment. This has the potential for accommodating different learning styles, by providing self paced interactive instruction.

Effective training must consist of a structure that allows for learning at three levels. Rasmussen (1983) proposed a model that describes learning behavior on three basic levels. Skill-based behavior refers to motor skill responses that require little conscious control. Rule-based behavior entails using sequences of routines based on the recognition of a familiar situation. Knowledge-based behavior is used when the situation does not have a familiar set of rules that can be used. This is a higher level or conceptual approach that is goal-oriented. Effective knowledge based behavior is dependent on the structure as well the information in that structure.

The effectiveness of models in training was examined by Mayer (1989). He compared the effectiveness of training with or without conceptual models in 8 domains. Topics included radar theory, Ohm's law, density, data bases, nitrogen cycle, camera operation, car brakes, and basic programming. His performance measures were conceptual recall, verbatim retention, and creative problem solving. Based on the results of this investigation, Mayer concluded learners with concrete models recalled more conceptual information, performed more poorly on verbatim retention, and generated more creative solutions.

Visual representations may enhance model development. Mayer & Gallini (1990) proposed that illustrations could enhance conceptual model development during learning. They showed that illustrations were effective for conceptual recall, and creative problem solving, but not for verbatim recall. The effectiveness was generally primarily obtained from novice, and not experienced individuals.

Improved quality in manufacturing is based on a better understanding of the manufacturing processes. Many of the non-traditional parameters are difficult to visualize, therefore conceptual visualization is key. Training is the facilitator for understanding. The scope of this study is to examine computer based training in non-traditional manufacturing processes, specifically electrical spark discharge machining. The objective is to compare alpha-numerical and graphical-bullet methods in computer based training.

METHODS

A study was conducted to evaluate tutorial effectiveness in two different interaction styles: (1). alphanumeric, and (2). Graphical bullet. Alpha-numerical screens provided information that was read through scrolling dialogue. Interaction was achieved through placing values in text entry fields and observing the output in another text field. Comparison between last observed results and current results were provided directly below the current value field. Graphical-bullet information was provided through sequential drawings which displayed the same concepts as alpha-numerical dialog. Cause and effect relationships were compared as before using bar graphs. Positive values were indicated upward and negative downward. The material on these screens was divided into six distinctive units of information about electro-discharge machining. Twelve experienced researchers, and twelve novice (undergraduate engineering students) subjects were tested. Half of each subject group used the alphanumeric and while the others used the graphical bullet style during the first session. They repeated the test with the style they had not used on a separate day within a week of the first session. The independent variables of this within subject experiment were the two levels of expertise, display style, and order they used the display styles. The dependent measures were three groups of three multiple choice questions. These groups were broken down into basic reading recall (Q1), cause-and-effect relationships (Q2), and extrapolation into higher level knowledge (Q3). A nine question set was asked after the learner was finished with each instructional unit (six total). The time they spent interacting with the unit, as well as the time they spent answering questions was recorded. Interaction was structured with twenty questions designed to create use of parameters. The whole exercise was conducted on a Macintosh 2ci, using a SuperCard.

RESULTS

Question responses and time for interaction data were analyzed using ANOVA. Expertise (Figure 1) was a significant main effect for total questions ($F(1, 47)=18.74$, 0.0001) and cause-and-effect relationship questions ($F(1, 47)=9.50$, 0.0001). A student Newman Keuls test indicated that experts performed better than novices with both of these types of questions. Expertise was also a significant main effect for the amount of time subjects spent interacting and answering questions ($F(1, 47)=70.41$, 0.0001). Experts spent nearly a minute less time interacting with each screen but spent about a minute more time answering questions. Display style was not a significant main effect for question or time performance. However, display style was a significant main effect for the number of times subjects recalculated parameters ($F(1,11)=9.71,0.0098$). A student Newman Keuls test indicated that subjects performed more recalculation interaction with the graphical versus the alphanumeric display style.

Figure 2 illustrates the order effect on both recall and total question performance. Order is a significant main effect for total question performance ($F(1, 11)=6.41$, 0.0278). A student Newman Keuls test indicated that subjects who started with the graphical tutorial outperformed those who began with the alphanumeric tutorial. In addition, the first order interaction for order by display style is significant for recall ($F(1, 9)=12.27. 0.0067$) and total question performance. Subjects activated the help utility significantly less ($F(1, 11)=6.91$, 0.0235) when they started with the graphical tutorial versus starting with the alphanumeric tutorial.

DISCUSSION

The intent of this investigation was to determine the effectiveness of computer based training of electro-discharge machine operators, and to determine if differences exist between alphanumeric text based and graphical representations. A number of interesting results have been obtained. Firstly, computer based interactive training may be a viable training tool for processes such as electro-discharge machining, especially given the fact that these processes lack the perceptual cues of more traditional machining processing. Secondly, experts appear to be more concerned with accuracy, than with time, although some amount of speed error tradeoff was observed in the experiments. Perhaps, the most interesting finding of this study is the interaction between order and display style, it appears that better learning takes place when people start with graphical display than with alphanumeric display. Perhaps graphical display offers better visualization in such processes. More research is definitely is needed in this area.

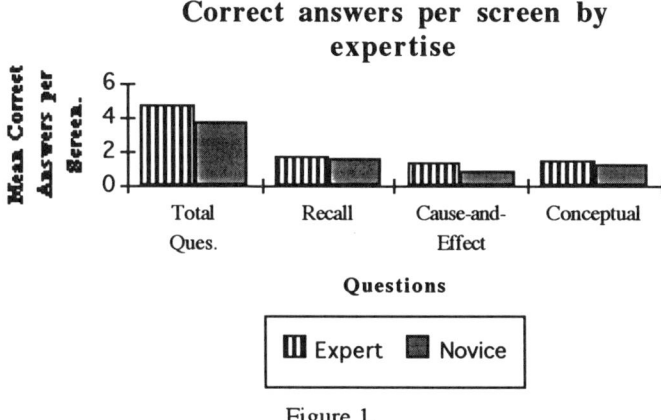

Figure 1.

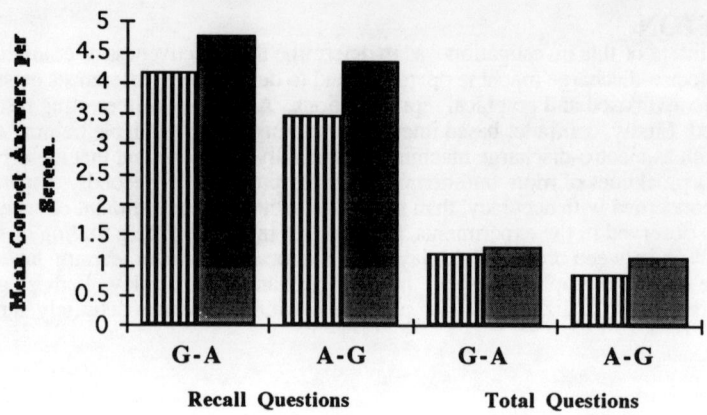

Figure 2.

REFERENCES

Mayer, R. E., (1989), "Models of Understanding", Review of Education Research, 59(1), pp. 43-64.

Mayer, R. E., and Gallini, J. K., (1990), "When is an illustration worth ten thousand words", Journal of Educational Psychology, 82(4), pp. 715-726.

Rasmussen J., (1983), Skills, Rules, and Knowledge: Signals, Signs, and Symbols, and Other Distinctions in Human Performance Models. IEEE Transactions on Systems, Man, and Cybernetics, 13(3), pp. 257-266.

Advances in Agile Manufacturing
P.T. Kidd and W. Karwowski (Eds.)
IOS Press, 1994

Shop Floor Control Systems
Supporting Teams

Dr.-Ing. Kai MERTINS, Dipl.-Ing. Martin CARBON

IPK-Berlin, Division Systems Planning, Pascalstrasse 8-9, D-10587 Berlin, Germany

Abstract. The increasing number of companies that change their shop floor organisation towards group and team work causes the need for new technical aids that are suitable for groups. Until today the cooperation and coordination of teams was supported by software only on office level by so called groupware. The short term scheduling seems to be a promising field of application for improved forms of decission support on shop floor level in order to increase productivity. Latest research about the tacit knowledge of workers for the task of shop floor control shows, that all members of a production team have their individual implicit knowledge. These pieces of knowledge have to be combined in order to optimise the production. A decision room with a new scheduling system is suggested to fulfill the demands mentioned above.

1. Introduction

Increasing globalisation of IT markets has resulted in competitors from all over the world selling their products in Europe. Many of them have the advantage of much lower labour costs. For this reason European industries must strengthen their competitiveness. The need for skilled and experienced workers on the shop floor becomes more and more important in order to handle complex tasks by using their knowledge and broad experience.

1.1. Organisational trends

Concerning the general trends, decentral concepts are very much discussed under names as segments, production islands or even fractals. In Germany there is an increase of implemented production islands by about 30% per year. Group work* is essential in this concept. It is also an important element of the virulent lean production discussion.

1.2. Gaining Experience and Cooperative Decision Making

Research about building and using of implicit knowledge in production is very new. The results of the just finished German project "Computer Supported Experience Guided Work in production - CeA" [1] dealing with the task shop floor control can be summarized by:
- All shop floor members have experience about the flow of orders
- Their implicit experience or tacit knowledge can be used in discussion processes by their experience guided argumentation
- Using the sum of experience of all shop floor members should enable the team to find an team-wide optimization of the flow of orders

---

* In Germany the term group work (Gruppenarbeit) means the positive form of cooperation within a group for a shared goal. Team work (Teamarbeit) often means the working together of different heterogenous profesions in remote locations. Angloamerican literature puts the term teams as a superior form of joint forces for a shared vision, against a group as an accidentally formed unit of people [2].

2. What technology is needed ?

The following description refers to the just started EWIG-project (Experience guided Shop Floor Control System Suitable for Groups, overall concept see figure 1), funded by the Federal Ministry of Research and Development within the programm "Arbeit und Technik". Beside researchers, the project incorporates user companies as well as vendors.

2.1. Cooperative Decision Making Suitable for Groups

Decentralized production concepts often incorporate group work and jointly taken group decisions on allocation of tasks to group members or to special machines or to special dates where they suit best the current circumstances. It is known that group work makes the best use of capacity and delivers the best decisions. On the other hand nearly no technical support is given for a joint discussion process in front of a screen or other medium of information where every team member can read and interact. Only some general attempts are made to enlarge screens or to explore the potential benefits of multimedia [3] to group decision processes.

2.2. Distributed Decision Making and Scheduling Tasks

Distributed decision making covers decisions by a remote located team at the same time or at different times being homogenious or interdepartmental and/or interprofessional. It may cover decisions such as shift of staff between groups related to a joint schedule or the primary use of material that is in shortage by one or another group or the change of a manufacturing process jointly with the quality and process planning department.

The success of cooperation in decision making often lies in the human interaction [4] that should be supported by new technical aids and not replaced. Research about the technical support for distributed decision making is under way in Europe [5] as well as in the USA [6] or in Japan [7]. The emphasis in these approaches is more adressed to IT systems design than to the improvement of work situations and is also mostly limited to office automation.

An increasing support of communication and coordination between production islands [8] is necessary to increase utilization. So the negative effects of new forms of organizations can be erased and combined with the positive aspects. Advanced manufacturing has been identified as a field suitable for computer supported cooperative work (CSCW) because some of the key issues in CIM systems design are identical to issues adressed by CSCW [9].

As well as facilitating cooperation within a work group, coordination between workgroups needs to be improved in order to reduce negative side effects such as lower utilization of some machines in production islands. To realize an optimized flow of order not only through one workshop, but through the whole company or even through different companies, new technical aids to support communication and distributed decision making are required. The supporting systems will be interconnected for that purpose.

2.3. Simulation

Simulation is available in a limited way in SFC systems. The input data for a schedule can be changed, a selection of machines or a limited time frame for the simulation can be made and the scheduling run will be performed. After that, the user can accept the results or refuse it and reset the system. It is not possible to receive results of different scheduling runs and compare them by having the different input data per version still active.

Alternative scheduling strategies are possible to achieve given factory objectives. Because of highly complex and interdependent scheduling parameters, simulation modules are required to build up experience by making it possible to investigate alternative scheduling. Simulation tools can also be used to handle the complex task of shop floor control without knowing in detail all interdependencies of all parameters. It is important that the results of alternative schedules can be compared easily, and that single planning steps can be safely undone to provide the user the possibility to explore schedules, without risking to loose data of previous and possibly better solutions.

2.4. Human Centered Approaches for Knowledge Based Modules

The new approach of building up and using implicit knowledge means that skilled staff incorporates the knowledge. Only a fraction of their knowledge is explicit and can be implemented in systems and then be used by the whole group and newcomers in the group. So any use of knowledge based modules should be handled with care and strictly as *decision support* system. Modern, human centered approaches to the development of knowledge based systems as computer assisted problem solving (CAPS) intend to provide effective user control instead of automating the decision making [10].

The use of the experience in the team can be supported further by providing tools for knowledge assimilation and representation. Using these modules, the team can be relieved from repetitive steps within the task of scheduling. The knowledge based modules will be used especially for storing information needed in scheduling, i.e. knowledge on technological sequences or special knowledge on resources.

2.5. Basic Modules of SFC-Systems and Feedback Data Handling

The advanced features such as cooperative decision support suitable for groups, simulation or representation of knowledge, being described before need basic Shop Floor Control modules. The division of systems planning of IPK-Berlin, is already working more than 10 years on concepts and realisations of short term scheduling systems [11], [12], [13]. Within several projects, shop floor control systems for different branches were developed.

The building up of tacit knowledge depends on the fact that the results of an action will be transparent to the decision making individual. Within the task of shop floor control, feedback data from the manufacturing process has to be provided to the persons in charge of scheduling. Important for this process is the balancing of the amount of data. Neither too many nor unsufficient data should be provided. Special modifiable statistic modules and aggregation modules will be developed to tackle this. Beside feedback data also current and broad data and cost calculation data has to be accessable for the scheduling team.

3. Expected Benefits

The EWIG-system will enhance workers abilities for decicion making, extend their experience and facilate coordination processes. Concept and software aims at the following economical, social and human benefits:
- Economic benefits will be visible by better reaction of companies towards customers and by minimized internal losses. This will be reached through the build-up and use of tacit knowledge on shop floor level. These benefits can be summarized by:
  + reducing due times and as a result, reducing inventory
  + shortening delivery times and reducing setup times
  + increasing utilization of machines

- Aspects of the social and human benefits are:
  + raised motivation of employees and their identification with their work
  + increased transparency of operational sequences, especially of planning sequences to support decision making on shop floor level
  + increased organizational flexibility based on better skills and experience of work groups and their advanced information infrastructure

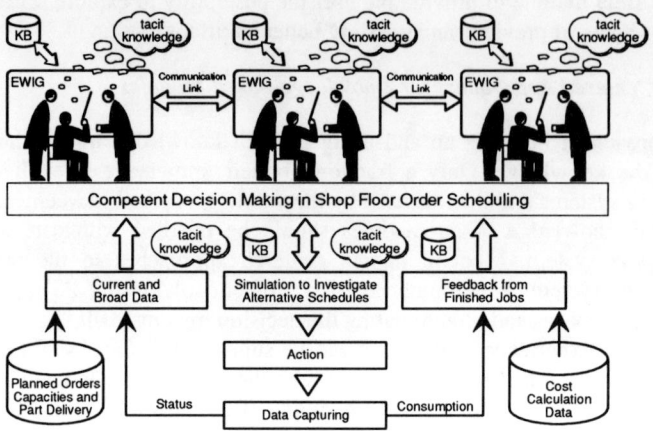

Figure 1: Overall EWIG system

4. References

[1] K. Mertins and M. Carbon: Computer Supported Experience Guided Work. In: P. Brödner and W. Karwowski (Ed.): Ergonomics of Hybrid automated systems III. Amsterdam, London: Elsevier, 1992, pp. 177-182.

[2] K. Mertins, B. Schallock and M. Carbon: Production Management Suitable for Group Work. In: M.J. Smith and G. Salvendy (Ed.): Human-Computer Interaction - Applications and Case Studies. Amsterdam, London, New York, Tokyo: Elsevier, 1993, pp. 96-101.

[3] K. Noro: Why video & hypermedia are needed as human interface in group works, its appropriate usage. In: Luczak et. al. (Eds.), "Work with display units", Abstract Book, TU Berlin, 1992.

[4] C. Savage: 5th generation management - Integrating Enterprises through human networking. Bedford MA: Digital Press, 1990.

[5] H. Fischer: Verteilte Planungssysteme zur Flexibilitätssteigerung der rechnerintegrierten Teilefertigung. München: Hanser, 1990.

[6] M. Shaw: Dynamic Scheduling in Cellular Manufacturing Systems. A Framework for Networked Decision Making. Journal of Manufacturing Systems. Vol. 7 (2) pp. 83-94.

[7] S. Ohsuga: How can Knowledge based Systems solve large scale Problems. In: H.J. Bullinger (Ed.): Human Aspects in Computing, Elsevier, 1991, pp. 801-806.

[8] M. Habich: Handlungssynchronisation autonomer, dezentraler Dispositionszentren in flexiblen Fertigungsstrukturen. LPS Bochum, diss., 1990.

[9] K. Schmidt: Computer Support for Cooperative Work in Advanced Manufacturing. The International Journal of Human Factors in Manufacturing. John Wiley & Sons, Inc., 1991, Vol.7 (4), pp. 303-320.

[10] J. Kirby: On the interdisciplinary design of human-centered systems. The International Journal of Human Factors in Manufacturing, John Wiley & Sons, Inc., 1992, Vol.2 (3), pp. 277-287.

[11] G. Spur, G. Seliger and A. Eggers: Dialogue Oriented Workshop Order Scheduling in Flexible Automated Manufacturing. In: B.J. Davies (Ed.): Proceedings of the 23rd International Machine Tool Design and Research Conference. Manchester, 1982, pp. 497-501.

[12] K. Mertins: Steuerung rechnergeführter Fertigungssysteme. Wien: Hanser, 1985.

[13] S. Reisch, F.W. Lutze, K. Mertins and R. Albrecht: Industrielle Softwareproduktion für die Fertigungsleittechnik. ZwF 86 (1991) 2, S. 65-69.

Advances in Agile Manufacturing
P.T. Kidd and W. Karwowski (Eds.)
IOS Press, 1994

Albert – A Decision Support Tool for Operators in Manufacturing Systems

Anna Johansson and Johan Stahre
Department of Production Engineering
Chalmers University of Technology
412 96 Göteborg, Sweden.
email: aj@pe.chalmers.se and jost@pe.chalmers.se

Abstract
A supervisory control model developed for continuous process industry has been
extended to include also the human/process interaction in discrete manufacturing. To
exemplify the different ways of interaction, an operator decision support tool has been
developed and implemented. This tool provides the operators with error detection and
suggestions for recovery procedures.

Introduction

A major problem in modern industry is low productivity caused by high failure rates of
advanced manufacturing systems. These failures must be corrected by the system operators, which means that the system is highly dependent on human intervention. System
failures can be blamed partly on the complexity of automation and encapsulated machines.
Therefore, the operators need tools to guide them in the understanding of the complex and
often computerized systems. Tools for situations like this have since long been used in the
continuous process industry. In order to find a general method for designing these tools, a
model that describes the situation is important.

Theoretical background

A modification of Sheridan's supervisory control model [1] for the human/process interaction in continuous process industry has been found to apply well in discrete manufacturing
[2].

The modified model called *Operator/process interaction model (OPIM)* , (figure 1),
consists of three parts. The left part is the *operator model*, which describes five roles into
which all the operators' work tasks, can be divided. The right part is the *process model*
where all possible processes are represented, i.e. manual, semiautomatic, and fully automatic processes, or combinations thereof. In the middle, the *Human/machine interface*
describes four different ways of interaction between the operator and the process.

In complex, discrete manufacturing all four of these interaction ways are represented.
Direct interaction (fig 1, interaction way 1) occurs whenever the operator needs to touch
machines or products by hand, either to adjust a work piece that is out of position, or to
repair a machine. The communication is reversed every time the operator obtains information about the system by directly looking at or listening to the process. In these situations
the operators sometimes need mechanical tools such as wrenches, screw drivers, powered
drills, measuring devices, etc. That way is marked in the model with communication
through a *tool*. (fig 1, interaction way 2).

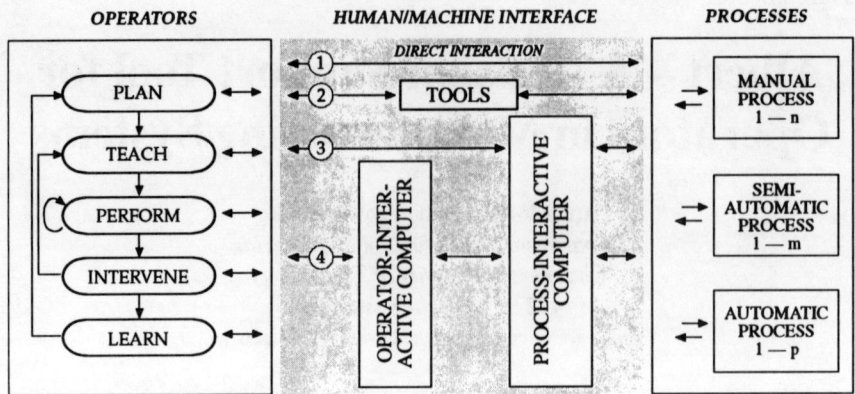

Figure 1: The Operator/Process Interaction Model

Most machines in advanced manufacturing systems are controlled by PLCs (Programmable Logic Controllers) and computers closer to the process, presented in the model by the *process-interactive computer (PIC)*. The operator quite often is forced to make adjustments directly through these control systems (fig 1, 3).

The more complex systems also have various computer-based support tools, marked in the model with the *operator-interactive computer (OIC)* , through which the operator interacts with the process (fig 1, 4 via PIC). In these, the information is collected, organized and presented in a user-friendly way. The more the interaction is made this way, the more automated and the more similar to process industry is the manufacturing system.

Result

To exemplify the benefits of the OPIM, a part of the OIC in the model has been developed and implemented in an experimental CIM cell at Chalmers University of Technology. The system is called *Albert* and supports the operator in error detection, suggestions of error recovery and restart procedures. These are work tasks that can be placed in three of the operator roles: perform, intervene, and learn. In order to give the operators tools that they will use and appreciate, the tasks that a decision support system, such as Albert, supports should be carefully chosen. This is done according to a data collection method that involves the operators who work in the system [2], [3]. The method is a way to select what tasks the operators find difficult and those tasks that often cause problems.

System description

In order to fulfil the functional requirement and to handle and support different ways of interaction presented in the OPIM, *Albert* consists of three branches [4].

The *manual search branch* is initiated by the operator and can be performed whether any of the manufacturing processes are in operation or not. The operator might see or hear something that seems wrong in the process, (fig1, 1, going left). The operator interacts with the system by answering questions about the situation, and is given more and more specific support in what he might do about it, (fig 1, first part of 4, both ways). The suggested action could be direct intervention, the use of a specific tool, or interaction with the

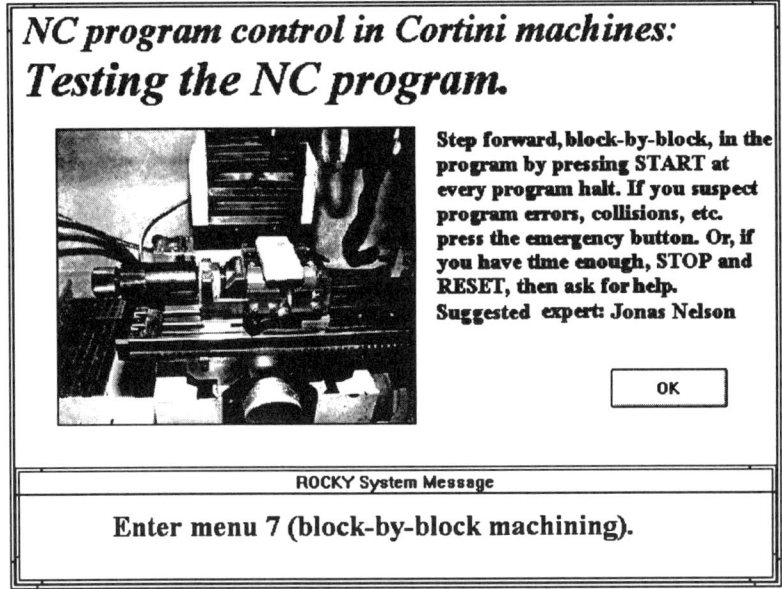

NC program control in Cortini machines:
Testing the NC program.

Step forward, block-by-block, in the program by pressing START at every program halt. If you suspect program errors, collisions, etc. press the emergency button. Or, if you have time enough, STOP and RESET, then ask for help.
Suggested expert: Jonas Nelson

OK

ROCKY System Message
Enter menu 7 (block-by-block machining).

Figure 2: An example of the operator interface in Albert

PIC. Albert does not, in itself, provide interaction directly with the process, but other support systems in the OIC or PIC can handle this way of communication (fig 1, 4, going right from OIC), e.g. a tool for planning or controlling the system .

In the *automatic branch* sensors in the process are continuously scanned, (fig 1, 4, going left). When an error occurs, a search tree automatically identifies it. The error is detected, but more information might be needed to find out what caused it and how to handle it. Therefore an error-specific subtree is called on and a manual search starts.

The third part is the *restart branch* which consists of sequentially organized instructions that can be used both for starting up the machines for a new work shift and as error recovery procedures. It gives support for the first three ways of communication.

In a complex manufacturing system many operators with different levels of skills will use the support system. In order to design Albert for skilled as well as novice operators, the information at every screen contains redundant information, so that it can be presented in several ways e.g. high resolution pictures, brief information, question or request, alternative answer buttons, and text with more detailed information (see figure 2).

Several cognitive aspects have been considered through sets of design guidelines, e.g. consistency in the placement of pictures and buttons, the amount of information on every screen, number of colors, etc. [5], [6].

Discussion

It is arguable that this kind of support tools, represented by Albert, provides good support in complex manufacturing systems. Things pointing in that direction are:
- The amount of information that has to be analyzed in a system, often exceeds what a human possibly can assimilate at one time, but can be handled by a computer.
- Some errors occur intermittently and so infrequently that it is difficult for an operator to remember what action to take. These situations can be stored in a structured way.

- The learning time to become a skilled operator, is in many cases longer than a year, which makes it important to provide not yet full-fledged operators with support. They will learn quicker and they will manage their work without frequently asking for help.

But there are aspects that have to be considered if such a support system for discrete manufacturing systems is to be successful:

- As much as 50% of the work time is spent with manual work, e.g. loading material, checking dimensions, setting up for an other product version, etc. [7]. This work can not be handled through an OIC, only supported.
- It might be inadvisable to let a system like Albert interfere with a control system by providing it directly with communication possibilities with the process (fig 1, index 4, going right from OIC).

It is important that all existing ways of interaction are represented, or at least supported, so that the operator can work and solve the problems in a natural way. This marks an important difference between support tools developed for manufacturing, where there are four different ways of interaction, and those developed for the continuous process or air/space industry where normally only one way is available.

An interesting aspect is whether the existing guidelines for development of support tools can handle this kind of interaction, or if they have to be adjusted to be more appropriate.

Conclusions

This paper has tried to show how human/machine interaction in advanced, discrete manufacturing systems can be described. As an example of one part of the theoretical model, a prototype decision support system has been developed and implemented. The prototype supports the four ways of interaction described in the model and also has shown that this kind of system provides the operators with good support.

Acknowledgements

The authors would like to thank Prof. N. Mårtensson, L. Mårtensson, the National Swedish Board of Technical Development and the Swedish Workhealth found.

References

[1] T.B. Sheridan, Supervisory Control. In: G. Salvendy, (Ed.) Handbook of Human Factors. Wiley, New York, 1987, pp 1243-1263.

[2] J. Stahre and A. Johansson, Decision Support for Flexible Manufacturing. To be presented at: Fourth Int. Conf. on Human Aspects of Advanced Manufacturing and Hybrid Automation. Manchester, IOS Press, 1994.

[3] J. Stahre, Evaluating Human/Machine Interaction Problems in Advanced Manufacturing. (Submitted for Publication) In: *Computer Integrated Manufacturing Systems*, ISSN 0951-5240, Butterworth – Heinemann.

[4] J. Stahre, and A. Johansson, Operator Decision Support in Manufacturing Systems – Error Detection and Restart Using an Expert System. In: Proceedings of Robotics Workshop, ISBN 91-7871-136-3, Linköping, 1993.

[5] B. Shneiderman, Designing the User Interface – Strategies for Effective Human-Computer Interaction. 2nd edition, ISBN : 0-201-57286-9, Addison-Wesley Publishing Co, Reading, MA, 1992.

[6] R.B. Bailey, Human/Computer Interface I and II. In: Human performance Engineering – Using Human Factors/Ergonomics to Achieve Computer System Usability. ISBN 0-13-446048-0, Prentice Hall inc., New Jersey, 1989.

[7] B. Edgren, G. Johansson, L. Mårtensson, The Job Content in Flexible Manufacturing Systems. *FMS, Manufacturing Systems*, 19:4 (1990) 303-308.

Advances in Agile Manufacturing
P.T. Kidd and W. Karwowski (Eds.)
IOS Press, 1994

Workshop-Oriented Operative Design Based on Manufacturing and Fixturing Features

B.E. Hirsch[1], K.D. Thoben[1], E. Hämmerle[2], H. Nordloh[1]
[1] BIBA at the University of Bremen, Hochschulring 20, 28359 Bremen, Germany
[2] University of Auckland, Private Bag 92019, Auckland, New Zealand

Abtract. Feature-based design as an approach to design for manufacturing has been the subject of research for many years. Feature is a generic term used by engineers to express the relation between design and manufacturing data. New attempts are leading to organisational integration of design, process planning and manufacturing. Manufacturing features allow an easy assignment of manufacturing processes to the part geometry. Fixturing features describe the basic principle of the part fixture. Part fixtures ensure that during the manufacturing process the raw material or the unmachined part is fixed in a defined position. In this paper the approach of strategic and operative design in a workshop-oriented environment and the concepts of manufacturing and fixturing features will be described.

1. Introduction

The current situation on the shopfloor is characterized by a number of unsatisfying conditions, e.g.:

- process plans not corresponding with the current reality,
- fixtures, tools and NC programs not being available in the required form,
- material not being available at the latest possible date and
- planning data in general (e.g. start and finish dates of operations, relationship between operation and resource) not being correct or complete.

These conditions are the result of process-oriented work organizations, the sequential sequence of order processing, central and deterministic planning approach, software systems that only support central planning, and the restriction allowing shopfloor personnel to perform machining tasks only.

However, in a workshop-oriented factory the production flow is organized according to the product families produced, often applying flexible manufacturing strategies and decentralized control procedures. It represents an intermediate alternative combining the flexibility of a functional factory with the efficiency and controllability of an automated factory for mass production. These characteristics have to be considered by appropriate software systems.

2.   Work Organization and Workshop-Oriented Systems

Most of the problems mentioned above are organizational-based and can therefore only be solved by a correct work organization. However, a strong relationship between work organization and the selection and design of technical systems exists. A specific work organization determines the technical system that should be used. On the other hand a particular technical system allows the development and application of a specific form of work organization.

In the past the complexity of manufacturing operations seemed to necessitate that shopfloor work be planned and controlled in central planning departments. The results were a very limited scope of decision and operation on the shopfloor. In some cases this has changed in recent years. It has been recognized that some answers to technological questions could only be given on the shopfloor. Replanning due to changed conditions can best be done by shopfloor personnel because of their manufacturing expertise [1]. Workshop-oriented systems support design, planning and control tasks performed on the shopfloor by workshop personnel. The systems make use of their knowledge and promote the enlargement of the workers' experience and knowledge. In order to do so the factory is best organized in a human-centred way. A human-centred work organization is briefly described by product orientation, a high degree of autonomy and the re-integration of design, planning and control tasks on the shopfloor for entire work contents. A workshop-oriented organization applies these human-centred principles.

3.   Systems Design Principles

Workshop-oriented systems have to provide a comprehensive functionality, an adequate data model and a reasonable user interface. Product-oriented organizations are mostly associated with the production of part families. Part family members have similarities in geometrical layout, common manufacturing processes and sequences as well as common fixturing principles. Furthermore, workshop-oriented systems should not emphasize a full automation of the planning logic. Instead the system should be a tool for information representation. Hence, an interactive procedure seems best suited where the decisions are made by the user and the system provides sufficient information. The analysis of process planning approaches has proven that a complementary way of variant process planning technique and partly generative implementation methods is favourable for workshop-oriented systems [3].

A key element of the data model is the technology-oriented product description in terms of manufacturing and fixturing features [2] [3] [4]. These also provide the basis for the planning procedures. Therefore, features will be explained in more detail in the following chapter.

3.1.   Manufacturing Features

The part model has been designed with respect to the requirements for manufacturing the part families. It utilizes a feature-based representation of part families and part instances. The feature model is based on an object-oriented modelling technique where feature types are represented in a hierarchical taxonomy of classes that can inherit attributes from their upper classes.

Complete parts are represented by a data structure containing all the features describing the part. It is a tree structure where the root of the tree is the billet node representing the unmachined part, possibly a raw material block or a casting. The billet has a number of surfaces, on which machinable features have been defined. Feature relationship is an important information needed by the process planning system to determine sequences of machining operations. Therefore, in a part family description the position, orientation and dimensions of manufacturing features should depend on the parameters of a part family. The feature definition contains a geometrical description of the feature as well as the definition of methods available for manufacturing the feature. Alternative methods can be defined for the features. A method comprises a number of work elements. A work element is the smallest unit in a process plan and represents a single machining cycle where a tool is applied to the feature. Hence, the work element description contains references to tools available and NC macros for generating the part program.

3.2. Fixturing Features

Fixturing features represent fixturing functions and a point/area of a workpiece where the fixture element and the workpiece are in contact. Fixturing features enlarge the product model with fixturing knowledge. This is useful for method planning, tool selection and collision detection. Fixturing features provide an easy way to select fixture elements. Fixturing features embody a parametric description because they are positioned relative to a surface or a feature. Whenever, for example, the workpiece dimensions are changed, then the fixturing feature position and the fixture element positions respectively are changed automatically as well [3].

4. Development of a Prototype Operative Design System

Figure 1 gives an architectural overview of the MCOES system implemented [5]; At the same time the MCOES main menu consisting of five main subsystems:

- feature-based design interface
- process plan preparation system for performing strategic process planning
- process plan generation system for performing operative process planning
- method editing interface
- factory modeller.

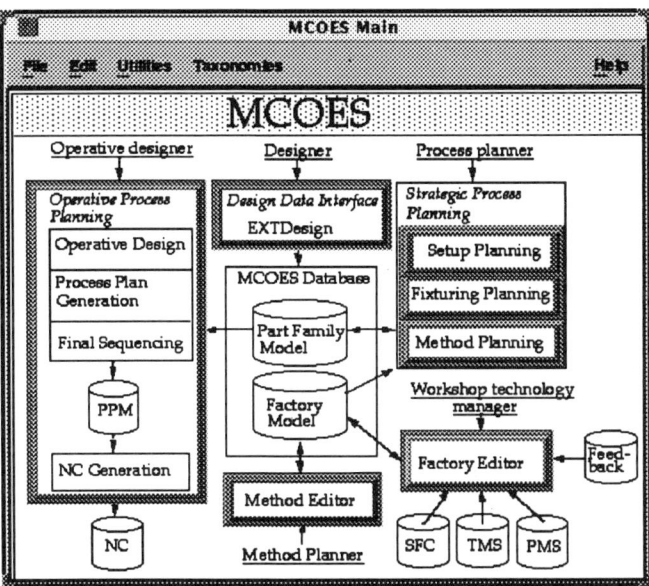

Figure 1: MCOES System Architecture [5]

With the MCOES system the generation of process plans and part programs takes place in three stages:

- The system setup stage, whereby a factory model representing factory facilities is generated and a collection of methods representing tested, proven manufacturing processes is created. At this stage, the planning focuses on a single process.
- Strategic process planning generates a global plan scheme for a given new product family. Because many decisions depend on the result of other decisions, there is no clear sequence of planning operations. Two examples might make this clear. Methods cannot be chosen without knowing the fixturing in order to avoid obstructing. Fixturing cannot be chosen without the knowledge of the processes chosen in order to consider the machining forces. Therefore, three assistant type systems — setup, method and fixture planning — have been developed for the strategic planning task. The modules can be executed relatively independent of each other and support quick user interaction with an evolving process plan.
- The overall goal of operative process planning is to generate correct process plans and NC programs in a few minutes. This high performance is a result of the preparation work done by the other components of the system. The first step of operative process planning is the "operative design". Using the part family description of the design data interface as a basis, the operator instantiates a predefined order type and modifies the default parameter values according to customer requirements. Pixmaps illustrate the meaning of the part family parameters. The result is an instantiated family model. The process plan is then generated automatically by evaluating the plan specification created during strategic planning. Based on the parametric part description the NC program is generated automatically and can be visualized and, if necessary, modified by the operator.

In reality, the three stages are all on-going parallel processes performed by product designers responsible for initial product family description, method planers defining available manufacturing processes, process planners and operative designers creating process plans and controlling the NC program simulations. The system setup and the part families are continuously changed according to the changes in the factory and the products.

5.   Conclusion

A workshop-oriented system for design and process planning has been introduced. Its integrated environment and the common data model will improve communication between design, process planning and machine operators, and it will make the decision process of the planning steps transparent. The concept of part families supports documentation and reuse of company specific manufacturing experience. The highly interactive character requires skilled users as acting nodes. Therefore, it aims to improve the performance of human resources instead of replacing them.

References

[1] H.-J. Bullinger, K.P. Fähnrich and H. Erzberger, Planen, Programmieren und Prüfen in der Werkstatt, *Technische Rundschau* 10 (1991) 26-34.
[2] J.J. Shah., Conceptual Development of Form Features and Feature Modelers, *Research in Engineering Design*, Vol 2 (1991) pp 93-108, Berlin, Springer
[3] E.Hämmerle, Werkstattorientierte Systeme zur Arbeitsplanung und kurzfristigen Fertigungssteuerung, Ph.D Thesis at the University of Bremen, (1993)
[4] J. Opas, Approaches for Computer-Aided Fixturing Planning, Analysis, Synthesis, and Implementation, Thesis for Licentiate of Technology, Helsinki University of Technology (1993)
[5] M. Mäntylä, BRITE/EURAM Project 3528 Manufacturing Cell Operator's Expert System (MCOES) Synthesis Report/Overview, Helsinki University of Technology (1993)

Advances in Agile Manufacturing
P.T. Kidd and W. Karwowski (Eds.)
IOS Press, 1994

Machine Fault Diagnosis: Operator Strategies and Performance Support

Nong Ye
Wright State University, Dayton, Ohio 45435, U. S. A.

Abstract. This paper discussed theories and techniques developed in four areas of research on fault diagnosis of advanced manufacturing systems: fault modeling and reasoning, human diagnostic strategies, diagnostic decision support by expert systems and neural networks, and on-line diagnostics. Theories and techniques in each area have their own strengths and weaknesses. A cost-effective solution to diagnostic problems relies on the integration of existing techniques.

1. Introduction

The increasing automation of manufacturing systems shifts the role of the human operator from a physical worker to an operation supervisor. When manufacturing systems run normally in an automative mode, the job of the human operator is merely to monitor production quality and detect abnormality of system operation such as system failures or malfunctions. When abnormality is detected, the human operator must quickly respond to system fault events by diagnosing the cause of fault events and carrying out remedial actions so that significant production loss can be avoided, and thus production quality is maintained at the minimum cost. However, fault diagnosis of highly automated manufacturing systems presents considerable difficulty to the human operator due to three factors: the inherent complexity of manufacturing systems, the unpredictability of fault events, and the limitation of human cognitive capacities.

To keep systems running productively, significant improvement in fault diagnosis technology is highly desirable. Researchers and engineers have been working in several areas: fault modeling and reasoning; human diagnostic strategies, diagnostic decision support by expert systems and neural networks, and on-line diagnostics. This paper discusses strengths and weaknesses of theories and techniques developed in these areas. A future direction is pointed out for a cost-effective solution to diagnostic problems. A hybrid intelligent system that follows this direction is presented.

2. Fault Modeling and Reasoning

Fault modeling is the basis of diagnostic reasoning. There are two main streams of fault modeling: the modeling of fault propagation and the modeling of system structure.

The modeling of fault propagation is to represent the cause-effect relationships between root faults, their propagation effects at different stages of system operation, and external symptoms. Root faults, their propagation effects, and external symptoms must be enumerated and pre-specified before modeling the fault propagation. Based on the modeling of fault propagation, faults can be reasoned out from observed symptoms. Many studies in this field are reported in the book edited by Rasmussen and Rouse[1]. The study by Chang, DiCesare, and Goldbogen[2] also falls in this field.

The most common representation of fault propagation is fault trees. Some variants of fault trees include event trees and cause-consequence diagrams. Many studies in fault propagation modeling[2, 3, 4, 5] focused on how to synthesize large fault trees for a system from mini-fault trees for individual system components.

The modeling of system structure and behavior is used mostly by so-called "model-based diagnosis" whose origin can be traced back to fault diagnosis of digital circuits. In general, digital circuits have straight-forward connections between components as well as simple quantitative relationships between inputs and outputs of components. The correct structure of digital circuits and the normal behavior of components can be described precisely. Since the abnormal behavior of circuit components is usually difficult to characterize, model-based diagnosis represents the correct structure of a system and the normal behavior of system components for diagnostic reasoning. The book edited by Hamscher, Console, and de Kleer[6] collects literature in model-based diagnosis.

Theories and techniques developed in the area of fault modeling and reasoning have limited applications to real-world diagnostic problems due to computational complexity. For example, if one tries to trace fault causes from fault symptoms along cause-effect paths, the number of possible paths may reach a combinatorial explosion as the size of systems increases.

3. Human Diagnostic Strategies

People often develop special skills and strategies to overcome difficulty in fault diagnosis and to achieve satisfactory performance. Many human factors studies have been carried out to explore and understand those special skills and strategies. Results of such studies may bring new insights into diagnostic problem solving.

Rouse and coworkers[7, 8, 9] studied many aspects of human performance in diagnostic problem solving, such as the transfer of context-free training to context-specific tasks and the effect of rule-based strategies. Su and Govindaraj[10] examined strategies used by experienced marine engineers in fault diagnosis. Their results indicated that subjects, who developed a good initial feasible set of possible faults and focused on a well-developed hypothesis, demonstrated better performance than other subjects who did not have a good initial feasible set of faults or kept several hypotheses at the same time. Cohen, Mitchell, and Govindaraj[11] made field observations on the fault diagnosis strategies used by operators in the test area of a printed-circuit-boards assembly line. To diagnose faults of printed-circuit-boards, operators first determined how familiar a problem appeared to them. If the symptom pattern was common, the testing information of circuit boards revealed a known problem, or a similar problem was recently experienced, then operators directly identified faults. Otherwise, operators followed generalized routes of test to locate faults.

Rasmussen[12] observed three types of strategies used by people: the symptomatic search by pattern recognition, the symptomatic search by hypothesis and test, and the topographic search. The symptomatic search by pattern recognition is to use a familiar pattern of symptoms as a template to retrieve a pair of symptoms-faults in human long-term memory. The symptomatic search by hypothesis and test is first to generate a rough hypothesis of faults first from the pattern of symptoms and then to test the hypothesis by simulating system operation to see if the hypothesis accounts for the symptoms. If not, a new hypothesis shall be generated. The topographic search is to use a topographic route of search and test to narrow down the fault area.

In general, existing studies on human diagnostic strategies provide insights into ingredients pertinent to a skilled performance of fault diagnosis. Focusing on individual aspects of human behavior, however, human factors experiments hardly reveal details that are essential to establish an operational model of human intelligence. Individual facts from existing studies need to be integrated into a coherent architecture of diagnostic problem solving.

4. Diagnostic Decision Support by Expert Systems and Neural Networks

Expert systems and neural networks represent the majority of artificial intelligent systems that were developed as human decision support. Diagnostic knowledge in expert systems can be categorized into two types: shallow knowledge that was obtained from human experience and deep knowledge that was obtained from fault trees. Among experts systems that used shallow knowledge are MYCIN[13], DART[14] and an expert system[15] for fault diagnosis of underground power distribution cables. Among expert systems that used deep knowledge are an expert system[16] for fault diagnosis of a chip-mounting machine and an expert system[17] for fault diagnosis of a nuclear auxiliary boiler feed system. However, the development of expert systems is often costly due to difficulty in knowledge transformation, knowledge transformation, knowledge generalization and knowledge update.[14, 18, 19, 20]

The architecture of neural networks overcomes the weaknesses of expert systems through parallel processing and leaning capacity. Ye, Zhao, and Salvendy[20] developed a neural network to diagnose faults of manufacturing systems using the back-propagation learning algorithm. Among neural networks using the back-propagation algorithm are ones developed by Knapp and Wang[21] and Watanabe and coworkers[22], and Venkatasubramanian and Chan[18]. Those neural networks can directly recognize faults from such symptoms as machine vibration signals and system state variables. However, neural networks that are based on the diagnostic strategy of pattern recognition are not able to distinguish among faults that share a common pattern of symptoms.

5. On-Line Diagnostics

Advanced manufacturing systems such as machine tools and electro-mechanical equipment usually adopt some form of on-line diagnostics. Two types of on-line diagnostics have been developed: pattern recognition of symptom data and state variable analysis.

For pattern recognition, symptom data (i.e., vibration signals, power consumption distributions, heating curves, etc.) under various abnormal conditions is usually collected from laboratory experiments and then used to classify symptom patterns of real fault events[23]. However, experiments of associating symptom patterns with faults are often expensive due to the amount of experimental effort and the cost of experimental materials and tools. Nevertheless, the result is often specific to tools and materials used in the experiments.

For state variable analysis, state variables are routinely measured and checked against their normal value at each step of machine operation or manufacturing process[16, 24]. Sensors for detecting operation parameters like motion, position, force, air pressure, and speed, are placed at various locations of systems. If some state variables do not have their desired value, fault trees that associate the values of state variables with faults are used to isolate problem areas. However, limited state information can help in isolating a problem area but not in identifying individual components of fault.

6. Hybrid Intelligent System for Diagnostic Decision Support

A practical approach to diagnostic decision support must be economically justiciable to promote marketing benefits from the perspectives of both the user and the maker of manufacturing systems. A full coverage of diagnostic functions by artificial intelligence is technically and economically infeasible at this stage of computing. The most promising approach is to develop a decision support system with a proper allocation of diagnostic functions between the computer and the human operator. The success of such decision support system relies on two ingredients: 1) advancement of artificial intelligent architecture by making use of cognitive-oriented intelligent models; and 2) integration of existing techniques for implementation of artificial intelligent architecture so that existing techniques can complement one another.

Ye and Zhao[19] developed a hybrid intelligent system that consisted of these two ingredients. The design of system architecture was based on a cognitive-oriented diagnostic strategy of hypothesis and test. Hypothesis generation and hypothesis testing were implemented using two separate neural networks. Decision making involved in hypothesis validation, hypothesis revision, and on-site verification of hypothesis was accomplished using a decision making algorithm. The hybrid intelligent system demonstrated highly reliable diagnosis performance for both single-fault and multiple-fault events. Its merits were also shown in knowledge generalization, fuzzy information processing, and common symptom handling.

Acknowledgments

This material is based upon work supported by the National Science Foundation under Grant No. DDM-9210523.

REFERENCES

[1] Rasmussen, J. and Rouse, W. B., 1981, *Human Detection and Diagnosis of System Failures* (Plenum Press, New York).

[2] Chang, S. J., DiCesare, F., and Goldbogen, G., 1991, Failure propagation trees for diagnosis in manufacturing systems, *IEEE Transactions on Systems, Man, and Cybernetics*, 21(4), 767-775.

[3] Fussell, J. B., 1973, A formal methodology for fault tree construction, *Nuclear Science and Engineering*, 52, 421.

[4] Lapp, S. A. and Powers, G., 1977, Computer-assisted generation and analysis of fault trees. In *Loss Prevention and Safety Promotion in the Process Industries* (Dechema, Frankfurt), pp. 377.

[5] Powers, G. J. and Lapp, S. A., 1976, Computer aided fault tree synthesis, *Chem. Engng Prog.*, 72(4), 89.

[6] Hamscher, W., Console, L., and de Kleer, J. (eds.), 1992, *Model-Based Diagnosis* (Morgan Kaufmann Publishers, San Mateo, CA).

[7] Rouse, W. B., 1983, Models of human problem solving: Detection, diagnosis, and Compensation for system failures, *Automatica*, 19(6), 613-625.

[8] Rouse, W. B., 1981, Experimental studies and mathematical models of human problem solving performance in fault diagnosis tasks. In *Human Detection and Diagnosis of System Failures* by J. Rasmussen and W. B. Rouse (eds.) (Plenum Press, New York), pp. 199-216.

[9] Rouse, W., B., Rouse, S. H., and Pellegrino, S. J., 1980, A rule-based model of human problem solving performance in fault diagnosis tasks, *IEEE Transactions on Systems, Man, and Cybernetics*, 10, 366.

[10] Su, Y.-L. and Govindaraj, T., 1986, Fault diagnosis in a large dynamic system: Experiments on a training simulator, *IEEE Transactions on Systems, Man, and Cybernetics*, 16(1), 129-141.

[11] Cohen, S. M., Mitchell, C. M., and Govindaraj, 1992, Analysis and aiding the human operator in electronics assembly. In *Design for Manufacturability* by M. Hellander and M. Nagamachi (eds.) (Taylor & Francis, London), pp. 361-376.

[12] Rasmussen, J., 1986, *Information Processing and Human-Machine Interaction* (North-Holland, New York).

[13] Buchanan, B. and Shortliffe, E., 1984, *Rule-Based Expert Systems: The MYCIN Experiments of the Stanford Heuristic Programming Project* (Addison-Wesley Publishing, Reading, MA).

[14] Bennett, J. S. and Hollander, C. R., 1981, DART: An expert system for computer fault diagnosis. In *Proceedings of International Joint Conference on Artificial Intelligence*, pp. 843-845.

[15] Kuan, K. K. and Warwick, K., 1992, Real-time expert system for fault location on high voltage underground distribution cables, *IEE Proceedings-C*, 139(3), 235-240.

[16] Naruo, N., Lehto, M., Salvendy, G., 1990, Development of a knowledge-based decision support system for diagnosing malfunctions of advanced production equipment, *International Journal of Production Research*, 28(12), 2259-2276.

[17] Adamson, M. S. and Ronerge, P. R., 1991, The development of a deep knowledge diagnostic expert system using fault tree analysis information, *The Canadian Journal of Chemical Engineering*, 69, 76-80.

[18] Venkatasubramanian, V. and Chan, K., 1989, A neural network methodology for process fault diagnosis, *AIChE Journal*, 35(12), 1993-2002.

[19] Ye, N. and Zhao, B., 1993, A hybrid intelligent system for fault diagnosis by hypothesis and test, *in review*.

[20] Ye, N., Zhao, B., and Salvendy, G., 1993, Neural-networks-aided fault diagnosis in supervisory control of advanced manufacturing systems, *International Journal of Advanced Manufacturing Technology*, 8, 200-209.

[21] Knapp, G. M. and Wang, H.-P., 1992, Machine fault classification: A neural network approach, *International Journal of Production Research*, 30(4), 811-823.

[22] Watanabe, K., Matsura, A. M., Kubora, M., and Himmelblau, D. M., 1989, Incipient fault diagnosis of chemical process via artificial neural networks, *AIChE Journal*, 35(11), 1803-1812.

[23] Hong, S. Y. and S. M. Wu, 1992, Preventive diagnosis of minor machine failure by DDS spectrum analysis. In *JAPAN/USA Symposium on Flexible Automation*, San Francisco, CA, pp. 1521-1530.

[24] Hong, S. Y., 1982, *Development of an End-Effector for Robotic Drilling with On-Line Sensing and Diagnosis* (Ph. D. Dissertation of Department of Mechanical Engineering, University of Wisconsin, Madison, Wisconsin, U. S. A.).

Advances in Agile Manufacturing
P.T. Kidd and W. Karwowski (Eds.)
IOS Press, 1994

Computer aided planning of structures and processing in autonomous working group networks including the use of simulation tools

Dipl.-Ing. Mathias Monjé
Special research centre 187, PO 102148, 44721 Bochum, Germany

Abstract. One innovative potential of the production concept of autonomous manufacturing islands (A.M.I.) lies in the flexibility and qualification of the human staff. A
human orientated job design and the codetermination of the future crews are important
assumptions for a successfull implementation of such organizational structures.
Additional to this the planning of such production systems consists a large variety of
design options (for example central or decentral NC-programming) and possible
technical solutions (for example the utilization of DNC or computer networks).
This paper shows a method for a successful planning of the structure and the
processing in A.M.I. including the use of simulation tools.

1. Introduction

At present many industrial companies try themselves to restruct their production. This
became necessary to remain competitive on the market. To be able to satisfy the demands of
the customers in time some new goals have to be reached:

- the flexibility has to be increased
- the processing time has to be decreased
- the duedate assurance has to be increased
- ...

In this context the production concept of autonomous manufacturing islands (A.M.I.) is of
great promise. The main ideas are:

- utilization of teamwork
- product orientated organization, decentralization
- restauration of the ability of the worker in product- and process-innovation
- ...

For a successful implementation of such oranizational structures a good and detailed
planning is needed. For the planer of A.M.I. the turning away from old organizational
structures results in new additional goals (such as flexibilty, qualification, e.g.). To get a
high acceptance in the future teams, human orientated job design and the integration of the
workers in the planning process are indispensable.

The concept of A.M.I. is variable in a wide area and has to be adapted to the respective conditions. The following figure shows the design options for autonomous manufacturing islands.

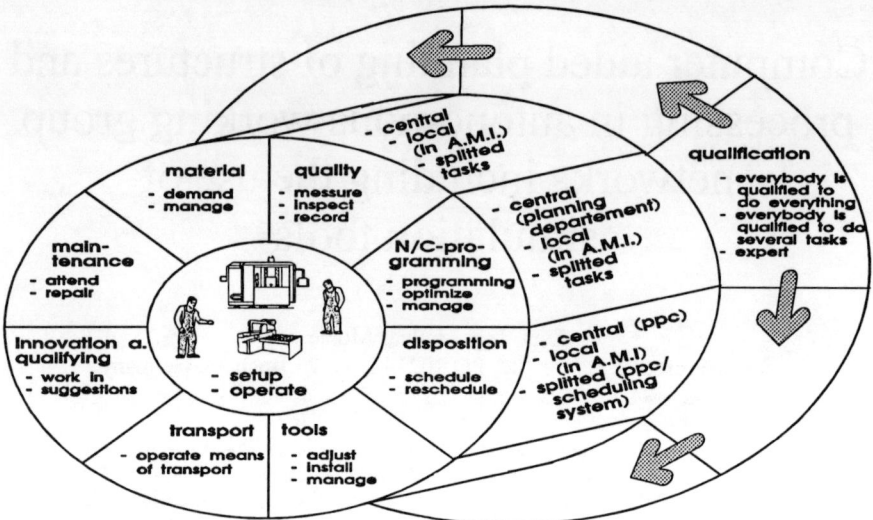

Fig. 1: Design options for autonomous manufacturing islands

All this goes to show that the planning of A.M.I. appears as a very complex problem.

Another point is that in small and mediumsized enterprises the planning competence and the planning engineers are probably not available. This circumstance requires a computer-aided planning system which supports the planning team without limiting its creativity.

The following explanation describes a planning system for the planning of such production structures developed at the University of Bochum within the special research centre 187.

2. Requirements on a planning system for A.M.I.

From the introductory comments the requirements on a planning system for A.M.I. can be specified. The following list shows the most important of them:

- the future crews should take part in the planning process
  To guarantee the acceptance of the new production structure from the future crew it is important to integrate the workers in the planning process.
- relation to the reality as accurately as possible
  To simplify the comprehension of the planning process the user should not have to abstract the reality more than necessary.
- the system must not influence the user
  If the system influences the planning team in a special direction it is questionable if a found solution is the best one.
- interim results should be used for further discussions
  The planning team should have the option to use interim results as a basis for further discussions in the planning process.

- structures of qualification must be considered
  One innovative potential of A.M.I. lies in the qualification of the human staff.
  Therfore it has to be taken into account in the planning process.
- comparison between variants of organizational structures has to be possible
  The system should be an aid to decision-making for the planner
- the planning system should be easy to use
  Because of the inhomogenous planning team (workers, planning engineers, etc.)
  the system should have an easy to use graphical desktop.

3. The modell centred and object orientated approach

To comply with the upper requirements a modell centred and object oriented approach was chosen. Object classes were built to modell a production system in the computer. These objects are very similar to the reality. The following figure shows the objects such a modell consists of.

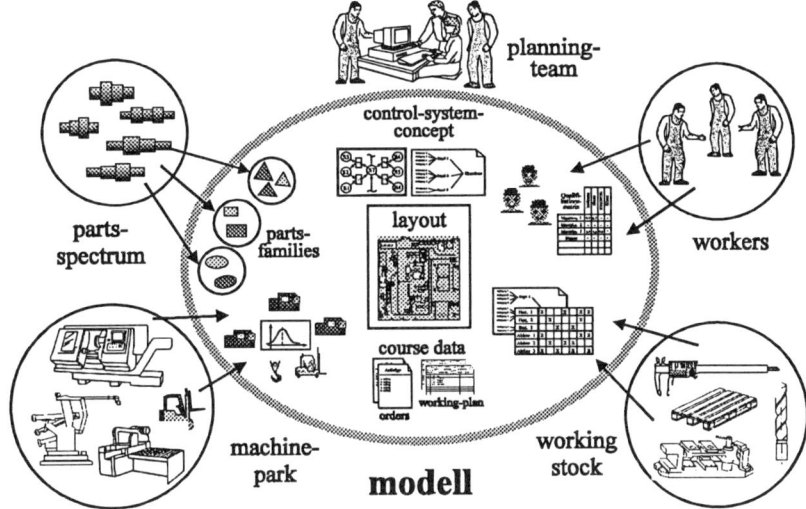

Fig. 2: Object world

This object world is the basis for the planning proces. The planning team can model several variants of production systems within the computer. This is easy because the objects are very realistic. Furthermore this case makes it easier to transmit solutions of the planning process to the reality.

Within the framework of the continuous research the new concept of properties was developed. This new concept allows the planning team to give the objects a specific behaviour. In the context of A.M.I. it is possible for example to give an object 'worker' a qualification for operating specific machines. As it was shown above this is a very important feature for planning A.M.I..

For the comparison of different model variants a simulation tool was developed. This tool is helpful to decide whether the one or another variant is the better one. With the simulation tool the planner is able to simulate the production process itself. So he gets characteristics of the modelled production system without having built it in reality. The following explanations gives more details of the planning process.

4. Objectorientated simulation as a planning instrument for A.M.I.

In the special research centre 187, Bochum an simulation system was developed which uses a modell centred and objectorientated approach. The implementaion of properties for the objects allows the planner to give machines or workers a special behavior. With these properties for example it is possible to model qualification structures of the personal staff. Therefore the system is very suitable for planning of A.M.I.. The working plans and jobs can be designed with a database editor or they can be taken from a ppc-system.

The system consists of several tools for modelling and simulation. The following text gives further details about these tools.

- structure editor
  The structure editor allows the user to design the organizational structure. The machines will be chosen here from a machine catalogue. The user can combine some machines to a group. Workers and means of transport can be assigned to machines or to machine groups. Furthermore the planner has the possibility to give the object special properties
- resource positioner
  This tool is useful for positioning resources such as machines or means of transport. Here the planner should design the layout of the factory.
- database editor
  The database editor is needed to create working plans and jobs. Also it is possible to get data from a ppc-system and to work them up with the editor.
- the simulator
  The simulator simulates the production process with the data from the other tools. As a result it delivers characteristics of the simulated production system such as processing time or duedate assurance.

5. Final conclusion

Using an objectorientated simulation tool as an instrument for factory planning gives the planning team new possibilities. For a restructuring of a production system it is possible to decide whether a concept is good or better than others without interfering the running production. For a new planning it is possibible to test different configurations without a large investment.

It may be asserted however that it is not possible to compare the simulation with a running process. To use the simulation for restructuring it is important to start with a reference simulation of the real process and to take this as a basis for the simulation of variants.

It is surely a matter of fact that simulation is one of the important techniques for the next years.

Advances in Agile Manufacturing
P.T. Kidd and W. Karwowski (Eds.)
IOS Press, 1994

Development of a Diagnosis Information System for Semi-Autonomous Production Islands

Bernhard ZIMOLONG and Udo KONRADT
Sonderforschungsbereich 187, University of Bochum, Universitaetsstr. 150,
44780 Bochum, Germany

Abstract. This paper describes the incremental development of a decision support system for maintenance tasks in semi-autonomous production islands. Based on the empirical analysis of maintenance tasks of the shop floor and user characteristics in failure diagnosis software demands are derived and implemented in an object oriented hypertext system. The design steps and the main features of the system are presented.

1. Introduction

In semi-autonomous production islands, tasks usually separated from the manufacturing processes, are partially or wholly integrated into a working group. Work design of autonomous production islands emphasizes the integration of production planning, time-scheduling of material, manufacturing, quality control, maintenance and repair tasks under the responsibility of a working group. To support the work of semi-autonomous working groups, a computer-based tool ('Diagnosis-Information-System', DIS) has been developed.

Knowledge-based decision support systems are not only potentive tools but also a precondition for the decentralisation of maintenance tasks. Support systems specifically promote a lot of aspects of semi-autonomous working groups, including the integration of workers with different qualifications, the reduction in initial time of semi-skilled workmen and the increase of qualification of the working group members through the accesss of the jointly acquired knowledge base. In order to guarantee and promote human and organizational potentials human, organizational and technical factors should be considered relatively early in design, which is a central issue in sociotechnical and human-centred systems design [1]. System design should not automatize and diminish human decision making such as in expert systems, but rather use and support the potentials of the human decision maker in order to increase overall system efficiency.

2. Empirical Analysis of Cognitive Requirements

The design of human centred software requires to take into account the user's professional competence in addition to the degree of experience and types of cognitive preferences. Therefore an analysis of tasks of operators at numerical controlled machines and a cognitive analysis of the diagnosis process of maintenance personnel was performed. Figure 1 shows the results of a representative survey of the SFB 187, which encompasses about 2.200 firms of the metal-manufactuturing industry in West Germany [2].

	Completely		To a greater part		To a lesser part		Not at all	
	1991	1992	1991	1992	1991	1992	1991	1992
Machine operation	86.4	86.7	12.4	12.5	1.0	0.5	0.2	0.3
Work piece handling	74.6	76.9	17.7	17.9	6.1	3.4	1.6	1.8
Machine preparation	66.1	69.9	25.9	22.7	7.5	6.1	0.5	1.3
Tool setting	37.6	44.6	29.0	28.6	22.4	18.8	11.0	8.0
Tool disposal	13.9	16.4	25.0	27.6	39.1	37.1	22.0	18.9
Programming	23.8	25.7	21.4	23.0	32.4	30.6	22.4	20.7
Program optimization	21.0	24.6	31.3	32.4	34.0	28.7	13.7	14.3
Quality proofing	24.6	28.8	44.2	44.2	25.4	23.5	5.8	3.5
Inspection	9.4	13.8	22.1	23.1	45.4	43.1	23.1	20.0
Cleaning	32.6	36.8	30.9	29.8	25.9	22.9	10.6	10.5
Diagnosis / Repair	2.3	3.2	5.9	8.6	38.1	41.4	53.7	46.8

Figure 1.   Task analysis of operators at numerical controlled machines in 1991 and 1992. Results of a representative survey of SFB 187

In 1991, 8.2 percent of the companies reported that operators at the shop floor were completely or to a greater part engaged in maintenance and servicing tasks. In 1992 there was even an increase of 3.7 to 11.9 percent. The results show the increasing significance of the integration of maintenance and repair tasks into the work of operators. In small and medium sized companies maintenance and repair tasks are usually performed by operators and not by specialists [3]. Operators are usually qualified to handle preventive maintenance, troubleshooting and in some cases even repair.

The cognitive and social requirements of maintenance and repair tasks were identified through the analysis of the work strategies of technicians performing electrical and mechanical maintenance of numerical controlled machine-tools. Particular interaction profiles and informational demands of beginners, experienced staff and experts were analyzed [4] [5]. In process control as well as in failure diagnosis, strategies are used for coping with task complexity. Results show that the most frequent troubleshooting strategies were 'Historical information', 'Least effort', 'Reconstruction', and 'Sensory check'. Historical information about types and frequencies of breakdowns of the machine-tools is used through maintenance records, failure statistics and contact with colleagues. Least effort means that easy checks are used first. Reconstruction requires that the operator or a collegue is asked about the failure course. By sensory check symptoms such as loose connections, odors or sounds are percepted. It is not affordable to use measuring instruments. In contrast, strategies such as 'Information uncertainty' and 'Split half', which lead to a binary reduction of the problem space only play a minor role in troubleshooting of machine tools.

In diagnostic search two basically different ways of maintenance and repair personnel were proposed [6]. 'Symptomatic search' compresses a set of symptoms collected by the operator and this set is matched to a set of symptoms under normal system state. If the search is performed in the actual system or physical domain it is called 'topographic search'. Our results suggest a further class of generalized strategies called 'case-based'. In case-based strategies, symptoms available in the diagnostic situation are collected and compared to those in similiar cases. In novel failure cases, topographic search dominates. In routine failures case-based strategies are most frequently used and the importance of topographical strategies is diminished. Finally, topographical and case-based strategies are dominant in familiar failure cases.

The analysis of strategies in failure diagnosis lead to the result that it is essential to store failure cases in a data bank and provide flexible data access. A support system for maintenance and repair should serve as an information and database. Additionally, the system should support the planning and decision processes by providing short if-then sequences for beginners as well as case-based and topographical information for experts.

3. Incremental Object Oriented Design Steps

The development of software systems is composed of different phases that creates the software life cycle. Although the models differ in types of classification, procedures and details, they usually contain affordance analysis, design, implementation and evaluation. Evolutionary software development is able to overcome a lot of psychological and organizational problems during systems development which results in more realistic cost management, decreased amount of error correction and confidence that a product will satisfy given requirements for quality.

Requirement analysis is a central point in the evolutionary software design, because it determines the decomposition of problems, the hierachical structure of the system, the specification of main features and the design of user interfaces. Therefore a spiral model of software development was taken as project model. Spiral models assume that software unterlies a process of further development. Because a single cycle is not sufficient to

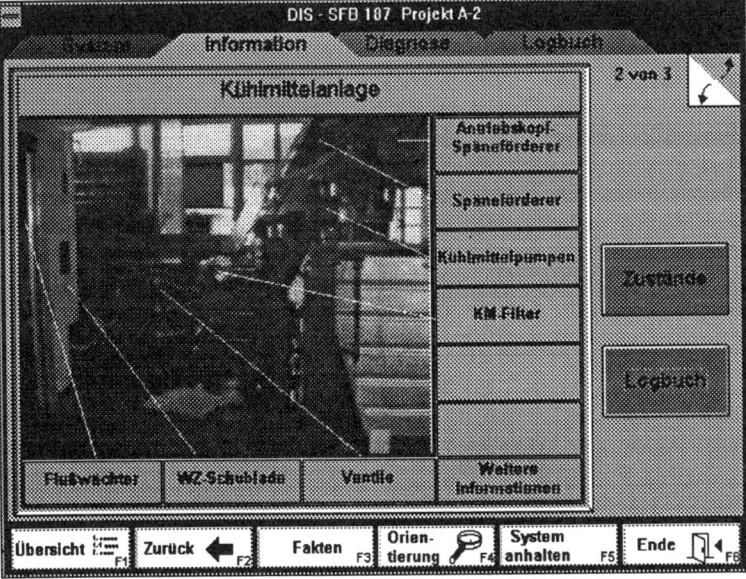

Figure 2.   Presentation of hydraulic components at a CNC-machine in the Diagnosis Information System

consider the dynamics and changes in system affordance complete cycles are repeated. This work is done through rapid prototyping, a technique that allows the user to participate by evaluating different system prototyps in relation to his task affordances. This leads to a continuous further development and a step-wise specification and adaption to user demands.

Object-oriented analysis and design (OOAD) is a methodology to support the translation of user demands more smoothly in prototypes [7]. In OOAD data structure are defined in user language and actions are matched. Domains of users are represented through specification of a set of semantic objects that are revealed in cognitive task analysis. In this case objects were derived from strategies of operators.

4. Main Features of the Diagnosis Information System

The Diagnosis Information System is object-oriented, contains multimedia-components and is easily expandible by the user. In case of a breakdown or an unallowed deviation within the worksystem, DIS informs the operator upon request about possible types and patterns of symptoms and failures, examines and explains their causes and reveals the ensuing failures. It also serves as a data base and offers the failure history at machines on the shopfloor. It provides the user with information about the most common causes of failure as well as with machine documentations and functional drawings. Components of the machine are represented in pictures which are identical to the real machine and are not displayed in abstract forms (Figure 2). Thereby context-relevant marks and information should be used. Through development process DIS was continuously refined by user participation. A final evaluation study concerning the proposed efficiency, the work immanent learning effects and user acceptance is under work.

References

[1]   C. Clegg and G. Symon, A review on human-centred manufacturing technology and a framework for its design and evaluation. *International Review of Ergonomics* 2 (1989) 15-47.

[2]   B. Ostendorf and B. Seitz, Alte und neue Formen der Arbeitsorganisation und Qualifikation - Ein Überblick. In: J. Schmid and U. Widmaier (Eds.), Flexible Arbeitssysteme im Maschinenbau. Opladen, Leske + Budrich, 1992, pp. 75-89.

[3]   U. Konradt and B. Zimolong, Arbeitstätigkeiten, Qualifikationsanforderungen und Organisationsformen an CNC-gesteuerten Werkzeugmaschinen und Bearbeitungszentren. *Zeitschrift für Arbeitswissenschaft* 47 (1993), 71-78.

[4]   B. Zimolong and U. Konradt, Interactive Support System for Maintenance and Repair in Production Islands. In: M.J. Smith and G. Salvendy (Eds.), Human-Computer Interaction. Applications and Case Studies. Elsevier, New York, 1993, pp. 86-89.

[5]   U. Konradt, Handlungsstrategien bei der Störungsdiagnose an flexiblen Fertigungseinrichtungen. *Zeitschrift für Arbeits- und Organisationspsychologie* 38 (1994) in press.

[6]   J. Rasmussen, Models of mental strategies in process plant diagnosis. In: J. Rasmussen and W. B. Rouse (Eds.), Human Detection and Diagnosis of System Failures. Plenum Press, New York, 1981, pp. 241-258.

[7]   D.E. Monarchi and G.I. Puhr, A Research Typology for Object-Oriented Analysis and Design. *CACM* 35(9) (1992) 35-47.

Advances in Agile Manufacturing
P.T. Kidd and W. Karwowski (Eds.)
IOS Press, 1994

An Experimental Evaluation of User Performance in CAD tasks

Lambros LAIOS and Maria ATHOUSSAKI
University of Piraeus, Karaoli & Dimitriou 80, 18534 Piraeus, Greece

Abstract. This paper investigates the performance of designers in two CAD tasks with different task uncertainty. Each task consists of two phases involving commands of the DRAW and EDIT category. An experimental evaluation was carried out with participation of 16 subject. The results on a number of performance measures subjected to statistical analysis and are discussed in terms of uncertainty involved within each phase and task.

1. Introduction

Today a large number of CAD systems is available in the market. The design of the user interface is of great importance for the performance and the acceptance of CAD systems. An insight of human performance when using existing CAD software could contribute in the design of improved CAD products.

Many areas of the human-computer interaction have developed satisfactory measurements of human performance which have proved effective for the evaluation of the usefulness of various system characteristics. The above measurements do not apply in the evaluation of CAD systems mainly because design tasks are ill-defined in terms of knowledge to define objectives and optimum processes required to reach a final solution.

The performance measurement methodology used in this study was based on the idea that the mental load of the information processing for a designer using a CAD system consists of two factors [1] :
1. The workload that derives from the task that has to be performed
2. The workload generated by the way the system functions.

To measure effectively the human performance both factors must be controlled. As the designer workload imposed by the system is determined by system characteristics, the study focuses on manipulation of the workload which is imposed by task aspects, specifically task uncertainty. We expect that the investigation of the interaction of the designers with the CAD systems will highlight functional deficiencies which adversely affect human performance.

2. Research Design

2.1. The Design Tasks

Two tasks of systematically varied uncertainty were developed on a 2D-CAD system [2]. The tasks consisted of two phases, one involving design of simple parts and another involving composition of the single parts into a more complex drawing. The two CAD tasks were developed with the same drawing content. The first was a well defined task with detailed information given to the user for execution and with complete specification of the parts. The second was an ill-defined task with no information given to the user and with no detailed part specification.

The first phase requested commands mainly from the "DRAW" category of AutoCAD and the second from the "EDIT" [2]. The DRAW category consisted of commands used to draw a shape right from the start, using simple lines, circles, e.t.c and the EDIT category consisted of commands used to edit or compose a shape from it existing parts. The above are important categories of the AutoCAD commands used in the design process. The task uncertainty and design phase were defined as independent variables.

2.2. Methodology - Dependent Variables

An experimental procedure was carried out and the results were subjected to statistical analysis using a repeated measures design with degree of task definition the between groups variable and phase the within group variable.

A method based on a model by [1] was used to transform the task procedure and the commands to quantifiable measurements. This method regards the movements that lead closer to the design target as forward movements represented by the number $+1$, the movements that have no effect to the design target as neutral movements represented by the number 0 and the movements that have negative result on the design target as backward movements represented by the number -1.

For each subject the following quantities were used as dependent variables :
1. The working time
2. The sum of the total movements
3. The sum of the forward movements
4. The sum of the neutral movements
5. The sum of the backward movements
6. A measure of subjective difficulty (1-5 interval scale).
   The hypotheses tested are :
Ho1 : The degree of task uncertainty does not affect the dependent variables
Ho2 : The design phase does not affect the dependent variables
Ho3 : There is no interaction between the independent variables on the dependent variables

2.3. Subjects - Data Gathering

Sixteen subjects participated in the experiment, eight in both phases of the first task, eight in both phases of the second task. All were engineers with a working experience in the use of AutoCAD. Data were gathered through :
a. observing of the activity by an expert during the experiment
b. recording of the commands used by subjects
c. aswering of a questionnaire after the completion of the task.

The data were analyzed through a 2x2 repeated measures research design. The ANOVA results and the mean values of performance measurements are shown in Table 1 and Figure 1, respectively.

Table 1. ANOVA Results

	Source	Sum of Squares	Degrees of Freedom	Mean Square	F	Sig. of F.
Time	Task	101.53	1	101.53	3.68	0.076
	Blocks w. Task	386.69	14	27.62		
	Phase	38.28	1	38.28	6.77	0.021
	Task by Phase	205.03	1	205.03	6.25	0.000
	Phase x Blocks w. Task	79.19	14	5.66		
Total Movements	Task	50323.78	1	0323.78	6.67	0.001
	Blocks w. Task	42269.94	14	3019.28		
	Phase	1046.53	1	1046.53	0.85	N.S
	Task by Phase	7969.53	1	7969.53	6.50	0.023
	Phase x Blocks w. Task	17177.44	14	1226.96		
Forward Movements	Task	21736.13	1	1736.13	5.01	0.000
	Blocks w. Task	12168.37	14	869.17		
	Phase	1326.12	1	1326.12	1.99	N.S
	Task by Phase	1860.50	1	1860.50	2.79	N.S
	Phase x Blocks w. Task	9334.37	14	666.74		
Backward Movements	Task	1770.13	1	1770.13	5.10	0.040
	Blocks w. Task	4857.75	14	346.98		
	Phase	1.12	1	1.12	0.01	N.S
	Task by Phase	1128.13	1	1128.13	3.54	0.002
	Phase x Blocks w. Task	1166.75	14	83.34		
Neutral Movements	Task	1339.03	1	1339.03	5.75	0.031
	Blocks w. Task	3262.69	14	233.05		
	Phase	1.53	1	1.53	0.01	N.S
	Task by Phase	205.03	1	205.03	1.47	N.S
	Phase x Blocks w. Task	1950.94	14	139.35		
Task Difficulty	Task	0.50	1	0.50	1.27	N.S
	Blocks w. Task	5.50	14	0.39		
	Phase	0.12	1	0.12	1.00	N.S
	Task by Phase	6.12	1	6.12	9.00	0.000
	Phase x Blocks w. Task	1.75	14	0.13		

N.S = No Significance

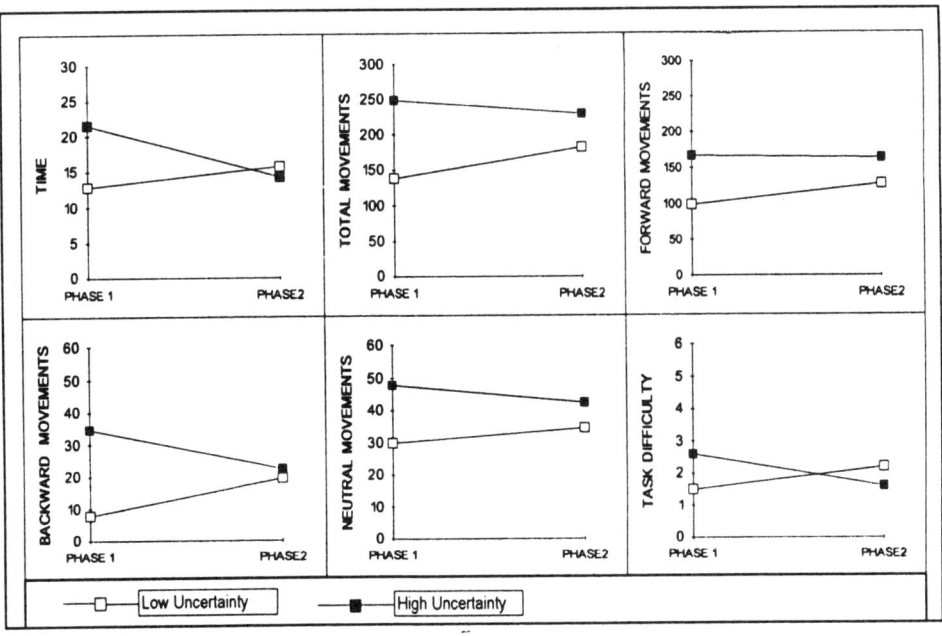

Figure 1. Time, movements and subjective difficulty rating for two CAD tasks

3. Results

From the ANOVA results we observe a number of statistically significant differences. In most cases task uncertainty affected the value of dependent variables. The effect of the phase was significant only in the time taken to complete a task. In all measures but forward and neutral movements, there was a statistically significant interaction between task uncertainty and phase.

Our results show that the differences in human performance due to uncertainty are larger in the first than in the second phase of the design tasks. Even in the case of forward and neutral movements where the interaction between uncertainty and phase is not significant, mean values indicate a similar tendency.

4. Discussion

This study indicates that ill-defined CAD tasks are more difficult to perform than well defined tasks, however, any existing differences are not same across task phases. This finding is supported by the presence of significant interactions between uncertainty and phase of a task in a number of performance measures such as working time, total and backward movements and subjective difficulty. Greater differences in the performance measures between the two tasks have been observed in the "DRAW" phase than in the "EDIT" phase of the design.

A likely explanation of these findings can be provided by considering the nature of uncertainty in CAD tasks. Uncertainty can be generated by both the degree of design specification and the number of possible courses of actions which are required to reach a design solution. In the well-defined CAD task uncertainty is minimal, because specifications are complete and the course of action is determined. In the ill-defined CAD task, uncertainty is present. However, in the "EDIT" phase, uncertainty is due mainly to presence of many different courses of actions and not due to lack of specifications. This explains the lack of difference in working time or the small differences in design movements between the two tasks in the "EDIT" phase. It is clear that the "EDIT" phase does not rely too much on specifications since designers compose drawings from parts prepared in the "DRAW" phase. The above findings are further supported by the measures of perceived task difficulty. Subjects rated the "DRAW" phase of the ill-defined task as more difficult than that of the well-defined task. However, ratings are reversed in the "EDIT" phase.

The main implication for developers and users of CAD software is that ill-defined design tasks should further be supported especially in the "DRAW" phase by providing aids such as special commands of standardized shapes or libraries of objects of frequent use. This is expected to reduce working time and design effort.

References

[1] J. Sharit & D.L. Cuomo, A Study of Human Performance in Computer - Aided Architectural Design, International Journal of Human - Computer Interaction, 1(1) 69-107, 1989.
[2] AutoCAD 10 Reference Manual : Autodesk AG.

Advances in Agile Manufacturing
P.T. Kidd and W. Karwowski (Eds.)
IOS Press, 1994

Computer Support for Engineering Design Tasks

through Skill-Oriented Technologies

Bernd GROEGER

Technologieberatungstelle TBS, Nikolaus-Dürkopp-Straße 17, 33602 Bielefeld - Germany

Ileana HAMBURG

Institut Arbeit und Technik, Florastraße 26-28, 45879 Gelsenkirchen - Germany

Abstract. This paper gives some proposals for skill-oriented technologies like the modelling of design objects and the use of a design methodology which can improve the engineering design process and make easier the computer aided individual and group work of design engineers. These technologies are implemented in a prototype for interactive design of machine components.

1. Introduction

The engineering design department of an enterprise has special responsibilities for the quality and costs of a product; the engineering design often is the first step in the chain of product development-manufacturing-marketing. In order to improve the quality of the designs produced, design engineers try to develop and to introduce new forms of work into the design process (like team work in multidisciplinary groups) and new skill-oriented aiding technologies into the cooperation with other technical specialists. One of these is the use of a design engineering methodology (see Section 2) together with suitable computer systems.

In this paper, some of such skill-oriented interactive computer technologies which support the complex process of engineering design of interconnected machine elements, both for individual and group work, are briefly described. The extension of the prototype which was developed within the FABER project [1] at the Institut Arbeit und Technik in Gelsenkirchen with these technologies (Section 3) represents the progressing work at this project.

2. Skill-Oriented Support for Engineering Design

The aim of the engineer's training is to convey both factual knowledge and a methodical working style. This refers not only to "hard" fields like car-manufacturing, mechanical engineering or aircraft construction, but also to "soft" fields like software development.

In Germany, the VDI-Richtlinie 2221 "Methodik zum Entwickeln und Konstruieren technischer Systeme und Produkte" [2] plays an important role in this training. It gives the fundamental principles for engineering design methodology and is also successfully applied in England, China and the USA. But not only in the university training an orientation to a reliable methodology is important. This way is recommended both from an academical and from a didactical point of view. For example in the engineering design process it facilitates a systematical treatment of the problem using the creative skills of the designer, increases the possibility to review more alternatives and to create a better documentation of the solutions.

Many computer aided systems which support engineering design tasks have been developed since 1970 [3], [4], [5]. Most of them only process data and facts; they are not oriented on an engineering design methodology. That is sufficient for a limited knowledge domain and for the application of a particular method or procedure, but not for solving complex tasks like the conceptual design of machine elements with the help of computer. The software systems aiding such complex processes should not force the user to work according to a fixed scheme. From the ergonomical point of view, they should be able to adapt themselves to the user's style of work and, additionally, to propose a skill-oriented solution using a suitable methodology to the user. According to these requirements, it is important to give the design engineer a tool which helps him to plan and to organize flexibly his design activity. This refers not only to the order of the engineering design steps but also to a proper handling of the work environment (e.g. hardware, software, existing knowledge and product information).

In order to support the new style of cooperative work in groups (which is characterised by common objectives and shared environment) such a tool is indispensable. It can also be used at the beginning of the design work in order to realise a suitable definition of design spaces which should be carried out concurrently by several specialists.

We briefly describe a tool which enables the design engineer (or a group member in charge) to develop skill-oriented plans which are updatable during the work according to the context. These plans are, for example, some methodically prepared pattern solutions which may be used for new, altered or variant designs [6]. In group work, each engineer can extend those parts of the plans which refer to his own design space with new steps (e.g. a new procedure, a rough or fine CAD-System modelling). Changes in the central plan should be discussed in group. The parameters mentioned below may be used (see FABER - Section 3) for a detailed specification of the steps of an engineering design task (design steps). We start from the structure a design plan consisting of n sections and a section consisting of m steps (Table 1).

Description of a design plan	Description of a section	Description of a design step
Name of the design plan	Section number	Design step number
Comments	Name of the section	Name of the design step
Person in charge	Comments	Comments
Final data of the plane	Person in charge	Preconditions
Plan version	Final data of the section	Results
Components	Section version	Aids
Software		

Table 1: Parameters for the description of engineering design plans

3. Example of Skill-Oriented Technologies for the Engineering Design of Machine Components

The FABER project is oriented on the principles of VDI-Richtlinie 2221. The object-oriented prototype of an interactive computer aided system developed within this project [] supports both individual and team work during the conceptual design of interconnected machine elements (with gears as an example) [7]. It runs in network environments (SUN workstations). We used algorithmical programming (for the calculation procedures), logical programming techniques for the processing of engineering knowledge and fuzzy logic-based methods to process uncertain knowledge and to assist individual and group decisions [8]. The group members can communicate (through electronic mail) and exchange data and information. They also have the possibility to create their own environments (with new objects, rules, properties and procedures). Using the CAD-system AUTOCAD and the programming language LISP, we recently implemented a library with the engineering design objects (wheels, bearings and shafts). They can be used skill-orientedly by the engineer to visualize the geometrical data of the gear which are calculated by the system.

The first version of the system demonstrates how various techniques, like algorithmic programming, knowledge based software techniques and fuzzy logic, can be usefully combined to support complex tasks of engineering design through an appropriate software system. But the tests and cooperative discussions with experienced and unexperienced engineers showed us that our system does not offer the user enough possibilities to decide and to control the design process himself. According to these requirements, the system was extended with the new function of "develop design plans" which helps the designer and the group members to organize their work more flexibly. Two pattern plans are proposed by the system: one for individual work and one for group work. They describe possible processing sequences of the design tasks with the interdependencies between the separate steps. The steps for gear design may be data input, a program for data calculation, a CAD makro command, a decision or a evaluation step, consultation of knowledge base, investigation of data base, etc. The user activates one of the pattern plans which appears on the screen; he chooses the desired design steps and gives the required parameters for each step. The system loggs these steps in a file, checks the logical consistence and serves them to the user; the step number (given by the user) determines the order of steps to be processed. The pattern plans are developed in the logical language PROLOG so that they can be easily extended or updated. The system can also give the user skill-oriented advice for the choice of a suitable system of tooth, of the number of teeth or a suitable module during the rough estimation process, for the selection of suitable bearings and of shuft-shub connections from special catalogues based on design knowledge stored in it. The engineer's decisions in these choices are very difficult because the objectives to be reached are multiple and often conflicting. For group work, the pattern plan includes a step of discussion and evaluation after the rough estimation phase. If some engineers develop more different alternatives (based on different experience), then the system evaluates and compares them and proposes a "group alternative" which can be accepted or modified by the group members.

4. Conclusions

In order to support a wide spectrum of design engineering tasks including calculation procedures, knowledge processing and object modelling as well as - more recently - multidisciplinary team work, the software engineers have developed and applied various software techniques. The design engineer needs tools which give him methodical support and facilitate his design work with computer systems using these techniques. We developed such skill-oriented tools for modelling the design objects and planning the design work. We are working at improving them and cooperative group work by using hypermedia.

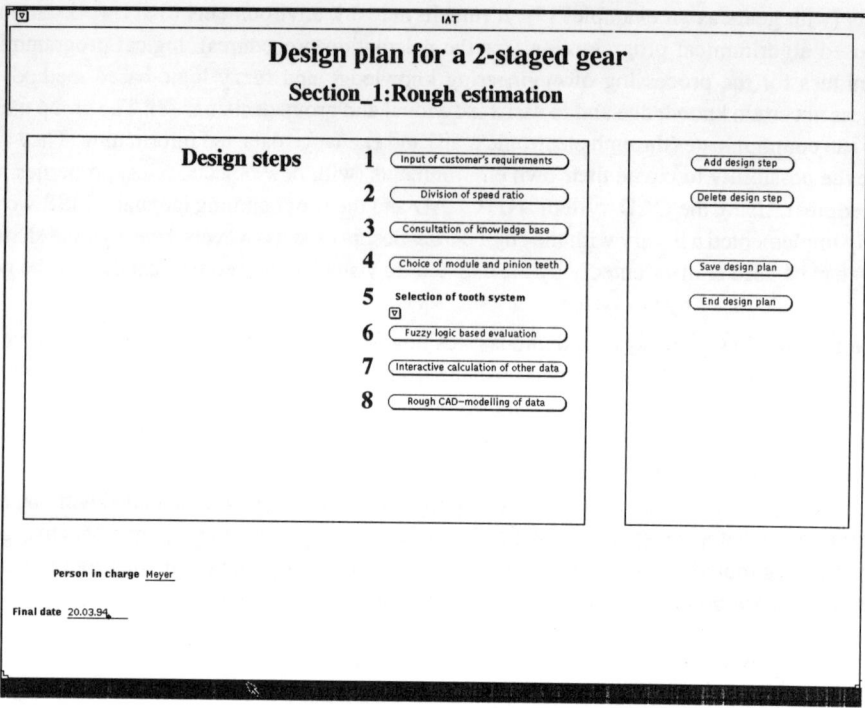

Figure 1: A design plan for the interactive engineering design of gear

References

[1] P. Brödner *et al.*, Arbeitsorientierte Gestaltung von DV-Systemen für Ingenieure. Information als Produktionsfaktor. Tagungsband zur 22. GI-Jahrestagung, Berlin, 1992, pp 629-639.

[2] N.N., VDI-Richtlinie 2221: Methodik zum Entwickeln und Konstruieren technischer Systeme und Produkte. VDI-Verlag, Düsseldorf, 1990.

[3] W. Beitz, Konstruktionsleitsystem als Integrationshilfe. VDI Berichte, Nr. 812: Rechnerunterstützte Produktentwicklung. VDI-Verlag, Düsseldorf, 1990.

[4] B. Groeger, Die Einbeziehung der Wissensverarbeitung in den rechnerunterstützen Konstruktionsprozeß. In: W. Beitz (Ed.), Schriftenreihe Konstruktionstechnik Technische Universität, 23, Berlin, 1992.

[5] I. Hamburg, Interactive calculation during the design phase. In: Ergonomics of Hybrid Automated Systems III. Proceedings of the 3rd International Conference on Human Aspects of Advanced Manufacturing and Hybrid Automation, Elsevier, 1992, pp. 325-331.

[6] B. Groeger *et al.*, Auslegung von Verbindungs elementen am Beispiel der Welle-Nabe-Verbindungen mit Hilfe der Wissenssverarbeitung, *Konstruktion* **44** (1992) pp. 145-153.

[7] G. Niemann and H. Winter, Maschinenelemente. Springer, Berlin, 1981.

[8] I. Hamburg, Using Fuzzy Logic in the Interactive Computer Support of Engineering Design. In: N. Roozenburg (Ed.), Proceedings of ICED 93, Den Hague. Heurista, Zürich, 1993, pp. 1343-1350.

Advances in Agile Manufacturing
P.T. Kidd and W. Karwowski (Eds.)
IOS Press, 1994

AN OBJECT-ORIENTED APPROACH IN BUILDING GRAPHICAL USER INTERFACES FOR CADCS

Costin Pribeanu
Research Institute for Informatics, Bd Averescu 8-10, 71316 Bucuresti, Romania

Abstract. An object-oriented graphical editor was designed as an intermediate layer between a graphical kernel and a class of object-oriented applications. Insulating a higher level functionality improves a clean separating of graphics from application specific. This approach is intended to create an appropriate framework for developing object-oriented applications which integrate powerful graphics tools.

1. Introduction

Human aspects in advanced manufacturing lead to reconsidering of operator limitations and abilities. Operators work with analogies and internal models which help them being aware of what happens in different situations [1]. This process of understanding can be speeded up by using visual presented information for process behaviour representation. Therefore, in recent years Computer Aided Design in Control Systems (CADCS) is more graphics oriented.

This paper is intended to present an object-oriented approach in building graphical user interfaces for control systems design. Designers can describe process behaviour by associating a set of visual schemes and describing their contents. Schemes are edited during the design process with an object-oriented graphical editor.

The graphical editor was designed as an intermediate layer between a graphical kernel and a class of object-oriented applications which could be developed on top. This approach leads to a cleaner separation of graphics from application specific and provides support for building better user interfaces.

2. Process Representation

Modelling of man-machine systems requires suitable models both for human operator and automated system. Human abilities and limitations need decision support in order to face the increasing complexity of technical systems. It was argued in [2] that adding new tasks makes structure and functions more transparent for the user.

Visual presented information can compress in a synthetic manner a great amount of knowledge and provides means for bridging with internal models. Thus, graphics tools make the understanding process of complex situations to be better supported.

Practice in control system design requires a sharing of responsibilities between designer and operator in order to provide decision support for dealing with critical situations. This means modelling of human error mechanisms starting from the earliest stages of design and using suitable models of representation.

In [3] is argued that design of control systems must include designing of man-machine interaction in order to provide a better understanding of the process under control. An approach applying this strategy, presented in [4], is based on designing a process control system by describing all possible scenarios which could be defined with process state variables.

This can be done by associating a process with a set of visual schemes. A process scheme can be defined as graphical, numerical and textual information associated with some physical system and having the capability to reflect in realtime its behaviour and state changes.

The basic entity of a process scheme is the block, defined as a distinct part of a scheme which can be defined, modified and displayed separately. A block has a finite number of visual states depending on some characteristic state variables. All blocks of a scheme, each one in the actual visual state, are visible on the screen.

3. User Interfaces

It is widely agreed that building graphical user interfaces (GUI) is a difficult and expensive task which involves serious investments. Portability of software is therefore an important goal. Independence from application program, device independence and portability of graphical information were the main reasons for developing the graphics standard GKS [5], adopted in 1985. Different requirements concerning data structure and dimensionality determined the proposal of graphics standard PHIGS [6].

A new generation of windows based graphics provide many facilities in building user interfaces. The 'look and feel' - a new paradigm for interface styles makes basic operations and concepts familiar along different applications enhancing the usability of graphical interfaces. Most of window based systems are object-oriented in order to increase software reusability.

These graphics tools were primarily developed as window management systems, their functionality being intended to support the WIMP interface style (Window, Icons, Mouse, Pull-down menus). Therefore Windows systems are rather resources managers than graphics systems.

However, portability of graphical interfaces requires more than a 'look and feel' feature. Interfaces of graphics systems are defined at different levels and should provide a language independent definition of functionality as well as device independence.

Some considerations on the opportunity of defining a new generation of object-oriented graphics standards, which must integrate advances in user interface development, are presented in [7]. Portability requirements of graphics systems and evolutions in user interface area lead for searching a convergency. This could be done by using a standard computer graphics model in an object-oriented framework.

4. Object-Oriented Graphical Editor

Process schemes generating is supported by a set of graphics tools for designing scenarios and control procedures. The structure of the associated schemes and the contents of each scheme are defined with a graphical editor.

This approach is based on defining the graphical interface as an intermediate layer between an existing implementation of GKS standard and an object-oriented application class, in order to provide a high level functionality for modelling a hierarchical data structure.

Functionality of the graphical editor is comprised of application specific functions which include means for defining schemes, blocks, visual states, variables, facilities for process behavior simulation and general purpose graphics functions such as primitive definition, attributes setting, viewing (panning, windowing, zooming), two-dimensional modelling of graphical entities.

Since GKS does not provide functions for high level editing (segmentation is a linear structure) building up a hierarchy for representing real objects was done on top of the kernel system using the GKS metafile. For each entity layer was designed an editing context as an abstraction of the interface functionality. Each editing frame corresponds to a GKS workstation.

Graphical data manipulation between different editing levels is achieved through some special functions. An window layer can be activated for describing a new entity, or for opening an existing, visible entity for editing. Closing the window leads to corresponding update of the previous layer.

Viewing functions (zooming, panning, windowing) and primitives specification were generalized as a superclass, since they only depend on graphical kernel. Each entity level corresponds to a subclass and inherits viewing and output specification methods from the general class. Editing operations are generic functions defined at the higherest level and implemented at each descendent level via virtual methods.

The graphical interface was implemented in two separate modules. The former includes superclass and higher level class (SymbolClass) implementation. It is closer to the graphical kernel and is provided by the interface designer. The second includes application defined classes (BlockClass and SchemeClass) implementation. This module can be designed by the application programmer.

Insulating a higher level functionality into an intermediate layer improves a clean separating of graphics from application specific. Since the interface model is layered itself, application-required hierarchical data structures can be built on top of the kernel system. This approach is intended to create an appropriate framework for developing object-oriented applications which integrate powerful graphics tools.

5. Conclusions

Evolution and actual trends in research and development related with control systems lead to conclude that understanding of human factor is an important goal in modelling man-machine systems. This can be done in a better way by reconsidering of operator limitations and abilities and providing suitable computer aids.

Graphics tools in control systems provide means for developing user-friendly interfaces. In [8] it is shown that a key concept for object-oriented paradigm is developing of standardized parts which could be integrated in higher level solutions. Using the object-oriented paradigm in computer graphics leads to a better structuring of graphics software and increases reusability of interface components.

References

[1] Wahlstrom, B. "Modeling of Man-Machine Systems - A Challenge for Systems Analysis," In: Utkin, V. and U. Taaksoo (eds.), *Preprints of the 11-th IFAC World Congress*, Vol.10, Tallinn, 1990.

[2] Johannsen, G. "Complexity in Man-Machine Systems," In: Utkin V. and U. Taaksoo(eds.), *Preprints of the 11-th IFAC World Congress*, Vol.10, Tallinn, 1990.

[3] Pribeanu, C. "Graphics Tools for CADCS, "*Advances in Modelling and Analysis, C*, vol.43, no.1, pp.11-38.

[4] Vasiliu, C. and C. Pribeanu. "A Realtime Graphics System - Paradigm for a Man-Machine Interaction Structure, "*Rumanian Review of Informatics and Automation*, no. 1, 1993, pp. 17-30.

[5] ISO 7942: 1985. Information processing systems-Computer graphics-*Graphical Kernel System (GKS), functional description*.

[6] ISO/IEC 9592-1: 1989 Information processing systems - Computer graphics-*Programmer's Hierarchical Interactive Graphics Systems (PHIGS),functional description*.

[7] Wisskirchen, P. and K. Kansy. "The new Graphics Standards - Object Oriented !," In: E.H.Blake and
     P. Wisskirchen (eds.), *Advances in Object Oriented Graphics I*, Springer-Verlag, 1991.
[8] Cox, B.J. "There is a Silver Bullet," *Byte*, vol. 15, no. 10, 1990, pp. 209-218.

Advances in Agile Manufacturing
P.T. Kidd and W. Karwowski (Eds.)
IOS Press, 1994

Human-Process Communication and its application in chemical industry, power plant control and mining

Dipl.-Ing. M. Arnold, Dipl.-Ing. M. Heim, Dipl.-Ing. N. Ingendahl, Prof. Dr. M. Polke
RWTH Aachen, Lehrstuhl für Prozeßleittechnik, 52056 Aachen, Germany

Abstract: With human-process-communication a new perspective on operating processes is introduced. Current machine-oriented concepts of man-machine-interaction *narrow the view*: they only deal with the plant equipment that covers the process. Man-process-communication makes this cover transparent to the staff. Operation tasks are hierachically structered. On each level the staff has to be set in a position to mesh into the process.

1. Introduction

Process Control Engineering examines methods and concepts to support humans in the task of controlling processes. The monitoring systems for process control by technical staff are located in control rooms. Nowadays, process monitoring may be characterized by the fact that the information presented is plant-oriented; typical for process monitoring are data organized in piping and instrumentation diagrams, block diagrams and PID Control Structures. This plant-oriented view, however, is to be understood as a narrow view which communicates information about plant-status rather than information about the covered process. The main purpose of process control, the production process, is covered by its technical means.

The control of technical processes always has to cope with the question of which tasks are solved directly by human operators and which tasks are to be delegated to technical systems. The degree of automation is the fraction of the process control tasks that are performed by the control system. Contemporary production may characterized by:

- increasing demands in product quality, process reliability, efficiency and ecology;
- basic tasks along with frequent intervention are intended to be automatically performed;
- complex tasks which require high cognitive skills, faced e.g. in process logistic questions, touch the limits of formalization;
- permanent optimization of process control during the whole life-cycle of plant.

2. Man-process-communication - the human role and demands

The technical staff at process control takes the responsibility for:

- the assurance of the designated operation;
- the detection of potentials for optimization;
- the verification of potentials for optimization;

Coming to terms with this responsibility necessitates the interaction of operators in the process itself. The complete interaction between human and process is the aim of *human-process-communication*. Here we should like to point out some guidelines for human-process-communication:

- the technical staff has to be enabled to take care of its tasks.
- the impact of intended and executed process operations has to be made clear to the technical staff (i.e. prediction).
- the significance of process information concerning aspects as quality, process reliability has to be shown.

3. Methods and concepts

The design of human-process-communication is based on a wide range of disciplines. Instead of dealing with these disciplines in detail some derived methods and concepts are presented in the following.

3.1. Information Structuring

A prerequisit to methods and concepts for human-process-communication is the information structuring of the process. During this analysis the relevant process and product properties are determined. The so called phase model of production is a graphical method for structuring process information [1]. The resulting information structure can be used as a design aid and as a tool for operation in human-process-communication systems.

3.2. Task-orientation

It is significant to arrange process information in relation to the tasks of process operation. In general, process information should be grouped into views under consideration of tasks and constraints as quality, process reliability, etc. A fundamental knowledge of the process and its behaviour is necessary in order to design these views.

3.3. Presentation and Interaction

The operator communicates with the process using presentation- and interaction-objects from the process-control-system. The method of information structuring and the concept of task-orientation, introduced above, supply the systems designer with helpful information for selecting and arranging different presentation- and interaction-objects. In order to support this design step, a taxonomy of presentation-objects has been developed at the institute of process control engineering. By defining so-called application attributes the "best suited" presentation-objects can be selected.

As an example, the presentation of a power plant's emissions is shown below. The process information may be presented in the form of a cumulative frequency chart, which will be actualized online (figure 1).

An interesting perspective for human-process-communication is introduced by model based prediction: while varying the model's independent variables and watching the dependent variables, the operator learns about the process behaviour and dynamics. With the help of model based prediction the operator has the ability to interactively explore his scope of action.

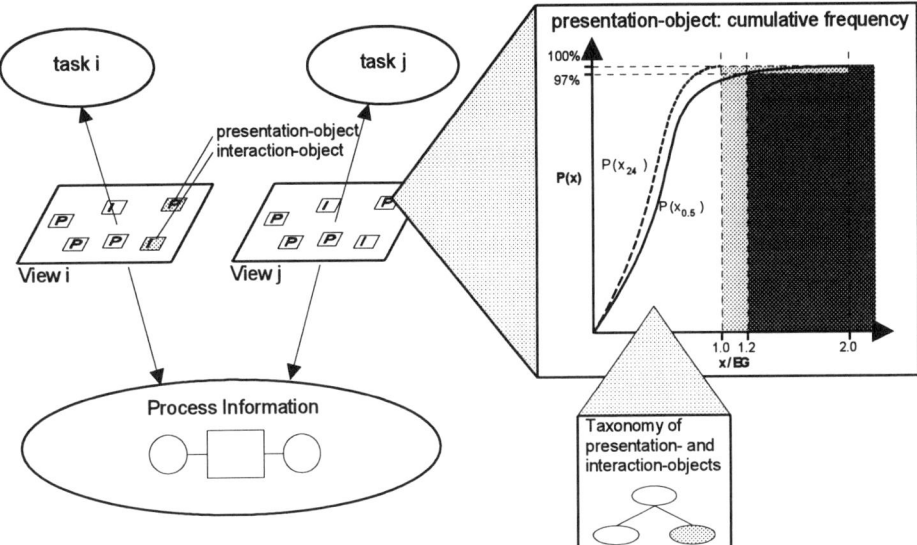

Figure 1: Task-orientation, presentation and interaction

3.4. State-orientation and Navigation

Tasks of process control are related to each other. The goal is not to get the optimal solution in executing a single task but to get the best overall solution while controlling the whole process. In general, tasks have to be planned, executed and controlled while applying an integrated view.

In order to meet these requirements, the concept of task-orientation is accompanied by the concept of state-orientation. In this context the expression *state* is related to objects of the process domain as well as to objects of the plant's machinery and devices. The object's behaviour is described by state models. State changes in the state model objects generate need of action for certain tasks. Implicit relations among state model objects should explicitly be presented to the operator as navigatory links between different tasks. These relations support the staff in determining a state dependent scope of action and consideration of actual goals and tasks. Examples are given in [2].

The presented task- and state-oriented human-process-communication may be illustrated by an operator navigating through a net consisting of task-nodes and flexible state-links.

4. System and design requirements of human-process-communication

4.1. Command mode

Operation tasks are hierarchically structured depending on their degree of abstraction. Due to responsibility at each level of abstraction, the technical staff has to be enabled to mesh into the process at any time. Consequently, a very important property of process-oriented functional units (see [1], Chap. 4.5) is an interface which allows to switch between manual command and automatic command mode. In principle, access to functional units must be first realized and made visible to the technical staff. The implementation (Hardware, Software) depends, among other things, on process reliability and availability requirements.

Further design aspects can be taken from software ergonomics and studies on socio-technical systems [3].

4.2. Configuration and process knowledge acquisition

Experience proves that process and plant behaviour may not be fully determined within project activities. Throughout the whole life-cycle of plant process knowledge is acquired. This fact necessitates the modification of the process monitoring system online by system designers and, more importantly, end users. Current systems do not support this feature sufficiently. System development should therefore consider features which particularly support:

- the acquisition of process knowledge e.g. by the product-oriented documentation of process states and
- the adaption or creation of new "views" by choosing process information and suitable presentation- and interaction- objects

during operating mode.

5. Summary

Process Control Engineering examines concepts, methods and models to support humans in the task of monitoring and controlling processes. Nowadays, machine-oriented concepts of process monitoring can be characterized as *narrow views* which communicate information about plant-status rather than information about the covered process. The main purpose of process control being the production process, is covered by its technical means. Human-process-communication intends to guide the operators to the process itself. A prerequisit to methods and concepts for human-process-communication is the information structuring of the process. In general, process information grouped into views under consideration of tasks and constraints are defined as task-oriented human-process-communication. The concept of task-orientation is accompanied by the concept of state-orientation and navigation.

6. References

[1] Polke, M. (Hrsg.): Prozeßleittechnik. Oldenbourg Verlag, München, Wien 2. Auflage 1994.

[2] Arnold, M.; Ingendahl, N. und Polke, M.: Umweltbewußte Führung prozeßtechnischer Anlagen. Chemische Technik 45 (1993) S. 363 - 374.

[3] Hartmann, E.A. und Fuchs-Frohnhofen, P.: Von Menschen und Handrädern. Technische Rundschau 85 (1993) S. 33 - 38.

Part VIII
Human Performance and Ergonomic Design Issues

Advances in Agile Manufacturing
P.T. Kidd and W. Karwowski (Eds.)
IOS Press, 1994

Stimulus-Response Compatibility Effects in Assembly Task: Component Assembly Time

Lynn A. Fish[1], Colin G. Drury[2], and Martin G. Helander[2]
Canisius College[1], Buffalo, NY 14208 USA
State University of New York at Buffalo[2],Buffalo, NY 14260 USA

Abstract. Stimulus-response compatibility (SRC) effects are expected to decrease as operators become practiced with a particular layout. SRC effects were tested using four different combinations of workstation layout and assembly sequence for a circuit board assembly. Experimental results show that workstation layout significantly effected the twenty-two components assembly time and the number of errors over three hundred assembly cycles.

1. Introduction

In an industrial setting, the impact of the workstation layout and assembly sequence on the development of the assembler's conceptual model of the task is not well known. The assembler's conceptual model may affect the assembly time and accuracy of an assembly [1]. As part of a larger modeling study, workstation layout and assembly sequence compatibility were analyzed for their effect upon speed and accuracy of an assembler over three hundred cycles [2]. The results described within this paper will discuss the SRC effects on component assembly time at three specific assembly intervals during the ten day experiment.

Traditional experimentation has emphasized a few trials to test compatibility effects. Compatible layouts have been assumed to yield faster assembly times and fewer errors [3]. A functional layout, where components are arranged by functional category, or a geographical layout, where components are arranged in the same relative position on the workstation as on the product, may facilitate different ways of assembly task chunking. The assembly sequence, which may also be in a functional or geographical order, may also facilitate chunking. These different chunking methods may lead to faster assembly times or decreased incidence of errors. Also, chunking abilities have been shown to develop as subjects move from novice performance to the skilled activities of an experienced assembler [4]. It was hypothesized that compatible workstation layouts and assembly sequences would yield significantly faster assembly times and decreased errors in comparison to incompatible workstation layouts and assembly sequences. Spatial memory and visual chunking, which are cognitive abilities, may impact upon the assembler's ability to perform the task and were factored into the analysis. The implications of chunking by layout and sequence could improve assembly times, decrease error and improve learning times in manufacturing. The end result of these changes would be decreased manufacturing costs and increased productivity.

2. Method

Eight male subjects inserted 22 components on each board under four treatments related to differences in workstation layout and assembly sequence. The four treatments consisted of subjects performing the assembly under a geographical sequence on a geographical workstation (GG), or geographical sequence on a functional workstation (FG), or a functional sequence on a functional workstation (FF), or a functional sequence on a geographical workstation (GF). Two subjects were assigned to each treatment. Subjects' visual chunking and spatial memory were also tested. The equation used to predict assembly time was:

$$\text{Assembly Time} = a \text{ Spatial memory covariate} + b \text{ Visual chunking covariate} + c \text{ Layout(i)} + d \text{ Sequence(j)} + f \text{ LayoutxSequence(ij)} + g \text{ Day(k)} + e$$

Each board consisted of 22 components - 10 resistors, 6 capacitors and 4 DIPS in a series of three horizontal rows. Each subject assembled 30 circuit boards per day over a period of ten consecutive days. Each subject was trained in the proper assembly sequence on the specified workstation. They were also trained and tested on the expected quality criteria and proper component orientation. They were asked to perform the task as quickly and as accurately as possible with equal emphasis on speed and accuracy. Subjects were given daily feedback regarding their assembly times and specific errors committed. Average assembly times, subject learning curves, and the number of errors were collected for all 30 boards per day by subject over the ten days of experimentation.

For analysis purposes, the assembly task was divided into the time to assemble the bare circuit board to the fixture (T1), the time to assemble all 22 components (T2), and the time to remove the completed circuit board (T3). The emphasis of the paper is on the SRC effects during T2.

3. Results

Based upon the learning curves for all subjects, a decision was made to gather more detailed data from day 1, day 4 and day 10. The first day was chosen to represent a novice assembler, the fourth day was chosen to represent the transition stage from novice toward experienced assembler, and the tenth day was chosen to represent an experienced assembler. As shown in Figure 1, the average time to assemble all 22 components per board ranged from .98 to 1.48 minutes and averaged 1.22 minutes. The number of errors decreased significantly (p=.000) from 39 to 11 over the ten days.

For each of the three days under analysis, covariance analysis on T2 indicated that visual chunking factor (p=.02) was significant. The spatial memory factor was also significant on the first (p=.004) and fourth days (p=.007). Figure 2 graphically demonstrates the relationship between subject visual chunking abilities and each subjects' average assembly time for all components (T2). The effect due to the treatments was significant on the fourth day (p=.048), but was not significant on either of the other two days. Figure 3 demonstrates the effect on T2 for each of the treatments on the fourth day. The FF and FG treatments (1.1 minutes per board) were

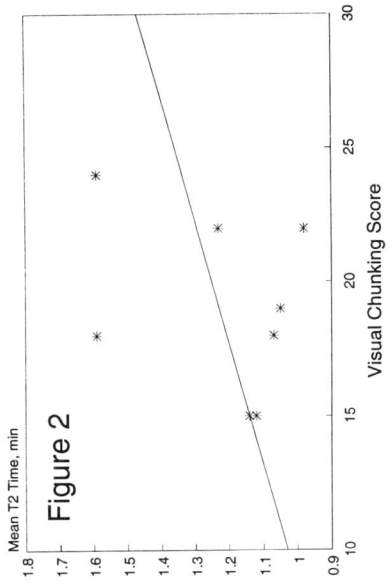

Figure 1

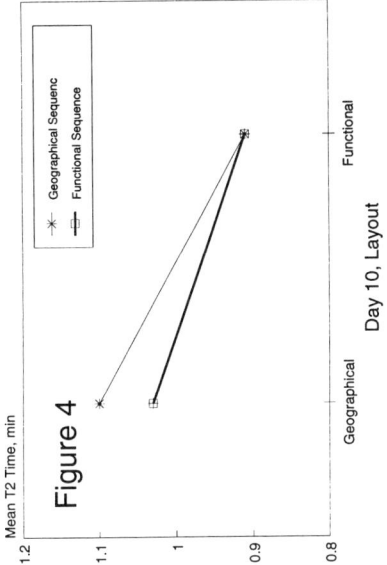

Figure 2

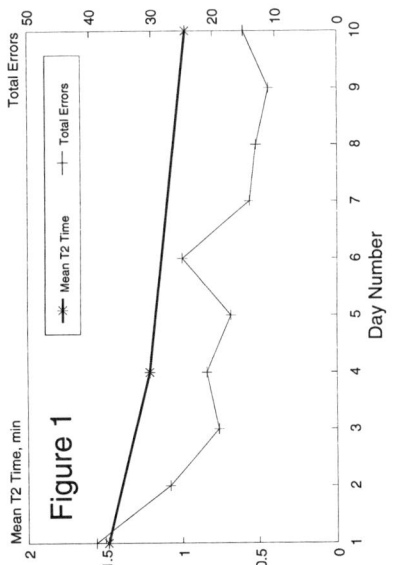

Figure 3

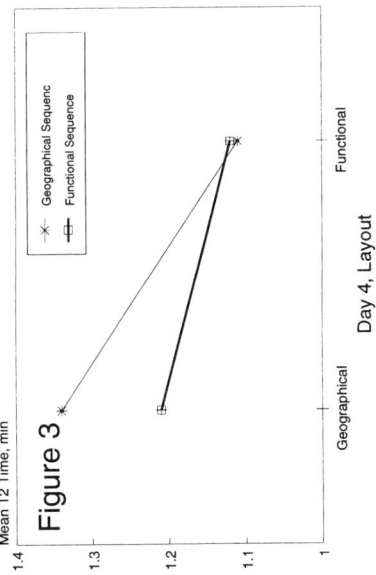

Figure 4

significantly faster than the other two treatments (1.34 minutes per board for GG and 1.21 minutes per board for GF). On the tenth day, the effect due to the layout was still significant (p=.000). As shown in Figure 4, T2 was significantly faster when the subjects utilized the functional layout (.91 minutes per board) than the geographical layout (1.065 minutes per board).

Over the ten days of experimentation, chi-square analysis on the number of errors revealed an insignificant difference between treatments (p=.07). However, chi-square analysis on the number of errors between workstation layouts was significant. The functional layout yielded fewer errors (84) than the geographical layout (117).

4. Discussion

As subjects became more experienced, SRC effects during the component assembly time increased in significance. Specifically, on the first day of experimentation, there were no effects which could be attributed to layout or sequence factors. However, after 90 cycles, the effect due to the layout and sequence was significant. The results favored the functional layout, regardless of sequence. This effect became more pronounced after 270 cycles, when only the effect due to layout was significant. Again, the functional layout yielded significantly faster component assembly times than the geographical layout. These results have implications for assembly as subjects using a functional layout had significantly faster component assembly times and fewer errors than subjects using a geographical layout. These results became more pronounced as subjects gained experience.

Hence, functional grouping of components by location improved the subjects' ability to chunking information. Subjects were able to chunk component location by function and reduce the information processing requirements. This resulted in improved assembly times and fewer errors. Therefore, as subjects gain experience, assembly speed and quality can be significantly impacted upon by workstation layout, regardless of the assembly sequence. This can impact upon manufacturing costs, training times, assembly times, quality and productivity in a modern small-batch assembly environment.

5. Acknowledgment

This work was partially supported by the Mark Diamond Research Foundation.

6. References

[1] R. Eberts and J. Posey, The Mental Model in Stimulus-Response Compatibility. Stimulus-Response Compatibility in R. Proctor and T. Reeve (Edts.) B.V. North-Holland: Elsevier Science Publications, 1990.
[2] L. Fish, C. Drury, and M. Helander, Stimulus-Response Compatibility in Assembly: An Overview. IEA '94 Conference, Toronto, Canada, August 1994.
[3] P. Fitts and C. Seeger, Stimulus-Response Compatibility: Spatial characteristics of stimulus-response codes. *Journal of Experimental Psychology* **46** (1953) 199-210.
[4] A. Newell, Unified Theories of Cognition. Harvard University Press: Cambridge, Massachusetts, 1990.

Advances in Agile Manufacturing
P.T. Kidd and W. Karwowski (Eds.)
IOS Press, 1994

Quantitative Identification for Catastrophe Model and its Application

Yasufumi Kume
Department of Industrial Engineering,Kinki University,3-4-1
Kowakae,Higashiosaka,Osaka 577,Japan

Abstract This paper pays attention to cusp catastrophe in the elementary catastrophe and the model is compared with stochastic cusp catastrophe model. Besides, in the case of given data on the state variable and control factors, the cusp catastrophe model by which the relationship between the state variable and control factors is possible to explain is discussed. As the result, it is clarified that the mode value of stochastic cusp catastrophe model is quite correspond to the deterministic cusp catastrophe model. Also, the identification method for cusp catastrophe model in the case of given the data of state variable and control factors is proposed. And, the application of this identification method to visual perception function test using the characteristic of human factors is performed and the cusp catastrophe model for every subjects is identified.

1. Introduction

The Automation of Production System, that is, FA·OA is progressed. As the result, worker has been saved. Also, the working form for labor in this factory changes from the manual work in which the waist or hand-arm is used to the brain work in which the five senses or brain are used. Therefore, the creativity has been much expected to the worker[1]. On the other hand, the characteristics of human factors have been needed for us to grasp . As the qualitative explanation for the unstable phenomenon of judgment of sensory test etc, the catastrophe model is suited and as the method of quantitative analysis for data, the stochastic catastrophe model is proposed[2]. But the relationship between state variable and control factors can not be explained by the data analysis using stochastic catastrophe model[4] quantitatively. This paper pays attention to cusp catastrophe in the elementary catastrophe[3] and the model is compared with stochastic cusp catastrophe model. Besides, in the case of given data on the state variable and control factors, the cusp catastrophe model by which the relationship between the state variable and control factors is possible to explain is discussed.

2. Quantitative Identification Method for Model

Cusp catastrophe model described by Thom is the typical one of the models with equivalent geometrical properties. So, the quantitative model with equivalent geometrical properties to Thom's equation is described by the following equation using simple increasing functions g, f.

$$-x^3 + f(\beta)x + g(\alpha) = 0 \quad (1)$$

As equation(1), functions g, f, depends on object model, the problem is how to identify the function that its structure is not clear. Therefore, by means the idea like GMDH[5], basic functions ϕ, φ for g, f stand for

$$\phi(\alpha) = a_1\alpha^3 + a_2\alpha + a_3\alpha^{1/3} = 0, \quad (2)$$

$$\varphi(\beta) = b_1\beta^3 + b_2\beta + b_3\beta^{1/3} = 0, \qquad (3)$$

and $g(\alpha)$ and $f(\beta)$ are identified by repeating the identification of basic function at the each rank, where $a_j \geq 0$, $b_j \geq 0$ (j=1,2,3). The model of i rank is given by

$$- x_i^3 + \varphi(\beta_i)x_i + \phi(\alpha_i) = 0. \qquad (4)$$

Also, substituting equation(2) and (3) for (4),

$$- x_i^3 + (b_1\beta_i^3 + b_2\beta_i + b_3\beta_i^{1/3})x_i + (a_1\alpha_i^3 + a_2\alpha_i + a_3\alpha_i^{1/3}) = 0. \qquad (5)$$

As equation(5) is implicit function, it is difficult to apply Least Square Method to obtaining a_j, b_j. So, in much same way as fuzzy regression [6], equation (5) is regarded as a system, and the method minimizing ambiguity is applied to. As equation (5) is linear, a_j and b_j are obtained by the linear programming problem.

Objective function

$$\min \lambda \qquad (6)$$

Constraints

$$\left. \begin{array}{l} - x_i^3 + \phi(\beta_i)x_i + \varphi(\alpha_i) \leq \lambda \\[2mm] x_i^3 - \phi(\beta_i)x_i - \varphi(\alpha_i) \leq \lambda \\[2mm] (i = 1,2,\cdots,m) \end{array} \right\} \qquad (7)$$

$$\lambda \geq 0. \qquad (8)$$

3. Numerical Example

3.1 Experiment and Data

As experimental specimen, the series of continuous figures which change from man face to girl shape are drawn by the imposed interpolate method[7]. As the method, sequential order experiments, that is, from man face to girl shape and *vice versa* are performed. Each series experiment is carried out eight times. When any figure in the series i is observed eight times and judgment of man face is z times, the degree of judgment for man face J_i is given by

$$J_i = J_{il} + (J_{iu} - J_{il}) \times \frac{z}{8}, \qquad (9)$$

where J_{il} and J_{iu} are lower limit and upper limit in the series i obtained by the experiment of scaling of sensation for the judgment degree for man face J_i, respectively[8].

Normal factor, splitting factor and state variable are determined using the data obtained by experiment as follows. Normal factor α is psychological quantity describing change from man face to girl shape and α is positive order in the case of $A_i \Rightarrow U_i$. Splitting factor β is psychological quantity describing the details of figures and β is positive order in the case of $A_1 \Rightarrow A_{15}$. State variable describes the degree of judgment to man face. Also, in order to use the identification method for model proposed in this paper, data are linear transformed so that the sequence of state variable x on α axis with fixed β is symmetric with respect to a point of origin of coordinate.

3.2 Identification of Model

The three subjects A,B,C are selected at random in the whole subjects and g and f are identified for each subject. As the result, the models obtained are shown in Fig.1, 2, 3.

4. Discussion

In usual deterministic catastrophe model, bifurcation set is shown by a curve on the control surface. But in the model identified by the method proposed in this paper, the bifurcation set may be described by the area of the curve with width. When control factors lie in this area, it is estimated that catastrophe phenomenon occurs. Also, in the models of subjects A,B,C, these models have peculiar features, and each of the models may be explained the criterion of judgment and the characteristics of organic function.

In addition, the ambiguity of judgment in the sensory test may be controlled by means of identifying and investigating the model at every subjects.

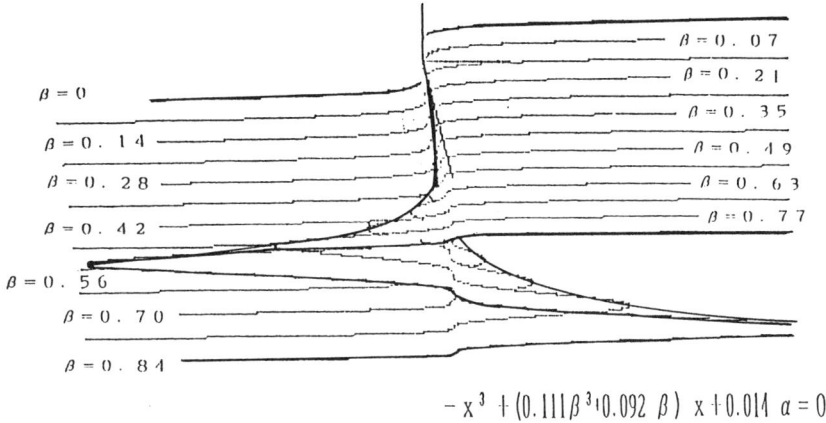

$$-x^3 + (0.111\beta^3 + 0.092\,\beta)\,x + 0.014\,a = 0$$

Fig.1 The Catastrophe Manifold for Subject A

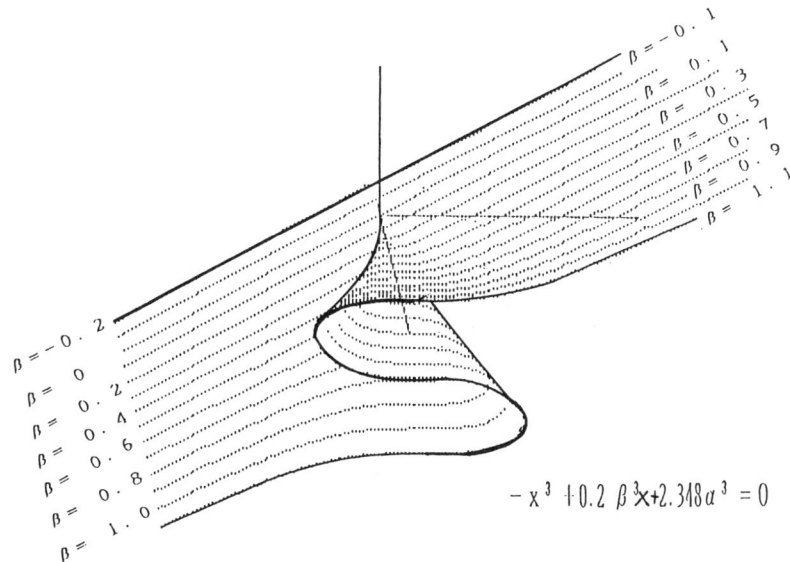

$$-x^3 + 0.2\,\beta^3 x + 2.348 a^3 = 0$$

Fig.2 The Catastrophe Manifold for Subject B

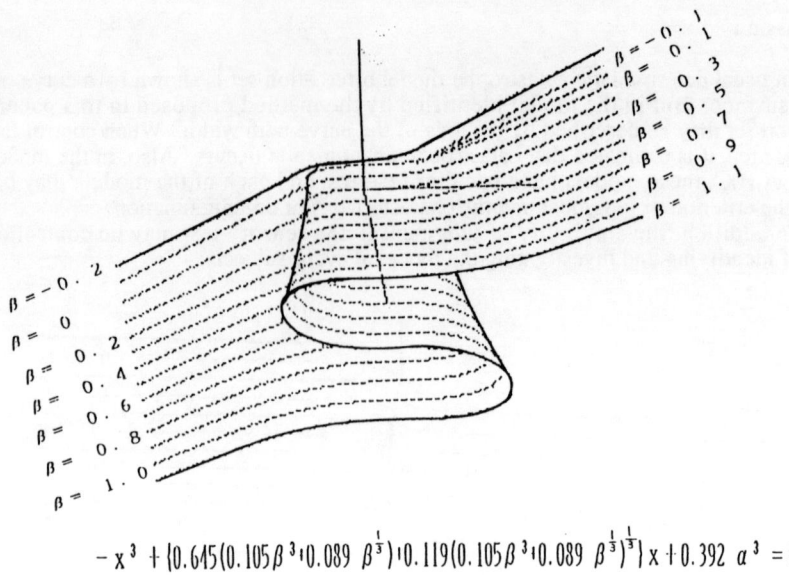

$$-x^3 + \{0.645(0.105\beta^3 + 0.089\ \beta^{\frac{1}{3}}) + 0.119(0.105\beta^3 + 0.089\ \beta^{\frac{1}{3}})^{\frac{1}{3}}\}x + 0.392\ a^3 = 0$$

Fig.3 The Catastrophe Manifold for Subject C

5. Conclusion

In this paper, the cusp catastrophe model is paid attention to and the comparison of the deterministic cusp catastrophe model and the stochastic cusp catastrophe model is performed and investigated. As a result, it is clarified that the mode value of stochastic cusp catastrophe model is quite correspond to the deterministic cusp catastrophe model. Also, the identification method for cusp catastrophe model is proposed in the case of given the data of state variable and control factors. And, the application of this identification method to visual perception test using the characteristic of human factors is performed and the cusp catastrophe model for every subject is identified.

References

1)R.G.King,A Structural Model of Creative Process for Improved Interface Design, Advances in CAD/CAM Workstations,Ed. by Peter C.C.Wang (1986) PP.126-143.
2)L.Cobb,Stochastic Catastrophe Model and Multi model Distributions, Behavior Science,23 (1978)PP.360-374.
3)H.Noguchi & T.Fukuda:Elemental Catastrophe, Kyoritu Shuppan(1980).
4)L.Cobb and S.Zacks, Application of Catastrophe Theory for Statistical Modeling into Biosciences, Journal of the American Statistical Association,80,392(1985)PP.793-802.
5)A.G.Ivakhnenko,The Group Method of Data Handling - A Rival of the Method of Stochastic Approximation, Soviet Automatic Control, Vol.13,No.3(1968) PP.43-55
6)H.Tanaka,A.Ueshima and K.Asai,Linear Regression Model, Journal of the Operations Research Society Japan,Vol.25,No.2(1982)PP.162-174.
7)A.Murata, Y. Kume and F.Hashimoto, Characteristics of Bistable Perceptive Phenomenon for Ambiguous Man/Girl Figure - Application of Catastrophe Model to Visual Organic Functions - , Journal of Japan Industrial Management Association, Vol.35, No.1 (1984) PP.44 - 49.
8)A.Murata,Y.Kume and F.Hashimoto,Cusp Catastrophe Phenomenon in Visual Organic Functions, Journal of Japan Industrial Management Association, Vol.35, No.3 (1984) PP.150-155.

Advances in Agile Manufacturing
P.T. Kidd and W. Karwowski (Eds.)
IOS Press, 1994

The Measurement of Mental Workload in Supervisory Systems

Ding-Yu Lin and Sheue-Ling Hwang
Department of Industrial Engineering
National Tsing-Hwa University, Hsin-Chu, Taiwan, R.O.C.

Abstract. The purpose of this study is to find some appropriate index to measure the mental workload in supervisory tasks.Seven elements relevant to the supervisory task were proposed after task analysis.The results revealed that span,discriminate,predict,recall and transfer attention were important elements,and then a multiple regression equation was acquired.We can use the equation to predict the mental workload of the human in an automated production system.

1. Introduction

As the automation system increases its complexity,more and more parameters may affect the efficiency and safety of the system.One of the important parameters is the task allocation between human and machine of the system.Usually the static task allocation is employed by analyzing the functions of the system in advance according to the merits and demerits of human and machine[1].The task contents of human and machine will not be changed no matter what the condition has been variated.It is obvious that the static task allocation is short of flexibility.Therefore,the concept of dynamic task allocation has been developed[2].The main advantage of dynamic task allocation is that system performance can be optimized .But the bottleneck of dynamic allocation is how to detect the workload of the human at any time and allocate the suitable task to them.So,the purpose of this study is to find some appropriate index to measure the mental workload in a supervisory task.

Some workload measurements have been proposed, such as physiological variable,secondary task methods,major task measurement ,subjective measurement , attention allocation ,etc.[3][4][5][6][7].The "Mento-Factor System" which was mentioned in Niebel's book "Motion and Time Study" can also be used to analyze the mental task of the human[8].Since the "Mento-Factor System" is suitable for analyzing the mental workload of the human in operating the automatic system, some elements from the "Mento-Factor System"will be applied to measure mental workload in the present study.These elements are response,span,discriminate,recall,convert, compute,sustain,transfer attention and predict.

2.Method

The main characteristic of the modern production is to produce less quantity and various products.While the Flexibile Manufacturing System(FMS) is a typical representation of the small batch production,the supervisory tasks of FMS were selected as the contents of the experiment in this study.Twenty-one junior college students and twenty-one graduate students participated in the experiment.Figure 1 is the monitoring display of the simulated FMS.The FMS includes vertical machines,horizontal machines,AS/RS,AGVs, automatic inspecting system and Robot which may be under normal / abnormal conditions.Two experiments were conducted in this study.

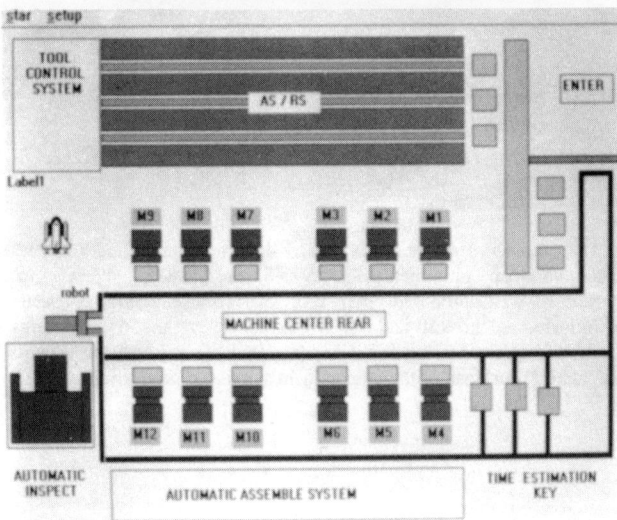

Fig. 1 The main screen of the simulated FMS

2.1 Experiment One

The subjects took some tests which were designed basing on the nine elements..These test problems were put together to form a "Test bank".The error rate was used as a performance index and the t-test was used to examine the difference of the ability between the junior college students and graduate students.The results of Experiment One showed that the performance between the college students and graduate students were significantly different on transfer attention ,predict,discriminate and span(α =0.05).The graduate students' ability is much better than the college students'.But the performance on sustain, recall,compute and response were not significantly different between these two groups.These results imply that the discriminative abilities of transfer attention,predict,convert,discriminate and span are higher than the other elements when measuring the mental workload.

2.2 Experiment Two

From the results of the Experiment One,some elements were significantly different between the college students and graduate students.To check whether these significantly different elements would lead to more or less mental workload and find which element would be more relevent to mental workload,some monitoring problems in a simulated FMS were designed basing on these elements.The primary task of the subjects was to deal with unnormal conditions of the machines in the FMS,and then the secondary task were to estimate ten seconds when some signals presenting in the screen.First of all,the subjects should read the experimental manual proficiently and pass a quiz.As long as his/her score was higher than 90 points,he/she practiced ten problems of simulated FMS.Then,the experimental task processed formally. It took one hour for every subject to complete the experiment.

3.Result

When the completion time of seven elements(span,discriminate,convert,predict, recall,response and transfer attention)are treated as the independent variables and time estimation (TE) as the dependent variable,the backward method of the multiple regression analysis can be applied to find out the important elements.The results revealed that span,discriminate,predict, recall and transfer attention were important elements.The multiple regression equation was as follows:

$$TE = -0.488 + 0.812 \ S + 0.106 \ D - 0.132 \ P + 0.164 \ R - 5.833 \ T \ ----(1)$$

where   TE:Time estimation (in seconds)
       S:span
       D:discriminate
       P:predict
       R:recall
       T:transfer atttention

Comparing the results of Experiment One to Experiment Two,one found out that most elements are important in both experiment except 'recall' and 'convert'.

4.Conclusion

In this study,we found out some important element to measure the mental workload of the human operator in the FMS,and this method can also be applied to other similar automated systems.As soon as the degree of mental workload is defined.the dynamic task allocation will be achieved and the whole performance of the human-machine system can be improved.In addition,these elements can be used as a reference index in the selection and training of the operator.For the further study,an expert system will be developed to decide the task allocation based on the degree of mental workload.Thus,the dynamic task allocation will be feasible and that can improve the efficiency of the automatic systems.

References

[1]N. Jordan, Allocation of Functions Between Man and Machines in Automated Systems, Journal of Applied Psychology, Vol.47,No.3,1963,pp. 161-165.

[2]J. S. Greenstein and S. T. Lam, An Experimental Study of Dialogue-based Communication for Dynamic Human-Computer Task Allocation,Int. J. Man-Machine Studies 23 (1985) 605-621.

[3]T. B. Sheridan and H. G. Stassen, Definitions, Models and Measures of Human Workload, Mental Workload,Neville Moray ,1979,pp. 219-233.

[4]Gunnar Johannsen, Workload and Workload Measurement, Mental Workload,Neville Moray ,1979,pp. 3-11.

[5]M. E. Revesman and J. S. Groenstein, Application of a Mathematical Model of Human Decisionmaking forHuman - Computer Communication,IEEE Transactions on Systems,Man and Cybernetics Vol.SMC-16.No.1,1986,pp. 142-147.

[6]J. Rasmussen, Reflections on the Concept of Operator Workload, Mental Workload,Neville Moray ,1979,pp. 29-40.

[7]W. W. Wierwille, M. Rahimi and J. G. Casali, Evaluation of Sixteen Measures of Mental Work Load Using a Simulated Flight Task Emphasizing Mediational Activity, Human Factors,Vol.27,No.5,1985, pp. 489-502.

[8] B. W. Niebel, Motion and Time Study, Irwin,1988,pp. 490-493.

Advances in Agile Manufacturing
P.T. Kidd and W. Karwowski (Eds.)
IOS Press, 1994

Lessons Learned from the Interaction of Health Care Professionals and Automation: Applications to General Industry

Lee T. Ostrom, Timothy J. Leahy, Steven D. Novack, and William R. Nelson
Idaho National Engineering Laboratory
P.O. Box 1625
Idaho Falls, ID 83415-3855

Abstract. This paper discusses some human issues associated with health care professionals interacting with automation. The information for this paper was obtained from the results of two events investigated during a study sponsored by the United States Nuclear Regulatory Commission that sought to determine the causes of misadministration events. These two events involved the use of high dose rate (HDR) brachytherapy remote afterloaders. Several human issues concerning the automation were determined. It was discovered that the HDR devices were not designed to be user centered and the health care professionals controlling the systems do not always understand: 1) how the system is designed; 2) how the system is programmed; 3) how the safety systems work; and 4) lacked sensitivity to indications of errors or problems. Also, the workplace layout has a bearing on the ability to monitor the system.

1. Introduction

This paper discusses the lessons learned from investigations into the causes of misadministration events involving highly automated radiation therapy medical devices. The lessons learned from the results of these investigations have applicability to all those who design automated devices.

In January of 1992, the United States Code of Federal Regulations (CFR) 10 CFR 35 was amended to require that all medical use licensees establish and implement a Quality Management Program (QMP) to provide high confidence that byproduct material or radiation from byproduct material will be administered as directed by the authorized user. On the basis of a review of therapy misadministrations, abnormal occurrences, and diagnostic misadministrations in the therapeutic range that had occurred between November of 1980 and December of 1988, the United States Nuclear Regulatory Commission (NRC) concluded that such a program could enhance patient safety. The NRC contracted with the Idaho National Engineering Laboratory (INEL) to perform detailed analyses of misadministration events. The objectives of these analyses were to identify the direct causes, contributing factors, actions the licensee took to mitigate the event, and the consequences of these events. Also, the INEL sought to determine the role of the Quality Management (QM) rule on the event. This project consisted of a series of team investigations of reported misadministrations and the analysis of the results of those team investigations. The general results from those investigations is found in Ostrom, Leahy, and Novack[1].

This paper focuses on two incidents investigated that involved high dose rate (HDR) brachytherapy afterloading devices. Brachytherapy is a cancer treatment modality in which

Work supported by the U.S. Nuclear Regulatory Commission Office of Nuclear Materials Safeguards and Safety, under DOE Idaho Field Office Contract DE-AC07-761D01570. Views expressed in this report are not necessarily those of the Nuclear Regulatory Commission.

the radiation source is placed close to the tumor. The HDR devices examined as a part of this study contain an approximately 3.7 E-19 Becquerel (ten Curie) Iridium-192 source that is attached to the end of a wire. A computer controlled stepping motor pushes the source out of its lead storage container (by means of the wire) through a catheter that has been implanted in the patient to its pre-assigned locations close to the tumor being treated for certain dwell times. The computer receives its inputs for dwell locations and times from a treatment file that is loaded into the computer either by a floppy disc or direct entry. The operator can stop, but not modify the treatment once it has started. Treatments using this device last between a few seconds and several minutes. The following presents a brief overview of the two events and the direct causes that pertain to the technology that lead to the events occurring.

2. Event 1

On the afternoon of November 27, 1991, the day before Thanksgiving holiday, a male patient scheduled to receive his fifth and final radiation therapy treatment for cancer of the nasal septum was placed in the HDR treatment room. A catheter was attached to the patient's nose. This catheter was attached to the HDR unit by a trained resident physician. When the patient was ready to be treated, a physicist was paged to operate the unit. The physicist who operated the HDR unit during this particular patient's first four treatments was not available. A second authorized physicist proceeded to the treatment area where he picked up a patient's chart located to the left of the HDR console and programmed the unit's computer with the treatment card taken from the chart. Entry of the information from the treatment card into the unit's console produces a printout of the treatment parameters (source dwell times and positions). The HDR unit was activated after the physicist and the resident physician verified that the treatment parameters on the chart corresponded with those on the printout. As the treatment began, one of the three observers standing near the console inquired about the length of the treatment. The resident physician indicated that the treatment would last about one and one-half minutes while the physicist indicated a time greater than 400 seconds. Based upon this disparity the resident physician directed the physicist to stop the treatment. Both the physicist and the resident physician reviewed the chart and discovered that it did not belong to the patient being treated. The appropriate patient chart had been placed to the right side of the console. The patient received 76 cGy to the lips as a result of the first treatment. The unit was reprogrammed with the correct information and the treatment progressed normally.

There were several contributing causes to this event. Those that pertain to the type of technology used were:

- The resident physician had not been trained about the HDR device and he was not aware that the numbers on the printout he was verifying represented source positions and dwell times. The printout from the device had a poor print quality and to an untrained individual it was not clear what the numbers on the printout represented. Had the resident physician known, he may have recognized that the information he was verifying was not correct for his patient.

- The part-time physicist operating the unit was not acquainted with the patient. He assumed that the nasal catheter represented an endobronchial treatment.

- The case constituted the first use of the HDR for this type of treatment. Gynecological treatments account for approximately 90% of the workload and endobronchial treatments account for most of the rest. Occasionally, a different type of treatment, such as the nasal septum case, occurred.

3. Event 2

On November 16, 1992, an 82 year old female patient was undergoing radiation therapy for an anal carcinoma. The radiation therapy was to be administered by a HDR afterloader with five connecting catheters. For that day's treatment, a dose of 6 Gy (600 rad) was to be administered through five catheters implanted as a single-plane perineal (rectal) implant encompassing the tumor. After a trial run through the five catheters with a dummy source, the Ir-192 source was easily placed in four of the five catheters. After several unsuccessful attempts to insert the source into the fifth catheter, the physician directed termination of the treatment. An area radiation monitor in the treatment room was observed in an alarm condition--flashing red light--at some point during the unsuccessful attempts to insert the source into the fifth catheter. Although three technologists and the physician were aware of the alarm, no one used the available portable survey meter to detect whether radioactivity was present. Believing that the area radiation monitor was malfunctioning, they reset the area radiation monitor and returned the patient to a local nursing home without performing any radiological surveys. The staff were unaware that the Ir-192 source had remained in the patient.

The patient was returned to the nursing home where she resided with four of the original five treatment catheters, one containing the Ir-192 source, in place. One loose catheter had been removed at the clinic. The source remained in the patient's body for almost four days. On the fourth day, the catheter with the source came loose, and early on the morning of November 20, 1992 the catheter fell out. The patient died on November 21, 1992.

On December 1, 1992, the Licensee's medical physicist notified NRC Region I that a 1.37 E+11 Bq (3.7-Ci) Ir-192 sealed source was missing from the licensee's HDR afterloader. The medical physicist believed that a radioactive source that was discovered by Browning-Ferris Industries (BFI) at their non-radioactive medical waste incinerator facility in Warren, OH, and later returned to another BFI facility in Carnegie, PA (BFI-Carnegie), could be the same source that was missing from the HDR afterloader at the licensee's facility.

The direct causes for this event that pertain to the technology used were:

- The device was designed so that if the emergency retract motor engaged, The source wire measuring device disengaged. In addition, the source park light illuminated once the end of the wire was detected passing into the entrance of the device's shielded container. Also, the error message indicating that the emergency retract motor had engaged was only displayed for a brief period of time on the screen. The device had no visible indications when it was in a faulted condition, for instance a trouble light or a message on the computer monitor that stayed on the monitor for a long period of time. No one at the facility was aware that the device was designed in this manner. An error message was printed by the device printer, but the technologists did not commonly read these printouts.

- The technologists, physician, and medical physicist assumed the source wire would not break. This assumption was instilled in them by the HDR device manufacturer. This displayed a "poor" safety culture because the individuals did not have a questioning attitude. They assumed that everything was fine when a radiation alarm sounded, thinking that it was the alarm that failed and not the source wire. This assumption was reinforced by the HDR device's display that showed the source being safely parked after the wire broke.

- Neither the physician or the technologists at the facility were adequately trained on using the HDR device for use under normal conditions. The clinic staff were not used to operating a device that contain a radioactive source. Most of the treatments the clinic performed utilized a linear accelerator, which does not contain any radioactive material.

- The clinic had a set of emergency operating procedures, but they did not specifically address the event and the staff had not been trained on them.

- The technologists had not been given radiation safety training since they began their employment at the facility. In some cases, this was several years.

- The workplace design prevented the HDR operator from watching the closed circuit TV patient monitor and the HDR computer monitor at the same time. Therefore, while watching the patient, the HDR operator may not have seen some of the important error messages presented on the HDR computer monitor.

4. Summary and Conclusions

From these examples it is clear that the medical devices were not designed to be user centered and the health care professionals involved did not understand: (1) how the systems were designed and functioned; (2) how the systems were programmed; (3) how the safety systems worked; and (4) lacked sensitivity to indications of errors or problems. Also, that work place layout affects the operator's ability to monitor the process. The findings from these investigations are not unique, but they further demonstrate the need to design devices that are user centered. Leveson and Turner[2] state in their study that the Therac-25 device relied on the software for safe operation. This is another example where the manufacturer and the operators relied on an automated device to be safe, rather than designing the device to be human centered. Billings[3] discusses the need to design automated devices using the principles of human centered automation. Billings states that the basic premise of human-centered automation is that the operator bears the ultimate responsibility for the safety of any operation, not the automation. The axiom is that the human operator must be in command.

In summary, automated devices should be designed to be human centered and not machine centered. It appears that if the HDR devices involved in the incidents above would have been designed to be human centered, the potential for the incidents to occur would have been far less.

5. References

1. Ostrom, L.T., Leahy, T.J., and Novack, S.D., *Summary of 1991 and 1992 Misadministration Event Investigations*, EG&G-2707, NUREG/CR-6088, 1994.

2. Leveson, N.G., and Turner, C.S., *An Investigation of Therac-25 Accidents*, University of California Technical Report #92-108, University of Washington Technical Report #92-11-05, 1992.

3. Billings, C.E., *Human-Centered Aircraft Automation Philosophy*, NASA-Ames Research Center, Moffett Field, CA., 1991.

Advances in Agile Manufacturing
P.T. Kidd and W. Karwowski (Eds.)
IOS Press, 1994

Emerging Automation Approaches in Roadway Traffic Management

Michael J. KELLY
Georgia Institute of Technology, Atlanta, Georgia, USA

Abstract. Roadway traffic congestion is a costly, world-wide problem. Congestion and accidents can be substantially reduced by "smart car" and "smart highway" technology. Automation of many functions of the driver and traffic management center operator can increase the functional capacity of the roadway; it can also introduce new systems and human factors challenges. Approximately two dozen operational control centers were visited in North America and Europe. Numerous automation approaches were identified and lessons learned during the implementation and operation of each were identified and documented.

1. Introduction

Roadway traffic congestion is a costly world-wide problem. It has been estimated that congestion-caused accidents, loss of resources, lost time while sitting in congested lanes, and added pollution cost $170 billion per year in the United States, alone [1]. One partial solution is to bring high technology to traffic management through the widespread use of the "smart cars" and "smart highways" of the intelligent vehicle-highway system (IVHS). It has been estimated that, by the year 2000, over 200 United States cities will have adopted some elements of IVHS technology; similar technology growth is being accomplished in Europe and on the Pacific Rim. Receiving major emphasis is the IVHS traffic management system (TMS) designed to facilitate the safe movement of persons and goods, with minimal delay, throughout the roadway network.

The approach of the IVHS Traffic Management System is to (1) maximize the functional capacity of the roadway network by optimally spacing vehicles in place and time, (2) reduce the frequency and impact of roadway incidents (e.g., stalls, spills, and accidents), (3) help identify, locate, and respond to roadway incidents and other emergencies, (4) influence drivers to select travel schedules, routes, and transportation modes that will optimize network traffic flow, and (5) maintain public confidence [2].

1.1 The Typical Traffic Management System

Current IVHS TMSs involve a complex network of traffic sensors, information processing and decision aiding systems, and effector systems. Traffic sensors typically consist of "loop detectors," wire coils embedded in the pavement that detect and signal the passage of ferrous metal in vehicles, and closed-circuit television (CCTV) cameras at critical points on the roadway. Effector systems most frequently include intersection traffic signals, freeway ramp meters, and variable message signs (VMS) that may warn drivers of traffic problems and suggest alternate routes. Within the traffic management center, computers fuse the data from, perhaps, thousands of loop detectors, display sensor and CCTV information to support the operators' situation awareness, and, in some cases, make and carry out decisions.

1.2 Automation Goals and Experiences.

Traditionally, the many data gathering, decisionmaking, and communication aspects of the TMS mission have been performed manually. New technologies in data sensing, data fusion, communication, automation, and decision support systems now allow many of these functions to be fully or partially automated, increasing the efficiency with which traffic can be controlled. Some existing centers have established a design goal of full automation, a lights-out TMS. Yet many attempts at automating tasks now performed by humans have been clumsy and not fully successful; other attempts have successfully automated major functions by reducing their complexity or by shedding parts of the task to other agencies. For the near future, full service TMSs will continue to be hybrid automated systems with most automation designed to support human operators rather than replace them.

2. Method

Approximately two dozen operational control centers in North America and Europe were visited. The majority of these were centers that monitored, controlled, or influenced automobile traffic on urban streets or major highways. Detailed interviews were conducted with managers, operators, and engineers to document system design, function allocation, operating procedures, and human factors lessons learned. Additional visits to hardware and software vendors provided insight into expected evolution paths for TMS automation and design difficulties the vendors sometimes experience in working with the IVHS community.

3. Results

Partial automation is becoming very common in traffic management systems. Automation of changes in traffic signal and ramp meter timing plans according to time of day or traffic demand is already common; future signal control software will be increasingly capable of automatic reaction. Fusion of sensor data into meaningful displays is well advanced. Automation of VMS sign control and message content is the next likely TMC function for extensive automation. Significant automation of other decisions and functions is lagging.

3.1 Three Sample Automation Approaches

The large differences in automation philosophy can be seen by comparing the philosophy and architecture of three relatively new IVHS TMSs. The three systems are all equally successful at performing their mission even though they perform them very differently.

The least automated of the three is in a city in the United States. The main function of the TMS is to monitor and control the city's traffic signals in response to predictable and unpredictable traffic. A network of approximately 1000 loop detectors and 20 CCTV cameras provides information on traffic flow and possible incidents. The traffic signal "timing plan" (the time devoted to each green, yellow, and red phase) is automatically controlled by the time of day. The software would also allow automated adaptive control according to traffic demand, but this feature is not used. Much of the operator's time is spent manually changing timing plans to optimize individual intersections. The operator also detects and reports incidents, monitors and troubleshoots TMS status, communicates with drivers using a network of VMS signs, and works closely with police in traffic control.

The second example, a system in Canada, monitors and controls traffic on one of the busiest freeways in North America. A network of loop detectors, computers, and VMS signs provides automatic measurement of traffic flow, selection of appropriate VMS messages, and display of congestion warnings to alert the drivers and support their rerouting decisions. Congestion levels that suggest incidents are indicated to the operators on a graphical display. Operators using CCTV then, verify that an incident is present, call up an appropriate data entry page on their computers, and enter the location and precise nature of the incident. Appropriate VMS incident messages are automatically selected according to a complex set of preestablished rules and displayed on the VMS network while operators communicate with other agencies to help coordinate any required emergency response.

The third example, being implemented on major roadways in large portions of The Netherlands, has the highest degree of automation. The Motorway Monitoring and Control System (MMCS) is a "lane control" system that can dynamically provide different speed limits or traffic restrictions for each motorway lane. Stations consisting of loop detectors in each lane, VMS signs for each lane, and a networked computer are placed approximately every 500 M. When traffic is freely flowing, no restrictions are displayed and speed limits remain at 120 KM/H. When slow or stopped traffic is detected in a lane, the station computer talks to upstream and downstream stations; upstream stations will lower the speed limit appropriately in that lane and, perhaps, advise traffic to move to other lanes. All sign changes are coordinated and approved by a central computer system. Operators can, but rarely do, override the automated VMS system. The primary role of operators is monitoring for MMCS maintenance problems and closing roadway lanes for maintenance or as requested by other agencies. Operators interact with MMCS through the central computer to compose appropriate sign patterns and transmit them to the roadway stations. In MMCS, the operators' incident detection and management responsibilities are nil.

3.2 Some Lessons Learned

Automation of traffic signals is straightforward but can create problems when operators enter the loop. In one center that employs automated signal timing changes at set times of day, the operator frequently makes extensive manual changes only to have them all erased, without warning, by an automated change to a new plan. In addition, the user interface to the automated system was clumsy. For example, the operators would type <1> to initiate a command; they would type <11> to shut down the automated system and all traffic signals.

Automated detection of roadway incidents is relatively ineffective. Numerous algorithms have been developed by traffic engineers based on measured variation in traffic flow. Use of these algorithms becomes a classic signal detection problem with high detection rates associated with even higher false alarm rates. In practice, operators usually chose to ignore graphical incident detection displays based on fused loop detector data and to rely on CCTV - even when this approach causes excessive visual workload.

In several traffic management systems the ultimate design goal is full automation. Some existing support systems we examined are capable of approaching that goal. Yet, it appears that the human operator must always have a role in the traffic management system due to:

- failures in the automated system logic due to unforseen circumstances (e.g., the support system that interpreted maintenance activity in a closed lane as a wrong-way driver and closed the roadway).

- events that can't be detected by sensors but that require response (e.g., the airplane crash near the freeway that required road closure to all but emergency vehicles).

-    support system hardware or software malfunctions that require full or partial degradation to manual mode.

Appropriate allocation of functions to human and machine components of the system is crucial to system effectiveness. For example, where routine decision and response functions can be described by an IF-THEN relationship, and where the IF contingency is easily detectable, a high level of automation would be indicated. These might be of the form:

IF traffic speed declines to 35 KM/HR, THEN display "Congestion" warning on the upstream VMS.

Where, on the other hand, the available sensors cannot determine the important decision parameters in a timely and reliable manner or where the decision process is equivocal, lower levels of automation are indicated. Some manual component is indicated, for example, for a response of the form:

IF the accident victim is seriously injured, THEN dispatch paramedics.

In one TMS, an automation scheme was implemented in which the support system would determine the "best" pattern of changeable signs to present; the operator would review and consent for each change. After implementation, it was found that sign changes were so frequent (200 per hour) that the operators' workload was unacceptably high. After determining that the support system approached 100% reliability, the function allocation was changed to give the support system initial response capability. The operator may manually change a pattern if a problem is noticed. Three lessons may be obtained from this example: (1) initial function allocations need to be validated empirically, (2) higher than expected reliability of a support system may allow change to higher levels of automation, and (3) partially automating a task does not necessarily result in decreased operator workload.

4. References

1. Constantino, J. (1993). Statement of Dr. James Constantino. *Technology Policy: Surface Transport Infrastructure R&D: Hearings before the Subcommittee on Technology, Environment and Aviation of the Committee on Space, Science and Technology. US House of Representatives.* Washington: US Government Printing Office. Pp.6-19.
2. Folds, D.J., Brooks, J.L., Stocks, D.R., Fain, W.B., Courtney, T.K., and Blankenship, S.M. (1993). *Functional Definition of an Ideal Traffic Management System.* Atlanta: Georgia Tech Research Institute.

This research was performed under Contract DTFH61-92-C-00094 between the Federal Highway Administration and the Georgia Institute of Technology. Dr. Michael Kelly is Principal Investigator and Ms. Nazemeh Sobhi is the FHWA Contracting Officer's Technical Representative.

Advances in Agile Manufacturing
P.T. Kidd and W. Karwowski (Eds.)
IOS Press, 1994

Integrative Planning of an Assembly System

Siegfried Bauer

Fraunhofer-Institute for Industrial Engineering, Stuttgart (Germany)

Abstract. The design of an assembly system for a complex automotive suppliers product, whilst taking especially human aspects into account, was the task for an interdisciplinary planning team. External researchers in close colaboration with the internal company engineers and the representatives of the personnel made the work system come true. Hence, the result was obtained from a simulateneous and integrative planning process. Already during the design phase of the product, requirements from the employees point of view and the necessary assembly techniques have both been taken into account. Organizational and human interests have been taken into consideration in the simulation of the technical system and led to a human adequate layout. An enlarged economical calculation was based as well on employee referred aspects, as on effects of flexibility while evaluating the concept alternatives.

1   Introduction

The company kabelmetal electro GmbH in Nürnberg led a project entitled "Designing of a Flexible Assembly System Regarding in Particular the Abilities and Requirements of Different Groups of Personnel". The German Federal Minister for Research and Technology supported this project for 2.7 years with 2.3 m DM. Two scientific research firms by the name of Fraunhofer-Institute for Industrial Engineering and University of Kassel, Department "Transmission Technics".also took part in this project.

2   Project Design

2.1   Starting Situation and Problems

Security products from suppliers of the electrical and automotive industries, such as the so-called Clock Spring, which is a component of the airbag system, are characterized by high quality and availability. They also have a large scale integration of functions, few components and an adequate structure and design for assembly and automation. These products are gaining importance in an increasing way and also have certain requirements in the design of the assembly systems, where they were manufactured. Linked with the development in the area of the products, process changes are also increasing: we find new assembly techniques, intelligent workpiece carriers, flexibility and stabilization of processes, monitoring and information systems.

At the same time the requirements of today`s personnel in their work contents and working conditions are changing in a hollistic manor: Independence, information, participation, individual working time, etc. are factors which are becoming more and more important. Heterogeneous groups of personnel, such as women, foreign employees, skilled and unskilled or even handicapped workers have different views concerning the aspects of design and organization of the work systems.

2.2 Objectives of the Project

The objectives of the project are deductive by the following interior and exterior developments and influences:

- High level of quality and overall improved quality consciousness, improved manufacturing safety, traceability of product defects, higher qualification level and new technical solutions (because it is a safety product with some flexible parts)
- Systematical synchronization of product und process innovation (because of shortening product working life and increasing product changes)
- Reasonable production, high flexibility of personnel and techniques related to growth of volume and growth of types an variants (because of pressure from the market prices, end producers and global competition)
- High performance of delivery (because of the JIT- situation of a supplier for the automotive industrie)
- Considering specific requirements of the personnel (mostly women, mainly foreign employees, partly handicapped)

3   Proceeding and Course of the Project

The project included many of these problems, which can be found in a row of middle-sized companies. Therefore, there was an ideal playground to develop and to practice new approaches to find competetive and transferable solutions:

Guidance by objectives in a hollistic project requires a sensible structure for the item of the planning process as well as for the strict organization of the project. Simultaneously all involved participants should receive and bring a high measure of engagement and responsibility. Therefore, the leading engineers, economists, a psychologist, managers and a work council member were represented in a coordinating and decision making team (steering committee). In addition, 6 Sub-teams under different headings i.e., "Organization", "Product", "Techniques", "Technical support", "Personnel" and "Quality" were founded, with each team having 4 to 6 members. The basis for the tasks of these sub-teams were, on one hand, the objectives given by the steering committee, and on the other hand the results from an analysis of problems and deficiencies in the existing area at the beginning of the project.

The results of all studies and concepts were developed and supported by new computer based instruments. One of these instruments is the PERSIMO simulation system, developed at the Fraunhofer Institute for Industrial Engineering which allows one to evaluate the interactions of techniques, organization and personnel. Another instrument is the the computer aided "Enlarged Consideration of Economy" which can be used simultaneously during the planning process to check costs, benefits or flexibility factors while changing design parameters.

Beside the technical and economical interests, one emphasis was on the human aspect in designing the flexible assembly system. The results of a wide spread quotation among the employees while using standard instruments and the science of labour based knowledge of the planning team members such as the medicine of the company, the work council, the manager of personnel department and a social scientist were concentrated upon and prepared for decisions. In form of of a so-called duty-sheet, these requirements took on immense influence on the alternative conceptions developed by the engineers.

There was also a well defined embedding of the project in an alliance of other firms with similar projects but different products. Regularly exchanges of experiences in workshops and mutually organized visits brought new ideas and constructive suggestions into the planning team.

4 Results

The integrative approach to design and to put the new assembly area into reality can do more than just save the working places for a partially manual manufactured product. By changing the traditional successive proceeding of planning assembly systems into the new simultaneous devoloping of product and production system, the technological level could be raised in a big step. Due to the demanding work contents work conditions referring to the needs of the heterogeneuos groups of personnel were improved and the qualification level of the workers raised. Alltogether, it can be concluded that the hollistic and interdisciplinary proceeding in this project is a very good example of one way to stabilise the international position of the German production. The following list reveals some of the main results:

- Innovative products and a new assembly system were simultaneously created. Two aspects, assembly and automation adequate features could be put into practice which translated into new market shares and strong competetiveness. Therefore, the chosen approach to conduct the project can be successfully transferred to other companies which have similar problems to solve.

- A new assembly structure with a flexible transfer system and with several automatic work stations was interdisciplinarily planned and finally installed. The personnell is working with overlapping work contents. This means that several operations situated next to each other can be fullfilled by the same person. There is also an organized job-rotation after

two hours. First experiences with a new kind of work organization could be made during
the project.

- A well adjusted hollistic qualifying concept with themes such as "Products, Functions
and Applications", "Company Facilities and Organization", "Work Organization within
the System" and "Work Station Referred Work Contents" was developed and put into
practice. The company exceeded the workstation related qualifying process by far and
showed clearly that by the meaning of a comprehensive understanding of work an
important contribution to the motivation and identification of the personnel was given. In
addition, the product quality and personal flexibility improved.

- Human factors had a strong influence on the design of the work system, especially in
ergonomics and work organization. The new system is now able to compensate the different performance levels of the manual workers. Therefore, the integration of handicapped
workers into the working group was possible. Without the intensive analysis, delivered
by the sub-team "personnel", most of the very detailed requirements of the employees
would not have been considered by the engineers.

- Experiences by practicing with new cooperation-focused proceedings and methods were
gained by the involved participiciants of the planning process. Today, all over the
company, these methods are being used by every planning team.

- Important strain factors and particular monotone visual inspection were completely abo-
lished by developing and introducing automatic vision control and monitoring systems.

- By the participation of a work council member in the designing and decision process any
interests of the concerned workers could be respected during the entire planning
phase.("steering committee", Sub-team "Personnel")

References

[1] Bauer, S.: "Wirtschaftlichkeitsbetrachtungen in der flexibel automatisierten Serienmontage"; Fortschritt-
    Berichte VDI-Reihe 2 Nr. 264;. Düsseldorf: VDI-Verlag 1992

[2] Bauer, S.: "Integration von Arbeits- und Gesundheitsschutzanforderungen in die Investitionsplanung", in
    Bullinger/Volkholz/Betzl/ Köchling/Risch,"Alter und Erwerbsarbeit der Zukunft", S.264-268, Springer-
    Verlag, Heidelberg, 1993

[3] Bauer, S./Götz, W. u.a.: "Abschlußbericht zum Vorhaben 01 HH 169/6", Projektträgerschaft Arbeit und
    Technik, Bonn, 1994

[4] Götz, W./Bauer, S.: "Integrative Montagestrukturierung"; im Tagungsband zum 2. FhG-IAO-Forum:
    "Kundenorientierte Produktion-Wettbewerbsfaktor Arbeitsorganisation", Stuttgart, 1993

[5] Warnecke, H.J./Bullinger,H.-J.: FhG-IAO-Montageforum '93: Integrative Gestaltung innovativer
    Montagesysteme, Forschung und Praxis T36; Springer-Verlag, Heidelberg, 1993

Advances in Agile Manufacturing
P.T. Kidd and W. Karwowski (Eds.)
IOS Press, 1994

Ergodynamics in Hybrid Automated Systems: Mutual Human-Machine Adaptation and Transformation Dynamics

Valery F. Venda*, Ilona V. Venda* and Oleg V. Shevyakov**

Department of Mechanical and Industrial Engineering, University of Manitoba, Winnipeg, R3T 2N2, Canada

**Russia State Committee on Higher Education, 33 Shabolovka, Moscow, Russia*

Abstract. The maximal work efficiency and safety during transformation dynamics is a main goal of the ergodynamics in hybrid automated and intelligence systems. Therefore ergodynamics is defined as a part of ergonomics on human-centered mutual adaptation in human-machine-environment systems during their transformation dynamics. Three ergonomic approaches to improve efficiency and expand acceptable work conditions of human-machine system are discussed. Ergodynamics approach is based on the mutual human-machine adaptation concept. It is the most effective for dynamic hybrid automated and intelligence systems.

1. Ergodynamics: human centered mutual adaptation in dynamic human-machine-environment systems

In accordance to the Ergodynamics Law #1, maximal work efficiency may be reached if work functional structure and work environment are mutually adapted [3, 6].

The maximal work efficiency and safety during transformation dynamics is a main goal of the ergodynamics in hybrid automated and intelligence systems. Therefore *ergodynamics is defined as a part of ergonomics on human-centered mutual adaptation in human-machine-environment systems (HMES) during transformation dynamics of their structures.*

Work efficiency is being understood here very widely, we consider not only productivity and quality during some short period but also during long time, so work efficiency reflects indirectly health, safety, work satisfaction and other human related aspects.

Some processes are so complex and dynamic that they are prone to catastrophic human error. Among these are fuel, chemical and power complexes, engineering and transportation facilities, and other systems needing skilled staff and advanced work tools (information displays, expert systems, computer hardware and software). Their complex and fluid nature complicates problem solving for scientists, researchers, and designers. Coordination of these systems' functional dynamic features can be baffling, and yet staff must respond promptly and correctly to a vast number of situations. Human error can lead to such disasters as those at the Chernobyl and Three Mile Island nuclear power plants.

An HMES is the ambient environment close to the worker, such as the cockpit, control center, or shop floor. Its borders divide it from the external, global, social or physical environment. The HMES mutually adapts with the global environment as an integral social-technical system studied in macroergonomics [1]. Optimal mutual adaptation between the HMES' system components (including its people and machines) and between the system as a whole and environment leads to maximal efficiency, safety, and reliability of the system.

2. Multi-level mutual adaptation in the hybrid automated systems

Functional structure of the hybrid automated system is the product of the processes of mutual multi-level adaptation between human, machine (automata) and environment. Within HMES ergonomic design these processes should be adapted to each other at surpassing rates to form general functional system structure. The multi-level concept also embraces evolution, job selection, instruction, and psycho-physiological adjustment. Machines also adapt to human needs at many levels, from components to their overall concept. Forming of new functional structure S_N of human work by transformation of old structure S_O takes more time, efforts, resources if new structure more differs from old one. The difference is calculated as $\Delta F_{ON} = F_{N\ opt} - F_{O\ opt}$. It is very essential also how low efficiency drops at the transformation state, what is a value of Q_{ON}. There are two pairs of the functional structures shown at Figure 1. Left pair includes more different structures, right pair includes closer structures: $\Delta F_{ON}' > \Delta F_{ON}''$; $Q_{ON}' < Q_{ON}''$. Intersect points of the characteristic curves have the coordinates $(Q_{ON}'; F_{ON}')$ for the left pair (Figure 1,a) and $(Q_{ON}''; F_{ON}'')$ for the right pair (Figure 1,b). Higher transformation efficiency $Q_{ON}'' > Q_{ON}'$ and bigger overlap of the right pair of the characteristic curves means that right pair contains *more associated functional structures* than left pair.

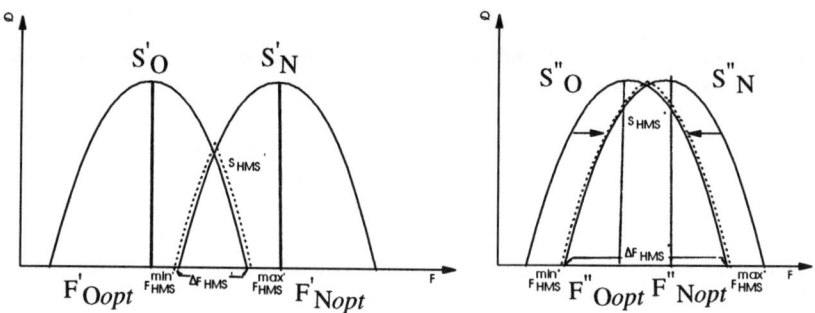

Figure 1. Two pairs of old S_O and new S_N functional structures: a) less associated (left pair, S_O', S_N') and b) more associated (right pair, S_O'', S_N''). The functional structures could be also interpreted as belonged to human S_O', S_O'' and machine S_N', S_N''; S_{HMS} - structure of interactive HMS that may work only at the interval of work efficiency-complexity factors ΔF_{HMS} common for the human and machine [6].

Mutual adaptation between human and machine creates a common human-machine system functional structure. This structure fits some certain range of environmental conditions. This conditions are limited by the value of the factor of efficiency-complexity F leading to the efficiency higher then minimal acceptable one: $Q(F_{min}) > Q^*$; $Q(F_{max}) > Q^*$. This interval $\Delta F = F_{max} - F_{min}$ could be easily found using characteristic curve of the functional structure $Q(F)$. If the conditions go outside this range (and this should be forecasted) the human-machine system must change its structure and strategy.

A functional structure of human-machine system is modeled graphically as the margins of the overlap of two characteristic curves: human and machine (read "designers' ") functional structures. Two pairs of the functional structures shown at Figure 1 could be also interpreted as the functional structures of human S_O' and machine S_N'. Left pair has a small overlap, thus maximal efficiency of human-machine system $Q_{ON\ max}'$ is low and a range of the conditions in which the system may work is very narrow. A functional structure of the human-machine system will be modeled as a broken bell-shaped curve S_{HMS}'. Right pair of functional structures of human (user) S_O'' and machine (hardware and software) S_N'' has much more overlap, thus human-machine system has higher maximal efficiency $Q_{HMS\ max}''>Q_{HMS\ max}'$, and it can work in wider range of conditions: $\Delta F_{HMS}''>\Delta F_{HMS}'$.

A main goal of ergonomic improvement of human-machine system may be simply expressed as changing left version at **Figure 1** to the right one.

We should note that we consider here only human-machine systems where interaction between human and machine is the only way and condition to get effective result. For example the MS Word for Windows does not work itself. Neither a user unfamiliar with this software installed into PC cannot get a result. User-computer common efficiency will depend on what part of the MS Word for Windows is user familiar with. Thus a functional structure of the system user-PC will be found as an overlap of the functional structures of the user and MS Word for Windows.

3. Three ergonomics approaches to the interactive system improvement

There are three ergonomic approaches could be used to improve efficiency and expand acceptable work conditions of human-machine system if we have unsatisfactory version (Figure 2,a):

1. Transform S_H to S_M. It is an unrealistic designers' dream: to select and train users so that they will completely follow designers' functional structure and reach maximal efficiency using the machine. Thus machine stays unchanged, S_M constant; users change their functional structures to fit machine: S_H = variable. Notice that a human structure should be changed very essentially (see Figure 2, b). Usually designers cannot count on this variant, the users are not so flexible. People are also very reluctant to transformations through very deep hollows of their work efficiency.

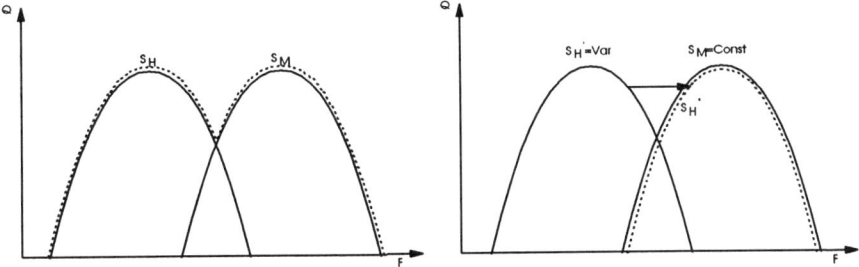

Figure 2. Unsatisfactory human-machine adaptation (a) and adaptation of human (S_H) to machine (S_M): S_H= variable, S_M= constant (b).

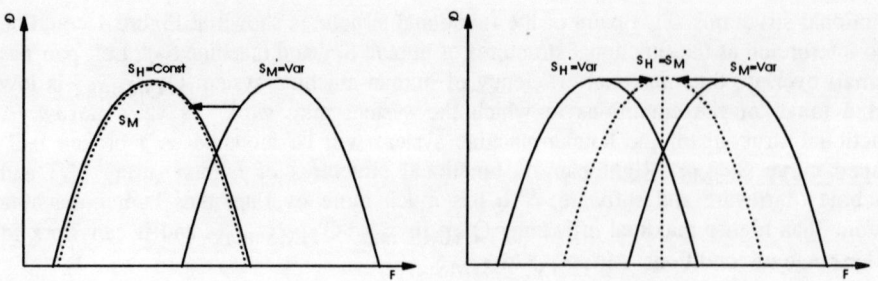

Figure 3. Adaptation of machine (S_M) to human (S_H): S_M = variable, S_H = constant (a) and mutual adaptation between human (S_H) and machine (S_M): S_H = variable, S_M = variable, with twice lesser changes of every system component (b).

2. Transform S_M to S_H (see Figure 3,a). This is not a way how technological progress should work: machines almost never return to old peoples' functional structures: car, airplane, electric mill did not follow human functional structures when previous transportation, water mills were used. Notice that in approach with one way adaptation of machine to human change in the machine functional structure will be very significant. One-way adaptation always demands too big changes of one partner to allow counter-partner to enjoy stability of functional structure. Thus one-way transformations are usually far ones (in a sense of distance between functional structures).

Far transformations of one human structure to another distant one are usually a difficult and rather painful process. Transformations always lead to decreasing efficiency, loss of previous experience revealed in that part of previous structure that should be dismantled. It is especially true for the far transformations [3-6].

3. Ergodynamics approach based on mutual adaptation and transformation dynamics concepts [6] considers two-way changes when human functional structures moves toward machine's one, and machine's functional structure moves toward human structure (see Figure 3,b).

References

[1] Hendrick, H.W. Organizational design. In: G.Salvendy (ed.), Handbook of Human Factors, New York: Wiley, 1986, 470-494.

[2] Karwowski, W. (ed.), Ergonomics in Hybrid Automated Systems, Amsterdam: Elsevier, 1988.

[3] Venda, V. F. On the laws of mutual adaptation in human-machine and other systems. In: Karwowski, W. (ed.), Trends in Ergonomics. Amsterdam: Elsevier, 1986.

[4] Venda, V. F., and Venda, Yuri V. Transformation dynamics in complex systems, *Journal of Washington Academy of Science*, #4, December, (1991).

[5] Venda, Yuri V., and Venda, V.F. Introduction to the Transformation Dynamics, In: Advances in Industrial Ergonomics and Safety-IV, London: Francis and Taylor, 1992.

[6] Venda, V.F. and Venda, Y.V. Dynamics in ergonomics, psychology, and decisions. Norwood, N.J.: Ablex, 1994.

Advances in Agile Manufacturing
P.T. Kidd and W. Karwowski (Eds.)
IOS Press, 1994

The effects of operator state and dialogue control on operator performance in automated systems

David G.WASTELL
Department of Computer Science, Manchester University, Oxford Rd., Manchester, U.K.

G. Robert J. HOCKEY and J. SAUER
Department of Psychology, Hull University, Hull, U.K.

Abstract. A laboratory test-bed has been developed to allow human computer interaction with complex systems to be investigated under controlled conditions. An experiment is described which investigated the influence of two factors on supervisory performance: operator fatigue and operator control. No strong effects of fatigue were found on performance. Although enhanced control led to more operator intervention, this increased activity was largely unproductive. These results illustrate the adaptive self-regulatory nature of the human cognitive system. They suggest that low operator control is not *per se* an undesirable design feature in automated systems; the key issue is to provide an appropriate division of labour between man and machine.

1. Introduction

Rasmussen has coined the phrase "complex work environment" [1] as a generic term for high technology work domains (in areas such as industrial process control and advanced manufacturing) which are characterised by two main features. First, they involve a challenging technical task, namely the control of a complex dynamic process made up of many tightly-coupled components whose interaction is difficult to prescribe. Second, they are characterised by a high degree of automation, i.e. the mode of operator control is supervisory rather than manual [2]. The role of the operator is to monitor a largely self-regulating system, intervening only in exceptional circumstances (e.g. to effect new production targets or to deal with fault conditions).

This paper reports a simulation study which examines two issues which are relevant to the performance of human operators in complex work domains. The first issue concerns the level of operator involvement. Although automation is intended to improve performance, there is evidence that these benefits are not always realised. Zuboff [3] observes that lodging intelligence in the "smart machine" can emasculate the human operator. Because operators are detached from the "control loop", they fail to develop the skills and understanding that are necessary to deal with problem situations [4]. Bainbridge [5] refers to this as an "irony of automation". The second issue addressed is operator state. We believe that any understanding of human behaviour in complex situations must take into account the psychological state of the subject. Negative states such as anxiety and fatigue are of high *prima facie* relevance to work in complex domains. Experimental studies of the effects of sleep loss, for instance, on performance [6] typically show a worsening of performance with sleep deprivation. The second objective of the present study was therefore to evaluate the effect of fatigue on operator performance.

2. The Air Management Task (AMT)

The study reported here was carried out as part of a larger programme of work commissioned by the European Space Agency (ESA). The general aim of this work is to deepen our understanding of the dynamics of human computer interaction in complex work domains and to evaluate the impact of key issues such as fatigue and display design. The target environment is manned space flight where many key functions are highly automated, e.g. navigation and life support. A simulated task environment (AMT) has been developed to allow operator behaviour to be studied under controlled laboratory conditions.

AMT (Air Management Task) simulates the control systems regulating air quality in the living quarters of a spaceship. It is described in detail elsewhere [7]. In brief, AMT involves two closely coupled sub-systems, one for air supply, the other for air purification. AMT incorporates an automatic control system which under normal circumstances automatically maintains critical parameters (e.g. oxygen and carbon dioxide levels) within prescribed ranges. AMT can also be operated in manual mode. Manual control, however, is not a trivial undertaking. The two subsystems are tightly coupled and their use has to be carefully synchronised. Problems will be encountered if the dependencies between the two subsystems and their dynamic interaction are not fully understood.

The interface to AMT is a direct manipulation interface based on a graphical mimic; the interface allows the status of key parameters to be inspected and manual control to be exerted when required. In order to study fault management performance, AMT allows a range of fault conditions to be programmed by the experimenter. To alert the operator to the possible presence of a failure, AMT incorporates an alarm subsystem.

3. Design and method

Two issues were investigated in the research: operator control and operator fatigue. Fatigue was induced by depriving subjects of one night's sleep. Subjects came into the laboratory the evening prior to testing. They were then free to use the time as they wished but they were obliged to stay awake. The vigil was supervised. Operator control was manipulated by creating two versions of AMT. In the "low control" version, there was little scope for manual intervention. Only during abnormal conditions (i.e. the triggering of an alarm) was the subject freely able to inspect system status and to operate the subsystem controls manually. In the "enhanced control" version, operators were free to examine all aspects of system status and to intervene whenever they wanted. It was expected that operators would play a more active role in this condition and that they would build up a more detailed and up-to-date "mental model" of the system. As a result, it was expected that their performance would be superior in situations (i.e. faults) where manual control was required.

15 subjects participated in the experiment. They were given a comprehensive training programme involving 8-10 sessions. Following training, the test phase of the experiment began. The two independent variables generated four treatment combinations. Subjects carried out four two-hour work sessions, one session for each treatment combination. The sequence of conditions was varied to balance out any order effects.Test sessions lasted for two hours and were always carried out at the same time of day (beginning at 9:00); they were separated by one week on average. Several fault conditions (e.g. a leak or a blocked valve) were programmed to occur during each work session. Faults lasted 10 minutes after which subjects were notified that the automatic systems were back in full working order.

Subjects were instructed to operate as supervisory controllers. Their primary job, under normal conditions, was simply to monitor the state of the system and to check out alarm conditions. If a fault was diagnosed then they were instructed to assume manual control until the automatic controllers became operational again.

4. Summary of results

The primary measure of operator performance was defined as the ability to maintain key system parameters within target ranges. Despite subjective reports of increased fatigue after sleep loss (measured using a set of simple analogue rating scales), no strong effects of sleep loss on primary task performance were found. Regarding control, again no clear-cut and consistent effects on primary task performance were found. AMT also incorporates a range of secondary tasks, e.g. operators were required to maintain a system log in which they recorded system status at prescribed intervals. Some aspects of secondary task performance were affected by sleep loss, although these effects were subtle. For the log-keeping task, there was evidence that subjects tended to make an increased number of early recordings when fatigued. This re-scheduling was seen as an adaptive coping strategy.

The intensity of operator interaction with the system was also assessed by counting the number of times operators intervened to inspect the value of a system parameter or to change the setting of a component. Both sleep loss and high control led to an increase in manual intervention and information gathering, although these effects were selective.

5. Discussion

The failure to find any strong effects of sleep loss is consistent with Hockey's Variable State Activation Theory[8] which provides a general conceptual framework for understanding the interaction between fatigue and performance. The key idea in VSAT is to see the cognitive apparatus as a self-regulating, adaptive system. When a mismatch occurs between task demands and performance, the cognitive system responds in a compensatory way, e.g. by working harder or by concentrating "cognitive resources" on high priority activities. In general terms, VSAT accounts for the present findings. No gross effects of sleep loss were obtained. Instead a subtle reconfiguration of patterns of activity across primary and secondary tasks was observed. Primary task performance was largely "protected" by a compensatory increase in work and an adaptive re-scheduling of lower priority tasks, such that there was almost no discernible outward change in performance.

High control led to a greater level of operator engagement. The unreliability of the system was presumably an important factor here. Gathering more information and taking more action presumably gave operators a greater feeling of control over a system that they knew was prone to failure. Other researchers have shown that lack of "trust" in unreliable systems will engender a strong propensity to intervene [9]. What is surprising is that this increased engagement did not lead to improved performance. What appeared to be happening was that operators were often taking control in situations (in the high control condition) where the automatic systems would have performed equally as well. We may conclude that although the high control interface enabled more intervention, this extra intervention was often inappropriate. This result confirms Norman's precept that there is nothing absolutely good or bad about the relative merits of manual vs. automatic control [10]. The issue is one of appropriate "allocation of function" between human and machine.

An idiographic analysis of the coping strategies of the weaker performing subjects was also carried out. In general, the weaker subjects exhibited a number of common failings. Typically, they showed rather reactive "closed-loop" responses, rather than the concerted, open-loop measures which are required for effective system control. There was an interesting polarisation in their coping strategies. Some subjects reacted by adopting a narrow focus on a single parameter, often over-relying on the automatic systems even though these were malfunctioning. Other responded by taking too much on; they threw themselves into excessive manual activity, attempting in a frantic fashion to run the whole system manually.

This dichotomy recalls the categories of maladaptive behaviour that have been observed in other studies of behaviour in complex situations. Dorner's elegant microworld studies [11] revealed two such dysfunctional strategies. Dorner uses the term "encystment" to refer to the tendency of some weaker performers to wrap themselves up in the details of minor issues whilst ignoring the overall situation. In other subjects, Dorner observed the opposite response, namely a tendency to flit erratically from one issue to another (thematic vagabonding). These two behaviour patterns show strong affinity with Janis's concepts of hyper and hypovigilance [12]. The present results, taken together with the work of Dorner and Janis, suggest that there are two general maladaptive styles for dealing with complex situations, of withdrawal on the one hand and hyperactivity on the other. These strategies are defensive-avoidance manoeuvres. They allay anxiety but the response is maladaptive in the sense that subjects avoid coming to terms with objective task demands.

6. Acknowledgments

This research was funded by the European Space Agency Long Term Project Office (grant reference: RFQ 13-7280/91/H/FL) through a grant awarded to Professor G.R.J. Hockey.

References

[1] Vicente, K and Rasmussen, J. The ecology of human-machine systems II: mediating direct perception in complex work domains. EPRL Report number 90-01, EPRL, University of Illinois: Illinois, USA, 1990.

[2] Sheridan, T. B. Task allocation and supervisory control. In Helander, M (ed.), *Handbook of human-computer interaction*. Elsevier, Amsterdam, 1988.

[3] Zuboff, S. *In the age of the smart machine: the future of work and power.* Heineman, Oxford, 1988.

[4] Wickens, C. D. *Engineering psychology and human performance.* Harper Collins, New York, 1992.

[5] Bainbridge, L. Ironies of automation. In J. Rasmussen *et al.* (eds.), *New technology and human error.* Wiley, New York, 1987.

[6] Dinges, D.F. and Kribbs, N.B. Performing while sleepy: effects of experimentally induced sleepiness. In Monk, T. (Ed.), *Sleep, sleepiness and performance.* Wiley, New York, 1991.

[7] Hockey, G.R.J., Sauer, J. and Wastell, D.G. Study of human skill maintenance and error management (SKERSI). Technical Report, Psychology Department, University of Hull, 1993.

[8] Hockey, G.R.J. A state control theory of adaptation and individual differences in stress management. In Hockey, G.R.J., Gaillard, A. and M. Coles (Eds), *Energetics and human information processing*, Nijhoff, Dortrecht, 1986.

[9] Lee, J and Moray, N. Trust, control strategies and allocation of function in human-machine systems. *Ergonomics*, 35 (1992) 1243-1270.

[10] Norman, D.A. The 'problem' with automation: inappropriate feedback and interaction, not 'over-automation'. In D.E. Broadbent, J. Reason and A Baddeley (Eds.), *Human Factors in hazardous situations.* Clarendon Press, Oxford, 1990.

[11] Dorner, D. On the difficulties people have in dealing with complexity. In J. Rasmussen *et al.* (eds.), *New technology and human error.* Wiley, New York, 1987.

[12] Janis, I.L. *Stress, attitudes and decisions: selected papers.* Praeger, New York, 1982.

Advances in Agile Manufacturing
P.T. Kidd and W. Karwowski (Eds.)
IOS Press, 1994

Towards a Modelisation of
Trust in a Teleoperation System

I. DASSONVILLE, D. JOLLY, A.M. DESODT
Centre d'Automatique de Lille,Bât. P2, USTL
59655 Villeneuve d'Ascq Cedex,France

Abstract :This works parts of a more vast frame which is the elaboration of a decision support system for teleoperation missions. The choice between the different control modes is really difficult and is function of a lot of parameters. It's the reason why an assistant is indispensable. The decision support system, in a phase of conception. (which was presented two years ago) tries to take into account the maximum of these parameters. and especially those who are typical to the man and to the man / machine relation, as vigilance, stress, tiredness and trust. This is the modelisation of the man / machine trust which is the subject of this paper.
We'll use for this study. the Dempster-Shafer theory as a tool to quantify trust.
First of all we 'll recall this theory.
After that. we'll speak about the theory which was developed by Bonnie MUIR, John LEE and Neville MORAY on the trust in man-machine system ,
We'll then explain our strategy to modelise trust in a teleoperation system, and the way used to translate the facts linked to trust into belief functions.

1. Overview of the Theory of Dempster-Shafer

The first person who presented basic ideas of this theory was DEMPSTER in 1967 [1]. The starting idea was to consider a measurement of probability (a sure and exact information), and to replace this probability by intervals of probability (for which the edges are known) which take into account the uncertainty. Then he has created the low and high probabilities.
SHAFER in his book "A Mathematical Theory of Evidence" [2], took again this idea, extended it and formalised it:
The first matter is to determine the value of a variable. The value of this variable is in a number of components of predetermined values : The frame (F). By different information, the likelihood of each solution A is estimated : a degree of belief that the good solution is in A is assigned to A : $Bel(A) = \sum_{C \subset A} m(C)$
m is a basic probability mass and all the m form the basic probability assignment.
C represents an element of the frame.
The most important property of m is this one : $\sum_{F \subset A} m(F) = 1$

The combination is a fundamental process for belief functions. Its target is to incorporate some information from different sources to create a general belief. Many combinations exist. The most well-known and oldest is the combination of Dempster [3]

2. Trust in a Man-Machine System

When we consider the definitions of trust (in relationships) in the literature ([4],[5],[6]) many different points enter in the elaboration of this parameter. These different aspects are explained by *the multidimensional character of trust.*

Two sociologists explain this character :

In 1983, BARBER [7] explains it telling that trust is a compromise between three kinds of hope linked to the three dimensions : (1) persistence of natural and moral laws. (2) technically competent performance. (3) fiduciary responsibility.

In 1985, REMPEL [8] explains the multidimensional character of trust by the three dimensions : (1) predictability (2) dependability (3) faith .

In fact, if those different dimensions are crossed, then we can count 15 different aspects of trust between human and machine (fig.1):[9]

Figure 1 :

Expectation	Predictability (of acts)	Dependability (of disposition)	Faith (in motives)
Persistence of laws			
Natural physical	Events conform to natural laws	Nature is lawful	Natural laws are survive
Natural biological	Human life has survived	Human survival is lawful	Human life will survive
Moral social	Humans and computers act 'decent'	Humans and computers are 'good' and 'decent' by nature	Humans and computers will continue to be 'good' and 'decent' in the future
Technical competence	J's behaviour is predictable	j has a dependable nature	j will continue to be dependable in the future
Fiduciary responsibility	J's behaviour is consistently responsible	j has a responsible nature	j will continue to be responsible in the future

This table was found (and verified) by Bonnie MUIR who has worked on trust between human and machine.[10]

Her work on trust in a supervisory context has demonstrated that the *variable trust is an authentic causal variable* in a man-machine system. So, we can also study this variable in a teleoperation context (which is different from a supervisory context, but which is too a man-machine system)

3. Approach to modelise Trust in a Teleoperation System

3.1 Trust in a Teleoperation system

A teleoperation system is composed by three important entities :

- *The master universe*, where there are a master arm, and a man (the operator)
- *The slave universe*, where sensors are found with the slave arm
- *The space between the two universes*, where transmission of information, a computer and an assistant are found.

The difficulty to study operator's trust in such a system comes from the fact the operator is himself a part of this system.

The trust he'll have in the system , will be then function of trust he'll give to each part of the system and of his self-confidence :

$$t(H \to TS)= F\left[\ t(H \to MS), t(H \to SS), t(H \to TI), t(H \to C), t(H \to H)... \ \right]$$

But, trust will also vary with reliability of the system, with its performance... The difficulty is to evaluate a personality variable and in the same time to evaluate the dynamic aspect of this variable. Many parameters enter in the composition of trust, and a choice of pertinent parameters is needed. The parameters chosen for our application, because they seem to be the most characteristic, are:

- *Selfconfidence*
- *Trust in other*
- *Trust in machine , in general*

- *Trust in parts of the system*
- *Performance.*

In order to modelise human's trust in a teleoperation system, we expected to use belief functions.

First of all, we must initialise those functions. For doing that, we use a questionnaire which is under development with psychologists.

The operators are bound by this questionnaire before working with the teleoperation station. Its principal aim is to know if the operator has trust in himself, has trust in others, or trust in machines, because his behaviour with the teleoperation system will be really different according to his affinities.

In fact, this questionnaire permits to qualify the *a priori trust* (the a priori trust is the static part of the variable trust for one person), and initialise belief functions.

In order to qualify the *dynamic part of the variable trust* of this operator, we'll use scales during the experimentation. : the operator must estimate his trust on different parts of the system, his selfconfidence..., and his confidence on his responses.

The study of those different scales'll allow to qualify the dynamic evolution, with the study of performance.

In fact, the responses will be converted to belief functions, and at each new response, the belief functions will be computed again.

At the end of the experimentation, the operator will again answer a questionnaire to know if the trust in the teleoperation system is different from the beginning of the experimentation.

3.2 Our Simulation

Before testing our approach in the real site, a first experiment in a simulation context is required.

We have taken the following choices, to simulate the different parts of the system:

- Simulation of the master universe : an operator which manipulates a joystick (simulation of the master arm)
- Simulation of space between the two universes : a PC type computer
- Simulation of the slave universe : a cursor which executes joystick's orders
- Simulation of a task : an operator must conduct with the joystick the cursor from a point A to a point B along a trajectory.

Before executing the task, a questionnaire is submitted to the operator (the task won't be executed if the operator won't answer it)

This questionnaire is necessary to evaluate the a priori trust. The questions are about - Selfconfidence - Trust in others -Trust in machines . The responses of this questionnaire are used to initialise the experimentation.

When the operator has answered it, the task begins. Then he must conduct the cursor from a point A to a point B, without making mistake and the faster as possible. A mistake is made when the cursor touches an edge of the trajectory. The screen of the PC has the following design :

In the right corner, a window for the profile of the trajectory is reserved. In fact, the trajectory isn't always plane.

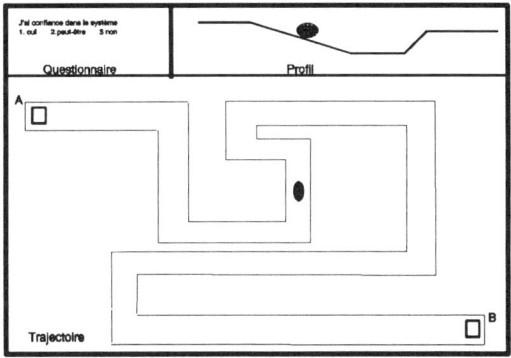

There are some slopes (up and down). So the cursor undergoes some acceleration and some deceleration. Theses changes of levels are been chosen to permit the execution of the three modes of the Teleoperation :
- In the automatic mode : the cursor executes alone the trajectory (the computer decides on the trajectory)
- In the manual mode : The joystick gives the orders to the cursor (the operator alone decides on the trajectory)
- In the semi-automatic mode : the computer takes into account the acceleration and deceleration of the cursor. To the operator, the trajectory seems to be plane.

In order to create a situation of distrust, the cursor doesn't always execute exactly joystick's orders. For example, in a slope, the cursor could have an abnormal acceleration, or could be inclined to go to the right (or the left) edge. The anomalies are pre-programmed and becomes visible always at the same time (from an operator to another one) .These are those anomalies which will incite the operator to change the driving mode, this information of change is a complementary observation of the scales concerning the evolution of trust.

4.Conclusion

At this moment, the experimental protocol and the questionnaire are realised. The first subjects are going to be submitted to the questionnaire and to the simulation. After these first trials, it will be necessary to quantify more precisely the parts of scales, and of change observation in the characterization of trust.

References

[1]   P.Smets *"What is Dempster-Shafer's model ?"* TR/IRIDIA/91-20
[2]   G.Shafer *"A Mathematical Theory of Evidence "* Princeton University Press, NJ,1976
[3]   I.Dassonville,D. Jolly, A.M. Desodt *"Modelisation of a teleoperation system using belief functions"* XII.European Annual Conference on Human Decisionn Making and Manual Control ,June 1993
[4]   M. Deutsch *"The resolution of conflict"* Theorical essays
[5]   F.Heider *"The psychology of interpersonal relations"* University of Kansas 1958-L63/2389 B.L.L.D.
[6]   J.B. Rotter *"Interpersonal Trust,Trust worthiness,and Gullibility"* University of Connecticut American Psychologist January 1980-1
[7]   B.Barber *"Logic and the limits of Trust "* New Brunswick, New Jersey:Rutgers University Press 1983
[8]   J.K.Remple J.G.Holmes M.P.Zamma *"Trust in close relationships"* Journal of Personality and social Psychology 1985 Vol 49,n°1,95-112
[9]   B.Muir *"Trust between humans and machine and the design of decision aides "* Internationnal Journal of Man Machine Studies 27,527-539
[10]   B.Muir *" Operators' Trust in and a percentage of time spend using the automatic controllers in a supervisory process control task"* Doctoral thesis. University of Toronto.

Advances in Agile Manufacturing
P.T. Kidd and W. Karwowski (Eds.)
IOS Press, 1994

511

Supervisory Control of a Computer Aided Teleoperation System

Hervé Le Bars, Philippe Gravez
CEA/DTA/DPSA/STR
Bat.38 92265 Fontenay-aux-Roses Cedex, France

Patrick MILLOT
LAIH - Université de Valenciennes et du Hainaut-Cambresis -
Le Mont-Houy - BP 311 59304 Valenciennes Cedex, France

Marie-Claude THOMAS
I3S - Université de Nice - Sophia Antipolis
Bat.4 Av.Albert Einstein Sophia Antipolis 06560 Valbonne Cedex, France

Abstract. This paper addresses the supervisory control of a Computer Aided Teleoperation (CAT) system. We want to help the operator in setting the system for the best with respect to the task. The assistance system we designed puts together several concepts as graphs representations, object-oriented mechanisms and meta-level module. The elements of this system, its architecture, its realisation and the experiments to be carried out are described.

Problematics and existing solution

Maintenance and dismantling of nuclear plants devices is a growing field of applications for telerobots. Here are some characteristics of this kind of applications :

- each mission is unique. They all have specificities that must be studied separately. Since there is generally no emergency, enough time can be spent on a preparation phase ;

- some elements can be reused from a mission to another. The work generally deals with operating a specialised tool and has two types of constraints : the ones related to the tool itself, and the ones related to the particular context of application. It is therefore possible to define generic tasks describing how to manipulate tools and to adapt them to each situation : how to grasp a tool in a rack, how to operate a plasma torch, how to use a drill. Formally, the preparation part of the mission can then be devided into two parts : a long term preparation to define generic tasks through laboratory experiments and a short term preparation to adapt "in situ" to the local conditions. In fact, a mission can be seen as the dynamic use of static tasks with parameters values for context arrangements ;

- the complexity and uncertainty of the remote environment forbids a total automation of the mission and calls for on line decision making and a close man-machine cooperation.

This last point explains why CEA chose a teleassistance approach for the use of computer in master-slave systems. This approach is to give the human operator the central role in the system to take advantage of his manipulative skill and initiative capabilities, and to assist him with sophisticated control modes and information feedback processing. At CEA, a software package called TAO-2 controls the interaction between the two arms, allowing manual, automatic and semi-automatic (control shared between the operator and the computer) control modes [1]. TAO-2 also provides assistance in information feedback, as force feedback tuning, possible use of proximity sensors informations, possible task-oriented informations, graphical simulation of the remote environment.

The operator uses the various capabilities of TAO-2 through a language called SPARTE [2]. SPARTE is dedicated to the description of CAT behaviours and can be used as a command language (the operator sets the system into the desired control mode with the desired information feedback settings) or as a programming language. In this case, the program written specifies the various behaviours the operator wants to go through to realise the task. The changing of behaviour, can be attached to special events connected with the task (operator switch, sensor information) : the program then adapts to the real conditions and to operator's acts that couldn't be foreseen. Even more, the operator can stop and resume the program very easily to adapt to unexpected problems.

SPARTE has been intensively experimented and has demonstrated the advantages of TAO-2 and the teleassistance approach. It's a good tool to communicate with the CAT system, but it is not fully satisfactory as a tool to assist the operator in choosing the right behaviour at the right moment :

- writing the program is not easy and must be done by a specialist ;
- the generic tasks defined out of the past experience can't be easily expressed and reus
- defining different possible strategies or changing strategy in the middle of the action is difficult ;
- no assistance is provided to the decision making.

All this calls for an assistance system taking into account the characteristics of the CAT system and missions to help the operator in planing and supervising the execution. The classical methodology of man-machine systems design must be adapted to this problem [3].

Basic concepts of the assistance system

The assistance system must store the generic knowledge on teleoperation tasks. We found it convenient to use a multilevel graph representation for this knowledge. The levels are the operation, which describes the utilisation of a tool (plasma cutting operation), the task, which expresses a local goal (grasp the drill) and the action, which refers to a CAT mode (go to the rack in manual mode). Each operation is represented as a graph of tasks and each task is represented as a graph of actions. The advantages of this representation are :

- genericity : the elements of the knowledge (operations, tasks or actions) are stored in a generic form in the database. If an element is used as the node of a higher level graph, all specific information is written in the link between it and the other nodes of the graph. This makes it possible to use any element separately as well as within a graph ;
- richness : the knowledge is explicitly expressed : it doesn't have to be compiled into rules or procedures to become usable. We can express things as they appear to us during experimentations, even if we can't really analyse them ;
- modularity : graphs are very easy to modify and can be used as part of a more sophisticated one.

However, this graph representation is somewhat too rigid and can become complex if knowledge grows : it has to be blended with other mechanisms.

To prevent the database from being to complex, a object-oriented structure is being used : all tasks are linked to objects that are related to them. For example, the task "grasp (Tool)" is stored into the object "Rack" Tool is into. We can therefore take advantage of the object mechanism

- polymorphism : the actual grasp task can be different from one rack to the other. The correct grasp task is automatically found when the right rack is selected ;
- inheritance : structured families can be built and objects can inherit general characteristics of their family and add their own specificity. Associated tasks can be built as the union of a general graph and specific branches.
- modularity : objects make the knowledge easy to retrieve and to modify.

The rigidity of graphs is another important problem. Graphs express sequences independent of the context : a (simple) representation of the operation "drill" is "grasp the drill" - "drill" - "release the drill". Obviously not all three tasks are at the same level : drill is the "heart" of the operation, grasp and release are making drill possible and are context dependent. Graphs can be simplified if the context dependent tasks are replaced by constraints on the fundamental ones. A constraint checking mechanism will take the generic graph and add tasks according to the current context.

Contrarily to most of robotics applications, there is no clear frontier between planing and executing in teleoperation. No comprehensive plan can be designed off line and the operator has to negotiate between the efficiency of a precise plan and the flexibility of a more general one. This feature introduces the problem of the temporal organisation of the different steps of the mission : how far must we go in the off line definition of the plan, what are the important decisions to be taken before the beginning of the realisation? These questions depend on generic knowledge as well as particular conditions, and no unique answer can be given. We therefore feel the need for a module specialised in the treatment of this problem : we must separate the planification process and the planification organisation process, which deals with reasoning on the planification knowledge and is situated at a meta level. An important place must be left in the system for a metaplanification module.

Based on these ideas, we designed and realised a supervisory control assistance system. A short description of this system is given in the next section.

A supervisory control assistance system

The knowledge involved in supervising a CAT mission has two main characteristics :

- it is poorly defined, since no expertise can be collected from an existing human supervisor ;

- it involves many different aspects and refers to different techniques : use of generic knowledge, context understanding, decision making, constraint evaluation, computation of parameters and so on.

A multi-agent approach is well suited to these characteristics. Its modular aspect supports evolution and modification, and it can implement different independent techniques. The architecture chosen to organise the agents is the blackboard [4]. The blackboard architecture clearly separates the resolution of the problem, done incrementally and opportunistically on a shared memory, and the control of the agents that help solving the problem :

This architecture suits particularly well to our need to independently express the processes of planification and metaplanification. Clearly, our metaplanification modules corresponds to the control part of a blackboard. To enable a intelligent process at this level, an original architecture has been implemented [5] (fig. 1).

The main advantages of this system are as follows :

- distribution of knowledge : each problem is clearly separated from the others. Building the system goes naturally with finding out what to do in each particular case. The system is fully modular and evolves with our knowledge ;

- justification of the decision making : at any moment, the state of the whole system can be available to the operator : current, past and future states of the plan, current, past and future steps in the mission, current and past contexts. All advice and directive to the operator is therefore justified. During experiments, all these informations can be recorded for analysis, and their significance can be evaluated.

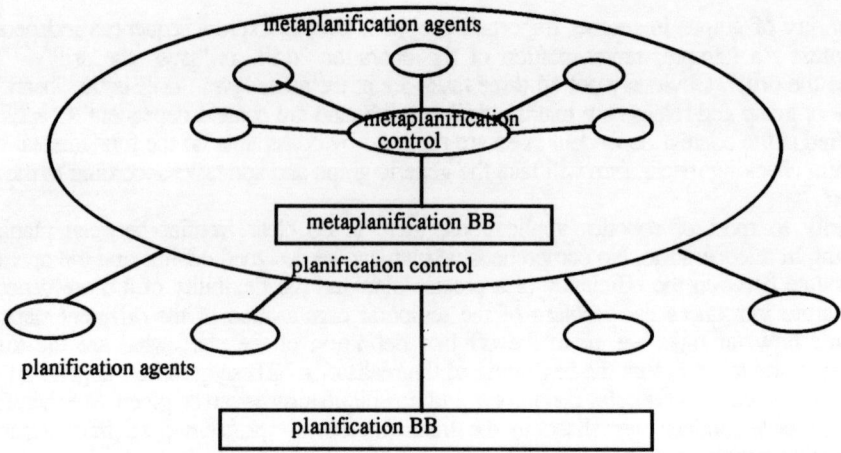

fig. 1 : blackboard architecture of the supervisory control system

Applications of the assistance system

This system is currently being connected to the TAO-2 package that drives MA-23 master and slave arms.

The system is tested on a drilling operation. The generic description of this operation takes a lot of different possibilities into account. One or several holes can be drilled : in the later case, they can belong to a predefined trajectory or not, and several methods can be used to define this trajectory (straight line defined by two points, curve, user-defined trajectory and so on). The drilling can be automatically, semi-automatically or manually done. In the two first cases, the orientation of the drilled surface must be learned. This learning can be achieve by different types of active or passive sensors. The grasping and release of the tools used can also be manually or automatically done. Depending on the tool rack, different techniques for learning and automatic grasping must be available. Moreover, incidents can happen during execution : drill jammed or broken, bad release or grasping of a tool, new objective set by the operator. Alternative strategies must be found and presented to the operator in each case.

Obviously, this system must show a much better ability to cope with the richness and flexibility of such an application than a SPARTE program can offer. The experiments aim at evaluating the performance of the system on those aspects and the validity of the proposed concepts. The system must allow not only a better supervisory control, but also a better global mission management, from planification to execution.

Adjustments in the human-computer interaction can also be made and a better understanding of the human supervisor role and needs has to be achieved.

References

[1] C. Terré, J. Vuillemey and Ph. Gravez, Remote task performing with the TAO-2 Computer Aided Telerobotics system. Proceedings of ORIA 1991, Marseille, France, pp 323-332.
[2] Ph. Gravez and R. Fournier, Symbolic control in Computer Aided Teleoperation : the SPARTE language and its basic principles. Proceedings of ICAR 1991, Pisa, Italy, pp 225-23
[3] P. Millot and E. Roussillon, Man-Machine cooperation in telerobotics : problematics and methodologies. Proceedings of the 2nd Franco-Israel Symposium on Robotics, Avril 1991, Saclay, France.
[4] M. Occello and M.C. Thomas, A parallel blackboard generic tool for intelligent robotics. Proceedings of IEEE Tools for Artificial Intelligence 1992, Arlington, USA. IEEE press.
[5] Ph. Gravez, H. Le Bars, M. Occello and M.C. Thomas, A distributed blackboard application to decision-making in Computer Aided Teleoperation. Proceedings of IEEE SMC 1993, Le Touquet, France, vol 3 pp 635-639.

Advances in Agile Manufacturing
P.T. Kidd and W. Karwowski (Eds.)
IOS Press, 1994

Vigilance In A Teleoperated Task

S. MESTIRI, D.JOLLY, J.M. JACQUESSON*, A.M. DESODT
Centre d'Automatique de Lille, Université des Sciences et Technologies de Lille
59655 Villeneuve d'Ascq Cedex
**Laboratoire de physiologie, Centre Hospitalier Universitaire Place de Verdun*
59000 Lille FRANCE

Abstract. In this article, we deal with the vigilance of human operator in a simulated teleoperation task. The task includes a manual control and an automatic control. The experiments already made, show some phenomena superposing on the measure of vigilance through performance : training, differences between subjects..., The measure of reaction times, the duration of reaction, and the surface of the response curve to steps perturbations allow us to put in evidence a decrease of vigilance.

1 - Vigilance

The vigilance is the capacity of the brain to respond actively to unexpected situations in relation with the attention capacity the subject is able to. It depends on the state of the central nervous system which is supposed to act on performance during a lengthly vigilance task(1), it is synonymous with sustained attention(2). Initially, the concept of vigilance has been introduced by the neurologist Head in 1923 (3) so as to define "the state of efficiency of the central nervous system" that insures quickness of adaptative responses. After clinical observations on patients either anorexic or under narcosis, he established a direct relation between the level of reactivity of the nervous system on one hand, and the accurateness and quickness of the comportemental responses on the other hand.

Then, and after a request coming from army, Mackworth(4) establishes an experimental protocol, putting in evidence the decrease of vigilance, that he defines as the deterioration of the operator's capacity to stay vigilant during a continuous stretch of time, as shown by the decrease of the number of signals detected then.

2 - Experimental Protocol

The decrease of vigilance during monotonous tasks is a psychophysiologic law(5), thus; the measure of physiologic indexes during vigilance tasks gives not only a descriptive complement but also an information useful to understand the phenomenon.

So as to evaluate the vigilance level we use objective and subjective measures.

Objective criteria : 1 - The results of task execution, the difference between the task to be executed, and the task really done.

2 - The recording of the operator's EEG during the task.

Subjective criteria : response of the subjects, at the end of the task, to a questionnaire derived from V.A.S.(6) concerning its state of vigilance, its performance, tiredness, stress and motivation.

3 Experiments and Results

During all the experiences, the subjects are seated in a confortable position, alone, in a dark room, at an 80 cm distance from the screen. The joystick is commanded in velocity. We began with the simulation of a teleoperation task before the experimentation with the telemanipulator of the laboratory.

The first experiments aim at the study of the phenomena that superimpose on the study of vigilance.

3.1 Experiment 1 : During the first experiments series, the subjects (7) had to cancel the displacement of a target on a screen. The target is driven both by the action of the subject on the joystick and by a sum of sinusoidal functions. For these experiments, the duration of the task is 15 mn, the performance is measured by the distance between the center of the screen and the position of the target.

Those experiments show at evidence :

* The learning effect : each subject having realised five trials scattered on the whole day, it has been possible to notice a significant increase of the performance (all subjects being miscued), this increase is the consequence of training.

* Inter individual differences : the comparison of the performances of subjects having realised the same number of trials show significant differences. This can be explained by differences such as age, sex, personality(10)...

3.2 Conclusion 1 : Each subject needs its proper reference for each kind of task, this reference moves during the time : if the performance increases (training) then, the reference changes while, if the performance decreases, we can consider that the importance of training is small in the face of the effect of the vigilance decrease.

3.3 Experiment 2 : The EEG is recorded on subjects in 4 situations, each of them lasting 5 mn : eyes open, eyes closed while calculation, eyes open while calculation, and reading aloud. This allowed us to characterise each situation and to put in evidence that some subjects generate alpha waves.

3.4 Conclusion 2 : We must take subjects that do not generate alpha waves and make trials either with closed or open eyes (alpha and mu waves are strongly linked to the state open or closed of the eyes).

3.5 Experiment 3 : So as to limit the effect of training and to get close to a teleoperation task, we realised a series of trials with a task composed of two different modes described on figure 1.

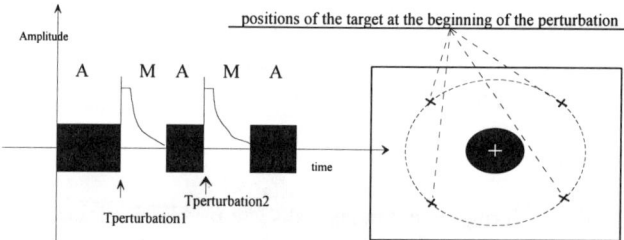

Figure 1 : Chronogram of a trial

During the automatic mode, the subject supervises the displacements of the target on a screen, he waits for the moment he must act and during the manual mode he tries to get the target back in the center of the screen. The transition between automatic and manual mode is observed at the time when the target goes outside its previous trajectory because of a step perturbation. When the target, during the manual mode, has come back to the center of the circle, the system works in a automatic way. The trials last 1 hour and 10mn and include about 70 random perturbations. Seven subjects have been studied.

Positions of the target according to vertical and horizontal axis, values send by the joystick, kind of control mode are recorded. Theses rough data allow, for each perturbation, to compute : reaction time, reaction duration, response time, and surfaces of reaction curves as shown in figure 2.

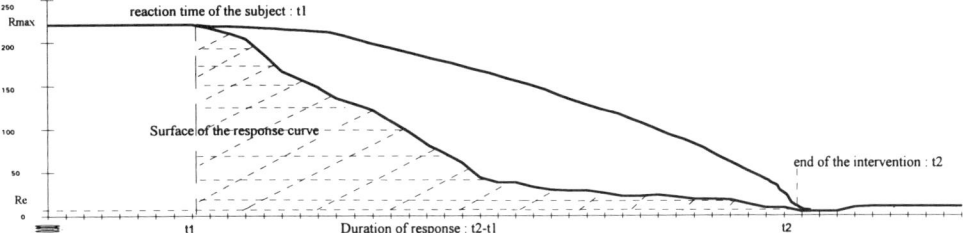

Figure 2. curves of response to a perturbation, t=0 is the time when the perturbation appears, the hachured part is the surface of the reaction curve (Scr). It is possible to have the same reaction time and the same reaction duration, but different reaction surfaces, so as to characterise the response quality.

It is important to know if secondary variables, such as direction of the perturbation (a) and duration of automatic mode(b), have an influence.

(a) The step perturbation has 2 possible directions (4 ways). A statistic study show that this variable has no influence : there are no significant differences in the responses of the subjects for the 4 kinds of perturbation.

(b) The perturbation appears at random, nevertheless, the automatic mode duration can be 10s, 20s, 30s or 40s, Fischer and Student tests show no-significant differences between the responses: responses can be compared even if the previous automatic mode lasts differently.

The values measuring the execution of the task by the subject, evolve in a instantaneous way. For clarity sake, we make an average on 10mn. The Student test shows a significant difference in the reaction time and reaction duration during the different periods. The detection of the target out going of the screen needs no training period on the contrary of the reaction duration. The training period is determined thanks to the values of the reaction times (Tr), for, in figure 3, we can notice three different periods :

In the first part of the this curve (representing the first 30mn of the trial), the reaction duration decreases, reaction time increases showing a decrease of vigilance but response time decreases due to the learning effect. In the second part of the reaction time and reaction duration(Dr), we consider that we are in mean vigilance. The evolution of the reaction duration is lessened by learning. In the last part of the curve, there is an increase of all the variable due to the loss of vigilance.

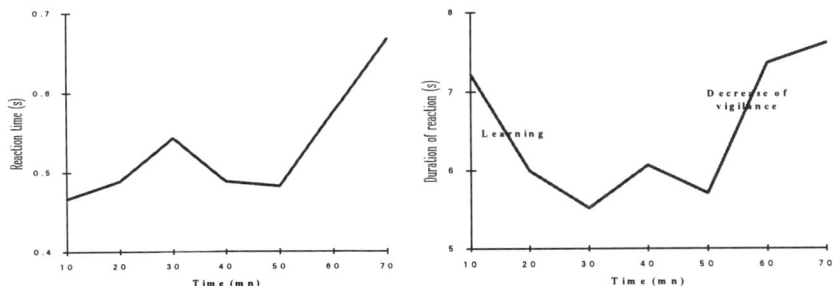

Figure 3 : Curves of the response times, reaction times. Each point is the average of the values on 10 mn

But, the values of reaction times indicate how quickly the subject reacts and not how efficiently he reacts. To characterise the quality of the response, we use Scr the surface of the response. This surface takes in account the strategy of the target on the screen. So, the computation of the vigilance level is made using the three components Tr, Dr, Scr with an

utility function defined as : $Vig(t) = a \times \left[\dfrac{min(Tr)}{Tr(t)} \right] + b \times \left[\dfrac{min(Dr)}{Dr(t)} \right] + c \times \left[\dfrac{min(Scr)}{Scr(t)} \right]$

The values a, b et c depends on the particular task realised, as well as the definition of the minimum vigilance threshold.

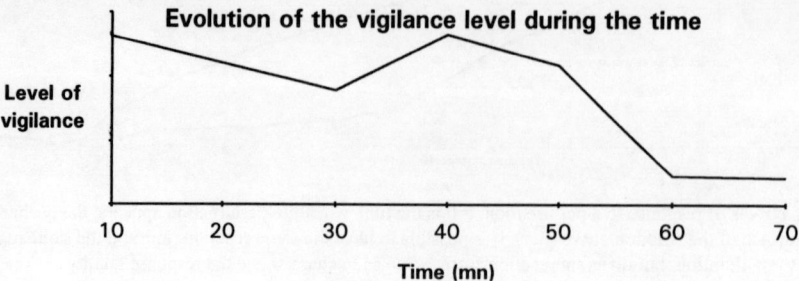

figure 4 : Curve showing the estimated level of vigilance

The level Vig(t) obtained at time t is a function of the best result obtained for each of the variables, for we use the minima of each variable as a reference. The training periods are so excluded of this curve.

4. Conclusion

The first experiences refered to in (7) show inter individual differences due to age, sex, and intra individual differences such as training, global state of the subject...This led us to change the task in a quick training one.

Because to the importance of inter individual differences, we compare the performance of the subject with respect to its own performances. When, during the trial, the performance gets better, then the reference values of the subject are changed.

The vigilance level computed is one of the criteria of the choice of the control mode in teleoperation.

Our next experiments will correlate measure of performance and EEG indexes such as evaluation of the ratios of alpha, beta and theta waves as a function of response time during all the task.

REFERENCES

(1)D. DAVIES,.R. PARASURANAM, "The psychology of vigilance", Academic Press, London, 1982
(2) R. PARASURAMAN, "Vigilance, monitoring and search", in Handboik of perception and human performance, Cognitive processes and performance, Vol. II chap 43,1986
(3)H.HEAD, "The conception of nervous and mentel energy, vigilance, a physiological state of the nervous system", 7th International Congress of Psychology, 1923
(8)N.H.MACKWORTH, "Research on the measurement of human performance", Medical Research Council HMSO, Special report n°268, London, 1950
(5)C.TARRIERE, A.WISNER, "Effets des bruits significatifs et non significatifs au cours d'une épreuve de vigilance", Le travail humain, 25, pp 1-28, 1962
(6)T.H.MONK, V.C.LENG, S.FOLKARD, E.D.WEITZMAN, "Circadian rythms in subjective alertness and core body temperature", Chronobiologia, 10, 49, 1983
(7)S.MESTIRI, D.JOLLY, J.M.JACQUESSON, A.M. DESODT, "Evaluation of the vigilance variations during a teleoperated task", ICARV, Singapore 92

Advances in Agile Manufacturing
P.T. Kidd and W. Karwowski (Eds.)
IOS Press, 1994

Combined Visual and Haptic Sensory System For the Identification of Remote Objects Using Teleoperation

Morris Driels

Department of Mechanical Engineering,
Naval Postgraduate School. Monterey, California 93943, USA

Abstract. It is proposed that a hybrid sensory feedback system comprising a visual peripheral component together with a haptic component corresponding to that of visual foveal information, is equivalent to that of full visual sensory feedback. Such a system is constructed and the ability of subjects to percieve objects using it is investigated by observing and classifying their search strategy.

1. Background

The focus of this work is directed towards the next generation of underwater remotely operated vehicles (ROV's) in which sensory information from the remote work head on the ocean floor is fedback to the human operator on the surface. If the quality of the feedback is high, the operator experiences a sense of telepresence, enabling him to perform the mission in a natural manner. Although visual feedback is obviously the most important sensory signal, divers confirm that the quality of direct visual contact with the object under study is at best poor, and often absent altogether. Human divers then rely only on their sense of touch to accomplish their mission, which they are able to do with surprising skill, considering the environment in which they work. It is important, therfore, that work be conducted into the role the haptic (touch) system plays in object identification of objects, particularly when a machine such as a telemenipulator is interposed between the master and slave units.

2. Previous Work

There are a number of related activities which may be brought to bear on the problem of exploring the potential correlation of haptic and visual search. Much work has been performed by Professor Stark and others [1] into visual search patterns. This concludes that visual search consists of a number of fixation points in an image with the observers eyes moving rapidly from one to the next.

These movements, or saccads, occupy about 10% of the total viewing time, and the mechanism is described as the scanpath process. Different observers have different scanpaths yet each observer repeats the scan-path in a cyclical manner. Further work by Stark and Ellis [2] led them to conclude that visual cognition was essentially a model based, or top down process in which observers attempted to match features to a stored model.

At this point, the haptic system is defined to be comprised of two major sensory systems. The first is the tactile system comprising the fingers and the palm, while the other is the kinesthetic (sometimes called the proprioceptive) system which uses information from the spatial position of the major limbs (wrist, forearm, upper arm) to determine the location of an object in space). Attention is focussed on the kinesthetic system since the original thrust of the work was to understand the mechanics of haptic probing strategies and the identification of objects in hazardous environments, such as in space or underwater, where much of the tactile (and possibly visual) sensory modes were absent.

The author has conducted research into the mechanics of haptic probing by decoupling the tactile sensory system from the kinesthetic system by using a telemanipulator as shown in figure 1. In this system the operator is visually and audiably masked and probes an object located on a task board. The manipulator reproduces his movements and is capable of full force reflectance. In studies of probing strategies [3], similarities with the visual scaning of objects was observed.

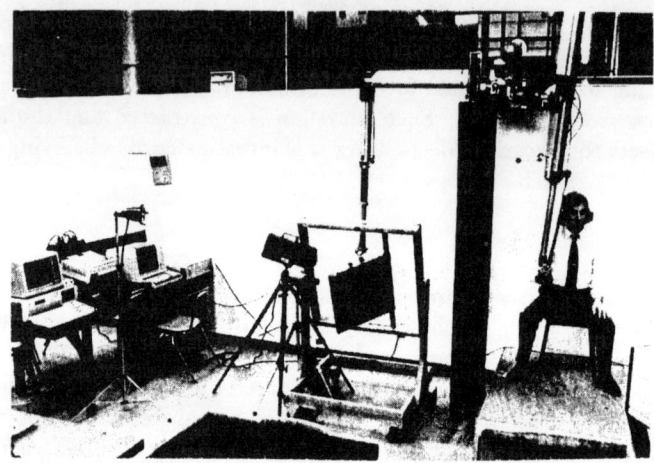

Figure 1: Force Reflecting Telemanipulator

The tests consisted of attempting to identify wooden letters of the alphabet using a single rigid probe located on the end of the remote part of the teleoperator. Operators provided verbal feedback on their actions and accurate probe motion was recorded on video tape. Because the probing took place over several minutes (compared to the few seconds of visual observation) more information was gathered about the mechanics of the search process. The qualitative observation from the tests prompted the hypothesis that searching begins as a feature based (or bottom up) activity until the operator thinks they know what the character is or suggests one of a small group that the character might be in. At this point the search transitions into the model based mode suggested by Stark as a means of testing the model. If successful the search ends, if not the operator returns to a feature based approach.

Just as the haptic system may be considered to be made up of the tactile and kinesthetic components, the visual system is often considered to operate in a similar manner. The foveal component is a small area of interest that subtends an angle of about one degree at the eye. In this restricted field of view, detailed observations may be made, and features extracted from the image. The other component of vision is the peripheral field, in which little detail of the observed object is available, but gross features (such as its boundaries or large internal holes) may be recognized.

If the above analogy between the haptic and visual search systems is correct, then haptic search may be considered to be equivalent to visual search without a peripheral component, in other words, a foveal search. Previous work comparing full visual search and haptic search of this particualr object set [4] confirmed that haptic search was very sequential in terms of the features explored, while visual search exhibited the traditional scan-path sequence.

3. Experimental Method

The research performed attempts to construct a hybrid sensory feedback system in which the operator is asked to explore objects using a peripheral visual sensory channel and a haptic sensory channel which may be considered foveal in nature. In this manner, it may be argued that a sensory system equivalent to full vision is available to the operator for object exploration. Whether this is in fact achieved is the subject of the research, and was investigated by conducting two sets of experiments.

(1) Operators used the force sensing manipulator described previously to explore randomly oriented, randomly selected wooden letters of the alphabet mounted on a task board. The exploratory procedure used was recorded on video tape and analysed. This experiment involves only the haptic senory channel, and may be expected to lead to a sequential search strategy.

(2) In the second experiment, the operator has the haptic sensory channel just described, but also a video image of the object being probed. In order to prevent the object from being rcognized directly from its visual image, and to make the image more representative of the blurred nature of peripheral vision, the image of the object was blurred by an image processing system.

4. Haptic Exploration

This was quantified by means of the sequence ratio S_r defined to be the ratio of the number of sequential features searched to the total number of features searched, and may be expressed as a percentage. Previous work by Acosta had noted that purely haptic probing was highly sequential in nature $(S_r \approx 95\%)$, while full visual search was relatively random with respect to object features $(S_r \approx 10\%)$. It was concluded that the sequence ratio was, therefore, a good discriminator between these two modes of search.

5. Haptic and Visual Search

In this series of tests, the subjects were allowed to view a blurred image of the object together with an overlayed probe position, in addition to the haptic sensory channel through the telemanipulator. The experimental data was recorded and analysed in the same manner as before, and the explorative procedure examined for how sequential it may or may not be.

6. Results and Discussion

The experimental results may be summarized in table 1. In this table, the number of reversals is defined to be the number of changes in direction of the haptic probe during the execution of an explorative procedure.

Six specific sets of data were compared for each subject's results. Three of these, The total number of features probed, time to recognition and number of probing reversals were considered as "raw" data which was heavily dependent upon both object selected and operator proficiency.

The total number of features searched and the number of reversals were greater in the haptic mode than in the hybrid mode for all subjects. The average time to recognize the object was almost identical in both modes for the first two subjects, and was greater in the haptic mode for the third subject. To obtain more normalized

	Dual Subject 1	Haptic Subject 1	Dual Subject 2	Haptic Subject 2	Dual Subject 3	Haptic Subject 3
S_r	93.77	93.59	98.20	96.39	92.24	89.82
Av # Features	87.09	122.0	42.86	52.17	41.40	68.43
Av Recog Time	122.5	121.6	88.86	86.00	49.70	68.57
Av Recog Rate	0.762	1.000	0.507	0.609	0.828	1.056
Av # Reversals	14.00	22.00	5.29	9.50	15.30	23.57
Av Reversal Rate	0.164	0.175	0.123	0.193	0.347	0.353

Table 1: Experimental Data

results between modes, two rate based sets of results were computed. The recognition rate (total number of features probed/time to recognition) and reversal rate (number of reversals/number of features probed). In all three subjects, the recognition rate was significantly higher (27%) in the haptic mode than in the hybrid mode. This indicates that the visual sensory channel does not increase the subjects ability to receive and process information at faster speeds.

The sequence ratio for purely haptic probing is generally in the region of 95%, confirming the previous results of Acosta. The average value of S_r for hybrid search was surprisingly high at 93.37% The importance of this result is that the peripheral visual component presented to the operator does not prompt him to alter the search strategy, and indicates significant reliance on haptically aquired data.

7. Conclusions

1. Inclusion of peripheral visual data does not significantly change the object search strategy, at least for the method used to present peripheral visual data.

2. Peripheral vision is a relatively weak source of information relative to haptic data.

3. The provision of peripheral visual data did provide some spatioal information, as indicated by fewer search reversals and fewer number of features needed for recognition.

8. References

[1] Noton D. and Stark L., "Eye Movements and Visual Perception", *Scientific American*, Vol 224, No 6, June 1971, pp 34-43.

[2] Stark L. and Ellis S., "Scanpaths Revisited: Cognitive Models Direct Active Looking", it Eye Movements: Cognition and Visual Perception, edited by D. Fisher, R. Monty and J. Senders, Lawrence Erlbaum Associates, New Jersey, 1981, pp 193-226.

[3] Driels M., "Haptic Recognition Through Remote Teleoperation", Ergonomics of Hybrid Automated Systems II, edited by W. Karwowski and M. Rahimi, Elsevier Science Publishers, Amsterdan, Holland 1990, pp 871-878.

[4] Acosta J., "Modeling Of Explorative Procedures for Remote Object Identification", MS Thesis, Naval Postgraduate School, Monterey, CA, 1991.

Advances in Agile Manufacturing
P.T. Kidd and W. Karowowski (Eds.)
IOS Press, 1994

Evaluation of the Criteria for a Decision Support System in Teleoperation

F. WAWAK, A.M. DESODT, D. JOLLY
Centre d'Automatique de Lille, Bât. P2, USTL
59655 Villeneuve d'Ascq Cedex, France

Abstract. The purpose of the paper is to underline methods to estimate fuzzy criteria putting together different kinds of information, in a decision support algorithm, this algorithm aiming to make easier the choice, for a task, of the driving mode in a teleoperation system. All the described methods conduct to a possibility distribution that is approximated to a trapezoidal form to be then compared to the fuzzy sets of the expert preferences evaluation. The fuzzy compatibility degree resulting from that comparison is the evaluation of the criterion.

1. Introduction

Our researches have been moved for several years now towards the elaboration of a decision support system for dynamic task allocation. Due to our application field that is teleoperation, we are working more especially on the study of a real time fuzzy decision support algorithm for the choice of control modes in teleoperation [1]. That fuzzy algorithm is based on the comparison of fuzzy appreciation of different criteria to the fuzzy preferences of the expert regarding the possible driving modes to achieve the task. The fuzzy compatibility degrees, results of the comparison, will be then aggregated to make up the global evaluation of the teleoperation system facing each driving mode, and thus to take the decision [2].

The paper aims to specify the build of the criteria evaluation setting out ideas on techniques using fuzzy tools put in the teleoperation context.

2. The Teleoperation Criteria

Nowadays, the development of Advanced Teleoperation systems multiplies driving modes. Indeed, to execute a task, it is possible to choose a large range of modes, from the manual to the automatic mode, depending on the task sharing between the man and the machine [3]. Hence, a choice among the modes has to be done aiming to obtain a good realisation of the task. The modelisation of a decision support system for teleoperation needs different kinds of criteria [4] coming from the task, from the man-machine system, from the link between the man and the machine, and from the man himself. So, the study of the teleoperation system state with respect to each control mode is made through different sets of criteria.

The first of these sets qualifies the task through its *required time* by the comparison of the available time with the required time for each driving mode, and qualifies its *execution simpleness* with each driving mode that could be expressed by an experienced operator.

The man-machine system is studied through its *performance* given by an historical account of the tasks execution success.

The criteria of the third set aim particularly to observe the links between the man and the machine that is to say the transmitted information. Therefore, we need to define the *quality* and the *quantity* of the information for the close past time.

The last set describes the operator of the man-machine system watching for his possible fickleness. Hence, we have chosen to study the man through his *vigilance* based on the analyse of his reaction times for the required task and through his physical *workload* that estimates his working part during the close past time.

According to the vagueness linked to the criteria described above, we have turned towards fuzzy tools to fulfil the evaluation of the criteria and of the preferences.

3. The Fuzzy Tools used to modelise the Criteria

The nature of the criteria being different from one another, we have to search for an homogeneity of the putting into form of the information bringing, the aim of the decision algorithm being to aggregate the information. Afterwards, we are looking for building possibility distributions to express the criteria real value and fuzzy sets to express the expert preferences.

There are two techniques to deduce a possibility distribution from data. In one hand, Zadeh suggests to build the possibility distributions from fuzzy sets by the association of a random variable to the fuzzy set, the possibility distribution taking the outline of the fuzzy set [5]. In an other hand, Dubois and Prade propose a statistical method based on the transformation of an histogram into a possibility distribution [6]. Thereby, the evaluation of the real value of the criteria is boiled down to the evaluation of a fuzzy set or of an histogram.

We can use several ways to construct a fuzzy set. The method that comes immediately in mind is the one which calculate the parameters of the fuzzy set outline, parameters depending on the outline functions. Yager puts emphasis the extension principle that allows to extend a crisp function to a fuzzy one if at least one of the inputs is fuzzy. The last method we are pointing out is the linguistic approach, the natural language being translated into fuzzy sets, for example, by the tools proposed by Zadeh with his method "PRUF" [7]. But, these evaluation to be realised need a continuous phenomenon that is measurable at each moment. The random phenomenon requires statistical estimations leaning on histograms.

An histogram is usually built by the count up of events classified into classes. That technique is expandable to a fuzzy count up. Thus, for the fuzzy case, we not just index the events into the classes but a membership degree is given to each event and we add up into the classes these membership degrees.

The different techniques are summed up below with the figure 1.

fig 1: paths to the possibility distribution

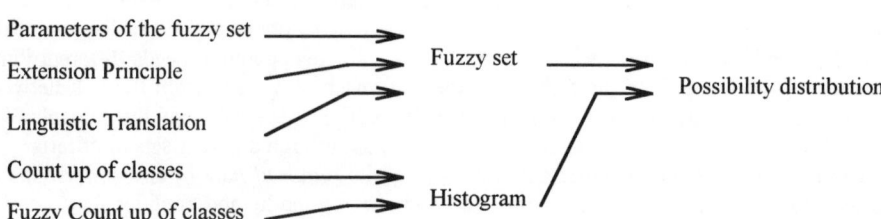

Parameters of the fuzzy set

Extension Principle

Linguistic Translation

Fuzzy set

Count up of classes

Fuzzy Count up of classes

Histogram

Possibility distribution

Now that the fuzzy tools have been exposed, we will discuss about the evaluation of the teleoperation criteria, keeping in mind the part 2 of the paper.

4. Evaluation of the Teleoperation Criteria

The determination of the criterion *required time* will use the technique based on the parameters of the fuzzy set outline. More precisely, concerning the real value of that criterion, we have to estimate the available time to do the task. For that case, the fuzzy set outline being the semi-trapezoidal function, the parameters we have to value are the start and the end of the slope (the two extremums of the intervals of membership zero and one).

The extension principle is applied to three criteria because of the introduction of a fuzzy window in their definition. These criteria, that are the *quantity*, the *quality* of the information and the *workload* of the operator, need to be defined through the close past time that is a vague notion, hence the fuzzy window of the time to approach the idea of the close past time for the expert. Now, starting from the definition of the criteria at a particular moment (mass of information, percentage of good information and percentage of working time are all classically defined for a certain period of time), we have to extend them for a fuzzy duration.

With the *execution simpleness* of the task, we need the intervention of an experienced operator that will estimate the facility of the execution of a task facing each driving mode. He will express his opinion that will be translated into a fuzzy set with the linguistic tools.

The last kind of criteria we are going to build are those based on random events like a value from a task which is required to be done at any moment. The *performance* of the man-machine system and the *vigilance* of the operator are from that kind. In fact, after the execution of a task by the teleoperation system, we define a successful degree calculated with the number of accomplished sub-task and we collect the reaction time of the operator when he take part in the task. Through a fuzzy window of time qualifying the close past time as above, we can now build two histograms with the technique described in the part 3, a first histogram with classes of successful degrees for the performance and a second with classes of the reaction times for the vigilance. At that step of the algorithm, we have to apply the method proposed by Dubois and Prade to transform the histogram into a possibility distribution. But, it appears the problem of the quantity of information contained in the histogram because the transformation needs a minimum of information to be validate, a minimum that is not ensured in that case. So, we are going to fit the method to our preoccupation introducing a degree of information mass defined by:

$\alpha = \min (1 , p/n)$
p being the mass of a class and n the minimum mass required for that class.

The build of a possibility distribution from the histogram is realised by the formula:

$$\Pi_i = 1 - \alpha_i \sum_j \max(0, p_j - p_i)$$

The criteria determinations are recapitulated below on the table 1.

Needless to say that the possibility distributions obtained with the different techniques shown above can not be used with their primary form that is too complicated for a real time algorithm. We have chosen to approximate them with a trapezoidal function. For this purpose, we first need a method to determine the modal intervals of each possibility distribution. These intervals will set up the small bases of the trapezoidal forms. Then, we deal with the method to estimate the slopes of the trapezium sides. Among a lot of possibilities, we have chosen a quick computed method that preserve the possibility mass of the distributions. That is to say, to calculate the modal intervals, we just use a threshold above which the values are modal values, and to preserve the information mass, the slopes are deduced from the area of the distribution.

Table 1: the criteria evaluation

CRITERION	REQUIRED INFORMATION	FUZZY TOOL
Required time	Estimation of the available time	Trapezoidal fuzzy set fixed by the parameters min. and max. of available time
Simpleness	Questionnaire to an experienced operator	Linguistic method to build fuzzy sets
Performance	Successful degrees of the tasks execution through the close past time	Successful degrees histogram of the member-ship degrees of the time fuzzy window
Quality	Rate of good information mass through the close past time	Extension principle to define a fuzzy rate
Quantity	Information mass transmitted through the close past time	Extension principle to define a fuzzy number
Vigilance	Reaction time rates through the close past time	R.T. rates histogram of the membership degrees of the time fuzzy window
Workload	Working part of the operator through the close past time	Extension principle to define a fuzzy percentage

5. Conclusion

The paper has developped ideas on the integration of different kinds of information in a decision system. This work is a part of a greater work about the decision in man-machine systems and many problems remain not solved in the decision making such as the integration of some dynamic aspects. But the treatment proposed here, associated with a pertinent choice of aggregation operators is a good incoming system. Works are actually under development to apply these methodologies to manufacturing systems where the presence of human operators with their own motivations influences the fabrication process.

References

[1] F. Wawak, D. Jolly and A.M. Desodt, Fuzzy logic for the choice of control mode in teleoperation, International Conference on Human Aspect of Advanced Manufacturing on Hybrid Automation, Gelsenkirchen Allemagne, 26-28 aôut 1992.

[2] F. Wawak, A.M. Desodt, D. Jolly, Fuzzy decision algorithm for Man-Machine systems, QUARDET'93, Univ. Politecnica de Catalunya, Barcelona, June 16 - 18, 1993, pp 745-751.

[3] Vertut, J. and Coiffet, Ph., Téléopération: Vers la Téléopération assistée par ordinateur, Les Robots, Tome 3B, Hermes, 1985.

[4] B.H. Kantowitz, R.D. Sorkin, Allocation of function, Chapter 3.3, Hand book of human factors,Ed G. Salvendy, Jwiney, 1987.

[5] L.A. Zadeh, Fuzzy sets as a basis for a theory of possibility, *Fuzzy sets and systems*, **1**, (1978) 3-28.

[6] D. Dubois and H. Prade, Fuzzy sets and statistical data, *European J. Operational Research*, **25**, (1986) 345-356.

[7] L.A. Zadeh, PRUF-a meaning represantation language for natural languages, *Int. J. Man-Machine Studies*, **10**, (1978) 395-460.

Advances in Agile Manufacturing
P.T. Kidd and W. Karwowski (Eds.)
IOS Press, 1994

Some Correlates of Workers Performance in a Hybrid Automated Industry

A. KHALEQUE
Psychology Department, Dhaka University, Dhaka, Bangladesh

Abstract. This paper reports the results of a study which was designed to assess and compare the multiple relationships between performance, job satisfaction, stress, fatigue and mental health of workers of a hybrid automated sanitary ware factory in Dhaka, Bangladesh. The study was conducted on a sample of 90 subjects, comprising 30 manual, 30 semi-automated and 30 automated workers. The results reveal that human aspects of work, such as job satisfaction, stress and fatigue, are related with both health and performance of the manual and semi-automated workers but those aspects are related only with the health of the automated workers and not with their performance.

1. Introduction

Automation as a process of technological development has many effects on the operator's health, well-being and quality of life [1]. However, it is indicated that the design of most automated equipments is primarily guided by the technical consideration, and the human aspects though important for improving performance and the quality of life of the operators, are generally ignored [2]. So the present study was designed to explore interrelationships between performance and some human aspects of operators in a hybrid automated manufacturing industry. In other words, the objectives of this study were to assess and compare multiple relationships between performance, job satisfaction, stress, fatigue and mental health of the manual, semi-automated and automated workers.

2. Method

2.1. Subjects

The study was conducted on a sample of 90 subjects, consisting of 30 manual, 30 semi-automated and 30 automated workers of a sanitary ware factory in Dhaka, Bangladesh. All the subjects were male. The mean ages of the manual, semi-automated and automated workers were 31, 32 and 35 years and their mean job experiences were 8, 10 and 12 years respectively. The educational qualification of the subjects of the three groups varied from primary to higher secondary levels.

2.2. Measuring Instruments

The following measuring instruments were used for assessing job satisfaction, stress, fatigue and mental health of the subjects.

2.2.1. The Job Satisfaction Scale. This scale was developed by Warr et al. [3]. It measures the degree of a person's self-reported satisfaction with the intrinsic and the extrinsic features of the job as a whole. The scale consists of 16 items of which 7 are concerned with the intrinsic job satisfaction, 8 with the extrinsic job satisfaction. Clegg and Wall [4] reported the alpha coefficient of 0.92 for the whole scale, 0.56 for the intrinsic job satisfaction items and 0.74 for the extrinsic job satisfaction items.

2.2.2. The Fatigue Scale. This scale was developed by Cooper and Marshall [5]. It consists of 22 items, the response categories of which range from "no stress at all" to "a great deal of stress", the scale values range from 0 to 5. The subjects are asked to choose the appropriate frequency according to their stress limit for each item of the whole scale. Cooper and Marshall [5] found this scale reliable and valid for measuring job stress.

2.2.3. The Fatigue Scale. This scale was developed by the Industrial Fatigue Research Committee of the Japanese Association of Industrial Health. It consists of 16 items about general and specific bodily and mental symptoms of fatigue to which the subject answers "yes" or "no". Several researchers have provided evidence of validity of this scale [6].

2.2.4. The General Health Questionnaire (GHQ). This Questionnaire was developed by Goldberg [7] and used for measuring mental health. It is a self-administered screening test for detecting minor psychiatric disorders in the general population. This questionnaire consists of 12 items with 4-point Likert type scaling. The scale value ranges from "definitely yes= 3" to "definitely no=0". The reverse scoring procedure is applied for negatively framed statements. Banks et al. [8] used it as an indicator of mental health in employment related and occupational problems. The development studies showed that the full scale exhibited high internal consistency (0.82) and good test-retest reliability (0.84).

3. Results

The results of the present study are presented in the following tables.

Table 1. Multiple correlations between job satisfaction, stress, fatigue, mental health and performance of the manual workers.

Variables	Job satisfaction	Stress	Fatigue	Mental health	Performance
Job satisfaction	-	-.07	-.46*	.64**	.20
Stress		-	.29	-.01	-.52
Fatigue			-	-.80**	.09
Mental health				-	.07
Performance					-

Note. N = 30; * = p<.01; ** =p<.001; the correlation coefficients without asterisk are not significant.

The results in table I show that there are significant positive correlations between job satisfaction and mental health and stress and performance. Moreover, there are significant negative correlations between job satisfaction and fatigue , and fatigue and mental health of the manual workers.

Table 2. Intercorrelations between job satisfaction, stress, fatigue, mental health and performance of the semi-automated workers.

Varibles	job satistsfaction	Stress	Fatigue	Mental health	performance
Job satisfaction	-	-.63**	-72**	.67**	.51*
Stress		-	-.38	-.43*	.11
Fatigue			-	-.53*	-.23
Mental health				-	.19
Performance					

Note. N= 30; * = p<.01; ** = p< 001; the correlation coeffients without asterisk are not significant.

The results of correlation matrix in table 2 show that there are significant positive correclations between job satistaction and performance, and mental health and job satistaction for the semi-automated workers. Moreover, there are significant negative correlations between job satisfaction, stress and fatigue; and mental health, fatigue and stress for this group of workers.

Table 3. Intercorrelations between the scores of job satisfaction. stress, fatigue, mental health and performance of the automated workers.

Varibles	Job satisfaction	Stress	Fatigue	Mental health	Performance
Job satisfaction	-	-.35	-.45*	.70**	.12
Stress		-	-.29	-.42*	.21
Fatigue			-	-.53*	-.11
Mental health				-	.20
Performance					-

Note. N= 30 * = p<.01; ** = p<.001; the correlation coeffieients without asterisk are not significant.

The results of correlation matrix in table 3 show that there is significant positive correlation between job satisfaction and mental health, and there are negative correlations between job satisfaction and fatigue, mental health and fatigue, and stress and mental health of the automated workers. However, their performance is not significantly related with any of these variables.

4. Conclusion

Thus the results of this study indicate that human aspects of work, such as job satisfaction, stress and fatigue are related with both health and performance of the manual and semi-automated workers but these aspects are related only with the health of the automated workers and not with their performance.

Finally, it may be concluded that this hybrid automated work should be redesign and reorganize to minimize the physical workload of the manual workers, and monotony and boredom of the semi-automated and automated workers for improving their performance, health, well-being and overall quality of life.

References

[1]   A. Khaleque and M. M. Hossain, Job Satistaction, Fatigue and Mental Health of Manual, Semi-automated and Automated Workers. In : S. Marras et al. (Eds.), The Ergonomics of Manual Works. Taylor and Francis, London, 1993.

[2]   P.T. Kidd, The Social Shaping of Technology : the Case of a CNC Lathe, Behaviour and Information Technology, 7(1988) 193-204.

[3]   P. Warr, J. Cook and T. Wall, Scales for the Measurement of some Work Attitudes and Aspects of Psychological Well-being, Journal of Occupational Psychology, 32 (1979) 129-148.

[4]   C.W. Clegg and T.D. Wall, Note on some Scales for Measuring Aspects of Psychological Well-being at Work, Journal of Occupational Psychology, 54 (1981) 221-225.

[5]   C.L. Cooper and J. Marshall, Occupational Sources of Stress: A Review of the Literature Relating to Coronary Heart Disease and Mental Health, Journal of Occupational Psychology, 49 (1976) 11-28.

[6]   H. Yoshitake, Relations between the Symptoms and Feelings of Fatigue, Ergonomics, 14 (1971) 175-186.

[7]   D.P. Goldberg, Manual of the General Health Questionaire, National Foundation for Educational Research, Windsor, 1978.

[8]   M. H. Banks et al., The Use of General Health Questionnaire as an Indicator of Mental Health in Occupational Studies, Journal of Occuptional Psychology, 53 (1980) 187-194.

Advances in Agile Manufacturing
P.T. Kidd and W. Karwowski (Eds.)
IOS Press, 1994

Operator Performance Requirements in an Advanced Traffic Management System

Dennis FOLDS and Deborah MITTA
ELSYS/CAD, Georgia Tech Research Institute, Atlanta, GA 30332-0800, USA

Abstract. A key element in any intelligent vehicle-highway system (IVHS) is an advanced traffic management system (ATMS), where the ATMS provides functionality for real-time traffic management and planning. In designing the ATMS from a human factors perspective, one of our primary objectives is to designate the appropriate human operator role (and corresponding level of automation) to each ATMS function. Operator role theory provides a framework for guiding function allocation. It prescribes a formal procedure for (1) specifying operator performance requirements for the ATMS and (2) designating the appropriate level of automation for each ATMS function.

1. Introduction

A key element in any intelligent vehicle highway system (IVHS) is an advanced traffic management system (ATMS). The ATMS provides functionality for real-time traffic management and planning. Its global mission is to facilitate the safe movement of individuals and goods (while minimizing delay) through a designated roadway system. The following five objectives, when satisfied, ensure accomplishment of this global mission: maximize available capacity of the designated roadway system, minimize the impact of incidents on delay and safety, assist in the provision of emergency services, contribute to the strategic regulation of demand, create and maintain public confidence in the ATMS [1].

In this research effort, the functions required to satisfy each of the five objectives have been defined in terms of four function types: input, throughput, output, and support [2]. Input functions receive information from sensors and external sources. Throughput functions process information and make decisions. Output functions disseminate information, control electronic devices, and issue requests. Support functions create and maintain the capabilities of other functions (e.g., store and retrieve information, train operators, and coordinate activities.

Ultimately, a total of 113 functions were specified [3]. Of this total, 32 input functions and 29 throughput functions were specified; 25 functions were designated as output functions, and 27 functions were designated as support functions.

1.1. Theoretical Framework

In this research, operator role theory provides a framework for guiding ATMS function allocation. Within this framework, a continuum of operator roles is defined such that at one end of the continuum, a function is allocated solely to a human operator, and at the opposite end, a function is allocated solely to a machine. Between the extremes, function performance is shared by human operator and machine components. Note that a *role* refers to the collection of activities an individual performs within a given context, rather than a specific task. An operator can have one role in a given function and another role in a second function. Within each role the operator may have several tasks.

Operator role theory was originally developed to describe human activities required for the operation of air defense systems [4]. Subsequently, operator role theory has been generalized to apply to all types of human-machine systems. It can be applied as a prescriptive (as well as descriptive) tool.

1.2. Research Objective

In designing the ATMS from a human factors perspective, one of our primary objectives has been to designate the appropriate human operator role for each function. As a consequence of such role designation, the level of automation associated with each function can be established.

2. Operator Role Theory

Design of the ATMS from a human factors perspective requires more than an assessment of individual display and control components and their arrangement in a set of workstations. Workstation design is critical; however, human factors inputs to the design process must impact high-level design philosophy. Specifically, the rationale underlying the allocation of functions among human operators and machines must be driven by human factors considerations.

Operator role theory was applied in the allocation of ATMS functions. According to this theoretical framework, a continuum of operator roles is defined such that at one extreme of the continuum a function is allocated solely to a human operator and at the opposite extreme a function is allocated solely to a machine. The continuum is divided into four regions, where each region indicates an operator role (and corresponding degree of automation). These regions are Direct Performer (no automation), Manual Controller (operator performs decision-making activities), Supervisory Controller (operator has the ability to override a machine-made decision), and Executive Controller (operator enables or disables a fully-automated function).

Note that three of the four roles are labeled with the term *Controller*, indicating that operator activities are focused on controlling machine components. The fourth role, labeled with the term *Performer*, indicates that direct performance of a function, rather than the control of machine components, is the focus of operator activities.

2.1. Fundamentals of Operator Role Theory

Operator roles are defined more precisely by considering the manner in which information is processed within a function. Each function is defined in terms of four stages of information processing: input, processing, response selection, and output. At the input stage, information is received from an external source by a sensor. At the processing stage, received information is manipulated by a processor. At the response selection stage, a controller decides what control actions are to be performed. At the output stage, an actuator executes control actions. In order to apply operator role theory to the function allocation process, the manner in which humans and machines accomplish each information processing stage must be specified.

In both the Direct Performer and Manual Controller roles, the controller is human. That is, a human performs response selection. In the Direct Performer role, however, sensors, processors, and actuators are human, while in the Manual Controller role, at least one of these three components is a machine.

In both the Supervisory Controller and Executive Controller roles, response selection is performed by a machine. In the Executive Controller role, sensors, processors, and actuators are machine components. Furthermore, except to terminate function execution, the human is unable to influence machine performance. In the Supervisory Controller role, however, the human can intervene during the response selection stage (e.g., override a machine-made decision). Additionally, the human may choose to modify performance of sensors, processors, or actuators. For a given operator role, human and machine components associated with the four information processing stages can be configured in a number of ways. These possible configurations are provided in Table 1.

Table 1. Human and Machine Configurations for Operator Roles[1]

Operator Role	Input	Processing	Response Selection	Output
Direct Performer	H or Hm	H	H	H
Manual Controller	H, Hm, or M	H, Hm, or M	H	H, Hm, or M
Supervisory Controller	Mh or M	Mh or M	Mh	M or Mh
Executive Controller	M	M	M	M

[1]The level of operator involvement at each information processing stage is described below.
H: The human is solely responsible for performing the processing stage.
Hm: The human (with machine assistance) performs the processing stage.
Mh: The machine (with human assistance) performs the processing stage.
M: The machine is solely responsible for performing the processing stage.

3. Applying Operator Role Theory to ATMS Function Allocation

In this section our procedure for allocating ATMS functions according to operator role theory is described. A team of human factors engineers, psychologists, and software engineers analyzed each lowest-level function. Appropriateness of assigning either of the two extreme operator roles (Direct Performer or Executive Controller) to a function was assessed. One of the two roles was considered appropriate if (1) its assignment to the function would satisfy that function's performance requirements and (2) a significant increase in capabilities could not be expected by the assignment of a different operator role to the function. If the team determined that a Direct Performer role would satisfy performance requirements and no significant gains in performance were expected by the assignment of some other role, it was assigned to the function. Similarly, if a function's performance requirements could be met by an Executive Controller role and no significant gains in performance were expected via some other role, an Executive Controller role was designated for that function.

In considering all remaining functions (those assigned neither Direct Performer nor Executive Controller roles), the analysts assessed the appropriateness of Manual Controller and Supervisory Controller roles. Again, appropriateness was evaluated in terms of the satisfactory achievement of performance requirements and the expectation of realizing significant performance gains through another role assignment. Once the team reached a consensus on operator role designations, further details of function allocation were completed with the four-stage information processing model. For each function, the analysts considered possible human/machine configurations associated with its respective operator role (Table 1) and identified the configuration that would best meet performance requirements.

4. Results

For each of the 113 ATMS functions, an appropriate operator role was designated. Additionally, a configuration of human and machine components was specified for each function's four information processing stages. The Direct Performer role was assigned to 40 of the 113 functions, the Manual Controller role to 28 functions, the Supervisory Controller role to 16 functions, and the Executive Controller role to 29 functions.

4.1. Example

Here, an example in which the function allocation process for a given throughput function (Anticipate Near-Term Traffic Conditions) is provided.

Function 2.1.1.2: Anticipate Near-Term Traffic Conditions

Operator Role: Supervisory Controller

Input: (M) Current load assessments and roadway system status, external reports (e.g., incident, weather, incident response, emergency response) are provided via software.

Processing: (Mh) Input data are provided to a near-term traffic model. From these data, the following road segment-specific information is predicted: traffic volume, density, speed.

Response Selection: (Mh) In some instances, the operator may have access to more relevant/accurate input data or may have better predictive capabilities than the software. Under these circumstances, the operator may choose to modify input information and override predictions (or, at a minimum, identify inaccurate/imprecise predictions).

Output: (M) Predictions derived by this function are sent via software to Function 2.1.2 (Develop Optimal Control Scheme) and Function 2.1.2 (Determine Special Vehicle Support Measures).

4.2. Summary

Our analysis yielded operator role designations for each ATMS function. In conjunction with these role designations, all functions were defined in terms of four processing stages (input, processing, response selection, output). Further analysis generated the operator performance (and automation) requirements essential for successful completion of each stage. Operator role theory prescribed a formal procedure for (1) specifying operator performance requirements for the ATMS and (2) designating the appropriate level of automation for each ATMS function.

5. Acknowledgment

This research was performed by the Georgia Tech Research Institute under the sponsorship of the Federal Highway Administration (Contract No. DTFH61-92-C-00094).[1]

References

[1] D. J. Folds, D. R. Stocks, H. F. Engler, and P. A. Parsonson, Operational Capabilities of an IVHS-Level Advanced Traffic Management System, Working Paper A-9309-A.2. Georgia Tech Research Institute, Atlanta, GA, 1993.
[2] D. J. Folds, D. R. Stocks, W. B. Fain, J. L. Brooks, J. B. Ray, and H. F. Engler, Functional Definition of an IVHS-Level Advanced Traffic Management System, Working Paper A-9309-B.2. Georgia Tech Research Institute, Atlanta, GA, 1993.
[3] D. J. Folds, A. D. Fisk, B. D. Williams, J. E. Doss, D. A. Mitta, W. B. Fain, A. C. Heller, and D. R. Stocks, Operator Roles and Automated Functions in an IVHS-Level ADvanced Traffic Management System, Working Paper A-9309-D.1. Georgia Tech Research Institute, Atlanta, GA, 1993.
[4] D. J. Folds, R. A. Beard, W. E. Sears, III, J. M. Gerth, and L. C. King, Operator Roles and Potential Vulnerabilities in Threat Air Defense Systems, Volume 1, AFWAS-TR-88-1161. Georgia Tech Research Institute, Atlanta, GA, 1989.

[1]Notice--This document is disseminated under sponsorship of the Department of Transportation in the interest of information exchange. The United States Government assumes no liability for its content or use thereof. The contents of this report reflect the views of the authors who are responsible for the facts and accuracy of the data presented herein. The contents do not necessarily reflect the official policy of the Department of Transportation.

Advances in Agile Manufacturing
P.T. Kidd and W. Karwowski (Eds.)
IOS Press, 1994

Cumulative Trauma Disorders in Advanced Manufacturing Environments

H. A. Romero and C. A. Wilhelmsen
Idaho National Engineering Laboratory, EG&G Idaho, Idaho Falls, ID, USA 83415-3855

Abstract. Repetitive motion injuries (RMIs) or cumulative trauma disorders will be an issue within advanced manufacturing environments just as they are within standard manufacturing environments. The reason is that the risk factors associated with RMIs are not always reduced by changing to advanced manufacturing environments. This paper examines the development of a work practices guideline equation for RMIs. The form recommended is analogous to the work practices guideline equation developed and used by the National Institute for Occupational Safety and Health (NIOSH) in the United States applied to manual lifting tasks. This paper recommends using the four factors associated with RMIs (force, repetition, fatigue breaks, and posture) and individual specific contributors (physical condition, and outside activities) as factors within the equation. This equation has yet to be quantified, but this is obviously the next step. The function of the equation will be to establish a Task Index for repetitive motion tasks. Task indices below an established Action Limit would be considered of low risk for developing a RMI. Tasks with indices between the Action Limit and an established Maximum Permissible Limit would be defined as requiring administrative controls to reduce the risk of developing a RMI. Tasks with indices greater than the calculated Maximum Permissible Limit would require engineering modifications to the task to reduce the likelihood of developing a RMI. The purpose of this paper is to demonstrate the need and recommend a methodology for quantification and testing of an equation.

1. Introduction

Human beings will still be called upon to perform physical tasks, often with high repetition, in advanced manufacturing environments (AMEs). AMEs are generally designed to reduce lead time, take advantage of batch production to increase quality, and introduce automation to increase efficiency. Yet, humans will still be required to perform repetitive work and, due to the increased throughput and precision, the repetitious nature of the work may worsen. However, quantification of the risks involved with these tasks has not been completed. Industries want an objective means to compare the risks of a task to the likelihood of developing a RMI. However, the question remains "What model or standard does industry have to measure or set guidelines on tasks to determine susceptibility to RMIs?" In 1992, RMIs accounted for 56 percent of gradual onset work-related illnesses according to the Occupational Safety and Health Administration (OSHA) [1]. NIOSH has guidelines for defining lifting hazards, but there is no standard for repetitive motion tasks. Industry needs a similar model for the determining the risk factors that increase the likelihood of a RMI.

A possible model could be a function of task contributors (i.e., force, posture, frequency of repetitions, and lack of fatigue breaks) and individual contributors (i.e., physical conditioning, and outside activities). Each of these factors would be applied to a particular body system being stressed at the job and a time weighted average approach could be used to recognize the amount of time on the task. These factors could be combined to produce an Acceptable Limit and Maximum Permissible Limit as already defined in the NIOSH Lifting Guide [3]. Specifically, the Acceptable Limit would be defined as a function value or task index below which the task presents no significant risk. A task index above the Acceptable Limit and below the Maximum Permissible Limit could be defined as requiring administrative modifications to the task to reduce the significant likelihood of a musculoskeletal injury. A task index above the Maximum Permissible Limit would require engineering modifications to reduce significantly the likelihood of developing a RMI. Once modifications have been introduced, the task index would be recomputed to ensure the modifications have been effective. Significant basic, applied, and empirical research must be conducted to quantify the form and function of the task index equation and the definition and values of the Acceptable and Maximum Permissible Limits for RMIs.

2. Defining terms

There are several terms that are recommended for inclusion in the model. To understand the reason for inclusion, these terms must first be defined as they relate to RMIs. These terms are: force, repetition, posture, fatigue breaks, physical conditioning, outside activities, Action Limit, and Maximum Permissible Limit [4]. *Force (F)* simply means that work must be performed to develop a RMI. Work is force times distance. The force does not have to be large. For example, typing usually does not exert much force in the wrist, yet the wrist is susceptible to a RMI from typing. *Repetition (R)* means performing the same activity more than once. Repetition is a key factor in developing a RMI. Repetition generally causes small microtraumas within the muscle-tendon-synovial sheath that accumulate and lead to a RMI. *Posture (P)* refers to the relative arrangement of body parts such as the orientation of the limbs, trunk, and head during a work task. Posture affects RMIs since there are several biomechanically neutral postures for the body depending on the body system being observed. Adopting neutral postures reduce the likelihood of developing a RMI. *Fatigue breaks (FB)* refer to the time spent on a significantly different task to alleviate physical fatigue in a specific body system. Physical fatigue is demonstrated by the buildup of lactic acid in the blood stream and the amount of reduction in dynamic strength capacity of the person. Two key elements of fatigue breaks are the duration and frequency. The *physical condition (PC)* of the worker is linked to susceptibility toward developing a RMI. Links have been shown between muscle tone, chronic diseases, high blood pressure, and a myriad of other factors. A program of stretching and flexing exercises combined with a program of overall wellness can reduce the susceptibility to RMIs. *Outside activities (OA)* that are hand intensive such as tennis, bowling, sewing, and playing the piano or violin can influence the susceptibility for developing RMIs. The *Action Limit (AL)* as defined in the NIOSH Work Practices Guideline for manual materials handling, is the amount of weight that requires some selection or training of potential employees or merit consideration of redesign to make the tasks suitable for people [3]. Therefore, the AL could be defined as the limit at which some action must be taken to reduce the likelihood of developing a RMI. *Maximum*

Permissible Limit (MPL) is defined in the NIOSH Work Practices Guide, as three times the AL that conforms to less than 15 percent of the work force having the capacity to do this type of lifting without increased risk of musculoskeletal injury. This definition could reflect RMIs by defining a level, perhaps three times the AL, beyond which most of the working population would be at a significant risk for developing a RMI.

3. Task Index equation

A recommended form of the equation is: Task Index $= \Sigma^m_{j=1}\Sigma^n_{i=1}\{(C_1*F + C_2*R + C_3*P + C_4*FB + C_5*PC + C_6*OA)*T_i\}S_j$. T_i refers to the time and S_j refers to the body system for which the risk factors are being evaluated. Low values of the Task Index would suggest that the combination of task specific and individual factors reflect a low likelihood for development of a RMI. Task indices greater than the AL and less than the MPL would require administrative interventions while task indices greater than the MPL would require engineering modifications to reduce the physical stressors. In either case, ethics demand that the company only make changes to the task contributors. It might be possible to limitedly affect physical conditioning by requiring all employees to participate in a wellness program. The coefficients (C_i) will reflect the relative weight of each variable in the Task Index equation. For example, physical conditioning is a minor contributor in the development of a RMI while repetition is a major contributor [5]. Therefore, the coefficient for physical conditioning (PC) might be $C_5 = 0.05$ while the coefficient for repetition (R) might be $C_2 = 0.7$. The coefficients would add to unity indicating the percent contribution for each task and individual contributor to the overall risk of developing a RMI. The range of values used for the task and individual contributors would best be derived from tabulated values reflecting the ranges of low, medium, and high risk contribution. For example, instead of inputting the exact number of repetitions per hour for R, a user would consult a table in which various ranges for R are listed. The user would then use the value listed in the table for R in the Task Index equation. The tabulated values for each contributor would have to be carefully selected to allow a minimum value to be calculated that would result in an index below the established AL. Additionally, the tabulated values would allow the maximum calculated index to be greater than the MPL. Additionally, the tabulated values would have to be derived on a per time basis to allow the time on the task to be considered. The AL and the MPL would be chosen to reflect a medium level of risk and a high level of risk as established within the tabulated values for the task and individual contributors. For example, the median value of the Task Index equation using the developed coefficients and the median value of the tabulated values for the contributors would be a reasonable estimate for the AL. A reasonable estimate for the MPL may be 90% of the maximum value possible for the Task Index, again using the developed coefficients and the maximum tabulated values for the contributors.

4. Quantifying the Task Index equation

Full quantification of the Task Index will require some basic and applied research. All the risk factors have not been identified, controversy exists over the risk factors currently identified, and there is little agreement on the contribution of these factors to the total risk of developing a RMI. However, industry needs this type of equation to protect its workers in the most economical manner. Therefore, empirical studies of the epidemiology of RMIs and a convened panel of experts should be able to develop a working equation with wide acceptance.

5. Intervention strategies

There are two important sets of intervention strategies. The first set is used when the task index is between the AL and the MPL. The second set is used when the task index is above the MPL. Task indices between the AL and the MPL will require, at least, administrative interventions to reduce the likelihood of a RMI. Examples of administrative interventions include: (1) establishing a program of fatigue breaks in which the person is required to perform physically different tasks every 1.0 to 1.5 hours for five to ten minutes; (2) training the person in task specific and individual contributors and how to reduce the combined contribution; (3) establish a program of stretching and flexing exercises; (4) encourage early recognition and reporting of RMI symptoms; and (5) train supervisors in recognizing RMI contributors. However, engineering modifications remain the most permanent manner of reducing the likelihood of a RMI and should be considered as an intervention method for Task Indices below the MPL and above the AL. Task Indices above the MPL require engineering interventions to reduce the task specific contributors to RMIs. Examples include: (1) redesigning the workplace to improve posture; (2) change the task to reduce the number of required repetitions; (3) redesign the task to reduce the amount of force required.

The purpose of the continued development and use of the Task Index equation is to provide industry with an understanding of the risk factors involved in RMIs and the relative contribution of these risk factors. Additionally companies could use the Task Index equation to determine which tasks contain a significant risk for developing a RMI and the level of intervention necessary to reduce this risk.

References

[1]   Eastman Kodak, (1986) *Ergonomic Design for People at Work, Volume 2*, Van Nostrand Rheinhold Company, New York.

[2]   Alexander, D. C., and Pulat B. M., (1985) *Industrial Ergonomics-A Practitioner's Guide*, Industrial Engineering and Management Press, Norcross, GA.

[3]   Putz-Anderson, V. & Water, T. R., (1991) Revisions in NIOSH guide to manual lifting, Paper presented at national conference entitles "A National Strategy for Occupational Musculoskeletal injury prevention-- Implementation Issues and research needs", University of Michigan, Ann Arbor, MI.

[4]   Putz-Anderson, V. (1988) *Cumulative Trauma Disorders: manual for musculoskeletal diseases of the upper limbs*, Taylor & Francis, Philadelphia, PA.

[5]   Sandler, H. M., (1993) *Are We Ready to Regulate Cumulative Trauma Disorders?*, Occupational Hazards, June, p. 51-53.

Part IX
Organisational and Cultural Change and Human Roles

Advances in Agile Manufacturing
P.T. Kidd and W. Karwowski (Eds.)
IOS Press, 1994

Continuous Improvement and Standardisation

Per Lindberg, Ph.D.
Anders Berger, Lic. of Engineering
Chalmers University of Technology / Gothenburg Center for Work Science
Chalmers Teknikpark, S-412 88 Gothenburg, Sweden

Abstract. This paper presents a framework that elaborates the impact of different levels of both work process and product standardisation with respect to three aspects of continuous improvements at the work group level. The three aspects in focus concern the differences in problem solving domain for work groups, the differences in learning levels in work groups as well as the differences in organisation and support activities implemented to generate and sustain continuous improvements. It is argued that product and work process standardisation are major determinants of the three aspects of continuous improvement activities at the work group level, and that previous research has chiefly been delimited to continuous improvement in systems characterised by highly standardised products and work processes. Continuous improvements in other systems need other organisation and type of support, and generates other types of learning and problem solving domains.

1. Introduction

Continuous improvements of processes, production and quality has always existed in different shapes and forms, i.e. day to day improvements. However, the concept "Continuous Improvement" (henceforth CI) has in recent years been given a more specific meaning and now designates a structured process with the aim of improving process and product quality. As such, CI is increasingly recognised as a valuable counterpart to more radical forms of innovation and renewal of manufacturing systems. The underlying principle of CI is to enable and encourage all personnel to actively use and enhance their skills through systematic problem solving. This means incremental small scale changes, through contribution of everyone, with the aim of enhancing learning through systematic problem solving. Thus, accompanying the debate of how to formally set up programmes to sustain these activities of incremental improvements, is the discussion relating to action learning, individual and organisational learning in the context of CI. As one determinant of organisational learning preconditions, Adler and Cole [1] uses the different levels of work process standardisation in the Volvo Uddevalla plant and the GM NUMMI plant to explain differences in the level of CI and learning, and the resulting productivity differences. They argue that a high degree of work process standardisation is required to enable an effective CI process. Berggren [7] argues that work process standardisation is of minor importance in explaining differences in productivity between NUMMI and Uddevalla.

A primary determinant of the selection of production system, and also of the standardisation of work processes, is the level of product standardisation [2]. This may be true for relatively standardised products as automobiles, and in a mass production environment. However, a company strategy based on producing customer-specific designs generally imply lowly standardised work processes. A unique product with unique features and functionality generally require both order-specific designs, production sequences, processes and work loads. This means that CI must be adapted to the lower level of standardisation in products and work processes.

2. Research purpose and design

The purpose in this article is to discuss how CI based on setting and improving standards are made possible in different product and work process environments and to elaborate the impact of different levels of product and work process standardisation on the establishment of CI. A two by two matrix is used to define four combinations of high and low degrees of standardisation of work processes and products. Continuous improvement activities in each work process/product combination are analysed and empirically supported through data from four case studies. Conclusions for the establishment of CI in the different environments are drawn.

3. Product and work process standardisation

Both Adler and Cole [1] and Robinson [3] points to standardisation of methods and procedures as a key ingredient for making CI on a systematic basis. Standardisation and maintenance of standards are "*activities to maintain the current condition, following predetermined procedures*"[4]. Standards and improvements are two sides of the same coin; standards are necessary to prevent regression, and improvements are needed to improve on standards. Standards may be set at different levels of aggregation; from detailed process operating procedures to system standards e.g. for education and training procedures. In this sense, both corporate policies and operator work instructions can be regarded as standards, although at different aggregation levels.

But as mentioned in the introduction, not all production processes are equally liable for standardisation, due to the fact that both products, work processes and procedures may vary. Table 1 shows the potential standardisation of work input and procedures, given product and work process standardisation.

Table 1. Product and work process standardisation vs. input and procedural standards (Note: SOP=Standard Operating Procedure)

Product standardisation	Operator work process standardisation	
	Low	*High*
Low	*Case 1 - Standards:* Input: Indirect/transformation through skills Procedure: Indirect/transformation through skills	*Case 2 - Standards:* Input: Indirect transformation through skills Procedure: Direct / SOP
High	*Case 3 - Standards:* Input: Direct / spec. by design Procedure: Indirect/transformation through skills	*Case 4 - Standards:* Input: Direct / spec. by design Procedure: Direct / SOP

In order to clarify the differences between these concepts, we will here comply to the following definitions. A work process consists of input (information and material) and work procedures to fulfil a task. The w*ork process standardisation* therefore means the degree of standardisation of inputs and procedures. *Product standardisation* here denotes the degree of brought about variability in input and work procedures from product variability. *Work procedure standardisation* is defined by Just-In-Time related criteria [5]; variability in work cycle length, sequences, number of process steps and level of Work-In-Process. *Input standardisation* refers to the degree of standardisation of predetermined criteria for the products to be produced.

A certain degree of standardisation of methods and procedures may be required in order to effectively make CI possible, but given the varying design of products and work processes, these standards are of different natures. Below is a brief description of 4 cases representing the 4 quadrants in table 1.

CASE 1. Low product standardisation and low work process standardisation. The case here is a manufacturer of advanced process automation equipment. Products are made to customer-specific design criteria, each product displaying unique functionality, configurations and sizes. Operators must interpret engineering drawings and specifications, and assemble and test each product in unique manner. The inputs (drawings, specifications etc.) must be interpreted and transformed by the operators own skills, and the work process set up accordingly. Detailed standards of methods and procedures are difficult to establish. The system standards are mainly skills and formal competence approval schemes.

CASE 2. Low product standardisation and high work process standardisation. Examples of this is in emergency hospital care, or in machinery maintenance. The input to the 'operators' is diffuse; symptoms must be transformed to diagnosis through skills and knowledge. However, as a diagnosis is made, the work procedure for treatment is relatively standardised. In some cases, diagnosis is organisationally separated from treatment.

CASE 3. High product standardisation and low work process standardisation. The case here is the Volvo Uddevalla car assembly system [6]. The design was deliberate; the details in the long assembly procedures and sequences was left to operators' discretion, with specified input in terms of detailed material supply and assembly instructions. Another example of this type of system may be found in engineering work; product functionalities are specified, but the procedures to establish functionalities are individual.

CASE 4. High product standardisation and high work process standardisation. This is the 'classic' car assembly line, as demonstrated in e.g. the NUMMI factory [1]. Standardised input in terms of specifications of quality and quantity requirements, and standardised work procedures (e.g. 60 second cycle times).

Table 2. Level of learning in the 4 cases

CASE 1	CASE 2
High learning for operators, individual, and craftsmanship-oriented. Low organisational learning.	Disconnected individual learning (diagnosis), and organisational (procedure) learning cycles.
CASE 3	CASE 4
Relatively high learning for the individual operator, but limited in the organisational sense.	Low individual learning, but the standardisation facilitates organisational learning.

As argued previously, continuous improvements are closely connected to the level of learning, i.e. the extent to which continuous improvements are used only to correct individual behaviour (*individual learning*), or as a base for *collective learning* to change system-wide procedures guiding the work [7]. The standardisation impact on the level of learning varies significantly in the four cases which is illustrated in table 2.

The objective of CI is to improve organisational effectiveness through individual and collective learning. These two learning levels are closely related since collective (organisational) learning can not take place without individual learning. On the other hand, individual learning should only be regarded as a necessary but not sufficient condition for collective learning [7]. The objective must therefore be to improve individual learning in cases 2 and 4 while sustaining the collective level. Correspondingly the objective in cases 2 and 3 should be improved collective learning while sustaining the individual level, given the preconditions of standardisation displayed in each situation.

4. Conclusion and discussion

In the design of CI systems, two determinants related to standardisation are paramount for the potential of CI and learning; *problem solving domain, PSD,* (i.e. the scope ,inclusion or exclusion, of continuous improvements in terms of work procedures, methods, work content and organisation, lay-out, product quality and redesign, administrative procedures and control) and *organisational setting and support, OSS,* (i.e. leadership, relations to staff, inter-functional teamwork, customer orientation, suggestion and incentive schemes, competence development, problem solving techniques, etc.). The following guidelines implicate how the *PSD* and *OSS* can be aligned with the degree of standardisation, in order to generate CI and learning at different levels as concluded in section 3.

Cases 1 and 3. The PSD for operators is wide; from interpretation of product specification to selection of appropriate procedures. PSD could be decreased and standardised through task specialisation, but a more viable solution in the long run would be to sequentially focus on prioritised themes through e.g. policy deployment and performance targets. The OSS for CI in this system should be standardised in terms of functional liaisons (e.g. quality, product design, engineering), systematic skill development and skills transfer between operators/groups, and incentive schemes for securing this. To sustain the high individual learning level while improving collective learning, standards should foremost concern the aggregate level of systems and procedures.

Cases 2 and 4. The PSD is relatively limited to the individual work process. PSD should be increased to include system-oriented problems through e.g. work enrichment. OSS should be directed to improvement of standards, by organised small group activities, and to increased individual learning through standardised competence development schemes.

References

[1] P. S. Adler and R. E. Cole, Designed for Learning: A Tale of Two Auto Plants, Sloan Management Review, Spring 1993, reprint 3436.
[2] T. Hill, , Manufacturing Strategy - The strategic management of the manufacturing function, Macmillan Education Ltd, London, 1985.
[3] J. Robinson, Continuous Improvement, Productivity Press, Cambridge, 1991.
[4] K. Suzaki, The New Shop Floor Management - Empowering People for Continuous Improvement, The Free Press, New York, 1993.
[5] Takahashi, K., JIT and Kaizen, Kaizen Institute of Europe, 1989.
[6]C. Berggren, NUMMI vs. Uddevalla, Sloan Management Review, Winter, 1994, Reprint 3523.
[7] J. Swieringa and A. Wierdsma, Becoming a Learning Organisation - Beyond the Learning Curve, Addison-Wesley, Cambridge, 1992.

Advances in Agile Manufacturing
P.T. Kidd and W. Karwowski (Eds.)
IOS Press, 1994

5steps - a strategy for change

Thomas Eriksson, M.Sc., Bernt Järneteg, M.Sc., Christer Johansson, Professor
Department of Mechanical Engineering, Linköping Institute of Technology, Sweden

Abstract. Dynamic, non-bureaucratic, organisation and continuous improvement have in general proved to be crucial for the competitiveness of the manufacturing company. This paper presents a methodology to start and, in the long term perspective secure, a process of continuous improvement within the manufacturing company. This methodology, 5steps, is based on the principles of TBM, Time Based Management. In 5steps time is used as a management tool and the main objective is to establish a work climate at all levels in the organisation where continuous improvement is a natural and important part of the daily work routines. Time is also the central factor on which all the improvement activities are focused.

1. Introduction

Many methods used for initiation of change focus on the organisation of the company. Methods based on establishing a team based organisation often give remarkable increases in productivity in a short time [2]. A possible drawback with methods that focus on reformation of the organisation is that it is difficult for small decentralised teams to improve the overall manufacturing process. There is also a risk that the team organisation itself obstructs structural change at process level in the system. On the other hand, if the overall processes have to be reformed before improvement at operation level is possible, then this can pose a serious threat against the team. The purpose of the 5steps strategy is to provide a strategy based on Time Based Management that avoids this drawback.

The philosophy of the 5steps strategy is that a process of continuous improvement is initiated in a clearly defined project that deals with manufacturing problems in the overall manufacturing process [2]. This approach has several advantages; staff are trained in problem solving and team work, competence is built up at all levels in the organisation, efficiency is increased regarding overall manufacturing process and a long term strategy is established upon which improvement is based. It is to a large extent a matter of building a foundation of knowledge and experience for an organisation based on team. This methodology may also be seen as a way of establishing a vision for improvement activities based on production technology instead of on organisation.

The first activity of the 5steps method is to focus on practical problems in the production system as quickly as possible. When the most significant problem areas are identified, the organisation needs tools to solve these problems. This means that new tools and principles can be related to a real need: the learning curve will be steeper due to higher motivation [1][4]. This will gradually build the competence and awareness of the nature of the production process needed for the effective operation of a team-based organisation [3].

2. 5steps - the model

As the name implies the process of change is handled in five steps or phases: definition, initiation, maturity, transformation and stabilisation. The first three phases are performed in a project organisation, functioning in parallel with the ordinary line organisation. In phase four this project organisation should be disbanded and the ordinary organisation reformed. The fifth step leads to a stable, team-based organisation. An outline of the model is shown in figure 1. The outline is to be read from the bottom upwards.

2.1 Step one: Definition

In this phase, the improvement project should be defined. This initial phase includes activities where the goals and objectives of the project are committed at senior management and the company board. A formal steering group for controlling and managing the process of change is also formed. In parallel with these activities an analysis of the current situation in the production system is carried out. The most important part of this activity is to locate primary problems and inefficient manufacturing processes. 5steps is based on the philosophy that all operative work in the project has the current situation in the production system as its starting point. It is an advantage if the problems identified in this phase are of low complexity. It is then easier to create "success stories". These "success stories" are invaluable when marketing the project in the organisation in order to obtain a broad acceptance and commitment to its objectives.

2.2 Step two: Initiation

The purpose of phase two is to "put the whole organisation on its feet": that is, as quickly as possible to involve as many associates as possible in improvement projects. It is important to start working in the real problem areas defined in phase one as quickly as possible. These problem areas act as starting points for the practical project work. If a problem area proves to be a dead end (for example if a problem turns out to be difficult to solve before other problems are eliminated) it is better to drop the problem temporarily and to continue work on other problems. The main task for the steering group in this step is to keep the project together and to define the long term strategies. Only the overall questions and work methods are to be discussed in the steering group. Questions of detail should be handled in the different improvement projects.

2.3 Step three: Maturity

The third step is the main part of the project as regards time and effort. This step is taken when the project organisation is established and is working systematically on problem solving. The organisation usually reaches this step when the improvement projects start working and the steering group has found its form to secure a process of continuous improvements. The associates are to be trained in team-work at operative level and it is the responsibility of the steering group to train and educate when there is a need. The practical work in each sub-project is also aimed at building up competence regarding production technology, through learning-by-doing. This will gradually increase the ability of associates to handle systematic improvement activities autonomously. These activities result in a simplified and systematised production system, and they provide experience of improvement activities and well-documented strategies for the future development of the company. The organisation is considered mature for phase four when these criteria are fulfilled.

2.4 Step four: Transformation

When the organisation is ready for the fourth step is a matter of judgement. This judgement must be made by the steering group, senior management and the company board. Is there

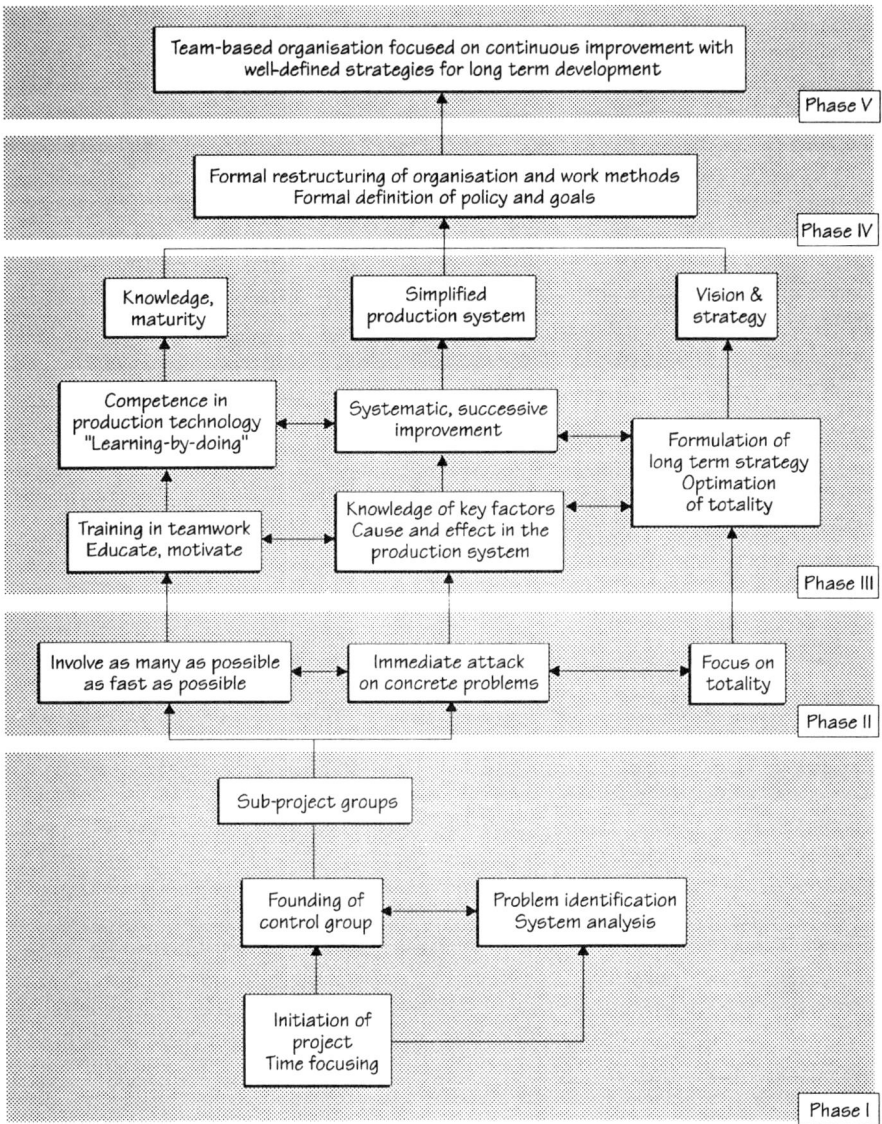

Figure 1. Outline of the 5steps-strategy. The left shows the development of experience of problem solving in small groups; the middle the development of the production system; and the right the work of the steering group. The steering group is responsible for forming a long-term strategy for the company and for systematising results in the different sub-projects. The information flow in each phase is horizontal; that is, between the sub-project groups and the steering group. An intensive and open flow of information is vital for satisfactory results.

enough knowledge and maturity among the associates? Does the production system work in such way that it can be used as a stable platform for continued, formalised improvement? Are our strategies consistent and well-defined? It is very important that the transformation be made at the right moment. The associates, the manufacturing system and the strategies are to have reached sufficient maturity to withstand the strain that the transformation process puts on the company. At the same time it is important that the transformation is not delayed unnecessarily. If the organisation is judged to have the right competence profile it is now ready for formal restructuring of the organisation as a whole. The organisational pyramid is turned upside-down. The project and project organisation is to be abandoned when the new organisation is finalised. The operative improvement work will be taken over by the formal teams. New improvement projects are executed in the new organisation by multi-disciplinary groups consisting of members from various expert functions or teams.

The forth phase has, however, not created a stable organisation. All experience points to the fact that it takes a lot of time before such a revolutionary change in work methods finds its forms in a new organisation. The fourth phase is not a traditional restructuring of the organisation but rather a change of corporate culture.

2.5 Step five: Stabilisation

When the basic principles and methods are fully established in the organisation as a natural part of the everyday work, the fifth step can be considered to have been reached. This means that a team-based organisation working with continuous improvement based on a long term strategy is established. Gaining stability is a matter of endurance and of interpreting the formulated goals consistently.

3. Conclusion and evaluation of the strategy

The 5steps model should be seen as a strategy which can be used to prepare an organisation for a new corporate culture in a systematically manner. It should be noted that the process of change is not completed within in the 5steps strategy. On the contrary, it is at this point that the major challenges arise: the system has to be improved in small steps on a daily basis in order to keep the forces for change alive and to increase competitiveness and productivity. It is important to be aware that transformation of an organisation and work methods in no way are painless. However, the transformation can be made more smoothly if the 5steps method is used instead of merely turning an unprepared organisation upside-down in one single movement.

The 5steps-strategy has been developed in the Eureka/Famos project known as SLIM (Short Lead-time In Manufacturing). Development, evaluation and improvement of the model have been carried out in interaction with experience from several case studies. Evaluation and improvement of the strategy is still in progress. Current results from case studies indicate that the 5steps is a successful strategy for starting and securing, a process of continuous improvement within a manufacturing company.

4. References

[1] T. Dahl, Peak Performance - The Role of Satisfaction, Stress and Control, A Presidental Address, The Eight Productivity Congress, Stockholm, 1993
[2] T. Eriksson, B. Järneteg, Med tiden som vapen - huvudrapport (Time as a weapon - main report) IVF-report, Linköping 1993.
[3] G. Hall, J. Rosenthal, L. Wade, How to Make Reengineering Really Work, Harvard Business Review pp. 119-131, November-December 1993.
[4] SAF, Kompetens är den enda resurs som växer när man använder den, Idéunderlag i serien "ständiga förbättringar"(Competence is the only resource that grows when you use it, In the series "continual improvements"), SAF 1994

Advances in Agile Manufacturing
P.T. Kidd and W. Karwowski (Eds.)
IOS Press, 1994

A Training Model for Integrated Use of Technology

- towards the organization as "a learning community"

Lise Busk Kofoed

Department of Development and Planning, Aalborg University

Fibigerstraede 13, 9220 Aalborg, Denmark. Phone: +45 9815 8522

Abstract. Generally it is a vast problem for SME (Small and Medium-sized Enterprises) to make their employees more qualified and to use present offers for training. This paper describes an attempt to consider SME's need for a flexible, integrated training when new technology is introduced. A model for the training is established, it builds on qualification analyses, periods of training at school and in the enterprise. Employees put together in interdisciplinary groups receive a problem and project oriented training, and distance learning when the training takes place in the enterprise.

1. Introduction

New market situations and technological changes, including more integration in the enterprises, make new demands to training of employees at all levels. New ways of acquiring knowledge become central challenges to the enterprises and the employees in the future. Learning as well as development of qualifications as a continuous process become an important condition when the enterprises want to optimize their efficiency, and at the same time it opens up possibilities of creating a more satisfying job and improved work conditions.

Development of a training model for integrated use of information technology, is an initiative that partly has to strengthen the overall training effort for SME, and partly has to result in a concrete model where both form and content match the training demands of the enterprises and the employees. The background for this paper is a research project in the area: "Integrated training and upgrading of employees in SME." The precondition is the wish for technical and organizational integration in the enterprises' production. (7 companies participated in the project.)

2. The enterprises' training requirements

Due to increasing demands for flexibility the companies must, to a higher degree than before, be able to adjust; and the ongoing changes also demand more from the technical production systems, the organization, and the employees. It requires far more *integration* of techology, organization, and knowledge.

If a strategy for integration is to function, then development of the human resources must be incorporated in the strategy.

Thereby the cornerstones in a flexible production are technical development and development of the employees, which requires an overall good working environment that develops and supports involvement, motivation, and the employees' flexibility. In relation to education this means that there is a need for new qualifications at all levels in the company, and that these qualifications must be gained alternatively. There is a demand for employee qualifications that emphasize the technical, organizational, and the social dimension where the integration aspect is central.

Specifically there is a demand for a joint and integrated training of employees who must be able to cooperate and communicate regarding a joint solution of tasks. There is a need for educations that qualify the employees to work with projects as well as tasks, and the content of the education must be related to the company's internal tasks. Furthermore the education must vitalize the attitude to steady training, as it is necessary when the qualification demands keep changing. In order to consider the dimension of integration, the training must be collective for a number of employee-groups and it must cut across functions, ie the education in the company has to be vertical as well as horizontal.

Evolution of the training model was planned in three phases: 1. Qualification analyses in the companies, 2. Planning of three courses, 3. Accomplishment of the courses. The following is build on the experiences gained from these courses.

3. Phase 1 - Analysis of Qualifications in the Companies

In general, planning of education in SME in Denmark is not part of a complete company strategy. Therefore it was important to initiate a strategy for education in SME. It was to be carried out be means of a qualification analysis. The qualification analysis is supposed to activate some processes regarding education strategies, likewise it should uncover the qualification demands. However, it requires financial resources and manhours to work with qualification analyses, whereas what characterises SME is short planning perspectives and shortage of resources. Consequently the teachers from the school, in collaboration with the companies' education managers carried out an 'amputated' qualification analysis[1]. It meant that thoughts and discussions were activated in the companies, regarding future educational needs, but the possibility of proving to the companies what a 'real' qualification analysis would mean, was lost.

Nevertheless, in general the companies were pleased with the qualification analyses. Either they have inspired an already started process, or they have started a process. There are many indications that the companies' education managers are aware of the value of an education strategy, but it is difficult for them to plan far ahead, and most of all there is a lack of plain methods and tools.

The qualification analysis formed basis for the content of the courses. Subjects for 3 courses had to be chosen. The qualification demands were specifically technical as well as "non-technical" ie.: cooperation, communication, method for project work and solving of assignments, as well as interdisciplinary comprehension and dialogue. Each course lasted for 37 hours. It was the course organizers' job to organize 3 courses that matched with the above mentioned demands.

The courses had the following broad titles:

1. SMD integrated production   2. EDP and quality   3. EDP based production

4. Phase 2 - Development and Planning of Courses

The concept for the course was based on the analysis of the companies' demands. Ordinary teaching took place at the school (Sønderborg Technical School), and project work and teaching took place in the company. The teaching in the companies was carried out as distance learning supported by computer based communication. The teachers had to be available in the in-company part of the education.

Each enterprise had to form a group of participants of 4-7 people including employees from different skill groups, representing general workers, skilled workers, technicians, and engineers. One of the participants had to be answerable to the company and was called tutor or company consultant. This was due to a wish that in all SME there should be a person who gradually acquires a further education, which could support the 'in-company training activities'. Before the start of the course, the group had to draft the project task to be worked with during the course, and the project needed approval by the company. The course schedule should be: 2½ days at school, in-company project work spread over approx. 1 month, out of which at least 4x4 hours were predetermined for project work. The entire course was concluded with a presentation and an evaluation of the project.

5. Phase 3 - Holding of the 3 Courses

Employees from 2 to 3 companies participated in each course. It was common to the participants that they were very inadequately informed about the form and the content of the course, and most of them did not have any knowledge of the project assignment. Consequently much time was spent, in the first school-part of the course, on giving the participants a common understanding of the project.

The content of the course at school had to shift between the technical subjects that conformed with solving of the project assignment, and project systematic and psychologic topics. The participants were very opposed to the so-called 'soft qualifications'. They did not consider them 'genuine' qualifications. It also appeared to be a problem, to connect the teaching in the subjects that should be the project assignment's technical part, with cooperation methods and the participants' daily job. Also it was a problem to teach an non-homogeneous group. The teachers tried to hit the central level, this made some find that it was too banal and others found it too difficult. Finally it seemed difficult for the participants to leave their daily position, where some, by means of education and functions had a dominating role (high confidence) and others had the opposite role (low confidence).

This became a hindrance for the non-dominating participants' learning and development as well as a barrier for the dominating with the 'know-all attitude' - as they did not listen to the other participants.

The in-company part of the course mainly went on without teachers, and the computer based communication was hardly used, either because it was too difficult to use and the participants not familiar with a PC, or because they were too busy with their project work. However, the participants would have appreciated that the teachers had contacted them more frequently during the in-company period.

6. Conclusion

The managing directors as well as the participants find that the training model is very good. They all made a very good project assignment, they have resources/ideas for improving the general performance of the company as well as for the specific task.

The employees at all levels in the company, find it very positive to cooperate with colleagues from other functions in the company. The learning process itself seems to open a broader perspective of the possibility of learning in an interdisciplinary way, but the teachers have to make this learning process visible. There is a general need to learn how to formulate a problem, eg problem analysis (how to work problem and project oriented). There is a lack of knowledge and experience about how to cooperate with colleagues belonging to other groups in the company - this counts for the highly educated as well as for the unskilled employees. It could be learned via roleplayings and simulation. There are problems concerning technical subjects as the "real thing to learn" and the more psychological subjects like cooperation and communication as "soft subjects of less importance" - but at the end of the course the latter subjects were estimated to be the most interesting for the employees. There are technical and pedagogical problems with the distance learning part of the project. Furthermore the management does not know how to organize and support in-company training.

The conclusion is that the structure of the training model is very good, but it is necessary to adjust it. The teacher's role must be developed technically and pedagogically. The teacher must be capable of teaching in an integrated and differentiated way. It is an advantage to support differentiated technical education in the electronic part of the distance learning, where the participants can have individual assignments. In order to make the distance learning function in the company, the teacher must play an active role as process consultant. Besides, the participants must be introduced to the use of a PC before the course commences, and each participant must have access to a PC that should be placed in the company in a room with free access. To use a PC must become as ordinary as the use of pen and paper. A collective ability of using the PC during the study will leave the participants in a more equal position. Plays (roleplaying and computer simulation plays) must be developed for the linking between theory and practice, and the teachers must assist the participants in finding their resources, and in their change of opinion regarding roles.

Change of attitudes has to be worked with at school, as well as in the company, so that the qualifications necessary for interdisciplinary cooperation are regarded as important as the technical qualifications.

The companies must be responsible for information regarding the education and provide the necessary organizational frames to make way for education and learning processes as an important and natural element in daily working life.

The education model started a process in the companies, which could be the foundation for "the learning community", if the companies follow up on the process and evolve implementation strategies for the employees' potentials and the qualifications acquired.

---

[1]   The qualification analysis used, called the AIDA-method, is rooted in the Organization of Executive and Managerial Staff in Denmark. The target group for AIDA is managers who are to participate in uncovering own and the employees' qualification demands in relation to the company's evolution.

Advances in Agile Manufacturing
P.T. Kidd and W. Karwowski (Eds.)
IOS Press, 1994

Computer Modelling of the Learning Organization

A.E.Kiv, V.G.Orischenko, I.A.Polozovskaya, I.G.Zaharchenko
Pedagogical Institute 26 Staroportofrankovskya str. Odessa 270020 Ukraine

Abstract. The new testing method of creative thinking is based on a comparison of results obtained by children during the particular type of computer logical problems solving and the corresponding theoretical dependencies obtained as a solution of specific differential equations. As a result of this comparison the coefficients of equations are determined by the software. These coefficients correspond to the certain characteristics of the creative thinking process. Our testing method has applications to the learning process correction in schools of the different level.

1. Introduction

The problem of the creative thinking modelling has many aspects. First of all the possibility of creative thinking model working out is a question of principle. There are different concepts that are used for creative thinking processes simulation. But the general view-point consists in assumption of the certain model structures arising in the head when the complex intellectual problems are deciding [1-3]. The examination of the way in which different authors use the concept of the creative thinking modelling suggests that the usages seem to be many and varied, and it is difficult to see how one relates to the other. Therefore it is important to find the method which can allow to confirm the efficiency of the concrete model. The way which we propose is to construct on the basis of the concrete model the system of tests and to investigate their fitness for creative abilities testing [4]. The results were used for the learning process improvement.

2. Creative Thinking Process Simulation

In accordance to proposed model of the creative thinking process we postulate an existence of the thinking space (TS), which contains the discrete elements corresponding to certain steps of a person in the direction of the creative problem decision. The thinking steps (or thinking elements) arise in local regions of TS and they are subdivided in three groups: effective steps (ES), wrong steps (WS) and intermediate steps (IS) (see Fig.1).

ES form the trajectory of the person moving in the direction of the creative problem decision. If the critical number of ES is accumulated the creative problem is solved (see Fig.2). We can say that our mind accomplished a phase transition to the new state that corresponded to the problem decision. The last assumption is correct for the large number of logical problems. WS correspond to elements which arise in TS as a result of the incorrect choice on the certain stages of the problem solving. The elements of the last group (IS) produce ES if they unite in complexes.

Differential equations which describe the creative thinking process are of the form:

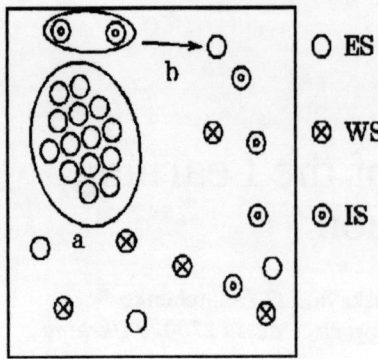

Fig. 1 The Illustration of Thinking Space
A critical number of ES is arised in
local region (a) IS produce ES (b)

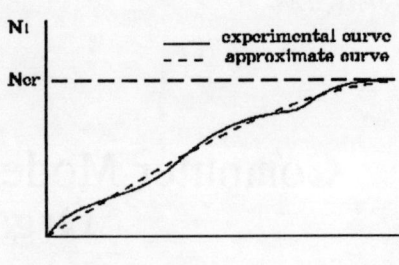

Fig. 2 The dependence of ES number on the
time

Approximated curve is the solution of
differential equations
N_{cr} is critical number of ES

$$\frac{dN_1}{dt} = I_1 + a_1 N_1 + b_1 N_3^k - c_1 N_2$$

$$\frac{dN_2}{dt} = I_2 + a_2 N_2 - b_2 N_3 - c_2 N_1$$

$$\frac{dN_3}{dt} = I_3 + a_3 N_3 + b_3 N_1 - c_3 N_2$$

where N_1, N_2 and N_3 are numbers of three types of thinking elements (ES, WS and IS), I_1, I_2 and I_3 are coefficients of the thinking intuitive component, a_1, a_2 and a_3 are coefficients of mastering accumulated information, b_1, b_2 and b_3 and also c_1, c_2 and c_3 are coefficients of the mutual influence of the different types of thinking elements, k is the order of "psychological reaction". In each concrete case the structure of these equations corresponds to the particular type of test problems which have computer realization. The coefficients in above mentioned equations can be used as characteristics of the creative thinking processes.

3. Computer Testing of the Creative Thinking of Children

The games contents must correspond to the structure of differential equations that describe the creative thinking process. It is significant that computer problems given to children do not require any preknowledge and look as games. We are working out the software that can give experimental dependencies $N_i(t)$. On the next stage the theoretical functions $N_i(t)$ obtained as solutions of the differential equations are compared with the experimental data. As a result of this comparison the coefficients of the creative thinking process I_i, a_i, b_i and c_i are determined by software. The procedure of testing is simple. For example, we can use the logical games: Hanoi Towers, River Crossing etc. When the children will receive the eventual result we can determine parameters of the creative thinking process by the special program ABILITY.EXE. These parameters (coefficients of differential equations) have the certain intervals of value change which were found on the basis of testing results investigations. Let us consider the testing data for groups of children in three age intervals.

This study simultaneously investigated the general indexes of the children abilities by Torrence method [5] and by our computer testing method. According to the sense given

Table 1 Parameters α and FT for Children

Age	Mean Parameters	Subjects									
		1	2	3	4	5	6	7	8	9	10
$\overline{A} = 6.9\ \overline{SD} = 0.5$	α	0.6	0.4	0.7	0.7	0.5	0.6	0.2	0.5	0.2	0.2
	$\overline{FT}$	0.4	0.4	0.5	0.6	0.6	0.4	0.4	0.4	0.1	0.2
$\overline{A} = 10.1\ \overline{SD} = 0.6$	α	0.7	0.7	0.8	0.3	0.8	0.4	0.6	0.7	0.6	0.6
	$\overline{FT}$	0.5	0.4	0.6	0.3	0.5	0.4	0.4	0.5	0.3	0.5
$\overline{A} = 12.7\ \overline{SD} = 0.5$	α	0.4	0.8	0.7	0.7	0.7	0.5	0.4	0.7	0.6	0.6
	$\overline{FT}$	0.5	0.6	0.6	0.7	0.5	0.6	0.3	0.5	0.6	0.4

Note: $\overline{A}$ is the mean age, $\overline{SD}$ is standart deviation.
to coefficients of differential equations in Part 2 we constructed auxiliary parameters so as they were comparable with Torrence characteristics of creative abilities. For instance, the parameter of searching activity in general we determine as

$$\alpha = \frac{I_1 + I_2 + I_3 + a_1 + a_2 + a_3 + b_1 + b_3}{b_2 + c_1 + c_2 + c_3}$$

We have compared this parameter with Torrence characteristic which may be called as the fluency of thinking (FT). The data for ten subjects (5 boys and 5 girls) of each from three age intervals in the elementary school of Odessa are given in Table 1. In given examples the choice of subjects was arbitrary. Parameters α and FT in the Table 1 are presented in relative units. The ratio to the maximum value of parameters for each group was used.

Similar comparison was fulfilled for other parameters (see Part 4). The correlation coefficients between Torrence parameters and received by our computer testing method are 0.7 - 0.9. The probability of received correlations 5 %. We state that our results are corresponded to those which are received by Torrence method. But our method allows to get the more precise information about creative thinking mechanisms. However further investigations are necessary.

4. Creative Thinking Testing and Learning Organization

The scheme of learning process organization includes the regular testing of creative abilities of children (see Fig.3). It was concluded on the basis of our results that the best variant is the semi-annual testing of children. We used the following procedure. Besides

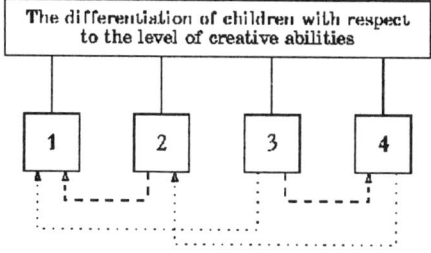

Fig. 3 Illustration of the Learning Process Correction
on the Basis of the Computer Testing of
Children Creative Abilities Evolution
(- - - - frequent events, ... rare events)

parameter α described in Part 3 three other parameters were constructed: parameter of the efficiency of mastering accumulated useful information

$$a = \frac{a_1 + b_1 + a_3 + b_2 + b_3 + c_2}{c_1 + c_3}$$

parameter of the efficiency of attention concentration

$$b = \frac{b_2 + c_2}{c_1 + c_3}$$

and parameter of the activity leading to wrong steps

$$\beta = \frac{I_2 + a_2}{b_2 + c_2}$$

On Fig.3 one can see the results received by the way of cluster analysis. There are four clusters of individuals with different creative abilities. These clusters characterized by following parameters:

1 $(\alpha - l, \beta - s, a - l, b - l)$,
2 $(\alpha - l, \beta - l, a - s$ (or $l), b - s)$,
3 $(\alpha - s, \beta - s, a - s, b - l)$,
4 $(\alpha - s, \beta - l, a - s$ (or $l), b - s)$.

Notation: l - large, s - small.

One can see the transitions of some individuals from one cluster to another which were proposed on the basis of our computer testing results during the learning process.

Each subgroup (1 - 4) of a whole group has the specific learning programs. The programs are corrected after each testing procedure. Therefore we realized by use of our testing methods the learning process which differ from usual static one and which can be considered as dynamic learning process.

5. Conclusion

The learning organization possibility is discussed on the basis of the new computer method of the creative thinking testing of children. The method is worked out by virtual of the mathematical model of creative thinking process and the description of this process by a set of differential equations. Computer programs allow to provide the abilities testing of children by way of logical computer games application. The dynamic learning process is suggested. It means that the learning process must be regularly corrected by taking into account the results of the creative thinking computer testing and the children abilities evolution.

References

[1] G.S. Altshuler, Creative Works as an Exact Science. Moscow, 1991.
[2] A.S. Maidanov, The Scientific Work Proces. Moscow, 1983.
[3] L. Bainbridge, Mental Models in Cognitive Skill. In Models in the Mind (Eds. A. Rutherford, Y. Rogers), N.Y. Academic Press, 1991.
[4] A.E. Kiv et al., Poster Sessions: Abridged Proceedings HCI International'93, Orlando, Florida, U.S.A., 1993.
[5] E.R. Torrance, Originality of Imagery in Identifying Creative Talent in Music // Gifted Child Quarterly, 1969.

Advances in Agile Manufacturing
P.T. Kidd and W. Karwowski (Eds.)
IOS Press, 1994

Performance Measurement in an Advanced Manufacturing Environment - Integrating Production Simulation and Shop-floor Kaizen

James JUNIPER

School of Economics, Finance and Property, Faculty of Business and Management, University of South Australia, North Tce., Adelaide, 5000.

Abstract. This paper reports on the progress of an on-going project on "Performance Measures for Lean Manufacturing". The chief collaborator is a South Australian-based automotive manufacturing company which is cooperating with academics from management accounting, mechanical engineering, organisation studies and computing fields.

The project's goal is to compare, link, and improve upon two systems of performance measurement which have been developed for the paint-line within the company's assembly plant. One of these is derived from a production simulation package that has been programmed to calculate the time, cost and financial benefits of potential modifications to the production-line. The simulation software has been developed in accordance with the writings of Dr. Eliyahu Goldratt on "*Theory of Constraints Management*". The second is based on the balanced-scorecard methodology developed by Beischel and Smith (from Ernst & Young), which enables the construction of a hierarchy of performance measures extending from the shop-floor to the highest level of corporate management.

1. Introduction

This paper reports on the results of a project on "Performance Measures for Lean Manufacturing which has been funded through an Australian Collaborative Research Grant. The industrial collaborator is a successful South Australian automotive manufacturer, while academic participants are drawn from management accounting, computing, mechanical engineering and organisation studies disciplines.

The objectives of the project are to develop two parallel systems of performance measurement which can be compared and enhanced interactively. The first engineering-based system builds upon a commercially financed production simulation of activities within the company's highly automated paintline. The completed simulation model is intended to be used, firstly, as an instrument for assessing various investment and up-grading proposals for the paintline; and secondly, as a system for performance measurement to support continuous improvement initiatives through-out the assembly plant and paint-line.

Under the direction of an advisory board, an engineering research group is to develop the performance measurement capabilities of the simulation software using the paint-line model to verify the feasibility and utility of such an exercise. A second, less detailed model of the assembly plant is also being constructed to examine how variations and volatility in up-stream activities can affect outcomes in the paint-line itself.

A second system of performance measurement is being developed by a management accounting research group, also operating under the control of the advisory board, but working under the day-to-day directions of a smaller committee. As the project proceeds research proposals are implemented, the organisational politics of the implementation process will be documented.

2. Production Simulation

Production activities have been modeled using Simview, a user-friendly simulation package whose development has been inspired by the writings of Eliyahu Goldratt, the Israeli physicist known for his application of fluid dynamics theory to problems of production scheduling and control [1] [2].

Simview replicates each process, including: arrival of materials and parts, assembly, processing, splitting of work-objects between different functions, conveyancing, inspection, rectification and rework, set-up, delay, break-down and repair. The user models the outcomes of each process in real-time by selecting an appropriate probability distribution from a wide range of possible functions. Appropriate parameter values are also chosen by the user for each distribution. Sources of volatility and variation (e.g. arising from supply shortages, inadequate quality control, insufficient capacity, or down-stream blockages) lead to the build up of inventory in front of critical bottlenecks in the production process. Buffer information is logged by the system and can be used to identify and remedy key factors which reduce throughput.

One of the most significant theoretical principles underlying the simulation philosophy is that average or plant-wide measures of performance can obscure key relationships which govern system dynamics and overall performance. The simulation process should uncover these key relationships and determine the system response to a range of possible process innovations. The system of performance measurement being developed by this research group has also been influenced by Goldratt's views on throughput accounting [3][4][5][6][7].

3. The Balanced Scorecard

The work of the management accounting group has been conditioned by recent writing which has adopted a critical stance towards finance-based approaches to performance measurement [8][9][10][11][12]. Traditional systems of standard costing, budgeting and variance analysis tend to focus on deviations between planned and actual financial outcomes. At upper levels within the firm, profitability is emphasised, while cost is the main focus at lower levels because in practice, revenues are more difficult to trace to individual operational units.

The critical literature has paid more attention to non-financial measures of performance. Following the philosophical outlook of Total Quality Management (TQM) - urging "bottom-up" empowerment of customers and workers to support continuous improvement- it has stressed that there are other sources of customer value than cost alone (e.g. quality, design, delivery time, after-sales service). Moreover, it is argued that cost measures are either too aggregated, or disseminated too late to provide adequate feedback. Additionally, the application of financial measures can often counter-productive. For example, they may "...encourage people to do better what should not be done at all." [8].

Representative of this new approach, Beischel and Smith's framework for performance measurement was adopted by the project team [13]. The first step in applying this framework is to determine the company's critical success factors, each of which is essential for its long-term survival.While these may vary by industry and market, five are deemed to be universal to all manufacturing companies: quality (meeting or exceeding customer needs and achieving customer satisfaction), customer service (external: meeting customer demand for end products; internal: meeting demands of customers internal to the firm), resource management (optimising outputs to inputs of people, inventory and fixed capital), costs (those that can be managed at the level reported) and flexibility (responsiveness to changing market, regulatory and environmental demands).

In the second step, measures defined at each management level must be able to be linked (inevitably defined in broader and more aggregated terms as measures move higher in the organisation) to a critical success factor at the top of the organisation. Each measure must also relate to the manufacturing process (i.e. it must be controllable and relevant).

Once the measures have been identified and linked, the third step involves the design of a series of score-cards for each success factor and organisational level. These define the appropriate measures, the frequency with which they should be measured (this will increase as you move down the organisation to the shop-floor) and the level of aggregation (measures should reflect a broader span of control and the more strategic character of decision-making at higher levels in the organisation). The final step requires the linkage of each scorecard with other scorecards for every measure and management level where particular measures at a specific level appear on more than one scorecard.

5. The Facility

The plant is vertically integrated (a stamping shop does most of the pressing of panels, fabrication of the dashboard units also occurs in-house, and many components are internally supplied) and the paint-line, itself, is highly automated. In one section, the primary coat is applied and hardened in a high temperature kiln. The first coat is sprayed on as a fine, electrostatically charged cloud. This section is followed by a rework and rectification area. Then the second coat is applied, similarly followed by rework and rectification, with a small section in between for painting the underside of each car. A buffer area at the end of the paint-line serves as a reservoir to assist in the matching of finished vehicles to distributor, and utlimately, customer demand, as reflected in the Master Production Schedule. Delays due to rework, rectification and other disruptions within the assembly area, and further delays within the paint-line itself, lead to increasing numbers of vehicles in buffer.

The company is one of the most successful manufacturers in Australia. Over the last five years, its corporate strategy has focused on improvement of quality through implementation of Total Quality Management, including: training in TQM, investment in automation, establishment of quality circles, and up-graded quality assurance and audit procedures. The Chief executive participates personally in weekly reviews by executive management of quality audit information. As a consequence, quality standards now equal those achieved by comparative assemblers in Japan, export volumes have increased significantly and the company's share of domestic market growing notably. Nevertheless, the company's financial performance, such as its return on investment and profitability has not improved adequately.

This has led to top-management support for review and evaluation of existing systems of performance measurement and control.

Although the company's quality control systems are quite sophisticated, performance measurement is driven by a traditional engineering outlook and by the standard costing system and variance analysis. A multitude of performance measures are routinely collected and fed up to higher levels in the organisation. However, as is common in many companies, the whole process has never been systematically reviewed or rationalised.

6. Initial Results

At this stage of the project, a survey of existing performance measures taken at each of the six levels of the management hierarchy within the paint-line has been completed. Over 63 different measures were identified through this survey. Work has also commenced on the development of balanced scorecards relating to the key success factors of manufacturing cost (profitability), including cost of quality, quality itself, and customer service (delivery time). The other key success factors - teamwork (encompassing flexibility and adaptability) and resource management - will be investigated over the next year.

The analysis of manufacturing cost has highlighted the importance of gathering information on the time spent on value-added or non-value-added activities (all activities not directly involved with application of paint to vehicles). Materials and energy costs per unit have also received emphasis, particularly the cost of consumables (solvents, cotton waste etc.) whose use has often been excessive, while the measurement of colour coat in litres per OK unit, should serve to reduce the cost of meeting paint thickness quality standards. The percentage of paint shop up-time coupled with information on the causes of stoppages should provide another measure of non-value added time due to maintenance work.

Two new manufacturing cost measures (which will obviously reappear on other scorecards relating to delivery time and resource management), are the percentage of daily schedule achievement (i.e. the percentage of cars completed on the day that were due to be completed on that day) and the average number of extra days per vehicle (above the expected completion time). Improvements in each of these simple measures should result in lower off-line inventory costs and reflect reduced on-line rework, delay and rectification costs. These measures will be supplemented by estimates of the number of bodies in the paint shop at any one time above the planned flow-through rate (an instantaneous efficiency measure) and the average vehicle days per unit in excess of planned time (an average efficiency measure).

Because managers are frequently castigated for holding large amounts of build material for obsolete models there is a tendency to delay presentation of disposal releases for as long as possible. This reduces opportunities to find alternative uses for materials and leads to excess inventory. Two new measures: identified potentially obsolete material; and, utilisation rates for potentially obsolete material should encourage more rational behaviour in this regard.

At the time of writing this paper, the project is still at a seminal stage and hard work on the development of new measures of performance is continuing. The next stage of implementation will be even more demanding and controversial. Hopefully, some useful observations and recommendations will be available for sharing with conference participants in July.

References

[1] Goldratt, E.M., and Fox, R.E., The Race, North River Press, New York, 1986.

[2] Goldratt, E.M., The Haystack Syndrome: Sifting Information Out of the Data Ocean, North River Press, New York, 1990.

[3] Galloway, D., & Waldron, D., Throughput Accounting, Part 1: The Need for a New Language for Manufacturing, *Management Accounting*, UK, January (1988).

[4] Galloway, D., & Wladron, D., Throughput Accounting, Part 2: Ranking Products Profitability, *Management Accounting*, UK, December (1988).

[5] Galloway, D., & Waldron, D., Throughput Accounting, Part 3: A Better Way to Control Labour Costs, *Management Accounting*, UK, January (1989).

[6] Galloway, D., & Waldron, D., Throughput Accounting, Part 4: Moving on to Complex Products, *Management Accounting*, UK, February (1989).

[7] Bakke, N. A. & Hellburg, R., Relevance lost? A critical discussion of different cost accounting principles in connection with decision making for both short and long term production scheduling. *International Journal of Production Economics*, 24, 1-18, Elsevier (1991).

[8] Johnson, H.T., Relevance Regained: From Top-down Control to Bottom-up Empowerment, The Free Press, New York, 1992.

[9] Cross, K., & Lynch, R., Accounting for Competitive Excellence, *Journal of Cost Management*, Spring, pp. 20-28 (1989).

[10] Kaplan, R.S., and Norton, D.P., The Balanced Scorecard Measures That Drive Performance, *Harvard Business Review*, Jan/Feb, pp. 71-79 (1992).

[11] Kaplan, R.S., and Norton, D.P., Putting the Balanced Scorecard to Work, *Harvard Business Review*, Sep/Oct, pp. 134-147 (1993).

[12] Maisel, L. S., Performance Measurement: The Balanced Scorecard Approach, *Journal of Cost Management* (1992).

[13] Beischel, M. E., & Smith, K. R., Linking the Shop Floor to the Top Floor, *Management Accounting*, UK, October (1991).

Team Based Manufacturing Cells: Inside the Black Box of Technology Implementation

Richard Badham and Paul Couchman,
MITOC, Department of Management, Wollongong NSW 2522 Australia

Abstract. Currently, many traditional socio-technical and human centred principles have become accepted parts of organisational restructuring to support product innovation, process flexibility and continuous improvement. Improving our understanding of how to manage this restructuring process, and ensure that potential benefits to both business and employees are realised, is an urgent task. In the past, there was a widespread inability of projects to effectively bring about supportive integrated technical and organisational change and make these changes 'stick' over time. There was also remarkably little documentation of how these change processes occurred, what contributed to failure, and what assisted success. This paper details how the Australian-German Smart Manufacturing Techniques project is addressing this task.

1. *Introduction*

A combination of Japan induced uncertainty about 'best practice' and the growth of the 'business school/consultancy' industry has promoted the rapid spread and universalisation of innovation 'fads and fashions' and 'management myths' about change to 'best practice'.[1] As a result, there has been a dramatic increase in 'airport text', 'management guru', or 'how we did it' discussions on technical and organizational change, offering universalistic principles as guides for how to innovate, and presenting broad generalities as 'recipes' for implementation. It is now 'conventional wisdom', for example, that of major importance is the development of an organisational 'mission', the gaining of 'senior management support', the effectiveness of project 'champions', 'involving everyone' in the process of change, developing clear 'performance measures' as a guide for change, and accepting the central yet lengthy process of 'cultural change' as part of successful innovation.

Such generalities are relatively unproblematic unless they are taken too seriously as either an adequate description of how change occurs or as useful heuristics sufficient to guide the change or implementation process. They say little, for example, about the artificial nature of formal strategies and their over-rationalistic view of organisational change, the need for different types of champions and their roles, what to do about creating senior management support and changing this into strong commitment, the impossibility of involving everyone equally in the change process, and the different types of involvement that are appropriate for different contexts, the constructed and negotiated nature of performance measures as an intermediary for organisational politics, and both the complex tacit nature of culture and the problematic nature of adopting culture change as a real focus of effort.

The importance of such issues, and the need for an improved understanding of technological change processes, is now becoming more widely accepted[2]. A key concern is, however, how these issues are best addressed. The present paper introduces the method and approach being used in a major Australian-German 'Smart Manufacturing Techniques: Team Based Cellular Manufacturing' (SMART) project. This project involves the two year funding by the Australian Department of Industry, Science and Technology of a research-industry network of four research and consultancy organisations and three firms to do two things. Firstly, to design and implement team based manufacturing cells in a press shop of an Australian white goods manufacturer, in the production and assembly area of an Australian small plastic irrigation products manufacturer, and in the assembly of instrument panels in an Australian automobile supplier company. Secondly, to accumulate knowledge and experience on techniques for designing and implementing team based cellular manufacturing (TBCM)*, and to present these in a form accessible to and able to assist other companies interested in initiating TBCM activities.

2. *Methodology*

One academic reaction to the use of 'illustrative case studies' used in many best seller management texts has been to argue for more rigourous quantitative methods. However, quantitative methods are inadequate as exploratory techniques for examining the previously untheorised micro-politics of change. In contrast, the use of more in depth, longitudinal and contextual case study methods have been advocated to provide a more rigourous case study approach.[3]. However, not only do such methods face a problem of generalisation, but they are dogged by the difficult, time consuming and costly process of establishing such longitudinal studies, and the dangers of the 'outsider' not really grasping the tacit cultural understandings and inner workings of the organisation/s under study. This is especially the case when there is pressure for 'quick' and 'relevant' results from heavily funded research.

---

* TBCM is a relatively clear and established technique for grouping similar parts to be fabricated or assembled into distinct sets, the creation of a number of machine or assembly operation groupings or cells to produce the different sets, the creation of a work team to run the machines or assembly operations within each cell, and the devolution of direct and indirect production functions relating to the individual cells (e.g. programming, scheduling, inspection, supervision, set up, maintenance etc.) to the respective teams. The significance of this technique is its combination and integration of technological and organisational design principles, and the role it plays in Europe as an exemplary expression of human centred systems design. This European 'production island' model contrasts strongly with the less autonomous cellular principles applied in Japanese total quality management. The Australian-German project is centrally concerned with the effect of taking central European human centred concepts and translating them into a different production culture.

The method adopted in the SMART project is to employ action research as a key research tool. The academic researchers are funded and conduct their research as change agents working nearly full time within industrial companies, and extra basic research funding is sought to provide time to reflect upon and systematise the insights gained as participant 'observers' in the change process. This method has three main benefits: firstly, it is successful in gaining funding for long term in depth research; secondly, by working closely with industry employees it provides access to the detailed change processes that are not amenable to even longitudinal interview series; thirdly, it clarifies and extends academic knowledge by the need to refine and extend change theory and methods in the process of attempting to create change; and, fourthly, it enables the researchers to act as more effective change agents lobbying for human centred system design principles and processes.

3. *Perspective*

The project began with a realisation that traditional human relations, ergonomic and socio-technical models of technical and organisational change have proved insufficient as a guide in the change process, and committed to the development of an improved model. After the first year of the project, the following configurational process model has proved useful as a guide for understanding and influencing techno-organizational change processes. The model is outlined in Figure 1 below. This model draws on 3 elements. In the case of technology: research on the way in which production technologies are more like malleable configurations than deterministic generic systems than was previously realised, and the influence of national production cultures and systems of innovation on technology transfer and incremental change . In the case of operators: UK research on user centred systems design, the configuration of the user within organisational processes, and the 'societal effect' of product and labour markets on operator characteristics. In the case of intrapreneurs as central components of the production system (or configuration): sociology of technology research on 'system builders', 'heterogeneous engineers' and 'actor networks', as well as organisational literature on project 'champions'. The resulting configurational process model provides a guide for integrating an understanding of micro-politics with broader macro-social characteristics. The understanding of production systems as such configurational processes then guides the action researcher in understanding the forces shaping the system that s/he is trying to change, and areas in which effective input is required. For example, a key role has been discovered to be the interaction between the change agent and a 'triad' of direct line manager, human resource intrapreneur and engineering intrapreneur in the change process, and the role of the change agent in relations between senior management and this triad.

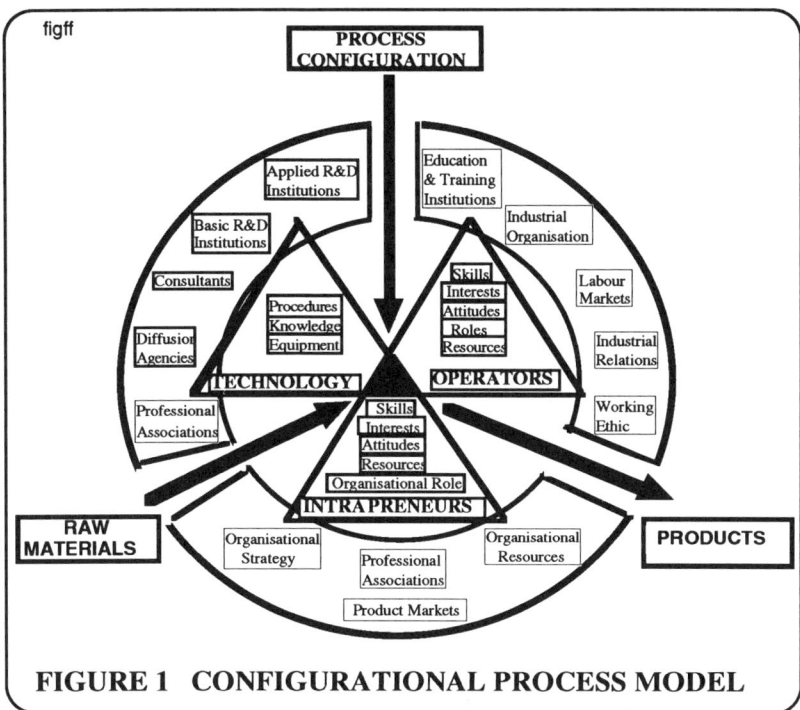

FIGURE 1   CONFIGURATIONAL PROCESS MODEL

4.   *Conclusion*

The effectiveness of action research using the configurational process model is currently being tested in the SMART project.   Initial indications are, however, that such a research model is remarkably effective as a change technique.   The close relationship developed with client companies has proven, so far, to be far superior to the more 'distant' relationship often adopted between academic or technical researchers and the corporate 'user'.   Moreover, the more 'politicised' view of the change process offered by the configurational model has enabled the researchers to develop and utilise this relationship to overcome obstacles to the change process.

[1] H.Ramsay, Management Myths, Paper presented to the *Department of Management,Seminar Series,* University of Wollongong, February, 1994; E.Abrahamson , Managerial Fads and Fashions: The Diffusion and Rejection of Innovations, *Academy of Management Review.* **16** (1991) 586-812

[2]. D.Buchanan, Approaches to Change and Changing Theories, Loughborough University Business School, *Working Paper Series,* 7 February, 1994

[3] P.Dawson, *Organizational Change,* Chapman Hall, London, 1994

Advances in Agile Manufacturing
P.T. Kidd and W. Karwowski (Eds.)
IOS Press, 1994

Improving Communication, Focus and Flexibility Through Restructuring

Amrik SOHAL
Syme Management, Monash University, Melbourne, Australia

Abstract. This paper describes the reorganisation that has taken place at Trico (Australia), a manufacturer of windscreen wiper assemblies for the automotive industry. The Trico Restructuring Improvement Program (TRIP) designed during 1990 and implemented in 1991 has resulted in considerable improvements throughout the organisation. The paper describes the design and implementation of TRIP and the resulting improvements.

1. Introduction

Since the beginning of the 1990s Australian manufacturing industry has been experiencing one of the worst recessions in its history. Closures and downsizing have been frequently reported. Remaining competitive has been difficult for most organisations and many have been introducing a variety of improvement programs based on Advanced Manufacturing Technologies (AMT), Total Quality Management (TQM) and Just-in-Time (JIT) methods [1,2]. However, these initiatives have not always been successfully implemented. In many companies these have faded away or simply died after few years [3]. One Australian company which has continued to grow throughout the recessionary period and sustained a competitive advantage is Trico (Australia) based in Melbourne, Victoria. This paper describes Trico's most recent restructuring program, the benefits achieved and the factors contributing to the company's continuing success.

2. The Company

Trico (Australia), currently employing 240 people, manufactures windscreen wiper assemblies for the automotive industry. It's products includes arms, blades and linkages. The Australian subsidiary was established in 1958 and operated very successfully with little competition from overseas until the early 1980s when it found itself in direct competition with the Japanese. For the first time in its history, the company made an operating loss. Factors contributing to this were identified as inefficient production, a high level of inventory, an inflexible manufacturing system and poor quality.

To improve its competitiveness the company adopted the JIT manufacturing strategy in mid-1984. The JIT program commenced with a focus on the reduction of set-up times on 200 ton high speed presses. The following five years involved the introduction of various elements of the JIT philosophy into Trico with significant improvements being achieved. For example, set-up time on the 200 ton press was reduced from seven hours in 1984 to half an hour in 1988, inventory turns improved from 2.5 in 1984 to 7.5 in 1989. Over the same period business volume increased 30-40 percent. Whereas the company was threatened with extinction from overseas competition in the first half of the 1980s, by 1988 Trico Australia was exporting 50,000 wiper assemblies per month to the USA.

The years 1988 and 1989 saw the JIT program focus on productivity gains which supported further improvements in the company. During 1990, the focus shifted from shop-floor operations to company-wide and specifically focussed on better use of support functions and resources involved in product development and introduction. Further, flexibility in the structure and in all administrative operations was sought.

3. Designing a New Program

Having implemented many of the elements of the JIT philosophy, Trico was looking at other alternatives to further improve it's responsiveness/flexibility and people skills. In early 1990, Trico's Managing Director, Operations Director and Company Secretary attended an external seminar on Social-technical Systems run by the Technology Transfer Council (TTC), a local consulting organisation. Trico quickly recognised the benefits of simultaneously optimising it's technical system and social systems. The company immediately set about designing a system similar to STS. A multi-disciplinary design team was formed in mid-1990 to evaluate the elements of STS and propose a direction for the implementation of teams across the structure of Trico. The 15 people involved in this design team were the Managing and Operations Directors; Company Secretary; Sales, Purchasing, Product and Quality Managers; Toolroom, Arm Manufacture and Blade Assembly Supervisors, Paint Plant and Arm Manufacture Leading Hands; Toolmaker and a Shop Steward.

The design team met on an average once a fortnight during the period August 1990 to October 1991 and discussed a variety of issues. During this period the design team completed three major projects: a survey of all Trico employees, the development of a technical matrix to link people to processes and employee training.

The employee survey was completed by October 1990. All of Trico's employees (totalling 200 in 1990) were surveyed to better understand what they liked and disliked about the company. In-depth interviews were conducted with 20 employees representing all the different work areas. In groups of around 10, the remaining employees were asked to complete a questionnaire. The analysis showed that people were generally satisfied with employment at Trico. This was mainly because of the working arrangements at that time - 9 hours working day, four days per week. Management was seen as "very positive and trying for improvements". The level of co-operative effort within natural work groups in handling day-to-day problems was considered very good however, there did not appear to be any knowledge of specific activity directed towards permanent problem elimination and performance measurement.

The technical matrix was developed for the generic products (arms, blades and linkages) to try and find areas where people could influence the processes of manufacture. The aim was to bring together people and processes across product families as teams.

In mid-1991, a two-day workshop on team working was developed with the assistance of a consultant from TTC. This included simulated games and exercises to teach employees the benefits of team work (by September 1993, all employees had participated in this workshop).

4. Launching the New Program

The design team was instrumental in preparing the groundwork for launching the new improvement program which they called TRIP - "Trico Restructuring Improvement Program". It's main objectives were to maximise output from the current site; consolidate job security; supply all people with relevant timely information and develop people.

The program was launched on 23 October 1991 at the Alexander Theatre, Monash University where all Trico employees were present. During the one and a half hour presentation the management team talked about the need for change, success of the earlier JIT program, vision of TRIP and the implementation timetable. Half an hour was set aside for questions and answers and the whole presentation was video taped. There were many questions from the floor which were mainly concerned with the TRIP program itself.

Restructuring and the formation of teams in the Blade manufacturing area started in November 1991. The management team took on the role of the Resource Group with four teams established in the Blade area by February 1992. Despite the careful preparation and attention to detail planning in the design stage, the launching foundered on rocky ground. The initial set back was the result of too high expectations of the teams and the *"too much too soon"* change syndrome. Management's expectations were too high in terms of what the teams would do, the decisions they would make and the benefits resulting. In fact, there was initially negative benefits, for example, there was a reduction in output. Direction and guidance from the Resource Group had been minimal during the first few months.

In May 1992, a new structure was established where Support Groups were put in place for each area between the Resource Group and the teams. The Support Group is a cross-functional team of 10 to 12 people consisting of team leaders from the area and other relevant staff from Engineering, Sales, Manufacturing, etc. The structure also calls for one member of the Resource Group to be part of the Support Group for each area.

The TRIP program has led to an organisational structure which enables improved communication, product and customer focus for all teams, and flexibility in all aspects of the organisation.

Overall, there has been considerable improvement in communication throughout the company. The new structure provides a minimum of two steps to transmit information either way between team members and the Resource Group. The communication process is further assisted by:

-       Quarterly reports direct from the Managing Director to all team members.
-       A quarterly magazine called "Trip-in Around Trico".
-       Weekly news sheets called "Trico Rumours".
-       Monthly lunches for each team. These are held in the company Dining Room with one hour allowed instead of the regular half hour.
-       Fortnightly Support Group meetings, held during working hours each lasting one hour.
-       Frequent team meetings, held during working hours. In March 1994, these meetings were being held once a week with the aim of developing these into five to ten minutes meetings at the start of each day.

The new structure has resulted in many more feedback loops leading to improved awareness of internal customer needs. By March 1994, 35 teams had been established with an average team membership of eight people. Eleven Support Groups were in place assisting these teams to better meet the needs of both external and internal customers.

The structure and changes initiated and which are continuing ensure flexibility in all areas of the organisation. The Support Teams are supplemented as required by working groups set up for specific tasks and disbanded when the project is completed. This flexibility achieved throughout the organisation has played a major role in securing the growth of the company. The much improved ability to react quickly to changes in production volumes, tooling requirements and design changes has enabled Trico to become a reliable and consistent supplier to the marketplace.

5. Improvements Resulting from the TRIP Program

The process which is now adopted with all teams is one of *"change by evolution"* rather than *"change by revolution"*. Trico's approach and current structure has been working very well since May 1992. Specific examples of results achieved to date include:

- Product introduction time reduced from 12 months in 1990 to 6 months in March 1994.
- The introduction of 30 new products in 1992 and 25 in 1993, against an historical average of 10 products per year.
- New roster for Engineers operating the 3D CAE system resulting in 20% extra output.
- Selection and installation of equipment by the teams. For example, the Press Shop Teams (a total of 20 employees working in three teams) re-organised their work area and purchased new racking. The Toolroom team leader and machine operators with the help of one engineer selected and recommended the purchase of a three-axis numerically-controlled machine worth $160,000. As part of this process four team members travelled to Adelaide to meet with the equipment supplier.

In most organisations restructuring generally involves retrenchments, closing plants and consolidating capacity. At Trico, TRIP have achieved extra employment, stable conditions, growth and the better use of resources. The continuing success of Trico can be attributed to:

- a management team which is totally committed to the company (i.e. the same senior management team as ten years ago). Top managers always looking for alternatives to improve company performance and enhance employee skills.
- employees readily accepting change. They recognise that structures and practices need to be adjusted as requirements change.
- the whole organisation working as a team with various types of support teams established for specific projects and disbanded as the projects are completed.

References

[1] P. Dawson, Organisational Change: A Processual Approach, Paul Chapman Publishing Limited, London, 1994.

[2]. D. Samson *et al.* Human Resource Issues in Manufacturing Improvement Initiatives: Case Study Experiences in Australia, *International Journal of Human Factors in Manufacturing*, 3/2 (1993) 135-152.

[3]. A.S. Sohal *et al.*, JIT Manufacturing: Industry Analysis and a Methodology for Implementation, *International Journal of Operations and Production Management*, 13/7 (1993) 22-56.

Advances in Agile Manufacturing
P.T. Kidd and W. Karwowski (Eds.)
IOS Press, 1994

Wheel of Change

Sue Holmes and David Weeks
33 Upper Belgrave Road BRISTOL BS8 2XL

Abstract. This paper applies the Just In Time principles from the manufacturing process to the decision-making process in people. The manufacturing industry has the opportunity to lead the world in a new way of thinking. It has learnt how to combine Western psychology of the individual and Eastern philosophy of the group with production schedules. The next step is Total Quality Self-Management.

1. Where Have We Been?

Do you remember the days when inventory was piled high and customer orders were chased only when someone shouted? The Economic Batch theory was typical of instant solutions to manufacturing problems. Then computers and automation were the answer to everything. What was really needed in the West was accepted JUST IN TIME: a step-by-step programme integrating men, machines and materials with Total Quality Management. [1].

But if someone had simply come into a factory and said, "Your inventory is too high, you need to change to Total Quality Management with Just in Time", did every thing change - just like that? Yet change is expected to happen exactly like that with something a million times more complex than the most advanced manufacturing system: People.

People are being asked to change:
* from one role to multi-skilling
* from doing it alone to teamwork
* from being told what to do to solving problems themselves.

To help them change, people as well as factories need Just In Time (JIT) : a step-by-step programme integrating minds, brains and ideas with Total Quality Self-Management.

2. What's Going On?

This paper asks the questions about people which a manufacturing consultant might ask when applying the JIT approach to a factory:
- What is the product?
- How is the factory laid out?
- What are the main constraints in the system?
- What are the keys to continuous improvement?

2.1. What Product Do People Make?

Every person in business ' makes' decisions, whatever their job. Creative thinking, based on best information, is the foundation of quality decision-making. Creative thinking offers new ways of looking at the lessons from the past, at the problems of the present, at the possibilities for the future; it also offers new ways of taking appropriate action.

A quality product begins with quality material. Ideas are the material of creative thinking, but it is the correct assembly of parts and the continuous flow of material which make a quality end product, whether in decision-making or in manufacturing. So what are the parts of the thinking process and how are they assembled?

2.2. How is the Human Ideas Factory Laid Out?

Fig. 1

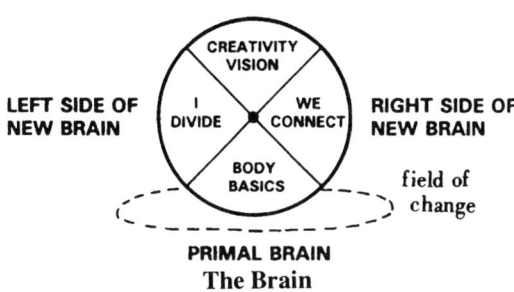

The Brain

The brain has 4 focus areas [Fig. 1]. Each can work alone or with other focus areas. Each can be at a different stage of development like the different parts of a factory. All 4 focus areas are essential to assemble the end product, i.e. quality decision-making. The primal brain [2], at the base of the neck, is the inherited survival kit; its focus is to balance the needs of the whole system. It is alerted when something is missing, it allocates resources and motivates change. The primal brain produces the raw material of ideas: images.

The new brain, or neocortex, is on either side of the head, with a central bridge linking the two parts. [3] The focus of the left side is division, the focus of the right side is connection. Concentration in the West on the left side focus of the new brain emphasises 'I' as a separate individual, whereas concentration on the right side focus in the East emphasises 'us' as connected parts of a group in teamwork. The left side works on the raw material of ideas in a similar way to piece part manufacturing, while the right side works more like an assembly shop fitting the pieces together.

The forebrain is behind the forehead and its focus is the way ahead. Like the directors of a company, it needs vision and the balanced flow of ideas from all parts of the human factory to think creatively and make quality decisions. This is the basic layout of the brain. Where are the blocks in the assembly of ideas?

2.3. What are the 3 Main Constraints in the System?

2.3.1. Brickwalls

Denial builds brickwalls to halt ideas: "What problem?" "It won't work". "No-one can teach me anything". Denying a problem does not make it go away. The companies who faced the Japanese challenge survived by learning a different way of thinking.[4]

2.3.2. Bottlenecks

Learning from the East can cause two kinds of bottlenecks. Firstly, the bridge between the two sides of the new brain may be too narrow to process efficiently all the ideas about teamwork stimulated in the right side. And secondly, learning about other people's ideas is not creative thinking, more like re-assembling the parts in re-manufacturing. Waiting for an original idea to be added to upgrade the product is another constraint.

2.3.3. Breakdowns

When the assembly of ideas is not synchronised between the two sides of the new brain, bottlenecks can turn into breakdowns, particularly if the bridge is not regularly used. For example, it is necessary for the left side to close down the flow from the right side in order to develop scientific detachment and logic. But the barrier has to be lifted from time to time to allow the right side to re-connect. Otherwise a temporary closure can become a fixture.

All these brickwalls, bottlenecks and breakdowns are constraints to the assembly of ideas. And they have the same effect as constraints in production: wasted time, effort and opportunities. The movement of material is the key factor in JIT manufacturing and so is the movement of ideas in the brain. How can the flow be improved?.

2.4. What are the 3 Keys to Continuous Improvement?

The primal brain holds the keys to improving the flow of ideas with its ability firstly, to be alerted by the will to change; secondly, to allocate resources through the wheel of change — [Fig. 2] and thirdly, to motivate the whole system to work with the prime agent of change - life itself.

2.4.1. The Will to Change

Do you *want* to change? The will can by-pass the brickwalls of denial, so that the primal brain receives the signal that it is missing out. Once it is alerted, the 'desire' button will seek out what is missing as automatically as the body seeks out food when hungry or rest when tired. It can be compared to the Kanban system, which automatically signals the need for more material.

2.4.2. The Wheel of Change

The primal brain protects the body with a field of emotional energy [Fig. 1], which flows through the Wheel of Change [seen from above Fig. 2]. There is a central hub, an outer rim which marks out a person's 'space', like a boundary fence, and the 4 spokes of anger, fear, grief and compassion [5]. All 4 emotions are essential for well-being.

A person is moved by e-motion when the wheel is touched by people or things outside, or by inner thoughts. Anger and compassion activate; fear and grief cause withdrawal. Anger and fear arouse tension; grief and compassion release it. The range of each emotion is very wide: from frustration to rage, from anxiety to terror, from letting go to devastation, from sympathy to being 'at one' with another person.

Regardless of which emotion is touched first, the flow of ideas has to pass through the whole circle for the wheel to be balanced and symmetrical. When the wheel of change is balanced, it can allocate resources to both sides of the new brain equally, and synchronise the flow between them to overcome bottlenecks on the bridge. The wheel's cycle is completed, for example, when anger and fear are recognised as part of the grieving process and support is given. Uncertainty about the future of manufacturing may arouse anger and fear, but letting go of the past is necessary to move into compassion.

Fig. 2

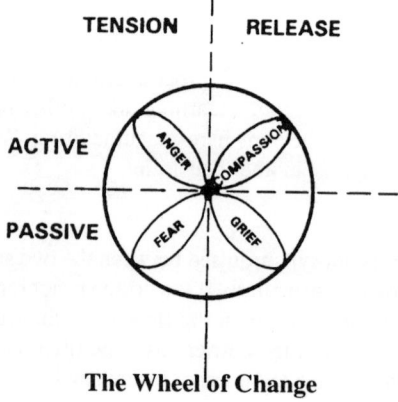

The Wheel of Change

It is through compassion that understanding of others develops into teamwork. The West can add value and originality to teamwork ideas from the East, with its own special development of "individuality". Individuality is that sense of adventure which makes the single-handed yachtsman, or the goal-scorer or the classic car owner stand out from the crowd. If fear of losing individual identity is causing a bottleneck on the bridge to 'us', it helps to remember Dr. Deming. He was the lone maverick who dared to penetrate the closed group of Japanese manufacturers and open up the exchange of ideas between Western left side and Eastern right side thinking. [6]

2.4.3. Life is Change

Mankind has survived because of its fitness to adapt to life's changes. The emotional response to something new fires the wheel of change into action. It is called motivation. Motivation can use the flow of ideas to go with the wind of change or to apply the brakes - but not for ever. Life is change and like time, it cannot be stopped. Critical change will come sooner or later, perhaps in divorce, or redundancy or bereavement. It your wheel is rusty and not fit for its purpose of adapting to change, then the breakdown could be in you.

Emotions, with intuition and common sense, form a vital part of the decision-making process. They give immediate feedback, like the shop-floor. Ignored, they cause problems. If that is hard to believe, look at yourself or others. Do you need a drink at the end of the day? Do they slump in front to the television or can never relax? Do you feel like hitting a ball - or the cat - instead of the boss? Do they ever feel they are overloaded and can't cope? These are signs that unreleased tension in the emotional wheel is causing a problem: it is called 'stress'.

If it is not released, the adrenalin of tension can stress the whole system to breaking point. If it is used positively, it can drive the flow of ideas round all parts of the brain to achieve creative thinking, quality decisions and world-class performance. The choice is yours.

> " You do not have to do this;
> survival is not compulsory". Dr. E Deming.

3. Where Next?

As the wheel of change drives a balanced flow of ideas to both sides of the new brain, then onto the forebrain, people *can* change Just In Time from one role to many, from doing it alone to teamwork, from being told what to do to making their own decisions. The next step is to integrate JIT decision-making with Total Quality Self-Management of the mind.

The manufacturing industry can show the rest of the world how to do it. They have the technology and the experience. Who says the spirit of adventure is dead?

References

[1] E. Goldratt, the Goal. Gower, 1984
[2] M. Odent, Primal Health. Century Hutchinson, 1986
[3] H. Adler, The Right Brain Manager. Piatkus, 1993
[4] D. Rowe, Living with the Bomb. Routledge, Kegan, Paul, 1985
[5] D. Connelly, Traditional Acupuncture: The Law of 5 Elements. Centre for TA, 1979
[6] F. Aguayo, Dr. Deming. Mercury, 1991

Advances in Agile Manufacturing
P.T. Kidd and W. Karwowski (Eds.)
IOS Press, 1994

Understanding and Coping with Resistance to Change

Bernard BURNES

Manchester School of Management, UMIST, Manchester, M60 1QD, UK

Abstract. There are many well-documented reasons why organisations fail to effectively manage change. These include lack of strategy, poor planning, absence of key skills and employee resistance. Many consider the latter, employee resistance, as the pivotal factor in the success or failure of a change project. It has been argued that human beings are by nature suspicious of and resistant to change. How true this is can be debated, what should not be at issue is that, when present, resistance can prove costly and in some cases make success all but impossible to achieve. Nevertheless, resistance is neither universal nor consistent. Even within the same organisation change can elicit different responses, even from the same people, depending up on the issues concerned, the approach taken and the circumstances of the organisation at that particular time. This paper will argue that applying the concept of Cognitive Dissonance to the management of change provides the key to understanding and overcoming resistance to change.

1. Introduction

There can be little doubt that the world we inhabit is becoming increasingly dynamic and unpredictable. Indeed, a number of writers now believe we have entered a new era of change altogether, one that is characterised by chaos rather than order [1,2]. The essence of the argument is that, "The long term future of a system in chaos is not simply difficult to foresee: it is inherently unknowable because the system can amplify out of all proportion changes that we can never hope to detect." [2:9].

The implications of this for organisations are significant, to say the least. In particular, it is likely that the pace and magnitude of organisational change will be greater than ever before. For some organisations, those committed to fostering a learning environment, change may become a normal everyday activity, and this in turn can be a source of competitive advantage [3]. However, for others, perhaps the majority, managing change will continue to be fraught with danger, and failure rather than success will be the outcome.

This paper seeks to address an issue which is seen as being a key factor in managing organisational change: employee resistance [4]. First of all, the paper draws upon the theory of cognitive dissonance in order to understand why resistance occurs. It then moves on to discuss approaches to avoiding or overcoming resistance. The paper concludes by arguing that it is not necessarily the magnitude or radicalness of a change initiative which should determine the approach to managing it. Rather, the approach should be selected in relation to the level of cognitive dissonance to which the change project might give rise.

2. Cognitive Dissonance

Cognitive dissonance states that people try to be consistent in both their attitudes and behaviour. When they sense an inconsistency either between two or more attitudes or between their attitudes and behaviour, people experience dissonance; that is, they feel frustrated and uncomfortable - sometimes extremely so - with the situation [5]. Therefore, individuals will seek a stable state where there is minimum dissonance. This latter point is important. It is unlikely that dissonance can ever be totally avoided, but where the elements creating the dissonance are relatively unimportant, the pressure to correct them will be low. However, where the issues involved are perceived by the individual to be significant, the presence of such dissonance will motivate the person concerned to try to reduce the dissonance and achieve consonance by changing either their attitudes or behaviour to bring them into line [6]. This may involve a process of cognitive restructuring which is unlikely to be free from difficulties for the individual concerned [7]. However, as Festinger [8] one of the originators of the concept points out, in addition to trying to reduce the dissonance, people will actively avoid situations and information which would be likely to increase the dissonance.

As an example, if a supervisor believes that tight control of those for whom she is responsible is required to make them work effectively she will be uncomfortable if required to give them a greater degree of autonomy. To reduce this discomfort (dissonance) she may either change her attitudes ("tight control is no longer effective/necessary in this day and age") or ignore/circumvent the new regime. However, if she cannot reduce the dissonance, for whatever reason, then this may result in stress, anger and resentment, none of which is likely to assist the move to more autonomous working arrangements.

Therefore, if an organisation embarks on a change project which is markedly out of step with the attitudes of those concerned, it will meet resistance unless those concerned change their attitudes; and this is only likely to occur if they believe that they have some choice in the matter [5, 6]. On the other hand, where the level of dissonance occasioned by proposed changes is low, attitudinal adjustments will be minor and potential resistance negligible. It follows that approaches to overcoming or avoiding resistance to change should be geared to the level of dissonance to which any proposed changes may give rise.

3. Avoiding and/or Overcoming Resistance

One of the major approaches to overcoming resistance to change is the involvement of those affected by change in its planning and execution [9, 4]. Some writers [10] have argued that those likely to be affected by change should be fully involved in the process. However, the work of Schmuck and Miles [11] challenges this and provides a bridge between levels of involvement and levels of cognitive dissonance. In examining the methods and techniques available for managing change, they did not see full involvement as necessary in all situations. Instead they saw involvement as a continuum related to the impact of the change taking place on the people concerned. This continuum runs from "acceptance", based on prescriptive/directive modes of intervention, to "theory and principle", which is where the change agent provides the change adopters with advice on which they can make their own free choice. Therefore, it is not necessarily the type, duration or magnitude of the change which determines the effectiveness of a particular approach. Rather it is the impact, or likely impact, on the people concerned.

Huse [12] developed this distinction further. Building on earlier work by Harrison [13], Huse categorises change interventions along a continuum based on the "depth" of intervention, ranging from the "shallow level" to the "deepest level". The greater the depth of the intervention, Huse argues, the more it becomes concerned with the psychological make-up and personality of the individual, and the greater the need for the full involvement of individuals if they are to accept the changes.

4. Conclusions

It follows from the above, that it is possible to tailor the methods and techniques for managing change to the level of dissonance (or potential resistance) that is caused by the proposed change. The key is that the greater the effect on the individual, especially in terms of psychological constructs and attitudes, the deeper the level of involvement required if successful change is to be achieved. Therefore, where a proposed change is in tune with the established norms of an organisation and the individual's own attitudes, that person will be more inclined to accept its legitimacy; ie the depth of involvement will be at the shallow end of the spectrum and may merely involve a passive acceptance.

This approach also helps to explain reported cases of radical or transformational change which have been achieved by directive rather than participative measures [14]. It might be assumed that the magnitude of such changes would give rise to considerable resistance if they were imposed rather than being negotiated. However, given that transformational change is most often attempted when an organisation is either in crisis or operating in a turbulent environment, this may not be the case. In such situations, instead of a directive approach causing resistance it could be welcomed by staff who recognise that difficult situations sometimes require extreme remedies. As Pettigrew and Whipp [15:174] have observed, "a crisis provides the space and legitimacy to effect major strategic reorientations."

Therefore, despite what "commonsense" might suggest, it may be that large-scale changes, in some situations at least, warrant lower levels of employee involvement than more limited change initiatives. However, where the proposed change is out of step with the dominant norms, or where these norms are in a state of flux, then a greater degree of involvement will be required; ie the depth of involvement may be at the deepest level, requiring such techniques as sensitivity training, personal counselling, and life and career planning.

It can be seen, therefore, that the concept of cognitive dissonance is potentially valuable in understanding the factors which can promote or hinder effective change.

References

[1] I, Nonaka, Creating Organizational Order Out of Chaos: Self Renewal in Japanese Firms. *California Management Review*, Spring, (1988).

[2] R. D. Stacey, Management into the 21st Century: The End of the Stable Equilibrium Paradigm? Paper presented to the Sixth Annual Conference of the British Academy of Management, September, (1992).

[3] D. A.Garvin, Building a Learning Organization. *Harvard Business Review*, July/August, (1993) 78-91.

[4] B. Burnes, Managing Change. Pitman, London, 1992.

[5] E. E. Jones, Interpersonal Perception. Freeman, New York, 1990.

[6] S. P. Robbins, Organizational Behavior: Concepts, Controversies, and Applications. Prentice-Hall, Englewood Cliffs, New Jersey, 1986.

[7] M. J. Mahoney, Cognition and Behavior Modification. Ballinger, Cambridge, Massachusetts, 1974.

[8] L. Festinger, The Theory of Cognitive Dissonance. Stanford University Press, Stanford, California, 1957.

[9] B. Burnes, Barriers to Employee Involvement in Technical Change: More Than a Case of the Good Guys and the Bad Guys. *Advanced Manufacturing Engineering*, 2, April, (1990) 69-75.

[10] C. Argyris, Intervention Theory and Method. Addison-Wesley, Reading, Massachusetts, 1970.

[11] R. Schmuck and M. Miles, Organizational Development in Schools. National Press, Palo Alto, California, 1971.

[12] E. F. Huse, Organization Development and Change. West: St Paul, Minnesota, 1980.

[13] R. Harrison, Choosing the Depth of an Organisational Intervention. *Journal of Applied Behavioural Science*, 6, (1970) 181-202.

[14] D. Dunphy and D. Stace, The Strategic Management of Corporate Change. *Human Relations*, (46) 1, (1993) 905-919.

[15] A. Pettigrew and R. Whipp, Managing Change for Competitive Success. Blackwell, Oxford, 1991.

Advances in Agile Manufacturing
P.T. Kidd and W. Karwowski (Eds.)
IOS Press, 1994

Four Cases for Improving Organizational Practices

Jyrki KIVINIITTY

Work Research Centre, University of Tampere, BOX 607, FIN-33101 Tampere, Finland

Abstract. *"Work, Culture and Technology"* is a combined operation carried out in machine workshops during the period 1992-94. The purpose of the research is the realization of techno-organizational changes and to discover inhibiting and facilitating factors. The operation includes four companies and a multidisciplinary research team.[1]

Views and premises on the development of productivity in companies vary from the tayloristic rationalization to the socio-technical and lean production models. For the change to be successful and in the intrest of the development under way in the companies, it is crucial that the company be able to derive benefit from (a) network-type mode of activity, which will be successful only if the personnel have adopted the (b) development tools introduced in the experimental phase. This is all part of the process of *organizational learning*, in which (c) professional skills are enhanced and cooperative work between groups make it possible to transcend boundaries between different functions in the company. To ensure that the results of the experimental phase are consolidated the company must have a strategic view of the necessity for continuous improvement. Each company involved in the research has proven, that it is capable of benefitting from developmental scope offered in the course of the combined operation.

1. Introduction

Many production organizations continue to be dominated by Tayloristic model of work organization developed at the beginning of the centry. Demarcations between the tasks of employees are strictly adhered to, and planning is separated from implementation. The task of the organization is to supervise the employees, whose roll is that of passive functionary. The thesis, revolutionary in its time, that "Science, not rule of thumb. Harmony, not discord. Cooperation, not individualism. Maximum output, in place of restricted output" [1; p. 140] was the driving force behind Taylorism.

Rapid realization that there were flaws in Taylorism (Hawthorn's studies in the 1930s) did nothing to impede its march to success. Also one of the main objectives of the socio-technical work design has been since the 1950s to eradicate the inconsistencies of the Tayloristic shaping of work. Similar "turning points" were the implementation of Total Quality Control (TQC) ja Just-In-Time (JIT) production in the 1980s [2].

2. Alternative Production Models

What was behind the development of new production models was the need to create organizational capable of meeting new demand for activity. What inhibited the

---
[1] Researchers include engineers, psychologists and sociolologists from the University of Tampere, the Technical Research Centre of Finland, Institute of Occupational health, and Ministry of Labour. The research was funded by Academy of Finland and the Finnish Work Environment Fund.

development of such models was Taylorism, and yet criticism has been expressed that scientific management was entranced in the "new" models too.[3] The most important prerequisites for the successful modern organization are expertise, network and group-based mode of activity and the creation of new knowledge. Behind innovations there is the "tacit" knowledge of certain individuals, and the organization must make use of this. [4]. At its weakest organizational ability to learn is system maintenance, and at its strongest it is institutional meta-learning.[5; p. 237] One may speak of organization's commitment to a novel *adaptable mode of activity*, which requires commitment to *organizational learning* [6][7][8].

Among the new models proposed are the lean production model [9], originating in Japan, the Anthropocentric Production System (APS) [10], originating in Europe, which may also be concidered one of the new forms of the socio-technical development [11]. Their common objective is the enhancement of employee skills. In the goal-setting there are variations: In lean production goals are set from the point of view of production chain, whereas in socio-technical and APS models the point of depature is the welfare of the personnel [12].

2. Aims of the Study

The objectives for the study are: (1) to obtaine information on the development mechanism of the techno-organizational changes and on the organizational learning and innovation process, (2) to solve operational problems (e.g. quality, flexibility, productivity) in a way, which promotes employees welfare, and (3) consolidate and diffuse tools and procedures used in the experiments in the companies.

3. Research Approach

Research is intensive case-study, which aims to *theory development* [13; p. 36-37]. research progresses cyclicly in three stages (see fig.1). As in *action reserch* "learning by doing" has an important role in case studies, and also the changes in companies mode of activity usually take longer than the "research cycle" [14; p. 242-243].

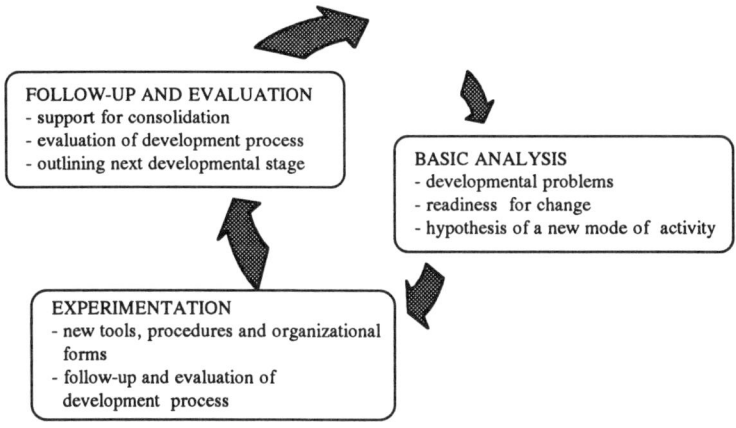

Figure 1. Research Cycle (and Change in Mode of Activity in the Company)

4. Material: Process of change in mode of activity in four companies

The research material consits of the phase of a techno-organizational change process under way in four machine shops (see table 1). It is the intention of the Paper and board machine company and the steel-cast foundry to carry through a "large-scale organizational change" [15; p. 145], while the bus coach assembly plant and the special machine tools shop aim at finding finding tools and procedures with which to overcome production problems.

5. Conclusions

The dominant paradigm in the companies' activity has so far been *craft-rationalized* more or less Tayloristic model. The non-functionality of the basic model is revealed in the form of functional problems between remaining rigid organization of labour and increasingly fluctuating enterprise environment. [3; p. 58] This former mode of operation has inhibited the enhancement of professional skills and problem solving. Without changing this dominant mode of operation it will not be possible to achieve such objectives. The change can be brought about if it is supported by methods and procedures, which are valid and sufficiently simple. The extent and depth of the change are dependent on the point from which the company sets out to make the change and on its ability to take on board new knowledge.

Table 1. The Research Sites

	PAPER AND BOARD MACHINE FACTORY	STEEL-CAST FOUNDRY	BUS COACH ASSEMBLY PLANT	SPECIAL MACHINE TOOLS FACTORY
ENTERPRISE SIZE	Big multi-branch group	Big multi-branch group	Middle-sized concern	Small enterprise
PLANT SIZE (EMPLOYEES)	Large (over 500 employees)	Large (over 500 employees)	Small (under 200 employees)	Small (under 100 employees)
PRODUCTION LAYOUT	6 focused factories: roll factory divided into 2 product shops	2 profit centres: steel foundry consists of 3 production lines	1 profit centre: 1 production line	1 profit centre: 3 product shops
AREA FOR EXPERIMENT	2 cells in a product shop & their support functions	1 production line & its support functions	2 work groups on the production line & staff functions	2 product shops & product development & their support functions
	Technical and organizational problems			
OBJECT OF CHANGE	Organization structure: shift to cell-based work organization	Organization structure: shift to product shop-based work organization	System of production management	Mode of activity and mutual cooperation of the product shops
	Tools and procedures for developmental activities			

All the companies taking part in the research proved themselves capable of utilizing elements of new mode of activity, whose guiding principle may be described as follows: (a) way of working based on networks and groups, which can be successful only if the personnel has at the experimental phase adobted (b) tools (methods of documentation, systematic progress etc.). In the interest of the continuity of the development work it is necessary, that the company take a strategic decision (c) to form a development organization. All these foregoing are part of the process of learning in the company, in which (d) professional skills are enhanced and the crucial nature of continuous development work is made apparent as a new area of expertise. The company should have organizational adaptability and a capacity for unlearning and error tolerance. What is more, development work demands slack resources.

At the end of the day it is the consolidation and evaluation phase, which determines how well the new mode of activity has gained a footing in the organization. In this respect the companies are currently in a stage which is crucial to the results. A great part of the results so far achieved will lose their strength if time needed for use of the tools and development work is cut down in the belief, that the methods have already been learned.[16; p.78] There is a danger sliding back into the old ways unless the critical waterched in the research cycle and the process of change is sucessfully surmounted.

References

[1] F. W. Taylor, The principles of scientific management, *In* Scientific management: comprising Shop management, The principles of scientific management, Testimony before the special house committee. ISBN: 0-8371-5706-4. Harper & Row, London, 1964.

[2] R. J. Schonberger, World class manufacturing: the lessons of simplicity applied. ISBN: 0-02-929270-0. Free Press, New York, 1986.

[3] D. Linhart, The shortcomings of an organizational revolution that is out of step, *Economic and Industrial Democracy* 1 (1993) 49-64.

[4] I. Nonaka, The knowledge-creating company, *Harvard Business Review* 6 (1991) 96-104.

[5] D. McKee, An Organizational Learning Approach to Product Innovation, *Journal of Product Innovation Management* 9 (1992) 232-245.

[6] P. S. Adler and R. E. Cole, Rejoinder, *Sloan Management Review* 35 (1994) 45-49.

[7] C. Berggren, NUMMI vs. Uddevala, *Sloan Management Review* 35 (1994) 37-45.

[8] D. Leonard-Barton, The factory as a learning laboratory, *Sloan Management Review* 33 (1992) 23-48.

[9] J.P. Womack, D. T. Jones and D. Roos, The machine that changed the world. ISBN: 0-06-097417-6. HarperPerennial, 1991.

[10] W. Wobbe, Anthropocentric production systems: a strategic issue for Europe. FAST/MONITOR Programme. APS Research Papers Series vol 1. Bruxells, 1991.

[11] F. M. v. Eijnatten, The Paradigm that Changed the Work Place. ISBN: 90-232-2805-7. Van Gorcum & Comp., AA Assen: The Swedish Center for Working Life, Stockholm, 1993.

[12] R.R. Rehder, Saturn, Uddevala and the Japanese lean systems: paradoxical prototypes for the twenty-first century, *The International Journal of Human Resource Management* 5 (1994) 1-31.

[13] R. K. Yin, Case study research: design and methods. ISBN: 0-8039-3471-8. Sage Publications, Newbury Park, 1989.

[14] P. H. Engelstad and B. Gustavsen, Swedish Network Development for Implementing National Work Reform Strategy, *Human Relations* 2 (1993) 219-248.

[15] G. E. Ledford, Jr and S. A. Mohrman, Self-design for high involement: a large-scale organizational change, *Human Relations* 2 (1993) 143-173.

[16] D. A. Garvin, Building a learning organization, *Harvard Business Review* 4 (1993) 78-91.

582

Advances in Agile Manufacturing
P.T. Kidd and W. Karwowski (Eds.)
IOS Press, 1994

INDUSTRIAL ROBOTS, WORK ORGANISATION AND WORKING CONDITIONS- AN EVALUATION OF DANISH INDUSTRY AND THE TREND TOWARDS ANTROPOCENTRIC SYSTEMS

Christian Koch (Assistant Professor)
The Unit for Technology Assessment
Danish Technical University
DK 2800 Lyngby
Denmark

tel:45 31474715

ABSTRACT:This evaluation of danish small and mediumsized enterprises utilizing industrial robots demonstrates that a technocentric concept still seems to dominate this part of danish industry. Some examples of an "nonideologic" concept were found, while a conscious anthropocentric concept was missing.

The robots weren't primarily introduced to improve working conditions. The working conditions at industrial robots are closely linked to the rhythm, speed and cycletime of the robot. The peripheral equipment also plays an important role. Most of the investigated work is characterized by loading and onloading of objects for the robot.

The working conditions has improved in some fields and deteriorated in others after the industrial robot was introduced. The latter applies to the psychological environment and to some aspects of the physical loads.

1.Background

This paper describes some results of a project made for the Danish Working Environment Fund (ref.[1]).

Most of the data for the project was collected in 1988-1990, covering 9 danish and 1 swedish company. The danish companies were SMEs with 17-700 employées. 7 companies were in the metalworking industry, one in plastics and one in the furniture-industry. The companies were chosen in order to collect a typical sample. Since some of the companies used robots for several purposes and with different characteristics the investigation was focused on *12 different concepts*, that is, specific combinations of tasks, robot(s), peripheral equipment, operators, work organisation and working conditions.

We carried out a series of semistructured interviews and job observations covering the 12 concepts in the companies using industrial robots.

In march 1993 we did a supplementary study of a spearpoint enterprise with 2200 employées. This was meant to be in contrast with the rest.

2.What could be expected ?

Many people have had expectations of the possibilities of industrial robots to improve some of the existing problems of work environment. The vision is that the robot will solve tasks that involves human strain.

Parallel to this, the findings of social scientists and the discussions and visions amongst them underlined the possibilities of and the emergence of a new concept for production; the socalled Post-fordistic (ref.[2]), Anthropocentric (ref.[3]) Humancentered or "Non empirical-ideologic" (ref.[4]).

With Wobbe (ref[3]) the anthropocentric concept is defined by

> *Advanced computeraided manufacturing which depend upon human skills, collaborative work organisation and adapted technologies. They are based on the new principles of work and factory organisation, reduced division of labour, the unity of conception and execution of work, decentralized decision making, skill enhancing job design as well as human centred technology design.*

One could therefore expect to find a tendency in danish industry along these lines. Was danish industry starting to abolish taylorism?

3.Results

This investigation demonstrates, however, that robots are not primarily introduced in enterprises with the intention to improve working conditions or to realize an anthropocentric concept.

The introduction of industrial robots in the enterprises covered by the investigation has not decisively led to a lift of work environment, qualitatively. The work environment in connection with industrial robots is in many ways similar to that of traditional industrial work. In some areas improvements have occured, in others it has not been the case.

The work conditions are closely linked to the rhythm, speed and cycle-time of the robot. Furthermore it is linked to the characteristics of the peripheral equipment and to the possibilities of building up a buffer stock at the workplace. These conditions highly influence the degree of freedom within a timeframe and consequently also to the risk of strain and stress in work.

Working with industrial robots is characterized by loading and unloading objects during one total work session, having a cycle in average of less than 10 minutes. Most of the jobs are organized in a way that includes very few different tasks. The possibilities of improving these jobs by enlarging them with for instance production-planning, programming or fixture manufacture, are very small due to the small extension of these tasks.

The investigation illustrates the use of various types of work organisation in connection with industrial robots. In the figure the three main types are grouped in narrow, middle and broad work organisation:

WORK ORGANISATION

TYPE	TASKS	WHO?
Narrow (technocentric)	planning details producing fixtures	others (superviser)
	programming preparing handle failures maintenance	preparetor
	loading/unloading monitoring	operator
Middle	planning details producing fixtures	others
	programming preparing maintenance	preperator
	handle failures/programadjustment maintenance (smaller tasks) loading/unloading monitoring	operator
Broad (anthropo-centric)	all the mentioned tasks	operator

The first model are represented by 5 concepts, the second by 3 and the third by 4 in the research. The middle type is still dominated by the loading and unloading of objects. Hence the dominance of technocentric concepts is evident. And even the broadest organized danish robot jobs can be developed. We visited a swedish work-place with group-organisation and robots that clearly illustrated this.

The label "anthropocentric" should be interpreted with caution since none of the cases explicitly worked with this concept. With Kern and Schumann we prefer to use the notion "non empirical-ideologic" (ref.[4]). It was for example the presence of a qualified, willing unskilled worker who made management integrate the programming and planning tasks into the operator's job, thus overruling traditional division of labour.

4. Types of robots

In the cases four types of robots are represented: paint, welding, assembly and handling robots. These types of robots and the models of work organisation in use show no connection. But the work environment at the different types of robots shows very distinguished characteristics. Working with welding robots is characterized by the operator being tied to the robot's working cycle, while the work with painting robots is declutched. This is due to the use of conveyer-systems for carrying forward the objects. However, at the spearpoint enterprise, the welding robots with pheripheral equipment resulted in a declutched work process as well.

5.Working Conditions

The psychological work environment of robots is especially connected with the contents of the work, that vary in accordance to the type of work organisation chosen. But many jobs typically are characterized by monotonously repeated work, constrained by the cycle of the robot. In these situations there is a great risk of monotonous reactions, and this is also experienced by the operators. Yet other situations show in addition a time   pressure which may prove stress provoking. Shift work and other types of abnormal working hours prevail among the robotinstallations and many robot operators suffer from psychosomatic diseases or they find trouble in combining family life and leasure time.

The physical work environment does not differ substantially from the general industrial work environment. However, improvements, of the layout and of the workplace design, have reduced a series of awkward, psysically straining postures. On the other hand the amount of physical handling has increased because of the large amount of objects processed by the robot. Due to this fact the total physical load grows comparatively in a number of jobs, and symptoms in the musculosceletal system, especially back and legs are very common.

The chemical exposition has been reduced by the robots. This is due to an effective exsuction or encapsulation of the robots, but especially because the robot operators are only for short periods in close contact with the robots' work processing. Thus the finding is that in some fields the work environment has improved and in other fields it has deteriorated. The latter applies to the physical environment and to certain aspects of physical loads. The examples of "good jobs" figuring in the investigation show that it is possible to improve the work environment through better planning. But even the best jobs cannot be claimed to be "good jobs". This will demand a more radical reorganization of production schemes and work organization in the firms.

6.Conclusion

This analysis shows that the utilization of industrial robots in SME in danish industry still follows the technocentric concept. There are only weak tendencies following an anthropocentric concept. Thus the task is to promote these tendencies.

REFERENCES

[1] C.Koch and A.Richter : Psychosocial and Physical Working Environment at Industrial Robots. The Danish Work Environment Fund 1991 (in danish)
[2] R. Piore and C.Säbel, The Second Industrial Divide,New York, Basic Boods,1984
[3] W.Wobbe, Advanced Manufacturing and Anthropocentric Production Systems in the European Community. Ergonomics of Hybrid Automated Systems III, P.Brödner and W. Karwowski (Editors) Elsevier 1992
[4] H.Kern and M.Schumann : Das Ende der arbeitsteilung ? Verlag C.H. BECK Frankfurt, 1985

Advances in Agile Manufacturing
P.T. Kidd and W. Karwowski (Eds.)
IOS Press, 1994

A Study of the Practicalities of Human-Centred Implementation in a British Manufacturing Company

Dr. I.S. Fan and Mr. Robert Gassmann
The CIM Institute, Cranfield University, Bedford, UK

Abstract: A series of pilot studies were made in a medium sized manufacturing company which makes braking systems for heavy goods vehicles. The study looked at the practicalities of adopting human-centred concepts in three cells at different stages of development. The study highlighted the difficulties of overcoming traditional working practice, and the importance of sustaining the programme of continuous improvement. This is true irrespective of the level of development of cell working practices and the skills of the cell members.

1. Introduction

In the past, the company has seen an increase in the demand of its products because of the increase in heavy goods traffic. Recently, however, intense competition has saturated this market, and customers are now buying fewer products and making them last much longer. As a consequence of this, and from the constraint of having a very static product range, they found themselves having to compete for orders on the bases of product quality, lead time and delivery performance. The company adopted a cellular manufacturing approach in an attempt to become world leaders in their industry.

Table 1: Chosen cells X, Y and Z, and selection criteria.

CELL FEATURES	AGE OF CELL		
	< SIX MONTHS	SIX MONTHS TO ONE YEAR	ONE TO TWO YEARS
	CELL X	CELL Y	CELL Z
Machines	2 Fanuc NC Tape Drills, 1 Precom 1 Takisawa CNC Lathe	2 Fanuc NC Tape Drills 2 Takisawa CNC Lathes 1 Washer, 1 Precom	1 NC Kingsbury, 3 Mills 1 Chamfer m/c, 1 Borer, 2 Washers
Cycle Times	192 seconds	180 seconds	10 minutes
Number of People	One person. Job rotation between four people; fortnightly and day / night shifts.	Two people. Job rotation between four people on day and night shifts.	Two people. Job rotation between four people on day and night shifts.
Size of Part Family	One part with four variations	Two parts with four variations each	Two parts with three variations each
Demand for Components	Steady, continuous.	Steady, continuous.	Currently very slow, few orders.
Skills / Training	Operating machines	Programming and operating machines.	Highly autonomous

2. Background

It was a company policy to consciously adopt a human-centred, participatory approach for the design, implementation and management of their cells. The research was undertaken to observe the practicalities of adopting human-centred concepts, and how the relationship between the cell members and management was affected. To establish a broad base of analysis during the pilot study cells were chosen according to their age. Three cells were chosen (cells X, Y and Z) differing in age and exhibiting different cell features (Refer to Table 1).

3. Summary of Results

The practicalities and problems of taking a more human-centred approach are summarised for each cell. The following aspects are encompassed: planning and control procedures, the role and purpose of meetings, scope of participation and involvement of operators and skill development policies.

3.1 Cell X

The weekly work load on the cell is specified by management. This was communicated to the operators by the superintendents at the start of each week. Informal communication between the superintendents and the operators occured if any disruptions were liable to influence the performance of the cell. Because of the dedicated nature of the cell and the stability of the demand, there was limited scope for the operators to make any decisions.

A charting method for recording the performance of the cell was used by the operators. The accumulated data was used for highlighting continuous improve-ment targets. A great deal of suspicion existed because any down time had to be logged and accounted for, and the operators were unsure of the way the data would be interpreted.

Continuous improvement meetings were held each month but they did not completely fulfil their intended purpose. The initiative to suggest any changes had to come from the operators. Improvements were often proposed in front of quite a confrontational audience of superintendents and line management, so that the operators found the idea of speaking out quite daunting. All the *final* continuous improvement decisions were made without the operators present, with the superintendents acting as intermediaries between management and the operators. Management did not yet feel confident of the abilities of the operators to include them in this process. The operators felt little sense of ownership because changes to the working conditions or cell procedures were imposed from above.

To increase production flexibility, the company offered free training courses to all operators. Many expressed the desire not to go on these training courses because passing the course was perceived to be of little value. The desirability of any reward was judged upon financial criteria. The benefits to the company of a more flexible workforce were not communicated sufficiently to the operators.

This symbolises the partial teamworking relationship between the shop-floor workers and management.

3.2 Cell Y

Cells X and Y shared the same managers and superintendents. The operators had a close relationship with the superintendents. This had evolved to the extent that the operators were consulted in all scheduling matters. As well as being used for manufacturing a family of parts, the cell was used to manufacture late or over-spill work from other cells. The opinions of the operators in these conditions are therefore very important. Within the cell, the operators decided when to rotate jobs and had learnt the skills to programme the machines, set offsets and replace tools. This made the cell members more autonomous which facilitated continuous improvement practices within the cell. They did not, however, want to become full setter-operators because the work involved was not reflected in the financial reward.

The operators in cells X and Y shared the same views regarding the purpose and role of the continuous improvement meetings. Charting was also used for gathering cell performance data, and final decisions were made in the absence of the operators. The operators, however, felt more confident in proposing changes because the managers and superintendents listen to them more. The reason for this is they had more skills, the ability to be semi-autonomous and they had gained a lot of experience working in the cell for about a year. This level of trust and respect has taken a long time to cultivate and the operators were looking forward to being included in all cell decision making meetings.

3.3 Cell Z

The cell members were fully trained as operators, setters, riggers and inspectors, and carry out front line maintenance. The cell was treated as a company within a company. The cell members liaised with 'customers' (other cells) down the line to determine the production schedule for the week, and liaised with 'customers' up the line to discuss what needed to be pulled off the cell during the week. The superintendents provided customer and organisational information for the cell members to help with scheduling. They decided on all aspects of production and met every month with management to discuss progress.

The cell operated under a kanban system with no charting. Charting was not required since the procedures and the skills of the people had been developed to the extent that a great deal of trust existed between management and the workers. Improvement decisions were made democratically between the cell members and the superintendents.

For a period of eight months, however, there had been a reduction in the demand for the cell components, which had been accompanied by the temporary shelving of the continuous improvement programme. The overall motivation levels of the cell members did not reflect the associated intrinsic benefits of the human-centred approach. Because of the reduction in communication, the inactivity on the shop-floor and the diminishing importance placed on the continuous improvement programme, the cell members felt insecure and unmotivated.

4. Discussion

The most striking feature emerging from the investigation was that the performance of the cell depended greatly on who was working in the cell. This was true even in the more technologically dominated cells and was as source of constant concern for the company. Management found it very difficult to accommodate the individual differences and needs of the workers, and tended to make assumptions about what people needed or why people behaved in a particular way. As a consequence, human-centred concepts were very difficult to implement and put into practice. The more technocentric, 'engineering' problems were tackled first because they were more tangible.

The fundamental problems involving changing traditional working practices and values were tackled from the beginning of the transition to cellular manufacturing primarily through establishing regular cell meetings, a company wide training programme, forming cell teams and encouraging skills development. In the case of cells X and Y, management were, at the time, not willing to discard the traditional decision making and information sharing practices. This would be done when they had sufficient confidence in the abilities of the operators. As a result, the process of discarding the traditional approaches for the more developed cell Y, had progressed further than for cell X.

Cell Z was the most mature and developed cell. Because of the down-turn in orders, and the subsequent shifting of management priorities, the continuous improvement programme was halted. As a result, the communication link between the shop-floor and management was severely restricted, and eventually many of the intrinsic benefits of the human-centred approach were lost. The cell members were tired with the levels of responsibility that other workers did not have to put up with.

5. Conclusion

The social problems involved with sustaining the performance of a cell, meeting the needs of individual people, accommodating individual differences and trying to continuously improve the working environment are enormous. They may only be understood fully, and tackled successfully after a great deal of learning and time. Nothing can be taken for granted. It is important for managers to be realistic, and quite often, to 'make do' for a period of time. The managers, superintendents and cell members recognised the need for improvement in the following areas: communication, sharing of information, trust, honesty and teamworking skills. Everybody expressed the notion that they were in the middle of a continuous learning process where mistakes were being made; but which was vital for gaining experience to support the programme of continuous improvement.

(The authors with to thank Bendix Heavy Vehicle Systems - Europe for their assistance with the research.)

Advances in Agile Manufacturing
P.T. Kidd and W. Karwowski (Eds.)
IOS Press, 1994

Advanced Manufacturing Systems and the Changing Nature of Work

Paul R Jackson, Sean Mullarkey, Sharon K Parker & Toby D Wall
MRC/ESRC Social and Applied Psychology Unit
University of Sheffield
England

Abstract. Integrated manufacturing practices are spreading, yet there is a high failure rate among organisations not achieving expected benefits. This paper argues that the main reason for failure is in the form of work design adopted. Details of two studies are presented, where the jobs of operators were changed to give them enhanced control over AMT systems. Results showed that substantial performance benefits were achieved, largely because workers learned to manage complex systems more effectively when given the opportunity to do so. Implications for other applications of integrated manufacturing are presented.

1. The Nature of Integrated Manufacturing

Manufacturing industry is undergoing fundamental changes in order to maintain competitiveness within an environment where market demands and technology are changing rapidly. Within the last ten years, the role of manufacturing within Britain's GDP has gone down systematically, and Japan's share of world GDP has risen to more than 10 percent. Meanwhile, Britain has been the major European recipient of Japanese inward investment.

Early explanations for Japan's success focussed on cultural or technological factors; but, more recently, it is becoming obvious that it is differences in management practice that are the major contributor to Japanese superiority. Companies such as Nissan, for example, use the same technology and workers as other companies, yet manage to achieve huge differentials in productivity and quality.

Changes in management practices have been grouped together under the single label 'integrated manufacturing' [1]: advanced manufacturing technology, which exploits computer technologies offering flexibility & potential for integration; just-in-time inventory control, which includes lead-time and inventory reduction, kanban systems, reduced set-up & changeover times, and supplier partnering; and total quality management, doing things right "first time, on time, every time", continuous improvement and meeting customer needs.

2. Work Design and Performance

While integrated manufacturing systems can be very effective, surveys show a high failure rate with organisations not achieving benefits they expected. We argue that the major reason for these shortcomings lies not in the technology itself but in the associated forms of work design. The dominant form of work design within western organisations has been variously described as 'specialist control' [2], 'command and control' [3] and the 'control-oriented approach' [4]. Under this form of work design, the jobs of shopfloor workers are precisely defined, and variances in in the production process are controlled by management or by technical experts. An alternative approach is one that devolves control to as low a level as possible so that shopfloor workers themselves are given the authority and the capability to manage production variances. This form of work design is described as 'operator control' [2], 'continual improvement' [3] and the 'high-involvement approach' [4].

Integrated manufacturing systems are often characterised by high levels of uncertainty, with operating problems which require human intervention. This uncertainty arises from two sources: *technological* uncertainty which refers to variability in the performance of the technology of the system itself; and *environmental* uncertainty which refers to variability in the production environment surrounding the technology. Whatever its source, as uncertainty increases, so does the need for more flexible work structures and practices, with greater skill and decision-making at lower levels. This follows the socio-technical systems principle of control of variance at source and is consistent with the philosophy of high-involvement management, based upon 'giving individuals at the lowest levels of organisations more information, knowledge, power, and rewards' [4, p. xi].

In this paper, we present evidence from two longitudinal studies in contrasting organisations of performance benefits when the jobs of operators are changed to give them enhanced control over AMT systems. The first study compared systems of contrasting levels of technological uncertainty, while the second study focussed on a highly uncertain robotics line.

3. Study 1 - Automatic Component Insertion

The first study [5] took place within the printed-circuit board assembly department of a manufacturer of mainframe and micro computers. The focus of the study was the performance of CNC insertion machines following work redesign to enhance levels of operator job control. The existing form of work design involved specialist control. Seven insertion machines were examined, three of which were characterised by higher operational uncertainty than the others. Operators were given authority over tasks such as recalibrating machines, adjusting mechanisms, inserting bypass steps for partial product builds and preventative maintenance. System downtime was taken as an index of machine performance and data were collected for 50 days before and 50 days after the redesign of jobs.

Little change was found for the low uncertainty systems, while the high uncertainty systems showed a substantial reduction in overall downtime and in the incidence of downtime. There was an initial reduction in overall downtime of 20 percent (from 50 minutes to 37 minutes per machine per 8-hour shift), and a much larger delayed reduction of 70 percent (from 37 minutes to 9 minutes per machine per 8-hour shift). In total, a reduction of over 80 percent in amount of downtime was achieved. There was also a reduction of over 75 percent in the number of stoppages per day (from 1.10 to 0.27 per machine); and no change in the average length of downtime per incident. The detailed pattern of findings is consistent with a skill-based explanation, under which operators develop and exercise informal knowledge and skills which lead to fault prevention, rather than a logistical explanation, under which operators simply respond more rapidly to technological faults and thus reduce waiting time [6].

4. Study 2 - Robotics Line

The second study [7] took place on a robotics line within a manufacturer of general purpose concrete breaker bits for heavy-duty hand-held hammer drills used in mining. The line performs forming, shaping and hardening of drill bits, with robots being used to transfer bits between stages in the manufacturing process. The system as a whole was a highly uncertain one, with downtime before the study started averaging about 40 percent. Stoppages were recorded on a tachograph, and no human intervention was possible without first stopping the system. As in study 1, the initial work design was an example of specialist control, since operators were not sanctioned to perform maintenance and problem-solving: in the event of a system breakdown, they were required to call in an engineer.

Operator control was increased by a change in work role allocations, such that operators were no longer forbidden to perform maintenance and fault-finding, as long as they called an engineer in the event of a serious system breakdown. Once again, the main indicators of system performance were total downtime and the number of incidents of downtime; and data were obtained on 101 shifts before and 127 shifts after the job change.

The effect of the job change was a reduction of 25 percent in overall downtime (from 268 minutes to 204 minutes per 12 hour shift), and there was an immediate reduction in downtime due to running repairs and adjustments, with no change in downtime due to major component failures. The data on incidence of downtime showed three main findings: (1) an immediate increase in the number of very short incidents (operator interventions) from 9 to 16 per 12 hour shift; (2) a delayed reduction in number of medium-length stoppages for running repairs and adjustments of 50 percent (from 4 to 2 per 12 hour shift); and (3) no change in the number of very long stoppages.

Taking these findings together, we see that there is an immediate quick-response effect producing shorter downtime per incident due to operators using their existing knowledge (and thereby avoiding the delays which often occurred waiting for engineers to arrive). This immediate effect was followed by a delayed elimination of stoppages through operators developing new knowledge that allows them to anticipate the occurrence of faults (perhaps by being more sensitive to the antecedents of a breakdown) and to act to prevent faults occurring.

5. Wider Implications

There are very few studies within the literature on integrated manufacturing which offer precise comparisons between alternative forms of work organisation or report on the effects of deliberate interventions. Most accounts are anecdotal or rely on impressionistic evidence. The two studies described here are valuable therefore since they combine longitudinal research designs, objective performance criteria and the use of a range of dependent variables in order to test competing explanations of the effects observed. The findings are clear and their implications are wide-ranging.

Where there is high technological or environmental uncertainty, specialist control of AMT systems harms performance, while devolving control to shopfloor workers themselves allows them to manage uncertainty effectively and leads to substantial performance benefits. Operator control of high-uncertainty systems benefits performance because active system management by operators allows them to generate new knowledge which they are sanctioned to use in their status as local experts. In order to support this high-involvement approach, it is necessary to: develop multi-skilled operators, define work roles which cut across traditional skill demarcation lines, reduce functional specialisation, and develop more decentralised organisational structures [8]. A broader work role for operators brings with it a requirement for training to enhance their diagnostic and rectification skills, and also necessitates a new role for technical specialists and supervisors as trainers, project managers and consultants. Finally, our research supports the view long espoused by TQM experts that the major barriers to workers performing effectively are those imposed by management.

References

[1]   J.W. Dean and S.A. Snell, Integrated manufacturing and job design: Moderating effects of organisational inertia. *Academy of Management Journal* **34** (1991) 776-804.

[2]   T.D. Wall, J.M. Corbett, C.W. Clegg, P.R. Jackson and R. Martin, Advanced manufacturing technology and work design: Towards a theoretical framework *Journal of Organisational Behavior* **11** (1990) 201-219.

[3]   R.H. Hayes, S.C. Wheelwright and K.B. Clark, Dynamic Manufacturing. Free Press, New York, 1988.

[4]   E.E. Lawler, The Ultimate Advantage: Creating the High Involvement Organisation. Jossey-Bass, San Francisco, 1992.

[5]   T.D. Wall, J.M. Corbett, R. Martin, C.W. Clegg and P.R. Jackson, Advanced manufacturing technology, work design and performance: A change study *Journal of Applied Psychology* **75** (1990) 691-697.

[6]   P.R. Jackson and T.D. Wall, How does operator control enhance performance of advanced manufacturing technology? *Ergonomics* **34** (1991) 1301-1311.

[7]   T.D. Wall, P.R. Jackson and K. Davids, Operator work design and robotics system performance: A serendipitous field study *Journal of Applied Psychology* **77** (1992) 353-362.

[8]   S.K. Parker, S. Mullarkey and P.R. Jackson, Dimensions of performance effectiveness in high-involvement work organisations *Human Resource Management Journal* in press.

Managerial Roles in Manufacturing Systems

Brian Trought

Management Research International, 59 Belton Lane, Great Gonerby, Grantham, Lincolnshire, U.K.

Abstract. Data has been collected from the work activities of a sample of manufacturing managers in the U.K. 59% of discrete activities are classified into a 5-roles taxonomy of work within and around manufacturing systems. The remaining 41% of activities are classified as manager-only activities. The data emphasises the frequent, overlapping work activities of managers and other system workers. This data together with earlier related data, indicates that managers in practice tend to be; only in partial control of product-flow, intuitive, non-rational, cooperative and lead a highly time-fragmented life at work. Some implications for these findings are discussed.

1. Introduction

The literature mostly fails to describe what manufacturing managers actually do at work, and how and why they do it in the way that they do. On the other hand the literature is rife with prescriptions on how to manage. It is normally not altogether clear on what these prescriptions are based. Surely, before we prescribe how managers should manage we ought to find out how managers do manage. This paper is a contribution to this dialogue.

2. The Research Base

The taxonomy of manufacturing work roles emerged from longitudinal, observational studies of the work activities in such systems, of people including managers, over the past nineteen years. The studies are continuing. This paper emphasises managerial activities and the data specified here has been collected more recently from nineteen managers. Observation time per manager was one full working week. The data records 4470 discrete managerial activities or acts of behaviour.

3. What Managers Actually Do

3.1. A Work Taxonomy

The 5-point, work roles taxonomy fitted the activities at work of all people who worked in and around the systems studied, including some activities by managers [1]. In the case of the nineteen managers, only 59% of their total, discrete work activities fitted easily into the taxonomy. However, all activities by non-managers fitted into the taxonomy. The 41% of managerial activities which did not fit the taxonomy were classified as manager-only activities because they were activities which, in a particular system, were normally carried out by a manager. The 5-point taxonomy which has been developed is:
a) System Enhancers - improvers of the initially installed design, activities which may be carried out by the original designers but more probably by production engineers and managers. (2.0%)

b) System Monitors - activities carried out by production engineers, system workers and managers. (22%)

c) System Facilitators - carried out in practice by maintenance engineers, by system workers and by managers. (19%)

d) Environmental Components - these are people who work in the system environment. Production planners and inspectors may fill this role, but also intermittently managers carry out activities in the environment of the system. (7%)

e) System Components - these are people who, together with the manufacturing plant, carry out activities which are within the line of product-flow. Obviously, some system workers fill this role, but it can also be shown that managers carry out this type of work activity, which is usually of a manual nature. (8.5%)

The percentage figure quoted in brackets at the end of each taxonomy point is the percentage of discrete, total managerial activities attributed by the observer to that system feature. A typical managerial activity which was recorded was when Manager B went onto his own shop-floor and checked the progress of some jobs with his chargehand. In this case the manager was acting as a System Monitor.

3.2. Manager-only Activities

What are the 41% of activities or behaviour which managers perform or exhibit that sets them apart from other people in manufacturing systems? In this sample such activities, for example, are report writing; paperwork in respect of wages, despatch, scrap notes, orders and batch cards; dealing with secretarial help; discipline; training; workers being set on or leaving and attending managerial meetings. There is also a small number of discrete managerial activities which can be called dual roles - 0.5% of total activities. They are dual because in them, two of the items of the taxonomy were being carried out simultaneously. The number of these duals is so small that, for the purpose of analysis here, they can be ignored.

4. Implications

4.1. Overlapping Work Roles

The data collected for this paper suggests that manufacturing managers perform the full range of work activities which are also the normal work activities of other system workers, including manual work. These overlapping activities constitute 59% of total managerial activities. Hence, how can we speak of the rational organization of work in these circumstances? Are sharper job descriptions required? Furthermore, the data provides strong evidence that there is no one best way to manage. The other 41% of managerial activities are manager-only activities. These are activities which are normally only carried out by managers. An important question then becomes: 'Why are managers carrying out a similar range of tasks (in 59% cases) in the system as other people who are organizationally inferior to them?' A common answer from managers is that they do what needs to be done, as they see it. This answer fits the reactive, 'firefighting' style adopted by all managers who I have studied. Progress chasing is a good example. Even where progress chasers are employed as relatively lowly members of a manufacturing system, managers frequently progress chase urgent jobs themselves. Perhaps weak work disciplines are the problem in these cases.

4.2. Managerial Performance

How do we assess managerial performance in this complexity of interlocking and overlapping activities. Surely not by considering individual activities and behaviour, as is frequently done. Better assessment of managerial performance is likely to be achieved by consideration of total system performance. System change or rectification will eventually become necessary. It is at this stage that individual, managerial activities and behaviour are likely to become more important. This later type of knowledge about managers is rarely gathered in any systematic way. The time surely has come for companies to do just this.

4.3. Outcomes of Uncertainty

Primarily because of endemic system uncertainties in U.K. manufacturing companies, the data also indicates that managers are only in partial control of product-flow [2], make decisions intuitively and non-rationally on short time scales [3] and hence lead highly fragmented but peer-cooperating working lives [4] which they do not necessarily view as complex. The reality of managerial activities and behaviour in and around manufacturing systems does not reflect most text book prescriptions - managers are still 'firefighting' [5].

These outcomes, only some of which have previously been noted in the literature by other managerial researchers, have often unplanned implications for manufacturing systems. In general terms the seeming rationality of the systems, as academically defined, is difficult to observe. Rather is there a manifestation under observation of a jumbled, hotch-potch of disparate sub-systems which only remain operational by continuous, varied and variable intervention by humans, both managerial and non-managerial. The human intervention can be frequently said to be non-rational [6]. We should not automatically assume that if common sense rationality in the manufacturing system does not prevail then irrationality does. Research suggests that people in manufacturing systems today, particularly managers, have many shared values within a common company culture and perform work according to what is often known as shop practice. The working out of this joint way of doing work is a type of rationality, as viewed from the inside of the organization. It makes rational sense to members of a particular organization, but it is company specific. This does not mean that disagreements, disruptions and conflict do not occur; they do. However, their occurrence is very low when compared with the consensus, cooperation and general coherence in doing work which epitomises much of the U.K. manufacturing industry [7]. So, if it is not altogether appropriate to use the term irrational for the way most systems operate and decisions are taken, what term can be used? A more appropriate term is, I believe, 'consensus rationality', which may be considered non-rational but not irrational.

The implication which appears to be consistent with system uncertainty and 'consensus rationality', based as the latter is on shared values, a common culture and company or shop practice, is that objectives and decisions should not be imposed. Rather they should evolve from a common consensus and preferably from all organizational levels. In this way system ownership is encouraged. The assumption here is that the consensus is the rationality. A prime example of this idea in practise is the self-managed work cell - the learning organization. Perhaps, a high incidence of formalized 'consensus rationality' within an organization is another definition of the learning organization. In companies where management sustains status differences , constantly voices their right to manage and imposes objectives and decisions, worker empowerment and 'consensus rationality' are still a distant dream. Few companies in the world have ventured totally down the empowerment path. One which has is Semco the Brazilian company. The fascinating book written by Ricardo Semler of Semco [8] sets out in stark detail the many pitfalls and colossal devotion of time which is likely to be necessary to implement worker empowerment for the greater good of the company and it's employees.

4.4. The Leadership Question

This research data also raises the question of whether there are implications from these findings for leadership in this setting, such as, do the multi, overlapping and duplicating roles carried out by workers, supervisors and managers enhance or detract from managerial leadership? The answer to this specific question appears to be that the system-integrated way in which the present sample of manufacturing managers spend their time is positively advantageous to the system. ' Firefighting' is still required and managers are best placed to do this. The practical aspects of most managerial jobs is shown by the 8.5% of their activities which are as working components in their systems. Managers can be seen to drive fork-lift trucks, repair machinery, unload lorries and even make up teams on production lines, whilst some managers have been observed working at their previously acquired craft skills. In many cases this willingness to turn a hand to practical tasks is a useful status raiser with the work-force. In these circumstances it is not difficult to visualise that managerial leadership - viewed here as a definite relationship between leader and led - may be frequently enhanced by any joint, overlapping work activities between managers and workers. The down side, of managerial activities which could perhaps be considered as being more correctly within the worker's domain, may be that other manager-only activities are being neglected. However, only careful consideration of individual cases can answer this point.

5. Conclusions

In a satisfying research sense what manufacturing managers actually do at work is still not widely understood. Systematic efforts to find out are still unfortunately few. The plea must therefore be that before we prescribe how managers should manage, we ought to find out how they do manage.

References

[1] B. Trought, Actual Managerial Jobs in and Around Advanced Manufacturing Systems. Proceedings of the conference, International Operations: crossing borders in manufacturing and Service, Eds. R.H. Hollier, R.J. Boaden and S.J. New. Held at UMIST, Manchester, U.K. June 1992. Publisher, Elsevier Science, Amsterdam, The Netherlands.

[2] B. Trought, Control: the ultimate barrier. Proceedings of the 10th International Conference on Production Research, Ed. P.F. McGoldrick. Held at University of Nottingham, Nottingham, U.K. August 1989. Publisher, Taylor and Francis, London, U.K.

[3] B.Trought, Manufacturing Decision-Making: Irrationality Rules - OK! Proceedings of the 6th National Conference on Production Research, held at, and published by, the University of Strathclyde, Glasgow, U.K. Eds. A. Carrie and I. Simpson. August 1990.

[4] B. Trought, Concurrent Engineering: the role of managerial peer relationships. Proceedings of the 12th International Conference on Production Research, Eds. V. Orpana and A. Lukka. Held at the University of Lappeenranta, Finland, August 1993. Publisher, Elsevier Science, Amsterdam, The Netherlands.

[5] B. Trought, Training the Firefighters. Article in the *Manufacturing Engineer,* Vol. 70, No. 8, October 1991. Publisher, Institution of Electrical Engineers, Stevenage, Herts. U.K.

[6] B. Trought, Model Collapse: the people effect. Proceedings of the 9th National Conference on Manufacturing Research, at and Published by the University of Bath, Bath, U.K. Eds. A. Bramley and A. Mileham. September, 1993.

[7] B. Trought, An Analysis of Lateral Relations in Manufacturing Management and of their Contribution to Product-flow. Unpublished Ph.D. thesis, University of Nottingham, U.K. 1984.

[8] R. Semler, Maverick. Published by Century, an imprint of Random House, Ltd. London, U.K. 1993.

Advances in Agile Manufacturing
P.T. Kidd and W. Karwowski (Eds.)
IOS Press, 1994

Human Resource and Automation Management in Company with Handicap People. Case of SILMET Co-operative

Stefan TRZCIELINSKI
Technical University of Poznan, Management Engineering Department
ul.Strzelecka 11, 60-965 Poznan, Poland
Andrzej JAWORSKI
SILMET Spoldzielnia Inwalidow
ul. Kujanska 10a, 77-400 Zlotow, Poland

Abstract. Since 1991 things went wrong in SILMET Co-operative and at turn of 1992/1993 the danger of bankruptcy was quite real. To change this situation a new president decided to invest in people, development and quality. Now SILMET is one of the best company in its region. In this paper the strategy, policy and methods which have been applied in SILMET to manage the human resource and automation are presented.

1. Introduction

SILMET is a co-operative of handicap people located in Zlotow a town placed in beautiful countryside north-west of Poland. For over 40 years SILMET has manufactured not complicated metal goods basing on simple plastic forming and machining technology. The co-operative did not managed to cope with the market economy and at the beginning of 1993 was in deep crisis. For 1991 and 1992 the salaries had not been changed although the inflation was 70% and 45% a year. About 50% of employees was paid less than the minimal salary established by the government while others were paid a little bit more. An egalitarism was common accepted. These are some features of a situation which a new president of SILMET decided to change.

One of the first things which was done was SWOT analysis [3]. It has resulted in working out a strategy of SILMET which can be concluded as it follows:
- The future of SILMET should be connected with commutator low power electric motors (called PRMO) for cars' wipers, blower, fan of radiator and for TV-sat antenna turning mechanism.
- The co-operative has to increase the market of PRMO users.
- The company should keep product and production flexibility in the area of its manufacture experiences (TV-sat antenna turning mechanism, wide range of small parts for bicycles, ducts for air conditioning systems, etc.).
- SILMET should stay a co-operative of handicap people.
According to the strategy a policy of managing people and automation has been developed.

2. Forcing People to Do Better

In the difficult market, social and financial situation of SILMET, it was a kind of individual rationality of all level managers to avoid decision making [1, p.299]. To change this the main functions which should be performed in the management system according to the accepted strategy was defined and a new organizational structure as well as the general areas of each department responsibility were designed [4, p.167]. Next management by exception has been applied. This effects in not including the top managers into operations management and enables them to concentrate on reaching the strategic goals.

In many cases it is very difficult for both the middle and the lowest level managers to be independent in decision making. Because of this two foremen had to be substituted. But the candidates which were chosen among the best workers to substitute them refused to be promoted although their salaries was going to be increased. They were afraid of taking the responsibility of the post. Therefore a trick was used. The candidates were informed that they would occupy the positions for one month. During this time they were slightly suggested haw the production cells and their workplace might be organized. They felt they could improve both the organization and the efficiency of their teems. Finally they agreed to keep their new positions. After three months they started to compete each other and their performance was so good that their salaries were proposed to be increased. This time they did not accepted the proposal. The egalitarian way of thinking had won.

When the lowest managers was able to manage the production cells effectively the system of piece wages was replaced with daily wages [5, p.500]. The reason of the change was that the system of piece wages demoralized the workers. For instance the following bad events took place:
- some sly workers always got the better paid work
- it used to happen that the defective parts were mixed with the good ones
- as the salary of disable people who receive also the annuity have been limited, to earn more, some illegal transactions were practiced among them.

As the lowest level managers are able to organize their work well, there is no problem with productivity of squads' and the system of daily wages is accepted.

3. To Turn Weaknesses into Strengths

There are 240 employees in SILMET including 160 handicap people. The most common kinds of disability are the following:
- mental defect (11%),
- deafness (9%),
- hypertension (8%).

Particular when the unemployment is high, the handicap people have less chance to get a job. They are thought to be less worth workers. To improve their chance, there are some tax and financial privileges in Poland for companies which employ at least 40% of disable employees including at least 10% of people with second degree of disability (in SILMET the figures are 68% and 14%). The privileges depend on both the right to keep a big part of VAT and income tax in the company and the possibility of obtaining a low interest loan to create or modernize a work-place for handicap people. Basing on disable workers SILMET has turned the weakness into its strength. Not only 28 new handicap people have been employed but more advanced technology of manufacturing PRMO electric motors has been introduced as well.

With an exception of a few high qualified jobs there is no problem to get needed workers because of the high rate of unemployment. The selection process involves the following steps [2, p.275]:
- testing the application form,
- initial interviewing by personnel manager,
- interviewing by the president,
- final selection by the immediate supervisor and acceptance by the president.

In very hard cases the accepting decision is made basing on humanitarian aspects. It occurs very often that the employees who had not have work for a long time are very ambitious and efficient.

Recruiting disable workers implicates a special requirements for training and rehabilitation. SILMET delivers the medical and rehabilitation service for its workers. This is necessary not only to ensure the needed quality and efficiency of their performance but also because of the humanitarian aspects.

4. Automated Versus Skill Based Manufacturing

According to the concept of marketing-mix one of the factors which must be respected to enter and increase the market and stay on it is a good quality product. To ensure the demanded quality an automation is introduced and this takes place also in SILMET. But particular in a case of company with handicap people a dilemma between implementing a higher level of automation and human resource management occurs. In short time the exemplification of the conflict is: automation benefits versus unemployment, social problems and loss of some tax privileges.

The policy of SILMET which is applied to resolve this conflict depends on automation of crucial operations for the products quality. Mostly they are connected with manufacturing of electric motor's rotor. They participate in about 18% of total labour-consumption. The scope of automation is correlated with production scale and each investment decision is supported by analysis of return on investment. Although SILMET has captured the Polish market of users of electric motors for TV-sat antenna turning mechanism, the scale of production is big not enough to automate the whole process of manufacturing the electric motor or even the rotor. There is possible to integrate all the automatic machines by automatic loading units. However this will be reasonable if the production scale increases as a result of increasing the market share of PRMO users. It is expected as SILMET is going to introduce a new generation electric motor for fan of radiator which will be used in cars which are assembled in Poland. But this will cause also the increase of labour consumption and the same the number of workers.

5. Conclusions

People behave rational and this is the duty of managers to utilize their rationality in order to reach the goals of whole organization. In SILMET it was done by working out a strategy and defining the areas of each department responsibility. In this way an organizational framework for managerial activities has been created. As far as the workshops are concerned, this makes possible to apply a job enlargement and enrichment rather then money oriented kind of motivation. But this kind of motivation is possible when the managers are competent. Therefore it is important to have a good promotion and recruitment policy. In SILMET the system of promoting does not work well and in fact it must be build. Contrary to it the system of recruitment is well developed. The function of planning of human resource includes forecasting human needs, searching the labour market and selection of potential candidates.

Human resource needs are forecasted according to the planned sales volume, the development of product, technology and automation and the possibility of gaining financial resources. As SILMET bases on handicap people there are some extra sources of obtaining funds for modernizing and creating new work-places. The policy of creating new work-places is related to implementing more advanced automated processes which are crucial for the product quality.

References

[1] M. Crozier and E. Friedberg, Czlowiek i system. Ograniczenia dzialania zespolowego. PWE, Warszawa, 1982.
[2] A. Elkins, Management. Structures, Functions, and Practices. ISBN: 0 201 01517 X. Addison-Wesley Publishing Company, 1980.
[3] Z. Martyniak, Metoda refleksji strategicznej, *Ekonomika i Organizacja Przedsiebiorstwa* 4 (1990) 4-6.
[4] E. Pawlowski and S. Trzcielinski, Projektowanie struktury organizacyjnej przedsiebiorstwa. Podstawy rozwijania i kojarzenia funkcji zarzadzania, vol.1 and 2, TNOiK, Poznan, 1987
[5] R. G. Schroeder, Operations Management. ISBN: 0 07 055625 6. McGrow-Hill Book Company, 1985.

Advances in Agile Manufacturing
P.T. Kidd and W. Karwowski (Eds.)
IOS Press, 1994

Human Resources in CIM Adoption

Professor Milan MALY
University of Economics Prague, W. Churchill Square 4,
CZ-130 67   Prague 3, Czech Republic

Abstract. The main goal of this paper is to present the results of testing the hypothesis about the changing role of manpower in the process of CIM adoption in companies and enterprises. Our methodological approach was based on testing of alternative hypotheses, on the sample of 60 CIM systems already installed.

1. Introduction

The paper emerges from the idea that the choice by firms, industries, and governments of specific strategies to take advantage of CIM will depend on how well the relevant decision makers understand the mechanism through which CIM will benefit different firms and industries. The deeper the level of understanding, the more productive will their strategic choices be.

This paper aims at contributing to a better understanding of the mechanism of CIM adoption in the area of human resources.

The main goal is to test the hypothesis about the changing role of manpower in the process of CIM adoption in companies and enterprises.

Our methodological approach was based on gathering the data from 60 systems already adopted CIM and consequently on testing of alternative hypotheses by means of this particular sample.

2. Findings

The main hypothesis specified in our research and consequently verified on our sample are the following:

(1) *The development of skills of manpower* goes to multidisciplinary skilled shop floor personnel and to multiskilled manager.

The division of labor in conventional systems is characterized by a very narrow specification of operators, by a division between traditional crafts and skilled workers, and by division and specialization of labor. The other groups consist of semi-skilled and unskilled workers (mainly for loading and unloading operations). The number of direct operators (handling the machine-tools and other devices directly) and mechanical maintenance-men is relatively high. This is the typical Taylor-type of division and specialization of labor.

In course of the time, when a higher level of automation and integration is introduced in the production process, the nature of the process requires integration of skills: thus the number of direct operators is decreasing and, on the other hand, the number of indirect suport staff is increasing. The ratio of electrical to mechanical maintenance men is going up, as well as the number of system analysts and programmers. We can examine the gradual substitution of the best qualified production personnel in a system for unskilled workers, such as loaders and unloaders.

Multi-skilled maintenance personnel combine mechanical, electronic and software know-how.

Even today it seems that the skills of many managers, based mainly on law and economics, might not be sufficient to solve the problems of managing the new technology and they must be changed to a multi-skilled profile.

(2) *Work structuring* in production tends to job enlargement, job enrichment and job rotation, as well as to an integration of skills. We distinguish two basic strategies in this area: substitution and development strategies.

The substitution strategy corresponds to the traditional job structure, characterized by a relatively low level of personnel skills: even unskilled workers can be considered useful in this case. The production is divided among the different groups of workers: the internal manufacturing operations that require overall planning and management are allocated to the few skillled foremen or group leaders: less qualified groups of specialized workers are placed outside of the actual production process, and unskilled auxiliary workers take care of the loading and unloading tasks. The genuine planning and preparatory operations that require high special qualifications are allocated to FMS-external partners who program and pre-set the tools and also take care of maintenance and repair.

The development strategy corresponds to the alternative job structure distinguished by that extremely homogenous, relatively highly qualified production staff. The internal manufacturing operations can be carried out by each of the workers who are rotating in different tasks. Planning and preparing of production is realized inside the system, only some programming and major repair is done outside. The time consumption for the daily production-process planning is increasing distinctly compared with a conventional system. There is discretion to regulate as much as possible on the shop floor.

(3) The next hypothesis is closely related to the previous one and it is connected with the so-called *technocentric or anthropocentric approach* in the man-machine architecture in CIM systems.

The problem has two main angles. Some of the authors stress the qualitative angle and regard the technocentric approach as "technology controlling man" and the anthropocentric approach as "man controlling technology".

In the first case the technocentric approach leaves the human subordinate in the system and has no higher requirements on human qualifications. The anthropocentric approach places man in control of the system and needs the multi-skilled operator who, with a very wide, open-ended repertoire of skills, will manage the system despite unforeseen disturbances.

The second, quantitative angle of this problem stems from

the idea that man can be replaced by machines and control devices. The role of the system operator will be to fill the gaps with the thoroughness of a designer. On this basis it is logically possible to draw the conclusion that the more expansive and complicated the system is, the fewer people will the production system neeed.

(4) The next hypothesis expresses the notion that *higher stages of automation and integration improve* the working condition such as a reduction of the hard, dangerous and monotonous work places. This fact has to have logically such a consequence as a reduction of manpower turnover (fluctuation) and sick leave ratio.

(5) Further hypothesis are *the training/retraining costs and training/retraining forms*. It was confirmed that the higher levels of automation and integration call for a longer time for training/retraining of the personnel and for higher training/retraining costs.

"The new type of worker is not the traditional man trained on the job who gets into the position or machine operator without formal training only on the basis of years of experience. Much of the brainwork has been removed from this workplace in normal running, by steering, regulating and supervising functions taken over by computer systems. But instead, according to the new concept of use of labor, this workplace is alloted new tasks for dealing with exceptional situations. Coping with these tasks requires more than in previous times a theoretical competence to a greater extent than could be obtained only by learning by doing" [1].

The appropriate organizational forms for training/retraining are the following:
- on-the-job training,
- off-the-job training organized by:
training department of user company
special training institution
vendors.

Firms use all above-mentioned organizational forms for training/retraining. The most common policy is the on-the-job training. Using this form the firm reached the best results in manpower turnover (fluctuation) and non-adaptable operators ratio.

(6) The next hypothesis specifies *the organizational forms of recruitment of personnel for CIM*. The following possible forms of recruitment are taken into account:
- recruitment from manufacturing areas close to CIM,
- recruitees having worked with previous system (machining centers, NC-machines, conventional system),
- young recruitees,
- recruitment from the best operators (creaming-off policy).

Firms predominantly combine ways of recruiting employees, mainly employees having previously worked with the system and the adoption of a creaming-off policy.

(7) The following hypothesis is connected with *the reward system suitable under the changed conditions*, i.e. at the higher level of automation and integration. The following forms were taken into account:
- piecework system,
- individual wages,
- time wages,
- group wages,

- group premium payment.

The analysis of reward system has not confirmed a tendency towards group wage systems. The prevalent wage system is still based on diferent forms of individual wage.

3. Conclusions

Analysing all above-mentioned hypothesis we have come to the conclusion that there are also some interesting differencies whose reason may lie in the different economic, cultural and historical conditions - such as small vs. large countries, different regions like USA vs. Japan vs. Europe, different industrial sectors, centrally planned vs. market economies, etc. The decision-makers adopting CIM have to clarify the main factors for different environmental, economic, cultural and other conditions for the successful adoption of these particular systems.

References

[1] M. Schumann, The Future of Work, Training and Innovation. EEC Brussels, 1986.

Advances in Agile Manufacturing
P.T. Kidd and W. Karwowski (Eds.)
IOS Press, 1994

Creating a Developmental Structure within a Manufacturing Organization

TUOMO ALASOINI

Ministry of Labour, Research Unit, P.O.Box 524, SF-00101 Helsinki, Finland

Abstract. This paper presents premisses of the Finnish research programme entitled 'Work, Culture and Technology' (1991-94) as to consolidation and diffusion of a developmental mechanism within a manufacturing organization. The research consists of developmental experiments in four engineering workshops with a view to creating preconditions for a new mode of operation. It is suggested that continuous improvement of production calls for collaborative network-like work patterns, and sophisticated organizational tools and procedures for detecting and solving problems. Consequently, a radical redefinition in the job assignments of workers, supervisors and middle management is needed.

1. Introduction

Owing to the rapid changes of the product market and the technological basis of production, creating a mechanism for continuous improvement is for a company today a major prerequisite for gaining a competitive edge. Yet many manufacturers are still characterized by a craft-like or Tayloristic mode of operation, both of which constitute an obstacle to perpetual development. In craft-like mode of operation, developmental activity within a company is unsystematic and uncoordinated, and is generally based on isolated innovations made by skilled employees. Tayloristic mode of operation, on the other hand, is plagued by the rigid demarcation between planning and execution, which hampers collaborative problem solving and communication between the various groups of employees.

Most socio-technical experiments also have been ill-prepared to respond to the challenge of continuous improvement [1]. The accent of the semi-autonomous group is on self-management, planning autonomy and task enlargement. Taken to extremes, an organization built on semi-autonomous workgroups is not conducive to systematic development of production, and at its worst may even exacerbate territorial thinking, internal friction and, hence, constitute a major hindrance for organizational changes [2]. An apparent problem in the introduction of semi-autonomous workgroups has turned out to be their slow consolidation and diffusion within a company; a fact due in part to lack of support, or in some cases even to overt resistance from the other parts of the work organization such as supervisors and middle management.

'Work, Culture and Technology' starts out with the assumption that to cope efficiently with the above problems an entirely new kind of approach is needed. This approach, in our view, must draw on a fresh outlook of collaboration between the various activities within a company, and supporting new organizational tools and procedures. We call this approach as *an adaptable mode of operation*.

2. Factory as a Field for Developmental Experimentation

The goal of 'Work, Culture and Technology' is to create, test and consolidate the new mode of operation in four export-oriented Finnish engineering workshops. The model is constructed on two principles: it (i) is capable of quick response by the means of parallel and cooperative work patterns and (ii) has a built-in mechanism for continuous improvement. Researchers take part in redesigning product shops, production cells or manufacturing lines to become a kind of 'laboratories', in which new tools and procedures for solving problems or implementing organizational changes are tested in a real production environment [3].

In a paper machine factory, for instance, the research fosters the transition of one of its product shops to cell-based work organization. Our concept of a 'networking cell' is characterized by flexible job assignments, responsibility for detailed scheduling and the subsequent working arrangements, and close connections to various support activities with a view to efficient problem solving. The cell is provided with systematic tools (such as a record on machine downtime and disturbances) and methods (such as regular cell meetings) for running of its everyday activities and handling of its operational problems. The new organizational tools and methods are tested in the building-up and running-in stages of two pilot cells, after which the researchers prepare a kind of manual on the construction and function of a 'networking cell' for a wider use in Finnish industry.

In a bus coach assembly plant, on the other hand, the goal of the research is to renew the system of production management and to contribute to solving of operational problems in the manufacturing line. As to the latter goal, the researchers developed a new method called as 'network-creating problem solving' in the course of the experiment. This six-stage method is tested at two workstations with an eye to finding causes and solutions to problems emanating from material deficiences and lack of information on drawings and constructions, and to implementing measures to solving them. In this case also and based on the results of the experiments, the researchers prepare a kind of manual of the new problem-solving method to be applied in other companies and to operational problems of other kind as well.

The development experiments in the four companies proceed in a bottom-up manner. That is, the creation of the new tools and procedures takes place in deliberately chosen parts of the manufacturing process including also their most important staff and support functions. After a careful evaluation of the experiments, a decision will be made whether the renewed mode of operation will be consolidated and diffused within a company on a wider scale.

The experimental stage of the research is always preceded by a basic analysis, in which the developmental needs of a company are defined together with management and the various groups of employees. In addition, they must be conceptualized so that they are clearly connected to certain very concrete, precisely defined and jointly approved *problems in the production process* such as poor product quality, long throughput times, insufficient response flexibility, excessive work-in-progress inventories, materials delays, etc. [4]. The researchers, however, must also be able to define the above problems as *inadequacies in the current mode of operation*. Otherwise, the novel research-assisted developmental activity in problem solving is in danger of petering out as soon as solutions to the most pressing problems are found or as the company study comes to an end.

To avoid the problem of slow consolidation and diffusion, it is essential, in our view, *not to focus on emphasizing the autonomy of a team (or any other organizational sub-unit under experimentation) but to perceive any team as a mesh in a company-wide collaborative network.* At best, network-like cooperation provides a team with much more effective problem-solving capabilities than group organization, where semi-autonomous workgroups are often left too much on their own as to their developmental activities. The network-like way of working with its regular inter-team exchange of information also enables employees to obtain a wider perspective of the production process, its problems and bottlenecks. Besides more effective problem solving, an important reason for building supportive networks between manufacturing teams and the other activities such as production planning, product design, maintenance and purchasing is to contribute to the rooting of the new working practices on a wider scale within a company.

3. Consolidation of the New Mode of Operation

There are three criteria for evaluating the reach of changes aimed at a new mode of operation in the companies: the *size* of the experimental field, the *depth to which change penetrates* in the experimental field, and the *mechanisms causing the mode of operation to become consolidated* within the organization.

Depending on the research site some 20-40 persons are working in the experimental field, though the indirect effects of the experiment may make themselves felt over a larger area. What is important is not only the size of the experimental field, but also what kind of key personnel and functions it includes with a view to network connections. In each of the companies, therefore, the experimental field includes at least part of the manufacturing process and its immediate support functions. In the bus coach assembly plant, for instance, the new problem-solving method was tested in two groups, which, in addition to the supervisors and blue-collar workers in question, included members from work planning, product design, purchasing and sales, a total of 16 persons from the company.

The purpose of the experimental stage of the research is to create and try out the tools and methods needed by the organization. These can be used to solve certain production problems. However, owing to the major contribution of the researchers, the experimental stage is an artificial situation. The critical stage in the consolidation of new practices is that following the experimental stage, in particular the period when the monitoring and the evaluation by the researchers are over.

The common goal in the four research sites is *the creation of a development organization in parallel with the production organization within the company*. This is a prerequisite for the new mode of operation being consolidated in the experimental field and for its diffusion elsewhere in the company. Being consolidated and diffused requires the wholehearted support of management and groups of employees. Moreover, work design, manufacturing management, and the systems of payment and workplace industrial relations must support the new ways of doings things. The contribution of the researchers is of primary importance in the creation of viable tools and methods, i.e. building up elements of the development organization of the company.

The development organization should be light and must not give rise to unwieldy administrative practices. In fact, the development organization does not need to be any kind of 'organization proper' in the literal sense of the word, but should rather be understood as a *'structure'* which supports the new mode of operation. Creating a development organization (or 'structure') generally entails a redistribution of tasks in the company, as development work inevitably consumes time.

A well-implemented experimental stage is important, since, in addition to increasing motivation of the personnel and creating tools for a transition to new practices, it also solves problems taking many people's time. As certain problems are removed and a more efficient battery of tools for their handling is evolved, more resources are released for development work. If it is feasible, at the same time, to shift responsibility for certain supervisory, planning and support functions downwards to the manufacturing teams, the opportunity opens up for foremen, work planners and methods designers, in particular, to expand their fields of activity to be coordinators in the development work on products and production [5].

The company should not indiscriminately proceed to lighten its organization as problems decrease and problem solving improves in the hope of saving on labour costs. Here there is a danger that the organization will be made as light as possible with an eye to achieving only the most immediate production goals. This may mean that resources are disposed of which, if properly channelled, would be available for development work, i.e. to create what is necessary for competitive production over a longer period of time. In our view, a certain degree of 'organizational slack' is a major precondition for innovative behaviour in any manufacturing firm [6].

4. Evaluation of the Developmental Experiments

The success of the research-assisted experiments in the companies cannot be assessed at this point in time, because the experimental periods still have some time to run. Evaluation of the results, however, is ultimately one of the most important tasks of the research.

For the new mode of operation to gain a wider acceptance in the companies and Finnish industry as well, the experiments must result in improvements (i) in performance, (ii) in the amount and quality of developmental activity within the companies, and (iii) in the well-being of employees. Immediate enhancements in performance and well-being, in particular, are quite simple to measure, but to acquire sufficient information on *all* direct and indirect, longer-term effects of the research, the follow-up period should last up to several years. In this respect, the relatively short time reserved for the experiments and their evaluation (2 years on average) unfortunately limits the chances of drawing firm conclusions on the consolidation of the mechanism for continuous improvement and, thus, on the superiority of the new mode of operation.

References

[1] F.M. van Eijnatten, The Paradigm That Changed the Work Place. Swedish Center for Working Life/Van Gorcum, Stockholm/Assen, 1993.

[2] P.S. Adler and R.E. Cole, Designed for Learning: a Tale of Two Auto Plants, *Sloan Management Review* 3/1993, 85-94.

[3] T. Alasoini, R. Hyötyläinen, A. Kasvio, J. Kiviniitty, S. Klemola, K. Ruuhilehto, P. Seppälä, K. Toikka and E. Tuominen, Interim Report on the Research Programme 'Work, Culture and Technology'. Tampere University Work Research Centre, Working Papers, Tampere, forthcoming.

[4] R.E. Cole, The Leadership, Organization and Co-determination Programme and its Evaluation: a Comparative Perspective. In: F. Naschold, R.E. Cole, B. Gustavsen and H. van Beinum, Constructing the New Industrial Society. Van Gorcum/Swedish Center for Working Life, Assen/Stockholm, 1993, pp. 121-132.

[5] R.L. Harmon, Reinventing the Factory II. Free Press, New York, 1992.

[6] K. Kuitunen, Innovative Behavior and Organizational Slack of a Firm. Helsinki School of Economics and Business Administration, Acta Academiae Oeconomicae Helsingiensis A:87, Helsinki, 1993.

Advances in Agile Manufacturing
P.T. Kidd and W. Karwowski (Eds.)
IOS Press, 1994

How to Measure and Increase "Leanness" of a Company

J. Cordes, D. Stokic , U. Kirchhoff
Institute for Applied System Technology GmbH, Wienerstraße 1,
D-28359 Bremen, Germany

Abstract: This paper shows a successful way to increase the "Leanness" of companies by using a developed Lean Reference System in order to improve the strength and competitiveness of companies. The developed reference system consisting of strongly interrelated lean mechanisms and rules is explained in order to use it for a thorough assessment of the company's status compared to an ideal lean company taking into account specific cultural and social constraints and specific goals of the company. The results of the assessment are used to establish a clear guideline for the planning of measures to increase "Leanness" in a wide variety of industrial sectors.

1. Introduction

The success of Lean Production puts companies under pressure to enhance their production in order to maintain their competitiveness. These enhancements must be strongly coupled with a reconsideration of the company's production philosophy stressing an integrated approach to increasing productivity and quality in parallel to cost reduction in all areas of the enterprise. These objectives cannot be achieved solely by the introduction of specific technologies and tools but requires a basic reconsideration of the company's culture and production philosophy.

The basis for such an enhancement process is a systematic analysis of the present situation within all departments of a company in order to identify the weak points. To support such an analysis, a clear systematic has been developed which includes a structure of the so-called mechanisms and rules of Lean Production. The reference system serves as a systematic guideline for the measurement of the current practice of enterprises in reference to Lean Production and for the development of measures in order to increase the "Leanness" of the company. In summary, the developed reference system of Lean Production can be seen as an extension to the improvement loop for production processes (Fig. 1).

This paper explains the application of the reference system of Lean Production in order to use it as a guideline for the assessment and improvement of a company's current "Leanness". The ways to use the reference system and the results of an assessment are presented to demonstrate the power of the method as an excellent guideline for the establishment of measures to improve the operation of the company in quality, cost and scheduling.

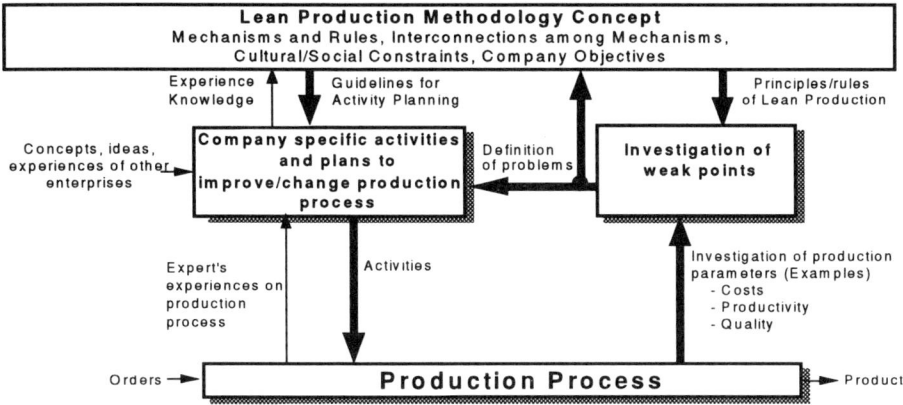

Fig. 1: The introduction of the reference system of Lean Production into the process improvement loop gives clear guidelines for improvement activities.

2. The Reference System of Lean Production

Basically the reference system of Lean Production consists of more than 50 so-called mechanisms which were identified by a thorough study of the Lean Production principles in literature ([1] to [4] and others) and by investigating a number of applications as well. Each of these mechanisms describes how a specific task of a company should be fulfilled. The level of detail of the mechanism's description can be adopted to the necessity of the company under assessment. The main mechanisms are listed in table 1.

Table 1: A selection of mechanisms of Lean Production

Integration of Servicing and Maintenance	Relation to Suppliers
Straightforward Research for Reasons of Defects	Teamwork / Job Rotation
Quality Assurance	Information System
Orientation to Customer Needs	Performance Tuning / Low Inventories
Flat Hierarchy	...

The description of each mechanism is structured into a number of rules and subrules. For example the mechanism "integration of servicing and maintenance" consists of the following rules:
1. Integration of maintenance and servicing into the production process
2. Reduced number of maintenance and servicing staff
3. Permanent preventive checking of the machines and devices
4. Technical devices for permanent checking for faulty parts
5. Total Productive Maintenance (TPM)

Each rule can be refined by a number of subrules.

The mechanisms of Lean Production tell how to operate an ideal lean company in order to achieve the numerous goals of lean producing companies. These goals are structured under the main goal of each company, that is 'profit optimisation'. The second main goal in operating a lean company is the establishment of a company's culture. This 'We' feeling serves as a basis for a number of subgoals, e.g. permanent improvement process (KAIZEN), cost reduction, customer orientation etc.

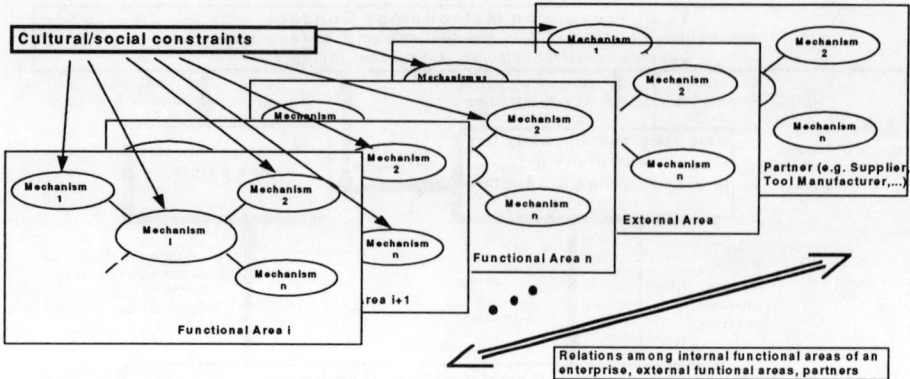

Fig. 2: The reference system of Lean Production represents a strongly interrelated net of lean mechanisms
and rules in a number of functional areas inside and outside of a company

The reference system of Lean Production is built upon the mechanisms in order to achieve the goals of the company. The reference system structures the mechanisms and shows clearly (Fig. 2)
- the mutual interrelations between mechanisms and their rules,
- the mutual interrelations between functional areas of a company concerning the valid mechanisms,
- the influence of cultural and social constraints upon the mechanisms, and
- the relation between the goals of Lean Production and the mechanisms to achieve each goal.

Putting all this together, the reference system serves as a consistent representation of an ideal lean company. The reference system was developed to assess, to monitor and to improve the operation of companies in a variety of different sectors of industry.

3. Assessment of Companies using the Reference System of Lean Production

The assessment to find out the distance of the company in question to an ideal lean company starts with an investigation of the weak points using the production parameters. Based on these parameters, measures are planned and carried out to improve the process.

The assessment of the company is carried out by using the lean mechanisms as a basis for interviews with the relevant company experts within the areas under examination. The employees are asked to rate the application of mechanisms and rules, their interrelations, etc. taking into account the specific environment of their company (e.g. culture).

The analysis of the results of the assessment creates a quantitative summary of the fulfilment of lean mechanisms, their interrelations, etc. Within this summary a list of mechanisms can be clearly identified by which the company lacks most when compared to an ideal lean company. The result of the assessment shows the distance between the 100% fulfilment of a lean mechanism and the actual implementation of the mechanism (Fig 3.).

The identified distances and the reference system together give management a tool to establish a strategy to increase "Leanness". Emphasis must be put upon the mechanisms which are most poorly applied and is indicated by the largest distances on the mechanism level (mechanisms at the bottom part of Fig. 3). The strategy for improvement consists of a series of measures based on the mechanisms and rules of the reference system to push the company along their way to increasing "Leanness". The goal is to achieve more "Leanness" under the constraints of company specific aspects and their social and cultural environment.

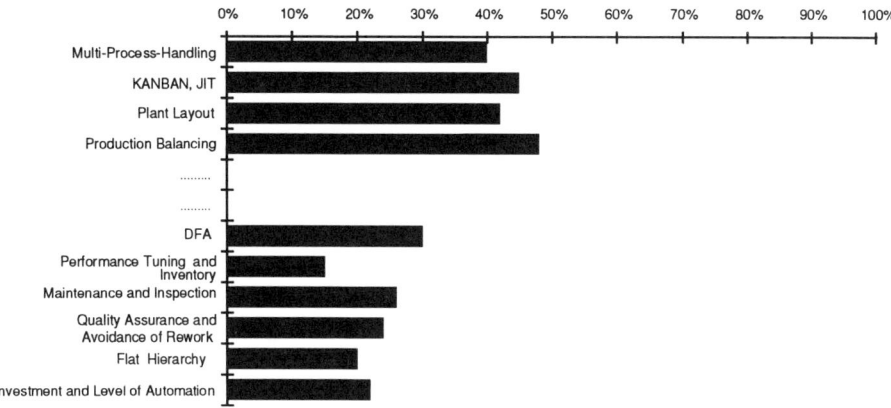

Fig. 3: The difference between 100% and the achieved percentage per mechanism is the current distance to an ideally applied lean mechanism.

While working out the activities to increase "Leanness" management must take into consideration a major change in the behaviour of the actors in all management and production processes in order to achieve a comprehensive company culture ('We'-feeling). The catalogue of measures must be put into operation in a very stepwise manner in order to maintain control of the effects due to the strong interrelations of mechanisms.

By using the reference system, management gets an overview of all effects which will occur in the different departments on the company's way to "Leanness". In addition to this, the reference system may be used to control the process to increase "Leanness" by subsequent assessments. Reduced distances indicate the success of the company on its way to improved competitiveness.

4. Conclusions

The application of the presented reference system of Lean Production gives the management of a company the basis for their decision as to where to start with changes in order to increase the "Leanness" of their company. The reference system supports them in controlling the process of change by monitoring the changes in the distances of the corresponding lean mechanism. The presented reference system can be easily adapted to the specific needs of companies on a broad spectrum of industrial processes.

The developed reference system has been successfully applied to a company in the automotive industry and is now used to measure and increase "Leanness" in a number of European small and medium sized enterprises (SME's).

References

[1] J.P. Womack, D.T. Jones, D. Roos, The Machine that changed the World, Rawson Associates, New York, 1990

[2] P. Wickens, The Road to NISSAN: Flexibility, Quality, Teamwork, MacMillan Press, England 1987

[3] S. Shingo, A Study of TOYOTA Production System from an Industrial Engineering Viewpoint, Productivity Press, 1989

[4] M. Imai, KAIZEN: The Key to Japan's Competitive Success, McGraw-Hill Publishing Company, 1986

Advances in Agile Manufacturing
P.T. Kidd and W. Karwowski (Eds.)
IOS Press, 1994

Lean Production
in Industrial Training

Liu You He
South China University of Technology, Guangzhou, China

Abstract

This paper introduces some ways to lead the industrial training into lean production which are testified to be successful in South China University of Technology. It is pointed out that great potentialities are existed in such a investigation.

1. Introduction

Basic industrial training is a required course for a great quantity of student in universities or colleges of technology. More than two thousand students take part in industrial training at our school-run practise factory each year. Further more, many enterprises and workshops send their workers to our practise factory for training. Because it is very important to improve the quality of labour for the industry, especially in Guangdong province in recent years.

Owing to the fact that all the trainees do not know anything about manufacturing technology and do not have any skill at the beginning. Many guiding personnel should be necessary to teach them. At the present, the ratio of the guiding personnel to the student in basic industrial training is about 1:6. The training efficiency is quite low. As students come to the training workshop batch by batch, some instructors have to repeat the same content day by day. This is a heavy and complicated work to a conscientious instructor. If he/she is slack in his/her work, the training effect will be bad. We have been looking for some better ways to improve the training quality and efficiency. In other words, we are going to lead the basic industrial training into lean production. The following are better ways we adopted in recent years.

2. Developing teach-yourself training

We try to give free scope to student's intelligence. We edit more detailed training instructions for them to read instead of giving more explaining. The student can learn in such a way more actively and flexibly. For example, in milling practice for two days, originally one instructor spent one hour to introduce the structure of the milling machine, the types of milling cutter and the operating steps to six students at the beginning. Now we issue a detailed training menu to each student. From this menu he/she can see the main structure of the milling machine, the function of each handle. Facing to a unknown machine, he/she can read and think a lot, finally he/she can try to operate the idling machine according to the instruction in the menu by themselves. One hours later they begin to read some new parts of the menu step by step---- to mount the cutter, to machine a slot and to use the dividing head. The total training time is still two days. But they can learn more. Because students understand that they will meet some new machines never seen before by them after they graduate. They must know how to learn to operate them by themselves. So this training model is very useful for them. They will try to do by themselves as possible as they can. If the training menu is a good one----easy to understand and easy to follow, questions asked by students will not be too more. The guiding personnel need not to give any lecture and demonstration at the first, but they should still stay there to monitor students' operation, to answer some students' questions. They can guide more students in such a way [1].

3. Teaching assistant

Generally, a student in the mechanical department should take part in the basic manufacturing training for six weeks in China. It is not enough to the student who want to learn more in the workshop. From 1987 We'd carried on a test for those students. We recruited some volunteers from them to train for more than 100 hours in one type of work in production in their spare time. They could learn more skills. If they can pass the examination for their operating ability and theoretical knowledge. They can be engaged to be the teaching assistant to help the guiding personnel to guide the beginner when they are free. This is a very good opportunity for them, as they can learn more and they can earn some money. They are very conscientious in their work and are better educated, they can play the good instructor [2]. The quantity of teaching assistant can be determined by the needs. So the quantity of the fixed staff can be reduced. For example, originally the number of instructors in machining workshop was 19, now 15 is enough. The wages of the formal staff is much more than that we pay for the assistants. So the total expense for training can be saved. Since 1987, we had trained more than 100 teaching assistants. Because these students are more familiar to production and operation. They are warmly welcome by the enterprises after graduating. This is of benefit to the society too.

4. Developing computer-aid instruction and audio-visual education programme

By means of computer-aid instruction, we can make some managing and testing programme automatically. We can give some self-paced CAI learning programmes to the student so that one instructor can guide more students. We can make some computer based presentation programmes to the instructor so that the trainees can build new knowledge efficiently.

One software we designed is a question bank for industrial training. It includes about

4000 questions covering all types of work in manufacturing technology. In the past, we could not often check the student during the training period. Students training in the workshop simultaneously are so more that the guiding personnel could not manage all this by himself. With this software, the student can check by himself. The results can be recorded automatically and can be collected by the guiding personnel later. As the software is very interesting, students like it very much and never take it as a heavy load.

With the audio-visual education programme we can lead the student to visit many workshops in front of the TV set. The training result is better and the quantity of guiding personnel can be cut down. This has been justified by our experience in recent years.

For example, we have made a serious of video tape for milling training.

a) the structure of the miller.   b) types and uses of the milling cutters.

c) how to machine a straight slot.   d) how to use the dividing head.

e) how to machine a helix slot and to bore a hole.

Each part sustained about 15-20 minutes. After watching one part, students do the practice about 3-4 hours. As the audio-visual education programmes have some advantages, especially in showing some internal details and some high-speed phenomenons which are difficult to understand in ordinary way.

Combining with the teach-yourself training menu mentioned above, we can arrange less instructors to guide more students and get better effect.

We are improving our work mentioned above by means of multimedia technique. combining the audio-visual education programmes and the computer. It can be predicted that better effected can be achieved.

5. Enhancing the quality of the guiding personnel

High qualified guiding personnel are necessary to perform the above tasks , for example, to edit the audio-visual education programme, to write the CAI software and the teach-yourself training instruction, to develop the new training item etc. This is the essential condition to reduce the quantity of guiding personnel. The head of the training base should be a professor who is proficient in industrial training and is willing to devote his energy in this cause. Under his leadership there must be a team which is small in number but highly trained.

6. Conclusion

It is beneficial to lead the industrial training into lean production. Not only the quantity of guiding personals can be decreased but also the training quality can be improved. Great potentialities are existed in such a investigation.

References

[1] Liu You He, Multiple-level training on student's manufacturing ability. Education research (S.C.U.T.) 1988, vol,11.

[2] Liu You He, Tan Shang Ming, Fang Huo Ling. Training assistants from students. Antholog of CMIT-II, Nanjing,1988.

Advances in Agile Manufacturing
P.T. Kidd and W. Karwowski (Eds.)
IOS Press, 1994

Technical Change Activities as Promoter of Operator Skills and Technical Innovation

Peter Friedrich
Division of Social and Organizational Psychology
National Institute of Occupational Health
S-171 84 Solna, Sweden

Abstract. The overall aim of the study was to contribute to the development of theoretical approaches to research on technical change from a work environmental perspective. The study focuses on technical change and skills development in relation to machine operators. The conclusions drawn from this analysis invite discussion of the suitability and limitations of the research approaches that are employed in the field of work and technology. An evident need for alternative approaches is shown. A series of critical questions is posed in relation to the theoretical foundations and methodological procedures of research on technical change and skills development at work. The empirical material on which this investigation is based has been obtained from case studies conducted in five production units with flexible-manufacturing-systems.

1. Technical change activity

Technology is shaped in many different social arenas; at a macro (global) level within the framework of international socio-economic structures, and at a micro (workplace) level consisting of individuals and organizations (1). Technical artefacts are primarily the product of *human activities* undertaken in the course of the artefacts' *development and use* .

One crucial function of human beings in technological systems, as well as the obvious role they play in inventing, designing and developing the systems themselves, is to *complete the feedback loop* between system performance and system goal, and thereby correct failures of system performance (2). By feeding back information on anticipated or actual deviations to professionals dealing with the technical system (e.g. production technicians, engineers maintenance workers) users can contribute to the design process.

The integration of the use and the development of technical artefacts forms a key arena in which discussion of the opportunities for *skills development* (3) available in operators' work can take place on a continuous basis. The primary focus of technology and work research has been on the use of technical equipment. The shift in perspective recommended here implies that technical change is treated as a task for the operator. From an action psychology perspective, it is shown that the potential for the skills development of operators should be sought not only in work directly related to production, but also within the framework of a company's development of its own technical resources. The latter arena offers hitherto unrealized opportunities.

The concept of *technical change activities* (4) denotes a new type of work which lies at the interface between the use and development of technical artefacts. Technical change activities are characterized as elements in an evolutionary developmental process, in which operators and technical specialists jointly participate. Technical change activities are *collective* by their very nature; i.e. the goal of such work (the further development of technical artefacts) can only be achieved dialectically, involving a process through which a variety of people contribute different skills and knowledge. Prevailing organizational preconceptions and attitudes towards the work of operators are often barriers to the realization of the potential possessed by operators to further technical development.

2. Experiences of technical change activities in the work of operators

Use and development of technical artefacts have been employed as analytical concepts referring to two qualitatively different *kinds of tasks,* which represent extreme points on a continuum of the tasks involved in the process of technical change; *knowledge development* takes place through alternation and interaction between the two. The continuum reflects the dimension "degree of active participation in developing the technical artefact". Given the lack of an already-developed classification of technical change activities, a typology (derived from an examination of our own empirical material) is proposed:

Production tasks: use of technical artefacts.
Maintenance tasks: work of a corrective nature to deal with acute deviations in the functioning of the technical artefact.
Improvement tasks: work undertaken to prevent deviations in the functioning of the technical artefact.
Renewal tasks: work initiated on the basis of experiences of use of the technical artefact which is designed to renew parts of that artefact.
Development tasks: work initiated by external demands made on the use of the technical artefact which is designed to develop that artefact.

Production, maintenance and improvement activities all have it in common that operators' experiences constitute a source of expertise applicable in work aimed at furthering the development of the technical artefact (technical change activities); i.e. they have relevance, at least an indirect one, to renewal and development. Through adopting such an approach, the activities of operators can be described as being aimed at the achievement of not only instantaneous targets (such as correcting disturbances) but also of longer term goals (such as the prevention of malfunctions through further technical development).

The results demonstrate that, in four out of five cases, operators' handling of deviations was not intended to achieve any permanent improvement in the conditions that prevailed at the installation; rather, it was corrective by nature, aimed at dealing with acute deviations in the production system. Analysis of operators' tasks shows that it was only in the production unit 'C' that opportunities were offered to operators to participate in technical change activities: 30% at most of the job tasks of the most highly-skilled operators were of this kind. In the other units, it is certainly the case that operators undertake simpler kinds of maintenance of existing equipment, and they also participate, at time of acquisition, in the planning and design of new equipment. However, these activities do not form part of a continuous process of change, and interchange of knowledge and skills, in which operators' experiences of the production process are seen as an essential element in integrated technical change and development.

From the analysis of the content of operators' work in the production units, it emerges that there was a "cooperative component" to tasks designed to achieve improvement, renewal and development. This is interpreted to mean that this type of work is collective by nature, i.e. it requires inputs of knowledge and skills from a variety of people, who mutually interact in a dialectical process that leads to the achievement of technical development.

3. An organizational perspective on deviation handling

Technical change and technical development in an organization are 'bottom-up' activities (from the perspective of the operator) and 'top-down' activities (from the point of view of engineers). The steady development of the technical artefact on the basis of operators' production experiences will scarcely take place automatically. Experiences are a source of development in themselves, but reaping the fruits of these requires conscious efforts. The integration of operators' production work and deviation handling as a means for developing the production system is a central element in the 'bottom-up' development process. Such a process, however, is highly sensitive to the decisions made and the opportunities offered by the 'top-down' procedures of management and other specialists. The company's *organizational infrastructure* is a factor of central importance in enabling an interactive process between *the effects of* and *the stimulation of* technical change to be initiated, to be perceived as meaningful by the individual, and to be sustained over time.

An empirical analysis of the personnel resources made available to five production units for maintenance and improvement on the one hand, and for renewal and development on the other showed that there was a satisfactory balance in only one of the production units investigated. In the others, resources were inadequate for either one of the two types of work, or for both. The production unit with satisfactory conditions for technical change activities ('C') had sufficient resources for different kinds of skills development (in relation to both product and production process). Technical change was not only initiated in production work, but also accelerated, kept alive, and led along the 'right' lines; both maintenance and improvement, and renewal and development workers followed up and decided upon the changes, and referred their initiatives to those higher up in the organization. At the same time, feedback to the operators provided stimulation for continued and renewed efforts in the field of technical change.

Whatever the type of work, the nature of the organization (centralized or decentralized) affects the opportunities for technical change activities. Remote physical location, long waiting times, unclear lines of responsibility and poor utilization of specialist skills are some examples of the consequences of a centralized organisation, which can hinder the interaction required. Having production units with centralized and geographically remote maintenance sections means lengthy waiting times for operators, and consequent dissatisfaction; there are often demands for immediate counter-measures to be taken by operators despite the repairer being on the site. This has a negative effect on the interaction between operators and maintenance personnel, and is a barrier to cooperation and joint problem-solving. The central location of maintenance staff also makes it impossible for them to maintain day-to-day contact with operators and the plant.

4. Paradoxes in "technology and work research"

The results of the study (3) and a survey of approaches adopted in social research on technology and work (5) give rise to the need for a discussion in principle on how social research into technical development can be conducted. Table 1 provides a comparative overview of central aspects of three different approaches. The definition of two of these is according to whether their focus is on the work environment or co-determination/-participation; the third, and contrasting, approach is that which considers "technical change at workplace level".

On the basis of this survey of the various approaches adopted within the field of technology and work research, and in the light of the empirical findings of this study, a series of critical questions can be posed in relation to the theoretical foundations and methodological procedures of research into technical change and occupational skills development:

1. With respect to *research into skills development*, the issue is whether existing, empirically-observable conditions can alone provide a basis on which the scope for action possessed by operators can be assessed. Is there a need to search for opportunities or potentials (*theoretical possibilities*) in order to promote the skills development of operators?
2. With respect to *research into technical development*, the conclusion that technical change is a *collective task* raises a series of questions concerning prevailing conceptions of the nature of technical development - its scientific underpinnings and its impact on the orientation of research.
3. Research from an occupational perspective into technical change within manufacturing industry has principally taken the form of studies of the use of technical artefacts at a specific point in time. Should not developments over time be taken into account?

The criticism can be raised that the principal approaches hitherto adopted improperly distinguish between technical artefacts (such as machines) on the one hand, and people and the social system on the other. The sharp line of demarcation between technology and techniques (technical development) on the one hand and work (job conditions) on the other is disputed. The making of such an analytical distinction has provided the basis for deterministic analyses which attempt to explain the way in which technology affects social conditions. It also explains the popularity of "social shaping" approaches, where studies are conducted of how given social conditions affect changing states of technology. If the theoretical distinction between technical artefacts and people were to be abandoned, this would provide for a better understanding of "work and technology" in an occupational context. The work of operators (production) and that of engineers (technical development) are both regarded as forms of action that belong to the same "arena" - i.e as involving skills development that can promote technical change. The work of operators is designed not only to meet production targets, but also to contribute to technical development, which then becomes an area of application for the experiences obtained in the field of production.

Table 1: Comparison between three General Approaches to Technology and Work Research

	Work environment	Co-determination/ participation	Technical change at workplace level
Perspective	Use of technique	The process of shaping and acquiring technology	*Technical development* and *use of technology*
Aim	Prevent negative consequences	Industrial democracy, prevent negative consequences	Scope for action, potential, *development of knowledge, skills development*
Object of research	Ill-health	Decision-making	Health
Perspective on change	Structural, preconditions	Infrastructural, process-oriented	Process-oriented
Contribution of personnel to change/development	Criticism	Experiences from work and work situation	*User experiences* related to production equipment, a production process and a product
Input/output relation of personnel's contribution	Hard to identify result; other people take over and transform the input in the best case	Results identifiable; other people transform the input	Input of personnel converted and implemented through action, *in cooperation with other specialists (collective work)*
Research focus	Use	User	User and designer
Understanding of the forces behind technical development	Not explicit	Social, democratic	*Social, interactive*
Conception of the nature of technical change	Radical, technological leap	Radical, technological leap	*Accretional, iterative, evolutionary*

References

[1] M. Hård. How the Diesel Engine Got Going: Local and Global Aspects of Development and Diffusion. Wissenschaftszentrum, Berlin 1991.
[2] P. Friedrich. Production Deviations: Opportunities for User-Based Technological Development. In: M.J. Smith and G. Salvendy. Human-Computer Interaction: Application and Case Studies. Proceedings of the Fifth International Conference on Human- Computer Interaction, Orlando, Florida, August 8-13, Vol 1, p 815-820.
[3] P. Friedrich. Kompetensutveckling vid lokal teknikförändring. En analys av fem fallstudier över operatörsarbete och datorstödd automatisering i verkstadsindustrin [Skills development and technical change activities at workplace level. Operators' work and computer-aided automation in manufacturing industry. Analysis of five case studies]. Akademitryck AB, Edsbruk, 1992 (In Swedish).
[4] P. Friedrich. Technical Change Activities: An Integrative Approach to Operators' Skills Development and Technical Change. Proceedings. IFAC Symposium on Automated Systems Based on Human Skill (and Intelligence). Madison, Wisconsin, September 22-25, 1992.
[5] P. Friedrich and M. Hård. Labor, Culture and R&D-policy: Technology-and-Society Studies in Sweden and Norway. In: M. Dierkes and U. Hoffmann (Eds). New Technology at the Outset. Forces in the Shaping of Technological Innovations. Campus Verlag, Frankfurt am Main, 1992.

Practical Tools to reorganize and support Work in Production

Matti VARTIAINEN

Laboratory of Industrial Psychology, HUT, Otakaari 4 A, 02150 Espoo, Finland

Abstract. New production paradigms present attractive models for future production work. The paradigms, however, fail to instrumentalize the road to their stated goal. Employees need various tools in order to be as autonomous and creative as the paradigms demand. This paper proposes some tools to help change and development processes on the shop floor. The paper concentrates on how to develop and support grouplike work.

1. "Goal" and "process" paradigms, and the "tool" approach

Several "goal" paradigms and models challenge the Taylorist work system and its principles to divide jobs into thinking (planning, designing, organizing, controlling) and doing tasks. *Lean production* [1] aims at scarce use of all resoures (time, costs, staff, etc.) compared with conventional enterprises. The main differences with traditional producers in organizing production work are in leadership, teamwork, communication, and simultaneous development. The *antropocentric production model* [2] stresses the combination of planning and executing, distributed decision-making, law hierarchies, cooperation between engineers, supervisors and workers, multiskilled work force, and a neat interaction of product design and production. The *fractal factory approach* [3] is based on the idea that each part of a 'fractal' contains the whole structure. For example, the idea of 'factories within a factory' means semi-autonomous production units and cells with a high variety of tasks, duties and responsibilities for employees.

As a result, enterprises organized according to the above-mentioned models should provide better products in wider variety at lower costs, and, simultanously, more challenging and fulfilling work for employees at every level of the enterprise.

"Process" paradigms and models try to help to understand how changes in organizations should be made and how to select developmental strategies for them. Generally speaking, we could define them as action research approaches emphasising the participation of employees in the change process. The process paradigms as well are often unable to instrumentalize the management of change. For example, the German evaluation team saw three structural limitations and weaknesses in the Swedish LOM Programme [4]. One of them [5] was that "the Programme ... appears to be seriously under-instrumentalised. Its armoury of instruments has proved effective in initiating development processes, but the lack of subsequent design and

process instruments is one of the main causes of the 'energy-drop' in the course of the project processes". In the LOM Programme, the researchers utilized clusters of enterprises, their conferences and principles of the democratic dialogue (communication) as their tools to organize and support change in organizations.

We propose theoretical and practical tools [6] as instruments to reach the future goals and avoid deficiencies during the change process. The *theoretical tools* include concepts and heuristic principles used during the whole change process, for example, the developmental cycle, work activity system as the object of change and development, sociotechnical work description, participation of the personnel, and models of 'good work'. The *practical tools* are more specific and technique-like instruments and procedures as shown in Figure 1. They include methods to analyse and model a work system, techniques for participative problem-solving and development, simulation games for the modelling and practising change, methods for work and job (re)design, and interactive information support for industrial work.

2. The challenge to reorganize and support work in production

2.1. Grouplike work as an organizational principle

There is general enthusiasm around Europe for such old semi-autonomous groups. Practically all the goal and process paradigms propose groups and teams as base-line organizational solutions. Our interests are in the organization transformations [7] which would result in jobs with temporal and procedural degrees of freedom, and possibilities to solve open problems. In practice, this means a 'hybrid' organization where small groups are both working and doing development tasks.

Our model of group organization is an open systems model: each group creates its own principles during its formation and development. These principles include values, norms, procedures and activities. The quality of the principles is expressed to the organization by the level of communication, productivity, creativity and well-being of the group members.

We have been working with a cluster of three enterprises in the field of light assembly. In each enterprise, there has been a steering committee and small groups to develop production. The steering committee from different organizations have also worked together in a conference to discover common solutions.

We have studied groups by questioning, discussing and observing them from the view points of well-being, work motivation, communication and procedural justice. The transformation of the production organization from individual work to group work has increased productivity in work, but not always increased work motivation or general job satisfaction. Interviews concerning procedural justice after four years of group work have revealed the following problems: communication between the supervisor and the group, lack in qualifications of group members concerning the more demanding tasks, production control and flexible functioning, pressure for conformity and a hard working pace. Group work was also sometimes regarded as a hindrance to promotion and personal development for group members.

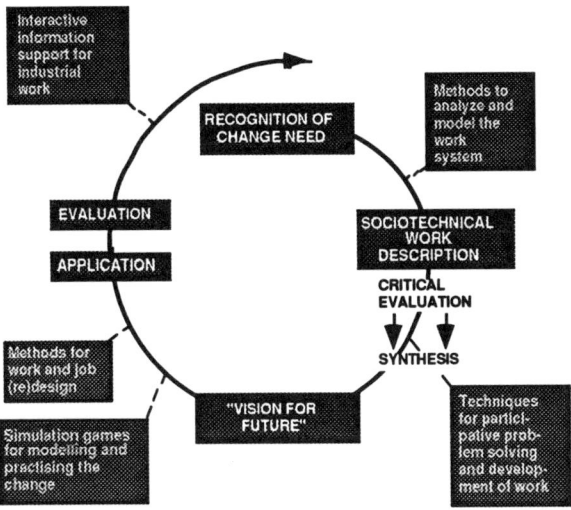

Figure 1. The developmental cycle and some practical tools.

2.2. Training of groups with simulation games

We developed the team-work game [8] to enhance group formation. The aims are to increase group members' abilities and skills to work in groups and to increase their functionality. The team-work game concentrates on certain group characteristics, some aspects of group-work, and on the developmental stages of group formation. The *group characteristics* are communication, common goals, cohesion, procedural and distributive justice, autonomy, and leadership in groups. The *aspects of group-work* are a group as a part of a larger work system, internal relationships in a group, and an individual in a group. The *developmental stages* dealt within the game are the formation of a group, conflicts in groups, the creation of activity principles, and the mature group. The simulation is realized as a board game. There are from three to eight participants plus a game leader in the game. A game lasts from two to four hours. Participants proceed by casting a dice, answering questions, discussing, and collecting points. The questions answered are organized into three categories: 'knowing each other', 'characteristics of group-work' and 'problems in groups'. The questions deal with the above-described group characteristics, aspects of group-work, and developmental stages of group formation. The debriefing during and after the game is used to conceptualize the actual and possible future problems in the group.

Experiences of the game gained from fifteen groups have been encouraging. Shop-floor and office employees have high esteem of the game compared with traditional training methods.

2.3. Interactive information support for groups

The need for information in group work has increased with the complexity of products. Traditionally, employees' competencies have been cultivated with special courses and task training. The information and training support of enterprises' information systems has been meagre. We [9] have developed an information support system called HOTS (HIPS On-the-job Support System) that utilises text, still video pictures and digitised speech. Experiences show the system be of great help in work.

3. Discussion

The role and costs/benefits of the tools described above are studied. Their basic aim is to develop means for personnel to participate fully in development work. It is not clear which of the tools are really useful. Some tools, for example methods of work analysis, are not easy to use without theoretical knowledge and wide training. One question is, what are the restrictions and possibilities to use group techniques on the shop floor? Simulation games are very attractive but what is their cost-benefit ratio and transfer ability to the real working contexts? Our data shows that employees do not always accept group work as an organisational arrangement. Why not? In principle, employees need more information on how to produce various goods in small batches in a flexible way. How should information be didactically organized? New tools may help and facilitate change but can they also become a hindrance?

References

[1] J.P. Womack et al., The machine that changed the world. Ranson Associate, New York, 1990.
[2] W. Wobbe, Antropocentric Production Systems: a strategic issue for Europa. FAST/MONITOR-Programme. APS Research Papers Series Vol. 1. Bruxelles 1991.
[3] H.J. Warnecke, H.J. (1993) The fractal company. Springer, Berlin, 1993.
[4] B. Gustavsen, Creating broad change in working life: the LOM programme. QWL Center, Ontario Ministry of Labour, Toronto, 1988.
[5] F. Naschold et al., Evaluation report commissioned by the board of the LOM Programme. Arbetsmiljöfonden, Stockholm, 1991.
[6] M. Vartiainen, Sociotechnical work description and other tools of participative work design. In: Y. Quéinnec and F. Daniellou (Eds.), Designing for everyone, Vol. 1. Proceedings of the 11th Congress of the International Ergonomics Association. Taylor & Francis, London, New York, Philadelphia, 1991, pp. 388-390.
[7] A. Pulkkis et al., From production groups to productive group work. In: V. Orpana and A. Lukka (Eds.), Production Research 1993, Proceedings of the 12th International Conference on Production Research, Lappeenranta, FInland, 16-20 August. Elsevier, 1993, pp. 661-662.
[8] M. Vartiainen and V. Ruohomäki, Simulation games as tools for work development, and their psychological bases. In: G. Bradley and H.W. Hendrick (Eds.) Human Factors in Organizational Design and Management - IV. Elsevier 1994. (In press)
[9] J.J.J. Kasvi et al., Developing a hypermedia authoring system for task training and information arrangement on the shop floor. In: V. Orpana and A. Lukka (Eds.), Production Research 1993, Proceedings of the 12th International Conference on Production Research, Lappeenranta, Finland, 16-20 August. Elsevier, 1993, pp. 647-648.

Advances in Agile Manufacturing
P.T. Kidd and W. Karwowski (Eds.)
IOS Press, 1994

Why Using Automation to Replace People Can be Wrong

Peter L. PRIMROSE
Total Technology Dept., UMIST, Manchester M60 1QD, U.K.

Abstract. The incorrect use of costing information has helped to exaggerate the importance of using automation to replace direct labour and transfer their skill to indirect workers. An understanding of the economics of automation and its role in manufacturing strategy shows that the objectives for investment have to change away from labour saving, resulting in a change in the required technology.

1. Introduction

In the late 1970's and early 1980's, when Advanced Manufacturing Technology (AMT) was being developed, there was a universal belief that manufacturing industry was over-manned. As a result, managers who were trying to reduce costs saw the introduction of AMT as being an ideal way to reduce the size of their direct labour force.

The emphasis on labour replacement was reflected in the literature describing AMT. For example, Robots were portrayed as machines that could replace people, the aim with Flexible Manufacturing Systems (FMS) was to achieve un-manned manufacture, and Computer-Aided Design (CAD) was able to reduce Drawing Office labour.

Because managers looked for applications that would replace labour, manufacturers responded by developing products that could optimise labour savings. The consequent publicity for these products helped to reinforce managers' belief that they were doing the right thing, thereby creating a vicious circle.

Unfortunately, not only did this emphasis on labour saving create industrial relations problems, it has now been realised that the incorrect use of costing information was giving a greatly exaggerated view of the importance of labour saving.

2. Cost Information

Most manufacturing companies have a cost management system whose primary purpose is to ensure that all the costs of manufacture are absorbed into the selling price of products. Because these systems are used to monitor the actual cost of manufacture against forecast, the information from them is widely available within a company and is often used to monitor the performance of departments.

At the time when the early cost systems were being developed, manufacturing companies were usually labour intensive and direct labour was often the largest single cost factor. As a result, systems were developed that used direct labour as the basis of measurement with all other costs being apportioned to this to establish a labour/hour rate.

Managers who use this information to help them improve the efficiency of their departments are likely to concentrate on projects aimed at reducing direct labour. For example, if labour cost is £5/hour plus 400 percent overhead allocation, giving a cost rate of £25/hour, reducing labour appears to save £25/hour. This suggests that replacing someone who works 2000 hours/year would save £50,000/year, therefore investment of £150,000 in automation to achieve this would provide a 3 year payback. However, if overheads and depreciation are unchanged, the saving may only be £10,000/year, making the investment in automation highly unprofitable.

Using costing information in this way can suggest that it is economical to employ two, or more, indirect workers to replace one direct worker, even if the wage of an indirect is greater than that of a direct. As a consequence, the balance between the number of direct and indirect workers has changed in many companies, resulting in a transfer of skills away from the shop floor.

The introduction of Computer Numerical Control (CNC) machines typifies this process. Managers often justified investment in CNC on the basis of reducing the number of operators required and also reducing the level of skill and rate of pay by changing the role from setter/operator to operator. The skill being transferred to a programmer who, as an indirect worker, was seen as being much cheaper than a direct worker.

Not only has the incorrect use of costing information exaggerated the importance of labour saving in the introduction of CNC machines, but the transfer of skills from operators to programmers may have been counter-productive in many cases.

In recent years, a number of manufacturers have been producing CNC machines with control systems that allow the operator to programme the machine while it is still producing the previous batch of components, eliminating the need for programmers.

Experience in several companies has shown that, provided operators are suitably trained and motivated, their programs can manufacture components more efficiently than those produced by office based programmers. The reason being that operators will normally have a much better knowledge of their machine and its capabilities than a programmer. Provided the cycle time and batch quantity are sufficiently large to avoid lost machine time, using operators to program their own machines can help to improve the economics of CNC machines.

3. Single Cost Rates

When early computerised cost systems were being developed in the 1960's, many companies were labour intensive and the capital cost of machine tools was relatively low. Because they were much simpler to use, many cost systems were based on having a single rate/hour for all direct workers. Surveys suggest that approx. 30% of companies (including some large companies in the aerospace industry) still use a single hourly labour rate.

Having a single rate means that operators of machines whose capital costs are low will have their time costed at an excessively high rate. On the other hand, operators of expensive equipment will have a rate that is too low.

To illustrate the consequences, a component may take ten minutes on a manual lathe, but only three minutes on a CNC machine. If both machines have a rate of £30/hour, using the CNC machine appears to reduce the cost from £5 to £1.50.

This leads to all components being planned for expensive high-technology machines which provide the shortest operation times, without reflecting their high capital costs. As a result, the most expensive machines become overloaded, and pressure builds up for investment in more of these expensive machines. If the procedures for investment appraisal are inadequate and do not identify what is happening, the company can end up making unnecessary and expensive investments, thereby increasing the replacement of labour by automation.

4. Investment Appraisal

As AMT developed it became more complex and expensive which made it more difficult for companies to justify investment on the grounds of cost reduction. Advocates of each area of technology therefore made considerable efforts to identify additional benefits, but were only able to describe these in general terms such as increased flexibility of production or better quality products.

This led to the belief that the benefits of AMT were intangible and unquantifiable and that investment could not be justified in financial terms but had to be made as a strategic decision (which sounds better than an act of faith). Unfortunately, the inability of advocates of AMT to define the benefits in quantifiable terms made it difficult for them to relate strategic objectives to technology selection. This in turn made it difficult for manufacturers to develop the technical features required to achieve strategic objectives. As a result, although investment in AMT has increasingly been made as an act of faith, the emphasis on direct savings such as labour replacement has continued.

Fortunately, the work done at UMIST [1] has shown that there is no such thing as an intangible benefit and that every benefit that can be identified can by redefined and quantified. The ability to quantify all the benefits of AMT allows us to relate technical specification to strategic objectives and compare the relative magnitude of benefits.

5. Manufacturing Strategy

There are four factors that influence a company's competitive ability, namely:-
*   Product function and features.
*   Delivery performance.
*   Product quality.
*   Selling price.
The aim with manufacturing strategy planning is to identify the changes that have to be made in these factors to meet market needs, and the way that changing manufacturing facilities can achieve this. Achieving these strategic objectives normally requires changes in technology, capacity, management philosophies and human resources. The major financial benefits of achieving the objectives come from increasing the volume and profitability of sales, rather than reducing operating costs.

Investing in AMT can make a significant contribution to

this but the applications selected will be different to those whose objective was primarily cost reduction. For example, one of the main benefits of investing in FMS is improved delivery performance, Robots and CNC can be used to improve product quality, and CAD to improve product design. Any emphasis on labour saving in such applications can be counter-productive.

Although selling price is a competitive factor, reducing the cost of the material and bought-in content of products will normally be much more important than reducing operating costs such as direct labour. Thus the use of CAD to help redesign products can not only improve the ability to meet customer needs for product function and features, it can also make a signifi-cant contribution to reducing material and bought-in costs. Any saving in drawing office labour from using CAD is likely to be insignificant by comparison.

The current literature relating to manufacturing strategy rarely considers human resources, except in the very limited context of the need to involve managers in the strategy planning process, and the perceived need to reduce labour. However, many of the changes identified in manufacturing strategy planning will have human resource management implications. Most people will need training for new skills and some will need to change their working conditions. Such changes may involve alterations in manning levels and shift patterns, increasing the flexibility of workers, or modification to payment systems.

As well as trying to increase customer satisfaction, companies have an obligation to keep trying to improve working conditions and eliminate unsafe working practices. While investing in automation can help them to do this, such invest-ments may also produce an increase in productivity. The danger of concentrating on labour savings in such cases is that the investment may not be made if the savings are inadequate to justify the cost.

6. Conclusions

Surveys of companies that have invested in AMT invariably show that the majority have not achieved the savings that were originally predicted and that AMT was not financially viable. This is not surprising because, in order to get projects approved, managers tend to underestimate costs, while being over-optimistic about the level of savings they can achieve.

Not only has this led to the incorrect belief that existing accountancy principles were inadequate for dealing with the complexities of AMT, it has also created unnecessary industrial relations problems. In some cases, especially with CAD, companies have refrained from making essential investments because they anticipated industrial relations problems.

For the introduction of AMT to be successful, everyone involved, both direct and indirect, has to be well motivated. By concentrating on applications where the aim is to replace labour it is difficult to get the enthusiasm needed to achieve the maximum savings. Being over-optimistic about the expected labour savings makes it harder to achieve this enthusiasm.

Reference

[1] P.L. Primrose, Investment in Manufacturing Technology, Chapman & Hall, 1991.

Advances in Agile Manufacturing
P.T. Kidd and W. Karwowski (Eds.)
IOS Press, 1994

Rehabilitation - an Industrial Economic Analysis

Per Dahlén[a] and Stig Wernersson[b]

[a]Lund University, Dept. of Production and Materials Engineering, 211 00 Lund, Sweden.
[b]Ericsson Telecom, Söderhamn, Sweden

Abstract. This paper produces a formula calculating the industrial profitability associated with the decision to rehabilitate an employee who is absent due to illness. The formula is exemplified in a case study from a Swedish engineering company. The results indicate that the frequent, short-term absence is usually profitable to eliminate through rehabilitation actions. Rehabilitation concerning long-term sick employees turned out to be difficult to justify if the worker does not regain almost full working ability after the rehabilitation. On the other hand, it is not difficult to find a number of good social economic reasons to rehabilitate a person so that he or she can continue a professional career. The indirect connection between sick leave exceeding 14 days and the social insurance contributions, in the Swedish system, seems to lower the incentives for the company to rehabilitate long-term absent personnel.

1. Introduction

The social benefit from the rehabilitation of personnel injured at work can't be questioned, neither the benefit for the individual. However there are reasons to analyse whether, rehabilitation is always profitable from an industrial economic point of view. Through experience [1] we know that the remaining decreased work ability and continued high absenteeism due to illness are common results of rehabilitation. This indicates that the economic result of rehabilitation will fluctuate depending on: the individual's working capacity after rehabilitation, wehether the individual will be able to work full time or part time, wheter the company has a suitable assignment to offer the employee after rehabilitation and whether the company has resources to allocate to rehabilitation.

The Swedish legislation (22 chap. 3 § in the law of public incurance) states that an employer has to investigate the possibilities to rehabilitate an employee who: 1 has been absent due to illness for more than four weeks, 2 has been absent due to illness on more than six occasions during a period of 12 months' or 3 if the employee demands an investigation. The Swedish legislation, however, leaves the company considerable freedom to chose the level of ambition and to weigh economic parameters concerning the rehabilitation [2]. The purpose of this paper is to produce a formula calculating the industrial profitability associated with the decision to rehabilitate an employee. The formula is to be exemplified in a case study from a Swedish engineering company.

2. Analysis of Costs and Revenues of a Decision to Rehabilitate

In the following text a number of costs and revenues that can be related to a rehabilitation decision will be defined. We assume that the company needs additional production capacity. The choice is between hiring new employees or rehabilitation of employees incapable of working full hours to full capacity. As a cost for the rehabilitation, all additional costs compared with the hiring of a new employee will be taken into account. The costs are to a large extent dependent on whether the rehabilitated person regains full working ability or not. Rehabilitation also has a revenue-side in the form of savings related to decreased costs for absent employees and a reduced need for hiring new employees.

A) Additional costs for labour hour per unit produced. The working capacity (W_c) and the working time (W_t) that the rehabilitated employee regains after rehabilitation decide if the labour costs per unit will increase compared to hiring a new employee. Notice that it is only the fixed part of the wage (W_f) that is to be taken into account in this case. In a piece wage system the costs per unit will not increase if a rehabilitated worker does not regain full capacity.

B) Variable costs due to the number of employees. More is required from a company than just to pay wages for an employee. The work must be administrated, the employee should have a work place, somewhere to change and so forth. A vast part of these costs do not vary due to the working time but with the number of employees. If the rehabilitated person does not regain full working capacity or working hours, the number of persons in work increases, generating the costs $(b_1, b_2..b_n)$.

C) Rehabilitation costs, representing the costs $(c_1, c_2..c_n)$ connected to the actual rehabilitation.

D) Decreased need for new employment. The rehabilitation might in some cases lower the costs for the company to hire new personnel. The savings $(d_1, d_2..d_n)$ [3] will be dependent on the work capacity and possible working hours received by the employee.

E) Decreased short-term absenteeism, (case 2 in the Swedish legislation). If the rehabilitation results in a decreased number of days lost to short-term absenteeism (A_s), it can be looked upon as a cost reduction $(e_1, e_2..e_n)$ [4] caused by rehabilitation .

F) Lowered long-term absenteeism due to illness (case 1 in the Swedish legislation). If rehabilitation results in decreased long-term absenteeism due to illness, it should be regarded as the cost reduction $(f_1 f_2..f_n)$ due to rehabilitation.

Formula 7 below illustrates the additional costs as a consequence of rehabilitation. The formula is a sum up of the parameters in table 1.

Table 1. Costs and Revenues Related to Rehabilitation

Costs	Formulas [SEK/year]	Revenues	Formulas [SEK/year]
Additional costs for labour hours per unit (A)	$= 1800 W_t W_f [1 - Wc]$ (1)	Decreased need for new employment (D)	$= W_c W_t \sum_{i=1}^{n} d_i$ (4)
Variable costs due to the number of employees (B)	$= [1 - W_c W_t] \sum_{i=1}^{n} b_i$ (2)	Decreased short-term absenteeism (E)	$= A_s \sum_{i=1}^{n} e_i$ (5)
Rehabilitation costs (C)	$= \sum_{i=1}^{n} c_i$ (3)	Lowered long-term absenteeism due to illness (F)	$= 1800 W_t \sum_{i=1}^{n} f_i$ (6)

Additional rehabilitation costs (ARC) = A+B+C-D-E-F or

$$ARC \approx 1800 W_f W_t [1 - W_c] + [1 - W_c W_t] \sum_{i=1}^{n} b_i + \sum_{i=1}^{n} c_i - W_c W_t \sum_{i=1}^{n} d_i - A_s \sum_{i=1}^{n} e_i - 1800 W_t \sum_{i=1}^{n} f_i \quad (7)$$

Case Study

To illustrate how the formula can be used a case study will be given. The study describes a production department where the work tasks are monotonous, leading to physical stress and absenteeism due to illness. The variables W_f, B, C and D in formula (7), vary, primarily due to the layout of the production system that the rehabilitated employee returns to. E and F are primarily dependent on political decisions. Finally W_c, W_t and A_s are parameters related to the individual and to the production system. If we choose to study a specific production system with given rules for E and F, only W_c, W_t and A_s will be variables related to the individual. Suppose that the company has made the following estimations of the costs for two employees, representing the two categories requiring a rehabilitation investigation according to Swedish legislation. One that has been absent due to illness for more than four weeks, and one employee frequently short-term absent due to illness:

1. Personnel absent for more than four weeks

$W_c = 0.2$ to 1.0 ($1.0 =$ full work capacity); $W_t = 0.2$ till 1.0 ($1.0 =$ full working hours);
$As = 0$ days; $W_f = 125$ SEK/hour (including payroll taxes).

$$\sum_{i=1}^{n} b_i = 30\ 000 \text{ SEK/year;} \qquad \sum_{i=1}^{n} c_i = 20\ 000 \text{ SEK;} \qquad \sum_{i=1}^{n} d_i = 120\ 000 \text{ SEK;}$$

$$\sum_{i=1}^{n} e_i = 1\ 500 \text{ SEK/day;} \qquad \sum_{i=1}^{n} f_i = 7 \text{ SEK/day}$$

2. Personnel absent more than six times during a 12 month period

$W_c = 0.2$ to 1.0; $W_t = 1.0$; $As = 0$ days; $W_f = 125$ SEK/hour.

$$\sum_{i=1}^{n} b_i = 30\ 000 \text{ SEK/year;} \qquad \sum_{i=1}^{n} c_i = 20\ 000 \text{ SEK;} \qquad \sum_{i=1}^{n} d_i = 120\ 000 \text{ SEK;}$$

$$\sum_{i=1}^{n} e_i = 1\ 500 \text{ SEK/day;} \qquad \sum_{i=1}^{n} f_i = 0 \text{ SEK/day}$$

The thick line in figure 1 illustrates the additional costs for the rehabilitation of an employee frequently absent for short periods, when returning to full-time work. The thinner lines represent the additional costs for the rehabilitation of a long-term absent employee, returning to full or part-time work. The figure shows that it is considerably more profitable to rehabilitate the short-term absent, compared with the long-term absent employee. If the short-term absent employee regains more than 50 percent of regular capacity, the calculations indicate that rehabilitation is very profitable. The cost reductions related to reduced sick pay and disruptions in production, apparently more than enough counter balance the additional costs for the rehabilitation of a short-term absent employee.

The profitability decreases when the long-term absent employee is studied. The additional costs are presumed to be higher, due to the resources required for the rehabilitation when the employee has been ill for a long period of time. Long-term absenteeism due to illness does not charge the company with high costs related to sick pay[1] and disruptions in production, as does the unplanned short period absenteeism due to illness. The opportunity costs for rehabilitation are therefore considerably lower for the long-term absent personnel lowering the profitability of rehabilitation.

---

[1] A Swedish company has to pay sick pay from the second to the fourteenth day of illness each sick period. The company also pays payroll taxes (approximately 40 percent of the sick pay).

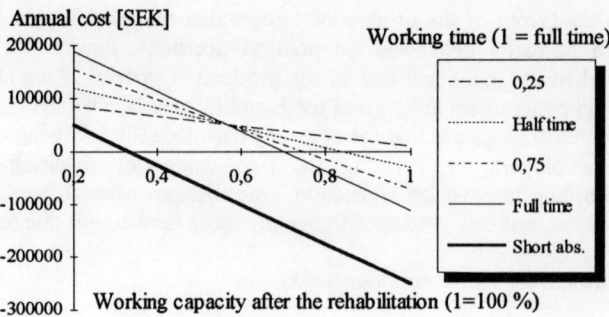

Figure 1. The additional cost during the first year after rehabilitation. Notice! (-)-sign turns the cost to a revenue.

The additional rehabilitation costs decrease when the regained work capacity increases. The figure also indicates that the costs increase with increased working time for the long-term absent employee if the work capacity is lower than X in the graph. If the working capacity is higher, the costs decrease with more working hours. It is also shown that the unprofitability is severe when the work capacity subsides 50 percent. When the working capacity increases to 70 - 80 percent, and the long-term absent employee can work at least 75 percent of full working hours, rehabilitation in this case study, seems to be an economical alternative to hiring new personnel.

3. Discussion

As already mentioned it is often less expensive to hire a new employee than to rehabilitate a long-term absent one. The unprofitability of rehabilitation is to a large extent a consequence of the low expences the company has to pay for absenteeism exceeding 14 days. The outcome of a calculation where the company also paid sickness benefits after the first 14 days of absence, should show incentives for the company to rehabilitate employees with working capacity so low that only ethical reasons exist today for rehabilitation.

In a production system frequently generates work injuries, the alternative of hiring new personnel is only a temporary solution. The risk is severe that the new employee will be affected by the work in the same way and also require expensive rehabilitation. The company has in this case reached a dead end and is again forced to choose - rehabilitate or hire new personnel. In the long run, the root of the evil must be eliminated e. g. a change in the production system generating personnel problems is necessary. It is possible to invest in automation or improve the organisation of work and thus break the monotonoy. Rehabilitation is thereby an economic incentive for an overhaul of both the automation level and the work organisation in the company.

References

[1] *Varannan rehabilitering (miss)lyckas, (Every second rehabilitation fails)*, Arbetsliv i utveckling-Tema rehabilitering, No., 3 pp., 4, 1993, (in Swedish).

[2] Iseskog, T., *Arbetsgivarens rehabiliteringsansvar*, (The employer's rehabilitation resposibility), Aktuell Juridik, 1991, (in Swedish).

[3] Bekiroglu, H., Gonen, T., *Labor turnover: roots, costs and some potential solutions*, Personnel Administrator, July, pp., 67-72, 1981.

[4] Mirvis, P., Lawler, E., *Measuring the Financial Impact of Employee Attitudes*, Journal of Applied Psychology, Vol., 62, No., 1, pp., 1-8, 1977.

Advances in Agile Manufacturing
P.T. Kidd and W. Karwowski (Eds.)
IOS Press, 1994

How greater operator control increases AMT performance

Peter Gardner, Nik Chmiel and Toby Wall
MRC/ESRC Social and Applied Psychology Unit, Department of Psychology, University of Sheffield, Sheffield, U.K.

Abstract. Research has shown that encouraging greater operator control in the management of advanced manufacturing technology (AMT) can produce benefits in terms of job satisfaction and decreased machine downtime. The reduction in downtime appears to be dependent on operators learning effective fault prevention strategies but the process by which this is done has received little attention. This paper brings together findings from the job design literature and psychological techniques for assessing knowledge and performance. We report a study of the ways in which inexperienced operators learn to diagnose faults in a simulated AMT system. We are able to show that the characteristics typical of an AMT system promote the acquisition of implicit knowledge, that is knowledge which is difficult to verbalise. Such knowledge has also been shown to be gained by 'hands-on' experience rather than formal instruction. Given that the traditional focus of job design is the motivation of the work force, we argue that there are also important implications for the training methods used for new operators of AMT systems.

1. Introduction

1.1 The Applied Context

Advanced manufacturing technology (AMT) has the potential to enhance competitiveness through improved quality, throughput and responsiveness to customer demand. Evidence suggests, however, that this potential is often not fully realised, not so much because of the limitations of the technology itself, but more because of inadequate supporting work organisation.

This evidence has fuelled interest in the role of job design in the effective use of AMT. Recent field studies have shown that redesigning shopfloor jobs to enhance the role of operators in system management not only promotes psychological well-being, but also substantially improves system performance. Wall et al. [1], for example, found that such increased operator control over CNC assembly machines resulted in reduced downtime, especially in the case of more complex systems. Jackson and Wall [2] showed that the reasons for these gains were not simply due to a quicker response to system errors, but reflected the acquisition of more fundamental fault management skills which accrued over time. This conclusion was supported by a subsequent study involving increased operator control over a robotics system (Wall, Jackson and Davids [3]). The more general implication is that a broader and more active involvement in system management provides an opportunity structure for learning how to operate AMT systems more effectively.

Clearly this conclusion prompts further inquiry. If the link between job design and the effectiveness of AMT systems is based on such learning processes then it is important to understand the nature of those processes and the conditions under which they arise. The indications from the above field studies, and the work organisation literature more generally, is that much of the skill and knowledge involved in managing AMT systems is implicit in nature. Wood [4], for example, observes that "the question of increasingly automated systems is often portrayed as being fraught with problems, and it is thought that workers' experience and tacit knowledge may be a better resource for overcoming these than any textbook formula or engineer's conceptions" (p. 170). This suggests a connection between this area of job design research and the currently separate literature on cognitive processes in the control of complex dynamic systems, which also highlights the importance of implicit knowledge.

1.2 A Cognitive Perspective

Research on cognitive processes in the control of complex systems has been conducted largely in the laboratory using computer-based tasks. The most significant theme to have emerged from a series of such experiments is the proposal of a distinction between two types of knowledge and methods of knowledge acquisition. A person is said to have learnt implicitly when they improve their performance over time (i.e. they learn to do the task) but do not correspondingly improve their verbal knowledge of how to do the task. Explicit learning is where performance and verbal explanation correspond to one another. The evidence demonstrates that different learning conditions promote the different types of learning. Whereas explicit knowledge can be passed on through traditional instructional methods, implicit learning depends more on 'hands-on' experience of the system. As discussed by Broadbent [5] and Chmiel & Wall [6], such findings have potentially very important applied implications, suggesting, for instance, differential roles for training and job design depending on the nature of the knowledge involved.

Therefore, it is important to understand the conditions under which explicit or implicit knowledge develops. It has been suggested that knowledge is most likely to be explicit under conditions where the key variables and their causal relationships are more obvious or 'salient', and implicit under conditions of lower salience (Berry and Broadbent [7, 8]; Berry [9]). The rather imprecise notion of 'salience' has been redefined more specifically with particular relevance to the characteristics of advanced manufacturing systems by Chmiel and Wall [10]. They propose that two key aspects of salience are *complexity* and *meaningfulness*. A cause and effect relationship is defined as complex where there are two or more contributory factors, neither of which is sufficient to produce the outcome by itself but together do so. Meaningfulness refers to the extent to which the relationship between cause and effect meets the prior expectation of the operator. The hypothesis is that explicit knowledge develops under conditions of low complexity and high meaningfulness; whereas implicit knowledge arises where cause and effect are more complex and less meaningful. Given that the latter conditions characterise many AMT systems, this would account for the suggestion that the knowledge development in this context is in large measure implicit, and that it is promoted through job design which provides for the 'hands on' experience upon which such learning depends.

1.3 The Present Study

Testing the above hypothesis in a field setting would be a very difficult task, since there is little opportunity to manipulate either complexity or meaningfulness, or to control extraneous variables. Since the issue is one concerned with the basic nature of learning and knowledge, and the key properties of dynamic systems can be replicated in the laboratory, this was the chosen context for the present experiment.

2. Method

An interactive computer simulation was produced that was based on the real-life robotics line studied by Wall, Jackson and Davids [3]. This line was used to manufacture hardened steel concrete breaker bits for pneumatic drills. The simulation was focused on the fault diagnostic component of the fault management requirement.

Using facilities provided in the computer simulation, participants in the study were asked to investigate the causes of a series of faults presented to them. The accuracy and efficiency of their diagnosis was measured. Accuracy was defined as the number of correct diagnoses, and efficiency scores were derived from the number of system components that were examined before a diagnosis was made. Each participant received three diagnosis sessions, each one consisting of 18 faults. After each session a questionnaire measuring verbal knowledge was administered.

There were four experimental conditions in which the causes of the faults were arranged to be salient or non-salient in terms of complexity and meaningfulness Thus, there was a 'salient' configuration in which the cause and effect relationships were both non-complex and meaningful. There was a 'non-salient' configuration in which the relationships were both complex and non-meaningful. There were two further groups, a 'complex' and a 'meaningfulness' group, in which non-salience was introduced on only one dimension. A fifth group did not take part in the fault diagnosis task on the computerised simulation but simply completed the questionnaire following an introduction to the system, in order to provide a comparison for prior knowledge. There were 12 subjects in each group.

The questionnaire was designed to measure verbalisable procedural knowledge and included six questions each based on one of the faulty products shown in the computer simulation. Participants were shown six printed pictures of the faults exactly as they appeared on the computer. They were then asked to decide the cause of each fault and indicate this on the answer sheet (they were given a list of possible faults to choose from). They were also asked to indicate, from a list, which system components they would have examined in order to make their diagnosis. Thus, verbalisable procedural knowledge could be directly compared to performance on the computer diagnostic task.

3. Results

With regard to task performance the findings from the study showed evidence of considerable learning over the three sessions for all four conditions. That is to say, participants were able to diagnose faults much more accurately by the end of the experiment, reaching levels of around 80-90% correct. Strikingly, the final performance for all four experimental groups was very similar by the end of the last session, demonstrating that differences in salience had limited impact on this outcome measure. With respect to efficiency all groups also improved across sessions; however participants in the non-salient conditions continued to seek more information from the system before arriving at a diagnosis.

There were substantial differences between the conditions regarding the extent to which task performance is reflected in verbal knowledge. The salient group recorded scores on the questionnaire which closely paralleled their performance on the task itself. This contrasted markedly with the questionnaire responses for all three low salience conditions, where verbal accuracy was well below task performance. Indeed, even the final questionnaire scores for these groups was little better than that of the control group who had no direct experience of the task.

Statistical analysis of the results demonstrated that the differences between the salient and non-salient conditions were significant in a number of respects, however the main finding was that the discrepancy between task and questionnaire performance was significantly greater for the non-salient conditions.

In summary, the study clearly demonstrates that salience, in terms of complexity and meaningfulness, affects the nature of fault diagnostic knowledge that developed, but not performance itself. Under salient conditions, that knowledge was largely explicit, in that actual performance was matched by the ability to express this verbally. Where cause and effect relationships were less salient, however, equivalent levels of performance were not matched by verbal report - the knowledge gained was to a large extent implicit.

4. Discussion

The results demonstrate the importance of meaningfulness and complexity in the development of knowledge about fault diagnosis. Implicit knowledge is promoted where cause and effect relationships are complex, low in meaningfulness, or both. In contrast highly salient relationships promote an association between performance and verbalisable, procedural knowledge. Further the learning observed suggests that implicit knowledge takes time to develop and supports high levels of diagnostic accuracy. The pattern of our results relating to efficiency accord with the theoretical suggestions about the nature of implict learning made by Berry & Broadbent [8]: that many contingencies between variables are noted and learned, even where they are irrelevant to the task at hand. In the present study this led to examining many more system components in the non-salient conditions than in the salient condition to achieve the same diagnostic accuracy.

From a practical perspective, the present study has some potentially important implications. To the extent that advanced manufacturing systems involve faults with complex causes and which are unpredictable, it seems likely that learning to deal with them effectively would require direct experience and result in knowledge which is not readily communicable to others. This suggests that operator job design will be a key consideration for the effective use of such systems, and that formal instruction may be of limited value. Those suggestions, however, require much closer examination.

5. References

[1]   Wall, T.D., Corbett, J.M., Martin, R., Clegg, C.W. and Jackson, P.R. (1990). Advanced manufacturing technology, work design, and performance: a change study. *Journal of Applied Psychology, 75,* 691-697.

[2]   Jackson, P.R. and Wall, T.D. (1991). How does operator control enhance performance of advanced manufacturing technology? *Ergonomics,34,* 1301-1311.

[3]   Wall, T.D., Jackson, P.R. & Davids, K. (1992). Operator work design and robotics system performance: a serendipitous field study. *Journal of Applied Psychology, 77,* 353-362.

[4]   Wood, S. (1990). Tacit skills, the Japanese management model, and new technology. *Applied Psychology : An International Review, 39,* 169-190.

[5]   Broadbent, D.E. (1990). Effective decisions and their verbal justification. *Philosophical Transactions of the Royal Society of London B, 327,* 493-502.

[6]   Chmiel, N.R.J. & Wall, T.D. (1993). New technology and job design: learning to manage computer controlled systems. *Paper presented at the BPS 1993 Occupational Psychology Conference.*

[7]   Berry, D.C. & Broadbent, D.E. (1987). The combination of explicit and implicit learning processes in task control. *Psychological Research, 49,* 7-15.

[8]   Berry, D.C. & Broadbent, D.E. (1988). Interactive tasks and the implicit-explicit distinction. *British Journal of Psychology, 79,* 251-272.

[9]   Berry, D.C. (1991). The role of action in implicit learning. *Quarterly Journal of Experimental Psychology, 43A,* 881-906.

[10]   Chmiel, N.R.J. & Wall, T.D. (in press). Fault prevention, job design, and the adaptive control of advanced manufacturing technology. *Applied Psychology : An International Review.*

Advances in Agile Manufacturing
P.T. Kidd and W. Karwowski (Eds.)
IOS Press, 1994

DESIGN STRATEGIES FOR HUMAN CENTRED MANUFACTURING SYSTEMS: CONCEPTS AND EXPERIENCE

Andrew Ainger
Human Centred Systems Limited
222 Maylands Avenue
Hemel Hempstead
Herts
HP2 4FE

ABSTRACT

This paper outlines a new approach to the development of software products. It draws on practical experience of employing: the traditional waterfall life-cycle model, the Spiral Model and concurrent engineering aspects of both small and large (Pan-European) software projects to present, discuss and propose a new life-cycle model; the Helical Life-Cycle Approach. The paper distinguishes between prototypes and models and postulates the need for the formalisation of a new software engineering job role which is focused around the design of human centered systems, i.e. a software engineering Model Maker.

BACKGROUND

In 1979 a report (Ref 1) (Figure 1) concluded that only 2% of software supplied was usable as delivered! A massive 47% of software was delivered but never used. These figures emanated from the USA's Department of Defence some 14 years ago, so it could be assumed that in the intervening years, with the advent of computer-aided everything (CAx),things would have changed. The new rigours of systems specification and design have changed the situation dramatically. From the 2% success rate in 1979 to, in 1991, a 99% failure rate! (Figure 2). It is realised that the comparisons are not exactly like with like. However, a general trend can be identified, that of the generally poor performance, as far as the user is concerned, of software systems.

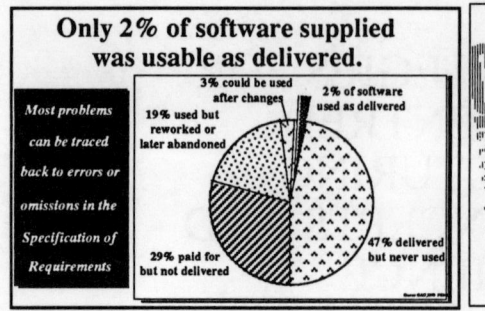

FIGURE 1   FIGURE 2

This general failure to meet user needs first time round accords with our own experience. Although the success rate of our projects (within a multi-million pound international organisation) appeared to be significantly higher than those experienced in the States, the overall performance was still felt to be low. As a result, an extensive internal survey of over 100 projects over a 10 year period was undertaken.

Regardless of hardware or software platforms, regardless of the qualifications of the development team and regardless of the financial size of the project, it appeared that the most successful projects had significantly more meetings (both formal and informal) with the users of the system than those that did not. There was another correlation that was even higher than the user meetings. It was found that all failed projects followed more rigorously the waterfall project life-cycle.

It was in the early 80's that work commenced on what is now termed the Helical Life-Cycle Approach. The initial work was conducted under a European Strategic Research and Development into Information Technology (ESPRIT) Project, (Ref 2). The Project was entitled "Human-Centred CIM Systems" and developed the idea that the system is designed around human beings and integrates human capabilities, skills, inventiveness, etc." The £5.6 million around human beings and integrates human capabilities, skills, inventiveness, etc." The £5.6 million Project involved 6 organisations from 3 European countries and is believed to be the first Pan-European Project to research Human Centred CIM Systems.

THE HELICAL APPROACH

The Helical Approach is focussed around the generation of models. To quote the Oxford Dictionary (1990) a model is "a representation of designed or actual object; design or style to be followed; give shape to, form".

In the Helical Approach (Figure 3) the generation of models assists both the user and the systems designer to communicate effectively. It is somewhat surprising to find that the use of models (as opposed to prototypes) is almost unheard of in the software field. However, in every other engineering discipline the generation of models can be a pre-requisite before the real or even prototype system/product/artifact is built. It is almost certain that if one walks into an architect's office the first thing that will be seen will be, under a glass dome, a "cardboard model" of the their latest mega project. This model is not expected to work. What is expected is that the overall view, the general picture that is given is as accurate as possible.

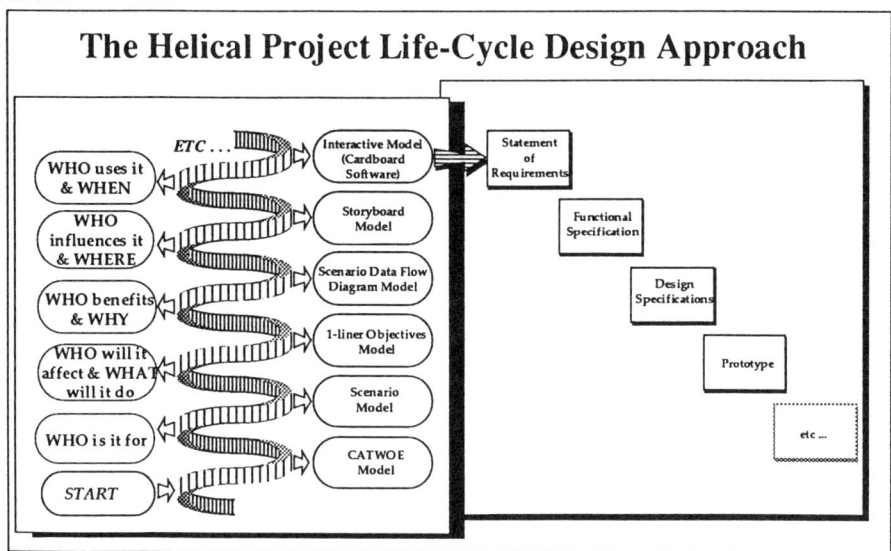

The Helical Project Life-Cycle Design Approach

FIGURE 3

It is relatively straightforward to conceive and build a model whether it be in cardboard or any other material of a block of flats. What is more problematic is the construction of a "cardboard model" of a proposed software product! However, just as one can have a model of a block of flats made of cardboard and many other materials, it is also possible to have many models of the proposed software system. It is only towards the end of the model generation sequence that "cardboard software" (a slide-show technique) is used. Many other model making "materials" have to be employed prior to the cardboard stage, such as CATWOE (soft systems), Scenarios and DFD's.

The model making "materials" of the Helix are not new. Some of the materials have existed for many years. The Helical Project Life-Cycle Approach employs and utilises concepts and insights from other disciplines to maximum benefit.

CONCLUSIONS

The benefits of adopting and using the Helical Project Life-Cycle have been substantial. Progress on projects employing the Helical Approach have been rapid and the success high (Ref 3 & 4). To date the Helix has been used successfully in four multi-million Pan-European development projects and numerous National ones.T he first product (ACiT - Ref

4) to result from using the Helical Approach was launched in London in 1990 and major companies such as BICC and ITT Cannon are now using the approach.

The biggest problems found when implementing the Helix have been mental barriers within the individuals concerned. It is almost second nature to go from specification to design and design to prototype. What we have to do is to breakdown those mental barriers. We have to think in the Helical way and to build models, to become multi-skilled, to think globally, to continually learn as we earn and to become part of the new Learning Earning culture.

REFERENCES

1.   USA D.O.D. (1979) GAO

2.   Hamlin M, April 1989 "Human-Centred CIM", Professional Engineer, UK.

3.   Ainger A, 1989 "Human-Centred Design of Human-Centred Systems", ESPRIT Conference, Brussels, Belgium.

4.   ACiT - Production Loading, Scheduling and Implications Software available from Human Centred Systems Limited +44 (0)442 219243.

Advances in Agile Manufacturing
P.T. Kidd and W. Karwowski (Eds.)
IOS Press, 1994

The Impact of Environment-Oriented Manufacture on Engineering Education - the Brunel University Programme

Rita VAN DER VORST
Department of Manufacturing and Engineering Systems, Brunel University,
West London, UB8 3PH, United Kingdom

Abstract. This paper gives a brief introduction to environmental engineering and its use to manufacturing. In the main part, a new initiative in environmental engineering education at Brunel University is described. The undergraduate programme called Special Environmental Engineering, SEE, is based in a Manufacturing Department and concentrates on Preventive Engineering and the consideration of Life Cycle Issues in the Design for Manufacture.

1. Introduction

"Engineers, as creators, promoters and implementors of technological solutions, bear a special responsibility to the future. In their work, they need to provide the technological growths essential to social, economic and cultural advance and, at the same time, ensure sustainability of development by conserving and enhancing the environment."

International Code of Environmental Ethics for Engineers

Although overall employment in UK manufacturing industry is decreasing, the demand for professional engineers has never been higher, and will continue to rise as we move towards the 21st century. This trend is reflected both within Europe and internationally.

Industry is trying to respond to ever more pressing environmental issues and legislation, as everyone begins to recognise the need to protect our world. Environmental Management Systems, Design for Recycling, Design for the Environment, Environmental Auditing, Environmental Impact Assessment are but a few examples for the new environmental tools, systems, guidelines etc. which need to be implemented in a cost-efficient way for the benefit of both company and environment. Engineers with a conventional education are often helpless in front of the complexities found in environmental problem solving. The engineer for the future, therefore, needs a broad multi- or even interdisciplinary engineering education to enable her to think in new ways and work with new concepts. These engineers need to know about environmental issues such as resources, energy and pollution as well as about technology, management, and management of technology to enable them to take over responsibility for their activities. Such system thinking and long term planning naturally develops to be preventive engineering and leads to the implementation of clean technologies.

The Engineering Council - the UK's national Engineering Governing Body - has recently recognised engineers' major new responsibilities in humanising our technology for use and understanding. Based on the International Code of Environmental Ethics for Engineers [1], it developed the Code of Professional Practice "Engineers and the Environment" which was introduced this year [2]. Following the Code, every registered engineer has to aim for greater awareness, understanding and effective management of environmental issues.

"By following the actions in this code, registrants will be able to give a lead in proposing and implementing sound engineering solutions to safeguard the future. They will also be able to progress the broader debate over how we achieve sustainable activity, and in so doing make a contribution to the stewardship of the earth." [2] In the discussion during the preparation of the Code it was suggested that, in future, about 10% to 15 % of every engineering degree programme should be devoted to environmental issues.

A further Engineering Council's development is a draft on "Guidelines on Environmental Issues". "The Guidelines are intended to provide a helpful source of reference for registrants on environmental issues" [3]. In the chapter on Education and Training is stated: " Environmental awareness should be a prime subject of education and training at all levels commencing at Primary School and continuing to Continuing Professional Development.

Clean technology based on life cycle considerations will be the order of the day. Thus the demand for environmentally aware and professionally educated design engineers has never been higher and will clearly increase. Especially, manufacturing companies will gain from engineers trained in both environmental design principles and manufacturing engineering. Further and Higher Education Institutions have to respond to those needs (see [4]).

2. The Special Environmental Engineering Programme at Brunel University

The Department of Manufacturing and Engineering Systems at Brunel University has developed an undergraduate engineering degree programme to meet new demands by industry for environmentally aware engineers. The Special Environmental Engineering Programme (SEE) is a broad based systems course, providing the option for every student to specialise in their final year in one of the engineering disciplines such as Mechanical Engineering, Power System, Control Engineering or Manufacturing Engineering.

The Programme aims to develop a sound generalists systems engineer having a good working knowledge and understanding of environmental issues. Engineering is treated as a broad based activity appropriately reflecting the human concerns of individuals, society and culture. Thus, reflecting societies' increased concern with the environment, the special environmental engineers will be able to design, manufacture, commission and manage clean, effective and efficient engineering artifacts and systems.

The Programme is based upon providing a deep working knowledge of technology, engineering principles and industrial application. This goes with an appropriate understanding of ethical, political, legal, management and business issues. The special environmental engineers will "know how" to design solutions as well as "knowing that" problems exit.

The speciality of the environmental Programme is found in a class of modules under the topic 'Environmental Principles' which introduce interdisciplinary topics related to environmental problem solving and allows space for an evaluative examination of solutions/responses to environmental problems.

Firstly the environmental engineering principle modules present factual information regarding the effects of man's activities, and secondly, they are devoted to the consideration of the values germane to the issues involved. Therefore, they are concerned with ethical considerations insofar as they relate to the environment. Both aspects, the factual and the evaluative, are treated in an integrated way.

Considering the demands on environmental engineers and their future positions in industry, the Department implemented new teaching/learning structures in the concept of the Special Environmental Engineering Programme. The opportunity to contribute to discussion groups, to work on individual as well as group projects and presentations, and in student-centred and student-led seminars will school the students to put in practice the factual knowledge gained through lectures and private study. Environmental projects and assignments as well as the seminars offered in the second year environmental principles courses are to be seen as tools to integrate an environmental approach into traditional engineering projects.

The new teaching structure chosen will facilitate awareness training and at the same time the students' personal development. The students' achievement will be the construction of an individual value systems, which will enable them to take decisions towards technical solutions considering both the economical and environmental/social consequences.

Life cycle considerations are employed as a case study to several of these issues. For example, based on a product's life cycle, the students learn to evaluate environmental impacts of a product not only during its manufacture, but throughout all stages. Social, regulatory as well as economical arguments resulting form the life cycle assessment have to be considered by the undergraduate environmental engineer when deciding for example about a design for recyclability.

In addition to lectures and activities on legislation, resources, life cycles, etc., Brunel's environmental engineering students are confronted with environmental issues through projects and assignments such as the Ecosystems Project, the Poster Project and Exhibition, the Philosophy assignment, and the Local Company Project.

1)	The introductory session of the Environmental Engineering Principles Module is devoted to the Ecosystem Project. The students' task is to model a complete ecosystem showing all the interactions between the organisms. In a final presentation to their peers, the students explain the interactions presented. This project is designed to help the student understand the complexities of the systems they are going to work with as graduate engineers.

2)   In the Poster Project, the students work on two corresponding posters, a general one to express what environmental engineering is, and a second poster to communicate the idea of environmental engineering to, for example, a group of sixth-form pupils. The students work in groups. One of the aims of this project it to learn about the different definitions, and ideas one can find about environmental engineering; another aim is to drive the student to think about their reasons for studying environmental engineering. In the final exhibition, the students will discuss their posters with academic staff and industrialists.

3)   In the assignment on the philosophy of technology the students discuss the value of ethical considerations to the environment, society and industry, based on philosophical literature, and their presentation to and discussion with the class.

4)   The Local Company Project offers the first opportunity for the students to work as "consultants" to industry. They might carry out a short energy audit, a feasibility study on a suggested heating system, or other small projects. A presentation and a scientific report are prepared for the use of the company.

Students studying on the Special Environmental Engineering Programme soon realise, that there is no easy solution to environmental problems, but that it is possible to work towards solutions through an understanding of the complex systems involved.

References

[1]   WFEO, FEANI, UNESCO, International Code of Environmental Ethics for Engineers. World Federation of Engineering Organisations, European Federation of National Engineering Associations, United Nations Education Scientific and Cultural Organisation, 1993

[2]   The Engineering Council, Code of Professional Practice, Engineers and the Environment. The Engineering Council, London, 1993

[3]   The Engineering Council, Guidelines on Environmental Issues, Draft, Revision A, 9 March 1994. The Engineering Council, London, 1994

[4]   Toyne, P., Environmental Responsibility, an agenda for further and higher education. Department for Education and the Welsh Office, HMSO, London, 1993

Advances in Agile Manufacturing
P.T. Kidd and W. Karwowski (Eds.)
IOS Press, 1994

INTEGRATION / FLEXIBILITY: AN AMBIGUOUS RELATION

CHRISTOPHE EVERAERE
Assistant Professor in Management
Institut d'Administration des Entreprises, Université Jean Moulin - Lyon 3
15 quai Claude Bernard B.P. 0638, 69239 Lyon cedex 02, France

Abstract

Within the research for the improvement of flexibility of industrial production systems, the word *integration* is often mentioned. This concept is namely associated with the implementation of new information technologies (CIM), but it is also linked with the organization of manufacturing systems, production activities and the design of products. However, it appears that integration does not always bear the same meaning. The purpose of this paper is to define and to precise the different meanings of integration and to see which ones are best suited to the simultaneous search for flexibility and productivity.

Introduction

Although the concept of integration is often linked with industrial modernization and the search for competitiveness [1], it turns out to bear different meanings that do not all suit the search for flexibility. Indeed, it is not rare to find in one text the word integration used with some obvious different meanings. Either, for example, robots are *integrated* into a manufacturing line; either functions are *integrated* between each other; either a subcontractor is *integrated* by a firm; either different units are organized in an *integrated* way.

With the help of the dictionary and relying on concrete examples, four definitions of integration are identified: *synchronization, made-in-house, merging, coordination* with for the last one three derivatives: *linkage, coherence, interaction*. Only the third one (merging) when applied to the product design and the fourth one (coordination) turn out to be compatible with the search for flexibility.

I focus on a wide definition of flexibility, that is *the ability of a system to adapt* and I put the emphasis of the *uncertainty* and *emergency* constraints which caracterize the economic model of reactivity.

1. INTEGRATION = (MECHANICAL) SYNCHRONIZATION

A first meaning can be identified in the field of automated manufacturing systems: an automated manufacturing line is said to be integrated when each element of the line is tightly coupled and synchronized with one another through the conveyor belt. This meaning has to do with the following definition of integration: "operating economically as a single coordinated physically interconnected unit..." (Webster, 1976)

Such an integrated manufacturing system requires a perfect <u>synchronization</u> in operations time and in space location of the machines. This requirement for synchronization has to do with integration. As a result, if any problem occurs somewhere in the line, the entire line stops at once or if the sequences or the speed of the line need to be modified, the entire line has to be modified and synchronized again. Thus, the adaptation ability of such an integrated system is very weak.

2. INTEGRATION = MADE-IN-HOUSE

Another meaning of integration is the economic one: an activity is said to be integrated when made-in-house, or in other words, when "a successive economic industrial process formely carried on independently becomes under a unified control" (Webster, 1976)

The relation between this definition of integration and flexibility appears once more problematic. Indeed, it has been pointed out that subcontracting - i.e. *dis-integration* or impartition of activities - is a source of flexibility since, despite transaction costs [2], subcontracting is seen as a way to share uncertainty and to push outside the cost and the management of production instability.

Let's make precise that the *dis-integration* of production activities is quite compatible with a very close cooperation between partners despite their different identities and their inner autonomy. M. Aoki [3] talks about "quasi-integration" to illustrate this intermediary level of relations between a simple market relation and a much closer relation between the units of a single company. All the literature about networks forms of organization confirms the tendency towards a dis-integration of production activities under more and more complex economic environment [4].

3. INTEGRATION = MERGING

According to a third meaning, something is said to be integrated when its components are merged or combined so as to be unified. As a result, the amount of components is reduced. This meaning has to do with the common definition of integration: "the condition of being formed into a whole by the addition or combination of parts or elements" (Webster, 1976).

The consequences of this meaning regarding the search for flexibility are different whether applied to the product design or to the organization of a manufacturing system.

When applied to the product design, that integration leads to a search for similarities between the components of different products so as to reduce their number and to have them resorted in a wider range of products. That is the way, we talk about the "modular conception of products" or "group technology". The integration or the merging of the components of a product enables to reduce production management complexity and to reach scale economies in the components so that they can be utilized in different products.

But the relationship with the search for flexibility is very different when the merging definition of integration is applied to the organization of a manufacturing system. As a result, tranformation stages are merged in more complex machines achieving several consecutive operations, which could be otherwise achieved by different specialized and less complex machines.

Having different transformation stages being merged or combined within one unified machine makes it more complex, more fragile and less turned to quick adaptation. On the contrary, several different machines can be more easily organized under the condition that the machines are not "bolted" between one another (first meaning).

4 INTEGRATION = COORDINATION

Within the problematic of flexibility, we brought out several meanings of the integration concept. Surprisingly, whereas integration is often associated with the so-called "modernization" of industrial production systems, the meanings we pointed out so far are little or not compatible with the search for flexibility.

Do we have then to give up using this concept in the context of adaptability under uncertainty and emergency constraints ? No, if integration has the following physiological meaning : <u>a system is said to be integrated when the activities of its units are coordinated in a harmonious way</u>. or in other words, "when a close cooperation among distinct entities is established" (Webster, 1976). I identify three derivatives of this definition of integration that follow different but complementary paths.

4. 1. *Integration = connexion*

A first application of the physiological meaning of integration is the MAP project (Manufacturing Automation Protocol). Launched by General Motors, MAP is an architecture of industrial communication systems that underlines the necessity of a *connexion* between the "units" so that they can communicate and cooordinate their activities [5].

The idea of a connexion is worth keeping since the adaptation ability under uncertainty and emergency constraints relies on quick information exchanges. It is all the more possible as the aim of the information technologies is to connect any dispersed areas. But let's prevent us from thinking that units will naturally communicate and have interactions only because they are physically connected. The technological dimension of communication does not suffice for the connected units of a production system tend to coordinate their activities in a harmonious way.

4.2. *Integration = coherence*

Another idea that has been pushed forward by an European project called CIM-OSA (Computer Integrated Manufacturing - Open System Access) is going futher and focuses on the *coherence* of the data the "units" exchange to one another.

It is also about "defining a computer architecture for integrated production systems" [6], but an important difference with the MAP project is that the emphasise shifts from the technological support of the information to the information itself: which information is worth sharing, for instance. The CIM-OSA project focuses on transparency, information circulation, on clear and compatible objectives, on decision making procedures, on coherence both in the content and the form of the information. Here integration means *coherence*. The usual example of this need for coherence is the difficulties to transmit information or plans from the design to the manufacturing departments because their information systems are not compatible or because they do not have common data bases.

However, the technological dimension keeps a too dominant priority. The C. in C.I.M. conveys the idea that computer remains the main stake to be dealt with. Yet, many other aspects are still in question, to reach a harmonious working that directly raise the issue of the organization itself and the conditions of innovation in the reactivity econonomic context.

4.3. *Integration = interaction*

Within a study of Japanese companies organization, H. Molet [7] talked about *integration* to emphasize the intensity of the relationships existing between departments, namely between the marketing, design and manufacturing departments. The problematic has to do with the innovation in new products with the requirement to have them developped more and more quiclkly to markets.

We talk about integration then to deal with this collective work between people from different departments within one firm or from different firms who design together new products in thinking at the same time about the contraints or the opportunities dealing with the commercialization, the design and the manufacturability of the new product.

Here integration means *interaction*.The experiences such as "project teams" or "transversal groups", that have to do with this integrated organizational working in the interactive meaning, show up that conflicts are not missing. But at least it is possible to talk about problems before, to have them raised up through the confrontation of points of view instead of finding out problems when it is too late or much more expansive to solve them. The "concurrent or simultaneous engineering" [8] terminology is nothing else but the applying of the integration concept in the interaction meaning.

The three ideas to reach the physiological meaning of integration are complementary. In a first time, there is *connexion*, against compartmen-talization and opacity in information exchanges. In a second time, there is *coherence* in the information exchanges, against specific and incompatible information systems. And then, there is *interaction* in the entire working and in the decision making to deal with adaptation under uncertainty and emergency constraints.

Conclusion

Four definitions of integration have been identified: *synchronization, made-in-house, merging, coordination* with for the last one three derivatives: *linkage, coherence, interaction.* Only the third one (merging) when applied to the product and the fourth one (coordination) seem to be suitable with the search for flexibility.

The dictionary has been helpful to understand a fashionable but rarely defined concept. Identifying the contexts in which the word is used, I hope to have clarified a highly staked concept since it has to do with the organization of a production system, its management and the designing of the products.

References

[1] DUDLEY G., HASSARD J., "Design Issues in the Development of Computer Integrated Manufacturing (CIM)", *Journal of General Management*, vol. 16 No. 1 Autumn, 1990.

[2] WILLIAMSON O., Markets and hierarchies. Analysis and antitrust implication,. The Free Press, New York, 1975.

[3] AOKI M., "Horizontal vs. Vertical information structure of the firm", *American Economic Review*, 1986, December 76(5): 971-83.

[4] POWELL, W., "Neither markets nor hierarchy : networks forms of organization", *Research in Organizational Behavior*, 12, 1990: 295-336.

[5] SAUNDERS D., "What MAP means to the production managers", *International Journal of Operations and Production Management*, v9n2, 1989.

[6] LORIMY B., "La recherche d'un nouveau concept d'entreprise intégrée. L'exemple du projet AMICE du programme Esprit", *Gérer et Comprendre*, Revue des Annnales des Mines, avril, 1988.

[7] MOLET H., "Gérer la complexité de la gestion de production", *Revue Française de Gestion Industrielle*, n° 3, 1987.

[8] HARTLEY, J., Simultaneous Engineering. J. Mortimer (eds.), second edition, Industrial Newsletters Ltd, Dunstable, UK, 1990.

People and Business Integration

Umit Sezer Bititci
DMEM, University of Strathclyde
Glasgow, UK

Abstract. This paper introduces a methodology for business integration which has been specifically developed to combine the strengths of the hard and soft approaches whilst eliminating their weaknesses. The methodology provides a structured and rigorous framework for business wide integration as well as ensuring maximum involvement of people. One of the key features of the methodology is the development and use of an integrated set of performance measures by people at all levels. This ensures that functions and individuals activities are streamlined towards maximum contribution to overall business objectives.

1. Introduction

Today worldwide competition between manufacturing enterprises is increasing significantly. Facing these challenges a manufacturing business can improve its competitive position through integration. Many organisations, large and small, are including within their strategic objectives specific goals focusing in achieving business excellence through integration.

In this context integration is defined as:-

> "*All functions of a business working together to achieve a common goal, using clearly defined objectives, benchmarks, disciplines, controls and systems in a flexible, efficient and effective way to maximize value added and to minimize waste.*"

The literature in this area indicates that the background to business integration is technology. The availability of technology has driven the integration effort which led to the concept of *Information Integrated Enterprise*. At present there are a number of tools, techniques and methods available to facilitate the integration of information systems within an organisation. These include I.CAM methodology [1], Information Engineering Methodology [2], GRAI Methodology [3], Strathclyde Integration Method [4]. On the other hand there is considerable emphasis being placed on streamlining of business activities for improvement of business performance. Work in this area focuses on the business needs and order winning criteria. The developments in Total Quality Management (TQM) concepts and principles also provide some guidelines, methodologies and frameworks for creating total harmony within a business.

It is now clear that what ever direction we approach from, Information Systems, Business Improvement or TQM, the overall objective is to achieve an integrated business platform.

It is also clear that technology is no longer the constraint for achieving business integration but methodologies are required which are independent from the technology and combine the rigour of the information systems oriented methods with the objectivity of the business oriented concepts. This conclusion is strongly supported by Self [5] and Hodgson & Waterlaw [6].

2. Total-I: A Methodology for Business Integration

With the above background in mind the Total-I methodology was developed and refined through a number of industrial applications. The methodology recognises that people are the most important aspect of any organisation, and integration as defined earlier can not be achieved unless people at all levels of an organisation are aware and committed. The methodology was developed with two objectives in mind, these were:-

-       to provide a structured process towards business wide integration.

-       to facilitate business wide integration through the provision of simple tools and techniques to be used by the people at all levels.

The Total-I methodology consists of five major phases and an optional phase for specification and selection of software modules. These six phases provide the basic structure behind the methodology - figure 1. Through out the core five phases the methodology encourages the development of an integrated set of performance measures (figure 2) which ensure that all parts of the business work towards common objectives.

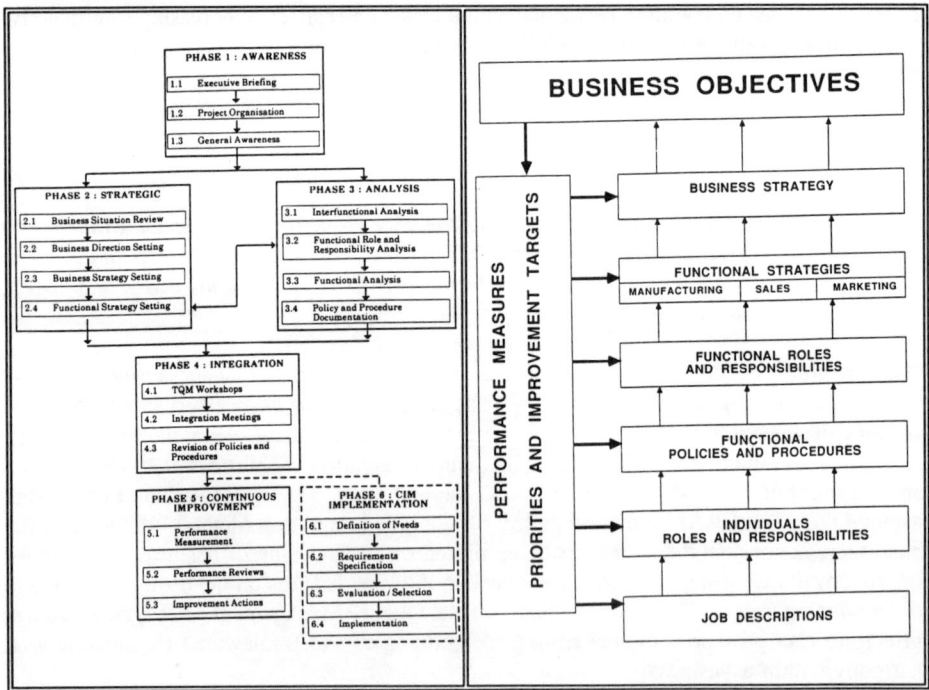

Figure 1. Structure of the Total-I methodology.   Figure 2. Integrated performance measures.

The *Awareness Phase* is designed to provide a focus at the start of the programme for application of the methodology. The methodology requires the involvement of people at all levels of an organisation. This phase ensures awareness of the programme and the methodology as well as establishing a structure for project management and setting out the critical mile stones for the programme.

The *Strategic Phase* is designed to ensure further senior management commitment and involvement. During this phase, through senior management workshops, the current situation of the business is reviewed, a realistic vision is set for the business and a business strategy is established to progress the business from its current position towards the business vision. At this stage, business level (strategic) performance measures are also identified to support this strategy.

The *Analysis Phase* provides a structured and a rigorous approach for analysis of the business processes and documentation of the current practice at operational levels. This is achieved through the involvement of people at operational levels. A modified/simplified version of the IDEF0 techniques is provided to facilitate the analysis. This is combined with some of the TQM techniques to facilitate the identification of the appropriate performance measures for each process. Although this is a separate phase from the Strategic phase. The interface points are clearly defined to ensure that the Strategic objectives and the operational objectives are common and complimentary.

The *Integration Phase* provides a structured environment and techniques to facilitate the integration and streamlining of all business activities through the operational personnel. This phase is driven by the strategic objectives of the business and the results of the analysis phase. Specific techniques are provided to ensure that the strategic objectives are correctly deployed at lower levels. This is achieved through the integration of Strategic and Operational performance measures. On completion of this phase an integrated business platform is achieved.

The *Continuous Improvement Phase* takes account of the organisational dynamics and recognises that the state of integration can be fragile unless it is maintained. The purpose of this phase is to install necessary mechanisms and disciplines to avoid disintegration over time and to effect further business improvements. The key element of this phase is the regular use of the integrated performance measures as a management tool.

The *CIM Implementation Subsystem* provides a sub-methodology for identification and implementation of appropriate software modules. This sub-methodology is based on the output from the integration phase and makes use of the streamlined, integrated and documented operational policies and procedures by converting these into a detailed statement of requirements.

3. Case Study Results

The methodology has been used in a number of industrial assignments. In each case the difficulty encountered was that the agreed performance measures, both strategic and operational, conflicted with the bonus criteria of some of the managers and directors. This required renegotiation of the incentive schemes to ensure that these fell in line with the strategic and functional performance measures.

To date all applications resulted in measurable improvements in business performance. In one specific case the following results were achieved:

- Increased inventory turns from 3 to 4.8   - Reduced scrap from £271K to £77K
- Reduced work-in-progress levels by 25%   - Improved internal quality by 5%
- Improved production throughput by 12% to 89.2%.

4. Concluding Remarks

In all applications the key for success was the commitment of the senior management and full participation of the people involved in the day to day management and execution of the business processes. In addition to specific benefits as illustrated above, the more general benefits include:-

- Focused and measurable business and functional strategies.
- Elimination of ambiguity and clarification of accountability.
- Improved communications and conformance to requirements.
- Proactive management style through monitoring of performance objectives.
- Relatively smoother implementation of changes due greater business awareness.
- Improved relationships between individuals and functions.
- The installation of a continuous improvement culture and increased motivation.
- Improved customer (internal and external) awareness by all personnel.

The application of the Total-I Methodology should not be seen as a stand alone project with a definite beginning and ending. It should be regarded as the start of a process of continuous change and improvement which will extend beyond the period of the project. The methodology provides the framework and facilitates business integration and performance improvement through people, systems and technology. The Total-I methodology is supported by a comprehensive Tool-kit and implementation guide which takes the user through the methodology step by step and provides useful hints on the use of the tool-kit.

References

[1]   CAM-I Architects Manual: ICAM Definition Method, Computer Integrated Manufacturing International Inc., Texas, USA.

[2]   W H Inmon, Information Engineering for the Practitioner: Putting Theory in to Practice, Prentice Hall, 1988.

[3]   G Doumeingts, GRAI Approach to Designing and Controlling Advanced Manufacturing Systems in a CIM Environment, Advanced Information Technology for Industrial Material Flow Systems, NATO ASI Series, Volume 53, Springer Verlag, 1989.

[4]   A S Carrie and U S Bititci, Tools for Integrated Manufacture, Proc. of MATADOR'90, 1990.

[5]   A Self, Factory 2000 - A Response to Order Winning Criteria, Computing and Control Engineering Journal, March 1992.

[6]   A Hodgson and G Waterlow, Integrating the Engineering and Production Functions in a Manufacturing Company, Computing and Control Engineering Journal, March 1992.

Advances in Agile Manufacturing
P.T. Kidd and W. Karwowski (Eds.)
IOS Press, 1994

Approaches to group working in two UK manufacturing firms: a comparative study

K.S. Ball* and C Baber [†]

* University of Bradford Management Centre, Emm Lane, Bradford, BD9 4JL

[†] Industrial Ergonomics Group, University of Birmingham, Birmingham, B15 2TT

Abstract: In this paper we consider group working in two very different U.K. firms. The firms appear to be polar opposites in terms of modern management and organisation theory, and we note how this difference in context accounts for ways in which each firm approaches group working. However, perhaps of most interest is the way in which the influence exerted by specific interest groups seems to have radical effects on the organisation and running of work groups. The development of group working is both helped and hindered as different groups shape the process, and. indeed by the very definition of "group" which the organisation adopts.

1. Introduction

Group working can facilitate a blend of knowledge, skills and abilities which can render it a productive and useful method of work organisation. Working in a group can also provide social and psychological support for its members (French, Caplan and Harrison, 1982). Campion and Medsher (1991) note that the relative performance of group working is highly dependent upon its organisational context (other factors, such as composition, structure, reward systems and task, seem to have a subsidary role in the group's success). There are, however, few published studies which specifically address the influence of contextual factors on group working.

In this paper, we identify contextual factors specifically associated with worker autonomy and skill which not only governed the composition and operation of the groups, but also shaped managerial perceptions of the groups' success or failure on these factors.

The desire for flexibility in modern firms can be associated with three perspectives on job design. These have been observed in many organisations, and, we submit, represent current practice:

(i) The humanist perspective is illustrated by Volvo's (now redundant) plants at Uddevalla and Kalmar. This can be characterised by high social contact, shared responsibility for quality and maintenance, autonomy, self allocation of tasks and encourages multiskilling;

(ii) The organisational choice perspective refers to the "empowerment" ideas commonly seen in Total Quality Management (TQM) and, more recently, Business Process Re-engineering. This approach encourages participation, although it might still refer to supervisory control and need not use multiskilling so much as multitasking (job enlargement rather than job enrichment);

(iii) The technological determinist perspective tends to allow litte social contact, no autonomy or devolved responsibility, little opportunity for multiskillng, although multitasking might be encouraged in manufacturing cells.

We also note differences in the definition of the word "group", depending on what one sees as being grouped: people, processes, technology or jobs. There is much scope for confusion and ambiguity in the use of the word 'group'.

2. The Study

Two U.K. manufacturing firms were studied over a three month period. Firm A was a small, young and growing automotive manufacturer with under 100 employees. Its production system was vertically integrated on site; the organisation's hierarchy was flat; management had low functional specialisation; the plant was not unionised (i.e. industrial relations were unitary and commitment based). Firm B was a large, mature firm with nearly 2000 employees. Its production system was partially integrated on site; the organisation's hierarchy had many levels and management had high functional specialisation. It was bureaucratic in nature, unionised and as such industrial relations were pluralist and control based. It was thus anticipated that, even though both firms asserted that they were using group working the actual outcomes of the process would be very different in terms of work organisation.

Data were gathered using semi structured interviews with production managers, and observation of the groups at work. It shouldbe noted that the production managers views are mediated to some extent by the relative power positions which they held in the organisation.

3. Findings

Firms A and B were very different in their approaches to group working. Firm A saw groups as the only way to produce effectively. Innovation was given a high priority and the use of production, consultation and quality groups was seen as best facilitating this.

Thus, in group A, we saw evidence of grouping based on process and production methods, on quality and consultation, and, to a lesser extent, on technology. Throughout these groupings, there is an opportunity for communication and consultation (although the avowed philosophy of 'permanent change' may have diluted the effective use of these channels).

Firm B, mainly because of its bureaucratic nature, viewed group working as merely experimental, and was reluctant to make any concessions to the production managers in respect of the groups' development. The current implementation of the 'group' is as a separate entity, partitioned from the rest of the factory floor, and with work flow designed in terms of process grouping. Consequently, the people working in the "group" have no opportunity for input to either the design or the working, and the 'payment by standard time' regime severely curtailed the possibility of 'group-working' (although ongoing research has produced changes in the design and in aiming towards more communication and group-working).

The 'groups' did exhibit some interesting similarities, with interest groups in each firm attempting to influence the work organisation. Firm A had attempted to adopt a Japanese approach to production. The result was a macrolevel set of organisational values and a structure that facilitated group working, but job designs with little evidence of multiskilling and autonomy. A commitment to continual change under a dominant managing director tended to limit to opportunity for participation and to restrict the success of communication.The work was line paced (albeit on a slow moving line) so the job design was still essentially determined by the technology, despite the highly skilled nature of the workforce. In this respect, firm A seemed to lie somewhere between the organisational choice and technological deterministic approaches, despite avowed aims of a humanistic group working regime; aims which were given lip service by many of the people to whom we spoke.

Firm B's 'group' can be viewed as an example of technological determinism. It was also seen, by other organisational departments, as 'politically motivated'. Consequently the other departments attempted to stunt the success of the group working programme by refusing to develop a group-based incentive scheme as an alternative to the current 'payment by standard time' system, and refusing to see work in progress as a cost, so rendering the programme unjustifiable on economic grounds. The causes of this opposition were bureaucratic, and resulted in modification of job and workspace design.

Thus, group working was seen as political because it was implemented outside the bureaucracy, and was tackling the bureaucracy's inflexibility through the use of conflict, and the manipulation of certain bureaucratic control tools (i.e. finance, wages and health and safety regulations). The tasks themselves were self paced, but had not changed from the line task. The only transformation was in the physical layout of the work.

The firms' respective definitions of "groups" differed vastly as a result of this. Firm A saw group work from a macro-perspective, with the organisation being the "group" frame of reference. Firm B saw the group as being more like a production cell - on a very micro level. Both firms recognised the innovative nature of group working, and both had very different ways of dealing with it. Control was still high on the agenda, but for firm A, who had instilled certain values in the workforce control was covert and internalised, i.e., via 'commitment', 'quality' and 'responsibility'. Firm B's cell system was characterised by a central surveillance point for the foremen, and other control mechanisms, such as the payment system. Current developments at firm B are based on 'just-in time', where control can be exercised through the work flow.

4. Conclusions

A number of lessons can be learnt from our observations. First, that the context of group working is vital to its success. It is significant that little evidence of autonomy and multiskilling was found. The preoccupation with the macro-environment prevented both firms from realising the potential of their workforces. Changes to group working must be planned and negotiated across the organisation, and taken seriously in terms of time and money if they are to succeed.

5. References

Campion, M A and Medsher, G J (1991) Job Design. In: Salvendy, G *Handbook of Industrial Engineering*. Wiley, New York.

French, J R P, Caplan, R D and Harrison, R V (1982) *The Mechanisms of Job Stress and Strain*. John Wiley and Sons Ltd, Chichester.

Janis, I (1972) *Victims of Groupthink*   Houghton Miffin, Boston

Advances in Agile Manufacturing
P.T. Kidd and W. Karowski (Eds.)
IOS Press, 1994

657

Work Organization in Swedish Industry: From Semi-Autonomous Groups to an Obscure Variety of Teams

Claudius H. RIEGLER

The Swedish Center for Working Life, Box 12670, S - 112 93 Stockholm, Sweden

Abstract: As a starting point, a short recapitulation is given of the literature on the development of collective forms of work organization in Swedish industry from the internationally well-known semi-autonomous groups in assembly and press line work in the car industry to team work in other branches as e.g. the electrical engineering and electronics industry. In the second part trends towards organizational renewal in the electrical engineering industry are presented. The third part describes the current situation of establishing team work and related vocational training in two big Swedish (multinational) companies by starting development projects that are partly financed by the Swedish Work Life Fund. Some consequences of the obscure variety of team organization for the future of co-operative work in industry are pointed out.

1. Swedish group work tradition in engineering industry

According to a trade union report on the position of white-collar employees in the ongoing transition of companies and work, "no survey has been conducted to establish how common teams or production groups are in the engineering industry in Sweden today" [8]. Inquiries in one province have shown that 70% of all the companies with more than 7 white-collar employees have introduced "some form of team system for parts or all of the metal trades". In a recently published sociological dissertation on group work in Swedish industry the author comes to the conclusion that "group-oriented work organization seems to be more and more common in industrial companies. It is, however, important to distinguish between different forms of group working, job rotation and control systems. The group-oriented organization can be an element of new management strategies and can also partly be a result of efforts made by employees and their unions" [7].

This ambiguity characterizes both traditional and newer concepts of co-operation in industrial work. But we should even make a distinction between concept and reality of group work in the auto industry and concept and reality of team work in other branches. An elder Swedish group work tradition is characterized by the establishment of permanent functional groups in greenfield plants where members have to co-operate to fulfill their tasks. The groups usually execute all tasks in a specific unit of production and not only in assembly. To varying extents, maintenance, quality control and material supply tasks are assigned to the groups. Their degree of autonomy is high. Supervisors are not members of the groups – the members themselves (but normally not all) become group speakers in a rotation system. Qualificational bottlenecks often originate from fluctuations in group membership. Therefore management has an interest in a stable workforce. In some cases, production workers display resentment towards additional and often strenuous training programmes [1].

Newer concepts of work organization in e.g. the electrical engineering and electronics industry are modifications of this tradition. Team work even here means that management tries to establish functional units, but these have more similarities with the team concept in the American car industry. Flexibility as a central goal means that employees – often by training programmes – must acquire qualifications making them able to rotate between more and more complicated work tasks, not to co-operate within a frame of relative autonomy.

Vocational training in the traditional group work organization often was dominated by collective learning by doing. Vocational training in the other case is separated into education and manufacturing. Even the production technique requires this school-type separation. The companies are arranging ambitious training programmes directed towards the individual employee, mainly blue-collar employees.

Group work has spread through Swedish auto industry since the early 70's. But in recent development work in car assembly, group work doesn't have the same high priority [5, 6]. Team work in the 90's is more common in other branches as the petro-chemical, chemical-technical and food industry [7] – and the electrical engineering and electronics industry [3, 4, 8].

2. Trends toward organizational renewal in the electrical engineering and electronics industry in Sweden

The electrical and electronics industry is an extremely heterogeneous industry where the work tasks range from research and development and skilled craft work to extremely taylorized assembly work. There is a polarization of occupational and skill profiles: the semi- and unskilled jobs on the assembly lines are becoming less in number whereas jobs which demand higher skills are increasing. There are important changes in organization and production concepts. Shorter technological product cycles are required in a situation with growing international market competition. This is why the goal for companies is to increase flexibility and product quality. The whole branch is in search for concepts and practices of flexible and "lean" overall organization (divisionalization, decentralization). The progressive implementation of new technology is partly followed by a restructuring of the labor force as indicated above. Blue-collar employees often get vocational training and take over work tasks for which formerly only the salaried employees were qualified. Management strategies aim at increasing the motivation and commitment of the employees. This results in evolving patterns of balanced interests and arrangements moving towards micro-corporatist modernization-alliances, aiming at consensual employee involvement and participation. Participation is an important and by management accepted precondition. Especially the blue-collar union with its concept of the "Good Work" elaborated in 1985 is a driving force behind organizational change, e.g. towards team work with upgraded work tasks partly transferred from white-collar employees – under the precondition that blue-collar employees can get solid and permanent vocational training in periods of massive reorganization.

3. Recent development projects in the Swedish electronics industry with team work as a crucial point

Our empirical material primarily consists of the evaluation of complex development projects in 7 Swedish plants in the electrical engineering and electronics industry (4 in ABB, 3 in Ericsson) which are partly financed by the Work Life Fund. The mandate of this Fund - financed by a special work environment tax (1,5% of the wage sum) in 1989/90 - is to create development in working life, with particular reference to reduction of absenteeism and improvement of rehabilitation processes, but also more general work environment improvements. With a five year running time, the Fund was established as a federal programme and is expected to cease its operations by the summer of 1994. In the meantime it will have generated more than 15 000 workplace development projects by spending 10 billion SEK and is in this respect one of the largest systematic efforts to change the working life in Sweden.

Example 1: An ABB plant (production of electrical transformators and inductors, 100 employees, situated in a small town in central Sweden)

In 1989 an enterprise-wide project was launched which aimed at reducing throughput times, mainly by getting rid of disturbances and delays. The initiative came from top management. Through the work of this project the need to create a broader and more deep going process of change became recognized. When the Work Life Fund was established, a so-called workplace programme was formed, building on the previous effort but putting more emphasis on health and safety issues. After this programme received support and was launched, the T 50 programme appeared, providing still further impulses to the efforts within this company. What came out of this combination of initiatives was a programme with the following main content. The first effort had implied some steps in a transformation from individual work-roles towards group work in the form of so-called flow groups. In the present effort this is carried further and the whole enterprise is restructured into groups. Along with this there has been a change of the tools and machine layout in the factory to improve on the conditions for group work. Major investments have also been done in training. The changes do not stop with the introduction of groups with a certain degree of autonomy in relation to production tasks. They continue in the form of a radical decentralization where production planning and several other former management tasks are allocated to the groups step by step. A new wage system is furthermore being developed and the administrative and managerial structure of the enterprise is being changed.

Example 2: An Ericsson plant (assembly of printed circuits and telephones, 1 000 employees, situated outside a medium-sized town in the south of Sweden)

Because of high rates of work load injuries and work accidents, one work group and later on four "vision groups" within this company tried to develop a programme of measures to avoid further problems. Through contacts with the ABB T 50 programme management and the unions realized that they had to make deeper transformations to cope with the situation. The Work Life Fund was contacted and discussed elements of an integrated so-called workplace programme in the sectors vocational training, wage system, production flows, work content, and work environment. These parts were implemented in parallel steps and led to a new work organization based on teams. By vocational training new responsibilities were taken over by employees. New tasks now are delegated to the team members who are even coordinate – through the elected (alternating) group speaker – the overall group effort. As a result work content and the psycho-social work climate at the workplace have improved, especially in the production teams.

A movement from individually defined work-roles towards group oriented work was carried out in some other plants within these two leading export companies. In Ericsson a major work organization project was carried out in a plant in Stockholm where a group work organization with cost centers led to an integration of white-collar employees work tasks in the production teams [8]. In another plant within the Telecom division in a small town, product work shops with flow groups were established and the planning and construction staffs were allocated near the production units. The Metalworkers' union is, despite the "Japanese" trends in management's reform work, after disputes positive to these changes because of the large amount of vocational training which almost all workers can get [3].

At two other plants within the Telecom division, production groups were established already in the 80's as an answer to a bad work environment and/or high absenteeism and turnover. They were combined with Japanese-inspired kanban-methods. On the island of Gotland, one plant – with financial aid from the regional Work Life Fund – established experimental teams at a small scale and provided much vocational training (a minimum 100 hours per blue-collar employee). Diffusion effects led to the establishment of new experimental teams in another production unit within the plant.

In a plant in a small industrial town in central Sweden with a majority of female workers (775 out of 1 000), production teams were built up in a so-called workplace programme partly financed by the Work Life Fund. The members decide who in the group is to be responsible and on training needs. Meanwhile white-collar tasks have been integrated in the teams which now have bigger responsibilities and rights and are more conscious of the importance of product quality.

In several companies in ABB production groups were established where work enlargement and vocational training for blue-collar workers is a main topic of development. Operators in teams of 7–10 persons assemble and test robotics, and the wage system is based on individual performance and ability to work in the group. Even in the switch gear production in a small town in central Sweden, the electrical apparatus assembly in a town on the east coast, and a retail center in another town in the province, team concepts often are solutions for deep organization and productivity problems. The unions – mainly the Metalworkers' union representing nearly all blue-collar employees – support these reforms as they offer an occasion to get substantial training in the company.

Many of the frequently used training programmes are arenas for participation in the sense that they are organized and run locally, often applying a dialogue oriented pedagogics and geared to meet the specific requirements of the concrete workplace programme. There are, of course, also more conventional training programmes to be found, including packages bought from outside to train people in more general themes, such as computer literacy.

The Work Life Fund has come to initiate a transition into concept driven change, often implying a movement from individually defined work-roles and into group oriented work organization. In a test survey of 300 companies made by the Statistical Bureau/SCB, group work is a prominent means of reorganizing the production process. In 25% of the cases people in project work were engaged more than half of the their time in introducing team work, and in 27% of the cases teams were introduced for more than half of the personnel after starting a workplace programme partly financed and thoroughly discussed with the Work Life Fund.

An element of transition is found in the ABB company case where the first development towards flow groups emerged out of productivity considerations rather than health and safey ones. Generally, however, the introduction of group organization on the shop floor has not been accompanied by more far-reaching changes upwards in the organization, in terms of decentralization of planning functions, of customer relations, and so on, creating a need for further development which the Work Life Fund has to bring about to some extent.

Group or team work in nearly all cases means that the demands from the production process promote the re-organization of work by teams of varying shape. Even the most fundamental work tasks require training which is offered mainly to blue-collar employees. This training often has very little to do with real qualification for co-operation. On the other side there are tendencies towards highly qualified vocational training given to "more or less independent teams with objective-oriented management" [8]. Training in these cases is used to integrate e.g. planning, data processing, but even programming in the work tasks of blue-collar employees. In the future, the proportion of traditional white-collar duties transferred to teams of shop floor operators will increase. This type of team work promotes the establishment of "grey-collar workshops" [8] which could lead to new forms of co-operative work in industry.

References

[1] P. Auer and C. H. Riegler, Post-Taylorism: The Enterprise as a Place of Learning Organizational Change. A Comprehensive Study on Work Organization Changes and its Context at VOLVO. ISBN: 91 87460 36 X. The Swedish Work Environment Fund and WZB, Stockholm and Berlin, 1990.

[2] P. Auer and C. H. Riegler, The Swedish Version of Group Work – The Future Model of Work Organization in the Engineering Sector? *Economic and Industrial Democracy* Vol. 11 No. 2 (1990) 291–299.

[3] G. Brulin, Från den 'svenska modellen' till företagskorporatism? Facket och den nya företagsledningsstrategin. ISBN 91 7924 0445. Arkiv avhandlingsserie 31, Lund, 1991.

[4] G. Brulin and A. Victorin, Improving the Quality of Working Life: The Swedish Model. In: OECD, New Directions in Work Organisation. The Industrial Response. ISBN 92 64 13667 3. Organisation for Economic Co-Operation and Development, Paris, 1992, pp. 149–165.

[5] K. Ellegård, T. Engström and L. Nilsson, Reforming Industrial Work – Principles and Realities in the Planning of Volvo's Car Assembly Plant in Uddevalla. ISBN 91 87460 53 X. The Swedish Work Environment Fund, Stockholm, 1991.

[6] K. Ellegård, T. Engström, B. Johansson, L. Nilsson and L. Medbo, Reflexiv produktion. Industriell verksamhet i förändring. ISBN 91 8761 604 1. AB Volvo, 1992.

[7] P. Sederblad, Arbetsorganisation och grupper. Studier av svenska industriföretag. ISBN 91 7966 250 1. Lund University Press, Lund, 1993.

[8] Swedish Union of Clerical and Technical Employees in Industry (SIF), White-Collar Employees Close to the Production Process. Companies and Work in Transition. International Metalworkers' Federation, Geneva, 1993.

Advances in Agile Manufacturing
P.T. Kidd and W. Karwowski (Eds.)
IOS Press, 1994

Weak Institutions And Strong Organizations ?

Dr. Hans-Joachim Braczyk and Dr. Gerd Schienstock
Center of Technology Assessment
Nobelstr. 15, 70569 Stuttgart, Germany
Phone: +49 - 711 - 6783-161, Fax -299

1. In recent debates on industrial paradigms, experts in organizational research increasingly paid attention to the institutional environment of organizations. Some are of the opinion that relevant institutions, such as industrial relations and the vocational training system, are essentially determined by national, cultural, and regional traditions. To a certain extent, those institutions are important or even indispensable for the operation and survival of organisations; they restrict and influence the development of organizational structures and work organization. Furthermore, institutions are interdependently linked to the institutional environment of organizations. Thus, transfer to and adaptation of a new industrial paradigm by organizations in a particular region appear to be strongly dependent on the institutions' compatibility to the paradigm in question. Regarding this, one could speak of the strength of institutions. Empirical evidence, however, leads us to just an opposite conclusion. We can observe considerable institutional weaknesses. Some implications for human centered organization models will be discussed in the following.

2. There is still a growing interest in so-called human-centered production concepts. Amongst others, the Swedish model (Berggren 1991), the so called new production model (Kern/Schumann 1984), and - most popular - the model of lean production (Womack et.al. 1991) have become increasingly attractive when firms strive for improving their competitiveness. There are several other concepts under discussion. The socio-technical approach (Eijnatten 1993) and the anthropocentric concept (Brödner 1985) are only two of several attempts to find convincing and promising alternatives to the presently dominating Tayloristic model. The Tayloristic paradigm appears to be significantly associated with technology-based production strategies.

Despite the fact that human-centered production concepts differ considerably concerning important views, they do have some aspects in common. First, these concepts, whether already realized or just a vision of the future, derive from fundamental criticism on the general Tayloristic or Fordistic production model. It has been assumed that under Tayloristic and Fordistic regimes in particular blue collar workers are restrained from using their skills by bureaucratic procedures, hierarchical structures, and institutionalized distrust. Second, there is the thesis that given skills of labor could be exploited much more efficiently, if firms could get rid of Fordistic structures and procedures and introduce human-centered production concepts instead. Third, under the regime of human-centered production concepts, the innovative capacity of firms would increase considerably. Fourth, technology serves as a tool; i.e. it supports human skills and qualifications. Neither does technology -as in the Fordistic model- contribute to a process of constantly lowering skill requirements, nor does it lead to job insecurity especially of blue collar workers. Fifth, - last but not least - human-centered production concepts allow full deployment of workers knowledge. It Therefore it is not surprising that not only managers and worker's representatives but also members of employers' associations and trade unions as well as politicians and social scientists are attracted by the prospects of human-centered production concepts and are fascinated by the idea of simultaneously meeting both the goal of economic efficiency and the need for higher standards of working life. However representatives of human centered production models make different and even divergent assumptions concerning the significance of the societal and institutional framework of organizations. This holds especially for the industrial relations system and the system of vocational training.

Some claim a strong and close relation between organizations and their institutional environment, the later being regarded as culturally formed and rooted in a national or regional context. Two consequences follow from this idea: first, the institutions relevant for economic organizations such as industrial relation system and vocational training system are strictly tailored for the specific needs and expectations that   derive from the organizations' commitment to a particular industrial paradigm. Second, in a given institutional environment only those production concepts can be realized which are well compatible with regulations and output of these institutions.

Others share the view that organizations necessarily depend on a strong and close relationship with their institutional environment. However, organizational changes are believed to be possible as soon as an appropriate institutional transformation takes place at the same time. This means that institutions have only weak roots in a certain country's or region's cultural soil. This implies, for example, that the adoption of the Japanese lean production concept by German firms is considered to be tied in with an institutional renewal, meaning that industrial relations and vocational training systems are to change into Japanese-like institutions.

A third view, too, confirms the idea of an interdependency between organizations and their institutioal environment. But it differs in so far as it assumes a rather weak and loose relationship. Therefore it is possible that institutions and organizations of various types and patterns can exist together. In other words: neither does a production model demand specific institutional arrangements nor does an existing institutional environment essentially prevent the organizations' change from one industrial paradigm to another. However, the relationship between organization and the institutional environment cannot be taken as being arbitrary; a certain affinity still exists.

3. In recent publications dealing with rationalization processes rather serious inconsistancies exist. On one hand, experts insist that for German firms, to bring their organization into line with the structure of the Japanese lean production model is almost impossible due to the specific characteristics of the culturally formed institutional arrangements in this country and their rigidity. On the other hand empirical results of various studies convincingly reveal changes in organizations as well as institutions. Our own research work on the adaptation process of the lean production concept Baden-Württemberg´s leading firms currently are involved in, confirms these empirical results. Simultaneously with an enforced introduction of the lean production model on firm level an institutional change within the industrial relation system has started and it is becoming more and more visible now.We also found that together with organizational restructuring the famous German vocational training system is coming under increasing pressure. This may justify to speak of the strength of organizations, as they play the most dynamic part in the transformation process, and we may also speak of rather weak institutions.

The German example illustrates that focal organizations do not really have a free choice concerning the remodeling of their principles for work organization. They still remain embedded in an institutional environment which itself is subject to changes. In the transformation process, neither organisation structures nor institutional arrangements can be granted causal priority. Instead, we have to assume a common co-evolution and mutual adaption. The mechanism here, triggering off the process of change, is the globalization of the market and a new world wide competition.

If we are to accept this argument, we have to be rather reluctant concerning recommendations of restructuring the production process in line with the idea of a human centered model. Both, organisational concepts and the institutional environment, do not only have to be related to each other but to meet the demands of global competition as well. Therefore, to conceive of a human centered production concept as been the one best way model can turn out to be a mistake; and in the end we are rather assuming wishful thinking than contributing to a realistic option for the actors concerned.

Advances in Agile Manufacturing
P.T. Kidd and W. Karwowski (Eds.)
IOS Press, 1994

Swedish experiences of
working in groups in industry

Annika Lantz
Division of Social and Organizational Psychology
National Institutet of Occupational Health
171 84 Solna, Sweden

Abstract. This paper provides a picture of the current state of knowledge of the effects of working-in-groups in Swedish industry. One central issue addressed by the review concerns the nature of what has been investigated in relation to the group form of working. One principal finding is that the majority of the studies focus on descriptions of job tasks, production technique and work-organizational aspects. Less attention, from a social-psychological perspective, has been paid to other important factors which influence the social interaction within work groups. How seperate frame factors affect social interaction and what characterizes this interaction is not consiedered systematically in empirical research on working-in-groups in industry. The results point to the need for methodological development and to the need for a crossdisciplinary approach to future research.

1. Background and aims

Swedish industry since the early 70s has promoted the development of new production concepts, not only for raising productivity but also for creating richer work content. Extensive research had demonstrated, among other things, the negative consequences of fragmented work tasks for health and wellbeing. Substantial resources have been put into the creation and study of new forms of work, often in conjunction with technical development, in which working-in-groups (task-oriented groups) has been an important component. The task-oriented group, as a means in the work process, would more easily satisfy the individual's psychological demands, while at the same time being an effective means for the implementation of the work. Despite the great attention paid to the new production concepts, both in Sweden and abroad, we know very little about the effects of their application. The aim of the current work is to provide a picture of the current state of knowledge from a social-psychological perspective. The results presented here, stem from a systematic review of the relevant literature that has appeared form 1970 onwards, which has treated working-in groups, regardless of how this phenomenon has been analyzed (1).

One central issue addressed by the review concerns the nature of what has been investigated in relation to the group form of working. In order to be able to systematize the variables/dimensions involved we take our point of departure in a model which treats the work group as a unit operating within a given organizational frame. The frame sets certain limits on and offers certain scope for the interaction that develops within the group, which in turn influences the work situation and conditions faced by the individual. In crude and simplified terms, two types of frame influences on the work of the group can be distinguished:

1. Factors that describe the organization solely in terms of its operative concept, goal, type of production (of goods or services), i.e. the overall task of the organization.
2. Factors that describe the organizational structure that has been built up to accomplish this task, e.g. the principles for the division and set-up of work, and the coordination of activities within the organization. Structural frame conditions are crudely placed on three dimensions: production technique, work organization and job tasks.

The effects of working in groups are understood as the result of interaction between frame conditions that describe job tasks, production technique, group composition and work organization; and further, the interaction that develops within the group.

2. What effects of working in groups have been demonstrated?

One way of discussing what we know about the effects of working-in-groups in industry is to relate the investigated variables and the study design to the conclusions drawn (table 1).

The bulk of the studies focused on descriptions of job tasks, production technique and aspects of the work organization. A few examined intra-group interaction empirically. Descriptions of the composition of groups generally consist in background information, and do not provide a basis for further analysis. We have not been able to find any research that empirically investigated how the overall task of the organization, i e operative concepts, goals etc influence the social interaction within work-groups.

Thus, if we study how conclusions on the effects of working-in-groups have been drawn, it is possible to separate out a series of comparative studies where the work organization in a common form of production set-up is compared with the group form of working. In these studies, conclusions on outcomes are drawn in terms of the interaction between production set-up and work organization . The results are not in unison, but suggest that working-in-groups has advantages over common forms of production. The investigations do not include a study design which makes it possible to examine the interaction effects between production set-up, work organization and group-processes; nor do they draw conclusions on which of the dimensions make the greatest contribution to the outcome.

Table 1. Conclusions on the effects of working-in-groups and the dimensions investigated (grouped according to the theoretical model).

Outcome variable	Dimensions studied (frame conditions and/or interaction)	Conclusions on the impact of working-in-groups
Direct effects on the organization		
Effectiveness measured in terms of the company's sick-leave statistics.	Not specified.	Social contact within the group leads to reduced sickness absenteeism.
Effectiveness measured in terms of the company's sick-leave statistics.	Production set-up. Intra-group interaction.	There are no differences in absenteeism which can be related to working-in-groups.
Effectiveness measured as labour productivity.	Description of the work organization provides the basis for the selection of cases.	There are no differences between line production and working-in-groups.
Productivity measured as production costs.	Description of the work organization provides the basis for the selection of cases.	Working-in-groups has advantages over common line production.
Productivity measured as production costs.	Description of the work organization provides the basis for the selection of cases.	No differences between common line production and working-in-groups.
Quality measured in terms of finished products.	Production set-up. Intra-group interaction.	Working-in-groups has advantages over common line production.
Quality measured in terms of finished product	Descriptions of the production set-up and work organization provide the basis for the selection of cases.	No differences between common line production and working-in-groups

Effects on the individual

Meaningfulness of work measured in terms of the individual's perception	Job tasks. Production set-up. Work organization.	Working-in-groups allows job rotation, greater task variety and longer cycles, which are regarded as favourable, but the work is still perceived as repetitive.
Opportunities for skills development.	Job tasks. Production set-up. Work organization.	Working-in-groups creates greater opportunities for skills development than a common production set-up.
Skills development measured in terms of individual competence.	Not specified.	Working-in-groups leads to increased skills development.
Skills development measured in terms of individual competence.	Work organization. Production set-up.	The opportunities offered by working-in-groups for increased skills development can be utilized to a greater or lesser extent depending on the nature of the frame conditions.
Physical workload measured as conditions in the physical work environment.	Job tasks. Production set-up. Work organization.	Via job rotation, working-in-groups leads in the longer term to a reduction in physical-strain injuries.
Physical workload measured as occupational injuries and diseases.	Description of the production set-up and work organization provides the basis for the selection of cases.	Difficult to determine whether working-in-groups has more favourable effects than a common work set-up because of methodological problems.
Physical workload measured in terms of the individual's perceptions and the researcher's judgement.	Job tasks. Production set-up. Work organization.	Working-in-groups does not reduce physical workload.
Job satisfaction measured as the individual's perception of work as a whole.	Overall judgement on intra-group interaction, job tasks, production set-up and work organization.	Working-in-groups leads to increased job satisfaction, primarily by providing greater social contact.
Influence measured as trade-union activities and individuals' judgements.	Production set-up. Work organization. Intra-group interaction.	Working-in-groups promotes increased trade-union activity and greater influence over the work.

A second group of studies draws conclusions on the conditions under which working-in-groups will have certain effects and also considers its potentials, but these (for a variety of reasons) are not investigated empirically. Example: In that the group form of working permits job rotation, the work becomes less monotonous and more varied — which, in the longer term, leads to increased job satisfaction.

A third group of studies draws conclusions on the effects of group-processes without investigating intra-group social interaction. The outcomes discovered are attributed to the organizational form, but it is unclear whether such conclusions are valid. An example: A reduction in absence through sickness is related to working-in-groups having led to "positive group pressure", without there being any investigation as to whether such "pressure" actually exists.

3. Discussion

One principal finding of our review is that the majority of the studies focus on descriptions of job tasks, production technique and work-organizational aspects. Less attention, from a social-psychological perspective, has been paid to other important factors which influence the interaction within work groups. This applies to both the studies that describe the characteristics of the group form of working and those designed to draw conclusions on its effects. How separate frame conditions affect interaction and what characterizes this interaction is not considered systematically in empirical research on working-in-groups in industry. The few, and rather restricted, studies in which an empirical approach is adopted describe interaction principally in terms of socio-emotional processes; the data tend to consist of responses to questions (in questionnaires or interviews) on how interaction is perceived.

The composition of the group is defined as an aggregate of individuals with certain attributes. The alternative way of defining the group, in terms of a system of functional relations, has not had any impact on empirical research on working-in-groups in industry. Research on working-in-groups in industrial companies differs at this point from that conducted within other kinds of organizations in Sweden. In this context, it is worth drawing attention to approaches governed by a perspective in which similarities between individuals are emphasized; such approaches lead to the neglect of questions on how dissimilarities (such as differences in occupational experience and training) can be utilized as resources in shared work.

Effects on the individual have largely been investigated on the basis of questions concerning whether working-in-groups leads to a reduction of negative consequences in the form of physical and mental workload. (Influence and task variation are then the central aspects.) A significant proportion of the research on the direct effects on the organization has also been motivated by a desire to legitimize new organizational forms.

In the light of the results, the question arises of what weight should be attributed to the interaction which develops within work groups. Extensive sociopsychological research ascribes great significance to both socio-emotional and task-related factors when questions are posed on whether working-in-groups in itself constitutes a means for the implementation of work, and on its intrinsic potentials for developing occupational activities. The latter issue is particularly interesting from the perspective of qualifications and skills development.

The different requirements for how research should be presented imposed by various publications have affected the analysis, and thus also its validity. Outcome research in itself involves methodological problems of a different kind. In addition to the design problem, there is, in this specific arena, a major need for methodological development. The incompatibilities in the results are probably (at least in certain cases) due to how concepts are variously defined, operationalized and examined in the different studies.

The results point to the need for a cross-disciplinary approach to future research on the effects of working-in-groups in industry. Social psychology should be able to make a contribution to describing and understanding the complex interaction between frame conditions and group processes. It should make it possible to bring to the surface phenomena other than those which have been investigated so far.

References

[1] A. Lantz and S. Sconfienza, Swedish experiences of Working In Groups in Industry. Research Report 1994:xx. National Institute of Occupatiunal Health, Solna 1994.

Part X
Quality and Maintenance Strategies

Advances in Agile Manufacturing
P.T. Kidd and W. Karwowski (Eds.)
IOS Press, 1994

Identification and Classification of Issues and Difficulties Associated with Quality Management Techniques and Tools.

Ruth E McQUATER, Barrie G DALE, Ruth J BOADEN, Mark WILCOX.
Quality Management Centre, Manchester School of Management, UMIST,
Manchester M60 1QD, UK

Abstract. There are many difficulties and issues associated with the use of quality management techniques and tools (QMT&T). This paper describes general and specific influences on their selection, use and application. It also describes briefly a triangulation methodology for measuring quantitatively and qualitatively the effects of these influences, and to determine the impact on QMT&T specifically. As an outcome of the research, a workbook will be produced, that will aid in the selection of the appropriate technique or tool for a particular function, and will help in the audit of QMT&T already *in situ*.

1. Introduction

Total quality advocates total involvement, an on going opportunity to be creative, and to make things happen. Whether TQM is viewed as a strategy or a philosophy, it suggests that individuals in an organisation have open communication channels to all levels. It further recommends that employee empowerment becomes the norm, to facilitate a process of continuous improvement (CI) [1]. In order to effect change and allow CI under the umbrella of Total Quality Management (TQM), quality management techniques and tools (QMT&T) are used, albeit with varying degrees of success. If they are not used, improvements are likely to be random and spontaneous rather than comprehensive and systematic [2].

This paper will discuss research being currently undertaken to determine specific issues and difficulties associated with the use of quality management techniques and tools. It will discuss the purpose of, and the objectives of, the ongoing project. Brief consideration will given to the methodology used and how it may be applied in practice.

2. Quality Management Techniques and Tools

Quality management techniques and tools are an essential part of the continuous improvement process. There are a variety of techniques and tools which have both specialist and universal applications [3,4]. Ishikawa is quoted as saying that 95% of all problems can be solved with the simple tools (ie, Pareto analysis, graphs, control charts) [2]. The specialist techniques (eg quality function deployment (QFD) and design of experiments (DOE) are most often used by a selectively small number of company employees as they demand high skill levels and specialist training. However the techniques or tools to work well they must be best suited to the process or function to which they are being applied.

Problems associated with the use and application of specific QMT&T are highlighted and discussed in the literature, and have been variously attributed to lack of management support [5], a lack of understanding of the process or the technique or both [6], and a lack of planning with regard to training and implementation of techniques [7]. An extensive review of the literature has indicated that little work has been undertaken to determine whether or not these issues and difficulties are unique to individual QMT&T or are found across all QMT&T. It is important to pinpoint these issues in order that the problems surrounding the selection, use and application of them are addressed to the satisfaction of the companies and organisations using them, and to help aid the introduction of new techniques and tools. The objective of the current research is to identify the association QMT&T in relation to the following propositions:

1)   The difficulties encountered in applying QMT&T are common across techniques.
2)   The effective use of QMT&T (in facilitating quality improvement) is related to the perception of their potential by the user.

These objectives are being pursued by consideration of the following aspects:

- The techniques and tools already used in companies.
- The techniques and tools used for particular applications and in what functions.
- The training for users of techniques and tools.

These particular research issues form part of a larger project funded by a SERC/ACME grant held by Dale and Boaden, Manchester School of Management, UMIST, (Total Quality Management: Integration and Development, GR/H21449).

3. Influences on the Selection, Use and Application of QMT&T

In addition to previous surveys and research undertaken, discussions with industrial collaborators in the research project have shown that there are many influences that affect the use and application of QMT&T. Although this research is looking at specific issues, the more general influences that impact on them cannot be ignored.

The Concise Oxford Dictionary describes an influence as being the effect a person or thing has on another [8]. A different definition by Hardy (1993) describes an influence as the process whereby one individual seeks to modify the attitudes or behaviour of another [9]. In the context of this investigation, this definition may be too narrow, since there may be many factors involved, and it is the cumulative effect of these that have a cause and effect relationship. This can be considered in the following context:- causal factor A has an effect on causal factor B, which then exerts influence on the process or individual C. The causal factors may not be obvious in isolation as having an effect on the narrow sphere of QMT&T, but when added together they cause a cascade effect.

Discussions with Quality Professionals, representative of a wide sector of manufacturing industries and literature searches, have enabled the following set of influences to be identified. They are by no means comprehensive, nor are they defined in terms of organisational theory. They are knowledge, education, training, experience, perceived benefits, customer demand, management style, environment and resources. Investigation of all possible factors, both overt and unforseen, is beyond the terms of the current project as it would entail creating a huge model. Therefore, more specific influences and causal factors must be considered in depth, with the more general ones being alluded to, for to ignore them would prevent consideration of the broader issues.

4. Investigation of the Specific Influences on Quality Management Techniques and Tools

The initial work already carried out has enabled a standard categorisation of issues and difficulties to be made [10]. These are under four general headings: role of tool in the quality improvement process; organisation and infrastructure; data collection; tool use and application. These are further sub-divided into specific topics. It had been proposed to develop this into an auditing matrix [11], to determine problems that individuals and/or companies are having with specific techniques or tools, in different functions. It consequently proved to be inflexible and did not take into consideration specific influences that affect the use and application of QMT&T directly.

One of the original matrices derived from this earlier work, however, has formed the foundation for the development of a semi-structured auditing tool. This has been constructed in such a way, to allow specific questions to be asked related to the selection, use and application of QMT&T, but also allows the researcher to ask more open-ended questions. This is particularly important when questioning the effect of specific influences (ie resources, management, training, education and experience) on the use and application of QMT&T.

To allow the more general influences to be taken into consideration, qualitative data may be gathered from semi-structured interviews and observations in order to find out how respondents view their environment. This, in addition to the quantitative data gathered from the audit, using methodological triangulation, will maximise the amount of data collected and add to its validity.

5. Fieldwork

It is proposed that an audit of the selection, use and application of QMT&T take place in four organisations. These will range from one company introducing quality management techniques and tools for the first time, one with only one year's experience of TQM; one with several years experience of QMT&T and one which has achieved an internal award (based on the Malcolm Baldridge Quality Award).

In this latter company a comparison is to be made with a German sister plant, the purpose of which is two-fold. The first, is to determine if the issues and difficulties that are apparent in the UK are unique, or if they are pan European. The second is to investigate the changing emphasis in post-compulsory education and training in the UK and how this compares to the German system. It is becoming evident that many of the issues and difficulties encountered in the use and application of QMT&T, in particular, those requiring the use of statistical methods, have their roots in education, with training a key issue in the selection and implementation of a technique or tool. It is hoped that analysis of the data obtained from the fieldwork will pinpoint particular areas of difficulty and with its interpretation provide a guide to possible solutions in order that they may be addressed.

The audit tool, though designed to carry out this particular research, will be developed into a self assessment workbook. This will proffer help in the selection of a particular technique or tool that will best suit a particular function. Since there is rarely one single choice it will offer a best fit scenario, with a selection of alternatives. The purpose is to produce a workbook with such a flexible audit capacity to enable measurement of QMT&T on several levels. For example, these could be an audit of individual techniques or tools for either a specific project, function or organisation wide; and audit of all techniques and tools used by an organisation; or for monitoring and evaluation of training of a new technique or tool.

6. Conclusion

Quality management techniques and tools are essential in the drive for Total Quality and continuous improvement. Difficulties have been highlighted that affect their efficient selection, use and application, particularly the more mathematically based techniques. These issues and difficulties that are becoming apparent are a microcosm of the fundamental issues that affect the drive for total quality management *per se*. By conducting detailed investigation in a range of industrial organisations, it is hoped that these issues and the influences on the use of QMT&T be identified, with particular emphasis on education, training, management, resources and experience. Education and training is being investigated rigorously because of the very great concern being voiced regarding changes in the UK's education system, particularly in post compulsory education and with the demise of company apprenticeship and in-house training schemes [12]. By addressing these issues and in helping industrial organisations become self-aware of the issues and influences on the uses of QMT&T, (by using the self-assessment audit tool), many of the problems will be overcome. It is possible that issues and difficulties will be highlighted where none were thought to exist, thereby aiding in the continuous improvement process.

References

[1] RE McQuater, Total Quality - Implications for Small Businesses. MSc Dissertation Dept of Engineering University of the West of England (1992).
[2] D Graves. Forget the Myths get on with TQM - Fast. *National Productivity Review*, Summer (1993) p301-311.
[3] Continual Inprovement Handbook, A Quick Reference Guide for Tools and Concepts. Executive Learning Inc.
[4] J Marsh .Hand Book of Quality Tools. IFS Ltd (1993)
[5] A Kochan (1990). SPC for Success. *Total Quality Management*, December (1990) p 351-353.
[6] Crossfield RT, Dale BG. Applying Taguchi Methods to the Design Improvement Process of Turbochargers. *Quality Engineering* 3 (4), 501-516 (1991).
[7] G Binney (1994). "What Goes Wrong", in Making Quality Work: Lessons from Europe's Leading Companies p 64. Special Report No P655. The Economist Intelligence Unit, Ashridge.(1994)
[8] The concise Oxford Dictionary of Current English. Eighth Edition, Ed R.E Allen. Clarendon Press, Oxford.
[9] C Handy. Understanding Organisations. Fourth Edition, Penguin Books Ltd, London (1993).
[10] BG Dale, RJ Boaden, M Wilcox. Difficulties Encountered in Quality Management Tools and Techniques. Working Paper No 3 ACME/SERC GR/H21499 (1993).
[11] BG Dale , RJ Boaden , M Wilcox. Quality Management Tool and Techniques Classification. Working Paper No 11 ACME/SERC GR/H21499 (1993).
[12] A Smithers and P Robinson. Beyond Compulsory Schooling, A Numerical Picture. The Council For Industry and Higher Education (1991).

Acknowledgments
The authors wish to express their thanks to the ACME Directorate, SERC for their financial support of the project "TQM: Integration and Development" (GR/H21499).

Quality and Work Organization with Advanced Automation in Portugal

António Brandão MONIZ

Industrial Sociology Group, Faculty of Sciences and Technology-UNL, Quinta da Torre, P-2825 Monte da Caparica, Portugal

Ilona KOVÁCS

Section of Sociology, Social Sciences Dept., Institute for Economy and Management (ISEG-UTL), Rua Miguel Lupi, 20 - Gab. 213, P-1000 Lisboa, Portugal

Zulema Lopes PEREIRA

Quality Engineering Group, Industrial Production Dept., Faculty of Sciences and Technology-UNL, Quinta da Torre, P-2825 Monte da Caparica, Portugal

Abstract. In this paper it is analysed the relationships between work organisation and quality systems in firms that uses some forms of advanced automation. Are characterised the existing quality control structures in the Portuguese industry, and the main factors that hidden or fosters the development of sociotechnical methods of quality control organisation strategies. Are analysed some industrial cases that explains more clearly the critical issues of the implementation of quality systems and work organisation systems. A few recommendations are given about the possibilities for the development of new forms of work organisation aith quality systems associated to automated manufacturing systems.

1. Introduction

The increasing market competitiveness felt all over the world has been perceived by Portuguese economic forces as a serious threat in the near future. The general belief is that, to become more competitive, Portuguese products and services have to be improved with regard to quality and productivity [1] . To tackle with the problem, a few initiatives have been carried out throughout the country, both at central and regional levels. Among them, one can cite the organisation of awareness seminars, the creation of infrastructures such as quality control laboratories and technological sectoral centres, the funding of new industrial equipment, the promotion of a national quality campaign and several other activities related to quality and productivity improvement under the umbrella of the program PEDIP.

As a result, there has been a slight change in managerial attitude and some firms have started the development of new management systems towards a total quality and productivity culture, which also imply new forms of work organisation. It is felt, however, that much more has to be done and achieved, especially with regard to small and medium-sized enterprises which form the great majority of Portuguese industry.

The present analysis is based on recent studies and results obtained from a survey which was undertaken in 1992 in Portuguese industrial firms. The survey was funded by the Ministry of Industry in the field of PEDIP [2]. About a thousand questionnaires were sent and 120 responses were received, 111 of which were analysed. This sample will be thereafter considered as A1 (sample 1). Sample 2 (A2) is composed by 41 firms (37% of sample 1), in which some type of quality control activity was found with advanced technologies. After this sample was analysed, it was concluded that only 11 of those firms, corresponding to 10% of sample A1, had or are introducing Quality systems. In this paper the sample formed by the above mentioned 11 firms will be designated as A3.

2. Data analysis

Portuguese firms seem to be really concerned with quality and productivity improvement, as Table 1 shows. In fact, these two features come first, in a scale from 1 to 5, among the objectives mentioned in the questionnaire. Results also show that, firms of A2 and A3 samples are more worried with quality, productivity, working conditions improvement and human resources than those of A1.

Table 1 Firms objectives

Firms objectives	A3	A2	A1
Increase of productivity	4.8	**4.9**	4.4
Quality improvement	4.8	**4.8**	4.4
Improvement of working conditions	3.7	**3.8**	**3.4**
Development of human resources	4.3	**4.0**	3.3
Improvement of Management Quality	4.7	**4.4**	3.8

As regards critical issues, work organisation is one of main concerns (Table 2). Lack of qualified workforce seems to be of great importance for firms in which integrated quality systems already exist (or are intended to be implemented), but the lack of motivation is not a special preocuppation in these enterprises.

Table 2 Critical issues

Critical issues	Value	(scale 1-5)	
	A3	A2	A1
Existing work organisation	3.4	**3.6**	3.5
Lack of personnel motivation	2.6	2.8	2.9
Lack of skilled personnel	**3.9**	3.5	3.4

Another important feature, since the late 80s, has been a more intensive use of new technologies within Portuguese industry. A sociological survey conducted in 1986 and interviews carried out in 1987-88 [3] showed that about 21% of industrial firms used some sort of advanced technological system (CAD, CAD/CAM, PPC, etc.). Computers were mainly used in administrative and financial management (37%), and in production management (12%). Use of CNC machine-tools or manipulators and robots was quite limited (2.5%). The same happened with the use of new quality and R&D techniques (4%). Recent results from the 1992 survey [2] show that the diffusion of new technologies has increased, especially in the late 80s, in the following areas:

- Administrative and financial management computerisation ,
- Computer aided design,
- Quality control,
- Machining.

Data also shows that there is a change in production strategies. In fact, for the past five years, companies have been directed towards new markets and specialisation in few products. In the mid term companies plan to be even more directed to new markets but, at the same time, they plan to introduce new products (table 3). On the other hand, priority given to the specialisation in few products decreased. This is true both for A2 and A1 samples. It is logical that to consider new markets and products as priorities will imply the development of appropriate advanced technologies and quality systems.

Table 3 Production strategies (%)

Production strategies	A2		A1	
	In the last 5 years	For the next 5 years	In the last 5 years	For the next 5 years
Specialization in few products	24	19	20	15
Market specialisation	20	15	18	16
Product diversification	14	11	17	11
Penetration in new market	25	29	22	30
Introduction of new products	12	17	16	20
Integration in networks	5	8	4	6
Others	1	1	2	2
Total of choices	100	100	100	100

Table 3 also shows that the integration in networks starts to be important for Portuguese industrial firms. One can then expect organisational aspects to improve, although quite slowly, due to the greater development of relationships between firms (including subcontract links), specially in those firms with quality systems and advanced technologies. As far as work organisation is concerned, it could be concluded that Taylor principles still exist in many of industrial enterprises.

Table 4 Characteristics of work organisation (%)

Characteristics of work organisation	Yes			No		
	A3	A2	A1	A3	A2	A1
Workers do simple tasks easily performed	36	83	89	9	17	5
Repetition of the same task by the same worker exists	45	83	84	9	17	11
Attribution of a job to each person	36	63	72	55	37	23
Each task has a pre-determined time and way to do it	73	85	74	18	15	20
Supervisors have as main function the control of orders execution	73	71	77	27	29	15
Only management and supervisors are responsible for the design and/or preparation and job control; workers do not take decisions about issues related to their job	18	61	68	55	37	26
Work is performed individually and not in group	18	73	77	55	27	19
Total of firms	11	41	111	11	41	111

There is, however, an important segment in which work organisation does not follow the classical principles (as shown above). It has to be noted that only in 36% of A3 firms simple tasks are easily performed by workers. On the other hand, there is a significant difference between the two samples as far as work planning and control and teamwork are concerned. As can be seen, there are much less cases in A3 in which "only management and supervisors are responsible for the design and/or preparation of job control" or in which "work is individually performed and teamwork does not exist".

It can be concluded that the tendency for flexible and multi-skilled working methods also increases, when concern about quality matters grows, which is according to the theory of TQM. In fact, as the broad literature on the matter can show, Taylor model of organisation has been one of the main obstacles to the implementation of appropriate quality systems.

Results on the work place where quality control is performed seem to confirm these remarks, as showed in the next table.

Table 5 Location of quality control

Quality control	A1 %	A2 %	A3%
In each job by the operator	37.8	73.2	72.7
By specialists	36.9	73.2	72.7
Materials control (laboratory)	54.1	56.1	81.8
Final product control (laboratory)	47.7	63.4	81.8

This data shows once again that there is, in A2 and A3 samples, a greater motivation towards a total quality philosophy and new methods of work organisation than in A1. Still related to work organisation, one can see in Table 6 that the most common forms have been multi-skilled working groups and job rotation.

Table 6 New forms of work organisation

Forms of work organisation	A2	A3
Job Rotation	56.1	52.3
Multi-skilled Working Groups	58.5	63.6
Self-managing Work Teams or production cells	32.0	45.4

Self-managing work teams, i.e. groups of workers who can plan, execute and control their work, thus contrasting with the traditional Taylor system, have greater expression in sample A3. Also in this sample the definition of task execution and planning is allocated in 45% of the cases to the working teams. As regards participation and representation of work force in problem solving, it can be concluded that legal representatives, such as trade unions, have been dominant in A1 (table 7). It is believed, however, that participation of unions and quality circles (or similar types of quality teams) will tend to grow in the near future.

Table 7 Participation instances

Participation/representation instances at firms	A1 (%)	A2 (%)	A3 (%)
Workers Committees	26	24	18
Union leaders/Union Committee	54	46	36
Committee for Working and Safety Conditions	50	54	55
Quality Circles, Progress Groups, Suggestions Systems	29	34	55

In Portugal, more than in other European countries, modernisation of companies is done without direct or indirect involvement of work force. Thus new technologies, mainly in SME's, are introduced without union or any other workers representatives intervention [4]. This can lead to conflicts and difficulties in the implementation of new methods.

4. Conclusions and recommendations

Portuguese industrial firms still present serious deficiencies related to productivity and quality improvement. There is, however, an important part in which new methods of organisation improvement are visible. Data showed that these firms export a great percentage of their products and, therefore, they need to be competitive in order to survive. Results from this study also showed that there is a straight relationship between advanced forms of work organisation and participation and efficient quality management systems. In fact, as companies move towards a total quality system, the need for implementing flexible models of organisation and autonomous working groups also grows.

The fact that work organisation has been pointed out as a major critical issue might reveal a certain "open mind" to new methods. It is crucial, however, that a few initiatives are taken with Government participation and commitment. Promotion of innovating experiences and discussion of their results can play an important role in achieving a change in attitudes. There has been, for the past few years, a great concern about acquisition of new equipment. This is not enough. In order to become more competitive, Portuguese industry must have organisational structures that allow and motivate workers to contribute for quality and productivity improvement.

To change attitudes, education and training will be essential. The inclusion, in training and educational programmes, of new forms of work organisation, employees participation and quality management will certainly contribute for a quicker development. These programmes should not be confined to technicians and engineers, but must also be extended to upper and middle management and shop-floor workers.

REFERENCES

1 PEREIRA, Z. L. (1993). The emergence of quality as an issue in Portugal. World Quality Congress, Helsinki, EOQ, pp. 113 - 119.
2 KOVÁCS, I.; MONIZ, A.B.; CERDEIRA, M.C. (1993). Mudança Tecnológica e Organizacional do Trabalho na Indústria Portuguesa. Lisboa, PEDIP-DGI-CGTP-CESO I&D.
3 MONIZ, A. B. (1989) Modernização da indústria portuguesa: análise de um inquérito sociológico. Economia e Sociedade. (1): pp. 117 - 160.
4 KOVÁCS, I. (1992). Novas tecnologias, organização e competitividade. KOVÁCS, I. et al: Sistemas flexíveis de produção e reorganização do trabalho. Lisbon, DGI-CESO I&D, pp. 59-60.
5 .PEREIRA, Z. L.; MONIZ, A. B.; KOVÁCS, I. (1994). Quality and Work Organisation in Portuguese Industry. European Congress fro Quality. EOQ, Lisbon.

Advances in Agile Manufacturing
P.T. Kidd and W. Karwowski (Eds.)
IOS Press, 1994

Quality Assurance in a Low Scale Industry

J. Zackrisson [1], M. Mellbin [2] and H. Shahnavaz [1]

[1] Ergonomic Department, Luleå University of Technology, S-951 87 Luleå, Sweden
[2] ABB Railcar AB, S-891 83 Örnsköldsvik, Sweden

Abstract. Management and utilisation of human resources are very different in low scale and large scale industries. In small and low scale industries the workers must possess several skills, be flexible and take responsibility for many functions in the organisation. Common statistical QC- tools are many times substituted by human related tools in quality assurance. The company at study, works with thin sheets aluminium welding. The company is reorganised and a quality assurance system according to the EN-29000/ISO 9000-system is prepared. Reorganisation and quality development have given important side-effects on productivity, worker's satisfaction and environmental factors, which are highlighted in this paper.

1. Introduction

Production, marketing, financial planning, work environment planning, human resources, human relations, and maintenance management are important elements of business and industrial organisation. The demand for higher productivity and the increasing request for fiting the product to consumers needs and expectations have caused creation of many new specialists in these areas. Today, total quality management is a concept, summarising recent competitive advantages in the industrial and service sectors [1]. The project, which partly is reported in this paper, is based on this view of modern quality management, successfully applied, for example, in the Japanese high tech manufacturing industry.[2] [3].

Management tools to control business and quality improvement are primarily intended for industries that have Tayloristic production system [4] and rarely, low scale industries are attained. The project discribed in this paper concernd about a low scale industry with few employees which is dependent on high skilled workers. To give an idea of the relationship between the company at study and other types of industries Figure 1 is presented.

Figure 1. Characteristics of the industries with respect to the number of employees and the number of produced items.[4]

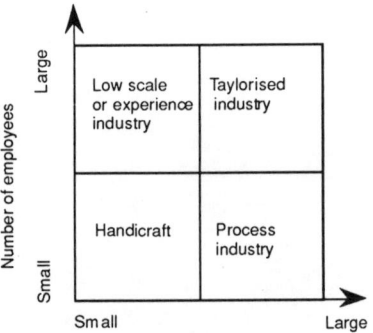

The company is highly motivated for introducing a quality assurance system and the research team (the authors) were acting as partners in that process.

2. The Aims of the Project

The main goal of any reorganisation of a company is increasing efficiency and profitability. From the employees´ view point the goal is formulated, for example, as higer expectation of security and safety in the job. The aim of this project was twofold. In the first step, the ambition was to maintain at least the quality and the productivity at the same levels as before the reorganization. The short-term effects on the employees and the customers were analysed to attain a solid base for a long term strategy. The operator´s performance and the subcontractors attitude to the new company were consequently analysed and compared to the time before the reorganisation. That implies, this step concerns about changes in productivity and in psycho-social characteristics of the peopele involved in relation to the working environment.

In a second step, not reported in this paper, the long term strategy, includes quality improvement and total quality management. More specifically, the research team attempts to affect the quality by introducing a quality assurance system to increase the profitability and the competitive position of the company. The intention was to certify the products, which is not the case today.

The human resources are the company's greatest and most valuable resources. The company's responsibility to improve the environment is obvious, but in the company at study, especial interest had to be focused on the individual health and safety audit system [5]. An early-warning information system to detect risk to the individual worker´s health in connection whith aluminium welding task is planned.

3. Methods and Concepts

This research and development project works on participation of workers in the improvement of the working conditions and the market competition. The project has been designed to comply with action research, in which the researchers continuously follow up the changing process [6].

The concept quality is defined in accordance to the proposed definition "quality is fitness to use" [7] and the Deming philosophy of quality control [2]. This implys, for example, that the quality concept .is focused on the products ability to satisfy the consumer's needs and expectations. Consequently, in this paper, quality of performance is defined as consumer's reaction to a product. Formally expressed:

Quality of performance = experience - expectation

Even if quality of performance is not possible to formulate in a single mathematical formula, its evidence and changing direction could be studied or measured by a set of quality indicators or characteristics such as: performance, features, reliability, conformance, durability, serviceability, aesthetics, and perceived quality.

Today, quality in modern industries is everyone's responsibility. A quality system starting from the top management and distributed to the operator level. This is a formal system to provide quality assurance and according to ANSI/ASQC Standard A3 (1987) quality assurance is defined as: "all those planned or systematic actions necessary to provide confidence that a product or service will satisfy given needs" [8].

Total Quality Management (TQM) enables every part of the organisation to work towards the same goal [9]. In the second step of the project the fundamental ideas behind this concept will be implemented in the quality development system.

The evaluation model is equivalent to an experimental design, when treatment or an invention is introduced in a natural setting. That means, a "before-after" experimental design is assumed to be a valid method for analysis of the empirical results [10]. The period before reorganisation is chosen so that the turbulence just before this event is eliminated as much as possible. Of course, primarily that concerns the hard data generation.

Quality characteristics are grouped into two broad classes namely, variables and attributes. Attitudes, such as beliefs, preferences, and intentions, are basic elements in the evaluation. In this study ,attitudes are measured and refereed to the workers and the customer. The Likert scaling method is used and five symmetric categories are labelled to indicate the degree of agreement or disagreement with a variety of statements related to the attitude object [11].

4. The Firm at Study

The research project has taken place at a small Swedish company. After a reorganisation of ABB Railcar AB, the company at study was established as a partner owned firm. In 1991 the market for new trams, the main product of ABB Railcar, was falling. To keep the well trained and high-potential employees, that were specialised in aluminium welding, the company started a new division and manufacturing plant. The division got an autonomous role and was a subcontractor to a shipyard. The product is a strategic ship component in aluminium.

However, the product was not profitable enough compared with the expectation and was not in line with the company's business idea. In that situation ABB Railcar decided to abandon the activity. Six workers and one foreman announced their interest to take over the production. ABB Railcar supported the employees to start the new firm "Örn-Alp AB". The firm bought the equipment in the plant and after some negotiations the firm signed contracts with the shipyard and its suppliers.

In 1993 the new company started the production in a new workshop. In that moment the project was initiated and all comparisons are refered to the time before the establishment of the new firm and thereafter.

The shipyard was very satisfied with the total quality that ABB Railcar supplied. That means, when the new firm succeeded the ABB Railcar division, the construction, the manufacturing process, the distribution system, and the cost level were under control. However, the future success was highly dependent on the improvement of the total quality.

From a more general point of view, the new company is located in an industrial area. The industrial culture is built on heavy engineering industry. The distance to the shipyard company is no problem. The supply of skilled workers is estimated to be good in the future. The relations to the local authorities are also confidential.

All this facts have inspired and motivated the workteam to succeed in their business as a partner-owned firm. Seven employees participate in the project. All had appointments in the origin company. One of them is mainly the firm's manager in the new situation. He is very motivated to build up a quality assurance system to be competitive in the growing market.

5. Results

The data collection is based on personal interviewing of the workers at Örn-Alp AB and the customers. The result of the survey is given in Table 1. All together 6 workers at Örn-Alp AB and 3 custumers were interviewed by the research team at the end of the studied period (six month).

Table 1. Short-term effects of the reorganisation on the workers' situation.

Characteristics of the work	Individual Statements				
	Decreased a lot	Decreased	No change	Increased	Increased a lot
Operator's control over his work pale				6	
Feeling content with the working team			3	3	
Motivation at work				5	1
Work committment				6	
Mental stress			5	1	
Future prospects			1	5	

The "objective" productivity (produced items/time unit) has increased by 36% during the observed period of six months. However, the workers' "subjective" estimates of their individual productivity increased only by 20% to 25%. The difference explains , for example, by smarter individual work planning and more efficient layout of the work shop. But, the primarily cause of the increased productivity seems to be the increased motivation. That interpretation is also verified by the interviews when the respondents answered motivation as the most important characteristics (5 of 6 workers). Of course, all observed positive reactions have contributed to the successful start of the new company.

The customer's reaction is that the technical quality is about the same, but the service quality is higher. The time of delivery is more reliable. Design changes is easier to execute as a result of a more effective communication. The customer's attitude to the total quality has increased.

Thus, the interviews reveals that the reorganisation of the production has got a considerable positive short-term effect on productivity and at the same time the effect on the workers' psycho-social situation was definitly positive. Today the incitement to increase the quality improvement and to plan for new products is a common goal. The development of the decided quality assurance system and the incentive to be certified according to EN 29002/ ISO 9002 is attained despite great costs. A negative effect may be that the workers have great expectations on rapid increasing dividends of their shares.

6. Discussion and Further Steps

The result of the study has shown that the first step toward the long term strategy of the company for quality assurance system has succeeded. The solid base for quality assurance is established and the system is running satisfacting. However, the risk effects of aluminium welding to operator´s health in the current contition is not well understood. That is a latent threat against the high motivation in the long run. An integrated health and safety audit system is designed by the authors to avoid any problems. The automation of aluminium welding is seemed not to be accomplished in the near future. Another problem, not yet solved, may appear, when the team need to be enlarged.

References

[1] J.S. Okland, (1989). *Total Quality Management.* Heineman, Oxford.

[2] W.E. Deming, (1982). *Quality, Productivity and Competitive Position.* MIT Press; Cambridge, Mass.

[3] K. Ishikawa, (1985). How to Apply Company-wide Quality Control in Foreign Countries. *Rep. Stat. Appl. Ress., JUSE*, Vol. 32, 4, 32-40.

[4] J. Zackrisson, M.Fransén, M. Mellbin and H.Shahnavaz, (1993). Quality by a Step-by-Step Program in Low Scale Industries. Presented at the 12th ICPR, Lappeenranta University of Technology, Finland.

[5] A.I. Glendon, A.J. Boyle and D.M. Hewitt, (1992). Computerised Health and Safety Audit Systems. *Computer Applications in Ergonomics, Occupational Safety and Health* / M. Mattila and W Karwowski (Eds). North Holland, Amsterdam.

[6] R.N. Rapoport, (1970). Three Dilemmas in Action Research. *Human Relations,* Vol 23, 6, 499-513.

[7] J.M. Juran, (Ed) (1974). *Quality Control Handbook*, 3rd edition. McGrow-Hill, New York.

[8] American Society for Quality Control (1987). ANSI/ASQC *Standard A3-1987, Quality Systems Terminology, ASQC*, Milwaukee, WI.

[9] A.V. Feigenbaum, (1983). *Total Quality Control.* MacGraw-Hill, New York.

[10] R.V. Hogg and J. Ledolter, (1989). *Engineering Statistics.* Macmillan Publishing Company, New York

[11] E.P. Cox, (1980) The Optimal Number of Response Alternatives for a Scale: A Review. *Journal of Marketing Research*, 17, 407-422.

Advances in Agile Manufacturing
P.T. Kidd and W. Karwowski (Eds.)
IOS Press, 1994

Development and Evaluation of Hybrid Inspection Systems

Tung-Hsu (Tony) Hou
Department of Industrial Management
National Yunlin Institute of Technology
Touliu, Taiwan, R.O.C.

Li Lin and Colin G. Drury
Department of Industrial Engineering
State Univeristy of New York at Buffalo
Buffalo, New York, U.S.A.

Abstract. Hybrid human-computer inspection systems, taking advantages of humans and computers, are candidates to improve inspection performance. In this paper, alternatives of designing a hybrid inspection system based on function analysis of human and automated inspection are described. Two case studies used to compare alternative inspection systems are discussed. These case studies consistently show that hybrid inspection systems have better performance than either human inspection or automated inspection systems.

1. Introduction

Since human and automated inspection systems have limitations in accuracy (Drury and Sinclair, 1983), hybrid human-computer inspection systems that take advantages of humans and computers are candidates to improve inspection performance. In this paper, alternatives of developing hybrid inspection systems are proposed based on a review of the functions of human visual inspection and automated inspection. In order to demonstrate the feasibility of hybrid systems and to show the effectiveness of humans in an inspection system, two case studies were conducted. Both have implications for inspection systems design.

2. Function Analysis of Human and Automated Inspection

Functions of humans in an attribute inspection task have been classified into four categories: presentation, search, decision-making, and action (Drury, 1992). Among these functions, search and decision-making are recognized as the two most important and most difficult components, with ``presentation" and ``action" typically being simple execution components.

Interestingly, the four functions have direct counterparts in an automated inspection system. Chin (1986) discussed a simple automated inspection system that consists of a transporter, a scanner, a processor, and a sorter. The transporter moves an object to be inspected into the scanning station, where the sensor collects the object's visual data and sends the information to the processor for analysis. After the analysis, classification about the status of the test product is made and the processor directs the sorter to sort the product. The sensing and analysis processes can be equated with the human search process while the classification is analogous to the human decision-making process. Although there is a close relationship between image analysis and classification procedures, these two processes are still separable. Therefore, the major functions of automated inspection can also be modelled as consisting of a search process and a decision-making process.

The computer search process involves illumination, preprocessing, processing, and feature extraction subtasks to segment an object or a flaw from its background (Jain, 1989).

Pattern recognition, usually used as a decision-making scheme in an automated inspection system, falls into three categories: template matching, syntactic approach, and decision-theoretic approach (Fu, 1982). Each method corresponds to one of the automated inspection algorithms: image reference, design rule checking, and a hybrid approach.

3. Alternatives of Developing a Hybrid Inspection System

Based on the function analysis of inspection, possible alternatives of developing hybrid inspection systems are (1) unaided human inspection, (2) computer-search human-decision inspection, (3) human-computer decision-sharing inspection, and (4) fully automated inspection, with the others being dominated by systems (2), (3) and (4) (Hou, et al., 1993).

For the first category of system design, the role of computer is only to provide high contrast images to the human inspector. As for the second category, the computer performs only the search task. Results of image analysis and processing are shown to the inspector, who then makes the decision about the status of the test image. In the third allocation, computers perform both search and decision-making. However, if the confidence level of the decision is low, the operator will take over the decision-making process. For the second and third system designs, the inspector is free from the tedious search task and can concentrate on the decision-making task. Alternative (4) is a fully automated system.

4. Evaluation of Hybrid Inspection Systems

4.1. Case Study 1

In this case study, human inspection, automated inspection, computer-search human-decision, and human-computer decision-sharing systems were compared (Hou, et al., 1993). The test object were pseudo 3-D images of surface mount devices (SMDs). Missing components, wrong-sized components, and misaligned components were used as the fault types. The number of components on the board, contrast level between components and background, and visual noise in the displayed image were used as experiment factors. Each factor was assigned three levels; therefore, a 3^3 factorial design was conducted. Batches of 20 printed circuit boards (PCBs) were tested for each of these 27 situations. In each batch of 20 PCBs, eight boards were good, four boards had a fault of missing component, four boards had a fault of wrong-sized component, and the other four boards had a fault of misalignment. The human inspection, computer-search human-decision system, and human-computer decision-sharing system were each assigned three subjects to test the system performance.

The overall system accuracy, in terms of sensitivity, is given in Table 1, which showed that accuracy of the two hybrid systems was significantly higher than that of the two automated inspection systems at the 95 % confidence level. Although there was no significant difference between the hybrid systems and the human inspection system, the computer-search human-decision system had a better sensitivity than the human visual inspection. The analysis of sensitivity demonstrated the advantage of using a combined human-computer inspection system.

System performance of these alternative systems was compared with different levels of number of components, contrast, and noise. It was found that the computer-search human-decision system had the best sensitivity in each situation, except for the 0.5 contrast level. Under the 0.5 contrast level, human inspection system was only marginally better than the computer-search human-decision system.

Table 1: Mean sensitivity values of alternative systems

System	Mean	Tukey Grouping
Computer-search human-decision	0.9573	A
Human system	0.9494	A
Human-computer decision-sharing	0.9291	A
Automated system 1	0.8949	B
Automated system 2	0.8330	C

4.2 Case Study 2

In this case study, a human inspection, a feature-based automated inspection, and a hybrid inspection systems were compared. Pseudo 3-D SMDs images containing 41 components were used. Missing components, wrong-sized components, and misaligned components were used as the experimental factors. Two hundred boards consisting of 27 faulty boards and 173 good boards were tested for each fault type. Five subjects were used in each of the pure human inspection and hybrid inspection tests.

The automated system, with a preset window around each component, used area, average gray level, and center of gravity in the preset window as features to detect a missing component, wrong-sized component, or misaligned component respectively. This automated system has been shown to have a significantly higher accuracy than a template matching based automated inspection system (Hou and Kuo, 1994), which used image subtraction (i.e., an exclusive OR operation between a standard image and a test image) and a preset threshold value to detect faulty components. The hybrid inspection system presented the results of image subtraction to the subject who then determined the status of the test SMD image, either to accept or to reject. Therefore, this hybrid system was a computer-search human-decision type inspection system.

Inspection accuracy in terms of hit rate, false alarm rate, and sensitivity for the three test systems are shown in Table 2. This case study showed again that hybrid inspection had a higher accuracy than either human or automated inspection systems. Table 2 also showed that the automated inspection system was better at detecting wrong-sized components while the hybrid inspection was better at detecting misaligned components. It seems that each inspection system has its advantages in detecting different types of faults.

Table 2: Inspection accuracy of alternative inspection systems

Fault Type	Human		Automated		Hybrid	
	Hit	FA	Hit	FA	Hit	FA
Missing	0.89	0	1	0	1	0
Wrong-sized	0.74	0.01	1	0	0.94	0
Misaligned	0.59	0	0.7	0	0.87	0
Total	0.74	0.005	0.9	0	0.94	0
Sensitivity	0.933		0.975		0.985	

FA = False Alarm

5. Conclusions

Hybrid human-computer inspection is a new design concept and is a candidate to improve inspection accuracy. Two case studies in this research consistently showed that hybrid inspection systems have better accuracy than either human inspection or automated inspection. In addition, it seems that a combination of computer searching and human decision-making has better performance. Therefore, computers are better suited to search and humans to decision-making.

This research also found that human inspectors were indeed adaptable to low contrast while a hybrid inspection system still had the same performance as the human inspection when the constrast level was fairly low. Therefore, humans maintained their performance, either alone or as part of a hybrid system. One of the values for humans in hybrid systems appears to be in the interpretation of noisy or degraded images.

Although hybrid inspection systems had better accuracy in general, case study 2 showed that each inspection system had its advantages in detecting different fault types. Therefore, an inspection station may be like an assembly system, which consists of more than two inspection subsystems, each being assigned to detect some dedicated fault types.

6. Acknowledgement

This work was partially supported by the National Science Council of the Republic of China under grant NSC 82-0113-E-028-T.

References

[1] Chin, R.T., ``Algorithms and Techniques for Automated Visual Inspection," in *Handbook of Pattern Recognition and Image Processing,* edited by T. Young and K.S. Fu, Academic Press, 1986.

[2] Drury, C.G., ``Inspection Performance," in *Handbook of Industrial Engineering,* edited by G. Salvendy, John Wiley and Sons, 1992.

[3] Drury, C.G. and Sinclair, M.A., ``Human and Machine Performance in an Inspection Task," Human Factors, 25(4), pp. 391-399, 1983.

[4] Fu, K.S., ``Pattern Recognition for Automatic Visual Inspection," IEEE Computer, pp.34-40, December, 1982.

[5] Hou, T.H. and Kuo, W. L., ``Performance Evaluation of Automated Inspection Systems" The Ninth National Conference On Technology and Vocational Education, March 1994 (to appear).

[6] Hou, T.H., Lin, L., and Drury, C.G., ``An Empirical Study of Hybrid Inspection Systems and Allocation of Inspection Functions," Journal of Human Factors in Manufacturing, Vol. 3, No. 4, pp. 351-367, 1993.

[7] Jain, J. K., *Fundamental of Digital Image Processing,* Prentice Hall, Englewood Cliffs, NJ 07632, 1989.

Advances in Agile Manufacturing
P.T. Kidd and W. Karwowski (Eds.)
IOS Press, 1994

Implementing New Working Practises in Manufacturing for Concurrent Quality and Maintenance Control

Ulf SANDBERG

Industrial Control Systems, Royal Inst. of Technology, S-100 44 Stockholm, Sweden

Abstract. This paper presents experiences and results from industrial projects in which manufacturing companies launched the implementation of a new, computer supported, working practise for work with process and product quality. A successful implementation of the working practise requires a change in how the work is performed as well as active use of a computer system for work support. As such this paper may serve as a general example of problems that occur when work dependent on the use of computer systems must be changed. Crucial factors for success are how the change is initiated and motivated, the possibilities of comparing the old working practise with the new one, and how the computer system is introduced. As changes are necessary in modern companies, lack of comparisons in these areas constitute a hindrance to reaching goals and obstruct a necessary development in the continuous improvements of the organisation and the work performed.

1. Introduction

Today´s flow oriented, lean production manufacturing systems are becoming more and more sensitive to process and product quality defects. When such deviations from normal production occur, quick responses are required to detect, identify, and correct a product or process quality deviation [1]. The work needed to correct and prevent quality deviations constitute a working practise [2] in which people closest to manufacturing are given tools, knowledge, and skills to be able to perform analysis and necessary corrective actions. Based on acquired knowledge, preventive measures are defined and carried out by the first line of workers or by experts within the company. The working practise includes use of a computer system for work support [3,4].

The experiences and results presented in this paper originate from industrial projects in which manufacturing companies have begun an implementation towards the new working practise. The existing working practise in the companies has consisted of a specific maintenance organisation responsible for the process quality and in some cases a quality organisation responsible for the product quality. In some companies, the working practises have been supported by computers. The computer support for the new working practise was intended to be purchased through a tender process and   delivered by an external vendor or by an internal organisation responsible for computer system development and operation. In principle, the existing old system could either be modified and upgraded or a completely new system could be bought. In the paper it is assumed that a decision already has been made to embark upon a change towards the new working practise. Even if the decision is made in consensus between management and workers, the principal problems discussed here remain valid.

The implementation of the working practise is dependent on an acceptance from the people who are going to carry out the new work. The working practise is built upon a two level hierarchy in competence and it covers two traditionally separate parts of the organisation. Hence, the implementation project must present, discuss, and motivate why the new working practice, including the computer support, is better than the existing one. For the individual this means acceptance, resistance, or a passive attitude towards the new situation.

If the change from an "old way of doing things" to a "new way" cannot be made in a smooth and systematic fashion, the company looses pace, time, and money. Irrelevant obstacles are used as criteria for resistance to the change. Insufficient methods and project steps carried out in the wrong order contribute to an "arbitrariness" when work is developed. This must be replaced by systematic procedures based on motivated methods and models understood and accepted by everybody involved.

2. Problem areas

2.1 From an old to a new working practise

Going from an old to a new working practise is a natural step in a company's ambition for continuous development and its aim towards as efficient methods and manners as possible. Irrespective of how the change is carried out, the individual judges what is presented to him and indicates a positive, negative, or passive response. The response can only be understood by examining those circumstances that were perceived at the time of the decision.

Most companies do not have any kind of account as to how work is actually carried out. Job descriptions can be found, but they are seldom up to date. A continuous evaluation of the performance of the work carried out is not often executed. This means that the circumstances motivating the existing job and the existing way of doing things are not consciously known. Things are just as they are. Goal orientation and an analysis of the quality of the work performed are more results of individual efforts than a pursued company venture.

Therefore, any changes from existing practises are difficult to realize. The chances of success depend mainly on the individual [5], his (or hers) perceived need for change combined with an evaluation of the result in terms of both a work related outcome and a personal outcome. This will affect the change and indirectly decide if the change is feasible at all. If it is pusued, this will also affect how quick a change, painful a change, and expensive a change will be.

2.2 Going from an existing to a new computer support system

The implementation of the computer system is normally executed in steps: specification, tender, acceptance of a vendor, development etc. The traditional development of computer systems severely treathens the implementation of a new working practise if the implementation is initiated to late. The system is already needed when the working practise is presented and introduced. If no system is available, other versions of the system must be used so that end-users understand their interaction with the system and how it is supposed to support the work tasks.

The life-cycle of a computer system, and also a working practise, include an establishment phase, a continuation phase, and a liquidation phase, see Figure 1. Changing the existing system to a new one normally implies a decision of the liquidation of the existing one. After that, an establishment phase of a new system starts. Since the new working practise requires access to the new computer system already when the new work is in its establishment phase, start-up of system development has to be initiated already when the existing work and computer systems are in their continuation phase, see Figure 1. This also means that most of the new working practise must be defined in its smallest work steps in order to be able to develop the computer system using traditional system development methods.

If people working with the existing system are not to become too confused and even question what is going on, the whole process of changing the working practise must be initiated and started well within the continuation phase. This challenges the change process since the existing way of doing things seems to be justified for some time ahead. Motivation of the coming change in working practise confronts well established routines and work. The justification of an investment in a new computer system must be initiated during the time period when the existing system seems to have a long life-time left. People comfortable with the existing system become threatened if a change comes about too prematurely.

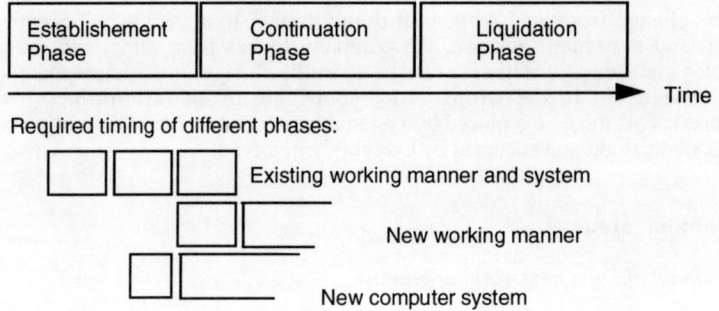

Figure 1 - Different life-cycle phases and their interdependencies

2.3 Bridging the constructional and behavioural domains

The gap between the behavioural domain, where work is performed in the manufacturing company, and the constructional domain, where the system is developed, has to be bridged early. A normal procedure, when computer systems in manufacturing are to be installed, is to choose a vendor after a tender. Before that, a specification of the computer system must be made. A complication here is that the company needs to choose a vendor before the new working practise is introduced in order to have a computer system ready for demonstration and presentation together with the introduction of the new working practise.

During the specification phase, the entire working practise must be defined. To be able to specify and continuously judge if the computer system is acceptable, the content of the functions in terms of man-machine interaction can be used [4]. As computer systems are judged mainly from their interfaces, and since no tradition to evaluate a system from its content exists, such a procedure may be dangerous. People tend to stick with their first impressions, and a new working practise with a corresponding computer system that does not give a positive first impression, will surely lengthen the implementation time.

Traditional system development makes investment decisions easy both for buyers and vendors. Since the working practise and computer support have to be developed and specified at the same time, people from the manufacturing company´s organisation have to be responsible for the knowledge about the behavioural domain. The bridge to the constructional domain can be made by some capable person within the company or someone brought in from outside. The ideal situation is certainly to have the intended vendor present already from the beginning, but that requires strong relationships between the buyer and the vendor. As the vendor must be present already at the beginning, the relationship must be built on trust. A chosen vendor using modern development tools, which in principle are available to all vendors, is looked upon as a production resource, similar as liasons established to other sub-contractors used in early phases of, e.g., product design.

If, however, the initiative to change an existing computer system to a new one comes from a vendor, the situation depicted in Figure 1 is rather complicated. Nevertheless, manufacturing companies that concentrate on their core businesses are assumed, to an increasing extent, to use ideas, products, and methods developed outside the company. Some hints to a correct approach in this case can be found in the material presented in this paper, but a full treatment of this problem complex is left to another publication.

3. Suggestions for improvements

The first step is to equalize the knowledge between all parties concerned:

- make goals clear; question if stated goals have any implications
- develop knowledge and descriptions about present work
- continuously define and analyse what affects present work and the results delivered

The response is a result of how the change is motivated. The old and new working practise must be comparable in terms of compliance with company goals, present and expected results from the work carried out, and what the new work requires from those who will perform it. It is important that figures showing present work results are realistic and not brushed up. Otherwise, all judgements about suggestions to new working practises will be subjective. Perhaps this is natural since companies have difficulites in fostering a systematic treatment of factors affecting present work and in distributing information about how, why, and with what result work is performed.

The whole change process must begin with a definition of the working practise. This must be described so that the inherent work tasks, activities, and work steps can be identified and understood. There is a lack of accepted methods and models for such a description. In practise, such descriptions seldom occur and the only way to understand how people work today is through interviews of all concerned. In different parts of an organisation, there may be some common views as to how, why, and with what result work is performed, but any cross-boundary knowledge is rare. This does not make integration or changes of work over organisational boundaries easy.

Knowledge about work, work performance and work results should be treated systematically, like successful people approach behaviour when applying work practises [6]. Theory about learning emphasise the individual´s near environment, the context of action, as well as the goals and models of the enterprise. These factors constitute the foundations for learning [7].

As the working practise involves use of a computer system, the work must be described and introduced using some kind of representation of the system. Using traditional system development methods require an early start-up of the entire change process. If the reason for the change is not known, the process may be severely threatened because of peoples resistance to it. The early start has to be motivated and the need for a system development before introducing the working practise. The relation with the intended vendor must be introduced before the liquidation phase of the existing system and the old working practise is reached.

If no complete new system is available, this normally means use of simple versions of it. A rational introduction of the system only requires a description of the content so that system support to different activities and work tasks can be understood. It is questionable if the users and vendors can reach this level. Looking at vendors and seeing how they describe their systems today is not encouraging in this respect.

References

[1] U.Sandberg, The Coupling Between Process and Product Quality, Proceedings of the EUROMAINTENANCE ´94, 26-28 April, Amsterdam, 1994.

[2] U. Sandberg, K. Franzén, An Approach to Work Design and Computer Support for Concurrent Maintenance and Quality Control in Manufacturing Industries, IFAC Symposium on Automated Systems Based on Human Skill and Intelligence, Sept. 23-25, Madison, 1992.

[3] U. Sandberg, K. Franzén, A Computer System for Concurrent Quality Assurance and Maintenance Control in Manufacturing Industries, IFAC Workshop on CIM in Process and Manufacturing Industries, Nov. 23-25, Helsingfors, 1992.

[4] U. Sandberg, An Activity Based Man-Machine Interface for Documentation and Analysis of Process and Product Quality in Manufacturing Industries, IFAC Symposium SAFEPROCESS´94, June 13-15, Espoo, 1992.

[5] J.L. Farr and M. Ford, Individual Innovation. In: M.A. West, J.L. Farr (Eds.), *Innovation and Creativity at Work*, Wiley, 1990, pp. 63-80.

[6] D. Dörner, et.al., *Vom Umgang mit Unbestimmtheit und Komplexität*, Huber, 1983.

[7] A.P. Sage, *Systems Engineering*, Wiley, 1992, pp. 539-546.

Advances in Agile Manufacturing
P.T. Kidd and W. Karwowski (Eds.)
IOS Press, 1994

Experience Guided Optimization of Preventive Maintenance Strategies - Experimental Evaluation of a Shop-Floor Assistance System

Dietmar Gude and Elena Psaralidis
*Abteilung Arbeitspsychologie, Institut für Arbeitsphysiologie an der Universität Dortmund, Ardeystraße 67, D-44139 Dortmund, Germany**

Abstract. This contribution presents the evaluation of a shop-floor assistance system for the optimization of preventive maintenance strategies, called Behavior-Outcome Feedback (BOF). In a simulation experiment the feedback converges to the actual optimal strategy with increasing size of the data base and exploratory user behavior. Moreover, the feedback was closer to the optimum than the strategy acquired by subjects operating under conventional conditions in a laboratory experiment. Thus, BOF seems to be both accurate and serviceable in supporting the execution of preventive maintenance functions.

1. Introduction

As response to the increasing demands for flexibility in the manufacturing process, a number of authors propose a human-centered approach, e.g., [1]. According to this approach the task spectrum of the shop-floor staff is to be reorganized by integrating, inter alia, preventive maintenance functions. In performing these functions, the shop-floor staff is confronted with an optimization problem. Typically, too frequent preventive maintenance and too many breakdowns affect the efficiency of the manufacturing process. Therefore, both short and long preventive maintenance intervals are suboptimal. Such an optimization problem is to be solved inductively on the shop-floor. This requires variation of the preventive maintenance strategy and recording of the consequences for the efficiency of the manufacturing process. Furthermore, to determine the efficiency of a given preventive maintenance strategy one has to precisely perceive, memorize, and integrate the data of a larger number of preventive maintenance activities and breakdown events. As these tasks are both strenuous and susceptible to human error, the shop-floor staff needs support.

On this background, an approach called Behavior-Outcome Feedback (BOF) has been proposed [2] and is now realized as an application program for flexible manufacturing cells [3]. In this program the shop-floor staff records its preventive maintenance and the breakdown events with their corresponding costs. Based on these data the program estimates the contingency between the preventive maintenance strategy and the efficiency. If required, the staff calls a diagram of the contingency, using it as a cue to the optimal strategy. The validity of this information can be checked by applying the suggested scheme and acquiring more precise feedback about the efficiency in this sector of the contingency.

Here, an evaluation of the key functionality of the program, the computation of the contingency, will be presented. For methodological reasons, this evaluation refers to its training

* The preparation of this paper was supported by a grant from the Deutsche Forschungsgemeinschaft (Sonderforschungsbereich 187, Teilprojekt A-4). We would like to thank Herbert Heuer, Christoph Koch, and Detlev Poweleit for helpful comments on an earlier draft of the manuscript, and Michaela Bruer, Britta Krämer, and Uwe Link for technical assistance.

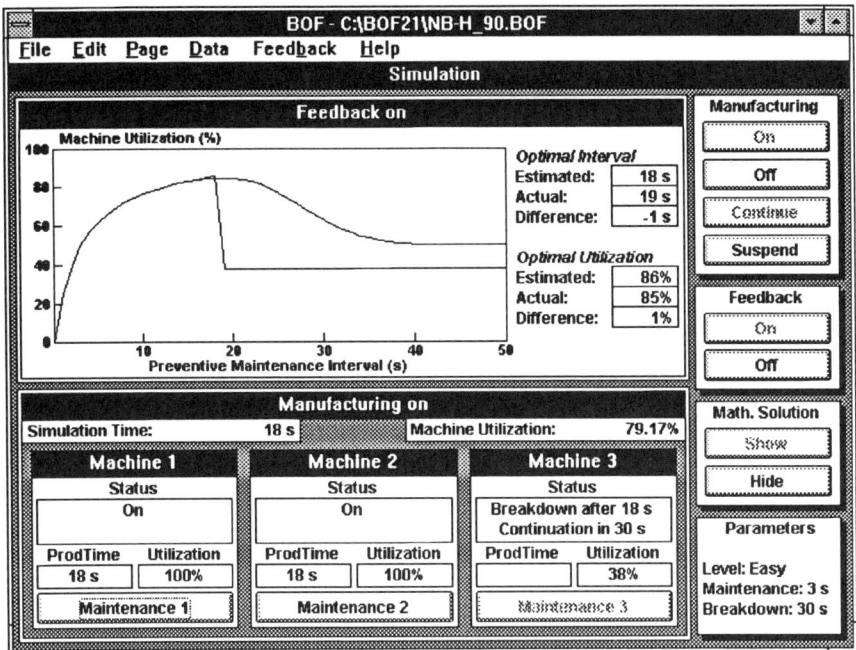

Figure 1. Interface of the simulation module.

facility, a simulation module consisting of a feedback window, a manufacturing window, and a control bar (Figure 1). The manufacturing window displays three machines, which work according to the same principles. Each machine breaks down after a certain amount of production time, resulting in a downtime of 30 s. The relationship between production time and probability of a breakdown is determined by a normal distribution with a mean of 30 s and a standard deviation of 5 or 15 s, depending on the selected level of the simulation. Breakdowns can be avoided by timely preventive maintenance, resulting in a downtime of 3 s.

Each machine is characterized by its status, production time, and utilization. The status signals whether the machine is currently operating, maintained, or broken down. The production time is equivalent to the time since the last breakdown or the last preventive maintenance. The utilization corresponds to the quotient of production time and availability time, which is to be maximized by applying an optimal preventive maintenance interval.

A solution of this optimization problem requires knowledge about the contingency between preventive maintenance interval and utilization. Figure 2 shows the contingency in the simulation, both for the distribution of breakdown events with a standard deviation of 5 s (high saliency) and 15 s (low saliency). According to this, the optimal strategy is a consistent preventive maintenance of the machines after 19 and 14 s of production time, respectively. In the low saliency condition the maximum is less prominent. Therefore, a variation of the strategy in this area results in a smaller variation of the utilization and, consequently, the maximum should be more difficult to determine.

The feedback window in Figure 1 shows two graphs representing the contingency between preventive maintenance interval and utilization in the high saliency variant of the simulation. The smooth graph corresponds to the actual relationship. The second, less

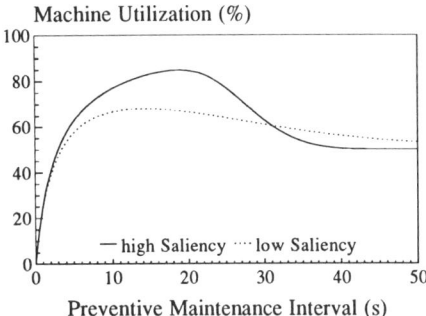

Figure 2. Contingency between preventive maintenance interval and machine utilization.

smooth graph represents the behavior-outcome feedback and is based on the single break-down event shown in the manufacturing window. This event implies a high utilization for the interval that would have prevented the breakdown and a low utilization for longer intervals. Furthermore, shorter intervals are suboptimal due to their increasing preventive maintenance time losses. Thus, the feedback is similar to the actual relationship, except that it predicts an abrupt decline in utilization for intervals that would not have prevented the breakdown.

Imagine a user who subsequently decides to investigate the hypothesis that the break-down occurred exceptionally early and, thus, adopts a preventive maintenance strategy of 25 s. A preventive maintenance event after 25 s of production time would be processed as indi-cating that the utilization of the intervals between 19 and 25 s is not as bad as expected and, therefore, the feedback would be more graded in this area than abrupt as in Figure 1.

This functionality was evaluated in a simulation experiment and a laboratory experiment. The simulation experiment investigated whether the feedback converges to the actual opti-mum. In the laboratory experiment maintenance behavior without support of the feedback was recorded, to determine whether the feedback is superior to the subjects´ optimization.

2. Method

The simulation experiment consisted of 80 runs of 1800 s each. During these runs, corre-sponding to the fed back optimal interval, preventive maintenance was launched automati-cally. In sum, four conditions of 20 runs each were realized by manipulating two factors, the saliency of the contingency between preventive maintenance interval and utilization (high vs. low) and the exploration component of the launched preventive maintenance (with vs. with-out). In the high saliency condition the standard deviation of the breakdown events was 5 s, in the low saliency variant this value was 15 s. In the condition with exploration the ma-chines were maintained 3 s later than the fed back optimal interval, in the variant without exploration the interval was applied strictly. In addition, each run was divided into 9 simula-tion intervals of 200 s. As the dependent variable, for each simulation interval the strategy error was computed. That is, the mean feedback value of a simulation interval was deter-mined and then the absolute difference of this mean and the actual optimal interval was as-certained. Thus, the smaller the strategy error the more accurate was the feedback.

These strategy error values were then compared with the preventive maintenance behav-ior recorded in the laboratory experiment, in which 40 students, 20 female and 20 male, par-ticipated. The subjects´ mean age was 23.0 years, with a standard deviation of 2.8 years. During 1800 s they had to optimize, without feedback, the utilization of the simulation, which consisted in this case of the manufacturing window only. The subjects were assigned to two groups, maintaining the high and the low saliency variant of the simulation, respec-tively. Differentiated for 9 observation intervals, their strategy error was determined analo-gous to the procedure described above.

3. Results

The strategy error of the feedback in the simulation experiment is depicted in Figure 3. These values were analyzed in a three-factorial mixed-effects ANOVA with saliency (high vs. low), exploration (with vs. without), and simulation interval (1-9) as the independent variables. According to the analysis, generally the strategy error in the high saliency condi-tion was smaller than that in the low saliency variant ($F[1,76] = 47.41, p < 0.001$). Thus, the more prominent the optimum of the contingency between preventive maintenance interval and utilization the more accurate was the feedback. In addition, the strategy error in the condition with exploration was smaller than that in the variant without exploration ($F[1,76] = 15.74, p < 0.001$). That is, the exploration improved the accuracy of the feedback. Finally, there was a significant main effect of simulation interval ($F[8,608] = 7.50, p < 0.001$) as well as a significant interaction of this factor with exploration ($F[8,608] = 8.06, p < 0.001$). This implies that the accuracy of the feedback improved with increasing size of the data base, but only in the condition with exploration, without exploration the accuracy was constant.

Figure 4 shows the subjects´ strategy error in the laboratory experiment and the corre-sponding values of the conditions with exploration in the simulation experiment. These re-

Strategy Error (s)

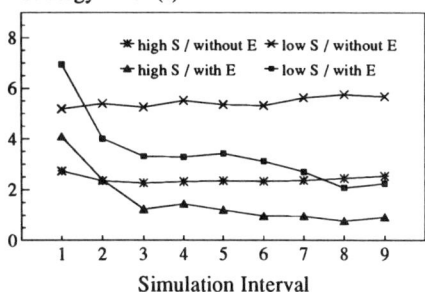

Simulation Interval

Figure 3. Strategy error of the behavior-outcome
feedback in the simulation experiment
(S = Saliency, E = Exploration).

Strategy Error (s)

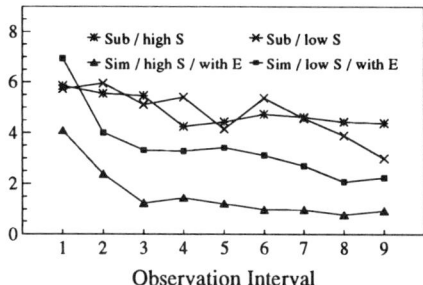

Observation Interval

Figure 4. Strategy error in the laboratory experiment
(Sub) and the simulation experiment (Sim)
(S = Saliency, E = Exploration).

sults were analyzed in a three-factorial mixed-effects ANOVA with source (subjects vs. simulation), saliency (high vs. low), and observation interval (1-9) as the independent variables. In general, the strategy error in the simulation was smaller than that in the laboratory experiment ($F[1,76] = 38.73, p < 0.001$). That is, the feedback was more accurate than the subjects´ performance. There was also a significant main effect of saliency ($F[1,76] = 6.05, p < 0.05$) and a significant interaction between source and saliency ($F[1,76] = 6.91, p < 0.05$). This implies that the strategy error in the high saliency condition was smaller than that in the low saliency variant only in the simulation condition, the subjects´ performance was not effected by this variable. Finally, the strategy error declined with the observation interval ($F[8,608] = 10.59, p < 0.001$), an effect which again was moderated by the source factor ($F[8,608] = 2.25, p < 0.05$). That is, the improvement of the feedback was larger than that of the subjects´ performance.

4. Discussion

According to the results of the simulation experiment, the behavior-outcome feedback converges to the actual optimal preventive maintenance strategy with increasing size of the data base and exploratory user behavior. As indicated by the preventive maintenance performance recorded in the laboratory experiment, this feedback is more accurate than subjects´ behavior operating under conventional conditions, underlining the utility of BOF.

However, the accuracy of the feedback depends on the saliency of the contingency between preventive maintenance strategy and utilization of the manufacturing system. At first sight, this feature seems to be inevitabe, due to the inductive nature of the data processing algorithms and their thence emerging sensitivity to the stochastic aspects of the manufacturing system. But on the other hand, the subjects were not affected by this factor and, therefore, their approach to solve the optimization problem seems to be qualitatively different. Investigating this difference might be helpful in improving the accuracy of the feedback for manufacturing systems with low saliency contingencies.

References

[1] P. Brödner, Design of Work and Technology in Manufacturing, *International Journal of Human Factors in Manufacturing* 1 (1991) 1-16.

[2] D. Gude and K.-H. Schmidt, Preventive Maintenance of Advanced Manufacturing Systems: A Laboratory Experiment and its Implications for the Human-Centered Approach, *International Journal of Human Factors in Manufacturing* 3 (1993) 335-350.

[3] D. Gude, C. Koch, and D. Poweleit, Behavior-Outcome Feedback - An Approach to Support the Inductive Optimization of Preventive Maintenance Strategies. In G. Bradley and H. W. Hendrik (Eds.), Human Factors in Organizational Design and Management - IV. Elsevier, Amsterdam, in press.

Author Index